Table F

The *t* Distribution

d.f.	Confidence Intervals	50%	80%	90%	95%	98%	99%
	ONE TAIL, α	0.25	0.10	0.05	0.025	0.01	0.005
d.f.	TWO TAILS, α	0.50	0.20	0.10	0.05	0.02	0.01
1		1.000	3.078	6.314	12.706	31.821	63.657
2		.816	1.886	2.920	4.303	6.965	9.925
3		.765	1.638	2.353	3.182	4.541	5.841
4		.741	1.533	2.132	2.776	3.747	4.604
5		.727	1.476	2.015	2.571	3.365	4.032
6		.718	1.440	1.943	2.447	3.143	3.707
7		.711	1.415	1.895	2.365	2.998	3.499
8		.706	1.397	1.860	2.306	2.896	3.355
9		.703	1.383	1.833	2.262	2.821	3.250
10		.700	1.372	1.812	2.228	2.764	3.169
11		.697	1.363	1.796	2.201	2.718	3.106
12		.695	1.356	1.782	2.179	2.681	3.055
13		.694	1.350	1.771	2.160	2.650	3.012
14		.692	1.345	1.761	2.145	2.624	2.977
15		.691	1.341	1.753	2.131	2.602	2.947
16		.690	1.337	1.746	2.120	2.583	2.921
17		.689	1.333	1.740	2.110	2.567	2.898
18		.688	1.330	1.734	2.101	2.552	2.878
19		.688	1.328	1.729	2.093	2.539	2.861
20		.687	1.325	1.725	2.086	2.528	2.845
21		.686	1.323	1.721	2.080	2.518	2.831
22		.686	1.321	1.717	2.074	2.508	2.819
23		.685	1.319	1.714	2.069	2.500	2.807
24		.685	1.318	1.711	2.064	2.492	2.797
25		.684	1.316	1.708	2.060	2.485	2.787
26		.684	1.315	1.706	2.056	2.479	2.779
27		.684	1.314	1.703	2.052	2.473	2.771
28		.683	1.313	1.701	2.048	2.467	2.763
(z) ∞		.674	1.282[a]	1.645[b]	1.960	2.326[c]	2.576[d]

Source: Adapted from Beyer, W. H., *Handbook of Tables for Probability and Statistics, 2nd Edition,* CRC Press, Boca Raton, Florida, 1986. With permission.

[a]This value has been rounded to 1.28 in the textbook.
[b]This value has been rounded to 1.65 in the textbook.
[c]This value has been rounded to 2.33 in the textbook.
[d]This value has been rounded to 2.58 in the textbook.

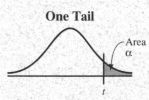

One Tail

Area α

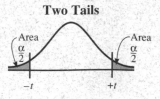

Two Tails

Area $\frac{\alpha}{2}$ Area $\frac{\alpha}{2}$

$-t$ $+t$

ELEMENTARY STATISTICS

A Step by Step Approach

second edition

ELEMENTARY STATISTICS

A Step by Step Approach

Allan G. Bluman

Community College of Allegheny County

 Wm. C. Brown Publishers

Dubuque, Iowa · Melbourne, Australia · Oxford, England

Book Team

Editor *Paula-Christy Heighton*
Developmental Editor *Daryl Bruflodt*
Designer *Jeff Storm*
Art Editor *Brenda A. Ernzen*
Photo Editor *Janice Hancock*

Wm. C. Brown Publishers
A Division of Wm. C. Brown Communications, Inc.

Vice President and General Manager *Beverly Kolz*
Vice President, Publisher *Earl McPeek*
Vice President, Director of Sales and Marketing *Virginia S. Moffat*
Vice President, Director of Production *Colleen A. Yonda*
National Sales Manager *Douglas J. DiNardo*
Marketing Manager *Julie Joyce Keck*
Advertising Manager *Janelle Keeffer*
Production Editorial Manager *Renée Menne*
Publishing Services Manager *Karen J. Slaght*
Permissions/Records Manager *Connie Allendorf*

Wm. C. Brown Communications, Inc.

President and Chief Executive Officer *G. Franklin Lewis*
Corporate Senior Vice President, President of WCB Manufacturing *Roger Meyer*
Corporate Senior Vice President and Chief Financial Officer *Robert Chesterman*

Copyedited by Carol I. Beal

Photo Credits
Page 13: The Bettmann Archive; p. 42: © Walter Bibikow/The Image Bank; p. 70: © Gabe Palmer/The Stock Market; p. 131: © ITTC Productions/The Image Bank; p. 165: © Sobel/Klonsky/The Image Bank; p. 266: © Alvis Upitis/The Image Bank; p. 327: © Steven Burr Williams/The Image Bank; p. 428: © Weinberg/Clark/The Image Bank; p. 464: © Gabe Palmer/The Stock Market; p. 553: © HMS Images/The Image Bank

Special acknowledgement is given to Minitab, Inc. for their assistance. MINITAB is a registered trademark of Minitab, Inc. For information about MINITAB Statistical Software Contact:
       Minitab, Inc.
       3081 Enterprise Drive
       State College, PA 16801 USA
       (814) 238–3280
       (814) 238–4383 FAX
       881612 Telex

Printed in the United States of America by Wm. C. Brown Communications, Inc., 2460 Kerper Boulevard, Dubuque, IA 52001

10 9 8 7 6 5 4 3 2

To my wife, Betty Claire,
To my parents, George and Eleanor,
and
To a special teacher and friend,
Dr. G. Bradley Seager, Jr.

CONTENTS

CHAPTER 10   Testing the Difference Between Means and Proportions

HYPOTHESIS-TESTING SUMMARY 1

CHAPTER 11   Correlation and Regression

CHAPTER 12   Chi-Square

CHAPTER 13   The F Test and Analysis of Variance

HYPOTHESIS-TESTING SUMMARY 2

CHAPTER 14   Nonparametric Statistics

HYPOTHESIS-TESTING SUMMARY 3

CHAPTER 15   Sampling and Simulation

CHAPTER 16   Quality Control

APPENDIX A   Algebra Review

APPENDIX B Tables

Speaking of Statistics

APPROACH

Elementary Statistics: A Step by Step Approach is a textbook for students in the beginning statistics course whose mathematical background is limited to basic algebra. The book uses a nontheoretical approach in which concepts are explained intuitively and supported by examples. There are no formal proofs in the book. The applications are general in nature, and the exercises include problems from agriculture, biology, business, economics, education, psychology, engineering, medicine, sociology, and computer science.

ABOUT THIS BOOK

The learning system found in *Elementary Statistics* provides the student with a valuable framework in which to learn and apply concepts.

- Each chapter begins with an introduction to provide the student with an overview and to show applications.

- Over 300 examples are provided, followed by the solutions to help students learn to solve problems. Examples are solved by using a step-by-step explanation.

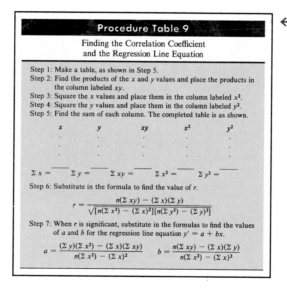

← Procedure tables summarize processes for the student.

Procedure Table 9

Finding the Correlation Coefficient
and the Regression Line Equation

Step 1: Make a table, as shown in Step 5.
Step 2: Find the products of the x and y values and place the products in the column labeled xy.
Step 3: Square the x values and place them in the column labeled x^2.
Step 4: Square the y values and place them in the column labeled y^2.
Step 5: Find the sum of each column. The completed table is as shown.

x	y	xy	x^2	y^2
.	.	.	.	.
.	.	.	.	.
.	.	.	.	.

$\Sigma x = \overline{\quad} \quad \Sigma y = \overline{\quad} \quad \Sigma xy = \overline{\quad} \quad \Sigma x^2 = \overline{\quad} \quad \Sigma y^2 = \overline{\quad}$

Step 6: Substitute in the formula to find the value of r.

$$r = \frac{n(\Sigma xy) - (\Sigma x)(\Sigma y)}{\sqrt{[n(\Sigma x^2) - (\Sigma x)^2][n(\Sigma y^2) - (\Sigma y)^2]}}$$

Step 7: When r is significant, substitute in the formulas to find the values of a and b for the regression line equation $y' = a + bx$.

$$a = \frac{(\Sigma y)(\Sigma x^2) - (\Sigma x)(\Sigma xy)}{n(\Sigma x^2) - (\Sigma x)^2} \qquad b = \frac{n(\Sigma xy) - (\Sigma x)(\Sigma y)}{n(\Sigma x^2) - (\Sigma x)^2}$$

Rules and definitions are set off for easy referencing by the student.

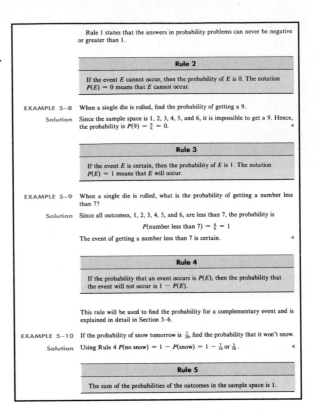

Rule 1 states that the answers in probability problems can never be negative or greater than 1.

Rule 2

If the event E cannot occur, then the probability of E is 0. The notation $P(E) = 0$ means that E cannot occur.

EXAMPLE 5–8   When a single die is rolled, find the probability of getting a 9.

Solution   Since the sample space is 1, 2, 3, 4, 5, and 6, it is impossible to get a 9. Hence, the probability is $P(9) = \frac{0}{6} = 0$.

Rule 3

If the event E is certain, then the probability of E is 1. The notation $P(E) = 1$ means that E will occur.

EXAMPLE 5–9   When a single die is rolled, what is the probability of getting a number less than 7?

Solution   Since all outcomes, 1, 2, 3, 4, 5, and 6, are less than 7, the probability is

$$P(\text{number less than 7}) = \frac{6}{6} = 1$$

The event of getting a number less than 7 is certain.

Rule 4

If the probability that an event occurs is $P(E)$, then the probability that the event will not occur is $1 - P(E)$.

This rule will be used to find the probability for a complementary event and is explained in detail in Section 5–6.

EXAMPLE 5–10   If the probability of snow tomorrow is $\frac{7}{10}$, find the probability that it won't snow.

Solution   Using Rule 4 $P(\text{no snow}) = 1 - P(\text{snow}) = 1 - \frac{7}{10}$ or $\frac{3}{10}$.

Rule 5

The sum of the probabilities of the outcomes in the sample space is 1.

Illustrations provide clear display of results for students.

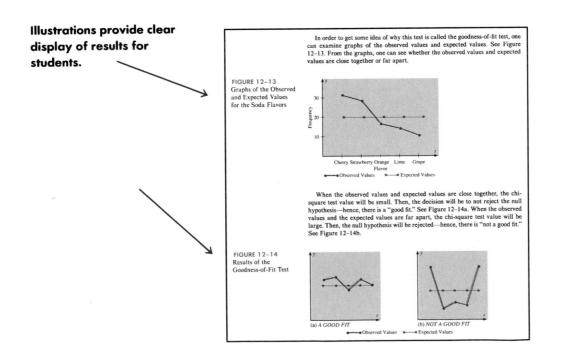

In order to get some idea of why this test is called the goodness-of-fit test, one can examine graphs of the observed values and expected values. See Figure 12–13. From the graphs, one can see whether the observed values and expected values are close together or far apart.

FIGURE 12–13
Graphs of the Observed and Expected Values for the Soda Flavors

When the observed values and expected values are close together, the chi-square test value will be small. Then, the decision will be to not reject the null hypothesis—hence, there is a "good fit." See Figure 12–14a. When the observed values and the expected values are far apart, the chi-square test value will be large. Then, the null hypothesis will be rejected—hence, there is "not a good fit." See Figure 12–14b.

FIGURE 12–14
Results of the Goodness-of-Fit Test

"Computer Applications" are found in each chapter to give students hands-on experience in using and writing statistical programs.

COMPUTER APPLICATIONS

The following data represent the number of customers using the drive-in services of a bank over a 50-day period.

90	135	162	151	148
99	150	107	132	99
127	97	122	92	97
126	150	132	143	143
119	135	105	136	127
142	96	123	144	103
127	114	136	145	132
108	150	129	105	125
132	112	105	116	149
106	128	99	132	96

1. Using the computer package of your choice, enter the data and find the mean, median, mode, range, variance, and standard deviation.

2. Write a program that will find the mean, variance, and standard deviation for the above data.

3. MINITAB can be used to find various descriptive measures of a data set and also to construct a box and whisker plot. To get descriptive measures, type in the following information for the above data.

```
MTB  > SET C1
DATA > 90 135 162 151 148
DATA > 99 150 107 132 99
DATA > 127 97 122 92 97
DATA > 126 150 132 143 143
DATA > 119 135 105 136 127
DATA > 142 96 123 144 103
DATA > 127 114 136 145 132
DATA > 108 150 129 105 125
DATA > 132 112 105 116 149
DATA > 106 128 99 132 96
DATA > END OF DATA
MTB  > DESCRIBE C1
```

A special section called **"Data Analysis: Applying the Concepts"** requires students to work with a data set to perform various statistical tests or procedures and then summarize the results.

√ DATA ANALYSIS Applying the Concepts

The Data Bank is located in the Appendix.

1. From the Data Bank, choose one of the following variables: age, weight, cholesterol level, systolic pressure, IQ, or sodium level. Select at least 30 values, and find the mean, median, mode, and midrange. State which measurement of central tendency best describes the average and why.

2. Find the range, variance, and standard deviation for the data selected in Exercise 1.

3. From the Data Bank, choose 10 values from any variable, construct a box and whisker plot, and interpret the results.

- "Speaking of Statistics" contains articles from newspapers, magazines, and books to show the student how statistics are used in real life.

- Exercises are located at the ends of sections and the ends of each chapter. With "Chapter Tests" over 1600 problems are available. Selected section exercises that contain more challenging solutions or additional information are designated with an asterisk (*).

- End-of-chapter "Summaries," "Important terms," and "Important formulas" give students a concise summary of the chapter topics and provide a good source for quiz or test preparation.

- An *algebra review* is found in the Appendix, and a "Glossary of Terms" and a "Glossary of Symbols" are provided for reference.

"Hypothesis-Testing Summaries" are found at the ends of Chapter 10 (z and t tests for testing means and proportions), Chapter 13 (the F test, analysis of variance, and the chi-square test), and Chapter 14 (nonparametric tests) to show students the different types of hypotheses and the types of tests to use.

HYPOTHESIS-TESTING SUMMARY 1

1. Comparison of a sample mean with a specific population mean.

    example: $H_0: \mu = 100$

    a. Use the z test when σ is known:

    $$z = \frac{\bar{X} - \mu}{\frac{\sigma}{\sqrt{n}}}$$

    b. Use the t test when σ is unknown:

    $$t = \frac{\bar{X} - \mu}{\frac{s}{\sqrt{n}}} \quad \text{with} \quad \text{d.f.} = n - 1$$

2. Comparison of two sample means.

    example: $H_0: \mu_1 = \mu_2$

    a. Use the z test when the population variances are known:

    $$z = \frac{(\bar{X}_1 - \bar{X}_2) - (\mu_1 - \mu_2)}{\sqrt{\frac{\sigma_1^2}{n_1} + \frac{\sigma_2^2}{n_2}}}$$

    b. Use the t test for independent samples when the population variances are unknown and the sample variances are unequal:

    $$t = \frac{(\bar{X}_1 - \bar{X}_2) - (\mu_1 - \mu_2)}{\sqrt{\frac{s_1^2}{n_1} + \frac{s_2^2}{n_2}}}$$

    with

    d.f. = the smaller of $n_1 - 1$ or $n_2 - 1$

    c. Use the t test for independent samples when the population variances are unknown and the sample variances are equal:

    $$t = \frac{(\bar{X}_1 - \bar{X}_2) - (\mu_1 - \mu_2)}{\sqrt{\frac{(n_1 - 1)s_1^2 + (n_2 - 1)s_2^2}{n_1 + n_2 - 2}} \sqrt{\frac{1}{n_1} + \frac{1}{n_2}}}$$

    with   d.f. $= n_1 + n_2 - 2$

d. Use the t test for dependent samples when the means are related:

    example: $H_0: \mu_D = 0$

    $$t = \frac{\bar{D} - \mu_D}{\frac{s_D}{\sqrt{n}}} \quad \text{with} \quad \text{d.f.} = n - 1$$

    where n = number of pairs.

3. Comparison of a sample proportion with a specific population proportion.

    example: $H_0: p = 0.32$

    Use the z test:

    $$z = \frac{X - np}{\sqrt{npq}} \quad \text{or} \quad z = \frac{X - \mu}{\sigma}$$

4. Comparison of two sample proportions.

    example: $H_0: p_1 = p_2$

    Use the z test:

    $$z = \frac{(\hat{p}_1 - \hat{p}_2) - (p_1 - p_2)}{\sqrt{\bar{p}\,\bar{q}\left(\frac{1}{n_1} + \frac{1}{n_2}\right)}}$$

    where $\bar{p} = \dfrac{X_1 + X_2}{n_1 + n_2}$ $\hat{p}_1 = \dfrac{X_1}{n_1}$

    $\bar{q} = 1 - \bar{p}$ $\hat{p}_2 = \dfrac{X_2}{n_2}$

CONTENT

Elementary Statistics provides excellent topic coverage, such as the following:

- Section 1–5 includes a history of probability and statistics and a brief discussion of some of the major contributors. Students will develop an appreciation of the background and development of statistics.
- Section 3–3 provides a detailed explanation of the difference between true variance and the unbiased estimator of the population variance.
- Counting techniques are covered in a full chapter (Chapter 4) rather than as part of the probability discussion. Probability immediately follows in Chapter 5.
- Section 5–7 illustrates how the counting rules are used with the probability rules.
- Chapter 9, "Hypothesis Testing," explains how to translate a stated problem and test a hypothesis. For example, *Elementary Statistics* provides a five-step plan to follow and test the student's hypothesis, including a summary of the results. Also included are clear explanations of the difference between the z test and the t test.
- In Section 13–5 the student is shown how to use the Scheffé and the Tukey tests, which follow the analysis of variance.
- Chapter 14, "Nonparametric Statistics," provides a clear explanation of when to use nonparametric tests and discusses the advantages and disadvantages of using these tests.

- Chapter 15 is a complete chapter on sampling techniques and provides many details and examples.
- Simulation techniques found in Sections 15–4 and 15–5 demonstrate how probability and random numbers can be used to create conditions similar to real-life situations.

NEW FEATURES

In response to the many users and reviewers, the following new material has been added to the second edition of the textbook.

Section 1–4: Data Collection and Sampling Techniques

This section has been expanded to include an explanation of the different ways in which data are collected (mail, telephone, etc.) and the four basic types of sampling techniques. For more in-depth coverage of sampling, instructors can refer to Chapter 15 at this time.

Section 2–3: Histograms, Frequency Polygons, and Ogives

Histograms, frequency polygons, and ogives using relative frequencies have been included in this section.

Section 3–3: Measures of Variation

Chebyshev's theorem and the empirical (normal) rule have been included in this section to help students better understand the concept of standard deviation.

Section 3–4: Measures of Position

An explanation of *outliers* has been added to this section to explain how they are identified and what to do with them when they are encountered in data collection and analysis.

Section 5–5: Conditional Probability and Bayes's Theorem

Conditional probability has been placed in this section and integrated with an explanation of Bayes's theorem.

Section 6–5: Other Types of Distributions

This section combines three subsections on the multinomial, Poisson, and hypergeometric distributions to give a smooth flow to the material.

Section 9–6: Additional Topics Regarding Hypothesis Testing

This new optional section includes a discussion of the *P*-value, type II error, power of a test, and practical significance; and it is intended to give students a more detailed knowledge of the concepts of hypothesis testing. Also, the relationship between confidence intervals and hypothesis testing is explained in this section.

Section 11–7: Other Types of Correlation Coefficients

This section gives a brief summary of other types of correlation coefficients, such as the biserial and tetrachoric, that are used in statistics. It is intended to show students that many different types of correlation coefficients are used and found in the statistical literature.

Section 13–5: The Scheffé and the Tukey Tests

The Tukey test has been included in this section as an additional technique to use after ANOVA.

Section 13–6: Two-Way Analysis of Variance

This new section explains the use of the two-way ANOVA technique and describes how the summary table can be used.

Section 15–3: Writing the Research Report

This new section has been included to explain to students how statistical research reports, theses, and dissertations are written.

In addition to the new sections, this edition of the textbook also includes a new chapter on quality control. Chapter 16 explains the purpose of quality control and how to construct and interpret the $\overline{X}$, R, $\overline{p}$, and $\overline{c}$ charts.

The chapter on confidence intervals and estimation (formerly Chapter 11) has been moved to precede the chapter on hypothesis testing (Chapter 9). However, it has been written so that the instructor can teach it before or after hypothesis testing.

The material on sampling and simulation was moved from Chapter 7 of the first edition to Chapter 15 of this edition. However, it can be taught with Chapter 1 or used anywhere that the instructor feels it would best suit his or her purpose.

Over 40 new "Speaking of Statistics" features have been included in this edition to show students how statistics are used in the real world. These sections can be employed as discussion topics to encourage critical thinking and divergent thinking about statistics students may find in newspapers and magazines. Critical-thinking problems can also be found in many of the exercises marked with asterisks (*).

More attention has been given in this edition, throughout the latter half of the textbook, to sample sizes and to the assumptions used in statistical tests. Furthermore, over 25% of the exercises are new.

Instructors can also help students to see the relationship between descriptive and inferential statistics by having the students do the * exercises that give individual data values, such as Exercises 9–28, 9–29, 9–31, 9–49, 9–50, 10–16, 10–32. These problems require students to compute the descriptive values, such as means and standard deviations, and then to test hypotheses involving these values.

Finally, the author believes that students have a better understanding of the nature of statistics if they actually see how various statistics are computed from data values rather than just reading them from computer printouts. Thus, in order to enable students to analyze and interpret data, the author has included "Data Analysis: Applying the Concepts" sections at the end of the chapters.

SUPPLEMENTS TO THE BOOK

Elementary Statistics provides valuable support material for the instructor and the student. An instructor may choose the *Instructor's Edition, Instructor's Solutions Manual,* and Esatest testing software. The student may choose the *Student Study Guide* by Pat Foard, the *Student's Solutions Manual,* or the Minitab Workbook by Lothar Dohse.

ACKNOWLEDGMENTS

I would like to thank the following people and companies for granting permission to reprint their statistical tables:

CRC Press, Inc.
Addison-Wesley Publishing Company, Inc.
Benjamin/Cummings Publishing Company
Prentice-Hall, Inc.
Institute of Mathematical Statistics

I am also grateful to the many authors and publishers who granted me permission to use their articles and cartoons.

Special thanks go to the staff of Wm. C. Brown Publishers, especially Earl McPeek, Publisher, Mathematics, and Jane Parrigin, Developmental Editor.

The following reviewers offered many helpful suggestions and comments:

Lisa Carrington, *Belleville Area College*
Nick Duerlinger, *Community College of Allegheny County*
Pam Jackson, *Oakland Community College*
Gael Mericle, *Mankato State University*
Ron Pierce, *Eastern Kentucky University*
Ronald Schwartz, *Wilkes University*
Jamal Shahin, *Salem State College*
Jim Walker, *American River College*
June Miller White, *St. Petersburg Junior College*

I would like to thank the following people for their accuracy check of the manuscript and the typeset galleys:

Paul Blankenship, *Lexington Community College*
Dr. Julia Brown, *Atlantic Community College*
Eugene Mastroianni
Ruth L. Mikkelson, *University of Wisconsin–Stout*
Barbara Van Fassen, *Jefferson Technical College*

I would like to acknowledge Minitab, Inc.'s cooperation in the preparation of this textbook. MINITAB is a registered trademark of:

Minitab, Inc.
3081 Enterprise Drive
State College, PA 16801
Phone: (814) 238–3280   Telex: 881612   FAX: (814) 238–4383

Finally, I would like to express my sincere appreciation to the many people who are in part responsible for this book. Special thanks to Betty Claire Bluman for her helpful suggestions, editorial assistance, and typing of the manuscript. Also, I would like to thank Dr. Louis A. Pingel of the University of Pittsburgh for his detailed review of the final manuscript of the first edition.

ACCURACY ASSURANCE

The author and the publisher acknowledge the fact that inaccuracies can cause frustrations for the instructor and students. Therefore, throughout the writing and production of this textbook we have worked diligently to eliminate errors. We would like to take this opportunity to describe the process used to assure the accuracy of this textbook.

Eugene Mastroianni checked the accuracy of all textual examples, solutions, and problems in the final manuscript. Dr. Julia Brown also verified the accuracy of the examples and solutions and worked all the odd-numbered problems to assure the accuracy of the answer appendix. Corrections were made to the manuscript before it was typeset.

Nick Duerlinger checked the accuracy of the *Instructor's Solutions Manual*. This served as another check of all the exercises and the answer appendix.

The galleys were proofread against the manuscript to allow for the correction of any errors introduced when the manuscript was typeset. The accuracy of all examples, solutions, odd-numbered problems, and true/false problems were again checked for accuracy after the manuscript was typeset. Paul Blankenship and Eugene Mastroianni checked the odd-numbered chapters while Barbara Van Fassen, Diana Kovalik, and Ruth Mikkelson checked the even-numbered chapters.

We are aware that it is not possible to guarantee that a textbook is error free. However, every effort has been made to minimize the number of errors in this textbook.

All examples and exercises in this textbook (unless cited) are hypothetical and are presented to enable students to achieve a basic understanding of the statistical concepts discussed. These examples and exercises should not be used in lieu of medical, psychological, or other professional advice. Neither the author nor the publisher shall be held responsible for any misuse of the information presented in this textbook.

CHAPTER

1

The Nature of Probability and Statistics

1-1 INTRODUCTION

Most people become familiar with probability and statistics through radio, television, newspapers, and magazines. For example, one may find the following statements in a newspaper or magazine.

"The median price of homes sold in February was $129,200 nationwide."[1]

"Nearly one in 10 people in the USA wears contact lenses."[2]

"Losing weight and cutting back on salt appear to be highly effective at lowering blood pressure—and ultimately cutting stroke and heart disease risk, a major study suggests."[3]

"Among older men, the mortality rate for smokers is twice the rate of those who have never smoked."[4]

"Half of the adults in a nationwide poll released this week said they would choose the same career if they could make choices in their life over again."[5]

"According to a Roper poll, 75% of American homes have a VCR; but only 58% of owners say they ever tape a show."[6]

Statistics is used in almost all fields of human endeavor. In sports, for example, a statistician may keep records of the number of yards a running back gains during a football game, or the number of hits a baseball player obtains in a season. In other areas, such as public health, an administrator would be concerned with the number of residents who contract a new strain of flu virus during a certain year. In education, the researcher might want to know if new methods of teaching are better than old ones. These are only a few examples of how statistics can be used in various occupations.

Furthermore, statistics is used to analyze the results of surveys and as a tool in scientific research to make decisions based on controlled experiments. Other uses of statistics include operations research, quality control, estimation, and prediction.

Students study statistics for several reasons:

1. The student as well as the professional person must be able to read and understand the various statistical studies performed in his or her field. To have this understanding, the person must be knowledgeable about the vocabulary, symbols, concepts, and statistical procedures used in these studies.
2. The student or professional person might be called upon to conduct research in his or her field; and since statistical procedures are basic to all research, this statistical knowledge will be necessary.
3. In doing statistical studies, the researcher must be able to summarize the data, draw general conclusions, and, if necessary, make reliable predictions or forecasts for future use.

It is the purpose of this chapter to introduce the student to the basic concepts of probability and statistics by answering some questions like the following:

What are the branches of statistics?
What are data?
How did probability and statistics originate?

[1]USA TODAY, March 27, 1992. Reprinted with permission.
[2]USA TODAY, March 4, 1992. Reprinted with permission.
[3]USA TODAY, March 4, 1992. Reprinted with permission.
[4]© 1992, *AARP Bulletin*. Reprinted with permission.
[5]AP September 3, 1992, *The Daily News*. Reprinted with permission.
[6]Reprinted with permission from *TV Guide*® magazine. Copyright © 1992 by News America Publications Inc.

Speaking of **Statistics**

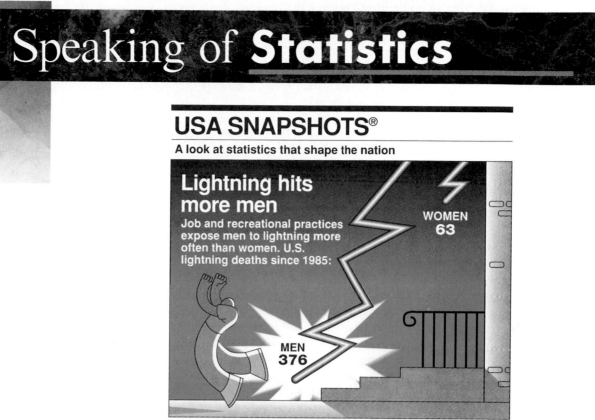

USA SNAPSHOTS®

A look at statistics that shape the nation

Lightning hits more men

Job and recreational practices expose men to lightning more often than women. U.S. lightning deaths since 1985:

WOMEN
63

MEN
376

Source: National Weather Service   By Keith Carter, USA TODAY

Source: Based on the National Weather Service. Copyright 1992, USA TODAY. Reprinted with permission.

The National Weather Service keeps records on weather from year to year. Discuss how the information might be used by an insurance company.

1-2 DESCRIPTIVE AND INFERENTIAL STATISTICS

In order to gain information about seemingly haphazard events, statisticians study random variables.

A variable is a characteristic or attribute that can assume different values. Height, weight, temperature, number of phone calls received, etc., are examples of variables. Variables whose values are determined by chance are called random variables

---

By studying random variables over a long period of time, statisticians are able to determine various patterns that enable them to gain information about the variables. Suppose that an insurance company studies its records over the past several years and determines that, on average, 3 out of every 100 automobiles the company insured were involved in an accident during a one-year period. Although there is no way to predict the specific automobiles that will be involved in an accident (random occurrence), the company can adjust its rates accordingly, since the company knows the general pattern over the long run. (That is, on average, 3% of the insured automobiles will be involved in an accident each year.)

The measurements or observations (values) for a variable are called **data,** and a collection of data values forms a **data set.** Each value in the data set is called a **data value** or a **datum.**

For example, suppose a researcher selects a specific day and records the number of calls a local office of the Internal Revenue Service receives each hour. The data set becomes

$$8, 10, 12, 12, 15, 11, 13, 6$$

since 8 calls were received the first hour, 10 calls the second hour, etc.

Data can be used in different ways. Depending on how data are used, the body of knowledge called statistics is sometimes divided into two main areas.

1. Descriptive statistics
2. Inferential statistics

In *descriptive statistics* the statistician tries to describe a situation. Consider the national census conducted by the United States government every ten years. Results of this census give the average age, income, and other characteristics of the U.S. population. In order to obtain this information, the Census Bureau must have some means to collect relevant data. Once data are collected, the bureau must organize and summarize the data. Finally, the bureau needs a means of presenting the data in some meaningful form, such as charts, graphs, or tables.

Descriptive statistics consists of the collection, organization, summarization, and presentation of data.

The second area of statistics is called *inferential statistics.* Here, the statistician tries to make inferences from *samples* to *populations.* Inferential statistics uses *probability*—i.e., the chance of an event occurring. Most people are familiar with the concepts of probability through various forms of gambling. People who play cards, dice, bingo, and lotteries are using the concepts of probability. In addition to being used in gambling, probability theory is also used in the insurance industry and other areas.

It is important to distinguish between a sample and a population.

A population is the totality of all subjects possessing certain common characteristics that are being studied.

In a statistical study the researcher must define the population being studied. If the researcher wishes to study the effects of smoking on cholesterol level, the researcher may define the population as all adult residents of Allegheny County who smoke at least one pack of cigarettes a day. The common characteristics then would be adults, residents of Allegheny County, and smokers who smoke at least one pack of cigarettes per day. In most cases, the population under study is usually very large, and it would be difficult and time-consuming to use all members; therefore, a sample must be selected.

Speaking of **Statistics**

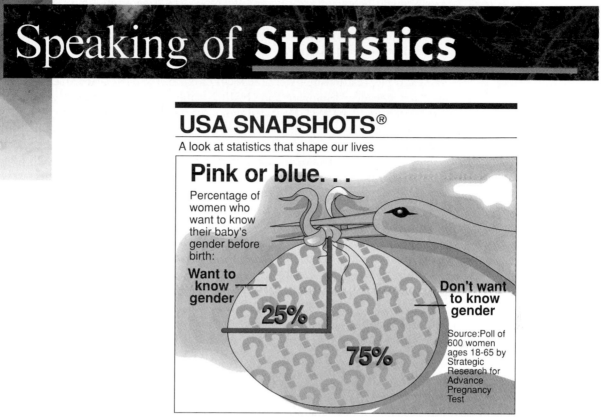

By Julie Stacey, USA TODAY

Source: Based on a poll of 600 women ages 18–65 by the Strategic Research for Advance Pregnancy Test. Copyright 1992, USA TODAY. Reprinted with permission.

Speculate about how the population would be defined. What about the respondents who were undecided?

A sample is a subgroup or subset of the population.

Since the population of adult smokers who reside in Allegheny County is large, the researcher may want to select a sample consisting of some reasonable number of smokers for the study. Statisticians use several methods to obtain samples. The four basic methods, random, systematic, stratified, and cluster sampling, are explained in Section 1–4.

After the smoking/cholesterol study is completed on the sample, the statistician may then infer, on the basis of the study results, that smoking either increases, decreases, or has no effect on the cholesterol levels of the smokers in the general population.

This area of inferential statistics is called hypothesis testing. **Hypothesis testing** is a decision-making process for evaluating claims about a population, based on information obtained from samples. In another instance, a researcher may wish to know if a new drug will reduce the number of heart attacks in men over 70 years of age. For this study, two groups of men over 70 would be selected. One group would be given the drug, and the other would be given a placebo (a substance with no medical benefits or damages). Later, the number of heart attacks occurring in each group of men would be counted, a statistical test would be run, and a decision would be made about the effectiveness of the drug.

Statisticians also use statistics to determine *relationships* among variables. For example, relationships were the focus of one of the most noted studies in the past few decades, "Smoking and Health," published by the surgeon general of the United States in 1964. He stated that after reviewing and evaluating the data, his group found a definite relationship between smoking and lung cancer. He did not say that cigarette smoking actually causes lung cancer but that there is a relationship between smoking and lung cancer. This conclusion was based on a study done in 1958 by Hammond and Horn. In this study, 187,783 men were observed over a period of 45 months. The death rate from lung cancer in this group of volunteers was ten times as great for smokers as for nonsmokers.

Finally, by studying past and present data and conditions, statisticians try to make predictions based on this information. For example, an automobile dealer may look at past sales records for a specific month in order to decide what types of automobiles and how many of each type to order for that month next year.

Inferential statistics consists of generalizing from samples to populations, performing hypothesis testing, determining relationships among variables, and making predictions.

1–3 VARIABLES AND TYPES OF DATA

As stated in the previous section, statisticians gain information about a particular situation by collecting data for random variables. This section will explore in more detail the nature of variables and types of data.

Variables can be classified as qualitative or quantitative. **Qualitative variables** are variables that can be placed into distinct categories, according to some characteristic or attribute. For example, if subjects are classified according to gender (male or female), then the variable "gender" is qualitative. Other examples of qualitative variables are religious preference and geographic locations.

Quantitative variables are numerical in nature and can be ordered or ranked. For example, the variable "age" is numerical, and people can be ranked in order according to the value of their ages. Other examples of quantitative variables are heights, weights, and body temperatures.

Quantitative variables can be further classified into two groups, discrete or continuous. *Discrete variables* can be assigned values such as 0, 1, 2, 3, and are said to be countable. Examples of discrete variables are the number of children in a family, the number of students in a classroom, and the number of calls received by a switchboard operator each day for one month.

Discrete variables assume values that can be counted.

Speaking of **Statistics**

Poll: Many Would Choose Same Career

NEW YORK (AP)—Half of the adults in a nationwide poll released this week said they would choose the same career if they could make choices in their life over again.

The 45 percent who said they would choose a different career were almost equally divided among those who would choose one more financially rewarding, more personally rewarding, or both. Five percent didn't know if they would choose differently.

Just 18 percent of the 1,296 people polled by the Roper Organization said they feel their careers are both personally and financially rewarding. Half those polled said the career they have chosen for themselves is more rewarding personally than financially.

Shearson Lehman Brothers, an investment banking company, sponsored the poll as part of a larger study of how people look at their lives, money and "the American dream."

Men were more likely to say they got more financial than personal rewards from their career, 25 percent compared with 18 percent of women.

Looking at the results by occupation, those most likely to want a different career are blue-collar workers, 62 percent, and those least likely are executives and professionals, 36 percent. Money may help explain this difference, since only 22 percent of those with family incomes of more than $100,000 would choose a different career if they could.

About three in 10 of those with weekday jobs said they often work on weekends, and that number rises to four in 10 among unmarried people. Asked to estimate the number of hours per week they work at their jobs, the midpoint among men is 45 hours, compared with 40 hours for women. Those who have more than $100,000 in savings and investments say they work a median of 50 hours.

Source: Associated Press. Reprinted with permission.

Here is an example of a poll taken by the Roper Organization concerning career choices. How would the information from the poll be useful in helping people select careers? Do you feel the information in the last sentence is accurate?

Continuous variables, by comparison, can assume all values between any two specific values. Temperature, for example, is a continuous variable, since the variable can assume all values between any two given temperatures.

Continuous variables can assume all values between any two specific values. They are obtained by measuring.

Since continuous data must be measured, rounding answers is necessary because of the limits of the measuring device. Usually, answers are rounded to the nearest given unit. For example, heights might be rounded to the nearest inch, weights to the nearest ounce, etc. Hence, a recorded height of 73 inches could mean any measure of 72.5 inches up to but not including 73.5 inches. Thus, the boundary of this measure is given as 72.5–73.5 inches. *Boundaries are written for convenience as 72.5–73.5 but understood to mean all values up to but not including 73.5.* Actual data values of 73.5 would be rounded to 74 and would be included in a class with boundaries of 73.5 up to but not including 74.5, written as 73.5–74.5. As another example, if a recorded weight is 86 pounds, the exact boundaries are 85.5 up to but not including 86.5, written as 85.5–86.5 pounds. Table 1–1 helps to clarify this concept. The boundaries of a continuous variable are given in one additional decimal place and always end with the digit 5.

Table 1–1

Recorded Values and Boundaries		
Variable	**Recorded Value**	**Boundaries**
Length	15 centimeters (cm)	14.5–15.5 cm
Temperature	86 degrees Fahrenheit (°F)	85.5°–86.5°
Time	0.43 second (sec)	0.425–0.435 sec
Weight	1.6 grams (g)	1.55–1.65 g

In addition to being classified as qualitative or quantitative, variables can also be classified by how they are categorized, counted, or measured. For example, can the data be organized into specific categories, such as area of residence (rural, suburban, or urban)? Can the data values be ranked, such as first place, second place, etc.? Or are the values obtained from measurement, such as heights, IQs, or temperature? This type of classification—i.e., how variables are categorized, counted, or measured—uses **measurement scales,** and four common types of scales are used: nominal, ordinal, interval, and ratio.

The nominal level of measurement classifies data into mutually exclusive (nonoverlapping), exhaustive categories in which no order or ranking can be imposed on the data.

A sample of college instructors classified according to subject taught (e.g., English, history, psychology, or mathematics) is an example of nominal-level measurement. Survey subjects classified as male or female is another example of nominal-level measurement. No ranking or order can be placed on the data. Residents classified according to ZIP codes is also an example of the nominal level of measurement. Even though numbers are assigned as ZIP codes, there is no meaningful order or ranking. Other examples of nominal-level data are political party (Democratic, Republican, Independent), religion (Lutheran, Jewish, Catholic, Methodist, etc.), and marital status (married, divorced, widowed, separated).

The next level of measurement is called the ordinal level. Data measured at this level can be placed into categories, and these categories can be ordered or ranked. For example, from student evaluations, guest speakers might be ranked as superior, average, or poor. Floats in a homecoming parade might be ranked as first place, second place, etc. *Note that precise measurement of differences in the ordinal level of measurement* does not *exist.* For instance, when people are classified according to their build (small, medium, or large), a large variation exists among the individuals in each class.

The ordinal level of measurement classifies data into categories that can be ranked; however, precise differences between the ranks do not exist.

Examples of ordinal-measured data are letter grades (A, B, C, D, F), rating scales, and rankings.

The third level of measurement is called the interval level. This level differs from the ordinal level in that precise differences do exist between units. For example, many standardized psychological tests yield values measured on an interval scale. IQ is an example of such a variable. There is a meaningful difference of one point between an IQ of 109 and an IQ of 110. Temperature is another example of interval measurement, since there is a meaningful difference of one degree between each unit, such as 72 degrees and 73 degrees. *One property is lacking in the interval scale: There is no true zero.* For example, IQ tests do not measure people who have no intelligence. For temperature, 0 degrees Fahrenheit does not mean no heat.

The interval level of measurement ranks data, and precise differences between units of measure do exist; however, there is no meaningful zero.

The highest level of measurement is called the ratio level. Examples of ratio scales are those used to measure height, weight, area, and number of phone calls received. Ratio scales have differences between units (1 inch, 1 pound, etc.) and a true zero. In addition, the ratio scale contains a true ratio between values. For example, if one person can lift 200 pounds and another can lift 100 pounds, then the ratio between them is 2 to 1. Put another way, the first person can lift twice as much as the second person.

The ratio level of measurement possesses all the characteristics of interval measurement, and there exists a true zero. In addition, true ratios exist between different units of measure.

There is not complete agreement among statisticians about the classification of data into one of the four categories. For example, some researchers classify IQ data as ratio data rather than interval. Also, data can be altered so that they fit into a lower category. For instance, if the incomes of all professors of a college were classified into three categories, low, average, and high, then a ratio variable becomes an ordinal variable. Table 1–2 gives some examples of each type of data.

Table 1-2

Examples of Measurement Scales			
Nominal-Level Data	**Ordinal-Level Data**	**Interval-Level Data**	**Ratio-Level Data**
Zip code Gender (male, female) Eye color (blue, brown, green, hazel) Political affiliation Religious affiliation Major field (mathematics, computers, etc.) Nationality	Grade (A, B, C, D, F) Judging (1st place, 2nd place, etc.) Rating scale (poor, good, excellent) Ranking of tennis players	SAT score IQ Temperature	Height Weight Time Salary Age

1-4 DATA COLLECTION AND SAMPLING TECHNIQUES

In research, statisticians use data in many different ways. As stated previously, data can be used to describe situations or events. For example, a manufacturer might want to know something about the consumers who will be purchasing his product, thus enabling him to plan an effective marketing strategy. In another situation, the management of a company might survey its employees to assess their needs in order to negotiate a new contract with the employees' union. Data can be used to determine whether the educational goals of a school district are being met. Finally, trends in various areas, such as the stock market, can be analyzed, enabling prospective buyers to make more intelligent decisions concerning what stocks to purchase. These examples illustrate a few situations where collecting data will help people make intelligent decisions for courses of action.

Data can be collected in a variety of ways. One of the most common methods of collecting data is through the use of surveys. Surveys can be done by using a variety of methods. Three of the most common methods are the telephone survey, the mailed questionnaire survey, and the personal interview survey.

Telephone surveys have an advantage over personal interview surveys in that they are less costly. Also, people may be more candid in their opinions since there is no face-to-face contact. A major drawback to the telephone survey is that some people in the population will not have phones or will not be home when the calls are made; hence, not all people have a chance of being surveyed.

Mailed questionnaire surveys can be used to cover a wider geographical area than telephone surveys or personal interviews since they are less expensive to conduct. Also, respondents can remain anonymous if desired. Disadvantages of mailed questionnaire surveys include a low number of responses and inappropriate answers to questions. Another drawback is that some people may have difficulty reading or understanding the questions.

Personal interview surveys have the advantage of obtaining in-depth responses to questions from the person being interviewed. One disadvantage is that interviewers must be trained in asking questions and recording responses, which makes the personal interview survey more costly than the other two survey methods. Another disadvantage is that the interviewer might be biased in his or her selection of respondents.

Data can also be collected in other ways, such as *surveying records* or *direct observation* of situations.

As stated in Section 1–2, researchers use samples in order to collect data and information about a particular variable from a large population. Using samples saves time and money and, in some cases, allows the researcher to get more detailed information about a particular subject. Samples cannot be selected in haphazard ways because the information obtained might be biased. For example, interviewing people on a street corner would not include responses from people working in offices at that time or people attending school; hence, all subjects in a particular population would not have a chance of being selected.

In order to obtain samples that are unbiased—i.e., giving each subject in the population a chance of being selected—statisticians use four basic methods of sampling: random, systematic, stratified, and cluster sampling.

Random Sampling

Random samples are selected by using chance methods or random numbers. One such method is to number each subject in the population. Then place numbered cards in a bowl, mix them thoroughly, and select as many cards as needed. The subjects whose numbers are selected constitute the sample. Since it is difficult to mix the cards thoroughly, there is a chance of obtaining a biased sample. For this reason, statisticians use another method of obtaining numbers, namely, random numbers. Random numbers can be generated by a computer or a calculator. Before the invention of computers, random numbers were obtained from a table of random numbers, similar to the one shown in Table 1–3. A more detailed explanation of random numbers is given in Chapter 15.

Table 1–3

Table of Random Numbers												
79	41	71	93	60	35	04	67	96	04	79	10	86
26	52	53	13	43	50	92	09	87	21	83	75	17
18	13	41	30	56	20	37	74	49	56	45	46	83
19	82	02	69	34	27	77	34	24	93	16	77	00
14	57	44	30	93	76	32	13	55	29	49	30	77
29	12	18	50	06	33	15	79	50	28	50	45	45
01	27	92	67	93	31	97	55	29	21	64	27	29
55	75	65	68	65	73	07	95	66	43	43	92	16
84	95	95	96	62	30	91	64	74	83	47	89	71
62	62	21	37	82	62	19	44	08	64	34	50	11
66	57	28	69	13	99	74	31	58	19	47	66	89
48	13	69	97	29	01	75	58	05	40	40	18	29
94	31	73	19	75	76	33	18	05	53	04	51	41
00	06	53	98	01	55	08	38	49	42	10	44	38
46	16	44	27	80	15	28	01	64	27	89	03	27
77	49	85	95	62	93	25	39	63	74	54	82	85
81	96	43	27	39	53	85	61	12	90	67	96	02
40	46	15	73	23	75	96	68	13	99	49	64	11

Systematic Sampling

Systematic samples are obtained by numbering each subject of the population and then selecting every kth number. For example, suppose there are 2000 subjects in the population and a sample of 50 subjects is needed. Since $2000 \div 50 = 40$, then $k = 40$, and every 40th subject would be selected; however, the first subject (numbered between 1 and 40) would be selected at random. Suppose subject 12 was the first subject selected; then the sample would consist of the subjects whose numbers were 12, 52, 92, etc., until 50 subjects were obtained. When using systematic sampling, one must be careful about how the subjects in the population are numbered. If subjects were arranged in a manner such as wife, husband, wife, husband, and every 40th subject was selected, the sample would consist of all husbands.

Stratified Sampling

Stratified samples are obtained by dividing the population into groups (called strata) according to some characteristic important to the study. For example, suppose a two-year college has a population of 2000 full-time students, with 60% freshmen and 40% sophomores. Suppose further that 50% of the freshman and sophomore students are males and 50% are females. The population then could be divided into strata of freshmen-sophomore and male-female, as shown in Figure 1–1.

FIGURE 1–1
Example of Stratified
Sampling

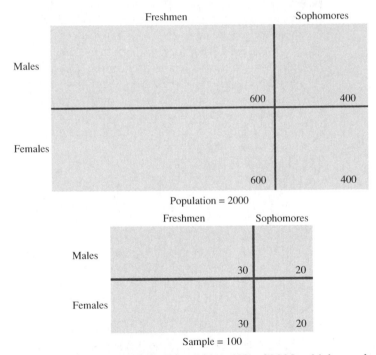

The numbers are obtained by taking 60% of 2000, which equals 1200 freshmen, and 40% of 2000, which is 800. Then half of each group is male and half is female; hence, there are 600 male freshmen, 600 female freshmen, 400 male sophomores, and 400 female sophomores.

If the researcher wanted to use a sample of 100 students, 30 would be selected at random from the freshmen males, 30 from the freshmen females, 20 from the sophomore males, and 20 from the sophomore females. Since these numbers are in the same proportion as in the population, the researcher would have a miniature population of two variables: freshmen-sophomore and male-female.

One advantage of stratified sampling is that it ensures that various strata of the population are represented in the sample. A disadvantage is that the process becomes very tedious if a large number of characteristics or strata are desired, for example, age, income, class rank, religion, or political affiliation.

Cluster Sampling

Cluster samples are selected by using intact groups called clusters. Suppose a researcher wishes to survey apartment dwellers in a large city. If there are ten apartment buildings in the city, the researcher could select at random two buildings from the ten and interview all the residents of these buildings. Cluster sampling is used when the population is large or when it involves subjects residing in a large geographic area. For example, if one wanted to do a study involving the patients in the hospitals in New York City, it would be very costly and time-consuming to try to obtain a random sample of patients since they would be spread over a large area. Instead, a few hospitals could be selected at random and the patients in these hospitals would be interviewed in a cluster.

A more detailed explanation of these sampling techniques is presented in Chapter 15.

1–5 HISTORY OF PROBABILITY AND STATISTICS

Probability and statistics have an interesting history of development. Probability theory had its origins in gambling. Although no one knows how old gambling really is, cubical dice made of bones and pottery have been found in ancient Greek and Egyptian tombs.

There is evidence that the Greek scientist Aristotle (384–322 B.C.) investigated simple probability concepts; however, not much else was done until the sixteenth century when Jerome Cardan (1501–1576) wrote the first book on probability, *The Book on Chance and Games.* Cardan was an astrologer, philosopher, physician, mathematician, and gambler. In his book on gambling, he included techniques on how to cheat and how to catch others at cheating. He also correctly listed the number of outcomes when two dice are rolled.

During the mid 1600s, a professional gambler named Chevalier de Méré made a considerable amount of money on a gambling game. He would bet unsuspecting patrons that in 4 rolls of a die he could get a 6 at least once. He was so successful at the game that soon people refused to play. He decided that a new game was necessary in order to continue his winnings. By reasoning, he figured he could roll two 6s in 24 rolls of a die; but his reasoning was incorrect, and he systematically lost. Unable to figure out why, he contacted Blaise Pascal (1623–1662), who was a famous mathematician at that time.

Pascal became interested and began studying probability theory. He corresponded with a French government official, Pierre de Fermat (1601–1665), whose hobby was mathematics. Together, the two formulated the beginnings of probability theory. Today, probability theory has applications in biology, physics, psychology, economics, and insurance, as well as many other areas.

The origin of descriptive statistics can be traced to the census taken by the Babylonians and Egyptians, 4500–3000 B.C. In addition, Roman Emperor Augustus (27 B.C.–17 A.D.) conducted surveys on births and deaths of the citizens of the empire as well as the amount of livestock each owned and the crops each harvested. In order to use this information, the Romans had to develop methods of collecting, organizing, and summarizing data.

During the fourteenth century, people began keeping records on births, deaths, and accidents in order to determine insurance rates. John Graunt (1620–1674) studied the number of males and females born and discovered that slightly more males were born than females, but that more males than females died during the first year of life. Gregor Mendel (1822–1884) used probability and statistics in his studies of heredity at a monastery in Brunn; Sir Francis Galton (1822–1911) did correlation studies using peas, moths, dogs, and humans. Sir Ronald Fisher

(1890–1962) developed statistical methods in inferential techniques, experimental design, estimation, and analysis of variance. Many new statistical methods have been developed since 1900, and new techniques in research and statistics are being studied and perfected each year.

With the advent of the computer, the statistician can process large amounts of data and do many complex calculations that were impossible a few years ago.

1-6 SUMMARY

The two major areas of statistics are descriptive and inferential. Descriptive statistics includes the collection, organization, summarization, and presentation of data. Inferential statistics includes making inferences from samples to populations, hypothesis testing, determining relationships, and making predictions. Inferential statistics is based on probability theory.

Since in most cases the populations under study are large, statisticians use subgroups called samples to get the necessary data for their studies. There are four basic methods used to obtain samples: random, systematic, stratified, and cluster.

Data can be classified as qualitative or quantitative. Quantitative data can be either discrete or continuous, depending on the values it can assume. Data can also be measured by various scales. The four basic levels of measurement are nominal, ordinal, interval, and ratio.

Finally, the applications of statistics are many and varied and are encountered in everyday life, such as in reading newspapers or magazines, listening to the radio, or watching television. In addition, statistics is used in almost every field of endeavor, and the educated individual should be knowledgeable about the vocabulary, concepts, and procedures of statistics.

LAFF - A - DAY

©1993 King Features Syndicate, Inc. World rights reserved.

"We've polled the entire populace, Your
Majesty, and we've come up with
exactly the results you ordered!"

Cluster sample
Continuous variables
Data
Data set
Data value or datum
Descriptive statistics
Discrete variables
Hypothesis testing

Inferential statistics
Interval level of measurement
Measurement scales
Nominal level of measurement
Ordinal level of measurement
Population
Probability

Qualitative variables
Quantitative variables
Random sample
Random variable
Ratio level of measurement
Sample
Stratified sample
Systematic sample
Variable

EXERCISES

1–1. Name and define the two areas of statistics.

1–2. What is probability? Name two areas where probability is used.

1–3. Suggest some ways statistics can be used in everyday life.

1–4. Explain the differences between a sample and a population.

1–5. Why are samples used in statistics?

1–6. (ANS.) In each statement that follows, tell whether descriptive or inferential statistics has been used.

a. The average age of the students in your psychology class is 20.1 years.
b. Of the students enrolled in Introduction to Engineering at Royal Oak Junior College this semester, 92% are male and 8% are female.
c. Teaching calculus by the computer method is more effective than teaching by the lecture method.
d. In the upcoming election, the pollsters have predicted that an incumbent will receive 63% of the vote and the new candidate will receive 37%.
e. During the Thanksgiving weekend last year, 73 people died in traffic accidents in New York.
f. From past figures, it has been predicted that 37% of registered voters will vote in the November election.
g. There is a relationship between smoking tobacco and an increased risk of developing lung cancer.
h. The chances of your being robbed next year are 3 out of 200.
i. The chances of your winning the Pennsylvania State Lottery on any given day are 1 out of 167,000.
j. According to insurance company figures, the chance of your living to age 83 is 62.8%.

(ANS.) means that the answer for the even-numbered exercise is included in the answer section.

1–7. Classify each as nominal-level, ordinal-level, interval-level, or ratio-level data.

a. Horsepower of motorcycle engines.
b. Ratings of newscasts in Houston (poor, fair, good, excellent).
c. Temperature of automatic popcorn poppers.
d. Time required by drivers to complete a course.
e. Salaries of cashiers of Day-Night grocery stores.
f. Marital status of respondents to a survey on savings accounts.
g. Ages of students enrolled in a martial arts course.
h. Weights of beef cattle fed a special diet.
i. Rankings of weight lifters.
j. Pages in the telephone book for the city of Los Angeles.

1–8. (ANS.) Classify each variable as qualitative or quantitative.

a. Colors of automobiles in a dealer's showroom.
b. Number of seats in movie theaters.
c. Classification of patients based on nursing care needed (complete, partial, or self-care).
d. Lengths of newborn cats of a certain species.
e. Number of complaint letters received by an airline per month.

1–9. Classify each variable as discrete or continuous.

a. Number of cartons of milk manufactured each day.
b. Temperatures of airplane interiors at a given airport.
c. Incomes of college students on work-study programs.
d. Lifetimes of transistors in a stereo set.
e. Weights of newborn calfs.
f. Number of tomatoes on each plant in a field.
g. Number of books in each professor's office on campus.
h. Capacity (in quarts) of water in automobile engines.
i. Bushels of wheat produced in the United States last year.
j. Miles driven by the Port Authority bus drivers each day.

1–10. Give the boundaries for each value.

a. 26.6 inches
b. 8 bushels
c. 0.39 gram
d. 13 milliliters
e. 138 ounces
f. 78 feet

1–11. Name the four basic sampling methods.

1–12. (ANS.) Classify each sample as random, systematic, stratified, or cluster.

a. All police officers of a small borough are interviewed to determine whether they feel the crime rate has changed over the past year.
b. Every fifth teenager entering an amusement park is asked to select his or her favorite ride.
c. School superintendents are selected by using random numbers in order to determine annual salaries.
d. Every hundredth bottle of shampoo manufactured is checked to determine its pH level.
e. Patients who enter an emergency room are divided into groups based on the nature of their injury. These groups are subdivided into age brackets and according to the gender of the individual. A proportional number of patients are selected from each group and are used for the sample.

1–13. Who wrote the first book on probability?

1–14. What two famous mathematicians studied probability theory in the 1600s?

1–15. What person used gambling to make a living and, as a result of incorrectly computing the probability of a game, caused mathematicians to develop formal probability theory?

1–16. What was the origin of probability theory?

1–17. How did descriptive statistics originate?

1–18. What famous monk used probability in his experiments on heredity?

1–19. Give three examples of nominal, ordinal, interval, and ratio data.

1–20. For each of the following statements, define a population and state how a sample can be obtained.

a. A researcher wishes to determine the preference of the voters in the upcoming presidential election.
b. A statistician wishes to find the unemployment rate for the state of Florida.
c. A survey is to be conducted to determine which brands of shampoo adult men use.
d. Police officers will be surveyed to see whether they think that a special defensive-driving course will help to reduce highway accidents.
e. The superintendent of a large school district wishes to determine the reading ability of the students in the school district who enter first grade.

*1–21. Select a statistical study from a newspaper or magazine article. Decide whether it uses descriptive or inferential statistics.

*1–22. Information from research studies is sometimes taken out of context. Explain why the claims of these studies might be suspect.

a. The average salary of the graduates of the class of 1980 is $32,500.
b. It is estimated that in Podunk there are 27,256 rats.
c. Only 3% of the men surveyed read *Cosmopolitan* magazine.
d. Based on a recent mail survey, 85% of the respondents favored gun control.
e. A recent study showed that high school dropouts drink more coffee than students who graduated; therefore, coffee dulls the brain.
f. Since most automobile accidents occur within 15 miles of a person's residence, it is safer to make long trips.

*1–23. Select a newspaper that carries local birth announcements. Keep a record of the ratio of female to male births for one month. Do the figures support John Graunt's findings? Does this "prove" he was correct or incorrect?

*Problems with an asterisk require more in-depth skills than those problems without it.

COMPUTER APPLICATIONS

Computers are used extensively in statistics as well as in most other areas of human endeavor. Throughout this textbook, there are sections entitled "Computer Applications." These sections give examples of how the computer can be used in statistics. There are two types of exercises. The first type provides problems to be used with MINITAB. Check with the computer center at your school to see whether this program is available.

The second type of exercise requires writing original programs to solve various exercises related to the topics discussed in the chapters.

√ DATA ANALYSIS Applying the Concepts

The Data Bank is located in the Appendix.

1. Identify the measurement level (nominal, ordinal, interval, ratio) for each variable in the Data Bank.

✎ TEST

Directions: Determine whether each statement is true or false.

1. Data that can be classified according to color is measured at the interval level.
2. Data such as the percentage of hearing loss a person suffers is measured on the ratio scale.
3. Making predictions based on present data is a part of descriptive statistics.
4. A subgroup of the population is called a small population.
5. The totality of all subjects under study is called a sample.
6. Hypothesis testing is part of inferential statistics.
7. Probability theory had its beginnings when mathematicians studied games of chance.
8. The number of leaves on various trees is an example of a discrete variable.
9. The lengths of trailer trucks that travel an interstate highway is an example of a continuous variable.
10. Descriptive statistics had its origin in the 1600s.

CHAPTER

2

Frequency Distributions and Graphs

2–1 INTRODUCTION

When conducting a statistical study, the researcher must gather data for the particular variable under study. For example, if a researcher wishes to study the number of people who were bitten by poisonous snakes in a specific geographic area over the past several years, he or she would have to gather the data from various doctors, hospitals, or health departments. Being bitten by a snake is a random event. That is, it is due to chance, since the person and the snake just happen to be in the same place at the same time.

In order to describe situations, draw conclusions, or make inferences about random events, the researcher must organize the data in some meaningful way. The most convenient method of organizing data is to construct a *frequency distribution.*

After organizing the data, the researcher must present the data so that it can be understood by those who will benefit from reading the study. The most useful method of presenting the data is by constructing *statistical charts and graphs.* There are many different types of charts and graphs, and each one has a specific purpose.

The purpose of this chapter is to explain how to organize data by constructing frequency distributions and how to present the data by constructing charts and graphs. The charts and graphs illustrated in this chapter include histograms, frequency polygons, ogives, pie graphs, bar graphs, time series graphs, pictographs, and stem and leaf plots.

2–2 ORGANIZING DATA

Suppose a researcher wished to do a study on the number of miles the employees of a large department store traveled to work each day. The researcher would first have to collect the data by asking each employee the approximate distance the store is from his or her home. When data are collected in original form, the data are called **raw data.** In this case, the data are as follows:

1	2	6	7	12	13	2	6	9	5
18	7	3	15	15	4	17	1	14	5
4	16	4	5	8	6	5	18	5	2
9	11	12	1	9	2	10	11	4	10
9	18	8	8	4	14	7	3	2	6

Since little information can be obtained from looking at raw data, the researcher will organize the data by constructing a frequency distribution. The **frequency** is the number of values in a specific class of the distribution. For this data set, a frequency distribution is shown as follows:

Class Limits (in Miles)	Tally	Frequency
1–3	||||‍ ||||‍	10
4–6	||||‍ ||||‍ ////	14
7–9	||||‍ ||||‍	10
10–12	||||‍ /	6
13–15	||||‍	5
16–18	||||‍	5
		Total 50

Now, some general observations can be obtained from looking at the data in the form of a frequency distribution. For example, the majority of employees live within 9 miles of the store.

A frequency distribution is the organization of raw data in table form, using classes and frequencies.

The classes in this distribution are 1–3, 4–6, etc. These values are called class limits, and the data values 1, 2, 3 can be tallied in the first class, 4, 5, 6 in the second class, etc.

There are three basic types of frequency distributions, and there are specific procedures for constructing each type. The three types are *categorical, ungrouped, and grouped frequency distributions*. The next three subsections show each type and the procedures for constructing each.

Categorical Frequency Distributions

The categorical frequency distribution is used for data which can be placed in specific categories such as nominal or ordinal level data. For example, data such as political affiliation, religious affiliation, or major field of study would use categorical frequency distributions.

EXAMPLE 2–1 Twenty-five army inductees were given a blood test to determine their blood type. The data set is as follows:

A	B	B	AB	O
O	O	B	AB	B
B	B	O	A	O
A	O	O	O	AB
AB	A	O	B	A

Construct a frequency distribution for the data.

Solution Since the data are categorical, discrete classes can be used. There are four blood types: A, B, O, and AB. These types will be used as the classes for the distribution.

The procedure for constructing a frequency distribution for categorical data is given next.

STEP 1 Make a table as shown.

A Class	B Tally	C Frequency	D Percent
A			
B			
O			
AB			

STEP 2 Tally the data and place the results in column B.

STEP 3 Count the tallies and place the results in column C.

STEP 4 Find the percentage of values in each class by using the formula

$$\% = \frac{f}{n} \cdot 100\%$$

where

f = frequency of the class
n = total number of values

For example, in the class of type A blood, the percentage is

$$\% = \frac{5}{25} \cdot 100\% = 20\%$$

Percentages are not normally a part of a frequency distribution, but they can be added since they are used in certain types of graphical presentations, such as pie graphs.

STEP 5 Find the totals for columns C and D (see the completed table that follows).

Class	Tally	Frequency	Percent
A	卄〃	5	20
B	卄〃 //	7	28
O	卄〃 ////	9	36
AB	////	4	16
	Total	25	100

Ungrouped Frequency Distributions

When the data are numerical instead of categorical, the procedure for constructing a frequency distribution is somewhat more complicated, as shown in the next two examples. Example 2-2 describes the procedure for constructing an ungrouped frequency distribution.

EXAMPLE 2-2 A psychologist administered the Grayback Test of Manual Dexterity to 25 third-grade students. The times, in minutes, required to complete the test are given below. Construct a frequency distribution for the data.

4	8	8	9	8
5	9	9	10	11
7	7	8	7	8
4	8	7	5	7
6	5	10	8	9

Solution **STEP 1** Find the range of the data. The **range** R is defined as

$$R = \text{highest value} - \text{lowest value}$$

For this data set, the range is $11 - 4$, or 7. Since the range is small, classes consisting of a single data value can be used, and they are 4, 5, 6, 7, 8, 9, 10, and 11.

STEP 2 Make a table as shown below.

STEP 3 Tally the data.

STEP 4   Complete the frequency column.

Class	Tally	Frequency
4	//	2
5	///	3
6	/	1
7	////	5
8	//// //	7
9	////	4
10	//	2
11	/	1

◄

The values of the data set are continuous (i.e., are minutes). Recall that in Chapter 1 a value of 4 minutes could theoretically mean any number from 3.5 up to 4.5 minutes; hence, boundaries for each class can be constructed by subtracting 0.5 from each class value and adding 0.5 to each class value, as shown next.

Class	Class Boundaries	Tally	Frequency
4	3.5–4.5	//	2
5	4.5–5.5	///	3
6	5.5–6.5	/	1
7	6.5–7.5	////	5
8	7.5–8.5	//// //	7
9	8.5–9.5	////	4
10	9.5–10.5	//	2
11	10.5–11.5	/	1

A cumulative frequency (cf) can be added to the frequency distribution shown above by adding the frequency in each class to the total of the frequencies of the classes above that class, as shown below.

Class	Class Boundaries	Frequency	Cumulative Frequency
4	3.5–4.5	2	$0 + 2 = 2$
5	4.5–5.5	3	$2 + 3 = 5$
6	5.5–6.5	1	$5 + 1 = 6$
7	6.5–7.5	5	$6 + 5 = 11$
8	7.5–8.5	7	$11 + 7 = 18$
9	8.5–9.5	4	$18 + 4 = 22$
10	9.5–10.5	2	$22 + 2 = 24$
11	10.5–11.5	1	$24 + 1 = 25$

Cumulative frequencies are used to show how many values are accumulated up to and including a specific class. For example, 18 students successfully completed the test in 8 minutes or less. Twenty-four students completed the test in 10 minutes or less.

Grouped Frequency Distributions

When the range of the data is large, the data must be grouped into classes that are more than one unit in width. For example, a distribution of the number of hours boat batteries lasted is as follows:

Class Limits	Class Boundaries	Tally	Frequency	Cumulative Frequency
24–30	23.5–30.5	///	3	3
31–37	30.5–37.5	/	1	4
38–44	37.5–44.5	⁺⁺⁺⁻	5	9
45–51	44.5–51.5	⁺⁺⁺⁻ ////	9	18
52–58	51.5–58.5	⁺⁺⁺⁻ /	6	24
59–65	58.5–65.5	/	1	25
			25	

The procedure for constructing the grouped frequency distribution above is given in the next example; however, several things should be noted. First, class limits are more than one unit in width. In this example, the limits of the first class are 24–30, meaning that all values from 24 through 30 will be tallied in this class.

Second, since the data are continuous and have been rounded, all measured values from 23.5 up to 30.5 will be tallied in this class; hence, the class boundaries are as shown in column 2. In the first class, the lower value 24 is called the **lower class limit,** and the upper value 30 is called the **upper class limit.** Also, 23.5 is called the **lower class boundary,** and 30.5 is called the **upper class boundary.**

Students sometimes have difficulty finding class boundaries when given the class limits. *The basic rule of thumb is that the class limits should have the same decimal place value as the data, but the class boundaries have one additional place value and end in a 5 digit.* For example, if the values in the data set are whole numbers, such as 24, 32, 18, the limits for a class might be 31–37, and the boundaries are 30.5–37.5. The boundaries are obtained by subtracting 0.5 from 31 (the lower class limit) and adding 0.5 to 37 (the upper class limit).

(lower limit)	$31 - 0.5 = 30.5$	(lower boundary)
(upper limit)	$37 + 0.5 = 37.5$	(upper boundary)

If the data are in tenths, such as 6.2, 7.8, 12.6, the limits for a class hypothetically might be 7.8–8.8, and the boundaries for that class will be 7.75–8.85. These values are obtained by subtracting 0.05 from 7.8 and adding 0.05 to 8.8.

Finally, the **class width** for a class in the frequency distribution is found by subtracting the lower class boundary from the upper class boundary:

class width = upper boundary − lower boundary

In the example above, the width of the first class is

$$30.5 - 23.5 = 7 \text{ units}$$

If the limits are subtracted (30 − 24 = 6 units), an incorrect answer is obtained, since all values, 24, 25, 26, 27, 28, 29, or 30, are tallied in the class 24–30.

The researcher must decide how many classes to use and the width of each class. For the construction of a frequency distribution, the rules given next should be followed.

1. *There should be between 5 and 20 classes.* A student would not be in error for having less than 5 classes or more than 20 classes; however, statisticians generally agree on these numbers.

2. *The class width should be an odd number.* The reason for making the class width an odd number is to ensure that the midpoints of each of the classes have the same place value as the data. The **class midpoint** X_m is obtained by adding the lower and upper boundaries and dividing by 2, or adding the lower and upper limits and dividing by 2:

$$X_m = \frac{\text{lower boundary} + \text{upper boundary}}{2}$$

or

$$X_m = \frac{\text{lower limit} + \text{upper limit}}{2}$$

For example, the midpoint of the first class is

$$\frac{24 + 30}{2} = 27 \qquad \text{or} \qquad \frac{23.5 + 30.5}{2} = 27$$

The midpoint is the numerical location of the center of the class. Midpoints are necessary for graphing (see Section 2–3) and are used in computation of the mean and standard deviation (see Sections 3–2 and 3–3). If the class width is an even number, the midpoint is in tenths. For example, if the class width is 6 and the boundaries are 5.5–11.5, the midpoint is

$$\frac{5.5 + 11.5}{2} = \frac{17}{2} = 8.5$$

3. *The classes must be mutually exclusive.* Mutually exclusive classes have nonoverlapping class limits so that data cannot be placed into two classes. Many times, frequency distributions such as

Age
10–20
20–30
30–40
40–50

are found in the literature or in surveys. If a person is 40 years old, into which class should he or she be placed? A better way to construct a frequency distribution is to use classes such as

Age
10–20
21–31
32–42
43–53

4. *The classes must be continuous.* Even if there are no values in a class, the class must be included in the frequency distribution. There should be no gaps in a frequency distribution. The only exception occurs when the class with a zero frequency is the first or last class. A class with a zero frequency at either end can be omitted without affecting the distribution.

5. *Classes must be exhaustive.* There should be enough classes to accommodate all the data.

6. *The classes must be equal in width.* The reason for having classes with equal widths is so that there is not a distorted view of the data.

One exception occurs when a distribution is **open-ended**—i.e., it has no specific beginning value or no specific ending value. Following are the class limits for two open-ended distributions.

Age	Minutes
10–20	Below 110
21–31	110–114
32–42	115–119
43–53	120–124
54–64	125–129
65 and above	130–134

The frequency distribution for age is open-ended for the last class, which means that anybody who is 65 years or older would be tallied in the last class. The distribution for minutes is open-ended for the first class, meaning that any minute values below 110 would be tallied in that class.

Example 2–3 shows the procedure for constructing a grouped frequency distribution, i.e., when the classes contain more than one data value.

EXAMPLE 2–3 The following data set was obtained from a survey of 20 sales representatives of the Memory Lane Card Company. Each value represents the distance, in miles, that the representatives traveled to meet their clients on a randomly selected day of the week. Construct a frequency distribution for the data, using 7 classes.

11	29	6	33	14
21	18	17	22	38
31	22	27	19	22
23	26	39	34	27

Solution The procedure for constructing a grouped frequency distribution for numerical data follows.

STEP 1 Find the highest value and lowest value: $H = 39$ and $L = 6$.

STEP 2 Find the range: $R = $ highest value $-$ lowest value.

$$R = 39 - 6 = 33$$

STEP 3 Select the number of classes desired (usually between 5 and 20); in this case, 7 is arbitrarily chosen.

STEP 4 Find the class width by dividing the range by the number of classes.

$$\text{width} = \frac{R}{\text{number of classes}} = \frac{33}{7} = 4.7$$

Round the answer up to the nearest whole number if there is a remainder: $4.7 \approx 5$. (Rounding up is different from rounding off. A number is rounded up if there is any decimal remainder when dividing. For example, $85 \div 6 = 14.167$ and is rounded up to 15. Also, $53 \div 4 = 13.25$ and is rounded up to 14.)

STEP 5 Select a starting point as the lowest class limit (this is usually the lowest score). Add the width to that score to get the lower limit of the next class. Keep adding until there are 7 classes, as shown.

  6
 11
 16
 21
 26
 31
 36

STEP 6 Subtract one unit from the lower limit of the second class to get the upper limit of the first class; then add the width to each upper limit to get all the upper limits.

$$11 - 1 = 10$$

So the first class is 6–10.

Class Limits
  6–10
 11–15
 16–20
 21–25
 26–30
 31–35
 36–40

STEP 7 Find the class boundaries by subtracting 0.5 from each lower class limit and adding 0.5 to the upper class limit, as shown.

Class Boundaries
  5.5–10.5
 10.5–15.5
 15.5–20.5
 20.5–25.5
 25.5–30.5
 30.5–35.5
 35.5–40.5

STEP 8 Tally the data.

STEP 9 Write the numerical values for the tallies in the frequency column.

STEP 10 Find the cumulative frequencies.

The completed frequency distribution follows.

Class Limits	Class Boundaries	Tally	Frequency	Cumulative Frequency
6–10	5.5–10.5	/	1	1
11–15	10.5–15.5	//	2	3
16–20	15.5–20.5	///	3	6
21–25	20.5–25.5	̶/̶/̶/̶/̶/	5	11
26–30	25.5–30.5	////	4	15
31–35	30.5–35.5	///	3	18
36–40	35.5–40.5	//	2	20

In this section three types of frequency distributions were shown. The first type, shown in Example 2–1, is used when the data is categorical (nominal), such as blood type or political affiliation. This type is called a **categorical distribution.** The second type of distribution is used for numerical data and when the range of data is small, as shown in Example 2–2. Since each class is only one unit, this distribution is called an **ungrouped frequency distribution.** The third type of distribution is used when the range is large and classes of several units in width are needed. This type is called a **grouped frequency distribution** and is shown in Example 2–3. All three types of distributions are used frequently in statistics and are helpful when one is organizing and presenting data.

The reasons for constructing a frequency distribution are as follows:

1. To organize the data in a meaningful, intelligible way.
2. To enable the reader to determine the nature or shape of the distribution.
3. To facilitate computational procedures for measures of average and spread (shown in Sections 3–2 and 3–3).
4. To enable the researcher to draw charts and graphs for the presentation of data (shown in Section 2–3).
5. To enable the reader to make comparisons between different data sets.

The procedure for constructing a grouped frequency distribution is summarized in Procedure Table 1.

Procedure Table 1

Constructing a Grouped Frequency Distribution

1. Find the highest and lowest values.
2. Find the range.
3. Select the number of classes desired.
4. Find the width by dividing the range by the number of classes and rounding up.
5. Select a starting point (usually the lowest value); add the width to get the lower limits.
6. Find the upper class limits.
7. Find the boundaries.
8. Tally the data.
9. Find the frequencies.
10. Find the cumulative frequency.

When one is constructing a frequency distribution, the guidelines presented in this section should be followed. However, one can construct several different but correct frequency distributions for the same data by using a different class width, a different number of classes, or a different starting point.

EXERCISES

2–1. List five reasons for organizing data into a frequency distribution.

2–2. How many classes should frequency distributions have?

2–3. Find class boundaries, midpoints, and widths for each class.

   a. 11–15
   b. 17–39
   c. 293–353
   d. 11.8–14.7
   e. 3.13–3.93

2–4. Why should the class width be an odd number?

2–5. Shown below are four frequency distributions. Each is incorrectly constructed. State the reason why.

	Class	Frequency
a.	27–32	1
	33–38	0
	39–44	6
	45–49	4
	50–55	2

	Class	Frequency
b.	5–9	1
	9–13	2
	13–17	5
	17–20	6
	20–24	3

	Class	Frequency
c.	123–127	3
	128–132	7
	138–142	2
	143–147	19

	Class	Frequency
d.	9–13	1
	14–19	6
	20–25	2
	26–28	5
	29–32	9

2–6. What are open-ended frequency distributions? Why are they necessary?

2–7. The following ZIP codes were obtained from the respondents to a mail survey. Construct a frequency distribution for the data.

15132	15130	15132	15130
15130	15131	15134	15133
15131	15133	15133	15133
15130	15131	15132	15130
15133	15134	15133	15133

2–8. At a college financial aid office, students who applied for a scholarship were classified according to their class rank: Fr = freshman, So = sophomore, Jr = junior, Se = senior. Construct a frequency distribution for the data.

Fr	Fr	Fr	Fr	Fr
Jr	Fr	Fr	So	Fr
Fr	So	Jr	So	Fr
So	Fr	Fr	Fr	So
Se	Jr	Jr	So	Fr
Fr	Fr	Fr	Fr	So
Se	Se	Jr	Jr	Se
So	So	So	So	So

2–9. A survey taken in a restaurant shows the following number of cups of coffee consumed with each meal. Construct an ungrouped frequency distribution.

0	2	2	1	1	2
3	5	3	2	2	2
1	0	1	2	4	2
0	1	0	1	4	4
2	2	0	1	1	5

2–10. In a survey of 20 patients who smoked, the following data were obtained. Each value represents the number of cigarettes the patient smoked per day. Construct a frequency distribution, using six classes.

10	8	6	14
22	13	17	19
11	9	18	14
13	12	15	15
5	11	16	11

2–11. The following list gives the ages of 40 college professors. Construct a frequency distribution, using eight classes. (The data for this exercise will be used for Exercise 2–21.)

37	41	41	47	62
27	44	43	40	58
62	43	50	61	53
65	58	45	50	27
36	65	43	41	30
42	29	32	48	31
63	38	37	47	26
50	35	31	49	34

2-12. In a study of 32 student grade point averages (GPA), the following data were obtained. Construct a frequency distribution, using seven classes. (*Note:* A = 4, B = 3, C = 2, D = 1, F = 0.) (The information in this exercise will be used again for Exercises 2-22 and 3-22.)

3.2	2.0	3.3	2.7	2.1	3.9
1.1	3.5	1.9	1.7	0.8	2.6
0.6	4.0	3.5	2.3	1.6	
2.8	2.6	1.6	1.6	2.4	
2.6	2.3	3.8	2.1	2.9	
3.0	1.7	4.0	1.2	3.1	

2-13. A supervisor wishes to check the miles her commuter buses are driven each day. Her findings are shown below. Construct a frequency distribution, using ten classes. (The data for this exercise will be used for Exercise 3-23.)

138	107	136	128	148	118	99
142	129	115	123	133	123	103
121	128	122	144	126	135	107
125	98	117	153	141	126	139
134	115	93	127	118	158	143

2-14. The following data represent the number of customers the Purple Mountain Restaurant served in a 30-day period. Construct a frequency distribution, using six classes. (The information in this exercise will be used for Exercises 2-24, 2-34, and 3-24.)

68	32	28	28	32	53	29
59	23	32	33	20	59	29
31	58	18	32	48	47	28
19	45	25	31	60	31	43
28	37					

2-15. In a study of 40 women, the following data of blood potassium levels, in milliequivalents per liter, were obtained. Construct a frequency distribution, using seven classes. (The information in this exercise will be used for Exercises 2-27 and 3-25.)

3.2	4.9	3.8	3.6	3.4
5.8	5.3	4.6	5.1	3.6
6.0	4.7	4.3	4.2	4.4
4.5	5.0	4.4	5.8	3.9
4.2	3.9	3.8	3.7	4.2
4.3	4.9	5.6	3.7	4.1
2.7	4.0	4.2	4.3	4.3
5.1	5.2	4.7	4.5	4.9

2-16. The blood glucose level, in milligrams per deciliter, for 60 patients is shown below. Construct a frequency distribution for the data set, using six classes.

55	115	118	114	59	109
63	97	90	59	105	81
84	81	82	61	103	77
82	76	68	86	97	80
77	85	69	62	101	83
58	83	101	86	84	78
59	92	88	97	87	92
70	86	72	84	82	84
101	80	93	56	65	91
75	78	100	74	74	90

*2-17. If the data is in units of 10—say 80, 230, 120, 160, etc.—and a hypothetical class is 150–200, what are the boundaries? The following data show the net worth of 20 new-car dealerships in Miami. The amounts are in millions of dollars. Construct a frequency distribution for the data.

130	70	180	160
120	60	150	140
160	80	140	130
210	80	130	140
90	110	120	150

*2-18. If the data are in units of hundreds—such as 300, 1200, 500, 900, etc.—what are the boundaries of a hypothetical class such as 800–1200?

2-3 HISTOGRAMS, FREQUENCY POLYGONS, AND OGIVES

After the data set has been organized into a frequency distribution, it can be presented in graphical form, because a person can more easily comprehend the meanings conveyed by charts and graphs. Here are the three most commonly used graphs in research:

1. The histogram
2. The frequency polygon
3. The cumulative frequency graph or ogive (pronounced ō–jive)

An example of each type of graph is shown in Figures 2–1a, 2–1b, and 2–1c, respectively. The data for each graph were obtained from the number of automobiles sold by a small-car dealer over a five-day period. The definitions and procedures for constructing these graphs are shown in this section.

FIGURE 2–1
Examples of Commonly Used Graphs

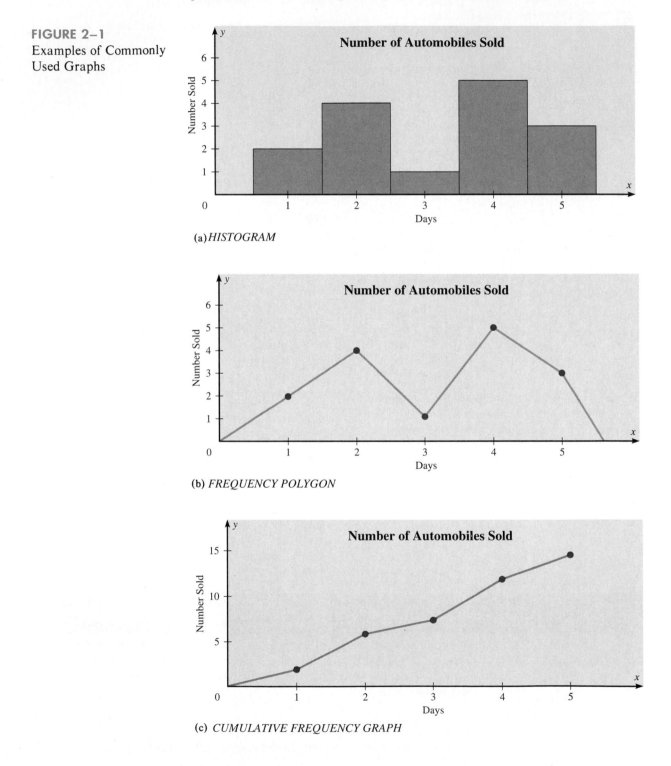

(a) *HISTOGRAM*

(b) *FREQUENCY POLYGON*

(c) *CUMULATIVE FREQUENCY GRAPH*

The Histogram

The histogram is a graph that displays the data by using vertical bars of various heights to represent the frequencies.

EXAMPLE 2–4 Construct a histogram to represent the data shown below for the miles driven by the sales representatives of the Memory Lane Card Company (see Example 2–3).

Class Boundaries	Frequency	Cumulative Frequency
5.5–10.5	1	1
10.5–15.5	2	3
15.5–20.5	3	6
20.5–25.5	5	11
25.5–30.5	4	15
30.5–35.5	3	18
35.5–40.5	2	20
	20	

Solution **STEP 1** Draw and label the x and y axes. The x axis will always be the horizontal axis, and the y axis will always be the vertical axis.

STEP 2 Represent the frequency on the y axis and the class boundaries on the x axis. See Figure 2–2.

FIGURE 2–2
Labeling the Axes for the Histogram for Example 2–4

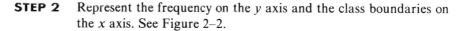
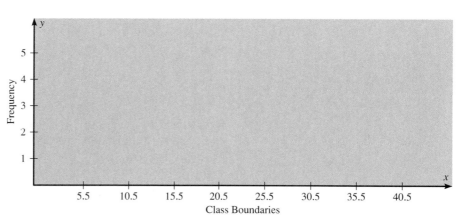

STEP 3 Using the frequencies as the heights, draw vertical bars for each class. See Figure 2–3.

FIGURE 2–3
Histogram for Example 2–4

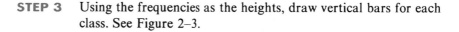
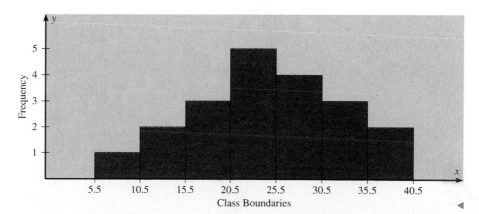

As the distribution in Figure 2–3 shows, the class with the greatest number of representatives is 20.5 to 25.5, with the frequencies clustering about the center.

The Frequency Polygon

Another way to represent the same data set is by using a frequency polygon.

The frequency polygon is a graph that displays the data by using lines that connect points plotted for the frequencies at the midpoints of the classes. The frequencies represent the heights of the midpoints.

The next example shows the procedure for constructing a frequency polygon.

EXAMPLE 2–5 Using the frequency distribution given in Example 2–4, construct a frequency polygon.

Solution **STEP 1** Find the midpoints of each class. Recall that midpoints are found by adding the upper and lower boundaries and dividing by 2.

$$\frac{5.5 + 10.5}{2} = 8 \qquad \frac{10.5 + 15.5}{2} = 13$$

And so on. The midpoints are listed next.

Class Boundaries	Midpoints	Frequency
5.5–10.5	8	1
10.5–15.5	13	2
15.5–20.5	18	3
20.5–25.5	23	5
25.5–30.5	28	4
30.5–35.5	33	3
35.5–40.5	38	2

STEP 2 Draw the x and y axes. Label the x axis with the midpoints of each class, and then use a suitable scale on the y axis for the frequencies, as shown in Figure 2–4.

FIGURE 2–4
Labeling the Axes for the Frequency Polygon for Example 2–5

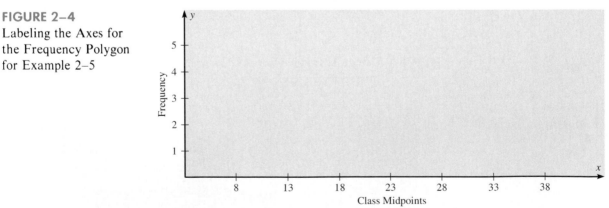

STEP 3 Using the midpoints for the x values and the frequencies as the y values, plot the points.

STEP 4   Connect adjacent points with straight lines. Draw a line back to the *x* axis at the beginning and end of the graph, at the same distance that the previous and next midpoints would be located, as shown in Figure 2–5.

FIGURE 2–5
Frequency Polygon for
Example 2–5

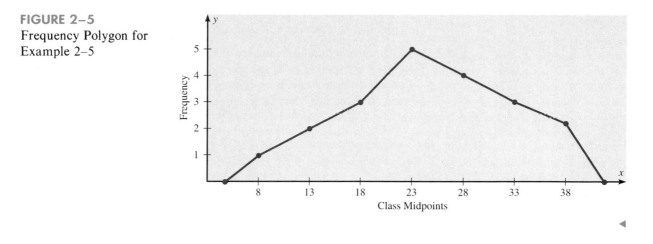

The frequency polygon and the histogram are two different ways to represent the same data set. The choice of which one to use is left to the discretion of the researcher. Still another type of graph that can be used is called the cumulative frequency graph or ogive.

The Ogive   The third type of graph represents the cumulative frequencies for the classes. This type of graph is called the cumulative frequency graph or ogive. The **cumulative frequency** is the sum of the frequencies accumulated up to the upper boundary of a class in the distribution.

The **ogive** is a graph that represents the cumulative frequencies for the classes in a frequency distribution.

Example 2–6 shows the procedure for constructing an ogive.

EXAMPLE 2–6   Construct an ogive for the frequency distribution described in Example 2–5.

Solution   **STEP 1**   Find the cumulative frequencies for each class.

Class Boundaries	Frequency	Cumulative Frequency
5.5–10.5	1	1
10.5–15.5	2	3
15.5–20.5	3	6
20.5–25.5	5	11
25.5–30.5	4	15
30.5–35.5	3	18
35.5–40.5	2	20

STEP 2 Draw the x and y axes. Label the x axis with the class boundaries. Use an appropriate scale for the y axis to represent the cumulative frequencies. (Depending on the numbers in the cumulative frequency columns, scales such as 0, 1, 2, 3, . . . , or 5, 10, 15, 20, . . . , or 1000, 2000, 3000, . . . , etc., can be used. Do *not* label the y axis with the numbers in the cumulative frequency column.) In this example, a scale of 0, 2, 4, 6, . . . will be used.

STEP 3 Plot the cumulative frequency at each upper class boundary, as shown in Figure 2–6. Upper boundaries are used since the cumulative frequencies represent the number of data values accumulated up to the upper boundary of that class.

FIGURE 2–6
Plotting the Cumulative Frequency for Example 2–6

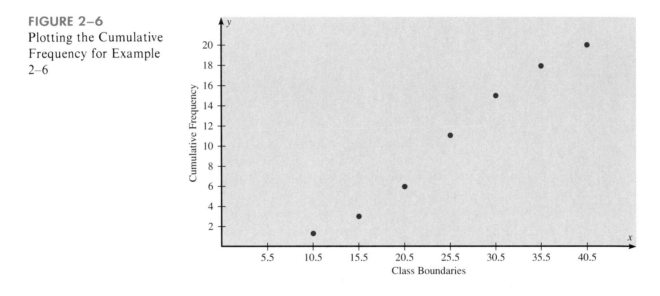

STEP 4 Starting with the first upper class boundary, 10.5, connect adjacent points with straight lines, as shown in Figure 2–7. Then extend the graph to the first lower class boundary, 5.5, on the x axis.

FIGURE 2–7
Ogive for Example 2–6

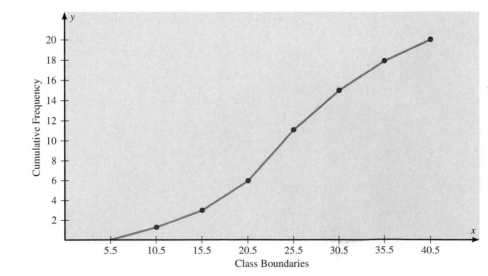

Cumulative frequency graphs are used to visually represent how many values are below a certain upper class boundary. For example, if a manager wanted to know how many sales representatives traveled less than 25.5 miles, the manager would locate 25.5 on the x axis, draw a vertical line up until it intersects the graph, then draw a horizontal line at that point to the y axis. The y axis value is 11, as shown in Figure 2–8.

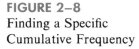

FIGURE 2–8
Finding a Specific
Cumulative Frequency

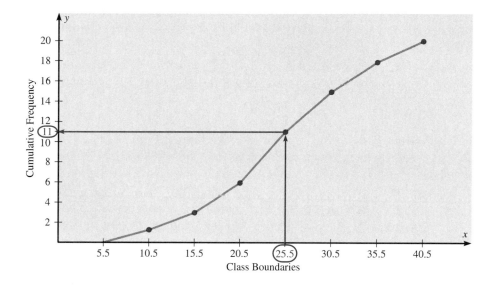

A summary for drawing the three types of graphs is shown in Procedure Table 2.

Procedure Table 2

Constructing Statistical Graphs

1. Draw and label the x and y axes.
2. Choose a suitable scale for the frequencies or cumulative frequencies and label it on the y axis.
3. Represent the class boundaries for the histogram or ogive or the midpoint for the frequency polygon on the x axis.
4. Plot the points.
5. Draw the bars or lines.

Relative Frequency Graphs

The histogram, the frequency polygon, and the ogive shown previously were constructed by using frequencies in terms of the raw data. These distributions can be converted into distributions using *proportions* instead of raw data as frequencies. These types of graphs are called **relative frequency graphs.**

In order to convert a frequency into a proportion or relative frequency, one must divide the frequency for each class by the total of the frequencies. The sum of the relative frequencies will always be 1. These graphs will be similar to the ones that use raw data as frequencies, but the values on the y axis will be in terms of proportions. The next example shows the three types of relative frequency graphs.

EXAMPLE 2-7 Construct a histogram, frequency polygon, and ogive using relative frequencies for the data in Example 2–3, which is repeated here.

Class Boundaries	Frequency	Cumulative Frequency
5.5–10.5	1	1
10.5–15.5	2	3
15.5–20.5	3	6
20.5–25.5	5	11
25.5–30.5	4	15
30.5–35.5	3	18
35.5–40.5	2	20
	20	

Solution **STEP 1** Convert each frequency to a proportion or relative frequency by dividing the frequency for each class by the total number of observations.

For class 5.5–10.5, the relative frequency is $\frac{1}{20} = 0.05$.
For class 10.5–15.5, the relative frequency is $\frac{2}{20} = 0.10$.
For class 15.5–20.5, the relative frequency is $\frac{3}{20} = 0.15$.
And so on.

STEP 2 Using the same procedure, find the relative frequencies for the cumulative frequency column. The relative frequencies are shown here.

Class Boundaries	Midpoints	Relative Frequency	Cumulative Relative Frequency
5.5–10.5	8	0.05	0.05
10.5–15.5	13	0.10	0.15
15.5–20.5	18	0.15	0.30
20.5–25.5	23	0.25	0.55
25.5–30.5	28	0.20	0.75
30.5–35.5	33	0.15	0.90
35.5–40.5	38	0.10	1.00
		1.00	

STEP 3 Draw each graph as shown in Figure 2–9. For the histogram and ogive, use the class boundaries along the x axis. For the frequency polygon, use the midpoints on the x axis. The scale on the y axis is in proportions.

FIGURE 2–9
Graphs for Example 2–7

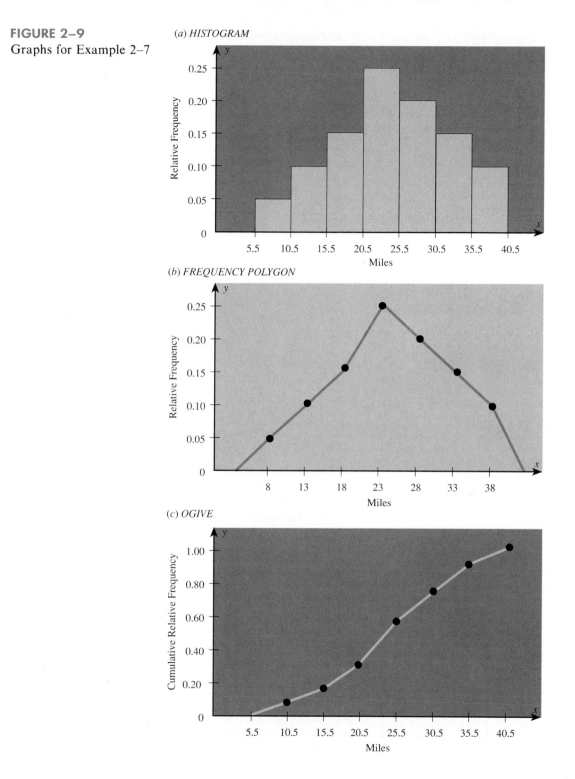

(a) HISTOGRAM

(b) FREQUENCY POLYGON

(c) OGIVE

EXERCISES

2–19. For 108 randomly selected college students, the following IQ frequency distribution was obtained. Construct a histogram, frequency polygon, and ogive for the data. (The data for this exercise will be used for Exercise 2–31.)

Class Limits	Frequency
90–98	6
99–107	22
108–116	43
117–125	28
126–134	9

2–20. For 75 employees of a large department store, the following distribution for years of service was obtained. Construct a histogram, frequency polygon, and ogive for the data. (The data for this exercise will be used for Exercise 2–32.)

Class Limits	Frequency
1–5	21
6–10	25
11–15	15
16–20	0
21–25	8
26–30	6

2–21. Construct a histogram, frequency polygon, and ogive for the data in Exercise 2–11.

2–22. Construct a histogram, frequency polygon, and ogive for the data in Exercise 2–12.

2–23. Thirty automobiles were tested for fuel efficiency, in miles per gallon (mpg). The following frequency distribution was obtained. Construct a histogram, frequency polygon, and ogive for the data. (The data for this exercise will be used for Exercise 2–33.)

Class Boundaries	Frequency
7.5–12.5	3
12.5–17.5	5
17.5–22.5	15
22.5–27.5	5
27.5–32.5	2

2–24. Construct a histogram, frequency polygon, and ogive for the data in Exercise 2–14.

2–25. In a class of 35 students, the following grade distribution was found. Construct a histogram, frequency polygon, and ogive for the data. (A = 4, B = 3, C = 2, D = 1, F = 0.) (The data in this exercise will be used for Exercise 2–35.)

Grade	Frequency
0	3
1	6
2	9
3	12
4	5

2–26. In a study of reaction times to a specific stimulus, an animal trainer obtained the following data, in seconds. Construct a histogram, frequency polygon, and ogive for the data.

Class Limits	Frequency
2.3–2.9	10
3.0–3.6	12
3.7–4.3	6
4.4–5.0	8
5.1–5.7	4
5.8–6.4	2

2–27. Construct a histogram, frequency polygon, and ogive for the data in Exercise 2–15.

2–28. Eighty randomly selected batteries were tested to determine their lifetimes. The following frequency distribution was obtained. The data values are in hours. Construct a histogram, frequency polygon, and ogive for the data.

Class Boundaries	Frequency
63.5–74.5	10
74.5–85.5	15
85.5–96.5	22
96.5–107.5	17
107.5–118.5	11
118.5–129.5	5

2–29. Construct a histogram, frequency polygon, and ogive for the data in Exercise 2–16.

2–30. In a study of reaction times to a specific stimulus, a psychologist obtained the following data, given in seconds. Construct a histogram, frequency polygon, and an ogive for the data.

Class Limits	Frequency
0.5–0.9	12
1.0–1.4	13
1.5–1.9	7
2.0–2.4	5
2.5–2.9	2
3.0–3.4	0
3.5–3.9	1

2–31. For the data in Exercise 2–19, construct a histogram, frequency polygon, and ogive, using relative frequencies.

2–32. For the data in Exercise 2–20, construct a histogram, frequency polygon, and ogive, using relative frequencies.

2–33. For the data in Exercise 2–23, construct a histogram, frequency polygon, and ogive, using relative frequencies.

2–34. For the data in Exercise 2–24, construct a histogram, frequency polygon, and ogive, using relative frequencies.

2–35. For the data in Exercise 2–25, construct a histogram, frequency polygon, and ogive, using relative frequencies.

*2–36. Use the histogram shown in Figure 2–10.

a. Construct a frequency distribution; include class limits, class frequencies, midpoints, and cumulative frequencies.
b. Construct a frequency polygon.
c. Construct an ogive.

FIGURE 2–10
Histogram for Exercise 2–36

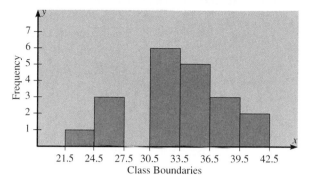

*2–37. Using the results from Exercise 2–36, answer the following questions.

a. How many values are in the class 27.5–30.5?
b. How many values fall between 24.5 and 36.5?
c. How many values are below 33.5?
d. How many values are above 30.5?

*2–38. Use the frequency polygon shown in Figure 2–11.

a. Construct a frequency distribution, including class limits, class boundaries, class frequencies, and cumulative frequencies.
b. Construct a histogram.
c. Construct an ogive.

FIGURE 2–11
Frequency Polygon for Exercise 2–38

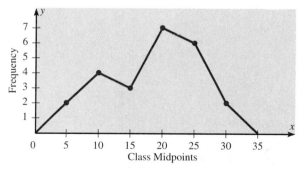

2–4 OTHER TYPES OF GRAPHS

In addition to the histogram, the frequency polygon, and the ogive, several other types of graphs are often used in statistics. They are the bar graph, the time series graph, the pie graph, and the pictograph. Figure 2–12 shows an example of each type of graph. Sometimes, a special graph called the stem and leaf plot is used in statistics; this graph is discussed at the end of this section.

FIGURE 2–12
Other Types of Graphs Used in Statistics

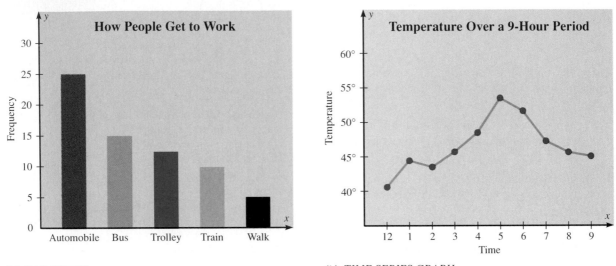

(a) *BAR GRAPH*   (b) *TIME SERIES GRAPH*

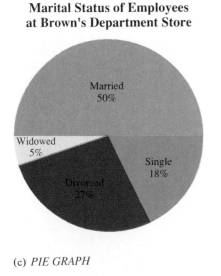

(c) *PIE GRAPH*

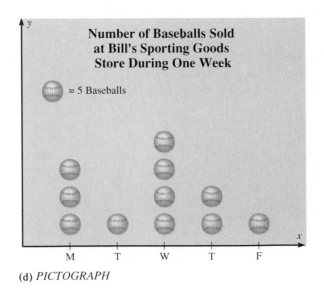

(d) *PICTOGRAPH*

The Bar Graph There are two types of bar graphs: the *vertical bar graph* and the *horizontal bar graph*. Figure 2–13a shows a vertical bar graph, and Figure 2–13b shows the horizontal bar graph for the same data.

 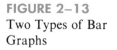
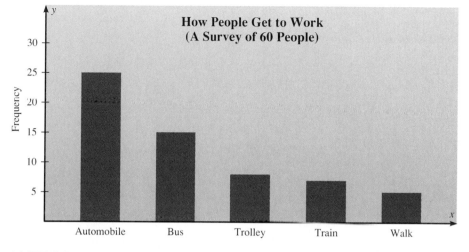

(a) *VERTICAL BAR GRAPH*

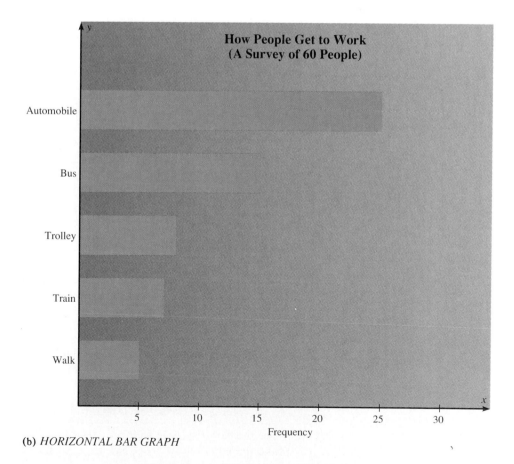

(b) *HORIZONTAL BAR GRAPH*

A bar graph uses vertical or horizontal bars to represent the frequencies of a distribution.

In order to draw the bar graph, one must have the data tabulated in a frequency distribution. The next example explains the procedure for constructing the bar graph.

EXAMPLE 2–8   The table below shows the number of fatal accidents on the Allegheny Expressway for the years shown. Construct a vertical bar graph for the data.

Year	Number of Accidents
1986	8
1987	7
1988	12
1989	15
1990	10

Solution   **STEP 1**   Draw and label the x and y axes.

STEP 2   Represent the years on the x axis and the frequency on the y axis.

STEP 3   Draw the bars corresponding to the data, as shown in Figure 2–14.

FIGURE 2–14
Bar Graph for Example 2–8

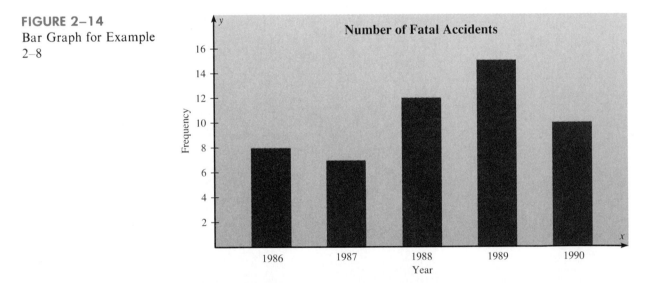

The graph shows that 1989 was the year in which the most fatal accidents occurred.   ◀

Horizontal bar graphs are constructed by placing the frequency on the x axis and the categories on the y axis, as shown in Figure 2–15.

Suggestions for Drawing a Bar Graph

1. Make the bars the same width.
2. Make the units on the axis that are used for the frequency equal in size.

FIGURE 2-15
Horizontal Bar Graph
for the Data Illustrated
in Figure 2-14

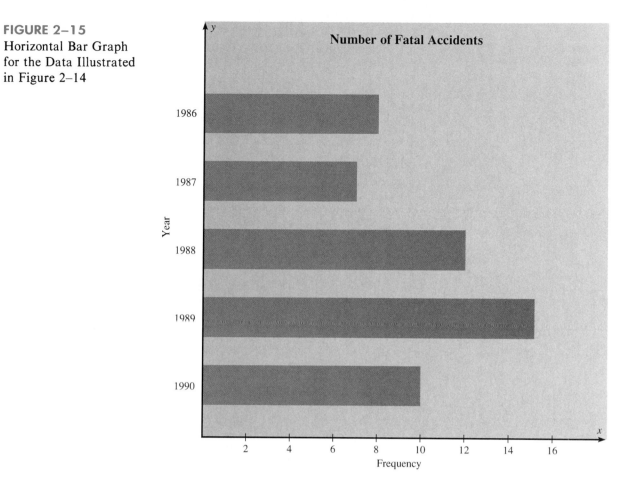

For the comparison of two sets of data, a *compound bar graph* can be used. Here, the two sets of data are represented on the same graph, but each of the bars is shaded differently for ease of comparison. See Figure 2-16.

FIGURE 2-16
Compound Bar Graph

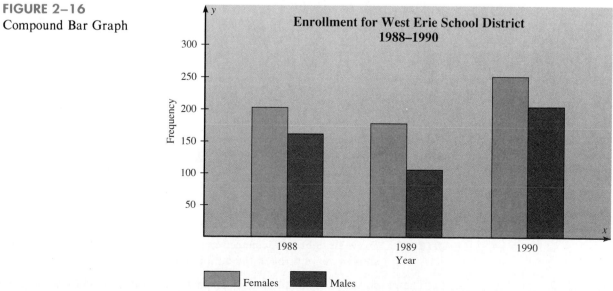

Occasionally, the bars will be placed above each other or end to end. This type of graph is called a *component bar graph* and is used to show the components that make up the whole. See Figure 2–17.

FIGURE 2–17
Component Bar Graph

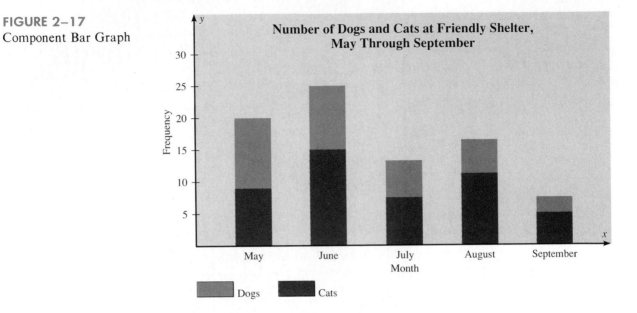

The Time Series Graph

Quite often, data can be shown by using a time series graph.

A time series graph represents data that occur over a specific period of time.

The next example shows the procedure for constructing a time series graph.

EXAMPLE 2–9 A mine safety engineer wishes to use the following data for a presentation showing how the number of deaths from surface mining has changed over the years. Draw a time series graph for the data.

Year	Number of Deaths
1930	46
1940	24
1950	32
1960	10
1970	8
1980	1
1990	3

Source: Mine Safety and Health Administration, U.S. Department of Labor; reprinted in the *Pittsburgh Press,* March 15, 1992. Used with permission.

Solution

STEP 1 Draw and label the *x* and *y* axes.

STEP 2 Label the *x* axis for years and the *y* axis for the number of deaths.

STEP 3 Plot each point according to the table.

STEP 4 Draw straight lines connecting adjacent points. Do not try to fit a smooth curve through the data points. See Figure 2–18.

The graph shows that the number of deaths have decreased over the years with the exceptions of 1940 to 1950 and 1980 to 1990.

FIGURE 2–18
Time Series Graph for
Example 2–9

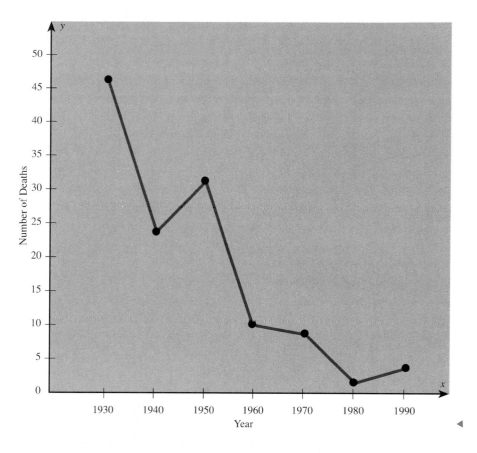

Speaking of **Statistics**

Pennsylvania Mining Deaths

Year	Underground	Surface
1930	701	46
1940	367	24
1950	158	32
1960	53	10
1970	28	8
1980	16	1
1990	2	3
1991	0	3

Source: Mine Safety and Health Administration, U.S. Department of
Labor; reprinted in the *Pittsburgh Press,* March 15, 1992. Used with
permission.

The deaths from mining work have been declining. Suggest some factors that might have
contributed to the decline in the number of deaths other than improved safety conditions.

As with the bar graph, two data sets can be compared on the same graph by using two lines, as shown in Figure 2–19.

FIGURE 2–19
Two Time Series Graphs
for Comparisons

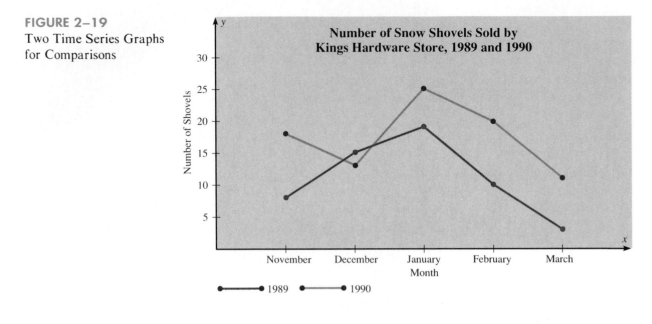

The Pie Graph

Pie graphs are used extensively in statistics.

A pie graph is a circle that is divided into sections or wedges according to the percentage of frequencies in each category of the distribution.

The next example shows the procedure for constructing a pie graph.

EXAMPLE 2–10 An automobile dealer reported information on the sales of cars for the month of July. Construct a pie graph to represent the data.

Type	Frequency
Convertibles	3
Station wagons	2
Compacts	5
Coupes	30
Sedans	20
	$n = 60$

Solution **STEP 1** Since there are 360° in a circle, the frequency for each class must be converted into a proportional part of the circle. This conversion is done by using the formula

$$\text{degrees} = \frac{f}{n} \cdot 360°$$

where

f = frequency for each class
n = sum of the frequencies

Speaking of **Statistics**

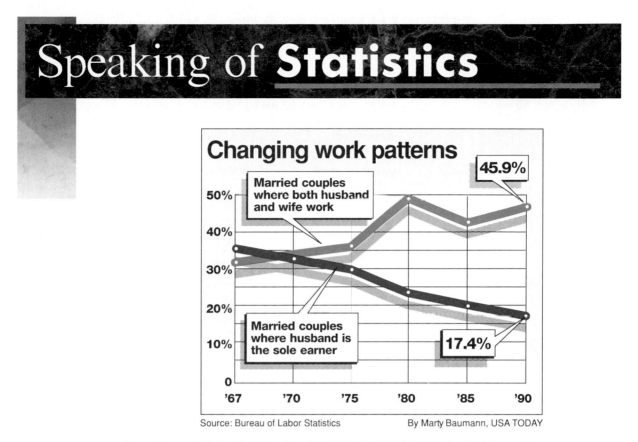

Changing work patterns

Married couples where both husband and wife work

Married couples where husband is the sole earner

45.9%

17.4%

50%
40%
30%
20%
10%
0

'67   '70   '75   '80   '85   '90

Source: Bureau of Labor Statistics   By Marty Baumann, USA TODAY

Source: Bureau of Labor Statistics. Copyright 1992, USA TODAY. Reprinted with permission.

This time series graph compares the percentages of married couples where both husband and wife work with the percentages of married couples where the husband is the sole earner. Comment on the patterns, and suggest what might happen in 1995 and 2000.

Hence, the following conversions are obtained.

convertibles $\dfrac{3}{60} \cdot 360° = 18°$

station wagons $\dfrac{2}{60} \cdot 360° = 12°$

compacts $\dfrac{5}{60} \cdot 360° = 30°$

coupes $\dfrac{30}{60} \cdot 360° = 180°$

sedans $\dfrac{20}{60} \cdot 360° = \underline{120°}$

total $= 360°$

FIGURE 2–20
Pie Graph for Example 2–10

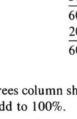

Types of Automobiles Sold

Convertibles 5%

Station Wagons $3\frac{1}{3}\%$

Compacts $8\frac{1}{3}\%$

Sedans $33\frac{1}{3}\%$

Coupes 50%

STEP 2 Each frequency must also be converted to a percentage. Recall from Example 2–1 that this conversion is done by using the formula

$$\% = \frac{f}{n} \cdot 100\%$$

Hence, the following percentages are obtained.

convertibles	$\dfrac{3}{60} \cdot 100\% =$	5%
station wagons	$\dfrac{2}{60} \cdot 100\% =$	$3\frac{1}{3}\%$
compacts	$\dfrac{5}{60} \cdot 100\% =$	$8\frac{1}{3}\%$
coupes	$\dfrac{30}{60} \cdot 100\% =$	50%
sedans	$\dfrac{20}{60} \cdot 100\% =$	$\underline{33\frac{1}{3}\%}$
	total =	100%

The degrees column should add to 360°, and the percentages column should add to 100%.

STEP 3 Next, using a protractor and a compass, draw the graph and label each section with the name and percentages, as shown in Figure 2–20.

In this case, then, the graph shows that most of the automobiles sold were coupes, followed by sedans. ◀

EXAMPLE 2–11 Construct a pie graph showing the blood types of the army inductees described in Example 2–1. The frequency distribution is repeated here.

Class	Frequency	Percent
A	5	20%
B	7	28%
O	9	36%
AB	$\underline{4}$	$\underline{16\%}$
	25	100%

Solution **STEP 1** Find the number of degrees for each class, using the formula

$$\text{degrees} = \frac{f}{n} \cdot 360°$$

For each class, then, one obtains the following results.

A $\frac{5}{25} \cdot 360° = 72°$
B $\frac{7}{25} \cdot 360° = 100.8°$
O $\frac{9}{25} \cdot 360° = 129.6°$
AB $\frac{4}{25} \cdot 360° = 57.6°$

STEP 2 Find the percentages. (This has already been done in Example 2–1.)

STEP 3 Next, using a protractor and a compass, graph each section and write its name and corresponding percentage, as shown in Figure 2–21.

The graph shows that in this case the most common blood type is type O.

FIGURE 2–21
Pie Graph for Example
2–11

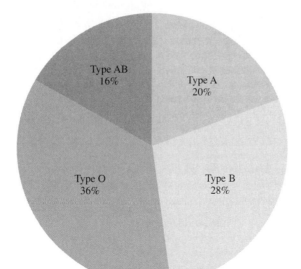

Speaking of **Statistics**

USA SNAPSHOTS®

A look at statistics that shape our lives

Toys galore

Nearly 5,000 new toys were exhibited during last week's International Toy Fair in New York. Toys sold in the USA;

Percentage of market

Games/puzzles — 8%
Activity toys — 9%
Infant — 9%
Vehicles — 11%
Dolls/stuffed animals — 15%
Other —

Video games 26%

22%

Source: Toy Manufacturers of America, 1990   By Marcia Staimer, USA TODAY

Source: Based on data from the Toy Manufacturers of America, 1990. Copyright 1992, USA TODAY. Reprinted with permission.

The data used here are presented in a pie chart. Comment on how toy manufacturers benefit from the information in this "USA SNAPSHOTS."

The Pictograph

The pictograph is similar to the bar graph; but instead of using bars to represent the frequencies, a researcher uses symbols or pictures. The researcher must, of course, define the unit of measure that each symbol represents. The symbols can be used vertically or horizontally, at the researcher's discretion.

A pictograph is a graph that uses symbols or pictures to represent data.

EXAMPLE 2–12

The following data represent the number of new houses built in White Oak Township for the given years. Construct a pictograph to represent the frequency distribution.

Year	Number of Houses
1987	253
1988	105
1989	164
1990	97

Solution

STEP 1 Decide the numerical value of each symbol according to the data. In this case, one 🏠 represents 50 houses.

STEP 2 Draw and label the x axis and the y axis.

STEP 3 Draw the correct number of houses for each year, as shown in Figure 2–22.

FIGURE 2–22
Pictograph for Example 2–12

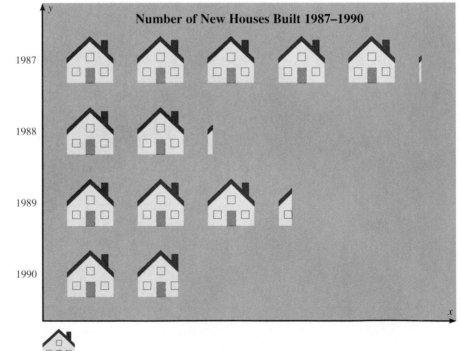

Number of New Houses Built 1987–1990

Each 🏠 Represents Approximately 50 Houses

The graph shows that more new houses were built in 1987 than in any other year, and more than twice as many were built in 1987 than were built in 1988 or 1990. ◄

If the symbol represents a large number of units, the researcher can use part of the symbol to represent a smaller number in the pictograph. For example, if

= 100,000 houses

then

= 50,000 houses

and

= 25,000 houses

etc.

Pictographs can be drawn horizontally or vertically. Figure 2–12d presented earlier shows a vertical pictograph.

Stem and Leaf Plots

The stem and leaf plot is a combination of sorting and graphing and looks very much like a frequency distribution. It has the advantage of retaining the actual data while showing it in a graphical form.

The stem and leaf plot is used in what is called *exploratory data analysis*. It was developed by John Tukey and presented in his book entitled *Exploratory Data Analysis* (Addison-Wesley, 1977). The purpose of exploratory data analysis is to enable the researcher to examine data in order to gain information about things such as unexplained patterns, the shape of the distribution, where data values cluster, and the existence of any gaps in the data that would not be apparent when using summary statistics (such as measures of average or variation; these measures are presented in Chapter 3). In addition to retaining the original data and showing it in graphic form, the stem and leaf plot is easier to construct than a grouped frequency distribution.

A stem and leaf plot is a data plot that uses part of a data value as the stem and part of the data value as the leaf to form groups or classes.

Example 2–13 shows the procedure for constructing a stem and leaf plot.

EXAMPLE 2–13   At an outpatient testing center, a sample of 20 days showed the following number of cardiograms done each day. Construct a stem and leaf plot for the data.

25	31	20	32	13
14	43	02	57	23
36	32	33	32	44
32	52	44	51	45

Solution   **STEP 1**   Arrange the data in order:

02, 13, 14, 20, 23, 25, 31, 32, 32, 32,

32, 33, 36, 43, 44, 44, 45, 51, 52, 57

Note: Arranging the data in order is not necessary, but it is helpful in the construction of the plot.

Speaking of **Statistics**

USA SNAPSHOTS®

A look at statistics that shape our lives

Who buys the most children's books

Retail store buyers:

41% Mothers

16% Teachers

11% Children

10% Grandparents

Source: *Publishers Weekly*   By Bob Laird, USA TODAY

Source: Based on a survey by *Publishers Weekly*. Copyright 1992, USA TODAY. Reprinted with permission.

This "USA SNAPSHOTS" is an example of a pictograph and uses percentages. Do the percentages add to 100? What information is missing from the pictograph? Explain.

STEP 2   Separate the data according to the first digit, as shown.

02   13, 14   20, 23, 25   31, 32, 32, 32, 32, 33, 36

43, 44, 44, 45   51, 52, 57

STEP 3   Since the values range from 2 to 57, a display can be made by using the leading digit as a *stem* and the trailing digit as the *leaf*. For example, for the value 32, the leading digit, 3, is the stem and the trailing digit, 2, is the leaf. For the value 14, the 1 is the stem and the 4 is the leaf; etc. Now a plot can be constructed, as shown.

FIGURE 2-23
Stem and Leaf Plot for
Example 2-13

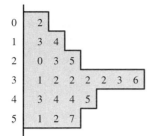

Leading Digit (Stem)	Trailing Digit (Leaf)
0	2
1	3 4
2	0 3 5
3	1 2 2 2 2 3 6
4	3 4 4 5
5	1 2 7

A graph can be drawn over the plot, as shown in Figure 2-23. ◄

The graph shows that the distribution peaks in the center and that there are no gaps in the data. For 7 of the 20 days, the number of patients receiving cardiograms was between 31 and 36. The plot also shows that the testing center treated from a minimum of 2 patients to a maximum of 57 patients in any one day.

EXAMPLE 2-14   An insurance company researcher conducted a survey on the number of car thefts in a large city for a period of 30 days last summer. The raw data are shown below. Construct a stem and leaf plot by using classes 50–54, 55–59, 60–64, 65–69, 70–74, and 75–79.

52	62	51	50	69
58	77	66	53	57
75	56	55	67	73
79	59	68	65	72
57	51	63	69	75
65	53	78	66	55

Solution   **STEP 1**   Arrange the data in order.

50, 51, 51, 52, 53, 53, 55, 55, 56, 57, 57, 58, 59, 62, 63, 65, 65, 66, 66, 67, 68, 69, 69, 72, 73, 75, 75, 77, 78, 79

STEP 2   Separate the data according to the classes.

50, 51, 51, 52, 53, 53   55, 55, 56, 57, 57, 58, 59
62, 63   65, 65, 66, 66, 67, 68, 69, 69   72, 73
75, 75, 77, 78 79

FIGURE 2-24
Stem and Leaf Plot for
Example 2-14

STEP 3   Plot the data as shown here.

Leading Digit (Stem)	Trailing Digit (Leaf)
5	0 1 1 2 3 3
5	5 5 6 7 7 8 9
6	2 3
6	5 5 6 6 7 8 9 9
7	2 3
7	5 5 7 8 9

The graph for this plot is shown in Figure 2-24. ◄

When the values are in the hundreds, such as 325, the stem is 32 and the leaf is 5. When data are grouped into classes, such as 20–24, 25–29, a stem and leaf plot might look like this:

Class	Leading Digit (Stem)	Trailing Digit (Leaf)	Value
20–24	2	0 1 3 4	20, 21, 23, 24
25–29	2	5 6	25, 26

Of course, the "Class" and the "Value" columns would be left out of the final plot.

In addition to the graphs shown in this section, there are other types of graphs, such as *logarithmic graphs* and *semilogarithmic graphs,* that are used in statistics. These graphs are studied in other statistical courses and will not be presented here.

Misleading Graphs

Graphs give a visual representation of data that enable readers to analyze and interpret the data more easily than they could simply by looking at numbers. However, inappropriately drawn graphs can misrepresent the data and lead the reader to false conclusions. For example, an automobile manufacturer's ad stated that 98% of the automobiles sold in the past ten years were still on the road. The ad then showed a graph similar to the one in Figure 2–25. The graph shows the percentage of the manufacturer's automobiles still on the road and the percentages of its competitors' automobiles still on the road. Is there a large difference? Not necessarily.

FIGURE 2–25
Graph of Automobile Manufacturer's Claim, Using a Scale from 95% to 100%

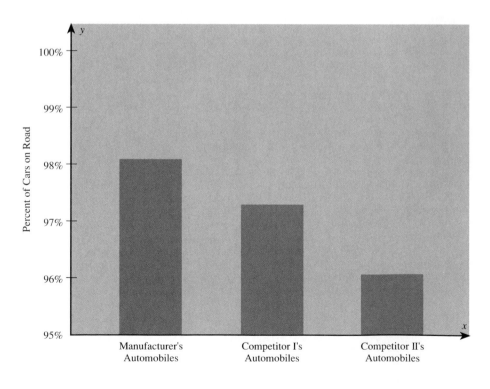

FIGURE 2–26
Graph in Figure 2–25
Redrawn, Using a Scale
from 0% to 100%

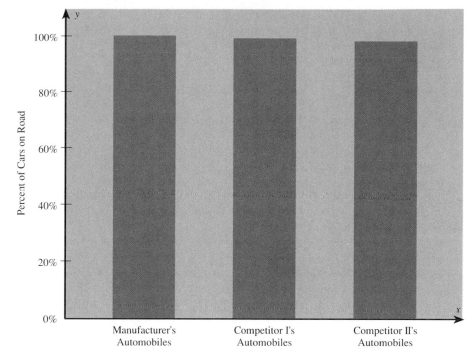

Notice the scale on the vertical axis in Figure 2–25. It has been cut off (or truncated), and it starts at 95%. When the graph is redrawn using a scale that goes from 0% to 100%, as in Figure 2–26, there is hardly a noticeable difference in the percentages. Thus, by changing the units at the starting point on the y axis, one can convey a very different visual representation of the data.

Let's consider another example. The percentage of the world's total motor vehicles produced by manufacturers in the United States declined from 25% in 1986 to 18% in 1991, as shown by the following data.

Year	1986	1987	1988	1989	1990	1991
Percent Produced in U.S.	25	23.8	23.3	22.1	20.3	18.0

Source: The World Almanac and Book of Facts, 1993 edition, copyright Pharos Books 1992, New York, NY 10166. Used with permission.

When one draws the graph, as shown in Figure 2–27a, a scale ranging from 0% to 100% shows a slight decrease. However, this decrease can be emphasized by using a scale that ranges from 15% to 25%, as shown in Figure 2–27b. Again, by changing the units or the starting point on the y axis, one can convey two different visual messages.

Another misleading graphing technique sometimes used is that of representing a one-dimensional increase or decrease by showing it in two dimensions,

FIGURE 2–27
Percent of World's
Motor Vehicles Produced
by Manufacturers in the
United States

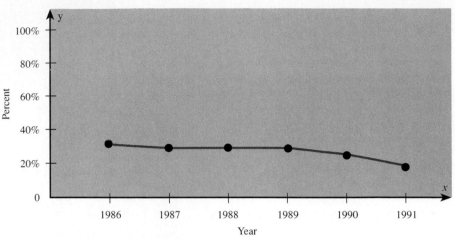

(a) *USING A SCALE FROM 0% TO 100%*

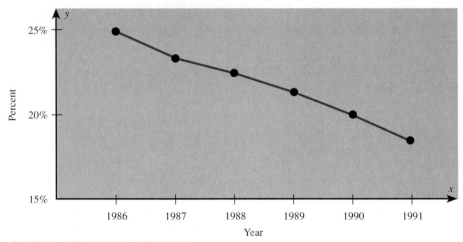

(b) *USING A SCALE FROM 15% TO 25%*

thus exaggerating it. For example, the average fee charged for a visit to a physician increased from $25.11 in 1983 to $46.43 in 1992. (*Source:* "USA TODAY, SNAPSHOTS" by Marcy Mullins, May 24, 1993; based on information from the AMA.) The increase shown by the bar graph in Figure 2–28a represents the change by a comparison of the heights of the two bars in one dimension. The same data are shown two-dimensionally by using circles in Figure 2–28b. Notice that the difference seems much larger because the eye is comparing the areas of the circles rather than the lengths of the diameters.

Note that it is not wrong to use the graphing techniques of truncating the scales or representing data by two-dimensional pictures. But when these techniques are used, the reader should be cautious of the conclusion drawn on the basis of the graphs.

FIGURE 2-28
Average Fees for a Visit to a Physician

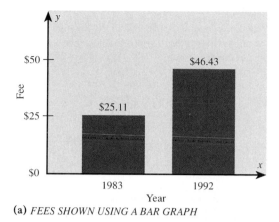

(a) *FEES SHOWN USING A BAR GRAPH*

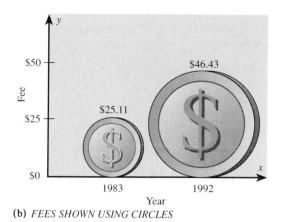

(b) *FEES SHOWN USING CIRCLES*

Speaking of **Statistics**

USA SNAPSHOTS®

A look at statistics that shape our lives

... like falling off a bicycle

About 600,000 bicycle riders are treated in emergency rooms each year.
Percent of all injuries by age group:

58.4%

8.9%   14.6%   16.8%   1.3%

0-4   5-14   15-24   25-64   65-up

Source: National Safe Kids Campaign   By Marty Baumann, USA TODAY

This bar graph shows the percentages of the total bicycle injuries for each age group treated in emergency rooms. What conclusion can be drawn from the graph? Can one conclude that the older a person is, the less likely the person is to have an accident while riding a bicycle? Comment on the age group classification.

EXERCISES

2–39. Construct a horizontal and a vertical bar graph for the lengths (in feet) of the five land tunnels.

Tunnel	Length
1. E. Johnson Memorial, Colo.	8959
2. Allegheny, Pa.	6070
3. Blue Mountain, Pa.	4339
4. Fort Pitt, Pittsburgh, Pa.	3560
5. Battery St., Seattle, Wash.	2140

Source: The World Almanac and Book of Facts, 1993 edition, copyright Pharos Books 1992, New York, NY 10166. Used with permission.

2–40. Construct a horizontal and a vertical bar graph for the areas (in square miles) of each of the Great Lakes.

Lake	Area
1. Superior	81,000
2. Michigan	67,900
3. Huron	74,700
4. Erie	32,630
5. Ontario	34,850

Source: The World Almanac and Book of Facts, 1993 edition, copyright Pharos Books 1992, New York, NY 10166. Used with permission.

2–41. The length of each river is given in miles. Construct a horizontal and a vertical bar chart to represent the data.

River	Length
Amazon	4000
Mississippi	2348
Nile	4160
Ohio	1310
Tennessee	652

Source: The World Almanac and Book of Facts, 1993 edition, copyright Pharos Books 1992, New York, NY 10166. Used with permission.

2–42. In a survey of 100 joggers asked to name the best surface on which to jog, the following data were obtained. Construct a vertical bar graph for the data.

Surface	Number of Joggers
Concrete	61
Asphalt	19
Grass	14
Artificial turf	6

2–43. The data shown below represent the average number of acres of six family-owned farms. Construct a horizontal bar graph for the data.

Farm	Acres
Johnston	17,500
Brown	14,900
Swenson	12,100
McCarthy	9,200
Bloom	8,900
Thomas	5,800

2–44. The data shown below represent the average acreage per farm in 1991 for seven selected states. Construct a horizontal bar graph for the data.

State	Acreage per Farm
New Jersey	166
Rhode Island	94
Massachusetts	99
Iowa	328
West Virginia	185
Texas	708
Wisconsin	222

Source: The World Almanac and Book of Facts, 1993 edition, copyright Pharos Books 1992, New York, NY 10166. Used with permission.

2–45. Draw a time series graph to represent the data for the number of U.S. airline fatalities for the given years.

Year	1985	1986	1987	1988	1989	1990	1991
No. of Fatalities	197	5	231	285	278	39	62

Source: The World Almanac and Book of Facts, 1993 edition, copyright Pharos Books 1992, New York, NY 10166. Used with permission.

2–46. The data below represent the personal consumption expenditures for transportation for the United States (in billions of dollars). Draw a time series graph to represent the data.

Year	1985	1986	1987	1988	1989	1990	1991
Amount	359.5	366.3	379.7	431.2	437.3	453.7	438.2

Source: The World Almanac and Book of Facts, 1993 edition, copyright Pharos Books 1992, New York, NY 10166. Used with permission.

2-47. The number of operable nuclear power reactors in the United States for the given year is shown below. Draw a time series graph to represent the data.

Year	1976	1979	1982	1985	1988	1991
No. Operable	61	68	77	95	108	111

Source: The World Almanac and Book of Facts, 1993 edition, copyright Pharos Books 1992, New York, NY 10166. Used with permission.

2-48. A bacteriologist charted the growth of a certain bacteria over a period of 8 hours. The data is shown below. Construct a time series graph to represent the data.

Hour	1	2	3	4	5	6	7	8
No. of Cells	2	5	8	13	17	22	30	38

2-49. In a survey of 100 male smokers concerning the use of tobacco, the following data were obtained. Construct a pie graph for the data.

Type of Smoker	Number of Users
Cigarettes	63
Cigars	22
Pipe	10
Chewing tobacco	5

2-50. A survey of 500 Philadelphia families—which asked the question "Where are you planning to vacation this summer?"—resulted in the following distribution. Construct a pie graph for the data.

Area	Number Vacationing
Great Lakes region	37
New England	104
East Coast	206
South	96
West Coast	57

2-51. The frequency distribution below shows the number of freshmen, sophomores, juniors, and seniors who attended a workshop on job opportunities in the news media. Construct a pie chart for the data.

Rank	Frequency
Freshmen	18
Sophomores	12
Juniors	6
Seniors	4
	40

2-52. In an insurance company study of the causes of 1000 deaths, the following data were obtained. Construct a pie graph to represent the data.

Cause of Death	Number of Deaths
1. Heart disease	432
2. Cancer	227
3. Stroke	93
4. Accidents	24
5. Other	224
	1000

2-53. The following data represent the number of hazardous waste sites in each state. Draw a pictograph to represent the data.

State	Number
Alabama	12
California	95
Maine	9
Mississippi	2
New Jersey	108
Pennsylvania	100

Source: The World Almanac and Book of Facts, 1993 edition, copyright Pharos Books 1992, New York, NY 10166. Used with permission.

2-54. The average rainfall for various locations is shown here. Construct a pictograph for the data.

City	Rainfall (Inches)
Allentown	43.6
Oakland	35.2
Hannahstown	56.0
Scottsboro	28.7
Bentleyville	17.1

2-55. The average annual snowfall for various locations in the United States is shown below. Construct a pictograph for the data.

City	Snowfall (Inches)
Boston	42
Sault Sainte Marie	113
Albany	65.2
Rochester	89.2
Juneau	105.8

Source: The World Almanac and Book of Facts, 1993 edition, copyright Pharos Books 1992, New York, NY 10166. Used with permission.

2–56. The number of passenger automobiles manufactured in 1991 is shown below. Construct a pictograph to represent the data.

Company	Number of Cars
Chrysler	510,147
Ford	1,171,680
General Motors	2,496,006

Source: The World Almanac and Book of Facts, 1993 edition, copyright Pharos Books 1992, New York, NY 10166. Used with permission.

2–57. Construct a stem and leaf plot for the following results of blood calcium level (in milligrams per deciliter) in 24 nurses.

7.2	10.2	11.2	10.3
9.0	10.7	7.3	9.1
8.1	9.6	9.6	9.2
9.7	8.2	9.8	9.9
7.3	9.3	9.4	8.5
10.0	11.1	8.3	10.3

2–58. Twenty salespeople reported the following number of calls completed last month. Construct a stem and leaf plot for the data.

72	102	88	103
93	107	100	93
82	97	91	73
81	119	83	89
82	86	102	106

2–59. The growth (in centimeters) of a certain variety of plant after 20 days is shown below. Construct a stem and leaf plot for the data.

20	12	39	38
41	43	51	52
59	55	53	59
50	58	35	38
23	32	43	53

*2–60. The number of successful space launches by the United States and Japan for the years 1986–1991 is shown below. Construct a compound bar graph for the data. What comparison can be made regarding the launches?

Year	1986	1987	1988	1989	1990	1991
U.S.	9	9	15	22	31	30
Japan	3	3	2	4	7	2

Source: The World Almanac and Book of Facts, 1993 edition, copyright Pharos Books 1992, New York, NY 10166. Used with permission.

*2–61. Meat production for veal and lamb for the years 1950–1990 is shown below. (Data are in millions of pounds.) Construct a compound time series graph for the data. What comparison can be made regarding meat production?

Year	1950	1960	1970	1980	1990
Veal	1230	1109	588	400	327
Lamb	576	769	551	318	358

Source: The World Almanac and Book of Facts, 1993 edition, copyright Pharos Books 1992, New York, NY 10166. Used with permission.

*2–62. The precipitation (in inches) for Juneau, Alaska, and Miami, Florida, for the months of January through August is shown below. Draw two time series graphs to represent the data. What comparisons can be made?

Month	Jan.	Feb.	Mar.	Apr.	May	June	July	Aug.
Juneau	3.7	3.7	3.3	2.9	3.4	3.0	4.1	5.0
Miami	2.1	2.1	1.9	3.1	6.5	9.2	6.0	7.0

Source: The World Almanac and Book of Facts, 1993 edition, copyright Pharos Books 1992, New York, NY 10166. Used with permission.

Source: Cartoon by Bradford Veley, Marquette, Michigan. Used with permission.

2–5 SUMMARY

When data are collected, they are called raw data. Since very little knowledge can be obtained from raw data, this data must be organized in some meaningful way. A frequency distribution using classes is the solution.

Once a frequency distribution is constructed, the representation of the data by using graphs is a simple task. The most commonly used graphs in research statistics are the histogram, frequency polygon, and ogive. Other graphs, such as the bar graph, time series graph, pictograph, pie graph, and stem and leaf plot, can also be used; and some of these graphs are seen frequently in newspapers, magazines, and various statistical reports. Two sets of data can be compared on the same graph by using either a compound bar graph or a component bar graph.

Important Terms

Bar graph
Categorical frequency
   distribution
Class midpoint
Class width
Cumulative frequency
Frequency
Frequency distribution
Frequency polygon
Grouped frequency distribution
Histogram
Lower class boundary
Lower class limit
Ogive
Open-ended distribution
Pictograph
Pie graph
Range
Raw data
Relative frequency graph
Stem and leaf plot
Time series graph
Ungrouped frequency
   distribution
Upper class boundary
Upper class limit

Important Formulas

Formula for the percentage of values in each class:

$$\% = \frac{f}{n} \cdot 100\%$$

where

f = frequency of the class   and n = total number of values

Formula for the range:

$$R = \text{highest value} - \text{lowest value}$$

Formula for the class width:

$$\text{class width} = \text{upper boundary} - \text{lower boundary}$$

Formula for the class midpoint:

$$X_m = \frac{\text{lower boundary} + \text{upper boundary}}{2}$$

or

$$X_m = \frac{\text{lower limit} + \text{upper limit}}{2}$$

Formula for the degrees for each section of a pie graph:

$$\text{degrees} = \frac{f}{n} \cdot 360°$$

Review Exercises

2–63. A questionnaire about how news is obtained resulted in the following information from 25 respondents. Construct a frequency distribution for the data (N = newspaper, T = television, R = radio, M = magazine).

N	N	R	T	T
R	N	T	M	R
M	M	N	R	M
T	R	M	N	M
T	R	R	N	N

2–64. Construct a pie graph for the data in Exercise 2–63.

2–65. A record store kept a record of sales for one randomly selected hour during a recent sale. The following data were obtained. Construct a frequency distribution for the data (LP = long-playing album, CD = compact disc, CAS = cassette, 45 = 45 rpm record).

LP	CD	45	45	CAS
LP	45	LP	CD	CD
LP	45	LP	LP	LP
45	CAS	CAS	CD	LP
LP	LP	CD	CAS	45

2–66. Draw a horizontal bar graph for the data in Exercise 2–65 showing the sales of each item.

2–67. When 20 randomly selected grocery bags were checked, the following number of items were found in each bag. Construct an ungrouped frequency distribution for the data.

17	18	13	14
12	17	11	20
13	18	19	17
14	16	17	12
16	15	19	22

2–68. Construct a histogram, frequency polygon, and ogive for the data in Exercise 2–67.

2–69. The temperature of Keystone Lake was recorded at noon each day for a month. Construct a frequency distribution for the data. Use five classes. (The data for this exercise will be used for Exercise 2–73.)

52	47	50	47	44
54	46	46	45	46
53	41	50	44	44
49	47	43	48	51
49	40	50	51	43
51	46	44	48	48

2–70. Construct a histogram, frequency polygon, and ogive for the data in Exercise 2–69.

2–71. During July, a private pool recorded the following number of swimmers. Construct a frequency distribution for the data. Use six classes. (The data for this exercise will be used for Exercise 2–74.)

143	156	156	163	167
142	171	170	169	164
138	158	160	162	164
173	157	158	159	160
138	172	166	166	159
120	125	165	136	168

2–72. Construct a histogram, frequency polygon, and ogive for the data in Exercise 2–71.

2–73. Construct a histogram, frequency polygon, and ogive by using relative frequencies for the data in Exercise 2–69.

2–74. Construct a histogram, frequency polygon, and ogive by using relative frequencies for the data in Exercise 2–71.

2–75. Shown below are the average running speeds of five animals (in miles per hour, mph). Construct a horizontal bar graph to represent the data.

Animal	**Speed**
Cheetah	70
Lion	50
Coyote	43
Fox	42
Elephant	25

Source: The World Almanac and Book of Facts, 1993 edition, copyright Pharos Books 1992, New York, NY 10166. Used with permission.

2–76. Shown below are the 1991 sales (in billions of dollars) for the top six U.S. corporations. Construct a vertical bar graph for the data.

Company	**Sales**
General Motors	$123,780
Exxon	103,242
Ford	88,963
General Electric	60,236
IBM	64,792
Mobil	56,910

Source: The World Almanac and Book of Facts, 1993 edition, copyright Pharos Books 1992, New York, NY 10166. Used with permission.

2–77. The given data represent the federal minimum hourly wage in the years shown. Draw a time series graph to represent the data.

Year	Wage
1960	$1.00
1965	$1.25
1970	$1.60
1975	$2.10
1980	$3.10
1985	$3.35
1990	$3.80

Source: The World Almanac and Book of Facts, 1993 edition, copyright Pharos Books 1992, New York, NY 10166. Used with permission.

2–78. The number of bank failures in the United States each year for the years 1983–1991 is shown below. Draw a time series graph to represent the data.

Year	Number of Failures
1983	48
1984	72
1985	120
1986	145
1987	184
1988	221
1989	207
1990	169
1991	127

Source: The World Almanac and Book of Facts, 1993 edition, copyright Pharos Books 1992, New York, NY 10166. Used with permission.

2–79. The following data represent the number of strikes (work stoppages) involving 1000 workers or more in the United States for the years shown. Draw a time series graph to represent the data.

Year	Number of Strikes
1960	222
1965	268
1970	381
1975	235
1980	187
1985	54
1990	44

Source: The World Almanac and Book of Facts, 1993 edition, copyright Pharos Books 1992, New York, NY 10166. Used with permission.

2–80. In a study of 100 working women, the numbers shown below indicate the major reason why each woman surveyed worked outside the home. Construct a pie graph for the data.

Reason	Number
To support self/family	62
For extra money	18
For something different to do	12
Other	8

2–81. A survey of the students in the school of engineering of a large university obtained the following data for students enrolled in specific fields. Construct a pie graph for the data.

Engineering Field	Number
Chemical	417
Electrical	389
Civil	506
Mechanical	201

2–82. The following data represent the number of orbits for manned space flights. Draw a pictograph for the data.

Space Flight	No. of Orbits
Apollo 18	136
Soyuz 19	90
Soyuz 11	360
Apollo-Saturn 7	163
Mercury-Atlas 9	22

Source: The World Almanac and Book of Facts, 1993 edition, copyright Pharos Books 1992, New York, NY 10166. Used with permission.

2–83. For the years shown, the following data represent the number of passenger cars sold in the United States. Construct a pictograph for the data. (Data are in thousands.)

Year	Sales
1960	7905
1970	8284
1980	8010
1990	9783

Source: The World Almanac and Book of Facts, 1993 edition, copyright Pharos Books 1992, New York, NY 10166. Used with permission.

2–84. The number of visitors to the Railroad Museum for 24 randomly selected hours is shown here. Construct a stem and leaf plot for the data.

67	62	38	73	34	43	72	35
53	55	58	63	47	42	51	62
32	29	47	62	29	38	36	41

2–85. The data set shown below represents the number of hours 25 part-time employees worked at the Sea Side Amusement Park during a randomly selected week in June. Construct a stem and leaf plot for the data.

16	25	18	39	25	17	29	14	37
22	18	12	23	32	35	24	26	
20	19	25	26	38	38	33	29	

2–86. A special aptitude test is given to job applicants. The data shown below represent the scores of 30 applicants. Construct a stem and leaf plot for the data.

204	210	227	218	254
256	238	242	253	227
251	243	233	251	241
237	247	211	222	231
218	212	217	227	209
260	230	228	242	200

COMPUTER APPLICATIONS

The following data represent the calories contained in 100 randomly selected hamburger sandwiches. Use the data in the exercises below.

310	339	362	389	367	327	352
321	352	357	381	339	360	343
387	368	359	372	320	325	377
390	371	390	359	372	311	319
351	327	344	312	331	321	388
362	308	327	313	366	343	313
368	362	332	307	353	331	383
347	333	361	361	353	352	341
333	341	317	337	362	312	390
322	346	319	371	377	309	386
315	352	316	352	367	361	331
354	355	349	328	350	350	382
327	359	369	343	319	352	366
330	332	361	351	386	333	317
314	335					

1. Using the computer package of your choice, enter the data.

   a. Construct a frequency distribution.
   b. Draw a histogram and a frequency polygon.

2. Write a program that will sort and print the data.
3. Write a program that will construct a frequency distribution for the data.

4. MINITAB can be used to construct a histogram and a stem and leaf plot, as shown in this exercise. For example, suppose one wants to construct a histogram and a stem and leaf plot for the following diastolic blood pressure readings of 25 patients.

68	80	69	81	72
100	101	73	102	93
91	92	93	88	82
83	75	75	89	96
103	83	84	85	89

The command SET C1 tells the computer that data will be entered. After the data have been entered, the END OF DATA statement is used. To construct a histogram, type HISTOGRAM OF C1. Finally, to construct a stem and leaf plot for the data, type STEM AND LEAF OF C1. The output will look like this.

```
MTB > SET C1
DATA > 68 80 69 81 72
DATA > 100 101 73 102 93
DATA > 91 92 93 88 82
DATA > 83 75 75 89 96
DATA > 103 83 84 85 89
DATA > END OF DATA
MTB > HISTOGRAM OF C1
```

The computer will print the following.

```
Histogram of C1 N = 25
Midpoint   Count
  70        3  ***
  75        3  ***
  80        3  ***
  85        4  ****
  90        5  *****
  95        3  ***
 100        3  ***
 105        1  *

MTB > SET C1
DATA > 68 80 69 81 72
DATA > 100 101 73 102 93
DATA > 91 92 93 88 82
DATA > 83 75 75 89 96
DATA > 103 83 84 85 89
DATA > END OF DATA
MTB > STEM-AND-LEAF OF C1
```

The computer will print the following.

```
Stem-and-leaf of C1      N = 25
Leaf Unit = 1.0
     2     6  89
     4     7  23
     6     7  55
    12     8  012334
   (4)     8  5899
     9     9  1233
     5     9  6
     4    10  0123
```

The computer will select the class boundaries and tally the data. It prints the midpoints 70, 75, 80, etc., the frequencies 3, 3, 3, 4, etc., and draws a histogram using asterisks. The class boundaries are 67.5–72.5, 72.5–77.5, etc.

For the stem and leaf plot, the stems are 6, 7, 8, 9, and 10. The leaves are given to the right of the stem. The column to the left is called a depths column. The depth of the line tells how many leaves (data values) lie on the line or beyond. The line with the ( ) contains the median, and the number in the parentheses tells how many leaves are in the line. In this example, there are four leaves. The cumulative frequencies start at the top and accumulate down to the line that contains the median. They also start at the bottom and accumulate up to the line that contains the median.

a. Use MINITAB to draw a stem and leaf diagram for Exercise 2–84.
b. Use MINITAB to construct a histogram for Exercise 2–69.

√ DATA ANALYSIS Applying the Concepts

1. From the Data Bank located in the Appendix, choose one of the following variables: age, weight, cholesterol level, systolic pressure, IQ, or sodium level. Select at least 30 values. For these values, construct a grouped frequency distribution. Draw a histogram, frequency polygon, and ogive for the distribution. Describe briefly the shape of the distribution.
2. From the Data Bank, choose one of the following variables: educational level, smoking status, or exercise. Select at least 30 values. Construct an ungrouped frequency distribution for the data. For the distribution, draw a bar graph, and describe briefly the nature of the graphs.
3. From the Data Bank, select at least 30 people and construct a categorical distribution for their marital status. Draw a pie chart, and describe briefly the findings.
4. From the Data Bank, select at least 30 values for one of the following: cholesterol level, IQ, or sodium level. Construct a stem and leaf plot for the data.

✎ TEST

Directions: Determine whether each statement is true or false.

1. In the construction of a frequency distribution, it is a good idea to have overlapping class limits, such as 10–20, 20–30, 30–40.

2. The midpoint of a class is found by adding the upper and lower limits and dividing by 2.

3. If the limits of a class are 10–19, then the width of the class is 9.

4. The ogive uses cumulative frequencies.

5. Histograms can be drawn by using vertical or horizontal bars.

6. If the limits of a class in a frequency distribution are 18–24, then the boundaries are 17.5–23.5.

7. When data is first collected, it is called raw data.

8. A frequency distribution should contain between 50 and 100 classes.

9. It is not important to keep the width of each class the same in a frequency distribution.

10. Frequency distributions can aid the researcher in drawing charts and graphs.

11. The type of graph used to represent data is determined by the type of data collected and by the researcher's purpose.

12. Data such as blood type (A, B, AB, O) can be organized into a categorical distribution.

13. In the construction of a frequency polygon, the class limits are used for the x axis.

14. Data collected over a period of time can be graphed by using a pie graph.

15. When the data is represented graphically by symbols or pictures, the graph is called a stem and leaf plot.

C H A P T E R

3

Data Description

3–1 INTRODUCTION

Variables that generate raw data occur at random. So that useful information can be obtained about these variables, as shown in the previous chapter, raw data can be organized by using frequency distributions. Furthermore, organized data can be presented by using various graphs. In addition to being used to organize and present data, statistical methods can be used to summarize data. The most familiar of these methods is finding averages. For example, one may read that the average salary for attorneys is $45,700, or the average salary for critical care nurses is $26,600.[1]

In the book *American Averages* by Mike Feinsilber and William B. Meed, the authors state:

"Average" when you stop to think of it is a funny concept. Although it describes all of us it describes none of us. . . . While none of us wants to be the average American, we all want to know about him or her.

The authors go on to give examples of averages:

The average American man is five feet, nine inches tall; the average woman is five feet, 3.6 inches.

The average American is sick in bed seven days a year missing five days of work.

On the average day, 24 million people receive animal bites.

By his or her 70th birthday, the average American will have eaten 14 steers, 1050 chickens, 3.5 lambs, and 25.2 hogs.[2]

In the above examples, the word *average* is ambiguous, since there are several different methods used to obtain an average. Loosely stated, the average means the center of the distribution or the most typical case. *Measures of average* are also called *measures of central tendency* and include the *mean, median, mode,* and *midrange.*

Knowing the average of a data set is not enough to describe the data set entirely. Even though a shoe store owner knows that the average size of a man's shoe is size 10, she would not be in business very long if she ordered only size 10 shoes.

As this example shows, in addition to knowing the average, one must know how the data values are dispersed. That is, do the data values cluster around the mean, or are they spread more evenly throughout the distribution? The measures that determine the spread of the data values are called *measures of variation* or *measures of dispersion.* These measures include the *range, variance,* and *standard deviation.*

Finally, another set of measures is necessary to describe data. These measures are called *measures of position.* They tell where a specific data value falls within the data set or its relative position in comparison with other data values. The most common position measures are *percentiles, deciles,* and *quartiles.* These measures are used extensively in psychology and education. Sometimes, they are referred to as *norms.*

The measures of central tendency, variation, and position are explained in this chapter.

[1]*Parade Magazine*, June 12, 1988. Reprinted with permission from *Parade*, copyright © 1988.

[2]Mike Feinsilber and William B. Meed, *American Averages* (New York: Bantam Doubleday Dell, 1980). Used with permission.

3-2 MEASURES OF CENTRAL TENDENCY

Chapter 1 stated that statisticians use samples taken from populations; however, when populations are small, it is not necessary to use samples since information can be gained by using the entire population. For example, suppose an insurance manager wanted to know the average weekly sales of all the company's representatives. If the company employed a large number of salespersons, say nationwide, he would have to use a sample and make an inference to the entire sales force. But if the company had only a few salespersons, say only 87 agents, he would be able to use all representatives' sales for a randomly chosen week and thus use the entire population.

Measures taken by using all the data values in the populations are called *parameters*. Measures obtained by using the data values of samples are called *statistics*. Hence, the average of the sales from a sample of representatives is called a statistic, and the average sales obtained from the entire population is called a parameter.

A statistic is a characteristic or measure obtained by using the data values from a sample.

A parameter is a characteristic or measure obtained by using all the data values for a specific population.

These concepts as well as the symbols used to represent them will be explained in detail in this chapter.

The Mean

The *mean,* also known as the arithmetic average, is found by adding the values of the data and dividing by the total number of values. For example, the mean of 3, 2, 6, 5, and 4 is found by adding $3 + 2 + 6 + 5 + 4 = 20$ and dividing by 5; hence, the mean of the data is $20 \div 5 = 4$. The values of the data are represented by X's. In the preceding data set, $X_1 = 3$, $X_2 = 2$, $X_3 = 6$, $X_4 = 5$, and $X_5 = 4$. To show a sum of the total X values, the symbol Σ (sigma) is used, and ΣX means to find the sum of the X values in the data set. The summation notation is explained in the Appendix.

The mean is the sum of the values divided by the total number of values. The symbol $\overline{X}$ represents the sample mean.

$$\overline{X} = \frac{X_1 + X_2 + X_3 + \cdots + X_n}{n} = \frac{\Sigma X}{n}$$

where n represents the total number of values in the sample.

For a population, the Greek letter μ is used for the mean.

$$\mu = \frac{X_1 + X_2 + X_3 + \cdots + X_N}{N} = \frac{\Sigma X}{N}$$

where N represents the number of values in the population.

In statistics, Greek letters are used to denote parameters and Roman letters are used to denote statistics. It will be assumed that the data are obtained from samples, unless otherwise specified.

EXAMPLE 3–1   The ages in weeks of six kittens at an animal shelter are 3, 8, 5, 12, 14, and 12. Find the mean.

Solution $$\overline{X} = \frac{\Sigma X}{n} = \frac{3 + 8 + 5 + 12 + 14 + 12}{6} = \frac{54}{6} = 9 \text{ weeks} \quad \blacktriangleleft$$

EXAMPLE 3–2   The miles-per-gallon fuel tests for ten automobiles are given below. Find the mean.

22.2, 23.7, 16.8, 19.7, 18.3, 19.7, 16.9, 17.2, 18.5, 21.0

Solution $$\overline{X} = \frac{\Sigma X}{n}$$

$$= \frac{22.2 + 23.7 + 16.8 + 19.7 + 18.3 + 19.7 + 16.9 + 17.2 + 18.5 + 21.0}{10}$$

$$= \frac{194}{10} = 19.4 \text{ miles per gallon} \quad \blacktriangleleft$$

EXAMPLE 3–3   A survey of 25 households showed the following number of phones per household. Find the mean for the data set.

1	2	1	0	1
3	1	2	2	2
4	1	3	1	1
1	2	1	1	3
1	5	2	4	0

Solution $$\overline{X} = \frac{\Sigma X}{n} = \frac{1 + 2 + 1 + \cdots + 0}{25} = \frac{45}{25} = 1.8 \text{ phones} \quad \blacktriangleleft$$

For data in an ungrouped frequency distribution, the mean can be found as shown in the next example.

EXAMPLE 3–4   The scores for 25 students on a 5-point quiz are shown below. Find the mean.

Score	Frequency
0	1
1	2
2	6
3	12
4	3
5	1
	$n = 25$

Solution   **STEP 1**   Multiply the score by the frequency for each class, and place the result in column C as shown here.

A Score (X)	B Frequency (f)	C $f \cdot X$
0	1	$0 \cdot 1 = 0$
1	2	$1 \cdot 2 = 2$
2	6	$2 \cdot 6 = 12$
3	12	$3 \cdot 12 = 36$
4	3	$4 \cdot 3 = 12$
5	1	$5 \cdot 1 = 5$
	$n = 25$	$\Sigma f \cdot X = 67$

STEP 2 Find the sum of column C, as shown above.

STEP 3 Divide the sum of the frequency column by *n,* as shown:

$$\overline{X} = \frac{67}{25} = 2.68$$

The formula is

$$\overline{X} = \frac{\Sigma f \cdot X}{n} \qquad \blacktriangleleft$$

EXAMPLE 3–5 Holmes Appliance reported the following number of television sets sold per month over a two-year period. Find the mean.

Number of Sets Sold	Frequency (Months)
5	2
6	3
7	8
8	1
9	6
10	4
	24

Solution

X	f	$f \cdot X$
5	2	10
6	3	18
7	8	56
8	1	8
9	6	54
10	4	40
	$n = 24$	$\Sigma f \cdot X = 186$

$$\overline{X} = \frac{\Sigma f \cdot X}{n} = \frac{186}{24} = 7.75 \text{ televisions} \qquad \blacktriangleleft$$

EXAMPLE 3–6 In a large hospital, a survey of 25 nurses showed the following number of patients per nurse on the afternoon shift. Find the mean.

Number of Patients	Frequency	$f \cdot X$
2	1	2
3	4	12
4	6	24
5	8	40
6	4	24
7	2	14
	$n = 25$	$\Sigma f \cdot X = 116$

$$\overline{X} = \frac{\Sigma f \cdot X}{n} = \frac{116}{25} = 4.64 \text{ patients} \qquad \blacktriangleleft$$

The procedure for finding the mean for grouped data is similar to that for ungrouped data, except that the midpoints of the classes are used for the X values.

EXAMPLE 3–7   Using the frequency distribution for Example 2–3 in Chapter 2, find the mean. The data represent the distances the sales representatives traveled to meet their clients on a randomly selected day of the week.

Solution   The procedure for finding the mean for grouped data is given here.

STEP 1   Make a table as shown.

A Class	B Frequency (f)	C Midpoint (X_m)	D $f \cdot X_m$
5.5–10.5	1		
10.5–15.5	2		
15.5–20.5	3		
20.5–25.5	5		
25.5–30.5	4		
30.5–35.5	3		
35.5–40.5	2		
	$n = 20$		

STEP 2   Find the midpoints of each class and enter them in column C.

$$X_m = \frac{5.5 + 10.5}{2} = 8, \qquad \frac{10.5 + 15.5}{2} = 13, \qquad \text{etc.}$$

STEP 3   For each class, multiply the frequency by the midpoint, as shown below, and place the product in column D.

$$1 \cdot 8 = 8, \qquad 2 \cdot 13 = 26, \qquad \text{etc.}$$

The completed table is shown here.

A Class	B Frequency (f)	C Midpoint (X_m)	D $f \cdot X_m$
5.5–10.5	1	8	8
10.5–15.5	2	13	26
15.5–20.5	3	18	54
20.5–25.5	5	23	115
25.5–30.5	4	28	112
30.5–35.5	3	33	99
35.5–40.5	2	38	76
	$n = 20$		$\Sigma f \cdot X_m = 490$

STEP 4   Find the sum of column D, as shown above.

STEP 5   Divide the sum by n to get the mean.

$$\overline{X} = \frac{\Sigma f \cdot X_m}{n} = \frac{490}{20} = 24.5 \text{ miles}$$

EXAMPLE 3–8 The following frequency distribution shows the amount of money (in dollars) 100 families spend for gasoline per month. Find the mean.

Class	Frequency
24.5–29.5	5
29.5–34.5	10
34.5–39.5	16
39.5–44.5	32
44.5–49.5	27
49.5–54.5	6
54.5–59.5	4
	$n = 100$

Solution **STEP 1** Make a table, as shown in Step 1 of Example 3–7.

STEP 2 Find the midpoints of each class, and place the values in column C.

$$X_m = \frac{24.5 + 29.5}{2} = 27, \quad \text{etc.}$$

STEP 3 Multiply the midpoint by the frequency for each class, and place the products in column D.

$$5 \cdot 27 = 135, \quad 10 \cdot 32 = 320, \quad \text{etc.}$$

STEP 4 Find the sum of column D.

$$\Sigma f \cdot X_m = 4200$$

STEP 5 Divide the sum by n.

$$\overline{X} = \frac{\Sigma f \cdot X_m}{n} = \frac{4200}{100} = \$42$$

These steps are summarized in the table below.

A Class	B Frequency (f)	C Midpoint (X_m)	D $f \cdot X_m$
24.5–29.5	5	27	135
29.5–34.5	10	32	320
34.5–39.5	16	37	592
39.5–44.5	32	42	1344
44.5–49.5	27	47	1269
49.5–54.5	6	52	312
54.5–59.5	4	57	228
	$n = 100$		$\Sigma f \cdot X_m = 4200$ ◄

The procedure for finding the mean for grouped data assumes that all of the raw data values in each class are equal to the midpoint of the class. In reality, this is not true, since the average of the raw data values in each class will not be exactly equal to the midpoint. However, using this procedure will give an acceptable approximation of the mean, since some values fall above the midpoint and some values fall below the midpoint for each class.

The procedure for finding the mean for grouped data is summarized in Procedure Table 3.

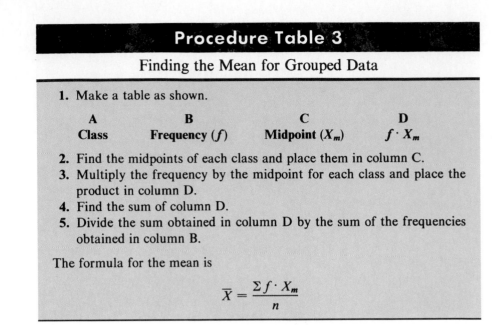

Procedure Table 3

Finding the Mean for Grouped Data

1. Make a table as shown.

A Class	B Frequency (f)	C Midpoint (X_m)	D $f \cdot X_m$

2. Find the midpoints of each class and place them in column C.
3. Multiply the frequency by the midpoint for each class and place the product in column D.
4. Find the sum of column D.
5. Divide the sum obtained in column D by the sum of the frequencies obtained in column B.

The formula for the mean is

$$\overline{X} = \frac{\Sigma f \cdot X_m}{n}$$

The Median   An article recently reported that the median income for college professors was $43,250. This measure of average means that half of all the professors surveyed earned more than $43,250, and half earned less than $43,250.

The *median* is the halfway point in a data set. Before one can find this point, the data must be arranged in order. When the data set is ordered, it is called a **data array.** The median either will be a specific value in the data set or will fall between two values, as shown in the following examples.

The median is the midpoint of the data array. The symbol for the median is MD.

Steps in Computing the Median of a Data Array

1. Arrange the data in order.
2. Find the midpoint of the array.

EXAMPLE 3–9   The weights (in pounds) of seven army recruits are 180, 201, 220, 191, 219, 209, and 186. Find the median.

Solution   **STEP 1**   Arrange the data in order.

180, 186, 191, 201, 209, 219, 220

STEP 2   Select the middle value.

180, 186, 191, (201,) 209, 219, 220

↑

median

Speaking of **Statistics**

USA SNAPSHOTS®

A look at statistics that shape your finances

Shop till you drop

Why do department stores have acres of women's clothes, while the men's section is tucked in a corner? What the average household spends a year on:

$607

$345

Women's clothes   **Men's clothes**

Source: *American Demographics*   By Elys A. McLean, USA TODAY

Source: Based on data from *American Demographics.* Copyright 1993, USA TODAY. Reprinted with permission.

This study uses the mean. How might the results have been obtained? What conclusions can be drawn from the data?

EXAMPLE 3–10   Find the median for the ages of seven preschool children. The ages are 1, 3, 4, 2, 3, 5, and 1.

Solution   1, 1, 2, ③ 3, 4, 5
                                              ↑
                                          median   ◄

Each of the above examples had an odd number of values in the data set; hence, the median was an actual data value. When there are an even number of values in the data set, the median will fall between two given values, as illustrated in the following examples.

EXAMPLE 3–11 The heights (in inches) of eight female police officers are given below. Find the median.

$$63, 64, 71, 66, 65, 66, 68, 62$$

Solution
$$62, 63, 64, 65, 66, 66, 68, 71$$
$$\uparrow$$
median

Since the middle point falls halfway between 65 and 66, the median is found by adding the two values and dividing by 2.

$$MD = \frac{65 + 66}{2} = 65.5 \qquad \blacktriangleleft$$

EXAMPLE 3–12 The ages of ten college students are given below. Find the median.

$$18, 24, 20, 35, 19, 23, 26, 23, 19, 20$$

Solution
$$18, 19, 19, 20, 20, 23, 23, 24, 26, 35$$
$$\uparrow$$
median

The median is

$$MD = \frac{20 + 23}{2} = 21.5 \qquad \blacktriangleleft$$

EXAMPLE 3–13 Six customers purchased the following number of magazines: 1, 7, 3, 2, 3, 4. Find the median.

Solution
$$1, 2, 3, 3, 4, 7 \qquad MD = \frac{3 + 3}{2} = 3$$
$$\uparrow$$
median $\qquad \blacktriangleleft$

For ungrouped frequency distributions, one finds the median by examining the frequencies to locate the middle value, as shown in the next example.

EXAMPLE 3–14 Holmes Appliance recorded the number of videocassette recorders (VCRs) sold per month over a two-year period. Find the median.

Number of Sets Sold	Frequency (Months)
1	3
2	8
3	5
4	4
5	2
6	1
7	$\underline{1}$
	$n = 24$

Solution To locate the middle point, divide n by 2, which is $24 \div 2 = 12$. Then locate the point where 12 values would fall below and 12 values would fall above. This can be done by looking at the cumulative frequency.

Class	Frequency	Cumulative Frequency	
1	3	3	
2	8	11	
3	5	16 ←	(This class contains the
4	4	20	12th through the 16th
5	2	22	values.)
6	1	23	
7	1	24	

The twelfth and thirteenth values fall in class 3. Hence, the median is 3. ◄

The procedure for finding the median for grouped data is shown in the next example.

EXAMPLE 3–15 Using the frequency distribution for Example 2–3 in Chapter 2, find the median.

Solution The procedure for finding the median for grouped data is shown here.

STEP 1 Make a table as shown next.

A Class Boundaries	B Frequency	C Cumulative Frequency
5.5–10.5	1	1
10.5–15.5	2	3
15.5–20.5	3	6
20.5–25.5	5	11·
25.5–30.5	4	15
30.5–35.5	3	18
35.5–40.5	2	20
	$n = 20$	

STEP 2 Divide n (the sum of column B) by 2 to find the halfway point.

$$\frac{20}{2} = 10$$

STEP 3 Find the class that contains the tenth value by using the cumulative frequency distribution. This class is called the **median class**; it contains the median.

Class	Frequency	Cumulative Frequency
5.5–10.5	1	1
⋮	⋮	⋮
15.5–20.5	3	6 ← cf
$L_m \rightarrow$ 20.5–25.5	5 ← f	11 ← Median class
⋮	⋮	⋮
	$n = 20$	

STEP 4   Substitute in the formula

$$MD = \frac{(n/2) - cf}{f}(w) + L_m$$

where L_m = lower boundary of the median class = 20.5

n = sum of frequencies = 20

cf = cumulative frequency of the class immediately preceding the median class = 6

f = frequency of the median class = 5

w = width of the median class = 25.5 − 20.5 = 5

STEP 5   Solve for the median.

$$MD = \frac{(20/2) - 6}{5}(5) + 20.5 = 24.5$$

The median is 24.5.   ◄

The reasoning behind this procedure is as follows: Since the halfway point falls in the class 20.5–25.5, and there are five values in this class, it is assumed that these five values are evenly distributed within the class, as shown in Figure 3–1.

FIGURE 3–1
Median Class

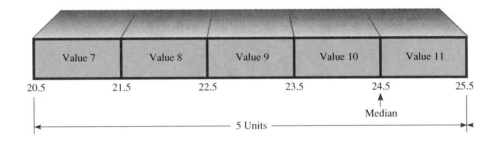

The middle point is between the tenth and eleventh value, which is 24.5. The formula

$$\frac{(n/2) - cf}{f}$$

gives the fractional part of the class where the median lies. In order to get the middle point, one must go $\frac{4}{5}$ of the way into the class. Since the class width is 5, $\frac{4}{5}$ of 5 is 4. This value (4) is added to the lower boundary (20.5) to get the median, 24.5.

EXAMPLE 3–16   The distribution shown in Step 1 of the solution for this example was constructed by asking the ages of 50 customers who purchased Spring Showers perfume. Find the median.

Solution   **STEP 1**   Construct a table as shown.

Age Class Boundaries	Frequency	Cumulative Frequency
13.5–22.5	3	3
22.5–31.5	9	12
31.5–40.5	12	24 ← cf
L_m → 40.5–49.5	20 ← f	44 ← Median class
49.5–58.5	3	47
58.5 and up	3	50
	n → 50	

STEP 2   50/2 = 25.

STEP 3   The median class is 40.5–49.5.

STEP 4
$$MD = \frac{(n/2) - cf}{f}(w) + L_m = \frac{(50/2) - 24}{20}(9) + 40.5$$

STEP 5   MD = 40.95.   ◄

One can compute the median for open-ended frequency distributions as long as the middle score does not occur in the open-ended class. If the median does fall in the open-ended class, the data probably have not been grouped correctly.

The procedure for finding the median for grouped data is summarized in Procedure Table 4.

Procedure Table 4

Finding the Median for Grouped Data

1. Make a table as shown below.

A Class Boundaries	B Frequency	C Cumulative Frequency

2. Divide n, the sum of the frequencies (column B), by 2 to get the halfway point.
3. Locate the median class in column C.
4. Substitute in the formula

$$MD = \frac{(n/2) - cf}{f}(w) + L_m$$

where n = sum of the frequencies
   cf = cumulative frequency of the class immediately preceding the median class
   f = frequency of the median class
   w = class width
   L_m = lower boundary of the median class

5. Solve for the median.

Speaking of **Statistics**

USA SNAPSHOTS®
A look at statistics that shape your finances

West leads in home prices
The median price of homes sold in February was $129,200 nationwide. Median prices by region:

Midwest
$93,900

Northeast
$162,900

West
$176,300

South
$112,900

Source: National Association of Realtors   By Elys A. McLean, USA TODAY

Source: Based on data from the National Association of Realtors. Copyright 1992, USA TODAY. Reprinted with permission.

The median is often used for a comparison of the price of homes in the United States in different geographic regions. Why might it be a better description of the average price of homes than the mean?

The Mode   The third measure of average is called the *mode*. The mode is the value that occurs most often in the data set and is sometimes said to be the most typical case.

The value that occurs most often in a data set is called the mode.

A data set can have more than one mode or no mode at all. These situations will be shown in some of the examples that follow.

EXAMPLE 3–17   At a fast-food restaurant, the following data were obtained for the number of cups of coffee ordered by each customer during the 10:00–11:00 A.M. hour on a randomly selected day of the week. The data have been arranged in order. Find the mode.

$$0, 0, 0, 1, 1, 1, 1, 1, 1, 1, 2, 2, 2, 2, 3, 3, 4, 5, 5$$

Solution   Since the most customers ordered one cup of coffee, the mode for the data set is 1. That is, the value 1 occurred more than any other value.   ◄

EXAMPLE 3–18   The number of calories per serving for eight different brands of cereal were recorded. Find the mode.

110, 113, 116, 118, 118, 118, 121, 130

Solution   Since 118 occurs most often, it is the mode. ◄

EXAMPLE 3–19   Six strains of bacteria were tested to see how long they could remain alive outside their normal environment. The time, in minutes, is recorded below. Find the mode.

2, 3, 5, 7, 8, 10

Solution   Since each value occurs only once, there is no mode. ◄

Note: Do not say that the mode is zero, since that would be incorrect. In some data, such as temperature, zero can be an actual value.

EXAMPLE 3–20   Eleven different automobiles were tested at a speed of 15 miles per hour for stopping distances. The data, in feet, are shown below. Find the mode.

15, 18, 18, 18, 20, 22, 24, 24, 24, 26, 26

Solution   Since 18 and 24 both occur 3 times, the modes are 18 and 24. This data set is said to be *bimodal*. ◄

The mode for grouped data is the modal class. The **modal class** is the class with the largest frequency.

EXAMPLE 3–21   Find the modal class for the frequency distribution of miles traveled by sales representatives of Memory Lane Card Company, used in Example 2–3 in Chapter 2.

Class	Frequency
5.5–10.5	1
10.5–15.5	2
15.5–20.5	3
20.5–25.5	5 ← Modal class
25.5–30.5	4
30.5–35.5	3
35.5–40.5	2

Solution   The modal class is 20.5–25.5, since it has the largest frequency. Sometimes, the midpoint of the class is used rather than the boundaries; hence, the mode could also be given as 23. ◄

EXAMPLE 3–22   Find the modal class for the frequency distribution for the test scores of 50 students in a world cultures class.

Class	Frequency
59.5–68.5	2
68.5–77.5	4
77.5–86.5	6
86.5–95.5	25
95.5–104.5	13

Solution   The modal class is 86.5–95.5 since it has the largest frequency, 25. ◄

The mode is also useful in finding the most typical case when the data are nominal or categorical.

EXAMPLE 3–23   A survey showed the following distribution for the number of students enrolled in each field. Find the mode.

Business	1425
Liberal arts	878
Computer science	632
Education	471
General studies	95

Solution   Since the category with the highest frequency is business, the most typical case is a business major.   ◄

EXAMPLE 3–24   In a parking lot, a student observed the following colors of automobiles. Find the mode.

Color	Frequency
Red	6
Blue	15
White	32
Black	5
Brown	12
Gray	18

Solution   The most typical color is white; hence, the mode is white.   ◄

The next example shows how to find the mode for an ungrouped frequency distribution.

EXAMPLE 3–25   The following data were collected on the number of blood tests a local hospital conducted for a random sample of 50 days. Find the mode.

Number of Tests per Day	Frequency (Days)
26	5
27	9
28	12
29	18
30	5
31	0
32	1

Solution   Since 29 tests were given on 18 days (the number of tests that occurred most often), the mode is 29.   ◄

For a data set the mean, median, and mode can be quite different. Consider the following example.

EXAMPLE 3–26   A small company consists of the owner, the manager, the salesperson, and two technicians, whose salaries are listed here. (Assume that this is the entire population.)

Staff	Salary
Owner	$50,000
Manager	20,000
Salesperson	12,000
Technician	9,000
Technician	9,000

Find the mean, median, and mode.

Solution

$$\mu = \frac{\Sigma X}{N} = \frac{50{,}000 + 20{,}000 + 12{,}000 + 9{,}000 + 9{,}000}{5} = \$20{,}000$$

mean = $20,000   median = $12,000   mode = $9,000   ◄

Thus, researchers and statisticians must know which measure of average is being used and when to use each measure of average. The properties and uses of the three measures of average are summarized here.

Properties and Uses of Averages

The Mean

1. The mean is computed by using all the values of the data.
2. The mean varies less than the other two measures when samples are taken from the same population and all three measures are computed for these samples.
3. The mean is used in computing other statistics, such as the variance.
4. The mean for the data set is unique.
5. The mean cannot be computed for an open-ended frequency distribution.
6. The mean is affected by extremely high or low values and may not be the appropriate average to use in these situations.

The Median

1. The median is used when one must find the center or middle value of a data set.
2. The median is used when one must determine whether the data values fall into the upper half or lower half of the distribution.
3. The median is used to find the average of an open-ended distribution.
4. The median is affected less than the mean by extremely high or extremely low values.

The Mode

1. The mode is used when the most typical case is desired.
2. The mode is the easiest average to compute.
3. The mode can be used when the data is nominal, such as religious preference, gender, or political affiliation.
4. The mode is not always unique. A data set can have more than one mode.
5. The mode does not always exist for a data set.

Speaking of **Statistics**

USA SNAPSHOTS®

A look at statistics that shape your finances

Who's lazy?

A U.S. employee must work 30 years to get as much vacation time as the typical European gets the first year on the job:

Average vacation 1st year

France — 5 weeks
England — 5 weeks
Sweden — 4 weeks
Germany — 3 weeks
USA — 2 weeks
Japan — 2 weeks

Source: Wyatt Co.   By Suzy Parker, USA TODAY

Source: Based on information from the Wyatt Co. Copyright 1992, USA TODAY. Reprinted with permission.

In this study, the average used was probably the mode. Do you agree or disagree? Explain your answer.

Distribution Shapes

Frequency distributions can assume many shapes. The three most important shapes are positively skewed, symmetrical, and negatively skewed. Figure 3–2 shows histograms of each.

In a **positively skewed distribution,** the majority of the data values fall to the left of the mean and cluster at the lower end of the distribution; the "tail" is to the right. Also, the mean is to the right of the median, and the mode is to the left of the median.

For example, if an instructor gave an examination and most of the students did poorly, the scores of these students would tend to cluster on the left side of the distribution. A few high scores would constitute the "tail" of the distribution, which would be on the right side of the distribution. Another example of a positively skewed distribution is the incomes of the population of the United States.

FIGURE 3–2
Types of Distributions

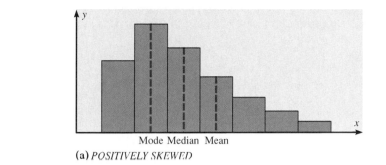

(a) *POSITIVELY SKEWED*

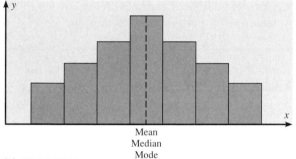

(b) *SYMMETRICAL*

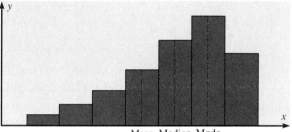

(c) *NEGATIVELY SKEWED*

Most of the incomes cluster about the low end of the distribution; those with high incomes are in the minority, and these incomes are in the "tail" at the right of the distribution.

In a **symmetrical distribution,** the data values are evenly distributed on both sides of the mean. In addition, when the distribution is unimodal, the mean, median, and mode are the same and are at the center of the distribution. Examples of symmetrical distributions are IQ scores and heights of adult males.

When the majority of the data values fall to the right of the mean and cluster at the upper end of the distribution, with the "tail" to the left, the distribution is said to be **negatively skewed.** Also, the mean is to the left of the median, and the mode is to the right of the median. As an example, a negatively skewed distribution results if the majority of students score very high on an instructor's examination. These scores will tend to cluster to the right of the distribution.

When a distribution is extremely skewed, the value of the mean will be pulled toward the tail, but the majority of the data values will be greater than the mean; hence, the median rather than the mean is a more appropriate measure of central tendency. An extremely skewed distribution can also affect other statistics.

A measure of skewness for a distribution is discussed in Exercise 3–88 in Section 3–3.

The Weighted Mean

Sometimes, one must find the mean of a data set in which all values are not equally represented. Consider the case of finding the average cost of a gallon of gasoline for three taxis. Suppose the drivers buy gasoline at three different service stations at a cost of $1.19, $1.27, and $1.32 per gallon. One might try to find the average by using the formula

$$\overline{X} = \frac{\Sigma X}{n}$$

$$= \frac{1.19 + 1.27 + 1.32}{3} = \$1.26$$

But all drivers did not purchase the same number of gallons. Hence, to find the true average cost per gallon, one must take into consideration the number of gallons each driver purchased.

The type of mean that considers an additional factor is called the *weighted mean*, and it is used when the values are not all equally represented.

The weighted mean of a variable X is found by multiplying each value by its corresponding weight and dividing the sum of the products by the sum of the weights.

$$\overline{X} = \frac{w_1 X_1 + w_2 X_2 + \cdots + w_n X_n}{w_1 + w_2 + \cdots + w_n} = \frac{\Sigma w X}{\Sigma w}$$

where w_1, w_2, . . . , w_n are the weights and X_1, X_2, . . . , X_n are the values.

EXAMPLE 3–27

Find the weighted mean of the cost of a gallon of gasoline when each driver purchases the number of gallons shown.

Driver	Gallons Purchased (w)	Cost (X)
1	5	$1.19
2	8	1.27
3	15	1.32

Solution

$$\overline{X} = \frac{\Sigma w X}{\Sigma w} = \frac{5(1.19) + 8(1.27) + 15(1.32)}{5 + 8 + 15} = \$1.28$$

A slightly higher value than $1.26 occurs since driver 3 purchased 15 gallons at a higher price. ◀

It is helpful to think of the X variable as one for which the mean is being found and the other variable as the one used for the weights, w. In the previous example, the mean of the cost of gasoline was desired; hence, cost is used as the X variable and gallons purchased is used for the w variable.

EXAMPLE 3–28

A parole officer wants to find the average number of prison terms (in months) his parolees served. The data follows. Find the weighted mean.

Number of Parolees (w)	Number of Months Served (X)
5	8
3	10
2	24

Solution $$\overline{X} = \frac{\Sigma\,wX}{\Sigma\,w} = \frac{5(8) + 3(10) + 2(24)}{5 + 3 + 2} = 11.8 \text{ months} \qquad \blacktriangleleft$$

The Midrange   The *midrange* is a rough estimate of the middle and is found by adding the lowest and highest values in the data set and dividing by 2. It is a very rough estimate of the average and can be affected by one extremely high or low value.

The midrange is defined as the sum of the lowest and highest values in the data set divided by 2. The symbol MR is used for the midrange.

$$MR = \frac{\text{lowest value} + \text{highest value}}{2}$$

EXAMPLE 3–29   Last winter, the city of Brownsville, Minnesota, reported the following number of water line breaks per month. Find the midrange.

$$2, 3, 6, 8, 4, 1$$

Solution $$MR = \frac{1 + 8}{2} = \frac{9}{2} = 4.5 \qquad \blacktriangleleft$$

If the data set contains one extremely large value or one extremely small value, a higher or lower midrange value will result and may not be a typical description of the middle.

EXAMPLE 3–30   Suppose the number of water line breaks were as follows: 2, 3, 6, 16, 4, and 1. Find the midrange.

Solution $$MR = \frac{1 + 16}{2} = \frac{17}{2} = 8.5$$

The value 8.5 is not typical of the average monthly number of breaks, since 16 breaks occurred in one month. $\qquad \blacktriangleleft$

In statistics there are several measures that can be used for an average. The most common measures include the mean, the median, the mode, and the midrange. Each has its own specific purpose and use. In Exercises 3–39, 3–40, and 3–41, examples of other averages—such as the harmonic mean, the geometric mean, and the quadratic mean—are shown. Their applications are limited to specific areas, as shown in the exercises.

EXERCISES

For Exercises 3–1 through 3–11, find (a) the mean, (b) the median, (c) the mode, and (d) the midrange.

3–1. Twelve secretaries were given a typing test, and the times (in minutes) to complete it were as follows:

$$8, 12, 15, 9, 6, 8, 10, 9, 8, 6, 7, 8$$

3–2. Ten fiction novels were randomly selected, and the number of pages were recorded as follows:

$$415, 398, 402, 399, 400, 405, 395, 401, 412, 407$$

3–3. Twelve members of the high school cross-country team were asked how many minutes each ran during practice sessions. Their answers are recorded here.

$$32, 28, 35, 37, 43, 51, 61, 39, 48, 51, 53, 49$$

3–4. Nine physicians were selected and given a blood pressure check. Their systolic pressures are recorded below.

$$135, 120, 116, 119, 121, 125, 132, 136, 124$$

3–5. The manager of a sports shop recorded the number of baseball caps he sold during the week. The data are shown here.

132, 121, 119, 116, 130, 121, 131

3–6. The heights (in inches) of six female police officer candidates are shown below.

62, 64, 63, 61, 62, 66

3–7. The grade point averages of ten students who applied for financial aid are shown below.

3.15, 3.62, 2.54, 2.81, 3.97,
1.85, 1.93, 2.63, 2.50, 2.80

3–8. The number of calls received by the Internal Revenue Service office in July are shown below.

18, 12, 25, 16, 27, 32, 25, 15, 23, 22, 37, 16,
25, 19, 16, 25, 19, 16, 29, 38, 29, 30, 21

3–9. The number of calories in 12 randomly selected microwave dinners are shown here.

560, 832, 780, 650, 470, 920,
1090, 970, 495, 550, 605, 735

3–10. The exam scores of 18 English composition students were recorded as follows:

78, 62, 98, 90, 88, 73, 79, 86, 81,
84, 93, 97, 63, 59, 78, 82, 87, 93

3–11. A traffic survey records the times (in seconds) of yellow lights at 12 intersections in a large city.

3.4, 3.6, 4.2, 4.8, 3.6, 3.7,
4.1, 4.0, 3.6, 4.1, 3.6, 4.9

For Exercises 3–12 through 3–21, find (a) the mean, (b) the median, and (c) the modal class.

3–12. For 108 randomly selected college students, the following IQ frequency distribution was obtained.

Class Limits	Frequency
90–98	6
99–107	22
108–116	43
117–125	28
126–134	9

3–13. For 50 antique car owners, the following distributions of the age of the cars was obtained.

Class Limits	Frequency
16–18	20
19–21	18
22–24	8
25–27	4

3–14. Thirty automobiles were tested for fuel efficiency (in miles per gallon). The following frequency distribution was obtained.

Class Boundaries	Frequency
7.5–12.5	3
12.5–17.5	5
17.5–22.5	15
22.5–27.5	5
27.5–32.5	2

3–15. The following number of books were read by each of the 28 students in a literature class.

Number of Books	Frequency
0	2
1	6
2	12
3	5
4	3

3–16. In a study of the time it takes an untrained mouse to run a maze, a researcher recorded the following data (in seconds).

Class Limits	Frequency
2.1–2.7	5
2.8–3.4	7
3.5–4.1	12
4.2–4.8	14
4.9–5.5	16
5.6–6.2	8

3–17. Eighty randomly selected lightbulbs were tested to determine their lifetimes (in hours). The following frequency distribution was obtained.

Class Boundaries	Frequency
52.5–63.5	6
63.5–74.5	12
74.5–85.5	25
85.5–96.5	18
96.5–107.5	14
107.5–118.5	5

3–18. The following data represent the net worth (in millions of dollars) of 45 national corporations.

Class Limits	Frequency
10–20	2
21–31	8
32–42	15
43–53	7
54–64	10
65–75	3

3-19. Twenty-five typists were tested, and the data below represent their scores on a typing speed test (in words per minute).

Class Limits	Frequency
54–58	2
59–63	5
64–68	8
69–73	0
74–78	4
79–83	5
84–88	1

3-20. The following frequency distribution represents the commission earned (in dollars) by 100 salespeople employed at several branches of a large chain store.

Class Limits	Frequency
150–158	5
159–167	16
168–176	20
177–185	21
186–194	20
195–203	15
204–212	3

3-21. This frequency distribution represents the data obtained from a sample of 75 copying machine service technicians. The values represent the days between service calls for various copying machines.

Class Boundaries	Frequency
15.5–18.5	14
18.5–21.5	12
21.5–24.5	18
24.5–27.5	10
27.5–30.5	15
30.5–33.5	6

3-22. Find the mean, median, and modal class for the data in Exercise 2–12, Chapter 2.

3-23. Find the mean, median, and modal class for the data in Exercise 2–13, Chapter 2.

3-24. Find the mean, median, and modal class for the data in Exercise 2–14, Chapter 2.

3-25. Find the mean, median, and modal class for the data in Exercise 2–15, Chapter 2.

3-26. Find the weighted mean price of three models of automobiles sold. The number and price of each model sold are shown in the following list.

Model	Number	Price
A	8	$10,000
B	10	$12,000
C	12	$8,000

3-27. Find the weighted mean of the number of hours 15 teenagers talked on the telephone over a three-day period.

Number of Teenagers	Hours Talked
7	2
5	3
3	4

3-28. A recent survey of a new diet cola reported the following percentages of people who liked the taste. Find the weighted mean of the percentages.

Area	% Favored	Number Surveyed
1	40	1000
2	30	3000
3	50	800

3-29. The costs of three models of helicopters are shown below. Find the weighted mean of the cost of the models.

Model	Number Sold	Cost
Sunscraper	9	$427,000
Skycoaster	6	365,000
Highflyer	12	725,000

3-30. An instructor grades as follows: exams, 20%; term paper, 30%; final exam, 50%. A student had grades of 83, 72, and 90, respectively, for exams, term paper, and final exam. Find the student's final average. Use the weighted mean.

3-31. Another instructor gives four 1-hour exams and one final exam, which counts as two 1-hour exams. Find a student's grade if she received 62, 83, 97, and 90 on the hour exams and 82 on the final exam.

3-32. For the following situations, state which measure of central tendency—mean, median, or mode—should be used.

a. The most typical case is desired.
b. The distribution is open-ended.
c. There is an extreme value in the data set.
d. The data is categorical.
e. There will be a need to do further statistical computations.
f. The values are to be divided into two approximately equal groups, one group containing the larger values and one group containing the smaller values.

3-33. Describe which measure of central tendency—mean, median, or mode—was probably used in each situation.

a. Half of the factory workers make more than $5.37 per hour and half make less than $5.37 per hour.
b. The average number of children per family in the Plaza Heights Complex is 1.8.
c. Most people prefer red convertibles over any other color.

d. The average number of times a person cuts the lawn is once a week.
e. The most common fear today is fear of speaking in public.
f. The average age of college professors is 42.3 years.

3–34. What type of symbols are used to represent sample statistics? Give an example.

3–35. What type of symbols are used to represent population parameters? Give an example.

*3–36. If the mean of five values is 64, find the sum of the values.

*3–37. If the mean of five values is 8.2, and four of the values are 6, 10, 7, and 12, find the fifth value.

*3–38. Find the mean of 10, 20, 30, 40, and 50.

a. Add 10 to each value, and find the mean.
b. Subtract 10 from each value, and find the mean.
c. Multiply each value by 10, and find the mean.
d. Divide each value by 10, and find the mean.
e. Make a general statement about each situation.

*3–39. The **harmonic mean** is defined as the number of values divided by the sum of the reciprocals of each value. The formula is

$$HM = \frac{n}{\Sigma\, 1/X}$$

For example, the harmonic mean of 1, 4, 5, and 2 is

$$HM = \frac{4}{\frac{1}{1} + \frac{1}{4} + \frac{1}{5} + \frac{1}{2}} = 2.05$$

This mean is useful for finding the average miles per hour. Suppose a person drove 100 miles at 40 miles per hour and returned driving 50 miles per hour. The average miles per hour is *not* 45 miles per hour, which is found by adding 40 and 50 and dividing by 2. The average is found as follows.
Since

$$\text{time} = \text{distance} \div \text{rate}$$

then

$$\text{time 1} = \frac{100}{40} = 2.5 \text{ hours to make the trip}$$

$$\text{time 2} = \frac{100}{50} = 2 \text{ hours to return}$$

Hence, the total time is 4.5 hours and the total miles driven are 200. Now, the average miles per hour is

$$\text{rate} = \frac{\text{distance}}{\text{time}} = \frac{200}{4.5} = 44.44 \text{ miles per hour}$$

This value can also be found by using the harmonic mean formula:

$$HM = \frac{2}{\frac{1}{40} + \frac{1}{50}} = 44.44$$

Using the harmonic mean, find each of the following.

a. A salesperson drives 300 miles round-trip at 30 miles per hour going to Chicago and 45 miles per hour returning home. Find the average miles per hour.
b. A bus driver drives the 50 miles to West Chester at 40 miles per hour and returns driving 25 miles per hour. Find the average miles per hour.
c. A carpenter buys $500 worth of nails at $50 per pound and $500 worth of nails at $10 per pound. Find the average cost of a pound of nails.

*3–40. The **geometric mean** is defined as the *n*th root of the product of *n* values. The formula is

$$GM = \sqrt[n]{(X_1)(X_2)(X_3) \cdots (X_n)}$$

The geometric mean of 4 and 16 is

$$GM = \sqrt{(4)(16)} = \sqrt{64} = 8$$

The geometric mean of 1, 3, and 9 is

$$GM = \sqrt[3]{(1)(3)(9)} = \sqrt[3]{27} = 3$$

The geometric mean is useful in finding the average of percentages, ratios, indexes, or growth rates. For example, if a person received a 20% raise after one year of service and a 10% raise after the second year of service, the average percentage raise per year is not 15% but 14.89%, as shown.

$$GM = \sqrt{(1.2)(1.1)} = 1.1489$$

or

$$GM = \sqrt{(120)(110)} = 114.89$$

His salary is 120% at the end of the first year and 110% at the end of the second year. This is equivalent to an average of 14.89%, since 114.89% − 100% = 14.89%.

This answer can also be shown by assuming that the person makes $10,000 to start and receives two raises of 20% and 10%.

$$\text{raise 1} = 10,000 \cdot 20\% = \$2000$$
$$\text{raise 2} = 12,000 \cdot 10\% = \$1200$$

His total salary raise is $3,200. This total is equivalent to

$$\$10,000 \cdot 14.89\% = \$1489.00$$
$$\$11,489 \cdot 14.89\% = \underline{\$1710.71}$$
$$\$3199.71 \approx \$3200$$

Speaking of **Statistics**

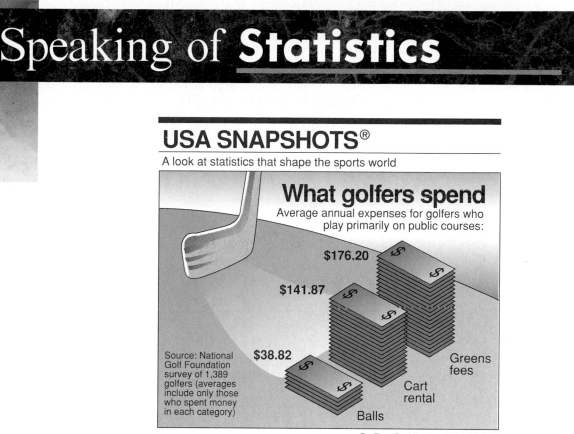

USA SNAPSHOTS®

A look at statistics that shape the sports world

What golfers spend

Average annual expenses for golfers who play primarily on public courses:

$176.20

$141.87

$38.82

Source: National Golf Foundation survey of 1,389 golfers (averages include only those who spent money in each category)

Balls

Cart rental

Greens fees

By Ron Coddington, USA TODAY

Source: Based on a survey done by the National Golf Foundation of 1389 golfers (averages include only those who spent money in each category). Copyright 1992, USA TODAY. Reprinted with permission.

Averages can be used to describe many sport situations, such as the amount spent by golfers for greens fees, cart rentals, and golf balls. Do you think the mean was used in this data summary? What was the sample size? Is it large enough to adequately describe the situation?

Find the geometric mean of each of the following.

a. The growth rate of the Living Life Insurance Corporation for the past three years was 35%, 24%, and 18%.
b. A person received the following percentage raises in salary over a four-year period: 8%, 6%, 4%, and 5%.
c. A stock increased each year for five years at the following percentages: 10%, 8%, 12%, 9%, and 3%.
d. The price increases, in percentages, for the cost of food in a specific geographic region for the past three years were 1%, 3%, and 5.5%.

*3–41. A useful mean in the physical sciences is the **quadratic mean,** which is found by taking the square root of the sum of the average of the squares of each value. The formula is

$$QM = \sqrt{\frac{\Sigma X^2}{n}}$$

The quadratic mean of 3, 5, 6, and 10 is

$$QM = \sqrt{\frac{3^2 + 5^2 + 6^2 + 10^2}{4}}$$
$$= \sqrt{42.5} = 6.52$$

Find the quadratic mean of 8, 6, 3, 5, and 4.

3-3 MEASURES OF VARIATION

In statistics, in order to accurately describe the data set, statisticians must know more than the measures of average. Consider the following example.

EXAMPLE 3-31 Two experimental brands of outdoor paint were tested to see how long they would last before fading. The results (in months) follow. Find the mean of each.

Brand A	Brand B
10	35
60	45
50	30
30	35
40	40
20	25

Solution The mean for brand A is

$$\mu = \frac{\Sigma X}{N} = \frac{210}{6} = 35 \text{ months}$$

The mean for brand B is

$$\mu = \frac{\Sigma X}{N} = \frac{210}{6} = 35 \text{ months} \qquad \blacktriangleleft$$

Since the means are equal in Example 3-31, one might conclude that both brands of paint last equally well. However, when the data sets are examined graphically, a somewhat different conclusion might be drawn. See Figure 3-3.

FIGURE 3-3
Examining Data Sets Graphically

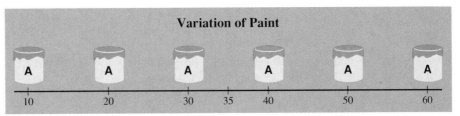

(a) *BRAND A*

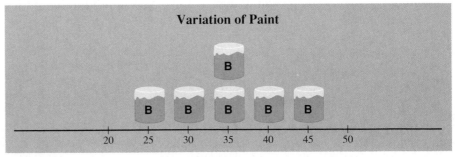

(b) *BRAND B*

As Figure 3–3 shows, even though the means are the same for both brands, the spread, or variation, is quite different. The figure shows that brand B performs more consistently; i.e., it is less variable. For measurement of the spread or variability of a data set, three measures are commonly used: *range, variance,* and *standard deviation.* Each measure will be discussed in this section.

Range   The range is the simplest of the three measures and is defined next.

The range is the highest value minus the lowest value. The symbol R is used for the range.

$$R = \text{highest value} - \text{lowest value}$$

EXAMPLE 3–32   Find the ranges for the paints in Example 3–31.

Solution   For brand A, the range is

$$R = 60 - 10 = 50 \text{ months}$$

For brand B, the range is

$$R = 45 - 25 = 20 \text{ months} \qquad \triangleleft$$

EXAMPLE 3–33   The data that follow represent the number of automobile inspections a garage performs per day for five randomly selected days. Find the range.

$$25, 16, 8, 15, 19$$

Solution $$R = 25 - 8 = 17 \qquad \triangleleft$$

One extremely high or one extremely low data value can affect the range markedly, as shown in the next example.

EXAMPLE 3–34   The salaries for the staff of the XYZ Manufacturing Company follow. Find the range.

Staff	Salary
Owner	$100,000
Manager	$40,000
Sales representative	$30,000
Workers	$25,000
	$15,000
	$18,000

Solution $$R = \$100{,}000 - \$15{,}000 = \$85{,}000 \qquad \triangleleft$$

Since the owner's salary is included in the data for Example 3–34, the range is a large number. In order to have an additional and more meaningful statistic to measure the variability, statisticians use measures called the variance and standard deviation.

Variance and Standard Deviation

Before the variance and standard deviation are defined formally, the computational procedure will be shown, since the definition is derived from the procedure.

EXAMPLE 3–35 Find the variance and standard deviation for the data set for brand A paint in Example 3–31.

$$10, 60, 50, 30, 40, 20$$

Solution **STEP 1** Find the mean for the data.

$$\mu = \frac{\Sigma X}{N} = \frac{10 + 60 + 50 + 30 + 40 + 20}{6} = \frac{210}{6} = 35$$

STEP 2 Subtract the mean from each data value.

$$10 - 35 = -25 \quad 50 - 35 = +15 \quad 40 - 35 = +5$$
$$60 - 35 = +25 \quad 30 - 35 = -5 \quad 20 - 35 = -15$$

STEP 3 Square each result.

$$(-25)^2 = 625 \quad (+15)^2 = 225 \quad (+5)^2 = 25$$
$$(+25)^2 = 625 \quad (-5)^2 = 25 \quad (-15)^2 = 225$$

STEP 4 Find the sum of the squares.

$$625 + 625 + 225 + 25 + 25 + 225 = 1750$$

STEP 5 Divide the sum by N to get the variance.

$$\text{variance} = 1750 \div 6 = 291.67$$

STEP 6 Take the square root of the variance to get the standard deviation.

$$\text{standard deviation} = \sqrt{291.67} = 17.08$$

It is helpful to make a table, as follows:

A Values (X)	B $X - \mu$	C $(X - \mu)^2$
10	−25	625
60	+25	625
50	+15	225
30	−5	25
40	+5	25
20	−15	225
		1750

Column A contains the raw data, X. Column B contains the differences obtained in Step 2, $X - \mu$. Column C contains the squares of the differences obtained in Step 3. ◄

By observing the preceding computational procedure, one notices several things. First, the square root of the variance gives the standard deviation; and vice versa, by squaring the standard deviation, one gets the variance. Second, the variance is actually the average of the square of the distance that each value is from the mean. Therefore, if the values are near the mean, the variance will be small. In contrast, if the values are further away from the mean, the variance will be large.

One might wonder why the squared distances are used instead of the actual distances. The reason is that the sum of the distances will always be zero. This result can be verified for a specific case by adding the values in column B of the table in Example 3–35. When each value is squared, the negative signs are eliminated.

Finally, why is it necessary to take the square root? The reason is that since the distances were squared, the units of the resultant numbers are the squares of the units of the original raw data. Taking the square root of the variance means that the standard deviation will be in the same units as the raw data.

The variance and standard deviation are formally defined next.

The variance is the average of the squares of the distance each value is from the mean. The symbol for the population variance is σ^2 (σ is the Greek lowercase letter sigma).

The formula for the population variance is

$$\sigma^2 = \frac{\Sigma (X - \mu)^2}{N}$$

where

$$X = \text{individual value}$$
$$\mu = \text{population mean}$$
$$N = \text{population size}$$

The standard deviation is the square root of the variance. The symbol for the population standard deviation is σ.

The corresponding formula for the standard deviation is

$$\sigma = \sqrt{\sigma^2} = \sqrt{\frac{\Sigma (X - \mu)^2}{N}}$$

EXAMPLE 3–36 Find the variance and standard deviation for brand B paint data in Example 3–31.

$$35, 45, 30, 35, 40, 25$$

Solution **STEP 1** Find the mean.

$$\mu = \frac{\Sigma X}{N} = \frac{35 + 45 + 30 + 35 + 40 + 25}{6} = \frac{210}{6} = 35$$

STEP 2 Subtract the mean from each value and place the result in column B of the table.

STEP 3 Square each result and place the squares in column C of the table.

A	B	C
X	$X - \mu$	$(X - \mu)^2$
35	0	0
45	10	100
30	−5	25
35	0	0
40	5	25
25	−10	100

STEP 4   Find the sum of the squares in column C.

$$0 + 100 + 25 + 0 + 25 + 100 = 250$$

STEP 5   Divide the sum by N to get the variance.

$$\sigma^2 = \frac{250}{6} = 41.67$$

STEP 6   Take the square root to get the standard deviation.

$$\sigma = \sqrt{41.67} = 6.455 \qquad \blacktriangleleft$$

Since the standard deviation of brand A is 17.08 (see Example 3–35), and the standard deviation of brand B is 6.455, the data are more variable for brand A. In summary, when the means are equal, the larger the variance or standard deviation is, the more variable the data are.

The Unbiased Estimator

When computing the variance for a sample, one would assume that the following formula would be used:

$$\frac{\Sigma (X - \overline{X})^2}{n}$$

where $\overline{X}$ is the sample mean and n is the sample size. This formula is *not* usually used, however, since in most cases the purpose of calculating the statistic is to estimate the corresponding parameter. For example, the sample mean $\overline{X}$ is used to estimate the population mean μ. The formula

$$\frac{\Sigma (X - \overline{X})^2}{n}$$

produces what is called a *biased estimate* of the population variance—an estimate that is different from the expected value of a population parameter. When the population is large and the sample is small (usually less than 30), the variance computed by this formula would underestimate the population variance. Therefore, instead of dividing by n, the variance of the sample is found by dividing by $n - 1$, giving a slightly larger value and an *unbiased* estimate of the population variance.

The **unbiased estimator** of the population variance is an estimator whose value approximates the expected value of a population variance. It is denoted by s^2, and its formula is

$$s^2 = \frac{\Sigma (X - \overline{X})^2}{n - 1}$$

where

$\overline{X}$ = sample mean
n = sample size

To find the standard deviation of a sample, one must take the square root of the sample variance, which was found by using the preceding formula.

Formula for the Sample Standard Deviation

The standard deviation of a sample (denoted by s) is

$$s = \sqrt{s^2} = \sqrt{\frac{\Sigma (X - \overline{X})^2}{n - 1}}$$

where
$$X = \text{individual value}$$
$$\overline{X} = \text{sample mean}$$
$$n = \text{sample size}$$

Shortcut formulas for computing the variance and standard deviation are presented next and will be used in the remainder of the chapter and in the exercises. These formulas are mathematically equivalent to the preceding formulas and do not involve using the mean.

Shortcut Formulas for s^2 and s

The shortcut formulas for computing the variance and standard deviation for data obtained from samples are as follows.

Variance	Standard Deviation
$s^2 = \dfrac{\Sigma X^2 - (\Sigma X)^2/n}{n - 1}$	$s = \sqrt{\dfrac{\Sigma X^2 - (\Sigma X)^2/n}{n - 1}}$

The next two examples explain how to use the shortcut formulas.

EXAMPLE 3–37 For a newly created position, the manager interviewed the following numbers of applicants each day over a five-day period: 16, 19, 15, 15, and 14. Find the variance and standard deviation.

Solution

STEP 1 Find the sum of the values.

$$\Sigma X = 16 + 19 + 15 + 15 + 14 = 79$$

STEP 2 Square each value and find the sum.

$$\Sigma X^2 = 16^2 + 19^2 + 15^2 + 15^2 + 14^2 = 1263$$

STEP 3 Substitute in the formulas and solve.

$$s^2 = \frac{\Sigma X^2 - [(\Sigma X)^2/n]}{n - 1} = \frac{1263 - [(79)^2/5]}{4} = 3.7$$
$$s = \sqrt{3.7} = 1.92$$

◄

Note that ΣX^2 is not the same as $(\Sigma X)^2$. The notation ΣX^2 means to square the values first, then sum; $(\Sigma X)^2$ means to sum the values first, then square.

EXAMPLE 3–38 The number of swimming pool sales for a randomly selected week in July is given below. Find the variance and standard deviation for the data, using the shortcut formulas.

$$2, 5, 8, 6, 3, 3, 1$$

Solution **STEP 1** $$\Sigma X = 2 + 5 + 8 + 6 + 3 + 3 + 1 = 28$$

STEP 2 $$\Sigma X^2 = 2^2 + 5^2 + 8^2 + 6^2 + 3^2 + 3^2 + 1^2 = 148$$

STEP 3 $$s^2 = \frac{\Sigma X^2 - [(\Sigma X)^2/n]}{n - 1} = \frac{148 - [(28)^2/7]}{6} = 6$$

$$s = \sqrt{6} = 2.45$$ ◀

Finding the Variance and Standard Deviation for Grouped Data

The procedure for finding the variance and standard deviation for grouped data is similar to that for finding the mean for grouped data, and it uses the midpoints of each class.

EXAMPLE 3–39 Find the variance and standard deviation for the frequency distribution of the data in Example 2–3 in Chapter 2. The data represent the distance (in miles) the sales representatives traveled to meet their clients on a randomly selected day of the week.

Class	Frequency	Midpoint
5.5–10.5	1	8
10.5–15.5	2	13
15.5–20.5	3	18
20.5–25.5	5	23
25.5–30.5	4	28
30.5–35.5	3	33
35.5–40.5	2	38

Solution **STEP 1** Make a table as shown, and find the midpoints of each class.

A	B	C	D	E
Class	Frequency (f)	Midpoint (X_m)	$f \cdot X_m$	$f \cdot X_m^2$
5.5–10.5	1	8		
10.5–15.5	2	13		
15.5–20.5	3	18		
20.5–25.5	5	23		
25.5–30.5	4	28		
30.5–35.5	3	33		
35.5–40.5	2	38		

STEP 2 Multiply the frequency by the midpoint for each class and place the products in column D.

$$1 \cdot 8 = 8 \qquad 2 \cdot 13 = 26 \qquad . . . \qquad 2 \cdot 38 = 76$$

STEP 3   Multiply the frequency by the square of the midpoint and place the products in column E.

$$1 \cdot 8^2 = 64 \qquad 2 \cdot 13^2 = 338 \qquad \ldots \qquad 2 \cdot 38^2 = 2888$$

STEP 4   Find the sum of columns B, D, and E. The sum of column B is n, the sum of column D is $\Sigma f \cdot X_m$, and the sum of column E is $\Sigma f \cdot X_m^2$.

The completed table follows.

A Class	B Frequency	C Midpoint	D $f \cdot X_m$	E $f \cdot X_m^2$
5.5–10.5	1	8	8	64
10.5–15.5	2	13	26	338
15.5–20.5	3	18	54	972
20.5–25.5	5	23	115	2,645
25.5–30.5	4	28	112	3,136
30.5–35.5	3	33	99	3,267
35.5–40.5	2	38	76	2,888
	$n = 20$		$\Sigma f \cdot X_m = 490$	$\Sigma f \cdot X_m^2 = 13,310$

STEP 5   Substitute in the formula and solve for s^2 to get the variance.

$$s^2 = \frac{\Sigma f \cdot X_m^2 - [(\Sigma f \cdot X_m)^2/n]}{n - 1}$$

$$= \frac{13,310 - [(490)^2/20]}{20 - 1} = 68.68$$

STEP 6   Take the square root to get the standard deviation.

$$s = \sqrt{68.68} = 8.29 \qquad \blacktriangleleft$$

EXAMPLE 3–40   Find the variance and standard deviation for the data of Example 3–8 in Section 3–2. The frequency distribution for the amount of money 100 families spend for gasoline per month follows.

Class	Frequency
24.5–29.5	5
29.5–34.5	10
34.5–39.5	16
39.5–44.5	32
44.5–49.5	27
49.5–54.5	6
54.5–59.5	4

Solution   **STEP 1**   Make a table and find the midpoints. The midpoints are

$$\frac{(24.5 + 29.5)}{2} = 27, \qquad \text{etc.}$$

Place the results in column C.

STEP 2   Multiply the frequency by the midpoint for each class,

$$5 \cdot 27 = 135, \qquad \text{etc.}$$

and place the results in column D.

STEP 3 Multiply the frequency by the square of the midpoint,

$$5 \cdot 27^2 = 3645, \quad \text{etc.}$$

and place the results in column E.

STEP 4 Find the sums of columns B, D, and E. The completed table follows.

A	B	C	D	E
Class	Frequency	Midpoint (X_m)	$f \cdot X_m$	$f \cdot X_m^2$
24.5–29.5	5	27	135	3,645
29.5–34.5	10	32	320	10,240
34.5–39.5	16	37	592	21,904
39.5–44.5	32	42	1344	56,448
44.5–49.5	27	47	1269	59,643
49.5–54.5	6	52	312	16,224
54.5–59.5	4	57	228	12,996
	$n = 100$		$\Sigma f \cdot X_m = 4200$	$\Sigma f \cdot X_m^2 = 181{,}100$

STEP 5 Substitute in the formula and solve for s^2.

$$s^2 = \frac{\Sigma f \cdot X_m^2 - [(\Sigma f \cdot X_m)^2 / n]}{n - 1}$$

$$= \frac{181{,}100 - [(4200)^2 / 100]}{99} = 47.4747$$

STEP 6 Take the square root to get the standard deviation.

$$s = \sqrt{47.4747} = 6.89 \qquad \blacktriangleleft$$

The procedure for finding the variance and standard deviation for grouped data is summarized in Procedure Table 5.

Procedure Table 5

Finding the Variance and Standard Deviation for Grouped Data

Step 1: Make a table as shown, and find the midpoints of each class.

A	B	C	D	E
Class	Frequency	Midpoint	$f \cdot X_m$	$f \cdot X_m^2$

Step 2: Multiply the frequency by the midpoint for each class, and place the products in column D.

Step 3: Multiply the frequency by the square of the midpoint, and place the products in column E.

Step 4: Find the sums of columns B, D, and E. (The sum of column B is n. The sum of column D is $\Sigma f \cdot X_m$. The sum of column E is $\Sigma f \cdot X_m^2$.)

Step 5: Substitute in the formula and solve to get the variance.

$$s^2 = \frac{\Sigma f \cdot X_m^2 - [(\Sigma f \cdot X_m)^2 / n]}{n - 1}$$

Step 6: Take the square root to get the standard deviation.

When the data are ungrouped, the same procedure is used, except that the class value is used instead of the midpoint. The next example illustrates this procedure.

EXAMPLE 3–41 Holmes Appliance reported the number of television sets sold per month over a two-year period. Find the variance and standard deviation for the data.

Number of Sets Sold	Frequency (Months)
5	2
6	3
7	8
8	1
9	6
10	4

Solution **STEP 1** Make a table as shown.

A Sets	B Frequency	C $f \cdot X$	D $f \cdot X^2$
5	2		
6	3		
7	8		
8	1		
9	6		
10	4		

STEP 2 Multiply the frequency by the values in column A, and place the products in column C.

$$2 \cdot 5 = 10 \qquad 3 \cdot 6 = 18 \qquad \text{etc.}$$

STEP 3 Multiply the frequency by the values squared, and place the products in column D.

$$2 \cdot 5^2 = 50 \qquad 3 \cdot 6^2 = 108 \qquad \text{etc.}$$

STEP 4 Find the sum of columns, B, C, and D. The completed table follows.

A Sets	B Frequency	C $f \cdot X$	D $f \cdot X^2$
5	2	10	50
6	3	18	108
7	8	56	392
8	1	8	64
9	6	54	486
10	4	40	400
	$n = 24$	$\Sigma f \cdot X = 186$	$\Sigma f \cdot X^2 = 1500$

STEP 5 Substitute in the formula to get the variance.

$$s^2 = \frac{\Sigma f \cdot X^2 - [(\Sigma f \cdot X)^2/n]}{n-1} = \frac{1500 - [(186)^2/24]}{23} = 2.54$$

STEP 6 Take the square root to get the standard deviation.

$$s = \sqrt{2.54} = 1.59 \qquad \blacktriangleleft$$

Uses of the Variance and Standard Deviation

1. As previously stated, the variances and standard deviations can be used to determine the spread of the data. If the variance or standard deviation is large, the data is more dispersed. This information is useful in comparing two or more data sets to determine which is more (most) variable.
2. The measures of variance and standard deviation are used to determine the consistency of a variable. For example, in the manufacture of fittings, like nuts and bolts, the variation in the diameters must be small or the parts will not fit together.
3. The variance and standard deviation are used to determine the number of data values that fall within a specified interval in a distribution. For example, Chebyshev's theorem (explained later) shows that for any distribution, at least 75% of the data values will fall within two standard deviations of the mean.
4. Finally, the variance and standard deviation are used quite often in inferential statistics. These uses will be shown in later chapters of this textbook.

Coefficient of Variation

Whenever two samples have the same units of measure, the variance and standard deviation for each can be compared directly. For example, suppose an automobile dealer wanted to compare the standard deviation of miles driven for the automobiles he received as trade-ins on new cars. He found that for a specific year the standard deviation for Buicks was 422 miles and the standard deviation for Cadillacs was 350 miles. He could say that the variation in mileage was greater in the Buicks. But what if a manager wanted to compare the standard deviations of two different variables, like the number of sales per salesperson over a three-month period and the commissions made by these salespersons?

A statistic that allows one to compare standard deviations when the units are different, as in the previous example, is called the coefficient of variation.

The coefficient of variation is the standard deviation divided by the mean, the result expressed as a percentage.

For samples,

$$\text{CVar} = \frac{s}{\bar{X}} \cdot 100\%$$

For populations,

$$\text{CVar} = \frac{\sigma}{\mu} \cdot 100\%$$

EXAMPLE 3-42 The mean of the number of sales of automobiles over a three-month period is 87 and the standard deviation is 5. The mean of the commissions is $5225 and the standard deviation is $773. Compare the variations of the two.

Solution The coefficients of variation are

$$\text{CVar} = \frac{s}{\bar{X}} = \frac{5}{87} \cdot 100\% = 5.75\% \qquad \text{sales}$$

$$\text{CVar} = \frac{773}{5225} \cdot 100\% = 14.79\% \qquad \text{commissions}$$

Since the coefficient of variation is larger for commissions, the commissions are more variable than the sales. ◄

EXAMPLE 3–43 The mean for the number of pages of a sample of women's magazines is 132, with a variance of 23; the mean for a sample of men's magazines is 182, with a variance of 62. Compare the variations of the two magazines.

Solution The coefficients of variation are

$$\text{CVar} = \frac{\sqrt{23}}{132} \cdot 100\% = 3.63\% \qquad \text{women's}$$

$$\text{CVar} = \frac{\sqrt{62}}{182} \cdot 100\% = 4.33\% \qquad \text{men's}$$

The number of pages in the men's magazines is more variable than in the women's, since the coefficient of variation is larger for the men's magazines. ◄

Chebyshev's Theorem

As stated previously, the variance and standard deviation of a variable can be used to determine the spread or dispersion of a variable. That is, the larger the variance or standard deviation, the more the data values are dispersed. For example, if two variables measured in the same units have the same mean, say 70, and variable one has a standard deviation of 2 while variable two has a standard deviation of 10, then the data for variable two will be more spread out than the data for variable one. **Chebyshev's theorem,** developed by the Russian mathematician Chebyshev, specifies the proportions of the spread in terms of the standard deviation.

Chebyshev's theorem **The proportion of values from a data set that will fall within *k* standard deviations of the mean will be at least $1 - 1/k^2$, where *k* is a number greater than 1.**

This theorem states that at least $\frac{3}{4}$, or 75%, of the data values will fall within two standard deviations of the mean of the data set. This result is found by substituting $k = 2$ in the formula.

$$1 - \frac{1}{k^2} \qquad \text{or} \qquad 1 - \frac{1}{2^2} = 1 - \frac{1}{4} \qquad \text{or} \qquad \frac{3}{4} = 75\%$$

In the previous example, since variable one has a mean of 70 and a standard deviation of 2, at least $\frac{3}{4}$ or 75% of the data values will fall between 66 and 74. These values are found by adding two standard deviations to the mean and subtracting two standard deviations from the mean, as shown.

$$70 + 2(2) = 70 + 4 = 74$$

and

$$70 - 2(2) = 70 - 4 = 66$$

For variable two, at least $\frac{3}{4}$ or 75% of the data values will fall between 50 and 90. Again, these values are found by adding and subtracting, respectively, two standard deviations to and from the mean.

$$70 + 2(10) = 70 + 20 = 90$$

and

$$70 - 2(10) = 70 - 20 = 50$$

Furthermore, the theorem states that at least $\frac{8}{9}$, or 88.89%, of the data values will fall within three standard deviations of the mean. This result is found by letting $k = 3$ and substituting in the formula.

$$1 - \frac{1}{k^2} \quad \text{or} \quad 1 - \frac{1}{3^2} = 1 - \frac{1}{9} = \frac{8}{9} = 88.89\%$$

For variable one, at least $\frac{8}{9}$ or 88.89% of the data values fall between 64 and 76, since

$$70 + 3(2) = 70 + 6 = 76$$

and

$$70 - 3(2) = 70 - 6 = 64$$

For variable two, at least $\frac{8}{9}$ or 88.89% of the data values will fall between 40 and 100.

The next two examples illustrate the application of Chebyshev's theorem.

EXAMPLE 3–44   The mean price of houses in a certain neighborhood is $50,000, and the standard deviation is $10,000. Find the price range for which at least 75% of the houses will sell.

Solution   Chebyshev's theorem states that $\frac{3}{4}$ or 75% of the data values will fall within two standard deviations of the mean. Hence,

$$\$50,000 + 2(\$10,000) = \$50,000 + \$20,000 = \$70,000$$

and

$$\$50,000 - 2(\$10,000) = \$50,000 - \$20,000 = \$30,000$$

Hence, at least 75% of all homes sold in the area will have a price range from $30,000 to $70,000.   ◄

Chebyshev's theorem can be used to find the minimum percentage of data values that will fall between any two given values. The procedure is shown in the next example.

EXAMPLE 3–45   A survey of local companies found that the mean amount of travel allowance for executives was $0.25 per mile. The standard deviation was $0.02. Using Chebyshev's theorem, find the minimum percentage of the data values that will fall between $0.20 and $0.30.

Solution   **STEP 1**   Subtract the mean from the larger value.

$$\$0.30 - \$0.25 = \$0.05$$

STEP 2   Divide the difference by the standard deviation to get k.

$$k = \frac{0.05}{0.02} = 2.5$$

STEP 3   Use Chebyshev's theorem to find the percentage.

$$1 - \frac{1}{k^2} = 1 - \frac{1}{2.5^2} = 1 - \frac{1}{6.25} = 1 - 0.16 = 0.84 \quad \text{or} \quad 84\%$$

Hence, at least 84% of the data values will fall between $0.20 and $0.30.   ◄

The Empirical or Normal Rule

Chebyshev's theorem applies to any distribution regardless of its shape. However, when a distribution is *bell-shaped* (or what is called *normal*), the following statements, which make up the **empirical rule,** are true.

Approximately 68% of the data values will fall within one standard deviation of the mean.

Approximately 95% of the data values will fall within two standard deviations of the mean.

Approximately 99.7% of the data values will fall within three standard deviations of the mean.

For example, suppose that the scores on a national achievement exam are normally distributed with a mean of 480 and a standard deviation of 90. If these scores are normally distributed, then approximately 68% will fall between 390 and 570 (480 + 90 = 570 and 480 − 90 = 390). Approximately 95% of the scores will fall between 300 and 660 (480 + 2 · 90 = 660 and 480 − 2 · 90 = 300). Approximately 99.7% will fall between 210 and 750 (480 + 3 · 90 = 750 and 480 − 3 · 90 = 210). More will be said about the empirical or normal rule in a later chapter.

EXERCISES

3-42. Name three measures of variation.

3-43. What is the relationship between the variance and standard deviation?

3-44. Why might the range *not* be the best estimate of variability?

3-45. What are the symbols used to represent the population variance, and standard deviation?

3-46. What are the symbols used to represent the sample variance and standard deviation?

3-47. Why is the unbiased estimator of variance used?

3-48. The three data sets below have the same mean and range, but is the variation the same? Prove your answer by computing the standard deviation. Assume the data are obtained from samples.

a. 5, 7, 9, 11, 13, 15, 17
b. 5, 6, 7, 11, 15, 16, 17
c. 5, 5, 5, 11, 17, 17, 17

For Exercises 3-49 through 3-59, find the range, variance, and standard deviation. Assume the data represent samples, and use the shortcut formula for the unbiased estimator to compute the variance and standard deviation.

3-49. Twelve students were given an arithmetic test, and the times (in minutes) to complete it were as follows.

10, 9, 12, 11, 8, 15, 9, 7, 8, 6, 12, 10

3–50. Ten used trail bikes are randomly selected, and the odometer reading of each is recorded as follows.

1902	103	653	1901	788
361	216	363	223	656

3–51. Fifteen students were selected and asked how many hours each studied for the final exam in statistics. Their answers are recorded here.

8, 6, 3, 0, 0, 5, 9, 2, 1, 3, 7, 10, 0, 3, 6

3–52. The weights of nine football players are recorded here.

206, 215, 305, 297, 265, 282, 301, 255, 261

3–53. The manager of a dress shop recorded the number of dresses she sold during the week. The data are shown here.

18, 16, 15, 18, 17, 19, 32

3–54. The heights (in inches) of nine male army recruits are shown here.

78, 72, 68, 73, 75, 69, 74, 73, 72

3–55. The grade point averages of ten students who applied for financial aid are shown here.

3.15, 3.62, 2.54, 2.81, 3.97,
1.85, 1.93, 2.63, 2.50, 2.80

3–56. The number of calls received by the Internal Revenue Service office in July appears here.

18, 12, 25, 16, 27, 32, 25, 15, 23, 22, 37, 16,
25, 19, 16, 25, 19, 16, 29, 38, 29, 30, 21

3–57. The number of calories in 12 randomly selected microwave dinners is shown here.

560, 832, 780, 650, 470, 920,
1090, 970, 495, 550, 605, 735

3–58. The number of patrons visiting a local theater during a randomly selected week is shown here.

156, 209, 361, 223, 216, 291, 187

3–59. The number of words printed in each of 12 randomly selected third-grade storybooks is listed below.

502	213	335	197	414	469
497	367	409	297	309	414

For Exercises 3–60 through 3–69, find the variance and standard deviation.

3–60. For 108 randomly selected college students, the following IQ frequency distribution was obtained.

Class Limits	Frequency
90–98	6
99–107	22
108–116	43
117–125	28
126–134	9

3–61. For 75 employees of a large department store, the following distribution for years of service was obtained.

Class Limits	Frequency
1–5	21
6–10	25
11–15	15
16–20	0
21–25	8
26–30	6

3–62. Thirty automobiles were tested for fuel efficiency (in miles per gallon). The following frequency distribution was obtained.

Class Boundaries	Frequency
7.5–12.5	3
12.5–17.5	5
17.5–22.5	15
22.5–27.5	5
27.5–32.5	2

3–63. In a class of 29 students, the following grade distribution was recorded (A = 4, B = 3, C = 2, D = 1, F = 0).

Grade	Frequency
0	1
1	3
2	5
3	14
4	6

3–64. In a study of reaction times to a specific stimulus, a psychologist recorded the following data (in seconds).

Class Limits	Frequency
2.1–2.7	12
2.8–3.4	13
3.5–4.1	7
4.2–4.8	5
4.9–5.5	2
5.6–6.2	1

3-65. Eighty randomly selected lightbulbs were tested to determine their lifetimes (in hours). The following frequency distribution was obtained.

Class Boundaries	Frequency
52.5–63.5	6
63.5–74.5	12
74.5–85.5	25
85.5–96.5	18
96.5–107.5	14
107.5–118.5	5

3-66. The following data represent the net worth (in millions of dollars) of 50 businesses in a large city.

Class Limits	Frequency
10–20	5
21–31	10
32–42	3
43–53	7
54–64	18
65–75	7

3-67. Twenty-five typists were tested, and the following data represent their scores (in words per minute) on a typing speed test.

Class Limits	Frequency
54–58	2
59–63	5
64–68	8
69–73	0
74–78	4
79–83	5
84–88	1

3-68. The following frequency distribution represents the commission earned (in dollars) by 100 salespersons employed at several branches of a large chain store.

Class Limits	Frequency
150–158	5
159–167	16
168–176	20
177–185	21
186–194	20
195–203	15
204–212	3

3-69. This frequency distribution represents the data obtained from a sample of word processor repairpersons. The values represent the days between service calls on 80 machines.

Class Boundaries	Frequency
25.5–28.5	5
28.5–31.5	9
31.5–34.5	32
34.5–37.5	20
37.5–40.5	12
40.5–43.5	2

3-70. The average IQ of the students in one calculus class is 110, with a standard deviation of 5; and the average IQ of students in another class is 106, with a standard deviation of 4. Which class is more variable in terms of IQ?

3-71. The average price of the Panther convertible is $40,000, with a standard deviation of $4000. The average price of the Suburban station wagon is $20,000, with a standard deviation of $2000. Compare the variances of both models.

3-72. The average score on an English final examination was 85, with a standard deviation of 5; and the average score on a history final exam was 110, with a standard deviation of 8. Which class was more variable?

3-73. The average age of the accountants at Three Rivers Corporation is 26, with a standard deviation of 6; and the average salary of the accountants was $31,000, with a standard deviation of $4000. Compare the variations of age and income.

3-74. Using Chebyshev's theorem, solve the following problems for a distribution with a mean of 80 and a standard deviation of 10.

a. At least what percentage of values will fall between 60 and 100?

b. At least what percentage of values will fall between 65 and 95?

3-75. The mean of a distribution is 20 and the standard deviation is 2. Answer each. Use Chebyshev's theorem.

a. At least what percentage of the values will fall between 10 and 30?

b. At least what percentage of the values will fall between 12 and 28?

3-76. In a distribution of 200 values, the mean is 50 and the standard deviation is 5. Answer each. Use Chebyshev's theorem.

a. At least how many values will fall between 30 and 70?

b. At most how many values will be less than 40 or more than 60?

3-77. A sample of the hourly wages of employees who work in restaurants in a large city has a mean of $5.02 and a standard deviation of $0.09. Using Chebyshev's theorem, find the range in which at least 75% of the data values will lie.

3–78. A sample of the labor costs per hour to assemble a certain product has a mean of $2.60 and a standard deviation of $0.15. Using Chebyshev's theorem, find the values in which at least 88.89% of the data will lie.

3–79. A survey of a number of the leading brands of cereal shows that the mean content of potassium per serving is 95 milligrams, and the standard deviation is 2 milligrams. Find the values in which at least 88.89% of the data will fall. Use Chebyshev's theorem.

3–80. The average score on a special test of knowledge of wood refinishing has a mean of 53 and a standard deviation of 6. Using Chebyshev's theorem, find the range of values in which at least 75% of the scores will lie.

3–81. The average of the number of trials it took a sample of mice to learn to traverse a maze was 12. The standard deviation was 3. Using Chebyshev's theorem, find the minimum percentage of data values that will fall in the 4-to-20 range.

3–82. The average cost of a certain type of grass seed is $4.00 per box. The standard deviation is $0.10. Using Chebyshev's theorem, find the minimum percentage of data values that will fall in the $3.82-to-$4.18 range.

*3–83. For the following data set, find the mean and standard deviation of the variable. The data represent the serum cholesterol level of 30 individuals. Count the number of data values that fall within two standard deviations of the mean. Compare this number with the number obtained from Chebyshev's theorem. Comment on the answer.

211	240	255	219	204
200	212	193	187	205
256	203	210	221	249
231	212	236	204	187
201	247	206	187	200
237	227	221	192	196

*3–84. For the following data set, find the mean and standard deviation of the variable. The data represent the ages of 30 customers who ordered a product advertised on television. Count the number of data values that fall within two standard deviations of the mean. Compare this number with the number obtained from Chebyshev's theorem. Comment on the answer.

42	44	62	35	20
30	56	20	23	41
55	22	31	27	66
21	18	24	42	25
32	50	31	26	36
39	40	18	36	22

*3–85. Using Chebyshev's theorem, complete the table to find the minimum percentage of the data values that fall within k standard deviations of the mean.

k	1.5	2	2.5	3	3.5
Percent					

*3–86. Use the following data set: 10, 20, 30, 40, 50.

a. Find the standard deviation.
b. Add 5 to each value, and then find the standard deviation.
c. Subtract 5 from each value, and find the standard deviation.
d. Multiply each value by 5, and find the standard deviation.
e. Divide each value by 5, and find the standard deviation.
f. Generalize the results of parts b through e.
g. Compare these results with those in Exercise 3–38.

*3–87. The **mean deviation** is found by using the following formula:

$$\text{mean deviation} = \frac{\Sigma \, |X - \overline{X}|}{n}$$

where
$$X = \text{value}$$
$$\overline{X} = \text{mean}$$
$$n = \text{number of values}$$
$$|\ | = \text{absolute value}$$

Find the mean deviation for the following data.

$$5, 9, 10, 11, 11, 12, 15, 18, 20, 22$$

*3–88. A measure to determine the skewness of a distribution is called the **Pearson coefficient of skewness**. The formula is

$$\text{skewness} = \frac{3(\overline{X} - \text{MD})}{s}$$

The values of the coefficient usually range from -3 to $+3$. When the distribution is symmetrical, the coefficient is zero; when the distribution is positively skewed, it is positive; and when the distribution is negatively skewed, it is negative.

Using the formula, find the coefficient of skewness for each distribution, and describe the shape of the distribution.

a. Mean = 10, median = 8, standard deviation = 3.
b. Mean = 42, median = 45, standard deviation = 4.
c. Mean = 18.6, median = 18.6, standard deviation = 1.5.
d. Mean = 98, median = 97.6, standard deviation = 4.

3-4 MEASURES OF POSITION (OPTIONAL)

In addition to measures of central tendency and measures of variation, there are also measures of position or location. These measures include standard scores, percentiles, deciles, and quartiles. They are used to locate the relative position of a data value in the data set. For example, if a value is located at the 80th percentile, it means that 80% of the values fall below that value in the distribution and 20% of the values fall above that value. The *median* is the value that corresponds to the 50th percentile, since half of the values fall below it and half of the values fall above it. This section discusses these measures of position as well as a special graph called a box and whisker plot.

Standard Scores

There is an old saying that states, "You can't compare apples and oranges." But with the use of statistics, it can be done to some extent. Suppose that a student scored 90 on a music test and 45 on an English exam. Direct comparison of raw scores is impossible, since the exams are not equivalent in terms of number of questions, value of each question, etc. However, a comparison of a relative standard similar to both can be made. This comparison uses the mean and standard deviation and is called a standard score or *z* score.

A standard score or z score for a value is obtained by subtracting the mean from the value and dividing the result by the standard deviation. The symbol for a standard score is z. The formula is

$$z = \frac{\text{value} - \text{mean}}{\text{standard deviation}}$$

For samples, the formula is

$$z = \frac{X - \overline{X}}{s}$$

For populations, the formula is

$$z = \frac{X - \mu}{\sigma}$$

EXAMPLE 3-46 A student scored 65 on a calculus test that had a mean of 50 and a standard deviation of 10; she scored 30 on a history test with a mean of 25 and a standard deviation of 5. Compare her relative positions on each test.

Solution First, find the *z* scores. For calculus the *z* score is

$$z = \frac{X - \overline{X}}{s} = \frac{65 - 50}{10} = 1.5$$

For history the *z* score is

$$z = \frac{30 - 25}{5} = 1.0$$

Since the *z* score for calculus is larger, her relative position in the calculus class is higher than her relative position in the history class.   ◀

Note that if the z score is positive, the score is above the mean. If the z score is 0, the score is the same as the mean. And if the z score is negative, the score is below the mean.

EXAMPLE 3–47 Find the z score for each test and state which is higher.

Test A	$X = 38$	$\overline{X} = 40$	$s = 5$
Test B	$X = 94$	$\overline{X} = 100$	$s = 10$

Solution For test A,

$$z = \frac{X - \overline{X}}{s} = \frac{38 - 40}{5} = -0.4$$

For test B,

$$z = \frac{94 - 100}{10} = -0.6$$

The score for test A is relatively higher than the score for test B. ◄

When all data for a variable are transformed into z scores, the resulting distribution will have a mean of 0 and a standard deviation of 1. A z score, then, is actually the number of standard deviations each variable is from the mean for a specific distribution. In Example 3–46, the calculus score of 65 was actually 1.5 standard deviations above the mean of 50.

Percentiles

Percentiles are position measures used in educational and health-related fields to indicate the position of an individual in a group.

In many situations, the graphs and tables showing the percentiles for various measures such as test scores, heights, or weights have already been completed. Figure 3–4 shows the percentile ranks for scaled scores on the Test of English as a Foreign Language. If a student had a scaled score of 58 for Section 1 (Listening and Comprehension) on the TOEFL, that student would have a percentile rank of 81. Hence, that student did better than 81% of the students who took Section 1 of the exam.

Figure 3–5 shows percentiles in graphic form of weights of girls from ages 2 to 18. In order to find the percentile rank of an 11-year-old who weighs 82 pounds, start at the 82-pound weight on the left axis and move horizontally to the right. Find the 11 on the horizontal axis and move up vertically. The two lines meet at the 50th percentile curved line; hence, an 11-year-old girl who weighs 82 pounds is in the 50th percentile for her age group. If the lines do not meet exactly on one of the curved percentile lines, then the percentile rank would have to be approximated.

Percentiles are also used to compare an individual's test score with the national norm. For example, tests such as the National Educational Development Test (NEDT) are taken by students in ninth or tenth grade. The student's scores

FIGURE 3–4

Percentile Ranks and Scaled Scores on the Test of English as a Foreign Language

Table 2. Percentile Ranks for TOEFL Scores — Total Group*

Scaled Score	Section 1 Listening Comprehension	Section 2 Structure and Written Expression	Section 3 Vocabulary and Reading Comprehension	Total Scaled Score	Percentile Rank
68	99	98			99
66	98	96	98	660	99
64	96	94	96	640	97
62	92	90	93	620	94
60	87	84	88	600	89
58	81	76	81	580	82
56	73	68	72	560	73
54	64	58	61	540	62
52	54	48	50	520	50
50	42	38	40	500	39
48	32	29	30	480	29
46	22	21	23	460	20
44	14	15	16	440	13
42	9	10	11	420	9
40	5	7	8	400	5
38	3	4	5	380	3
36	2	3	3	360	1
34	1	2	2	340	1
32		1	1	320	
30		1	1	300	
Mean	51.5	52.2	51.4	Mean	517
S.D.	7.1	7.9	7.5	S.D.	68

Source: Reprinted by permission of Educational Testing Service, the copyright owner.

*Based on the total group of 1,178,193 examinees tested from July 1989 through June 1991

are compared with those of other students locally and nationally by using percentile ranks. A similar test for elementary school students is called the California Achievement Test.

Percentiles are not the same as percentages. That is, if a student gets 72 correct answers out of a possible 100, she obtains a percentage score of 72. There is no indication of her position with respect to the rest of the class. She could have scored the highest, the lowest, or somewhere in between. On the other hand, if a raw score of 72 corresponds to the 64th percentile, then she did better than 64% of the students in her class.

Percentiles are symbolized by

$$P_1, P_2, P_3, \ldots, P_{99}$$

and divide the distribution into 100 groups. Percentile graphs can be constructed as shown in the next example.

FIGURE 3–5
Weights of Girls by Age
and Percentile Rankings

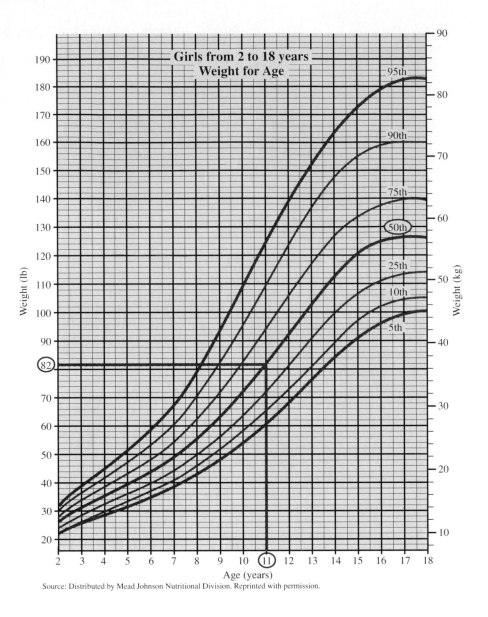

Source: Distributed by Mead Johnson Nutritional Division. Reprinted with permission.

EXAMPLE 3–48   The frequency distribution for the systolic blood pressure readings (in millimeters of mercury, mmHg) of 200 randomly selected college students follows.
Construct a percentile graph.

A Class Boundaries	B Frequency	C Cumulative Frequency	D Cumulative Percent
89.5–104.5	24		
104.5–119.5	62		
119.5–134.5	72		
134.5–149.5	26		
149.5–164.5	12		
164.5–179.5	4		
	200		

Solution **STEP 1** Find the cumulative frequencies and place them in column C.

STEP 2 Find the cumulative percentages and place them in column D. To do this step, use the formula

$$\text{cumulative \%} = \frac{\text{cumulative frequency}}{n} \cdot 100\%$$

For the first class,

$$\text{cumulative \%} = \frac{24}{200} \cdot 100\% = 12\%$$

The completed table is shown next.

A	B	C	D
		Cumulative	Cumulative
Class Boundaries	Frequency	Frequency	Percent
89.5–104.5	24	24	12
104.5–119.5	62	86	43
119.5–134.5	72	158	79
134.5–149.5	26	184	92
149.5–164.5	12	196	98
164.5–179.5	4	200	100
	200		

STEP 3 Graph the data, using class boundaries for the x axis and the percentages for the y axis, as shown in Figure 3–6.

FIGURE 3–6
Percentile Graph for
Example 3–48

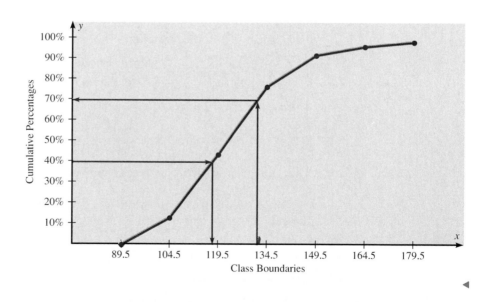

Once a cumulative percentage graph has been constructed, one can find the approximate corresponding percentile ranks for given blood pressure values and find approximate blood pressure values for given percentile ranks.

For example, to find the percentile rank of a blood pressure reading of 130, find 130 on the x axis and draw a vertical line to the graph. Then move horizontally to the value on the y axis. See Figure 3–6. A blood pressure of 130 corresponds to approximately the 70th percentile.

If the value that corresponds to the 40th percentile is desired, start on the *y* axis at 40 and draw a horizontal line to the graph. Then draw a vertical line to the *x* axis, and read the value. See Figure 3–6. The 40th percentile corresponds to a value of approximately 118. Thus, if a person has a blood pressure of 118, he or she is at the 40th percentile.

Note that finding values and the corresponding percentile ranks by using a graph yields only approximate answers. Several mathematical methods exist for computing percentiles for data. These methods can be used to find the approximate percentile rank of a data value or to find a data value corresponding to a given percentile. The next several examples show these methods.

Percentile Formula

The percentile corresponding to a given value (*X*) is computed by using the following formula:

$$\text{percentile} = \frac{\text{number of values below} + 0.5}{\text{total number of values}} \cdot 100\%$$

EXAMPLE 3–49 A teacher gives a 20-point test to ten students. The scores are shown below. Find the percentile rank of a score of 12.

$$18, 15, 12, 6, 8, 2, 3, 5, 20, 10$$

Solution Arrange the data in order from lowest to highest.

$$2, 3, 5, 6, 8, 10, 12, 15, 18, 20$$

Then, substitute in the formula.

$$\text{percentile} = \frac{\text{number of values below} + 0.5}{\text{total number of values}} \cdot 100\%$$

Since there are six values below a score of 12, the solution is

$$\text{percentile} = \frac{6 + 0.5}{10} \cdot 100\% = 65\text{th percentile}$$

Thus, a student whose score was 12 did better than 65% of the class. ◄

Note: One assumes that a score of 12 in Example 3–49, for instance, means theoretically any value between 11.5 and 12.5.

EXAMPLE 3–50 Using the data in Example 3–49, find the percentile rank for a score of 6.

Solution There are three values below 6; hence,

$$\text{percentile} = \frac{3 + 0.5}{10} \cdot 100\% = 35\text{th percentile}$$

If a student scored 6, the student did better than 35% of the class. ◄

The procedure illustrated in the next two examples shows how to find a value corresponding to a given percentile.

EXAMPLE 3–51   Using the scores in Example 3–49, find the value corresponding to the 25th percentile.

Solution   **STEP 1**   Arrange the data in order from lowest to highest.

$$2, 3, 5, 6, 8, 10, 12, 15, 18, 20$$

STEP 2   Compute

$$c = \frac{n \cdot p}{100}$$

where n = total number of values
p = percentile

Thus,

$$c = \frac{10 \cdot 25}{100} = 2.5$$

STEP 3   If c is not a whole number, round it up to the next whole number; in this case, $c = 3$. (If c is a whole number, see the next example.) Start at the lowest value and count over to the third value, which is 5. Hence, the value 5 corresponds to the 25th percentile.   ◄

EXAMPLE 3–52   Using the data set in Example 3–49, find the value that corresponds to the 60th percentile.

Solution   **STEP 1**   Arrange the data in order, smallest to largest.

$$2, 3, 5, 6, 8, 10, 12, 15, 18, 20$$

STEP 2   Substitute in the formula.

$$c = \frac{n \cdot p}{100} = \frac{10 \cdot 60}{100} = 6$$

STEP 3   If c is a whole number, use the value halfway between the c and $c + 1$ values when counting up from the lowest value—in this case, the 6th and 7th values.

$$2, 3, 5, 6, 8, 10, 12, 15, 18, 20$$

6th value   7th value

The value halfway between 10 and 12 is 11, and it is found by adding the two values and dividing by 2.

$$\frac{10 + 12}{2} = 11$$

Hence, 11 corresponds to the 60th percentile. Thus, anyone scoring 11 would have done better than 60% of the class.   ◄

Deciles and Quartiles

Deciles divide the distribution into ten groups. **Quartiles** divide the distribution into four groups. Deciles are denoted by

$$D_1, D_2, D_3, \ldots, D_9$$

and they correspond to

$$P_{10}, P_{20}, P_{30}, \ldots, P_{90}$$

Quartiles are denoted by

$$Q_1, Q_2, Q_3$$

and they correspond to

$$P_{25}, P_{50}, P_{75}$$

The median is the same as

$$P_{50} \quad \text{or} \quad Q_2$$

Box and Whisker Plots

When the data set contains a small number of values, a **box and whisker plot** is used to graphically represent the data set, since a histogram would be inappropriate. These plots, also called box plots, involve five specific values:

1. The lowest value of the data set
2. The lower hinge
3. The median
4. The upper hinge
5. The highest value of the data set

These values are called a **five-number summary** of the data set.

The lower hinge is defined as the median of all values less than or equal to the median when the data set has an odd number of values, or as the median of all values less than the median when the data set has an even number of values. The symbol for lower hinge is LH.

The upper hinge is defined as the median of all values greater than or equal to the median when the data set has an odd number of values, or as the median of all values greater than the median when the data set has an even number of values. The symbol for upper hinge is UH.

EXAMPLE 3–53 Find the lower hinge and the upper hinge for the following data set.

$$3, 8, 10, 12, 15, 18, 20$$

Solution Since the median is 12, the data set is divided into two subsets:

$$3, 8, 10, 12 \quad \text{and} \quad 12, 15, 18, 20$$

The median 12 is included in both subsets, since there are an odd number of data values. Next, the median of the lower subset is found as shown.

$$3, 8, 10, 12$$
$$\uparrow$$
$$9$$

And the median of the upper subset is found as shown.

$$12, 15, \underset{\uparrow}{} 18, 20$$

$$16.5$$

Hence, the lower hinge is 9 and the upper hinge is 16.5. ◄

In the preceding example, there were an odd number of values in the data set. When there are an even number of values in the data set, the lower hinge is the same as Q_1, and the upper hinge is the same as Q_3, as shown in the next example.

EXAMPLE 3-54 Find the lower and upper hinges for the data set shown here.

$$18, 20, 25, 31, 33, 38, 42, 50$$

Solution The median is 32, the lower hinge is 22.5, and the upper hinge is 40, as shown.

$$
\begin{array}{ccccccccc}
18 & 20 & 25 & 31 & 33 & 38 & 42 & 50 \\
 & & \uparrow & & \uparrow & & \uparrow & \\
 & & 22.5 & & 32 & & 40 & \\
 & & \text{LH} & & \text{median} & & \text{UH} &
\end{array}
$$

$$\text{MD} = \frac{31 + 33}{2} = 32$$

$$\text{LH} = \frac{20 + 25}{2} = 22.5$$

$$\text{UH} = \frac{38 + 42}{2} = 40 \qquad ◄$$

The next two examples show the procedure for constructing a box and whisker plot.

EXAMPLE 3-55 A stockbroker recorded the number of clients she saw each day over an 11-day period. The data are shown below. Construct a box and whisker plot for the data.

$$33, 38, 43, 30, 29, 40, 51, 27, 42, 23, 31$$

Solution **STEP 1** Arrange the data in order.

$$23, 27, 29, 30, 31, 33, 38, 40, 42, 43, 51$$

STEP 2 Find the median.

$$23, 27, 29, 30, 31, \underset{\uparrow}{33}, 38, 40, 42, 43, 51$$

$$\text{median}$$

STEP 3 Find the lower hinge.

$$23, 27, \underset{\uparrow}{29}, 30, 31, 33$$

$$29.5$$

STEP 4 Find the upper hinge.

$$33, 38, \underset{\uparrow}{40}, 42, 43, 51$$

$$41$$

STEP 5 Draw a scale for the data on the x axis.

STEP 6 Locate the lowest value, the lower hinge, the median, the upper hinge, and the highest value on the scale.

STEP 7 Draw a box around the hinges, draw a vertical line through the median, and connect the upper and lower values, as shown in Figure 3–7.

FIGURE 3–7
Box and Whisker Plot for Example 3–55

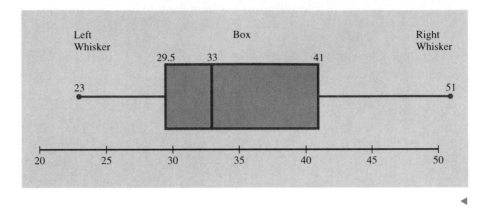

The box in the box and whisker plot represents the middle 50% of the data, and the whiskers represent the lower and upper ends of the data.

Information Obtained from a Box and Whisker Plot

1. a. If the median is near the center of the box, the distribution is approximately symmetric.
   b. If the median falls to the left of the center of the box, the distribution is positively skewed.
   c. If the median falls to the right of the center, the distribution is negatively skewed.
2. a. If the whiskers are about the same length, the distribution is approximately symmetric.
   b. If the right whisker is larger than the left whisker, the distribution is positively skewed.
   c. If the left whisker is larger than the right whisker, the distribution is negatively skewed.

The box and whisker plot for Example 3–55 indicates that the distribution is slightly positively skewed.

EXAMPLE 3–56 Another stockbroker recorded the number of clients he saw each day for ten days. The data follow. Construct a box and whisker plot for the data set.

$$11, 14, 18, 5, 16, 8, 19, 10, 17, 20$$

Solution **STEP 1** Arrange the data in order.

$$5, 8, 10, 11, 14, 16, 17, 18, 19, 20$$

STEP 2   Find the median and the hinges.

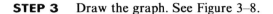

5, 8, 10, 11, 14, 16, 17, 18, 19, 20
  ↑   ↑   ↑
  LH   15 = MD UH

Note that in this case, the lower hinge and the upper hinge correspond to Q_1 and Q_3.

STEP 3   Draw the graph. See Figure 3–8.

FIGURE 3–8
Box and Whisker Plot
for Example 3–56

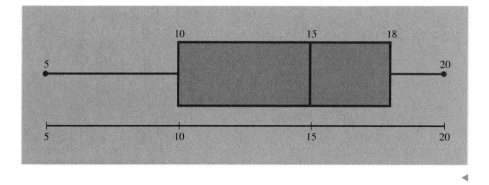

◄

The plot for Example 3–56 shows that the distribution is somewhat negatively skewed.

Outliers   A data set should be checked for extremely high or extremely low values. These values are called *outliers*.

An outlier is an extremely high or an extremely low data value when compared with the rest of the data values.

There are several ways to check for outliers. One method is shown in the next example.

EXAMPLE 3–57   Check the following data set for outliers.

5, 6, 12, 13, 15, 18, 22, 50

Solution   The data value 50 is extremely suspect. The steps in checking for an outlier follow.

STEP 1   Find Q_1 and Q_3. This is done by using the procedure shown in Examples 3–51 and 3–52. Q_1 responds to the 25th percentile; hence,

$$c = \frac{n \cdot p}{100} = \frac{8 \cdot 25}{100} = 2$$

Since c is a whole number, find the value halfway between the second and third values, counting from the left in the data set. This corresponds to 9 $\left(\text{i.e. } \dfrac{6 + 12}{2}\right)$, as shown below.

$$5, 6, \underset{\underset{Q_1 = 9}{\uparrow}}{12}, 13, 15, 18, 22, 50$$

Q_3 corresponds to the 75th percentile; hence,

$$c = \frac{n \cdot p}{100} = \frac{8 \cdot 75}{100} = 6$$

Find the value halfway between the 6th and 7th data values. In this case, it is 20 $\left(\text{i.e. } \dfrac{18 + 22}{2}\right)$, as shown.

$$5, 6, 12, 13, 15, 18, \underset{\underset{Q_3 = 20}{\uparrow}}{22}, 50$$

STEP 2 Find the **interquartile range**, which is $Q_3 - Q_1$.

$$Q_3 - Q_1 = 20 - 9 = 11$$

STEP 3 Multiply this value by 1.5.

$$1.5(11) = 16.5$$

STEP 4 Subtract the value obtained in Step 3 from Q_1 and add the value to Q_3.

$$9 - 16.5 = -7.5 \qquad \text{and} \qquad 20 + 16.5 = 36.5$$

STEP 5 Check the data set for any data values that fall outside the -7.5-to-36.5 range. The value 50 is outside this range; hence, it can be considered an outlier. ◄

There are several reasons to check a data set for outliers. First, the data value may have resulted from a measurement or observational error. Perhaps the researcher measured the variable incorrectly. Second, the data value may have resulted from a recording error. That is, it may have been written or typed incorrectly. Third, the data value may have been obtained from a subject that is not in the defined population. For example, suppose test scores were obtained from a seventh-grade class, but perhaps a student in that class was actually in the sixth grade and had special permission to attend the class. This student might have scored extremely low on that particular exam on that day. Fourth, the data value might be a legitimate value and occurred by chance, even though the probability is extremely small.

There are no hard and fast rules on what to do with outliers, nor is there complete agreement among statisticians on ways to identify them. Obviously, if they occurred as a result of an error, an attempt should be made to correct the error or else the data value should be omitted entirely. When they occur naturally by chance, the statistician must make a decision about whether to include them in the data set.

When a distribution is normal or bell-shaped, data values that are beyond three standard deviations of the mean can be considered as suspected outliers.

This section has explained how positions of specific data values can be located in a distribution by the use of z scores and percentiles. Also, when it is impractical to draw a histogram for a small data set, a box and whisker plot can be used to graphically represent the data. Finally, position measures can be used to check a data set for possible outliers.

EXERCISES

3–89. What is a z score?

3–90. Define *percentile rank*.

3–91. What is the difference between a percentage and a percentile?

3–92. Define *quartile*.

3–93. What is the relationship between quartiles and percentiles?

3–94. What is a decile?

3–95. How are deciles related to percentiles?

3–96. To which percentile, quartile, and decile does the median correspond?

3–97. If an IQ test has a mean of 100 and a standard deviation of 15, find the corresponding z score for each IQ.

a. 115   c. 93   e. 85
b. 122   d. 100

3–98. The reaction time to a stimulus for a certain test has a mean of 2.5 seconds and a standard deviation of 0.3 second. Find the corresponding z score for each reaction time.

a. 2.7   b. 3.9   c. 2.8   d. 3.1   e. 2.2

3–99. A final examination for an Introduction to Psychology course has a mean of 84 and a standard deviation of 4. Find the corresponding z score for each raw score.

a. 87   b. 79   c. 93   d. 76   e. 82

3–100. An aptitude test has a mean of 220 and a standard deviation of 10. Find the corresponding z score for each exam score.

a. 200
b. 232
c. 218
d. 212
e. 225

3–101. Which of the following exam grades has a better relative position?

a. A grade of 43 on a test with $\overline{X} = 40$ and $s = 3$.
b. A grade of 75 on a test with $\overline{X} = 72$ and $s = 5$.

3–102. A student scores 60 on a mathematics test that has a mean of 54 and a standard deviation of 3, and she scores 80 on a history test with a mean of 75 and a standard deviation of 2. On which test did she do better than the rest of the class?

3–103. Which score indicates the highest relative position?

a. A score of 3.2 on a test with $\overline{X} = 4.6$ and $s = 1.5$.
b. A score of 630 on a test with $\overline{X} = 800$ and $s = 200$.
c. A score of 43 on a test with $\overline{X} = 50$ and $s = 5$.

3–104. The following distribution represents the data for weights of fifth-grade boys. Find the approximate weights corresponding to each percentile by constructing a percentile graph.

a. 25th   c. 80th
b. 60th   d. 95th

Weight (Pounds)	Frequency
52.5–55.5	9
55.5–58.5	12
58.5–61.5	17
61.5–64.5	22
64.5–67.5	15

3–105. For the data in Exercise 3–104, find the approximate percentile rank of the following weights.

a. 57 pounds   c. 64 pounds
b. 62 pounds   d. 59 pounds

3–106. (ANS.) The data below represent the scores on a national achievement test for a group of tenth-grade students. Find the approximate percentile rank of the following scores by constructing a percentile graph.

a. 220   c. 276   e. 300
b. 245   d. 280

Score	Frequency
196.5–217.5	5
217.5–238.5	17
238.5–259.5	22
259.5–280.5	48
280.5–301.5	22
301.5–322.5	6

3–107. For the data in Exercise 3–106, find the approximate scores that correspond to the following percentiles.

a. 15th   c. 43rd   e. 80th
b. 29th   d. 65th

3–108. (ANS.) The following distribution represents the weights of 18-year-old males. Find the approximate values that correspond to the given percentiles by constructing a percentile graph.

a. 9th   c. 45th   e. 75th
b. 20th   d. 60th

Weight (Pounds)	Frequency
120.5–131.5	12
131.5–142.5	16
142.5–153.5	24
153.5–164.5	48
164.5–175.5	62
175.5–186.5	21
186.5–197.5	17

3–109. Using the data in Exercise 3–108, find the approximate percentile ranks of the following weights.

a. 125 pounds   d. 173 pounds
b. 138 pounds   e. 195 pounds
c. 160 pounds

3–110. Find the percentile ranks of each weight in the data set. The weights are in pounds.

78, 82, 86, 88, 92, 97

3–111. In Exercise 3–110, what value corresponds to the 30th percentile?

3–112. Find the percentile rank for each test score in the data set.

12, 28, 35, 42, 47, 49, 50

3–113. In Exercise 3–112, what value corresponds to the 60th percentile?

3–114. Find the percentile ranks for each test score in the data set.

5, 12, 15, 16, 20, 21

3–115. What test score in Exercise 3–114 corresponds to the 33rd percentile?

3–116. Construct a box and whisker plot for the IQ data set. Comment on the shape of the distribution.

95, 98, 102, 105, 115, 135

3–117. Construct a box and whisker plot for the heights (in inches): 54, 58, 62, 66, 67. Comment on the shape of the distribution.

3–118. Construct a box and whisker plot for the number of calculators sold during a randomly selected week. Comment on the shape of the distribution.

8, 12, 23, 5, 9, 15, 3

3–119. Using the procedure shown in Example 3–57, check each data set for outliers.

a. 16, 18, 22, 19, 3, 21, 17, 20
b. 24, 32, 54, 31, 16, 18, 19, 14, 17, 20
c. 321, 343, 350, 327, 200
d. 88, 72, 97, 84, 86, 85, 100
e. 145, 119, 122, 118, 125, 116
f. 14, 16, 27, 18, 13, 19, 36, 15, 20

*3–120. Another measure of average is called the **midquartile**; it is the numerical value halfway between Q_1 and Q_3, and the formula is

$$\text{midquartile} = \frac{Q_1 + Q_3}{2}$$

Using these formulas, find Q_1, Q_2, Q_3, the midquartile, and the interquartile range for each data set.

a. 5, 12, 16, 25, 32, 38
b. 53, 62, 78, 94, 96, 99, 103

3–5 SUMMARY

This chapter explains the basic ways to describe data. The three most commonly used measures of central tendency or average are the mean, median, and mode. The three most commonly used measures of variation or dispersion are the range, variance, and standard deviation. In addition, measures of position or location are also used in statistics. The measures of position include percentiles, quartiles, and deciles. Finally, when the data set is small and it is not convenient to draw a histogram, a box and whisker plot can be used to describe the data graphically.

Important Terms

Box and whisker plot
Chebyshev's theorem
Coefficient of variation
Data array
Decile
Empirical rule
Five-number summary
Interquartile range
Lower hinge
Mean
Median
Midrange
Modal class
Mode
Negatively skewed distribution
Outlier
Parameter
Percentile
Positively skewed distribution
Quartile
Range
Standard deviation
Statistic
Symmetrical distribution
Unbiased estimator
Upper hinge
Variance
Weighted mean
z score or standard score

Important Formulas

Formula for the mean for individual data:

$$\overline{X} = \frac{\Sigma X}{n} \qquad \mu = \frac{\Sigma X}{N}$$

Formula for the mean for grouped data:

$$\overline{X} = \frac{\Sigma f \cdot X_m}{n}$$

Formula for the median for grouped data:

$$\text{MD} = \frac{(n/2) - \text{cf}}{f}(w) + L_m$$

Formula for the weighted mean:

$$\overline{X} = \frac{\Sigma w \cdot X}{\Sigma w}$$

Formula for the midrange:

$$\text{MR} = \frac{\text{lowest value} + \text{highest value}}{2}$$

Formula for the range:

$$R = \text{highest value} - \text{lowest value}$$

Formula for the variance for population data:

$$\sigma^2 = \frac{\Sigma (X - \mu)^2}{N}$$

Formula for the variance for sample data (shortcut formula for the unbiased estimator):

$$s^2 = \frac{\Sigma X^2 - [(\Sigma X)^2/n]}{n - 1}$$

Formula for the variance for grouped data:

$$s^2 = \frac{\Sigma f \cdot X_m^2 - [(\Sigma f \cdot X_m)^2/n]}{n - 1}$$

Formula for the standard deviation for population data:

$$\sigma = \sqrt{\frac{\Sigma (X - \mu)^2}{N}}$$

Formula for the standard deviation for sample data (shortcut formula):

$$s = \sqrt{\frac{\Sigma X^2 - [(\Sigma X)^2/n]}{n - 1}}$$

Formula for the standard deviation for grouped data:

$$s = \sqrt{\frac{\Sigma f \cdot X_m^2 - [(\Sigma f \cdot X_m)^2/n]}{n - 1}}$$

(Continued)

Formula for the coefficient of variation:

$$\text{CVar} = \frac{s}{\overline{X}} \cdot 100\% \quad \text{or} \quad \text{CVar} = \frac{\sigma}{\mu} \cdot 100\%$$

Formula for Chebyshev's theorem: The proportion of values from a data set that will fall within k standard deviations of the mean will be at least

$$1 - \frac{1}{k^2}$$

where k is a number greater than 1

Formula for the z score or standard score:

$$z = \frac{X - \mu}{\sigma} \quad \text{or} \quad z = \frac{X - \overline{X}}{s}$$

Formula for the cumulative percentage:

$$\text{cumulative \%} = \frac{\text{cumulative frequency}}{n} \cdot 100\%$$

Formula for the percentile rank of a value X:

$$\text{percentile} = \frac{\text{number of values below} + 0.5}{\text{total number of values}} \cdot 100\%$$

Formula for finding a value corresponding to a given percentile:

$$c = \frac{n \cdot p}{100}$$

Formula for interquartile range:

$$\text{interquartile range} = Q_3 - Q_1$$

Review Exercises

3–121. The following temperatures were recorded in Pittsburgh for a week in July.

$$106, 98, 97, 101, 93, 90, 87$$

Find each of the following.

a. Mean
b. Median
c. Mode
d. Midrange
e. Range
f. Variance
g. Standard deviation

3–122. Ten motors were tested and the following data were obtained for the number of revolutions per minute the flywheel attached to the motors turned.

$$215, 308, 225, 236, 272,$$
$$300, 254, 260, 232, 216$$

Find each of the following.

a. Mean
b. Median
c. Mode
d. Midrange
e. Range
f. Variance
g. Standard deviation

3–123. Twelve batteries were tested after being used 1 hour. The output (in volts) is shown below.

Volts	Frequency
2	1
3	4
4	5
5	1
6	1

Find each of the following.

a. Mean   d. Range
b. Median   e. Variance
c. Mode   f. Standard deviation

3–124. A survey of 40 clothing stores reported the following number of sales that were held in the month of November.

Number of Sales	Frequency
5	3
6	5
7	12
8	9
9	8
10	3

Find each of the following.

a. Mean   d. Range
b. Median   e. Variance
c. Mode   f. Standard deviation

3–125. Shown below is a frequency distribution for the rise in tides at 30 selected locations in the United States.

Rise in Tides (inches)	Frequency
12.5–27.5	6
27.5–42.5	3
42.5–57.5	5
57.5–72.5	8
72.5–87.5	6
87.5–102.5	2

Find each of the following.

a. Mean   d. Variance
b. Median   e. Standard deviation
c. Modal class

3–126. A survey of 50 selected mail-order companies recorded the number of days it took to mail an order from the day it was received.

Days to Mail	Frequency
7–9	1
10–12	3
13–15	6
16–18	8
19–21	17
22–24	10
25–27	5

Find each of the following.

a. Mean   d. Variance
b. Median   e. Standard deviation
c. Modal class

3–127. In a dental survey of third-grade students, the following distribution was obtained for the number of cavities found. Find the average number of cavities for the class. Use the weighted mean.

Number of Students	Number of Cavities
12	0
8	1
5	2
5	3

3–128. An investor calculates the following percentages of each of three stock investments with payoffs as shown. Find the average payoff. Use the weighted mean.

Stock	Percent	Payoff
A	30%	$10,000
B	50%	$3,000
C	20%	$1,000

3–129. In an advertisement, a transmission service center stated that the average years of service of its employees was 13. The distribution is shown below. Using the weighted mean, calculate the correct average.

Number of Employees	Years of Service
8	3
1	6
1	30

3–130. The average number of textbooks in professors' offices is 16, and the standard deviation is 5. The average age of the professors is 43, with a standard deviation of 8. Which data set is more variable?

3–131. A survey of bookstores shows that the average number of magazines carried is 56, with a standard deviation of 12. The same survey showed that the average length of time each store was in business is 6 years, with a standard deviation of 2.5 years. Which is more variable, the number of magazines or the number of years?

3–132. (**OPT.**) The number of previous jobs each of six applicants held is shown here.

$$2, 4, 5, 6, 8, 9,$$

a. Find the percentile for each value.
b. What value corresponds to the 30th percentile?
c. Construct a box and whisker plot and comment on the nature of the distribution.

(**OPT.**) means that the exercise relates to a section marked "Optional."

3-133. (OPT.) On a comprehensive mathematics exam, the following distribution was obtained from 35 students.

Score	Frequency
40.5–45.5	7
45.5–50.5	6
50.5–55.5	12
55.5–60.5	5
60.5–65.5	5

a. Construct a percentile graph.
b. Find the values that correspond to the 35th, 65th, and 85th percentiles.
c. Find the percentile of each value.
   (i) 48   (ii) 54   (iii) 62

3-134. (OPT.) Check each data set for outliers.

a. 506, 511, 517, 514, 400, 521
b. 3, 7, 9, 6, 8, 10, 14, 16, 20, 12
c. 14, 18, 27, 26, 19, 13, 5, 25
d. 112, 157, 192, 116, 153, 129, 131

3-135. A survey of automobile rental agencies shows that the average cost of a car rental is $0.32 per mile. The standard deviation is $0.03. Using Chebyshev's theorem, find the range in which at least 75% of the data values will fall.

3-136. The average cost of a certain type of seed per acre is $42. The standard deviation is $3. Using Chebyshev's theorem, find the range in which at least 88.89% of the data values will fall.

3-137. The average labor charge for automobile mechanics is $54 per hour. The standard deviation is $4. Find the minimum percentage of data values that will fall within the $48-to-$60 range. Use Chebyshev's theorem.

3-138. For a certain type of job, it costs a company an average of $231 to train an employee to perform the task. The standard deviation is $5. Find the minimum percentage of data values that will fall in the $219-to-$243 range. Use Chebyshev's theorem.

3-139. The average cost of delivery charges for a refrigerator is $32. The standard deviation is $4. Find the minimum percentage of data values that will fall in the $20-to-$44 range. Use Chebyshev's theorem.

3-140. (OPT.) Which of the following exam grades has a better relative position?

a. A grade of 37 on a test with $\overline{X} = 42$ and $s = 5$.
b. A grade of 72 on a test with $\overline{X} = 80$ and $s = 6$.

COMPUTER APPLICATIONS

The following data represent the number of customers using the drive-in services of a bank over a 50-day period.

90	135	162	151	148
99	150	107	132	99
127	97	122	92	97
126	150	132	143	143
119	135	105	136	127
142	96	123	144	103
127	114	136	145	132
108	150	129	105	125
132	112	105	116	149
106	128	99	132	96

1. Using the computer package of your choice, enter the data and find the mean, median, mode, range, variance, and standard deviation.
2. Write a program that will find the mean, variance, and standard deviation for the above data.

3. MINITAB can be used to find various descriptive measures of a data set and also to construct a box and whisker plot. To get descriptive measures, type in the following information for the above data.

```
MTB  > SET C1
DATA > 90  135  162  151  148
DATA > 99  150  107  132  99
DATA > 127  97  122  92  97
DATA > 126  150  132  143  143
DATA > 119  135  105  136  127
DATA > 142  96  123  144  103
DATA > 127  114  136  145  132
DATA > 108  150  129  105  125
DATA > 132  112  105  116  149
DATA > 106  128  99  132  96
DATA > END OF DATA
MTB  > DESCRIBE C1
```

The command SET C1 tells the computer that the data will be entered. After the data have been entered, the END OF DATA statement is used. To obtain the descriptive measures, type DESCRIBE C1. The output will look like this.

```
        N    MEAN  MEDIAN  TRMEAN  STDEV  SEMEAN
C1     50  123.56  127.00  123.57  19.17   2.71

       MIN     MAX      Q1      Q3
C1  90.00  162.00  105.00  137.50
```

The sample size is 50. The mean is 123.56. The median is 127.00. The standard deviation is 19.17. The minimum value is 90.00, and the maximum value is 162. The first quartile is 105.00, and the third quartile is 137.50. The SEMEAN gives the standard error of the mean; this term will be explained later. The TRMEAN is a trimmed mean that is computed by omitting the smallest 5% and the largest 5% of the data values.

Try Exercise 3–121 using MINITAB.

For a box and whisker plot, type BOXPLOT C1. The output will look like this.

```
MTB > BOXPLOT C1


                  ---------------------
      ---------I               +     I-----------------
                  ---------------------
   ----+---------+---------+---------+---------+---------+--C1
      90        105       120       135       150       165
```

The minimum value is 90. The maximum value is 162. Q_1 is 105 and Q_3 is 137.5. The median is 127.

Here is another example for a box and whisker plot.

```
MTB  > SET C1
DATA > 33  38  43  30  29  40  51  27  42  23  31
DATA > BOXPLOT C1


               ---------------------
   -------------I       +          I-------------------
               ---------------------
   ----+---------+---------+---------+---------+---------+--C1
      25.0      30.0      35.0      40.0      45.0      50.0
```

Use MINITAB to construct a box and whisker plot for Exercise 3–132, part c.

√ DATA ANALYSIS Applying the Concepts

The Data Bank is located in the Appendix.

1. From the Data Bank, choose one of the following variables: age, weight, cholesterol level, systolic pressure, IQ, or sodium level. Select at least 30 values, and find the mean, median, mode, and midrange. State which measurement of central tendency best describes the average and why.

2. Find the range, variance, and standard deviation for the data selected in Exercise 1.

3. From the Data Bank, choose 10 values from any variable, construct a box and whisker plot, and interpret the results.

✎ TEST

Directions: Determine whether each statement is true or false.

1. A measure obtained from a sample is called a parameter.
2. Generally, Greek letters are used to represent parameters, and Roman letters are used to represent statistics.
3. When the mean is computed for individual data, all values in the data set are used.
4. The mean cannot be found for grouped data when there is an open class.
5. A single, extremely large value can affect the median more than the mean.
6. When the values in a data set are all different, there is no mode.
7. When data are in categories, such as place of residence (rural, suburban, urban), the most appropriate descriptive measure of average would be the median.
8. Half of all the values in the data set will fall above the mode and half will fall below the mode.
9. In a data set, the mode will always be unique.
10. One disadvantage of the median is that it is not unique.
11. The range and midrange are both measures of variation.
12. The standard deviation is the square root of the variance.
13. The unbiased estimator formula for the variance is used when samples are small, since the true variance formula tends to overestimate the population variance.
14. The symbol for the population standard deviation is s.
15. When a distribution is negatively skewed, the tail is to the right.
16. The coefficient of variation is used to compare standard deviations when the units of measure are different for the samples.
17. When the mean of the data set is 18 and the standard deviation is 3, the z score for a value of 12 is -2.
18. If a person's score on an exam corresponds to the 75th percentile, then that person obtained 75 correct answers out of 100 questions.
19. P_{50}, D_5, Q_2, and the median are the same for the values of a data set.
20. An alternative to the histogram when the data set contains a small number of values is the box and whisker plot.

CHAPTER

4

Counting Techniques

4–1 INTRODUCTION

Many problems in probability and statistics require a careful analysis of the outcomes of a sequence of events. For example, a sales representative may wish to select the most efficient way to visit several different stores in four cities. Sometimes, in order to determine costs or rates, a manager must know all possible outcomes of a classification scheme. Or an insurance company may wish to classify its drivers according to the following classes:

1. Gender (male, female).
2. Age (under 25, between 25 and 60, over 60).
3. Area of residence (rural, suburban, urban).
4. Distance driven to work (under 4 miles, between 4 and 10 miles, over 10 miles).
5. Value of the automobile (under $5000, between $5000 and $10,000, over $10,000).

In a psychological study, a researcher might attempt to train a rat to run a maze. In order to determine the rat's success, the researcher must know the number of possible choices the rat can make to traverse the maze. Then the researcher can differentiate between actual learning and chance successes.

On a game show, a contestant might be required to arrange four digits correctly to guess the exact price of a new car. To determine the probability of his guessing the correct answer, one must know the number of possible ways the four digits can be arranged.

The vice president of a company might wish to know the number of different possible ways four employees can be selected from a group of ten in order to be transferred to a new location.

These examples, as well as many others, illustrate the need to know the possible outcomes of situations. In order to determine the number of outcomes, one can use several rules of counting: the *multiplication rules,* the *permutation rules,* and the *combination rule.* Finally, when one must list the outcomes of a sequence of events, a useful device called a *tree diagram* can be used. These concepts will be explained in this chapter.

4–2 THE MULTIPLICATION RULES FOR COUNTING

This section discusses two rules used to determine the total number of outcomes of a sequence of events that are called the multiplication rules. These rules will be illustrated in examples and then formally defined.

EXAMPLE 4–1 A security supervisor of a large corporation wishes to issue each employee a parking permit with a four-digit number. How many different permits can be issued?

Solution Since there are four digits on each permit, there will be four spaces to fill, which can be represented as follows:

First Digit	Second Digit	Third Digit	Fourth Digit
☐	☐	☐	☐

There are ten ways to select each digit, since there are ten digits: 0, 1, 2, 3, 4, 5, 6, 7, 8, 9. Hence, the total number of permits that can be issued is

$$\boxed{10} \cdot \boxed{10} \cdot \boxed{10} \cdot \boxed{10} = 10,000 \qquad \blacktriangleleft$$

This example illustrates the first multiplication rule.

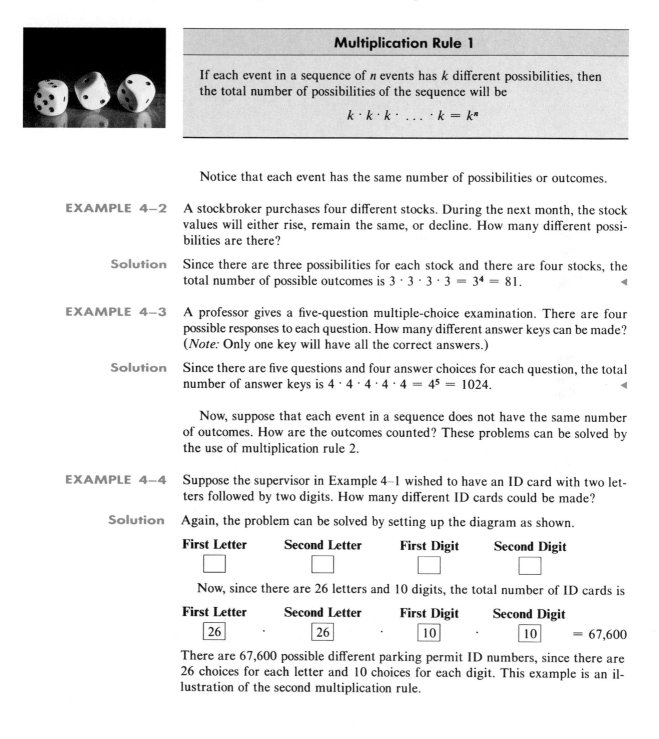

Multiplication Rule 1

If each event in a sequence of n events has k different possibilities, then the total number of possibilities of the sequence will be

$$k \cdot k \cdot k \cdot \ldots \cdot k = k^n$$

Notice that each event has the same number of possibilities or outcomes.

EXAMPLE 4–2   A stockbroker purchases four different stocks. During the next month, the stock values will either rise, remain the same, or decline. How many different possibilities are there?

Solution   Since there are three possibilities for each stock and there are four stocks, the total number of possible outcomes is $3 \cdot 3 \cdot 3 \cdot 3 = 3^4 = 81$. $\qquad \blacktriangleleft$

EXAMPLE 4–3   A professor gives a five-question multiple-choice examination. There are four possible responses to each question. How many different answer keys can be made? (*Note:* Only one key will have all the correct answers.)

Solution   Since there are five questions and four answer choices for each question, the total number of answer keys is $4 \cdot 4 \cdot 4 \cdot 4 \cdot 4 = 4^5 = 1024$. $\qquad \blacktriangleleft$

Now, suppose that each event in a sequence does not have the same number of outcomes. How are the outcomes counted? These problems can be solved by the use of multiplication rule 2.

EXAMPLE 4–4   Suppose the supervisor in Example 4–1 wished to have an ID card with two letters followed by two digits. How many different ID cards could be made?

Solution   Again, the problem can be solved by setting up the diagram as shown.

First Letter   **Second Letter**   **First Digit**   **Second Digit**

$\boxed{}$ $\boxed{}$ $\boxed{}$ $\boxed{}$

Now, since there are 26 letters and 10 digits, the total number of ID cards is

First Letter   **Second Letter**   **First Digit**   **Second Digit**

$$\boxed{26} \cdot \boxed{26} \cdot \boxed{10} \cdot \boxed{10} = 67,600$$

There are 67,600 possible different parking permit ID numbers, since there are 26 choices for each letter and 10 choices for each digit. This example is an illustration of the second multiplication rule.

Multiplication Rule 2

In a sequence of n events in which the first one has k_1 possibilities, the second event has k_2, the third has k_3, etc., the total possibilities of the sequence will be

$$k_1 \cdot k_2 \cdot k_3 \cdot \ldots \cdot k_n$$

◄

EXAMPLE 4–5   Paint is classified as follows:

Color	Red, blue, white, black, green, brown, yellow
Type	Latex, oil
Texture	Flat, semigloss, high gloss
Use	Outdoor, indoor

How many different kinds of paint are made?

Solution   Since there are seven color choices, two type choices, three texture choices, and two use choices, the total number of possible different paints is

Color		**Type**		**Texture**		**Use**	
7	·	2	·	3	·	2	= 84

◄

EXAMPLE 4–6   A nurse has three patients to visit. How many different ways can she make her rounds if she visits each patient only once?

Solution   She can choose from three patients for the first visit, choose from two patients for the second visit, since there are two left, and choose from one patient for the third visit, since there is only one patient left. Hence, the total number of different possible outcomes is

$$3 \cdot 2 \cdot 1 = 6$$

◄

EXAMPLE 4–7   There are four blood types, A, B, AB, and O. Blood can also be Rh+ and Rh−. And finally, a blood donor can be classified as either male or female. How many different ways can a donor have his or her blood labeled?

Solution   Since there are four possibilities for blood type, two possibilities for the Rh factor, and two possibilities for the gender of the donor, there are $4 \cdot 2 \cdot 2$, or 16, different classification categories.

◄

When determining the number of different possibilities of a sequence of events, one must also determine whether repetitions are permissible.

EXAMPLE 4–8   The digits 0, 1, 2, 3, and 4 are to be used in a four-digit identification card. How many different cards are possible if repetitions are permitted?

Solution   Since there are four spaces to fill and five choices for each space, the solution is

$$5 \cdot 5 \cdot 5 \cdot 5 = 5^4 = 625$$

◄

Now, what if repetitions are not permitted? For Example 4–8, the first digit can be chosen in five ways; but the second digit can be chosen in four ways, since there are only four digits left; etc. Thus, the solution is

$$5 \cdot 4 \cdot 3 \cdot 2 = 120$$

The same situation occurs when one is drawing balls from an urn or cards from a deck. If the ball or card is replaced before the next one is selected, then repetitions are permitted, since the same one can be selected again. But if the selected ball or card is not replaced, then repetitions are not permitted, since the same ball or card cannot be selected the second time.

EXAMPLE 4–9   An urn contains four balls whose colors are red, blue, black, and white. A ball is selected, its color is noted, and it is replaced. Then a second ball is selected, and its color is noted. How many different color schemes are possible?

Solution   There are four possible choices of colors for the first ball. And since the ball is replaced, there are four possible choices for the second ball. Hence, the solution is

$$4 \cdot 4 = 16$$

◄

EXAMPLE 4–10   If the first ball in Example 4–9 is not replaced, how many different outcomes are there?

Solution   There are four possible choices for the first ball and only three possible choices for the second ball. Hence, the solution is

$$4 \cdot 3 = 12$$

◄

These examples illustrate the multiplication rules. The next section will explain the permutation rules.

EXERCISES

4–1. A woman has three skirts, five blouses, and four scarves. How many different outfits can she wear, assuming that they are color-coordinated?

4–2. How many five-digit ZIP codes are possible if digits can be repeated? If there cannot be repetitions?

4–3. How many ways can a baseball manager arrange a batting order of nine players?

4–4. How many different ways can six speakers be seated in a row on the speakers' platform?

4–5. A store manager wishes to display six different kinds of laundry soap. How many different ways can this be done?

4–6. Seven dice are rolled. How many different outcomes are there?

4–7. There are eight different statistics books, six different geometry books, and three different trigonometry books. A student must select one book of each type. How many different ways can this be done?

4–8. At the Educational Film Festival eight films are to be shown. How many different ways can these eight films be shown?

4–9. The call letters of a radio station must have four letters. The first letter must be a *K* or a *W*. How many different station call letters can be made if repetitions are not allowed? If repetitions are allowed?

4–10. How many different four-digit identification tags can be made if the digits can be used more than once? If the first digit must be a 7 and repetitions are not permitted?

4–11. How many different ways can a six-volume set of books be arranged on a shelf?

4–12. If a baseball manager has five pitchers and two catchers, how many different possible pitcher-catcher combinations can he field?

4–13. There are three roads from city X to city Y and four roads from city Y to city Z. How many different trips can be made from city X to city Z?

4–14. A coin is tossed eight times. How many different outcomes are there?

4–15. A secret code word is made by using six different color tiles that can be placed in four slots. How many different words can be formed? Repetitions are allowed.

4–16. Given the digits 0, 1, 3, 5, and 6, how many different four-digit numbers can be made that are divisible by 4 if repetitions are allowed? If repetitions are not allowed? (*Hint:* A number is divisible by 4 if the last two digits form a number that is divisible by 4 or ends in 00.)

4–17. A corporation president must select a manager and an assistant manager for each of two stores. The first store has nine employees, and the second store has seven employees. How many different ways can the selection be accomplished if the employees remain at their respective stores?

*4–18. Pine Pizza Palace sells pizza plain or with one or more of the following toppings: pepperoni, sausage, mushrooms, olives, onions, or anchovies. How many different pizzas can be made? (*Hint:* A person can select or not select each item.)

*4–19. Generalize Exercise 4–18 for *n* different toppings. (*Hint:* For example, there are two ways to select pepperoni: either take it or not take it. For two toppings, a person can select none, both, or one of the two. Continue this reasoning for three toppings, etc.)

*4–20. How many different ways can a person select one or more coins if he has two nickels, one dime, and one half-dollar?

*4–21. A photographer has five photographs that she can mount on a page in her portfolio. How many different ways can she mount her photographs?

*4–22. In a barnyard there is an assortment of chickens and cows. Counting heads, one gets 15; counting legs, one gets 46. How many of each are there?

*4–23. How many committees of two or more people can be formed from four people? (*Hint:* Make a list using the letters A, B, C, and D to represent the people.)

4–3 PERMUTATIONS

Another set of rules used to determine the total number of possibilities of a sequence of events is the set of permutation rules. Again, the rules are illustrated by examples and then formally defined.

EXAMPLE 4–11 Suppose a photographer must arrange three people in a row for a photograph. How many different possible ways can the arrangement be done?

Solution Assume that the three people are Sue, Mary, and Bill. A listing of the possible ways can be shown as follows:

Sue	Mary	Bill	Mary	Sue	Bill
Mary	Bill	Mary	Sue	Bill	Sue
Bill	Sue	Sue	Bill	Mary	Mary

Hence, there are six different possibilities.　　　◄

An arrangement of *n* objects in a specific order is called a permutation of the objects.

In order to determine the number of different possibilities mathematically, one can use the multiplication rule as follows:

First Position　　**Second Position**　　**Third Position**

□　　　　　　□　　　　　　□

Now, one can fill the first position in three ways, the second position in two ways, and the third position in one way. Hence, there are

$$3 \cdot 2 \cdot 1 = 6$$

possibilities.

Although some permutation problems can be solved by multiplication rule 2, one must use the permutation rules in situations where the multiplication rules cannot be used. These situations will be shown later.

Permutation Rule 1

The number of permutations of n objects taken all together is $n!$

Note: For an explanation of the factorial notation and operations with factorials, see Appendix A–1. Table A in the Appendix gives the factorials for the numbers from 1 through 20.

Example 4–11 can also be solved by using permutation rule 1, as shown:

$$3! = 3 \cdot 2 \cdot 1 = 6$$

EXAMPLE 4–12   Suppose a businessman has a choice of five locations in which to establish his business. He decides to rank each location according to certain criteria, such as price of the store and parking facilities. How many different ways can he rank the five cities?

Solution   From permutation rule 1, there are

$$5! = 5 \cdot 4 \cdot 3 \cdot 2 \cdot 1 = 120$$

different possible rankings.   ◄

In both of the previous examples, all objects were used up. But what happens when all objects are not used up? The answer to this question is given in Example 4–13.

EXAMPLE 4–13   Suppose the businessman in Example 4–12 wishes to rank only the top three locations. How many different ways can he rank them?

Solution   Using the multiplication rule, he can select any one of the five for first choice, then any one of his remaining four locations for his second choice, and finally, any one of the remaining three locations for his third choice, as shown.

First Choice		**Second Choice**		**Third Choice**	
5	·	4	·	3	= 60 ◄

The solution in Example 4–13 is also a permutation. The computation is defined by permutation rule 2.

Permutation Rule 2

The arrangement of n objects in a specific order using r objects at a time is called a *permutation of n objects taken r objects at a time*. It is written as $_nP_r$, and the formula is

$$_nP_r = \frac{n!}{(n-r)!}$$

Example 4–13 can be solved by using the formula, as shown:

$$_5P_3 = \frac{5!}{(5-3)!} = \frac{5!}{2!} = 60$$

EXAMPLE 4–14 How many different arrangements of three boxcars can be selected from eight boxcars for a train? The order is important since each boxcar is to be delivered to a different location.

Solution
$$_8P_3 = \frac{8!}{(8-3)!} = \frac{8!}{5!} = 336 \qquad \blacktriangleleft$$

EXAMPLE 4–15 How many different ways can a chairperson and an assistant chairperson be selected for a research project if there are seven scientists available?

Solution
$$_7P_2 = \frac{7!}{(7-2)!} = \frac{7!}{5!} = 42 \qquad \blacktriangleleft$$

EXAMPLE 4–16 How many ways can four books be arranged on a shelf if they can be selected from nine books?

Solution
$$_9P_4 = \frac{9!}{(9-4)!} = \frac{9!}{5!} = 3024 \qquad \blacktriangleleft$$

A brief explanation is necessary for why the formula $_nP_r$ works. Given the letters a, b, c, d, and e, how many permutations of two letters can be made? The multiplication rule states that there are two positions to fill. The first one can be filled in five ways, and the second position can be filled in four ways, as shown:

$$5 \cdot 4 = 20$$

The formula $_5P_2$ shows that the numerator is 5! and the denominator is $(5 - 2)!$, or 3!, which can be shown as

$$_5P_2 = \frac{5!}{3!} = \frac{5 \cdot 4 \cdot 3 \cdot 2 \cdot 1}{3 \cdot 2 \cdot 1} = \frac{5 \cdot 4 \cdot \cancel{3} \cdot \cancel{2} \cdot \cancel{1}}{\cancel{3} \cdot \cancel{2} \cdot \cancel{1}}$$

Since the last three products in the numerator and denominator cancel, what is left is

$$5 \cdot 4 = 20$$

which is the same result as that given by the multiplication rule.

What happens if the objects are not all different? The answer is shown in the next example.

EXAMPLE 4-17 How many different permutations of the letters S E E M can be made?

Solution Since there are four letters, the total possible ways are 4!, or 24 different ways if each E is labeled differently, as in S, E_1, E_2, M; and there are 2! ways to permute E_1 and E_2. But since they are indistinguishable, these duplicates must be eliminated by dividing by 2!. So the final result is

$$\frac{4!}{2!} = 12$$ ◄

Example 4-17 illustrates permutation rule 3.

Permutation Rule 3

The number of permutations of n objects in which k_1 are alike, k_2 are alike, etc., is

$$\frac{n!}{k_1! \cdot k_2! \cdot \ldots \cdot k_m!}$$

EXAMPLE 4-18 How many different permutations can be made from the letters in the word *Mississippi?*

Solution Regroup the letters as M IIII SSSS PP. Then $n = 11$, $k_1 = 1$, $k_2 = 4$, $k_3 = 4$, and $k_4 = 2$. Hence, there are

$$\frac{11!}{1! \cdot 4! \cdot 4! \cdot 2!} = 34{,}650$$

different permutations. ◄

In this section, three permutation rules were explained. In the next section, another counting rule, the combination rule, will be explained.

EXERCISES

4-24. Evaluate each expression.

a. $_8P_2$ d. $_5P_3$ g. $_8P_0$ j. $_6P_2$
b. $_7P_5$ e. $_6P_0$ h. $_8P_8$
c. $_{12}P_4$ f. $_6P_6$ i. $_{11}P_3$

4-25. How many different four-letter permutations can be formed from the letters in the word *decagon?*

4-26. In a board of directors composed of eight people, how many ways can a chief executive officer, a director, and a treasurer be selected?

4-27. How many different ID cards can be made if there are six digits on a card and no digit can be used more than once?

4-28. How many ways can seven different types of soaps be displayed on a shelf in a grocery store?

4-29. How many different four-color code stripes can be made on a sports car if each code consists of the colors green, red, blue, and white? All colors are used only once.

4-30. An inspector must select three tests to perform in a certain order on a manufactured part. He has a choice of seven tests. How many ways can he perform three different tests?

4-31. How many different permutations of the letters in the word *statistics* are there?

4-32. The Anderson Research Company decides to test-market a product in six areas. How many different ways can three areas be selected in a certain order for the first test?

4-33. How many different ways can five radio commercials be run during an hour of time if each commercial must be run according to the following schedule?

Commercial	A	B	C	D	E
Number of Runs	3	2	1	1	3

4-34. How many different ways can 4 tickets be selected from 50 tickets if each ticket wins a different prize?

4-35. How many different ways can a researcher select 5 rats from 20 rats and assign each to a different test?

4-36. How many different signals can be made by using at least three distinct flags if there are five different flags from which to select?

4-37. An investigative agency has seven cases and five agents. How many different ways can the cases be assigned if only one case is assigned to each agent?

4-38. How many different code words can be made using the letters A, A, A, B, B, C, D, D, D, D if the word must contain ten letters?

4-39. If a teacher has five different assignments to be given to five different classes, how many ways can the assignments be accomplished?

4-40. How many different ways can a visiting nurse see six patients if she sees them all in one day?

4-41. A store owner has 50 items to advertise, and she can select one different item each week for the next six weeks to put on special. How many different ways can the selection be done?

*4-42. How many different ways can four people be seated in a circle?

*4-43. How many different ways can five people, A, B, C, D, and E, sit in a row at a movie theater if (a) A and B must sit together; (b) C must sit to the right of, but not necessarily next to, B; (c) D and E will not sit next to each other?

*4-44. How many different ways can six people be assigned to three offices if there will be two people in an office?

*4-45. In Exercise 4-44, if two people refuse to share an office with each other, how many possible ways can the assignments be made?

4-4 COMBINATIONS

Suppose a dress designer wishes to select two different colors of material to design a new dress, and he has on hand four different colors. How many different possibilities can there be in this situation?

This type of problem differs from previous ones in that the order of selection is not important. That is, if the designer selects yellow and red, this selection is the same as the selection red and yellow. This type of selection is called a combination. The difference between a permutation and a combination is that in a combination, the order or arrangement of the objects is not important, but order *is* important in a permutation. The next example illustrates this difference.

A selection of objects without regard to order is called a combination.

EXAMPLE 4-19   Given the letters A, B, C, and D, list the permutations and combinations for selecting two letters.

Solution   The listings follow.

Permutations				Combinations	
AB	BA	CA	DA	AB	BC
AC	BC	CB	DB	AC	BD
AD	BD	CD	DC	AD	CD

Note that in permutations AB is different from BA. But in combinations AB is the same as BA.   ◄

Combinations are used when the order or arrangement is not important, as in the selecting process. Suppose a committee of 5 students is to be selected from 25 students. The students represent a combination, since it does not matter who is selected first, second, etc.

Combination Rule

The number of combinations of r objects selected from n objects is denoted by $_nC_r$ and is given by the formula

$$_nC_r = \frac{n!}{(n - r)! \, r!}$$

EXAMPLE 4–20   How many combinations of four objects are there taken two at a time?

Solution   Since this is a combination problem, the answer is

$$_4C_2 = \frac{4!}{(4 - 2)! \, 2!} = \frac{4!}{2! \, 2!} = 6$$ ◄

This is the same result shown in Example 4–19.

Notice that the formula for $_nC_r$ is

$$\frac{n!}{(n - r)! \, r!}$$

which is the formula for permutations,

$$\frac{n!}{(n - r)!}$$

with an $r!$ in the denominator. This $r!$ divides out the duplicates from the number of permutations, as shown in Example 4–19. For each two letters there are two permutations, but only one combination. Hence, dividing the number of permutations by $r!$ eliminates the duplicates. This result can be verified for other values of n and r.

EXAMPLE 4–21   In order to survey the opinions of customers at local malls, a researcher decides to select 5 malls from a total of 12 malls in a specific geographic area. How many different ways can the selection be made?

Solution $$_{12}C_5 = \frac{12!}{(12 - 5)! \, 5!} = \frac{12!}{7! \, 5!} = 792$$ ◄

EXAMPLE 4–22   In a club there are 7 women and 5 men. A committee of 3 women and 2 men is to be chosen. How many different possibilities are there?

Solution   Here, one must select 3 women from 7 women, which can be done in $_7C_3$, or 35, ways. Next, 2 men must be selected from 5 men, which can be done in $_5C_2$, or 10, ways. Finally, by the multiplication rule, the total number of different ways is $35 \cdot 10 = 350$.   ◄

EXAMPLE 4–23   A committee of 5 people must be selected from 5 men and 8 women. How many ways can selection be done if there are at least 3 women on the committee?

Solution   The committee can consist of 3 women and 2 men, or 4 women and 1 man, or 5 women. To find the different possibilities, find each separately, and then add them.

$$_8C_3 \cdot {}_5C_2 + {}_8C_4 \cdot {}_5C_1 + {}_8C_5 = \frac{8!}{5! \, 3!} \cdot \frac{5!}{3! \, 2!} + \frac{8!}{4! \, 4!} \cdot \frac{5!}{4! \, 1!} + \frac{8!}{3! \, 5!}$$

$$= 56 \cdot 10 + 70 \cdot 5 + 56 = 966 \qquad \blacktriangleleft$$

In the type of problem demonstrated in Example 4–23, the word *and* means to multiply, and the word *or* means to add.

EXAMPLE 4–24   How many different 5-card flushes are there in a standard deck of cards?

Solution   Since a flush consists of 5 cards of the same suit, and there are 13 cards for each suit, there are $_{13}C_5 = 1287$ flushes for each suit, and there are 4 suits. Hence, the number of 5-card flushes is

$$4 \cdot 1287 = 5148 \qquad \blacktriangleleft$$

EXERCISES

4–46. Evaluate each expression.

a. $_5C_2$ d. $_6C_2$ g. $_3C_3$ j. $_4C_3$
b. $_8C_3$ e. $_6C_4$ h. $_9C_7$
c. $_7C_4$ f. $_3C_0$ i. $_{12}C_2$

4–47. How many ways can 3 cards be selected from a standard deck of 52 cards?

4–48. How many ways are there to select three bracelets from a box of ten bracelets?

4–49. How many ways can a student select five questions from an exam containing nine questions? How many ways are there if he must answer the first question and the last question?

4–50. How many ways can a committee of four people be selected from a group of ten people?

4–51. If a person can select three presents from ten presents under a Christmas tree, how many different combinations are there?

4–52. How many different possible tests can be made from a test bank of 20 questions if the test consists of 5 questions?

4–53. The general manager of a fast-food restaurant chain must select 6 restaurants from 11 for a promotional program. How many different possible ways can this selection be done?

4–54. How many ways can 3 cars and 4 trucks be selected from 8 cars and 11 trucks to be tested for a safety inspection?

4–55. In a train yard there are 4 tank cars, 12 boxcars, and 7 flatcars. How many ways can a train be made up consisting of 2 tank cars, 5 boxcars, and 3 flatcars?

4–56. There are 7 women and 5 men in a department. How many ways can a committee of 4 people be selected? How many ways can this committee be selected if there must be 2 men and 2 women on the committee? How many ways can this committee be selected if there must be at least 2 women on the committee?

4–57. Wake Up cereal comes in two types, crispy and crunchy. If a researcher has ten boxes of each, how many ways can she select three boxes of each for a quality control test?

4–58. How many ways can a dinner patron select three salads and two vegetables if there are six salads and five vegetables on the menu?

4–59. How many ways can a jury of 6 men and 6 women be selected from 12 men and 10 women?

4–60. How many ways can a foursome of 2 men and 2 women be selected from 10 men and 12 women in a golf club?

4–61. The state narcotics bureau must form a 5-member investigative team. If it has 25 agents from which to choose, how many different possible teams can be formed?

4–62. How many different ways can a computer programmer select 3 jobs from a possible 15?

4–63. The Environmental Protection Agency must investigate nine mills for complaints of air pollution. How many different ways can a representative select five of these to investigate this week?

4–64. How many ways can a person select eight videotapes from ten tapes?

4–65. A buyer decides to stock 8 different posters. How many different ways can she select these 8 if there are 20 from which to choose?

4–66. An advertising manager decides to have an ad campaign in which 8 special calculators will be hidden at various locations in a shopping mall. If he has 17 locations from which to pick, how many different possible combinations can he choose?

*4–67. Using combinations, calculate the number of each poker hand in a deck of cards. (The information in this exercise will be used again for Exercise 5–151.)

a. Royal flush   c. Four of a kind

b. Straight flush   d. Full house

*4–68. If a photographer has 4 photos to be arranged on a page and she can select any combination of 1 to 4, how many different possible ways can she choose the photos? (*Hint:* She can select combinations of 1, 2, 3, or 4.)

4–5 TREE DIAGRAMS

In the preceding sections, only the total number of possibilities was of interest; however, many times one wishes to list each possibility of a sequence of events. For example, if one wishes to list all possible outcomes of a seven-game World Series, the solution is a difficult task if done by guessing alone. Rather than do this listing in a haphazard way, one can use a tree diagram.

A tree diagram is a device used to list all possibilities of a sequence of events in a systematic way.

Tree diagrams are also useful in determining the probabilities of events, as will be shown in the next chapter.

EXAMPLE 4–25   Suppose a salesman can travel from New York to Pittsburgh by plane, train, or bus, and from Pittsburgh to Cincinnati by bus, boat, or automobile. List all possible ways he can travel from New York to Cincinnati.

Solution A tree diagram can be drawn to show the possible ways. First, the salesman can travel from New York to Pittsburgh by three methods. The tree diagram for this situation is shown in Figure 4–1.

FIGURE 4–1
Tree Diagram for New York–Pittsburgh Trips in Example 4–25

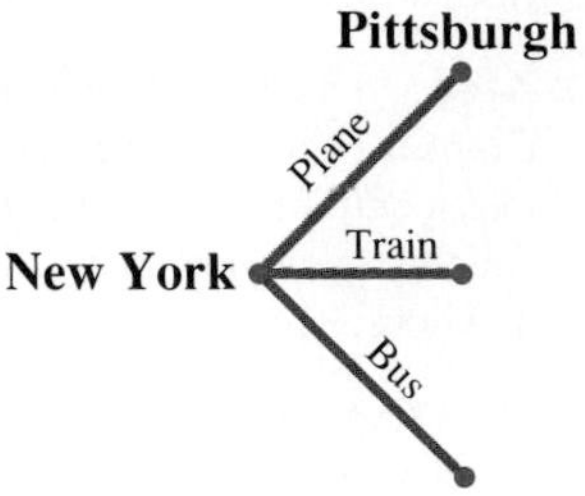

Then the salesman can travel from Pittsburgh to Cincinnati by bus, boat, or automobile. This tree diagram is shown in Figure 4–2.

FIGURE 4–2
Tree Diagram for Pittsburgh–Cincinnati Trips in Example 4–25

Next, the second branch is paired up with the first branch in three ways, as shown in Figure 4–3.

FIGURE 4–3
Complete Tree Diagram for Example 4–25

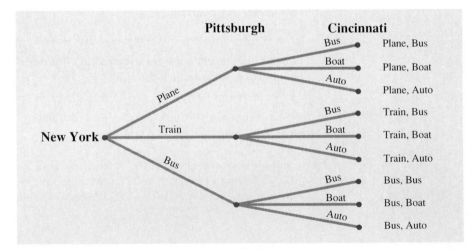

Finally, all outcomes can be listed by starting at New York and following the branches to Cincinnati, as shown at the right end of the tree in Figure 4–3. There are nine different ways. ◄

EXAMPLE 4–26 A coin is tossed and a die is rolled. Find all possible outcomes of this sequence of events.

Solution Since the coin can land either heads up or tails up, the outcomes can be shown as in Figure 4–4.

FIGURE 4–4
Tree Diagram
for Coin Tosses in
Example 4–26

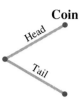

Since the die can land with any one of six numbers shown face up, the outcomes can be represented as shown in Figure 4–5.

FIGURE 4–5
Tree Diagram for
Die Rolls in
Example 4–26

Putting the two branches together, one gets the tree diagram shown in Figure 4–6.

There are 12 possible outcomes.

FIGURE 4–6
Complete Tree Diagram
for Example 4–26

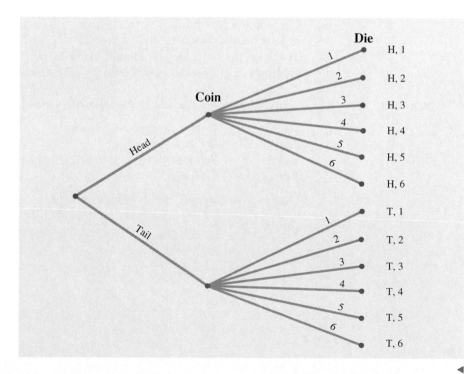

EXAMPLE 4–27   Sue and Tom play in a racquetball tournament. The first person to win two games out of three games wins the tournament. Find all possible outcomes.

Solution   Let S represent Sue's wins and T represent Tom's wins. A tree diagram can be drawn as shown in Figure 4–7.

There are six possible outcomes. Sue wins in three and Tom wins in three.

FIGURE 4–7
Tree Diagram for
Example 4–27

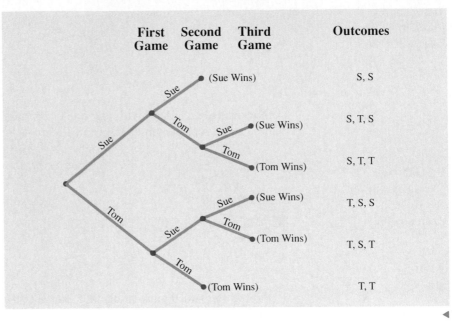

The tree's branches in Figure 4–7 are not all even in length, since each branch ends when a person has won two games.

One does not need to draw a tree in each and every situation; a systematic listing of the outcomes of simple problems may be sufficient. However, tree diagrams help eliminate the possibility of omitting outcomes.

EXAMPLE 4–28   A breakfast menu consists of the following items (one from each category):

Juice	Orange, grapefruit, cranberry
Toast	White, whole wheat
Eggs	Scrambled, fried, poached, hard-boiled
Beverage	Coffee, tea

Using a tree diagram, list all possible breakfast combinations.

Solution   See Figure 4–8.

FIGURE 4–8
Tree Diagram for
Example 4–28

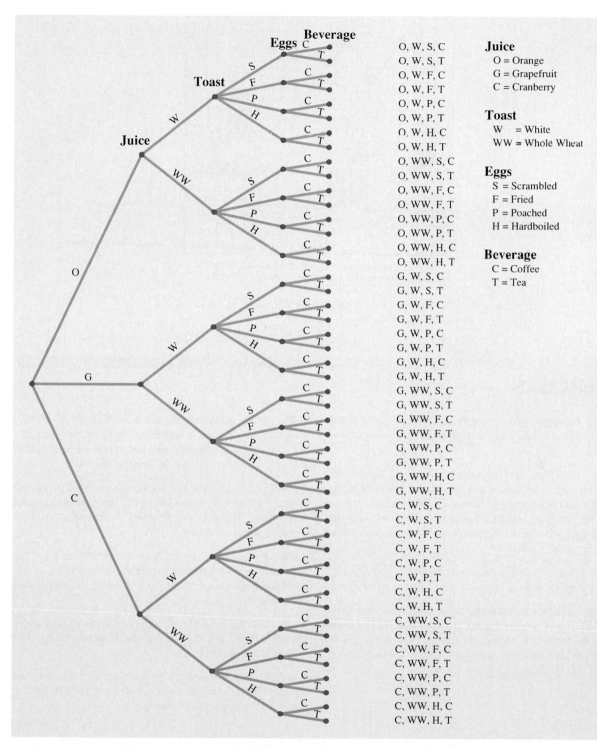

O, W, S, C
O, W, S, T
O, W, F, C
O, W, F, T
O, W, P, C
O, W, P, T
O, W, H, C
O, W, H, T
O, WW, S, C
O, WW, S, T
O, WW, F, C
O, WW, F, T
O, WW, P, C
O, WW, P, T
O, WW, H, C
O, WW, H, T
G, W, S, C
G, W, S, T
G, W, F, C
G, W, F, T
G, W, P, C
G, W, P, T
G, W, H, C
G, W, H, T
G, WW, S, C
G, WW, S, T
G, WW, F, C
G, WW, F, T
G, WW, P, C
G, WW, P, T
G, WW, H, C
G, WW, H, T
C, W, S, C
C, W, S, T
C, W, F, C
C, W, F, T
C, W, P, C
C, W, P, T
C, W, H, C
C, W, H, T
C, WW, S, C
C, WW, S, T
C, WW, F, C
C, WW, F, T
C, WW, P, C
C, WW, P, T
C, WW, H, C
C, WW, H, T

Juice
O = Orange
G = Grapefruit
C = Cranberry

Toast
W   = White
WW = Whole Wheat

Eggs
S = Scrambled
F = Fried
P = Poached
H = Hardboiled

Beverage
C = Coffee
T = Tea

"At this point in my report, I'll ask all of you to
follow me to the conference room directly
below us!"

Source: Cartoon by Bradford Veley, Marquette, Michigan.
Reprinted with permission.

EXERCISES

4–69. By means of a tree diagram, find all possible outcomes for the genders of the children in a family that has three children.

4–70. Bill's Burger Joint sells hot dogs and hamburgers plain or with a choice of any *one* of the following items: tomato, onion, or mustard. Draw a tree diagram to represent all possible selections he can serve. (*Hint:* The customer can either select an item or not. If she does not select the item, use "plain.")

4–71. A quiz consists of four true–false questions. How many possible answer keys are there? Use a tree diagram.

4–72. Students are classified according to eye color (blue, brown, green), gender (male, female), and major (chemistry, mathematics, physics, business). How many possible different classifications are there? Use a tree diagram.

4–73. A box contains a $1 bill, a $5 bill, and a $10 bill. Two bills are selected in succession, without replacing the first bill. Draw a tree diagram and represent all possible amounts of money that can be selected.

4–74. The Eagles and the Hawks play a hockey tournament. The first team to win two out of three games wins the tournament. Draw a tree diagram to represent the outcomes of the tournament.

4–75. An inspector selects three batteries from a lot. He tests each to see whether it is overcharged, normal, or undercharged. Draw a tree diagram to represent all possible outcomes.

4–76. List all possible outcomes of a five-game series played between the Pirates and the Yankees. (*Note:* The winner must win three games.)

4–77. A coin is tossed. If it comes up heads, it is tossed again. If it lands tails, a die is rolled. Find all possible outcomes of this sequence of events.

4–78. A person has a chance of obtaining a degree from each category listed below. Draw a tree diagram showing all possible ways a person could obtain these degrees.

Bachelor's	Master's	Doctor's
B.S.	M.S.	Ph.D.
B.A.	M.Ed.	D.Ed.
	M.A.	

4–79. If blood types can be A, B, AB, and O, and Rh+ and Rh−, draw a tree diagram and find the outcomes.

4–80. Shoppers in an area mall are classified as male or female, over 40 or 40 and under, and credit card or cash customers. Draw a tree diagram and find all classifications.

4–81. Patients are classified as male or female, diabetic or nondiabetic, and total, partial, or self-care. Draw a tree diagram and find the outcomes.

4–82. Urn 1 contains a blue ball and a red ball. Urn 2 contains a white ball and a black ball. If an urn is selected and a ball is drawn, find the outcomes, using a tree diagram.

4–83. Two arrows are shot at a target. They will either hit the bull's-eye (10 points), hit the outside (5 points), or miss the target (0 points). Draw a tree diagram and find the total possible scores.

4–84. Three balls numbered 1, 2, and 3 are placed in a box. The box is shaken and the balls are drawn out in sequence *without* replacement. Draw a tree diagram and find all possible three-digit numbers.

4–85. In Exercise 4–84, find all possible four-digit numbers if a ball numbered 0 is added.

4–86. Use the outcomes found in Exercise 4–85.

a. How many of the numbers are odd?
b. How many are even?
c. How many are greater than 3000?
d. How many are less than 2000?

*4–87. Solve each of the problems in Exercise 4–86 by using the multiplication rule instead of a tree diagram.

*4–88. Tree diagrams can be used to analyze various gambling games and in other areas where decisions must be made for outcomes based on chance. For example, suppose a company decides to run a taste test between two brands of colas, brand X and brand Y. A person is asked to taste a cola and identify it. The outcomes can be found by means of a tree diagram, as shown in Figure 4–9.

FIGURE 4–9

Tree Diagram for the Taste Test in Exercise 4–88

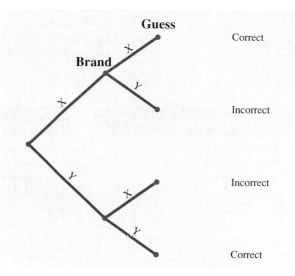

Suppose two people play a game as follows: Person A selects either a quarter or a half-dollar, and person B tries to guess the correct coin person A is holding. If B guesses the correct coin, he wins the coin; otherwise, he loses the amount he guesses. Analyze the game, using a tree diagram, and show the decisions.

*4–89. A video game manager has two locations, A and B. She can install either five or ten machines in either location, depending on the demand. If the demand is high, she should install ten machines; if it is low, she should install five machines. If she installs the wrong number of machines, she makes an incorrect decision and will lose money. Draw a tree diagram to represent the outcomes.

*4–90. There are three urns. Urn 1 contains two black balls, urn 2 contains a black and a white ball, and urn 3 contains two white balls. An urn is selected and a ball is drawn, and its color is noted. On the basis of the color, the person must guess the correct color of the other ball in the urn. Draw a tree diagram and show the outcomes and the decisions.

*4–91. A retailer decides to purchase snow shovels for a sale. If it snows, he will have a high demand, and he should stock 50 shovels. If it does not snow, he should stock 20 shovels. Draw a tree diagram to represent the outcomes and the decisions.

4–6 SUMMARY

In this chapter, the rules for counting the outcomes of events were shown. These rules included the multiplication rules, the permutation rules, and the combination rule. Using these rules, statisticians can find the solutions to a variety of problems in which they must know the number of possibilities that can occur. These rules will be used in the next chapter to determine the probabilities of events.

This chapter also described tree diagrams. When a listing of events is needed, the tree diagram can be used as a systematic way to find the solution.

Important Terms

Combination
Permutation
Tree diagram

Important Formulas

Multiplication rule 1: If each event in a sequence of n events has k different possibilities, then the total number of possibilities of the sequence will be

$$k \cdot k \cdot k \cdot \ldots \cdot k = k^n$$

Multiplication rule 2: In a sequence of n events in which the first one has k_1 possibilities, the second event has k_2 possibilities, the third has k_3 possibilities, etc., the total possibilities of the sequence will be

$$k_1 \cdot k_2 \cdot k_3 \cdot \ldots \cdot k_n$$

Permutation rule 1: The number of permutations of n objects taken all together is $n!$

Permutation rule 2: The number of permutations of n objects taken r objects at a time is

$$_nP_r = \frac{n!}{(n - r)!}$$

Permutation rule 3: The number of permutations of n objects in which k_1 are alike, k_2 are alike, etc., is

$$\frac{n!}{k_1! \cdot k_2! \cdot \ldots \cdot k_n!}$$

Combination rule: The number of combinations of r objects selected from n objects is

$$_nC_r = \frac{n!}{(n - r)! \, r!}$$

Speaking of **Statistics**

USA SNAPSHOTS®
A look at statistics that shape the nation

Playing to win
Number of $1 tickets someone would have to buy to cover every 6-number combination in selected lotto games. Tickets in millions:

Calif. (51 numbers) — 18.0
Fla., Mass. (49) — 13.9
Ill., N.Y., Lotto America (54)[1] — 12.9
Mich., Ohio (47) — 10.7
N.J. (46) — 9.4
Va., Conn., La. (44) — 7.0

1-Players get two plays for $1

Source: USA TODAY research   By Ron Coddington, USA TODAY

Source: Based on USA TODAY research. Copyright 1992, USA TODAY. Reprinted with permission.

The illustration shows the number of tickets one would have to purchase to cover every six-number combination for selected lotto games. Why are the numbers different for each state? Explain briefly how the results were computed.

Review Exercises

4–92. An identification tag consists of two letters followed by seven digits. How many different tags can be made if repetitions are allowed? If repetitions are not allowed?

4–93. How many different arrangements of the letters in the word *stubborn* can be made?

4–94. How many different three-digit combinations can be made by using the numbers 1, 3, 5, 7, and 9 without repetitions if the "right" combination can be used to open a safe? (*Hint:* Does a combination lock really use combinations?)

4–95. How many two-card pairs are there in a poker deck?

4–96. A quiz consists of six multiple-choice questions. Each question has three possible answer choices. How many different possible answer keys can be made?

4–97. How many different ways can seven different automobiles be selected from ten different automobiles?

4–98. How many different ways can a buyer select four television models from a possible choice of six models?

4–99. Draw a tree diagram to show all possible outcomes when a coin is flipped four times.

4–100. How many ways can three outfielders and four infielders be chosen from five outfielders and seven infielders?

4–101. How many different ways can eight computer operators be seated in a row?

4–102. How many ways can a student select two electives from a possible choice of ten electives?

4–103. There are 6 Republicans, 5 Democrats, and 4 Independent candidates. How many different ways can a committee of 3 Republicans, 2 Democrats, and 1 Independent be selected?

4–104. In Exercise 4–103, how many ways can a committee of four people be selected if they are all from the same party?

4–105. Employees can be classified according to gender (male, female), income (low, medium, high), and rank (assistant, instructor, dean). Draw a tree diagram and show all possible outcomes.

4–106. A disc jockey can select four records from ten to play in one segment. How many ways can this selection be done? (*Note:* Order is important.)

4–107. A judge is to rank six different brands of cookies according to their flavor. How many different ways can this ranking be done?

4–108. A vending machine serviceman must restock and collect money from 20 machines, each one at a different location. How many different ways can he select 5 machines to service in one day?

4–109. How many different ways can four paintings be arranged on a wall?

4–110. How many different ways can three fraternity members be selected from ten members if one must be president, one must be vice president, and one must be secretary/treasurer?

4–111. How many different ways can two balls be drawn from a bag containing five balls? Each ball is a different color, and the first ball is replaced before the second one is selected. How many different ways are there if the first ball is not replaced before the second one is selected?

4–112. A restaurant offers 3 choices of meat, 2 choices of potatoes, 4 choices of vegetables, and 5 choices of dessert. How many different possible meals can be made if a customer must select one item from each category?

4–113. If a girl has a blouse, a sweater, a pair of slacks, and a skirt, how many different outfits can she wear?

4–114. How many different computer passwords are possible if each consists of four symbols and if the first one must be a letter and the other three must be digits?

4–115. If a student has a choice of five computers, three printers, and two monitors, how many ways can she select a computer system?

4–116. A combination lock consists of the numbers 0 to 39. If no number can be used twice, how many different combinations are possible using three numbers?

4–117. There are 12 applicants for financial aid. The Seabest Scholarship allows 4 students to receive the money. How many different ways can these scholarships be awarded?

4–118. A candy store allows customers to select 3 different candies to be packaged and mailed. If there are 13 different varieties available, how many different possible selections can be made?

4–119. If a student can select 5 novels from a reading list of 20 for a course in literature, how many different possible ways can this selection be done?

4–120. If a student can select 1 of 3 language courses, 1 of 5 mathematics courses, and 1 of 4 history courses, how many different schedules can be made?

4–121. Find the number of different meals consisting of one or more of these cheeses, brick, Swiss, or Muenster, that can be fed to mice in an experiment.

COMPUTER APPLICATIONS

1. Write a computer program to do permutations and combinations. The program should ask the user to enter "P" for permutation or "C" for combination. Next, ask for the value of n and r. Finally, the program you write should print the answer.

TEST

Directions: Determine whether each statement is true or false.

1. A fast-food restaurant has 8 different sandwiches, 3 different kinds of french fries, and 5 different kinds of beverages. The number of different ways a person can select a sandwich, fries, and a beverage is 120.
2. If there are 5 contestants in a race, the number of different ways the first- and second-place winners can be selected is 25.
3. The number of permutations of 6 different objects taken all together is 6!.
4. If a true–false exam contains 10 questions, there are 20 different ways to answer all the questions.
5. The number of permutations of the letters in the word *moon* is 12.
6. Some permutation problems can also be solved by multiplication rule 2.
7. The arrangement A B C is the same as B A C for combinations.
8. The number of combinations when 8 objects are selected from 8 different objects is 8.
9. When objects are arranged in a specific order, the arrangement is called a combination.
10. A device that is helpful in listing the outcomes of a sequence of events is called a tree diagram.

CHAPTER

5

Probability

5–1 INTRODUCTION

A cynical person once said, "The only two sure things are death and taxes." This philosophy no doubt arose because so much in people's lives is affected by chance. From the time a person awakes until he or she goes to bed, that person makes decisions based on probability. For example, should I carry an umbrella to work today? Will my automobile battery last until spring? Should I accept that new position?

Probability as a general concept can be defined as the chance of an event occurring. Most people are familiar with probability from observing or playing games of chance, such as card games, slot machines, or lotteries. In addition to being used in games of chance, probability theory is used in the fields of insurance, investments, and weather forecasting, and in various other areas. Finally, as stated in Chapter 1, probability is the basis of inferential statistics. For example, predictions are based on probability, and hypotheses are tested by using probability.

The basic concepts of probability are explained in this chapter. These concepts include *probability experiments, sample spaces,* the *addition and multiplication rules,* and the *probabilities of complementary events.* Section 5–7 explains how the counting rules of Chapter 4 and the probability rules can be used together to solve a wide variety of problems.

5–2 SAMPLE SPACES AND PROBABILITY RULES

As noted in Chapter 1, the theory of probability was developed from the study of various games of chance using coins, dice, and cards. Since these situations lend themselves well to the application of concepts of probability, they will be used in this chapter as examples. This section begins by explaining some basic concepts of probability. Then the types of probability and probability rules are discussed.

Basic Concepts

Processes such as flipping a coin, rolling a die, or drawing a card from a deck are called probability experiments.

A probability experiment is a process that leads to well-defined results called outcomes.

An outcome is the result of a single trial of a probability experiment.

For example, when a coin is tossed, there are two possible outcomes: head or tail. (*Note:* We exclude the possibility of the coin landing on its edge.) In the roll of a single die, there are six possible outcomes: 1, 2, 3, 4, 5, or 6. In any experiment, the set of all possible outcomes is called the sample space.

A sample space is the set of all possible outcomes of a probability experiment.

Some sample spaces for various probability experiments are shown here.

Experiment	Sample Space
Toss one coin	Head, tail
Roll a die	1, 2, 3, 4, 5, 6
Answer a true-false question	True, false
Toss two coins	Head-head, tail-tail, head-tail, tail-head

It is important to realize that when two coins are tossed, there are *four* possible outcomes, as shown in the fourth experiment above. Both coins could fall heads up. Both coins could fall tails up. Coin 1 could fall heads up, and coin 2 could fall tails up. Or coin 1 could fall tails up, and coin 2 could fall heads up. Heads and tails will be abbreviated as H and T throughout this chapter.

EXAMPLE 5–1 Find the sample space for rolling two dice.

Solution Since each die can land in six different ways, and two dice are rolled, the sample space can be presented by using a rectangular array, as shown in Figure 5–1. The sample space is the list of pairs of numbers in the chart.

FIGURE 5–1
Sample Space for Rolling Two Dice (Example 5–1)

	Die 2					
Die 1	1	2	3	4	5	6
1	(1, 1)	(1, 2)	(1, 3)	(1, 4)	(1, 5)	(1, 6)
2	(2, 1)	(2, 2)	(2, 3)	(2, 4)	(2, 5)	(2, 6)
3	(3, 1)	(3, 2)	(3, 3)	(3, 4)	(3, 5)	(3, 6)
4	(4, 1)	(4, 2)	(4, 3)	(4, 4)	(4, 5)	(4, 6)
5	(5, 1)	(5, 2)	(5, 3)	(5, 4)	(5, 5)	(5, 6)
6	(6, 1)	(6, 2)	(6, 3)	(6, 4)	(6, 5)	(6, 6)

◄

EXAMPLE 5–2 Find the sample space for drawing one card from an ordinary deck of cards.

Solution Since there are four suits (hearts, clubs, diamonds, and spades) and 13 cards for each suit (ace through king), there are 52 outcomes in the sample space. See Figure 5–2.

FIGURE 5–2
Sample Space for Drawing a Card (Example 5–2)

A ♥	2 ♥	3 ♥	4 ♥	5 ♥	6 ♥	7 ♥	8 ♥	9 ♥	10 ♥	J ♥	Q ♥	K ♥
A ♦	2 ♦	3 ♦	4 ♦	5 ♦	6 ♦	7 ♦	8 ♦	9 ♦	10 ♦	J ♦	Q ♦	K ♦
A ♠	2 ♠	3 ♠	4 ♠	5 ♠	6 ♠	7 ♠	8 ♠	9 ♠	10 ♠	J ♠	Q ♠	K ♠
A ♣	2 ♣	3 ♣	4 ♣	5 ♣	6 ♣	7 ♣	8 ♣	9 ♣	10 ♣	J ♣	Q ♣	K ♣

◄

EXAMPLE 5–3 Find the sample space for the gender of the children if a family has three children. Use B for boy and G for girl.

Solution There are two genders, male and female, and each child could be either gender. Hence, there are eight possibilities, as shown here.

BBB BBG BGB GBB GGG GGB GBG BGG ◄

In the previous examples, the sample spaces were found by observation and reasoning; however, a tree diagram can also be used. In Chapter 4, the tree diagram was used to show all possible outcomes of a sequence of events. *The tree diagram can also be used as a systematic way to find all possible outcomes of a probability experiment.*

EXAMPLE 5–4 Use a tree diagram to find the sample space for the gender of three children in a family, as in Example 5–3.

Solution There are two possibilities for the first child, two for the second, and two for the third. Hence, the tree diagram can be drawn as shown in Figure 5–3.

FIGURE 5–3
Tree Diagram for
Example 5–4

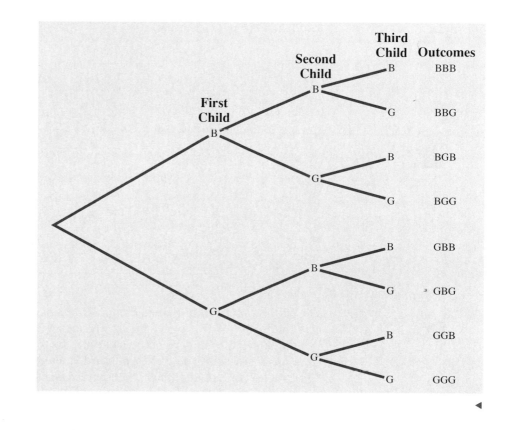

An outcome was defined previously as the result of a single trial of a probability experiment. In many problems, one must find the probability of two or more outcomes. For this reason, it is necessary to distinguish between an outcome and an event.

An event consists of one or more outcomes of a probability experiment.

An event can be one outcome or more than one outcome. For example, if a die is rolled and a 6 shows, this result is called an outcome, since it is a result of a single trial. This result is also called a **simple event.** If an odd number occurs, the result is a 1, a 3, or a 5. This result is called a **compound event,** since it consists of three outcomes or three simple events. In general, a compound event consists of two or more outcomes or simple events.

There are three basic types of probability:

1. Classical probability
2. Empirical or relative frequency probability
3. Subjective probability

Each type, along with the rules of probability for that type, is discussed in the sections that follow.

Classical Probability

Classical probability uses sample spaces to determine the numerical probability that an event will happen. One does not actually have to perform the experiment to determine that probability. Classical probability is so named because it was the first type of probability studied formally by mathematicians in the seventeenth and eighteenth centuries.

Formula for Classical Probability

The probability of any event E is

$$\frac{\text{number of ways } E \text{ can occur}}{\text{total number of outcomes in the sample space}}$$

This probability is denoted by

$$P(E) = \frac{n(E)}{n(S)}$$

This probability is called classical probability, and it uses the sample space S.

Classical probability assumes that all outcomes in the sample space are equally likely to occur. For example, when a single die is rolled, each outcome has the same probability of occurring. Since there are six outcomes, each outcome has a probability of $\frac{1}{6}$. When a card is selected from an ordinary deck of 52 cards, one assumes that the deck is shuffled, and each card has the same probability of being selected. In this case, it is $\frac{1}{52}$.

Equally likely events are events in the sample space that have the same probability of occurring.

EXAMPLE 5–5   For a card drawn from an ordinary deck, find the probability of getting a queen.

Solution   Since there are 4 queens and 52 cards, P (queen) $= \frac{4}{52} = \frac{1}{13}$. ◄

EXAMPLE 5–6   If a family has three children, find the probability that all the children are girls.

Solution   The sample space for the gender of children for a family that has three children is BBB, BBG, BGB, GBB, GGG, GGB, GBG, and BGG (see Examples 5–3 and 5–4). Since there is one way for all three children to be girls in eight possibilities,

$$P(\text{GGG}) = \frac{1}{8}$$ ◄

EXAMPLE 5–7   A card is drawn from an ordinary deck. Find these probabilities.

a. Of getting a jack.
b. Of getting the 6 of clubs.
c. Of getting a 3 or a diamond.

Solution   a. Refer to the sample space in Figure 5–2. There are 4 jacks and 52 possible outcomes. Hence,

$$P(\text{jack}) = \tfrac{4}{52} = \tfrac{1}{13}$$

b. Since there is only one 6 of clubs, the probability of getting a 6 of clubs is

$$P(\text{6 of clubs}) = \tfrac{1}{52}$$

c. There are four 3s and 13 diamonds, but the 3 of diamonds is counted twice in this listing. Hence, there are 16 possibilities of drawing a 3 or a diamond, so

$$P(\text{3 or diamond}) = \tfrac{16}{52} = \tfrac{4}{13}$$

See the sample space shown in Figure 5–2.   ◄

There are five basic probability rules. These rules will be helpful in solving probability problems, in understanding the nature of probability, and in checking answers.

Rule 1
The probability of any event E is a number between and including 0 and 1. This rule is denoted by $$0 \le P(E) \le 1$$

Rule 1 states that the answers in probability problems can never be negative or greater than 1.

Rule 2
If the event E cannot occur, then the probability of E is 0. The notation $P(E) = 0$ means that E cannot occur.

EXAMPLE 5–8   When a single die is rolled, find the probability of getting a 9.

Solution   Since the sample space is 1, 2, 3, 4, 5, and 6, it is impossible to get a 9. Hence, the probability is $P(9) = \tfrac{0}{6} = 0$.   ◄

Rule 3
If the event E is certain, then the probability of E is 1. The notation $P(E) = 1$ means that E will occur.

EXAMPLE 5–9   When a single die is rolled, what is the probability of getting a number less than 7?

Solution   Since all outcomes, 1, 2, 3, 4, 5, and 6, are less than 7, the probability is

$$P(\text{number less than 7}) = \tfrac{6}{6} = 1$$

The event of getting a number less than 7 is certain.   ◄

Rule 4

If the probability that an event occurs is $P(E)$, then the probability that the event will not occur is $1 - P(E)$.

This rule will be used to find the probability for a complementary event and is explained in detail in Section 5–6.

EXAMPLE 5–10   If the probability of snow tomorrow is $\tfrac{7}{10}$, find the probability that it won't snow.

Solution   Using Rule 4 $P(\text{no snow}) = 1 - P(\text{snow}) = 1 - \tfrac{7}{10}$ or $\tfrac{3}{10}$.   ◄

Rule 5

The sum of the probabilities of the outcomes in the sample space is 1.

For example, in the roll of a fair die, each outcome in the sample space has a probability of $\tfrac{1}{6}$. Hence, the sum of the probabilities of the outcomes is as shown.

Outcome	1	2	3	4	5	6
Probability	$\tfrac{1}{6}$	$\tfrac{1}{6}$	$\tfrac{1}{6}$	$\tfrac{1}{6}$	$\tfrac{1}{6}$	$\tfrac{1}{6}$
Sum	$\tfrac{1}{6} + \tfrac{1}{6} + \tfrac{1}{6} + \tfrac{1}{6} + \tfrac{1}{6} + \tfrac{1}{6} = \tfrac{6}{6} = 1$					

Probabilities can be expressed as fractions, decimals, or percentages. If one asks, "What is the probability of getting a head when a coin is tossed?" typical responses can be any of the following three.

"One-half."
"Point five."
"Fifty percent."

These answers are all equivalent. In most cases, the answers to examples and exercises given in this chapter are expressed as fractions or decimals, but percentages are used where appropriate.

Empirical Probability   The difference between classical probability and empirical probability is that classical probability uses sample spaces to determine numerical probability, and **empirical probability** uses frequency distributions based on observation. Suppose, for example, that a researcher asked 25 people if they liked the taste of a new soft drink. The responses were classified as "yes," "no," or "undecided." The results were categorized in a frequency distribution, as shown.

Response	Frequency
Yes	15
No	8
Undecided	2
Total	25

Probabilities now can be compared for various categories. For example, the probability of selecting a person who liked the taste is $\frac{15}{25}$, or $\frac{3}{5}$, since 15 out of 25 people in the survey answered "yes."

Formula for Empirical Probability

Given a frequency distribution, the probability of an event being in a given class is

$$P(E) = \frac{\text{frequency of the class}}{\text{total frequencies in the distribution}} = \frac{f}{n}$$

This probability is called empirical probability and is based on observation.

EXAMPLE 5–11   In the survey described above, find the probability that a person responded "no."

Solution
$$P(E) = \frac{f}{n} = \frac{8}{25}$$ ◄

EXAMPLE 5–12   In a recent study, 6 people were found to have type A blood, 7 people were found to have type B blood, 3 were found to have type AB blood, and 14 were found to have type O blood. Set up the frequency distribution and find each of the following probabilities.

a. A person has type O blood.
b. A person has type A or type B blood.
c. A person has neither type A nor type O blood.
d. A person does not have type AB blood.

Solution

Type	Frequency
A	6
B	7
AB	3
O	14
Total	30

a. $P(\text{O}) = \dfrac{f}{n} = \dfrac{14}{30} = \dfrac{7}{15}$

b. $P(\text{A or B}) = \dfrac{(6+7)}{30} = \dfrac{13}{30}$

(Add the frequencies of the two classes.)

c. $P(\text{neither A nor O}) = \dfrac{(7 + 3)}{30} = \dfrac{10}{30} = \dfrac{1}{3}$

(Neither A nor O means that a person has either type B or type AB blood.)

d. $P(\text{not AB}) = 1 - P(\text{AB}) = 1 - \dfrac{3}{30} = \dfrac{27}{30} = \dfrac{9}{10}$

(The probability of not AB can be found by subtracting the probability of type AB from 1.) ◄

EXAMPLE 5–13 Hospital records indicated that maternity patients stayed in the hospital for the number of days shown in the distribution.

Number of Days Stayed	Frequency
3	15
4	32
5	56
6	19
7	5
	127

Find these probabilities.

a. A patient stayed exactly 5 days.   c. A patient stayed at most 4 days.
b. A patient stayed less than 6 days.   d. A patient stayed at least 5 days.

Solution   a. $P(5) = \dfrac{56}{127}$

b. $P(\text{less than 6 days}) = \dfrac{(15 + 32 + 56)}{127} = \dfrac{103}{127}$

(Less than 6 days means 3, 4, or 5 days.)

c. $P(\text{at most 4 days}) = \dfrac{(15 + 32)}{127} = \dfrac{47}{127}$

since "at most 4 days" means 3 or 4 days.

d. $P(\text{at least 5 days}) = \dfrac{(56 + 19 + 5)}{127} = \dfrac{80}{127}$

since "at least 5 days" means 5, 6, or 7 days. ◄

Subjective Probability

The third type of probability is called subjective probability. **Subjective probability** uses a probability value based on an educated guess or estimate, employing opinions and inexact information.

In subjective probability, a person or group makes an educated guess at the chance that an event will occur. This guess is based on the person's experience and evaluation of a solution. For example, a sportswriter may say that there is a 70% probability that the Pirates will win the pennant next year. A physician might say that on the basis of her diagnosis, there is a 30% chance the patient will need an operation. A seismologist might say that there is an 80% probability that an earthquake will occur in a certain area. These are only a few examples of how subjective probability is used in everyday life.

All three types of probability (classical, empirical, and subjective) are used to solve a variety of problems in business, engineering, and other fields.

EXERCISES

5–1. What is a probability experiment?

5–2. Define *sample space*.

5–3. What is the difference between an outcome and an event?

5–4. What are equally likely events?

5–5. What is the range of the values of a probability event?

5–6. When an event is certain to occur, what is the probability?

5–7. If an event cannot happen, what value is assigned to its probability?

5–8. What is the sum of the probabilities of all of the outcomes in a sample space?

5–9. If the probability that the sun will shine in your town tomorrow is 0.60, what is the probability that the sun will not shine in your town tomorrow?

5–10. A probability experiment is conducted; which of the following cannot be considered a probability of an outcome?

a. $\frac{2}{3}$ e. -0.23 i. 75%
b. $-\frac{1}{2}$ f. 0   j. 250%
c. .98   g. 1.6
d. 1.65   h. 1

5–11. Classify each statement as an example of classical probability, empirical probability, or subjective probability.

a. The probability that a person will watch the 6:00 evening news is 0.15.
b. The probability of winning at a chuck-a-luck game is $\frac{5}{36}$.
c. The probability that a bus will be in an accident on a specific run is about 6%.
d. The probability of getting a royal flush when five cards are selected at random is 1/649,740.
e. The probability that a student will get a C or better in a statistics course is about 70%.
f. The probability that a new fast-food restaurant will be a success in Chicago is 35%.
g. The probability that interest rates will rise in the next six months is 0.50.

5–12. (ANS.) If a die is rolled one time, find these probabilities.

a. Of getting a 4.
b. Of getting an even number.
c. Of getting a number greater than 4.

d. Of getting a number less than 7.
e. Of getting a number greater than 0.
f. Of getting a number greater than 3 or an odd number.
g. Of getting a number greater than 3 and an odd number.

5–13. If two dice are rolled one time, find the probability of getting these results.

a. A sum of 6.
b. Doubles.
c. A sum of 7 or 11.
d. A sum greater than 9.
e. A sum less than or equal to 4.

5–14. (ANS.) If one card is drawn from a deck, find the probability of getting these results.

a. An ace.
b. A diamond.
c. An ace of diamonds.
d. A 4 or a 6.
e. A 4 or a club.
f. A 6 or a spade.
g. A heart or a club.
h. A red queen.
i. A red card or a 7.
j. A black card and a 10.

5–15. A box contains five red, two white, and three green marbles. If a marble is selected at random, find these probabilities.

a. That it is red.
b. That it is green.
c. That it is red or white.
d. That it is not green.
e. That it is not red.

5–16. If four students are seated on a bench, find the probability that the students are seated in alphabetical order from left to right, according to each student's last name. Assume that all last names are different.

5–17. If there are 50 tickets sold for a raffle and one person buys 7 tickets, what is the probability of that person winning the prize?

5–18. In an office there are seven women and nine men. If one person is promoted, find the probability that the person is a man.

5–19. If the probability that it will snow tomorrow is 60%, find the probability that it will not snow.

5–20. A certain brand of grass seed has an 86% probability of germination. If a lawn care specialist plants 9000 seeds, find the number that should germinate.

5–21. A couple plans to have three children. Find each probability.

a. Of all boys.
b. Of all girls or all boys.
c. Of exactly two boys or two girls.
d. Of at least one child of each gender.

5–22. In the game craps using two dice, a person wins on the first roll if a 7 or an 11 is rolled. Find the probability of winning on the first roll.

5–23. In a game of craps, a player loses on the roll if a 2, 3, or 12 is tossed. Find the probability of losing on the first roll.

5–24. When 350 students were interviewed, 186 stated that they preferred the lecture presentation to a group discussion. Find the probability that if a student were randomly selected, he or she would prefer the lecture presentation method.

5–25. A roulette wheel has 38 spaces numbered 1 through 36, 0, and 00. Find the probability of getting these results.

a. An odd number.
b. A number greater than 25.
c. A number less than 15 not counting 0 and 00.

5–26. A recent survey indicated that in a city of 2200 households, 1600 had a microwave oven. If a household is randomly selected, find the probability that it has a microwave oven.

5–27. A baseball player's batting average is 0.331. If she is at bat 53 times during the season, find the approximate number of times she gets to first base safely. Walks do not count.

5–28. A certain sterilizing agent is 89% effective. If a person uses this agent, find the probability that it will not be effective.

5–29. If three dice are rolled, find the probability of getting triples—i.e., 1, 1, 1; 2, 2, 2; etc.

5–30. Among 100 students at a small school, 50 are mathematics majors, 30 are English majors, and 20 are history majors. If a student is selected at random, find the probability that she is neither a mathematics major nor an English major.

5–31. A record store receives orders from the following age groups.

Age	Percentage
Under 20	30
20–29	13
30–39	20
40–49	27
50 and over	10

If a customer is selected at random, find the probability that he or she is in the following age groups. Use percentages.

a. Between 30 and 39 years of age.
b. Under 30 years of age.
c. Over 29 and under 50 years of age.
d. Under 20 or over 49 years of age.

*5–32. A person flipped a coin 100 times and obtained 73 heads. Can the person conclude that the coin was unbalanced?

*5–33. A medical doctor stated that with a certain treatment, a patient has a 50% chance of recovering without surgery—that is, "Either he will get well or he won't get well." Comment on his statement.

*5–34. The wheel spinner in Figure 5–4 is spun twice. Find the sample space, and then determine the probability of the following events.

FIGURE 5–4
Spinner for Exercise 5–34

a. An odd number on the first spin and an even number on the second spin. (*Note:* 0 is considered even.)
b. A sum greater than 4.
c. Even numbers on both spins.
d. A sum that is odd.
e. The same number on both spins.

*5–35. Roll a die 180 times and record the number of 1s, 2s, 3s, 4s, 5s, and 6s. Compute the probabilities of each, and compare these probabilities with the theoretical results.

*5–36. Toss two coins 100 times and record the number of heads (0, 1, 2). Compute the probabilities of each outcome, and compare these probabilities with the theoretical results.

5-3 THE ADDITION RULES FOR PROBABILITY

Many problems involve finding the probability of two or more events. For example, at a large political gathering, one might wish to know, for a person selected at random, the probability that the person is a female or is a Republican. In this case, there are three possibilities to consider:

1. The person is a female.
2. The person is a Republican.
3. The person is both a female and a Republican.

Consider another example. At the same gathering there are Republicans, Democrats, and Independents. If a person is selected at random, what is the probability that the person is a Democrat or an Independent? In this case, there are only two possibilities:

1. The person is a Democrat.
2. The person is an Independent.

The difference between the two examples is that in the first case, the person selected can be a female and a Republican at the same time. In the second case, the person selected cannot be both a Democrat and an Independent at the same time. In the second case, the two events are said to be mutually exclusive; in the first case, they are considered not mutually exclusive.

Two events are mutually exclusive if they cannot occur at the same time.

In another situation, the events of getting a 4 and getting a 6 when a single card is drawn from a deck are mutually exclusive events since a single card cannot be both a 4 and a 6. On the other hand, the events of getting a 4 and getting a heart on a single draw are not mutually exclusive, since one can select the 4 of hearts when drawing a single card from an ordinary deck.

EXAMPLE 5-14   Determine which events are mutually exclusive and which are not when a single die is rolled.

a. Getting an odd number and getting an even number.
b. Getting a 3 and getting an odd number.
c. Getting an odd number and getting a number less than 4.
d. Getting a number greater than 4 and getting a number less than 4.

Solution   a. The events are mutually exclusive, since the first event can be 1, 3, or 5, and the second event can be 2, 4, or 6.
b. The events are not mutually exclusive, since the first event is a 3 and the second can be 1, 3, or 5. Hence, 3 is contained in both outcomes.
c. The events are not mutually exclusive, since the first event can be 1, 3, or 5, and the second can be 1, 2, or 3. Hence, 1 and 3 are contained in both outcomes.
d. The events are mutually exclusive, since the first event can be 5 or 6, and the second event can be 1, 2, or 3.   ◄

EXAMPLE 5–15   Determine which events are mutually exclusive and which are not when a single card is drawn from a deck.

    a. Getting a 7 and getting a jack.
    b. Getting a club and getting a king.
    c. Getting a face card and getting an ace.
    d. Getting a face card and getting a spade.

Solution   Only the events in parts a and c are mutually exclusive.   ◄

The probability of two or more events can be determined by using the *addition rules*. The first addition rule is used when the events are mutually exclusive.

Addition Rule 1

When two events A and B are mutually exclusive, the probability that A or B occurs is

$$P(A \text{ or } B) = P(A) + P(B)$$

EXAMPLE 5–16   A drawer contains three pairs of red socks, two pairs of black socks, and four pairs of brown socks. If a person in a dark room selects a pair of socks, find the probability that the pair will be either black or brown. (*Note:* The socks are folded together in matching pairs.)

Solution   Since there are nine pairs of socks,

$$P(\text{black or brown}) = P(\text{black}) + P(\text{brown}) = \tfrac{2}{9} + \tfrac{4}{9} = \tfrac{6}{9} = \tfrac{2}{3}$$

The events are assumed to be mutually exclusive.   ◄

EXAMPLE 5–17   At a political rally, there are 20 Republicans, 13 Democrats, and 6 Independents. If a person is selected, find the probability that he or she is either a Democrat or an Independent.

Solution $$P(\text{Democrat or Independent}) = P(\text{Democrat}) + P(\text{Independent})$$
$$= \tfrac{13}{39} + \tfrac{6}{39} = \tfrac{19}{39}$$ ◄

EXAMPLE 5–18   A day of the week is selected at random. Find the probability that it is a weekend day.

Solution $$P(\text{Saturday or Sunday}) = P(\text{Saturday}) + P(\text{Sunday}) = \tfrac{1}{7} + \tfrac{1}{7} = \tfrac{2}{7}$$ ◄

When two events are not mutually exclusive, one must subtract the probabilities of the outcomes that are common to both events, since they have been counted twice. This technique is illustrated in the next example.

EXAMPLE 5–19   A single card is drawn from a deck. Find the probability that it is a king or a club.

Solution   Since the king of clubs is counted twice, the probability must be subtracted, as shown.

$$P(\text{king or club}) = P(\text{king}) + P(\text{club}) - P(\text{king of clubs})$$
$$= \tfrac{4}{52} + \tfrac{13}{52} - \tfrac{1}{52} = \tfrac{16}{52} = \tfrac{4}{13}$$ ◄

When events are not mutually exclusive, addition rule 2 can be used to find the probability of the events.

Addition Rule 2

If A and B are *not* mutually exclusive, then

$$P(A \text{ or } B) = P(A) + P(B) - P(A \text{ and } B)$$

Note: This rule can be used when the events are mutually exclusive, since $P(A \text{ and } B)$ will always equal 0. However, the author feels that it is important to make a distinction between the two situations.

EXAMPLE 5-20   In a hospital unit there are eight nurses and five physicians. Seven nurses and three physicians are females. If a staff person is selected, find the probability that the subject is a nurse or a male.

Solution   The sample space is shown below.

Staff	Females	Males	Total
Nurses	7	1	8
Physicians	3	2	5
Total	10	3	13

The probability is

$$P(\text{nurse or male}) = P(\text{nurse}) + P(\text{male}) - P(\text{male nurse})$$
$$= \tfrac{8}{13} + \tfrac{3}{13} - \tfrac{1}{13} = \tfrac{10}{13} \qquad \blacktriangleleft$$

EXAMPLE 5-21   On New Year's Eve, the probability of a person driving while intoxicated is 0.32, the probability of a person having a driving accident is 0.09, and the probability of a person having a driving accident while intoxicated is 0.06. What is the probability of a person driving while intoxicated or having a driving accident?

Solution
$$P(\text{intoxicated or accident}) = P(\text{intoxicated}) + P(\text{accident})$$
$$- P(\text{intoxicated and accident})$$
$$= 0.32 + 0.09 - 0.06 = 0.35 \qquad \blacktriangleleft$$

The probability rules can be extended to three or more events. For three mutually exclusive events A, B, and C,

$$P(A \text{ or } B \text{ or } C) = P(A) + P(B) + P(C)$$

For three events that are *not* mutually exclusive,

$$P(A \text{ or } B \text{ or } C) = P(A) + P(B) + P(C) - P(A \text{ and } B) - P(A \text{ and } C)$$
$$- P(B \text{ and } C) + P(A \text{ and } B \text{ and } C)$$

See Exercises 5-60, 5-61, and 5-63.

EXERCISES

5–37. Define *mutually exclusive events*.

5–38. Give an example of two events that are mutually exclusive and two events that are not mutually exclusive.

5–39. Determine whether the following events are mutually exclusive.

a. Roll a die: Get an even number, and get a number less than 3.
b. Roll a die: Get a prime number (2, 3, 5), and get an odd number.
c. Roll a die: Get a number greater than 3, and get a number less than 3.
d. Select a student in your class: The student has blond hair, and the student has blue eyes.
e. Select a student in your college: The student is a sophomore, and the student is a business major.
f. Select any course: It is a calculus course, and it is an English course.
g. Select a registered voter: The voter is a Republican, and the voter is a Democrat.

5–40. A ski shop decides to select a month for its annual sale. Find the probability that it will be November or December. Assume that all months have an equal probability of being selected.

5–41. In a large grocery store, there are two managers, three department heads, five stock clerks, and eight cashiers. If a person is selected at random, find the probability that he or she is either a cashier or a manager.

5–42. At a convention there were 7 mathematics instructors, 5 computer science instructors, 3 statistics instructors, and 4 science instructors. If an instructor is selected, find the probability of getting a science instructor or a mathematics instructor.

5–43. In a fish tank there are 27 goldfish, an angelfish, and 3 guppies. If a fish is selected at random, find the probability of getting a guppy or an angelfish.

5–44. On a small college campus, there are 5 English professors, 4 mathematics professors, 2 science professors, 3 psychology professors, and 3 history professors. If a professor is selected at random, find the probability that the professor is the following.

a. An English or psychology professor.
b. A mathematics or science professor.
c. A history, science, or mathematics professor.
d. An English, mathematics, or history professor.

5–45. The probability of a California teenager owning a surfboard is 0.43, of owning a skateboard is 0.38, and of owning both is 0.28. If a California teenager is selected at random, find the probability that he or she owns a surfboard or a skateboard.

5–46. The probability of a tourist visiting Indian Caverns is 0.80 and of visiting the Safari Zoo is 0.55. The probability of visiting both places on the same day is 0.42. Find the probability that a tourist visits Indian Caverns or visits Safari Zoo.

5–47. A single card is drawn from a deck. Find the probability of selecting the following.

a. A 4 or a diamond
b. A club or a diamond
c. A jack or a black card

5–48. In a statistics class there are 18 juniors and 10 seniors; 6 of the seniors are females, and 12 of the juniors are males. If a student is selected at random, find the probability of selecting the following.

a. A junior or a female
b. A senior or a female
c. A junior or a senior

5–49. In a large department store there are 500 employees. Three hundred fifty are females and 200 of them are under the age of 25. There are 75 males under 25. If an employee is selected for promotion, find the probability that the employee will be the following.

a. Under 25 or a female
b. Over 24 or a female
c. Male or over 24

5–50. A women's clothing store owner buys from three companies: A, B, and C. The most recent purchases are shown here.

Product	Company A	Company B	Company C
Dresses	24	18	12
Blouses	13	36	15

If one item is selected at random, find the following probabilities.

a. It was purchased from company A or is a dress.
b. It was purchased from company B or company C.
c. It is a blouse or was purchased from company A.

5-51. In a recent study, the following data were obtained in response to the question, "Do you favor the college building a swimming pool?"

Class	Yes	No	No Opinion
Freshmen	15	8	6
Sophomores	25	3	2

If a student is selected at random, find these probabilities.

a. The student has no opinion.
b. The student is a freshman or is against the issue.
c. The student is a sophomore or favors the issue.

5-52. A grocery store employs cashiers, stock clerks, and deli personnel. The distribution of employees according to marital status is shown next.

Marital Status	Cashiers	Stock Clerks	Deli Personnel
Married	8	12	3
Not married	5	15	2

If an employee is selected at random, find these probabilities.

a. The employee is a stock clerk or married.
b. The employee is not married.
c. The employee is a cashier or is not married.

5-53. In a certain geographic region, newspapers are classified as being published daily morning, daily evening, and weekly. Some have a comics section and some do not. The distribution is shown next.

Have Comics Section	Morning	Evening	Weekly
Yes	2	3	1
No	3	4	2

If a newspaper is selected at random, find these probabilities.

a. The newspaper is a weekly publication.
b. The newspaper is a daily morning publication or has comics.
c. The newspaper is published weekly or does not have comics.

5-54. Three cable channels, 6, 8, and 10, have quiz shows, comedies, and dramas. The number of each is shown below.

Type of Show	Channel 6	Channel 8	Channel 10
Quiz show	5	2	1
Comedy	3	2	8
Drama	4	4	2

If a show is selected at random, find these probabilities.

a. The show is a quiz show or it is shown on channel 8.
b. The show is a drama or a comedy.
c. The show is shown on channel 10 or it is a drama.

5-55. A local postal carrier distributes first-class letters, advertisements, or magazines. For a certain day, she distributed the following number of each type of item.

Delivery to	First-Class Letters	Advertisements	Magazines
Home	325	406	203
Business	732	1021	97

If an item of mail is selected at random, find these probabilities.

a. The item went to a home.
b. The item was an advertisement or it went to a business.
c. The item was a first-class letter or it went to a home.

5-56. The frequency distribution shown here illustrates the number of medical tests conducted on 30 randomly selected emergency patients.

Number of Tests Performed	Number of Patients
0	12
1	8
2	2
3	3
4 or more	5

If a patient is selected at random, find these probabilities.

a. The patient has had exactly 2 tests done.
b. The patient has had at least 2 tests done.
c. The patient has had at most 3 tests done.
d. The patient has had 3 or fewer tests done.
e. The patient has had 1 or 2 tests done.

5-57. The following distribution represents the length of time a patient will spend in a hospital.

Days	Frequency
0-3	2
4-7	15
8-11	8
12-15	6
16+	9

If a patient is selected, find these probabilities.

a. The patient spends 0-3 days in the hospital.
b. The patient spends less than 8 days in the hospital.
c. The patient spends 16 or more days in the hospital.
d. The patient spends a maximum of 11 days in the hospital.

5–58. A sales representative who visits customers at home finds she will sell 0, 1, 2, 3, or 4 items according to the following frequency distribution.

Items Sold	Frequency
0	8
1	10
2	3
3	2
4	1

Find the probability that she sells the following.

a. Exactly 1 item
b. More than 2 items
c. At least 1 item
d. At most 3 items

5–59. A recent study of 300 patients found that of 100 alcoholic patients, 87 had elevated cholesterol levels, and 43 of the 200 nonalcoholic patients had elevated cholesterol levels. If a patient is selected at random, find the probability that the patient is the following.

a. An alcoholic with elevated cholesterol level.
b. A nonalcoholic.
c. A nonalcoholic with normal cholesterol level.

5–60. If one card is drawn from an ordinary deck of cards, find the probability of getting the following.

a. A king or a queen or a jack.
b. A club or a heart or a spade.
c. A king or a queen or a diamond.
d. An ace or a diamond or a heart.
e. A 9 or a 10 or a spade or a club.

5–61. Two dice are rolled. Find the probability of getting the following.

a. A sum of 6 or 7 or 8.
b. Doubles or a sum of 4 or 6.
c. A sum greater than 9 or less than 4 or a 7.

5–62. An urn contains six red balls, two green balls, one blue ball, and one white ball. If a ball is drawn, find the probability of getting a red or a white ball.

5–63. Three dice are rolled. Find the probability of getting the following.

a. Triples   b. A sum of 5

*5–64. The probability that a customer selects a pizza with mushrooms or pepperoni is 0.55, and the probability that the customer selects mushrooms only is 0.32. If the probability that he or she selects pepperoni only is 0.17, find the probability of the customer selecting both items.

*5–65. In building new homes, a contractor finds that the probability of a home buyer selecting a two-car garage is 0.70 and of selecting a one-car garage is 0.20. Find the probability that the buyer will select no garage. The builder will not build houses with three-car garages.

*5–66. In Exercise 5–65, find the probability that the buyer will not want a two-car garage.

LAFF-A-DAY

"I know you haven't had an accident in thirteen years. We're raising your rates because you're about due one."

Source: Reprinted with special permission of King Features Syndicate, Inc.

5–4 THE MULTIPLICATION RULES FOR PROBABILITY

The previous section showed that the addition rules are used to compute probabilities for mutually exclusive and not mutually exclusive events. In this section, two more rules, the multiplication rules, are introduced. These rules can be used to find the probability of two or more events that occur in sequence. For example, if a coin is tossed and then a die is rolled, one can find the probability of getting a head on the coin *and* a 4 on the die. These two events are said to be independent since the outcome of the first event (tossing a coin) does not affect the outcome of the second event (rolling a die).

Two events *A* and *B* are independent if the probability of *A* occurring does not affect the probability of *B* occurring.

Here are other examples of independent events:

Rolling a die and getting a 6, and then rolling a second die and getting a 3.
Drawing a card from a deck and getting a queen, replacing it, and drawing a second card and getting a queen.

In order to find the probability of two independent events that occur in sequence, one must find the probability of each event occurring separately and then multiply the answers. For example, if a coin is tossed twice, the probability of getting two heads is $\frac{1}{2} \cdot \frac{1}{2} = \frac{1}{4}$. This result can be verified by looking at the sample space, HH, HT, TH, TT. Then $P(\text{HH}) = \frac{1}{4}$.

Multiplication Rule 1

When two events are independent, the probability of both occurring is

$$P(A \text{ and } B) = P(A) \cdot P(B)$$

EXAMPLE 5–22   A coin is flipped and a die is rolled. Find the probability of getting a head on the coin and a 4 on the die.

Solution $$P(\text{head and 4}) = P(\text{head}) \cdot P(4) = \frac{1}{2} \cdot \frac{1}{6} = \frac{1}{12}$$

Note that the sample space for the coin is H, T; and for the die it is 1, 2, 3, 4, 5, 6.   ◄

The problem in Example 5–22 can also be solved by using the sample space:

H1   H2   H3   H4   H5   H6   T1   T2   T3   T4   T5   T6

The solution is $\frac{1}{12}$, since there is only one way to get the head-4 outcome.

EXAMPLE 5–23   A card is drawn from a deck and replaced; then a second card is drawn. Find the probability of getting a queen and then an ace.

Solution   Since the card is replaced, the probability of getting a queen is $\frac{4}{52}$, and the probability of getting an ace is $\frac{4}{52}$. Hence, the probability of getting a queen and an ace is

$$P(\text{queen and ace}) = P(\text{queen}) \cdot P(\text{ace}) = \frac{4}{52} \cdot \frac{4}{52} = \frac{16}{2704} = \frac{1}{169} \quad ◄$$

EXAMPLE 5–24   An urn contains 3 red balls, 2 blue balls, and 5 white balls. A ball is selected, its color noted, and then it is replaced. A second ball is selected, and its color is noted. Find the probability of each of the following.

a. Selecting 2 blue balls.
b. Selecting a blue ball and then a white ball.
c. Selecting a red ball and then a blue ball.

Solution   a. $P(\text{blue and blue}) = P(\text{blue}) \cdot P(\text{blue}) = \dfrac{2}{10} \cdot \dfrac{2}{10} = \dfrac{4}{100} = \dfrac{1}{25}$

b. $P(\text{blue and white}) = P(\text{blue}) \cdot P(\text{white}) = \dfrac{2}{10} \cdot \dfrac{5}{10} = \dfrac{10}{100} = \dfrac{1}{10}$

c. $P(\text{red and blue}) = P(\text{red}) \cdot P(\text{blue}) = \dfrac{3}{10} \cdot \dfrac{2}{10} = \dfrac{6}{100} = \dfrac{3}{50}$ ◀

Multiplication rule 1 can be extended to three or more independent events by using the formula

$$P(A \text{ and } B \text{ and } C \text{ and } \ldots \text{ and } K) = P(A) \cdot P(B) \cdot P(C) \cdot \ldots \cdot P(K)$$

EXAMPLE 5–25   A survey found that 46% of people say they were sunburned at least once during the summer. If three people are selected at random, find the probability that all three will have been sunburned.

Solution   Let S denote sunburn. Then

$$P(S \text{ and } S \text{ and } S) = P(S) \cdot P(S) \cdot P(S)$$
$$= (0.46)(0.46)(0.46) = 0.097336$$ ◀

EXAMPLE 5–26   The probability that a specific medical test will show positive is 0.12. If four people are tested, find the probability that all four will show positive.

Solution   Let T be the symbol for a positive test result. Then

$$P(T \text{ and } T \text{ and } T \text{ and } T) = P(T) \cdot P(T) \cdot P(T) \cdot P(T)$$
$$= (0.12)(0.12)(0.12)(0.12) = 0.0002074$$ ◀

In the previous examples, the events were independent of each other, since the occurrence of the first event in no way affected the outcome of the second event. On the other hand, when the occurrence of the first event affects or changes the probability of the occurrence of the second event, the two events are said to be *dependent*. For example, suppose a card is drawn from a deck and *not* replaced, and then a second card is drawn. What is the probability of selecting an ace on the first card and a king on the second card?

Before an answer to the question can be given, one must realize that the events are dependent. The probability of selecting an ace on the first draw is $\frac{4}{52}$. But since that card is *not* replaced, the probability of selecting a king on the second card is $\frac{4}{51}$, since there are 51 cards remaining. The outcome of the first draw has affected the outcome of the second draw.

Dependent events are formally defined next.

When the outcome or occurrence of the first event affects the outcome or occurrence of the second event in such a way that the probability is changed, the events are said to be dependent.

Here are some examples of dependent events:

Drawing a card from a deck, not replacing it, and then drawing a second card.
Selecting a ball from an urn, not replacing it, and then selecting a second ball.
Being a lifeguard and getting a suntan.
Having high grades and getting a scholarship.
Parking in a no-parking zone and getting a parking ticket.

In order to find probabilities when events are dependent, one uses the multiplication rule with a modification in notation. For the problem discussed above, the probability of getting an ace on the first draw is $\frac{4}{52}$, and the probability of getting a king on the second draw is $\frac{4}{51}$. By the multiplication rule, the probability of both events occurring is

$$\frac{4}{52} \cdot \frac{4}{51} = \frac{16}{2652} = \frac{4}{663}$$

The event of getting a king on the second draw *given* that an ace was drawn the first time is called a conditional probability. The **conditional probability** of an event B in relationship to an event A is the probability that event B occurs after event A has already occurred. The notation for conditional probability is $P(B|A)$. This notation does not mean that B is divided by A; rather, it means the probability that event B occurs given that event A has already occurred. In the card example, $P(B|A)$ is the probability that the second card is a king given that the first card is an ace, and it is equal to $\frac{4}{51}$ since the first card was *not* replaced.

Multiplication Rule 2

When two events are dependent, the probability of both occurring is

$$P(A \text{ and } B) = P(A) \cdot P(B|A)$$

EXAMPLE 5-27   In a shipment of 25 microwave ovens, 2 are defective. If 2 ovens are randomly selected and tested, find the probability that both are defective if the first one is not replaced after it has been tested.

Solution   Since the events are dependent,

$$P(D_1 \text{ and } D_2) = P(D_1) \cdot P(D_2|D_1) = \frac{2}{25} \cdot \frac{1}{24} = \frac{2}{600} = \frac{1}{300} \qquad \blacktriangleleft$$

EXAMPLE 5-28   The World Wide Insurance Company found that 53% of the residents of a city had homeowner's insurance with its company. Of these clients, 27% also had automobile insurance with the company. If a resident is selected at random, find the probability that the resident has both homeowner's and automobile insurance with the World Wide Insurance Company.

Solution $$P(H \text{ and } A) = P(H) \cdot P(A|H) = (0.53)(0.27) = 0.1431 \qquad \blacktriangleleft$$

This multiplication rule can be extended to three or more events, as shown in the next example.

EXAMPLE 5-29   Three cards are drawn from an ordinary deck and not replaced. Find the probability of the following.

a. Getting three jacks.
b. Getting an ace, a king, and a queen in order.
c. Getting a club, a spade, and a heart in order.
d. Getting three clubs.

Solution   a. $P(3 \text{ jacks}) = \frac{4}{52} \cdot \frac{3}{51} \cdot \frac{2}{50} = \frac{24}{132{,}600} = \frac{1}{5525}$

b. $P(\text{ace and king and queen}) = \frac{4}{52} \cdot \frac{4}{51} \cdot \frac{4}{50} = \frac{64}{132{,}600} = \frac{8}{16{,}575}$

c. $P(\text{club and spade and heart}) = \frac{13}{52} \cdot \frac{13}{51} \cdot \frac{13}{50} = \frac{2197}{132{,}600} = \frac{169}{10{,}200}$

d. $P(3 \text{ clubs}) = \frac{13}{52} \cdot \frac{12}{51} \cdot \frac{11}{50} = \frac{1716}{132{,}600} = \frac{11}{850}$ ◄

Tree diagrams can be used as an aid to finding the solution to probability problems when the events are sequential. The next example illustrates the use of tree diagrams.

EXAMPLE 5–30   Box 1 contains 2 red balls and 1 blue ball. Box 2 contains 3 blue balls and 1 red ball. A coin is tossed. If it falls heads up, box 1 is selected and a ball is drawn. If it falls tails up, box 2 is selected and a ball is drawn. Find the probability of selecting a red ball.

Solution   With the use of a tree diagram, the sample space can be determined as shown in Figure 5–5. First, assign probabilities to each branch, as shown in Figure 5–5. Next, using the multiplication rule, multiply the probabilities for each branch, as shown in Figure 5–5.

FIGURE 5–5
Tree Diagram for
Example 5–30

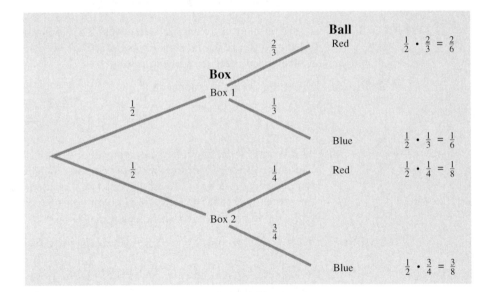

Finally, use the addition rule, since a red ball can be obtained from box 1 or box 2.

$$P(\text{red}) = \tfrac{2}{6} + \tfrac{1}{8} = \tfrac{8}{24} + \tfrac{3}{24} = \tfrac{11}{24}$$

(*Note:* The sum of all final probabilities will always be equal to 1.)   ◄

Tree diagrams can be used when the events are independent or dependent, and they can also be used for sequences of three or more events.

EXERCISES

5–67. State which events are independent and which are dependent.

a. Tossing a coin and drawing a card from a deck.
b. Drawing a ball from an urn, not replacing it, and then drawing a second ball.
c. Getting a raise in salary and purchasing a new car.
d. Driving on ice and having an accident.
e. Having a large shoe size and having a high IQ.
f. A father being left-handed and a daughter being left-handed.
g. Smoking excessively and having lung cancer.
h. Eating an excessive amount of ice cream and smoking an excessive amount of cigarettes.

5–68. If a student guesses on all the ten questions on a true-false exam, find the probability that he or she will get them all right.

5–69. If 18.6% of all professors of higher education in Pennsylvania have a Ph.D., find the probability that if three are selected, all will have a Ph.D.

5–70. A national study of patients who were overweight found that 56% also had elevated blood pressure. If two overweight patients are selected, find the probability that both have elevated blood pressure.

5–71. If 52% of the families surveyed have both spouses employed, find the probability of selecting three families with both spouses employed.

5–72. An automobile saleswoman finds that the probability of making a sale is 0.23. If she talks to four customers today, find the probability that she will sell four automobiles.

5–73. If 3% of all computers manufactured are defective, find the probability of selecting 2 computers in a row that are defective.

5–74. Find the probability of selecting two people at random who were born in the same month.

5–75. If two people are selected at random, find the probability that they have the same birthday (both month and day).

5–76. If three people are selected, find the probability that all three were born in March.

5–77. One-third of all deaths are caused by heart attacks. If three randomly selected deaths are chosen, find the probability that all resulted from heart attacks.

5–78. What is the probability that a husband, wife, and daughter have the same birthday?

5–79. A flashlight has six batteries, two of which are defective. If two are selected at random without replacement, find the probability that both are defective.

5–80. In Exercise 5–79, find the probability that the first battery tests good and the second one is defective.

5–81. If a certain drug is administered to mice, there is a 5% probability of an adverse reaction. If three mice are selected at random, find the probability that they will all have adverse reactions.

5–82. In a department store there are 120 customers, 90 of whom will buy at least one item. If 5 customers are selected at random, one by one, find the probability that all will buy at least one item.

5–83. Three cards are drawn from a deck *without* replacement. Find these probabilities.

a. All are jacks.
b. All are clubs.
c. All are red cards.

5–84. In a scientific study there are eight guinea pigs, five of which are pregnant. If three are selected at random without replacement, find the probability that all are pregnant.

5–85. In Exercise 5–84, find the probability that none are pregnant.

5–86. In a class consisting of 15 men and 12 women, two homework papers were selected at random. Find the probability that both papers belonged to women.

5–87. In Exercise 5–86, find the probability that both papers belonged to men.

5–88. A manufacturer makes two models of an item, model I products, which account for 80% of unit sales, and model II products, which account for 20% of unit sales. Because of defects, the manufacturer has to replace (or exchange) 10% of its model I products and 18% of its model II products. If a product is selected at random, find the probability that it will be defective.

5–89. An automobile manufacturer has three factories, A, B, and C. They produce 50%, 30%, and 20%, respectively, of a specific model of the automobile. Thirty percent of the automobiles produced in factory A are white, 40% produced in factory B are white, and 25% of those produced in factory C are white. If an automobile produced by the company is selected at random, find the probability that it is white.

5–90. An insurance company classifies drivers as low-risk, medium-risk, and high-risk. Sixty percent of those insured are low-risk, 30% are medium-risk, and 10% are high-risk. After a study, the company finds that during a one-year period, 1% of the low-risk drivers have an accident, 5% of the medium-risk drivers have an accident, and 9% of the high-risk drivers have an accident. If a driver is selected at random, find the probability that the driver will have an accident during the year.

5–91. In a certain geographic location, 25% of the wage earners have a college degree and 75% do not. Of those who have a college degree, 5% earn more than $100,000 a year. Of those who do not have a college degree, 2% earn more than $100,000 a year. If a wage earner is selected at random, find the probability that the wage earner earns more than $100,000 a year.

5–92. Urn 1 contains 6 white balls and 4 red balls. Urn 2 contains 3 white balls and 2 red balls. Urn 3 contains 4 white balls and 1 red ball. If an urn is selected at random and a ball is drawn, find the probability it will be a red ball.

*5–93. In an experiment, there are ten female and five male mice. Three are selected to receive an injection. Find the probability that all are males or all are females.

*5–94. At a medical center, there are 8 doctors, 6 of whom are males. There are 10 nurses, 4 of whom are males. If a person is selected at random, find the probability that the person is a female and a doctor.

*5–95. In Exercise 5–94, find the probability of selecting a male nurse.

5–5 CONDITIONAL PROBABILITY AND BAYES'S THEOREM (OPTIONAL)

In the previous section, the conditional probability of an event B in relationship to an event A was defined as the probability that event B occurs after event A has already occurred. The formulas for conditional probability and Bayes's theorem are given in this section.

Conditional Probability

The conditional probability of an event can be found by dividing both sides of the equation for multiplication rule 2 by $P(A)$, as shown:

$$P(A \text{ and } B) = P(A) \cdot P(B|A)$$

$$\frac{P(A \text{ and } B)}{P(A)} = \frac{\cancel{P(A)} \cdot P(B|A)}{\cancel{P(A)}}$$

$$\frac{P(A \text{ and } B)}{P(A)} = P(B|A)$$

Formula for Conditional Probability

The probability that the second event occurs given that the first event has occurred can be found by dividing the probability that both events occurred by the probability that the first event has occurred. The formula is

$$P(B|A) = \frac{P(A \text{ and } B)}{P(A)}$$

The next two examples illustrate the use of this rule.

EXAMPLE 5–31 A box contains black chips and white chips. A person selects two chips without replacement. If the probability of selecting a black chip *and* a white chip is $\frac{15}{56}$, and the probability of selecting a black chip on the first draw is $\frac{3}{8}$, find the probability of selecting the white chip on the second draw, *given* that the first chip selected was a black chip.

Solution   Let

$$B = \text{selecting a black chip} \qquad W = \text{selecting a white chip}$$

Then

$$P(W|B) = \frac{P(B \text{ and } W)}{P(B)} = \frac{\frac{15}{56}}{\frac{3}{8}}$$

$$= \frac{15}{56} \div \frac{3}{8} = \frac{15}{56} \cdot \frac{8}{3} = \frac{\overset{5}{\cancel{15}}}{\underset{7}{\cancel{56}}} \cdot \frac{\overset{1}{\cancel{8}}}{\underset{1}{\cancel{3}}} = \frac{5}{7}$$

Hence, the probability of selecting a white chip on the second draw given that the first chip selected was black is $\frac{5}{7}$.   ◄

EXAMPLE 5–32   The probability that Sam parks in a no-parking zone *and* gets a parking ticket is 0.06, and the probability that Sam cannot find a legal parking space and has to park in the no-parking zone is 0.20. On Tuesday, Sam arrives at school and has to park in a no-parking zone. Find the probability that he will get a parking ticket.

Solution   Let

$$N = \text{parking in a no-parking zone} \qquad T = \text{getting a ticket}$$

Then

$$P(T|N) = \frac{P(N \text{ and } T)}{P(N)} = \frac{0.06}{0.20} = 0.30$$

Hence, Sam has a 0.30 probability of getting a parking ticket, given that he parked in a no-parking zone.   ◄

The conditional probability of events occurring can also be computed when the data is given in table form, as shown in the next example.

EXAMPLE 5–33   A recent survey asked 100 people if they thought women in the armed forces should be permitted to participate in combat. The results of the survey are shown in the table.

Gender	Yes	No	Total
Male	32	18	50
Female	8	42	50
Total	40	60	100

Find these probabilities.

a. The respondent answered "yes," given that the respondent was a female.
b. The respondent was a male, given that the respondent answered "no."

Solution   Let M = respondent was a male Y = respondent answered "yes"
          F = respondent was a female N = respondent answered "no"

a. The problem is to find $P(Y|F)$. The rule states

$$P(Y|F) = \frac{P(F \text{ and } Y)}{P(F)}$$

The probability $P(F \text{ and } Y)$ is the number of females who responded "yes" divided by the total number of respondents:

$$P(F \text{ and } Y) = \frac{8}{100}$$

The probability $P(F)$ is the probability of selecting a female:

$$P(F) = \frac{50}{100}$$

Then

$$P(Y|F) = \frac{P(F \text{ and } Y)}{P(F)} = \frac{8/100}{50/100}$$

$$= \frac{8}{100} \div \frac{50}{100} = \frac{\overset{4}{\cancel{8}}}{\underset{1}{\cancel{100}}} \cdot \frac{\overset{1}{\cancel{100}}}{\underset{25}{\cancel{50}}} = \frac{4}{25}$$

b. The problem is to find $P(M|N)$.

$$P(M|N) = \frac{P(N \text{ and } M)}{P(N)} = \frac{18/100}{60/100}$$

$$= \frac{18}{100} \div \frac{60}{100} = \frac{\overset{3}{\cancel{18}}}{\underset{1}{\cancel{100}}} \cdot \frac{\overset{1}{\cancel{100}}}{\underset{10}{\cancel{60}}} = \frac{3}{10}$$

◄

Bayes's Theorem

Given two dependent events, A and B, the previous formulas for conditional probability allow one to find $P(A \text{ and } B)$, or $P(B|A)$. Related to these formulas is a rule developed by the English Presbyterian minister Thomas Bayes (1702–1761). The rule is known as **Bayes's theorem.**

Knowing the outcome of a particular situation, one can find the probability that the outcome occurred as a result of a particular previous event. In Example 5–30 of Section 5–4, there were two boxes, each containing red balls and blue balls. A box was selected and a ball was drawn. The example asked for the probability that the ball selected was red. Now, a different question can be asked: "If the ball is red, what is the probability it came from box 1?" In this case, the outcome is known, a red ball was selected, and one is asked to find the probability that it is a result of a previous event, that it came from box 1. Bayes's theorem can enable one to compute this probability, and the theorem can be explained by using tree diagrams.

The tree diagram for the solution of Example 5–30 is shown in Figure 5–6, along with the appropriate notation and the corresponding probabilities. In this case, A_1 is the event of selecting box 1, A_2 is the event of selecting box 2, R is the event of selecting a red ball, and B is the event of selecting a blue ball.

In order to answer the question, "If the ball selected is red, what is the probability that it came from box 1?" the two previous formulas,

$$P(B|A) = \frac{P(A \text{ and } B)}{P(A)} \tag{1}$$

$$P(A \text{ and } B) = P(A) \cdot P(B|A) \tag{2}$$

FIGURE 5–6
Tree Diagram for
Example 5–30

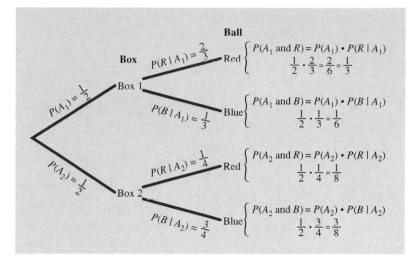

can be used. The notation that will be used is that of Example 5–30, shown in Figure 5–6. Finding the probability that box 1 was selected given that the ball selected was red can be written symbolically as $P(A_1|R)$. By Formula 1,

$$P(A_1|R) = \frac{P(R \text{ and } A_1)}{P(R)}$$

Note: $P(R \text{ and } A_1) = P(A_1 \text{ and } R)$.

By Formula 2,

$$P(A_1 \text{ and } R) = P(A_1) \cdot P(R|A_1)$$

and

$$P(R) = P(A_1 \text{ and } R) + P(A_2 \text{ and } R)$$

as shown in Figure 5–6. $P(R)$ was found by adding the products of the probabilities of the branches in which a red ball was selected. Now,

$$P(A_1 \text{ and } R) = P(A_1) \cdot P(R|A_1)$$
$$P(A_2 \text{ and } R) = P(A_2) \cdot P(R|A_2)$$

Substituting these values in the original formula for $P(A_1|R)$, one gets

$$P(A_1|R) = \frac{P(A_1) \cdot P(R|A_1)}{P(A_1) \cdot P(R|A_1) + P(A_2) \cdot P(R|A_2)}$$

Refer to Figure 5–6. The numerator of the fraction is the product of the top branch of the tree diagram, which consists of selecting a red ball and selecting box 1. And the denominator is the sum of the products of the two branches of the tree where the red ball was selected.

Using this formula and the probability values shown in Figure 5–6, one can find the probability that box 1 was selected given that the ball was red, as shown.

$$P(A_1 | R) = \frac{P(A_1) \cdot P(R|A_1)}{P(A_1) \cdot P(R|A_1) + P(A_2) \cdot P(R|A_2)}$$

$$= \frac{\frac{1}{2} \cdot \frac{2}{3}}{\frac{1}{2} \cdot \frac{2}{3} + \frac{1}{2} \cdot \frac{1}{4}} = \frac{\frac{1}{3}}{\frac{1}{3} + \frac{1}{8}} = \frac{\frac{1}{3}}{\frac{8}{24} + \frac{3}{24}} = \frac{\frac{1}{3}}{\frac{11}{24}}$$

$$= \frac{1}{3} \div \frac{11}{24} = \frac{1}{\cancel{3}_1} \cdot \frac{\cancel{24}^{8}}{11} = \frac{8}{11}$$

This formula is a simplified version of Bayes's theorem.

Before Bayes's theorem is stated, another example is shown.

EXAMPLE 5–34 A shipment of two boxes, each containing six telephones, is received by a store. Box 1 contains one defective phone and box 2 contains two defective phones. After the boxes are unpacked, a phone is selected and found to be defective. Find the probability that it came from box 2.

Solution **STEP 1** Select the proper notation. Let A_1 represent box 1 and A_2 represent box 2. Let D represent a defective phone and ND represent a phone that is not defective.

STEP 2 Draw a tree diagram and find the corresponding probabilities for each branch. The probability of selecting box 1 is $\frac{1}{2}$, and the probability of selecting box 2 is $\frac{1}{2}$. Since there is one defective phone in box 1, the probability of selecting it is $\frac{1}{6}$. The probability of selecting a nondefective phone from box 1 is $\frac{5}{6}$.

Since there are two defective phones in box 2, the probability of selecting a defective phone from box 2 is $\frac{2}{6}$, or $\frac{1}{3}$; and the probability of selecting a nondefective phone is $\frac{4}{6}$, or $\frac{2}{3}$. The tree diagram is shown in Figure 5–7.

FIGURE 5–7
Tree Diagram for
Example 5–34

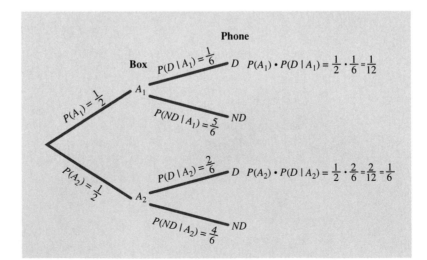

STEP 3 Write the corresponding formula. Since the example is asking for the probability that, given a defective phone, it came from box 2, the corresponding formula is as shown.

$$P(A_2|D) = \frac{P(A_2) \cdot P(D|A_2)}{P(A_1) \cdot P(D|A_1) + P(A_2) \cdot P(D|A_2)}$$

$$= \frac{\frac{1}{2} \cdot \frac{2}{6}}{\frac{1}{2} \cdot \frac{1}{6} + \frac{1}{2} \cdot \frac{2}{6}} = \frac{\frac{1}{6}}{\frac{1}{12} + \frac{2}{12}} = \frac{\frac{1}{6}}{\frac{3}{12}}$$

$$= \frac{1}{6} \div \frac{3}{12} = \frac{1}{\underset{1}{\cancel{6}}} \cdot \frac{\overset{2}{\cancel{12}}}{3} = \frac{2}{3}$$

◄

Bayes's theorem can be generalized to events with three or more outcomes and formally stated as in the next box.

Bayes's theorem For two events, *A* and *B*, where event *B* follows event *A*, event *A* can occur in $A_1, A_2, \ldots, A_n$ mutually exclusive ways, and event *B* can occur in $B_1, B_2, \ldots, B_m$ mutually exclusive ways,

$$P(A_1|B_1) = \frac{P(A_1) \cdot P(B_1|A_1)}{P(A_1) \cdot P(B_1|A_1) + P(A_2) \cdot P(B_1|A_2) + \cdots + P(A_n) \cdot P(B_1|A_n)}$$

for any specific events A_1 and B_1.

The numerator is the product of the probabilities on the branch of the tree that consists of outcomes A_1 and B_1. The denominator is the sum of the products of the probabilities of the branches containing B_1 and A_1, B_1 and A_2, . . . , B_1 and A_n.

EXAMPLE 5–35 On a game show, a contestant can select one of four boxes. Box 1 contains one $100 bill and nine $1 bills. Box 2 contains two $100 bills and eight $1 bills. Box 3 contains three $100 bills and seven $1 bills. Box 4 contains five $100 bills and five $1 bills. The contestant selects a box at random and selects a bill from the box at random. If a $100 bill is selected, find the probability that it came from box 4.

Solution **STEP 1** Select the proper notation. Let B_1, B_2, B_3, and B_4 represent the boxes and 100 and 1 represent the values of the bills in the boxes.

STEP 2 Draw a tree diagram and find the corresponding probabilities. The probability of selecting each box is $\frac{1}{4}$, or 0.25. The probabilities of selecting the $100 bill from each box, respectively, are $\frac{1}{10} = 0.1$, $\frac{2}{10} = 0.2$, $\frac{3}{10} = 0.3$, and $\frac{5}{10} = 0.5$. The tree diagram is shown in Figure 5–8.

FIGURE 5–8
Tree Diagram for
Example 5–35

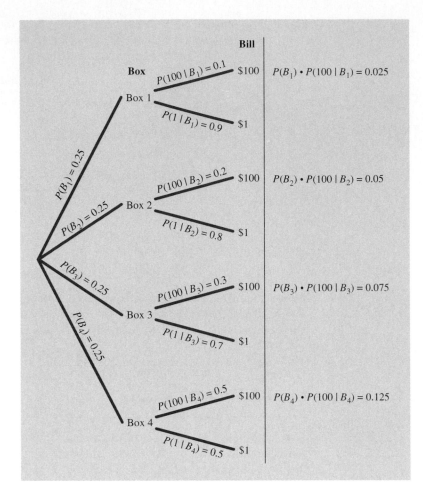

STEP 3   Using Bayes's theorem, write the corresponding formula. Since the example asks for the probability that box 4 was selected, given that $100 was obtained, the corresponding formula is as follows:

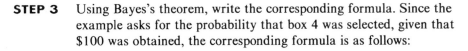

$$P(B_4 \mid 100) = \frac{P(B_4) \cdot P(100 \mid B_4)}{P(B_1) \cdot P(100 \mid B_1) + P(B_2) \cdot P(100 \mid B_2) + P(B_3) \cdot P(100 \mid B_3) + P(B_4) \cdot P(100 \mid B_4)}$$

$$= \frac{0.125}{0.025 + 0.05 + 0.075 + 0.125} = \frac{0.125}{0.275} = 0.455 \qquad \blacktriangleleft$$

In Example 5–35, the original probability of selecting box 1 was 0.25. However, once additional information was obtained—and the condition was considered, that a $100 bill was selected—the revised probability of selecting box 1 became 0.455.

Bayes's theorem can be used to revise probabilities of events once additional information becomes known. Bayes's theorem is used as the basis for a branch of statistics called Bayesian decision making, which includes the use of subjective probabilities in making statistical inferences.

EXERCISES

5–96. At a small college, the probability that a student takes physics and sociology is 0.092. The probability that a student takes sociology is 0.73. Find the probability that the student is taking physics, given that he or she is taking sociology.

5–97. In a certain city, the probability that an automobile will be stolen and found within one week is 0.0009. The probability that an automobile will be stolen is 0.0015. Find the probability that a stolen automobile will be found within one week.

5–98. A circuit to run a model railroad has eight switches. Two are defective. If a person selects two switches at random and tests them, find the probability that the second one is defective, given that the first one is defective.

5–99. At the Coulterville Country Club, 72% of the members play golf and drink beer, and 80% play golf. If a member is selected at random, find the probability that the member drinks beer, given that the member plays golf.

5–100. At a large university, the probability that a student takes calculus *and* is on the dean's list is 0.042. The probability that a student is on the dean's list is 0.21. Find the probability that the student is taking calculus, given that he or she is on the dean's list.

5–101. In Friendly Acres, 35% of the homes have a family room and a fireplace. If 70% of the homes have a family room, find the probability that a home has a fireplace, given that it has a family room.

5–102. In a pizza restaurant, 95% of the customers order pizza. If 65% of the customers order pizza and a salad, find the probability that a customer who orders pizza will also order a salad.

5–103. At an exclusive country club, 68% of the members play bridge and drink champagne, and 83% play bridge. If a member is selected at random, find the probability that the member drinks champagne, given that the member plays bridge.

5–104. Eighty students in the school's cafeteria were asked if they favored a ban on smoking in the cafeteria. The results of the survey are shown in the table.

Class	Favor	Oppose	No Opinion
Freshman	15	27	8
Sophomore	23	5	2

If a student is selected at random, find these probabilities.

a. Given that the student is a freshman, he or she opposes the ban.
b. Given that the student favors the ban, the student is a sophomore.

5–105. In a large shopping mall, a marketing agency conducted a survey on credit cards. The results are shown in the table.

Employment Status	Owns a Credit Card	Does Not Own a Credit Card
Presently employed	18	29
Presently unemployed	28	34

If a person is selected at random, find these probabilities.

a. The person owns a credit card, given that the person is employed.
b. The person is unemployed, given that the person owns a credit card.

5–106. A study of graduates' average grades and degrees showed the following results.

Degree	Grade		
	C	B	A
B.S.	5	8	15
B.A.	7	12	8

If a graduate is selected at random, find these probabilities.

a. The graduate has a B.S. degree, given that he or she has an A average.
b. Given that the graduate has a B.A. degree, the graduate has a C average.

5–107. A store owner purchases telephones from two companies. From company A 350 telephones are purchased, and 2% are defective. From company B 1050 telephones are purchased, and 4% are defective. Given that a phone is defective, find the probability that it came from company B.

5–108. Two manufacturers supply food to a large cafeteria. Manufacturer A supplies 2400 cans of soup, and 3% are found to be dented. Manufacturer B supplies 3600 cans, and 1% are found to be dented. Given that a can of soup is found to be dented, find the probability that it came from manufacturer B.

5–109. A test for a certain disease is found to be 95% accurate, meaning that it will correctly diagnose the disease in 95 out of 100 people who have the disease. For a certain segment of the population, the incidence of the disease is 9%. If a person tests positive for the disease, find the probability that the person actually has the disease. The test is also 95% accurate for a negative result.

5–110. Using the test in Exercise 5–109, if a person tests negative for the disease, find the probability that the person actually has the disease. Nine percent of the population have the disease.

5–111. A corporation has three methods of training employees. Because of time, space, and location, it sends 25% of its employees to location A, 35% of its employees to location B, and 45% of its employees to location C. Location A has an 80% success rate. That is, 80% of the employees who complete the course will pass the licensing exam. Location B has a 75% success rate, and location C has a 60% success rate. If a person has passed the exam, find the probability that the person went to location B.

5–112. In Exercise 5–111, if a person failed the exam, find the probability that the person went to location C.

5–113. A store purchases baseball hats from three different manufacturers. In manufacturer A's box, there are 12 blue hats, 6 red hats, and 6 green hats. In manufacturer B's box, there are 10 blue hats, 10 red hats, and 4 green hats. In manufacturer C's box, there are 8 blue hats, 8 red hats, and 8 green hats. A box is selected at random and a hat is selected at random from that box. If the hat is red, find the probability that it came from manufacturer A's box.

5–114. In Exercise 5–113, if the hat selected is green, find the probability that it came from manufacturer B's box.

5–115. A driver has three ways to get from one city to another. There is an 80% probability of encountering a traffic jam on route 1, a 60% probability on route 2, and a 30% probability on route 3. Because of other factors, such as distance and speed limits, the driver uses route 1 50% of the time and routes 2 and 3 each 25% of the time. If the driver calls the dispatcher to inform him that she is in a traffic jam, find the probability that she has selected route 1.

5–116. In Exercise 5–115, if the driver did not encounter a traffic jam, find the probability that she selected route 3.

5-6 COMPLEMENTARY EVENTS

Another important concept in probability theory is that of *complementary events*. When a die is rolled, for instance, the sample space consists of the outcomes of 1, 2, 3, 4, 5, and 6. The event E of getting odd numbers consists of the outcomes of 1, 3, and 5. The event of not getting an odd number is called the complement of event E, and it consists of the outcomes 2, 4, and 6.

The complement of an event E is the set of outcomes in the sample space that are not included in the outcomes of event E. The complement of E is denoted by $\bar{E}$ (read "E bar").

The next example further illustrates the concept of complementary events.

EXAMPLE 5–36   Find the complement of each event.

a. Rolling a die and getting a 4.
b. Selecting a letter of the alphabet and getting a vowel.
c. Selecting a month and getting a month that begins with a J.
d. Selecting a day of the week and getting a weekday.

Solution   a. Getting a 1, 2, 3, 5, or 6.
b. Getting a consonant (assume y is a consonant).
c. Getting February, March, April, May, August, September, October, November, or December.
d. Getting Saturday or Sunday.   ◀

Two important things should be noted. First, the event and its complement are mutually exclusive events; that is, an outcome cannot be both in the event and in its complement at the same time. Second, the outcomes of an event and the outcomes of the complement make up the entire sample space. For example, if two coins are tossed, the sample space is HH, HT, TH, and TT. The complement of "getting all heads" is not "getting all tails," since the event "all heads" is HH, and the complement of HH is HT, TH, and TT. Hence, the complement of the event "all heads" is the event "getting at least one tail."

Since the event and its complement make up the entire sample space, it follows from probability rule 5 (Section 5–2) that the sum of the probability of the event and the probability of its complement will equal 1. That is, $P(E) + P(\overline{E}) = 1$. In the previous example, let E = all heads, or HH, and let $\overline{E}$ = at least one tail, or HT, TH, TT. Then $P(E) = \frac{1}{4}$ and $P(\overline{E}) = \frac{3}{4}$; hence, $P(E) + P(\overline{E}) = \frac{1}{4} + \frac{3}{4} = 1$.

EXAMPLE 5–37 A die is rolled. Show that the sum of the probability of the event of getting a number less than 3 and the probability of its complement is equal to 1.

Solution The sample space is 1, 2, 3, 4, 5, 6. Each event has a probability of $\frac{1}{6}$. Hence,

$$P(1 \text{ or } 2) + P(3 \text{ or } 4 \text{ or } 5 \text{ or } 6) = \frac{2}{6} + \frac{4}{6} = \frac{6}{6} = 1 \qquad \blacktriangleleft$$

The rule for complementary events can be stated algebraically in three ways.

Rule for Complementary Events

$$P(\overline{E}) = 1 - P(E) \quad \text{or} \quad P(E) = 1 - P(\overline{E}) \quad \text{or} \quad P(E) + P(\overline{E}) = 1$$

Stated in words, the rule is: *If the probability of an event or the probability of its complement is known, then the other can be found by subtracting the probability from* 1. This rule is important in probability theory because at times the best solution to a problem is to find the probability of the complement of an event and then subtract from 1 to get the probability of the event itself.

EXAMPLE 5–38 A game is played by drawing four cards from an ordinary deck and replacing each card after it is drawn. Find the probability of winning if at least one ace is drawn.

Solution It is much easier to find the probability that no aces are drawn (i.e., losing) and then subtract from 1 than to find the solution directly, because that would involve finding the probability of getting one ace, two aces, three aces, and four aces and then adding the results.

Let E = at least one ace is drawn and $\overline{E}$ = no aces drawn. Then

$$P(\overline{E}) = \frac{48}{52} \cdot \frac{48}{52} \cdot \frac{48}{52} \cdot \frac{48}{52}$$

$$= \frac{12}{13} \cdot \frac{12}{13} \cdot \frac{12}{13} \cdot \frac{12}{13} = \frac{20{,}736}{28{,}561}$$

Hence,

$$P(E) = 1 - P(\overline{E})$$

$$P(\text{winning}) = 1 - P(\text{losing}) = 1 - \frac{20{,}736}{28{,}561} = \frac{7825}{28{,}561} \approx 0.27$$

or about 27% of the time. $\qquad \blacktriangleleft$

EXAMPLE 5-39 A coin is tossed five times. Find the probability of getting at least one tail.

Solution It is easier to find the probability of the complement of the event, which is "all heads," and then subtract the probability from 1 to get the probability of at least one tail.

$$P(E) = 1 - P(\overline{E})$$
$$P(\text{at least 1 tail}) = 1 - P(\text{all heads})$$
$$P(\text{all heads}) = (\tfrac{1}{2})^5 = \tfrac{1}{32}$$

Hence,

$$P(\text{at least 1 tail}) = 1 - \tfrac{1}{32} = \tfrac{31}{32} \quad \blacktriangleleft$$

EXAMPLE 5-40 A die is rolled three times. Find the probability of getting at least one 6.

Solution The complement of "at least one 6" is "getting no 6s." Since the probability of getting a 6 is $\tfrac{1}{6}$, the probability of not getting a 6 is $1 - \tfrac{1}{6}$, or $\tfrac{5}{6}$. Hence,

$$P(\text{at least one 6}) = 1 - P(\text{no 6s}) = 1 - \left(\frac{5}{6}\right)^3 = 1 - \frac{125}{216} = \frac{91}{216} \quad \blacktriangleleft$$

Using the rule for complementary events to solve problems can save time and effort in many cases.

EXERCISES

5-117. Find the complement of each event.

a. Selecting a defective resistor from a lot of resistors.
b. Selecting a letter of the alphabet from A through J.
c. Selecting a three-digit number from 000 to 333.
d. Selecting a month of the year that begins with J.
e. Selecting a female student in a statistics class.
f. Selecting a day of the year on which it snows.
g. Selecting an opponent that a basketball team beat last season.
h. Selecting a day of the year that is a national holiday.
i. Selecting a president who is a male.
j. Selecting a red card from a deck of cards.

5-118. In Exercise 5-117, find the probability of the event and its complement for parts b, c, d, and j.

5-119. In a lab there are eight technicians. Three are male and five are female. If three technicians are selected, find the probability that at least one female is selected.

5-120. There are five chemistry instructors and six physics instructors at a college. If a committee of four instructors is selected, find the probability of at least one physics instructor being selected.

5-121. On a surprise quiz consisting of five true-false questions, an unprepared student guesses each answer. Find the probability that he gets at least one correct.

5-122. A lot of condensers contains 12 good ones and 4 defective ones. If three are selected and tested, find the probability that there will be at least one defective condenser.

5-123. If a family has four children, find the probability that there is at least one girl among the children.

5-124. A car pool contains three kindergartners and five first-graders. If two children are ill, find the probability that at least one of them is a kindergartner.

5-125. If four cards are drawn from a deck and not replaced, find the probability of getting at least one club.

5-126. At a local clinic there are eight men, five women, and three children in the waiting room. If three patients are randomly selected, find the probability that there is at least one child among them.

5-127. It has been found that 6% of all automobiles on the road have defective brakes. If five automobiles are stopped and checked by the state police, find the probability that at least one automobile will have defective brakes.

5-128. A medication is 75% effective against a bacterial infection. Find the probability that if 12 people take the medication, at least one person's infection will not improve.

5-129. A coin is tossed six times. Find the probability of getting at least one tail.

5-130. If three digits are randomly selected, find the probability of getting at least one 7. Digits can be used more than once.

5-131. If a die is rolled three times, find the probability of getting at least one 6.

5-132. At a teachers' conference, there were four English teachers, three mathematics teachers, and five science teachers. If four teachers are selected for a committee, find the probability of selecting at least one science teacher.

5-133. If a die is rolled three times, find the probability of getting at least one even number.

5-134. At a faculty meeting, there were 7 professors, 5 associate professors, 6 assistant professors, and 12 instructors. If 4 people are selected at random to attend a conference, find the probability of selecting at least one professor.

⁺5-135. Odds are used in gambling games to make them fair. For example, if a person rolled a die and won every time he or she rolled a 6, then the person would win on the average of once every 6 times. So that the game is fair, the odds of 5 to 1 are given. This means that if the person bet $1 and won, he or she could win $5. On the average, the player would win $5 once in 6 rolls and lose $1 on the other 5 rolls—hence, the term *fair game*.

In most gambling games, the odds given are not fair. For example, if the odds of winning are really 20 to 1, the house might offer 15 to 1 in order to make a profit.

Odds can be expressed as a fraction or as a ratio, such as $\frac{5}{1}$, 5:1, or 5 to 1. Odds are computed in favor of the event or against the event. The formulas for odds are

$$\text{odds in favor} = \frac{P(E)}{1 - P(E)}$$

$$\text{odds against} = \frac{P(\overline{E})}{1 - P(\overline{E})}$$

In the die example,

$$\text{odds in favor of a 6} = \frac{\frac{1}{6}}{\frac{5}{6}} = \frac{1}{5} \text{ or } 1:5$$

$$\text{odds against a 6} = \frac{\frac{5}{6}}{\frac{1}{6}} = \frac{5}{1} \text{ or } 5:1.$$

Find the odds in favor and against each event.

a. Rolling a die and getting a 2.
b. Rolling a die and getting an even number.
c. Drawing a card from a deck and getting a spade.
d. Drawing a card and getting a red card.
e. Drawing a card and getting a queen.
f. Tossing two coins and getting two tails.
g. Tossing two coins and getting one tail.

5-7 PROBABILITY AND COUNTING TECHNIQUES (OPTIONAL)

The counting rules in Chapter 4 can be combined with the probability rules in this chapter to solve many types of probability problems. By using the multiplication rules, the permutations rules, and the combination rule, one can compute the probability of outcomes of many experiments, such as getting a full house when five cards are dealt or selecting a committee of three women and two men from a club consisting of ten women and ten men.

EXAMPLE 5-41 Find the probability of getting four aces when five cards are drawn from an ordinary deck of cards.

Solution There are $_{52}C_5$ ways to draw 5 cards from a deck. There is only 1 way to get 4 aces, but there are 48 possibilities to get the fifth card. Therefore, there are 48 ways to get 4 aces and 1 other card. Hence,

$$P(4 \text{ aces}) = 1 \cdot \frac{48}{_{52}C_5} = \frac{48}{2,598,960} = \frac{1}{54,145} \qquad \blacktriangleleft$$

EXAMPLE 5-42 A box contains 24 transistors, 4 of which are defective. If 4 are sold at random, find the following probabilities.

a. Exactly 2 are defective.   c. All are defective.
b. None are defective.   d. At least 1 is defective.

Solution   There are $_{24}C_4$ ways to sell 4 transistors, so the denominator in each case will be 10,626.

a. Two defective transistors can be selected as $_4C_2$ and 2 nondefective ones as $_{20}C_2$. Hence,

$$P(\text{2 defectives}) = \frac{_4C_2 \cdot {_{20}C_2}}{_{24}C_4} = \frac{1140}{10,626} = \frac{190}{1771}$$

b. The number of ways to choose no defectives is $_{20}C_4$. Hence,

$$P(\text{no defectives}) = \frac{_{20}C_4}{_{24}C_4} = \frac{4845}{10,626} = \frac{1615}{3542}$$

c. The number of ways to choose 4 defectives from 4 is $_4C_4$, or 1. Hence,

$$P(\text{all defective}) = \frac{1}{_{24}C_4} = \frac{1}{10,626}$$

d. To find the probability of at least 1 defective transistor, find the probability that there are no defective transistors, and then subtract that probability from 1.

$$P(\text{at least 1 defective}) = 1 - P(\text{no defectives})$$
$$= 1 - \frac{_{20}C_4}{_{24}C_4} = 1 - \frac{1615}{3542} = \frac{1927}{3542} \quad \blacktriangleleft$$

EXAMPLE 5–43   A store owner receives 12 computers: 9 are Model 900s and 3 are Model 600s. If two computers are sold at random, find the probability that one of each model was sold.

Solution $$P(\text{1 Model 900 and 1 Model 600}) = \frac{_9C_1 \cdot {_3C_1}}{_{12}C_2} = \frac{9 \cdot 3}{66} = \frac{9}{22} \quad \blacktriangleleft$$

EXAMPLE 5–44   A combination lock consists of the 26 letters of the alphabet. If a 3-letter combination is needed, find the probability that the combination will consist of the letters ABC in that order. The same letter can be used more than once.

Solution   Since repetitions are permitted, there are $26 \cdot 26 \cdot 26 = 17,576$ different possible combinations. And since there is only one ABC combination, the probability is $P(\text{ABC}) = 1/26^3 = 1/17,576$. $\quad \blacktriangleleft$

EXAMPLE 5–45   There are 8 married couples in a tennis club. If 1 man and 1 woman are selected at random to plan the summer tournament, find the probability that they are married to each other.

Solution   Since there are 8 ways to select the man and 8 ways to select the woman, there are $8 \cdot 8$, or 64, ways to select 1 man and 1 woman. Since there are 8 married couples, the solution is $\frac{8}{64} = \frac{1}{8}$. $\quad \blacktriangleleft$

Speaking of Statistics

Owners Face Busy Two-Day Agenda

By Gordon Forbes
USA TODAY

NO BET: Los Angeles Rams offensive tackle Jackie Slater has spent 17 years in the NFL, but he has failed to master one thing—the coin toss.

Slater is the Rams' designated caller in the pregame ritual. In four exhibitions and six regular-season games, the Rams have lost the toss every time.

Rams officials consulted a computer specialist, who said the odds against that happening are 1,023-1.

Source: Copyright 1992, USA TODAY. Reprinted with permission.

In 1992, according to the news article, the Los Angeles Rams lost the coin toss in 10 out of 10 games. In order to compute the odds, the author states that they consulted a computer specialist and found that the odds were 1023 to 1 against this happening. The probability obtained from the odds is 1 chance in 1024. Explain how this probability can be computed without using odds. (For more on odds, see Exercise 5–135.)

As indicated at the beginning of this section, the rules in this chapter and the previous chapter can be used to solve a large variety of probability problems found in business, gambling, economics, biology, etc.

EXERCISES

5–136. Find the probability of getting two face cards (king, queen, or jack) when two cards are drawn from a deck without replacement.

5–137. A parent-teacher committee consisting of 4 people is to be formed from 20 parents and 5 teachers. Find the probability that the committee will consist of the following. (Assume that the selection is random.)

a. All teachers
b. Two teachers and 2 parents
c. All parents
d. One teacher and 3 parents

5–138. Find the probability of getting a full house (three cards of one denomination and two of another) when five cards are dealt from an ordinary deck.

5–139. A committee of four people is to be formed from six doctors and eight dentists. Find the probability that the committee will consist of the following.

a. All dentists
b. Two dentists and two doctors
c. All doctors
d. Three doctors and one dentist
e. One doctor and three dentists

5–140. In a company there are seven executives: four women and three men. Three are selected to attend a management seminar. Find these probabilities.

a. All three selected will be women.
b. All three selected will be men.
c. Two men and one woman will be selected.
d. One man and two women will be selected.

5–141. A city council consists of ten members. Four are Republicans, three are Democrats, and three are Independents. If a committee of three is to be selected, find the probability of selecting the following.

a. All Republicans
b. All Democrats
c. One of each party
d. Two Democrats and one Independent
e. One Independent and two Republicans

5–142. In a class of 18 students, there are 11 men and 7 women. Four students are selected to present a demonstration on the use of the calculator. Find the probability that the group consists of the following.

a. All men
b. All women
c. Three men and 1 woman
d. One man and 3 women
e. Two men and 2 women

5–143. A package contains 12 resistors, 3 of which are defective. If 4 are selected, find the probability of getting the following.

a. No defective resistors
b. One defective resistor
c. Three defective resistors

5–144. If 50 tickets are sold and 2 prizes are to be awarded, find the probability that one person wins 2 prizes if that person buys 2 tickets.

5–145. An insurance sales representative selects three policies to review. The group of policies he can select from contains eight life policies, five automobile policies, and two homeowner's policies. Find the probability of selecting the following.

a. All life policies
b. Both homeowner's policies
c. All automobile policies
d. One of each policy
e. Two life and one automobile policies

5–146. Find the probability of getting a triple-digit number on a lotto that consists of selecting a three-digit number.

5–147. Find the probability of selecting three freshmen and two sophomores from a group of students consisting of ten freshmen and five sophomores.

5–148. When three dice are rolled, find the probability of getting a sum of 7.

5–149. Find the probability of selecting two mathematics books and three physics books from four mathematics books and eight physics books.

5–150. Find the probability that if five different-sized washers are arranged in a row, they will be arranged in order of size.

*5–151. Using the information in Exercise 4–67, Chapter 4, find the probability of each poker hand.

a. Royal flush
b. Straight flush
c. Four of a kind

5–8 SUMMARY

In this chapter, the basic concepts and rules of probability are explained. The three types of probability are classical, empirical, and subjective. Classical probability uses sample spaces, and empirical probability uses frequency distributions and is based on observation. In subjective probability, the researcher makes an educated guess about the chance of an event occurring.

A probability event consists of one or more outcomes of a probability experiment. Two events are said to be mutually exclusive if they cannot occur at the same time. Events can also be classified as independent or dependent. If events are independent, the probability that the first event occurs does not affect or change the probability of the next event occurring. If the probability of the second event occurring is altered or changed by the occurrence of the first event, then the events are dependent. The complement of an event is the set of outcomes in the sample space that are not included in the outcomes of the event itself. Complementary events are mutually exclusive.

When an outcome has occurred as a result of a particular previous event, Bayes's theorem allows one to compute the revised probability of the previous event.

Probability problems can be solved by using the addition rules, the multiplication rules, and the complementary event rule. Finally, when the number of outcomes of the sample space is large, probability problems can be solved by using the counting rules explained in Chapter 4.

Important Terms

Bayes's theorem
Classical probability
Complement of an event
Compound event
Conditional probability
Dependent events
Empirical probability
Equally likely events
Event
Independent events
Mutually exclusive events
Outcome
Probability
Probability experiment
Sample space
Simple event
Subjective probability

Important Formulas

Formula for classical probability:

$$P(E) = \frac{\text{number of ways } E \text{ can occur}}{\text{total number of outcomes in the sample space}} = \frac{n(E)}{n(S)}$$

Formula for empirical probability:

$$P(E) = \frac{\text{frequency of the class}}{\text{total frequencies in the distribution}} = \frac{f}{n}$$

Addition rule 1, for two mutually exclusive events:

$$P(A \text{ or } B) = P(A) + P(B)$$

Addition rule 2, for events that are not mutually exclusive:

$$P(A \text{ or } B) = P(A) + P(B) - P(A \text{ and } B)$$

Multiplication rule 1, for independent events:

$$P(A \text{ and } B) = P(A) \cdot P(B)$$

Multiplication rule 2, for dependent events:

$$P(A \text{ and } B) = P(A) \cdot P(B|A)$$

Formula for conditional probability:

$$P(B|A) = \frac{P(A \text{ and } B)}{P(A)}$$

Formula for Bayes's theorem:

$$P(A_1|B_1) = \frac{P(A_1) \cdot P(B_1|A_1)}{P(A_1) \cdot P(B_1|A_1) + P(A_2) \cdot P(B_1|A_2) + \cdots + P(A_n) \cdot P(B_1|A_n)}$$

Formula for complementary events:

$$P(\overline{E}) = 1 - P(E) \quad \text{or} \quad P(E) = 1 - P(\overline{E})$$
$$\text{or} \quad P(E) + P(\overline{E}) = 1$$

Review Exercises

5–152. When a die is rolled, find the probability of getting the following.

a. A 5
b. A 6
c. A number less than 5

5–153. When a card is drawn from a deck, find the probability of getting the following.

a. A heart
b. A 7 and a club
c. A 7 or a club
d. A jack
e. A black card

5–154. In a survey conducted at a local restaurant during breakfast hours, 20 people preferred orange juice, 16 people preferred grapefruit juice, and 9 people preferred apple juice with breakfast. If a person is selected at random, find the probability that a person prefers grapefruit juice.

5–155. What is the difference between independent and dependent events? Give an example of each.

5–156. What is meant by mutually exclusive events? Give an example.

5–157. During a sale at a men's store, 16 white sweaters, 3 red sweaters, 9 blue sweaters, and 7 yellow sweaters were purchased. If a customer is selected at random, find the probability that he bought the following.

a. A blue sweater.
b. A yellow or a white sweater.
c. A red, a blue, or a yellow sweater.
d. A sweater that was not white.

5–158. At a swimwear store, the managers found that 16 women bought white bathing suits, 4 bought red suits, 3 bought blue suits, and 7 bought yellow suits. If a woman is selected at random, find the probability that she bought the following.

a. A blue suit.
b. A yellow or a red suit.
c. A white or a yellow or a blue suit.
d. A suit that was not red.

5–159. When two dice are rolled, find the probability of getting the following.

a. A sum of 5 or 6.
b. A sum greater than 9.
c. A sum less than 4 or greater than 9.
d. A sum that is divisible by 4.
e. A sum of 14.
f. A sum less than 13.

5–160. The probability that a person owns an automobile is 0.80, that a person owns a boat is 0.30, and that a person owns both an automobile and a boat is 0.12. Find the probability that a person owns either a boat or an automobile, but not both.

5–161. There is a 0.39 probability that John will purchase a new car, a 0.73 probability that Mary will purchase a new car, and a 0.36 probability that both will purchase a new car. Find the probability that neither will purchase a new car.

5–162. If 82% of all flight attendants are female and five attendants are selected at random, find the probability that all are female.

5–163. Twenty-five percent of the engineering graduates of a university received a starting salary of $25,000 or more. If three of the graduates are selected at random, find the probability that all had a starting salary of $25,000 or more.

5–164. Three cards are drawn from an ordinary deck *without* replacement. Find the probability of getting the following.

a. All black cards
b. All spades
c. All queens

5–165. A coin is tossed and a card is drawn from a deck. Find the probability of getting the following.

a. A head and a 6
b. A tail and a red card
c. A head and a club

5–166. A box of candy contains six chocolate-covered cherries, three peppermint patties, two caramels, and two strawberry creams. If a piece of candy is selected, find the probability of getting a caramel or a peppermint patty.

5–167. A manufacturing company has three factories: X, Y, and Z. The daily output of each is shown below.

Product	Factory X	Factory Y	Factory Z
TVs	18	32	15
Stereos	6	20	13

If one item is selected at random, find these probabilities.

a. It was manufactured at factory X or is a stereo.
b. It was manufactured at factory Y or factory Z.
c. It is a TV or was manufactured at factory Z.

5–168. A vaccine has a 90% probability of being effective in preventing a certain disease. The probability of getting the disease if a person is not vaccinated is 50%. In a certain geographic region, 25% of the people get vaccinated. If a person is selected at random, find the probability that the person will contract the disease.

5–169. A manufacturer makes three models of a television set, models A, B, and C. A store sells 40% of model A sets, 40% of model B sets, and 20% of model C sets. Three percent of model A sets have stereo sound; 7% of model B sets have stereo sound; and 9% of model C sets have stereo sound. If a set is sold at random, find the probability that it has stereo sound.

5–170. (OPT.) The probability that Sue will live on campus and buy a new car is 0.37. If the probability that she will live on campus is 0.73, find the probability that she will buy a new car, given that she lives on campus.

5–171. (OPT.) The probability that a customer will buy a television set and buy an extended warranty is 0.03. If the probability that a customer will purchase a television set is 0.11, find the probability that the customer will also purchase the extended warranty.

5–172. (OPT.) Forty-three percent of the members of the Blue River Health Club have a lifetime membership and exercise regularly (three or more times a week). If 75% of the club members exercise regularly, find the probability that a randomly selected member is a life member, given that he or she exercises regularly.

5–173. (OPT.) The probability that the weather is bad and the bus arrives late is 0.023. John hears the weather forecast, and there is a 40% chance of snow tomorrow. Find the probability that the bus will be late, given that the weather is bad.

5–174. (OPT.) A number of students were grouped according to their reading ability and education. The table shows the results.

| Education | Reading Ability | | |
	Low	Average	High
High school graduate	6	18	43
Did not graduate	27	16	7

If a student is selected at random, find these probabilities.

a. The student has a low reading ability, given that the student is a high school graduate.
b. The student has a high reading ability, given that the student did not graduate.

5–175. (OPT.) At a large factory, the employees were surveyed and classified according to their level of education and whether or not they smoked. The data are shown in the table.

| | Educational Level | | |
Smoking Habit	Less Than 4 Years High School	High School Graduate	College Graduate
Smoke	6	14	19
Do not smoke	18	7	25

If an employee is selected at random, find these probabilities.

a. The employee smokes, given that he or she graduated from college.
b. Given that the employee did not graduate from high school, he or she is a smoker.

5–176. (OPT.) A pregnancy test is 98% accurate in detecting pregnancy. That is, if a woman is pregnant, it will show positive 98% of the time and show negative 2% of the time. Furthermore, if a woman is not pregnant, it will show negative 98% of the time and positive 2% of the time. Assume that there is a 50% probability that a woman who uses the test is pregnant. Find the probability that if the test shows positive, the woman is not pregnant.

5–177. (OPT.) In Exercise 5–176, if the test shows negative, find the probability the woman is pregnant.

5–178. In a certain high-risk group, the chances of a person getting infected with a virus are 60%. If four people are selected, find the probability that at least one person is infected.

5–179. A coin is tossed five times. Find the probability of getting at least one tail.

5–180. If 15% of all people are left-handed, and five people are selected at random, find the probability that at least one of them is left-handed.

5–181. (OPT.) A person has six bonds, three stocks, and two mutual funds. If three investments are selected, find the probability that one of each type is selected.

5–182. (OPT.) A newspaper advertises five different movies, three plays, and two baseball games for the weekend. If a couple selects three activities, find the probability that they attend two plays and one movie.

5–183. (OPT.) In an office there are three secretaries, four accountants, and two receptionists. If a committee of three is to be formed, find the probability that one of each will be selected.

✎ TEST

Directions: Determine whether each statement is true or false.

1. The set of all possible outcomes of a probability experiment is called the sample space.
2. Subjective probability has little use in the real world.
3. Classical probability uses a frequency distribution to compute the probability of an event.
4. In classical probability, all outcomes in the sample space are considered equally likely.
5. The probability of an event can be any number between -1 and $+1$.
6. When an event cannot occur, its probability is assumed to be -1.
7. The sum of the probabilities of all outcomes in the sample space is 1.
8. If the probability that an event will occur is 0.60, then the probability that the event will not occur is -0.60.
9. When two events cannot occur at the same time, they are said to be mutually exclusive.
10. If the probability of an event is 0.42, then the probability of its complement is 0.58.
11. When two events are not mutually exclusive, $P(A \text{ or } B)$ will equal $P(A) + P(B)$.
12. When two events are dependent, they must have the same probability of occurring.
13. The complement of the event of guessing five answers correctly on a true-false test is guessing five answers incorrectly on the test.
14. When two events A and B are dependent, $P(A \text{ and } B) = P(A) \cdot P(A|B)$.
15. An event and its complement can occur at the same time.

CHAPTER

6

Probability Distributions

6-1 INTRODUCTION

Many decisions in business, insurance, and other real-life situations are made by assigning probabilities to all possible outcomes pertaining to the situation and then evaluating the results. For example, a saleswoman can compute the probability that she makes 0, 1, 2, or 3 or more sales in a single day. An insurance company might be able to assign probabilities to the number of automobiles a family owns. A self-employed speaker might be able to compute the probabilities for giving 0, 1, 2, 3, or 4 or more speeches each week. Once these probabilities are assigned, statistics such as the mean, variance, and standard deviation can be computed for these events. With these statistics, various decisions can be made. The saleswoman will be able to compute the average number of sales she makes per week; and if she is working on commission, she will be able to approximate her weekly income over a period of time, say monthly. The public speaker will be able to plan ahead and approximate his average income and expenses. The insurance company can use its information to design special computer forms and programs to accommodate its customers' future needs.

This chapter explains the concepts and applications of what is called a *probability distribution*. In addition, special probability distributions, such as the *binomial, multinomial, Poisson,* and *hypergeometric* distributions, are explained.

6-2 PROBABILITY DISTRIBUTIONS

Before *probability distribution* is defined formally, the definition of *variable* is reviewed. In Chapter 1, a *variable* was defined as a characteristic or attribute that can assume different values. Various letters of the alphabet, such as X, Y, or Z, are used to represent variables. Since the variables in this chapter are associated with probability, they are called *random variables*.

For example, if a die is rolled, a letter such as X can be used to represent the outcomes. Then the values X can assume are 1, 2, 3, 4, 5, or 6, corresponding to the outcomes of rolling a single die. If two coins are tossed, a letter, say Y, can be used to represent the number of heads, in this case 0, 1, or 2. As another example, if the temperature at 8:00 A.M. is 43° and at noon it is 53°, then the values (T) the temperature assumes are said to be random, since they are due to various atmospheric conditions at the time the temperature was taken.

A random variable is a variable whose values are determined by chance.

Also recall from Chapter 1 that variables can be classified as discrete or continuous by observing the values the variable can assume. If a variable can assume only a specific number of values, such as the outcomes for the roll of a die or the outcomes for the toss of a coin, then the variable is called a *discrete variable. Discrete variables have values that can be counted.* For example, the number of joggers in Riverview Park each day and the number of phone calls received after a television advertisement is aired are examples of discrete variables since they can be counted.

Variables that can assume all values between any two given values are called *continuous variables.* For example, if the temperature goes from 62° to 78° in a 24-hour period, it has passed through every possible number from 62 to 78. *Continuous random variables are obtained from data that can be measured rather than counted.* Examples of continuous variables are heights, weights, temperatures, and time. In this chapter only discrete random variables are used; in Chapter 7 continuous random variables are explained.

The procedure shown here for constructing a probability distribution for a discrete random variable uses the probability experiment of tossing three coins. Recall that when three coins are tossed, the sample space is represented as TTT, TTH, THT, HTT, HHT, HTH, THH, HHH, and if X is the random variable for the number of heads, then X assumes the values 0, 1, 2, or 3.

Probabilities for the values of X can be determined as follows:

No Heads	One Head			Two Heads			Three Heads
TTT	TTH	THT	HTT	HHT	HTH	THH	HHH
$\frac{1}{8}$	$\frac{1}{8}$	$\frac{1}{8}$	$\frac{1}{8}$	$\frac{1}{8}$	$\frac{1}{8}$	$\frac{1}{8}$	$\frac{1}{8}$
$\frac{1}{8}$		$\frac{3}{8}$			$\frac{3}{8}$		$\frac{1}{8}$

Hence, the probability of getting no heads is $\frac{1}{8}$, one head is $\frac{3}{8}$, two heads is $\frac{3}{8}$, and three heads is $\frac{1}{8}$. From these values, a probability distribution can be constructed by listing the outcomes and assigning the probability of each outcome, as shown.

Number of Heads, X	0	1	2	3
Probability, $P(X)$	$\frac{1}{8}$	$\frac{3}{8}$	$\frac{3}{8}$	$\frac{1}{8}$

A probability distribution consists of the values a random variable can assume and the corresponding probabilities of the values. The probabilities are determined theoretically or by observation.

EXAMPLE 6–1   Construct a probability distribution for rolling a single die.

Solution   Since the sample space is 1, 2, 3, 4, 5, 6, and each outcome has a probability of $\frac{1}{6}$, the distribution is as follows:

Outcome, X	1	2	3	4	5	6
Probability, $P(X)$	$\frac{1}{6}$	$\frac{1}{6}$	$\frac{1}{6}$	$\frac{1}{6}$	$\frac{1}{6}$	$\frac{1}{6}$
◄

EXAMPLE 6–2   Construct a probability distribution for the number of girls a family with two children has. Let X be the number of girls.

Solution   The sample space can be shown as follows:

Outcome	BB	BG	GB	GG
Probability	$\frac{1}{4}$	$\frac{1}{4}$	$\frac{1}{4}$	$\frac{1}{4}$
	$\frac{1}{4}$		$\frac{1}{2}$	$\frac{1}{4}$

Hence, the distribution is as shown.

Number of Girls, X	0	1	2
Probability, $P(X)$	$\frac{1}{4}$	$\frac{1}{2}$	$\frac{1}{4}$
◄

Probability distributions can be shown graphically by representing the values of X on the x axis and the probabilities $P(X)$ on the y axis.

EXAMPLE 6–3 Represent graphically the probability distribution for the sample space for tossing three coins.

Number of Heads, X	0	1	2	3
Probability, $P(X)$	$\frac{1}{8}$	$\frac{3}{8}$	$\frac{3}{8}$	$\frac{1}{8}$

Solution The values X assumes are located on the x axis, and the values for $P(X)$ are located on the y axis. The graph is shown in Figure 6–1.

FIGURE 6–1
Probability Distribution
for Example 6–3

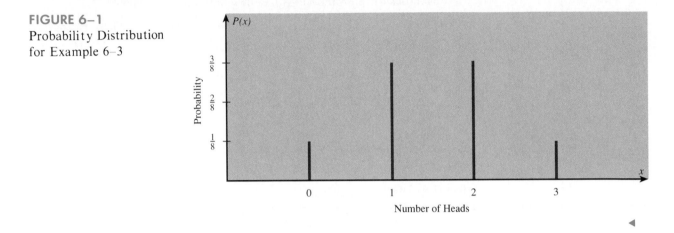

Note that for visual appearances, it is not necessary to start with 0 at the origin.

The preceding examples are illustrations of *theoretical* probability distributions. One did not need to actually perform the experiments to compute the probabilities. In contrast, actual probability distributions can only be constructed by observing the variable over a period of time. They are empirical, as shown in the next example.

EXAMPLE 6–4 During the summer months, a rental agency keeps track of the number of chain saws it rents each day for a period of 90 days. The number of saws rented per day is represented by the variable X. The results are shown below. Compute the probability $P(X)$ for each X, and construct a probability distribution and graph for the data.

X	Number of Days
0	45
1	30
2	15
Total	90

Solution The probability $P(X)$ can be computed for each X by dividing the number of days X saws were rented by the total days.

for 0 saws: $\frac{45}{90} = 0.50$ for 1 saw: $\frac{30}{90} = 0.33$ for 2 saws: $\frac{15}{90} = 0.17$

The distribution is as follows:

Number of Saws Rented, X	0	1	2
Probability, P(X)	0.50	0.33	0.17

The graph is shown in Figure 6–2.

FIGURE 6–2
Probability Distribution
for Example 6–4

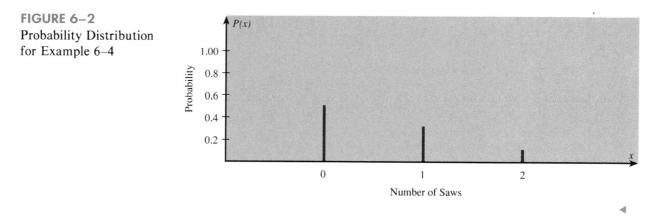

Two Requirements for a Probability Distribution

1. The sum of the probabilities of all the events in the sample space must equal 1; that is, $\Sigma\, P(X) = 1$.
2. The probability of each event in the sample space must be between or equal to 0 and 1. That is, $0 \le P(X) \le 1$.

EXAMPLE 6–5   Determine whether each distribution is a probability distribution.

a.

X	0	5	10	15	20
P(X)	$\frac{1}{5}$	$\frac{1}{5}$	$\frac{1}{5}$	$\frac{1}{5}$	$\frac{1}{5}$

c.

X	1	2	3	4
P(X)	$\frac{1}{4}$	$\frac{1}{8}$	$\frac{1}{16}$	$\frac{9}{16}$

b.

X	0	2	4	6
P(X)	−1.0	1.5	0.3	0.2

d.

X	2	3	7
P(X)	0.5	0.3	0.4

Solution   a. Yes, it is a probability distribution.
b. No, it is not a probability distribution, since $P(X)$ cannot be 1.5 or -1.0.
c. Yes, it is a probability distribution.
d. No, it is not, since $\Sigma\, P(X) = 1.2$.

Many variables in business, education, engineering, and other areas can be analyzed by using probability distributions. In the next section, methods for finding the mean and standard deviation for a probability distribution will be shown.

EXERCISES

6–1. Define and give three examples of a random variable.

6–2. Explain the difference between a discrete and a continuous random variable.

6–3. Give three examples of a discrete random variable.

6–4. Give three examples of a continuous random variable.

6–5. What is a probability distribution? Give an example.

For Exercises 6–6 through 6–11, determine whether the distribution represents a probability distribution. If it is not, state why.

6–6.

X	1	3	5	7	9	11
$P(X)$	$\frac{1}{6}$	$\frac{1}{6}$	$\frac{1}{6}$	$\frac{1}{6}$	$\frac{1}{6}$	$\frac{1}{6}$

6–7.

X	15	20	25	30	35
$P(X)$	$\frac{1}{12}$	$\frac{2}{12}$	$\frac{3}{12}$	$\frac{4}{12}$	$\frac{2}{12}$

6–8.

X	3	6	8
$P(X)$	-0.3	0.6	0.7

6–9.

X	1	2	3	4	5
$P(X)$	$\frac{3}{10}$	$\frac{1}{10}$	$\frac{1}{10}$	$\frac{2}{10}$	$\frac{3}{10}$

6–10.

X	6	12	18
$P(X)$	$\frac{3}{4}$	$\frac{1}{16}$	$\frac{3}{16}$

6–11.

X	5	10	15
$P(X)$	1.2	0.3	0.5

For Exercises 6–12 through 6–18, state whether the variable is discrete or continuous.

6–12. The speed of a jetliner.

6–13. The number of cups of coffee a fast-food restaurant serves each day.

6–14. The number of people who play the state lottery each day.

6–15. The weight of a sumo wrestler.

6–16. The time it takes to complete an exercise session.

6–17. The number of English majors in your school.

6–18. The blood pressures of all patients admitted to a hospital on a specific day.

For Exercises 6–19 through 6–26, construct a probability distribution for the data, and draw a graph for the distribution.

6–19. The probabilities that a patient will have 0, 1, 2, or 3 medical tests performed upon entering a hospital are $\frac{6}{15}$, $\frac{5}{15}$, $\frac{3}{15}$, and $\frac{1}{15}$, respectively.

6–20. The probabilities of a return on an investment of $1000, $2000, and $3000 are $\frac{1}{2}$, $\frac{1}{4}$, and $\frac{1}{4}$, respectively.

6–21. The probabilities of a machine manufacturing 0, 1, 2, 3, 4, or 5 defective parts in one day are 0.75, 0.17, 0.04, 0.025, 0.01, and 0.005, respectively.

6–22. The probabilities that a shopper will purchase 0, 1, 2, or 3 items are 0.20, 0.40, 0.20, and 0.20, respectively.

6–23. A die is loaded in such a way that the probabilities of getting 1, 2, 3, 4, 5, and 6 are $\frac{1}{2}$, $\frac{1}{6}$, $\frac{1}{12}$, $\frac{1}{12}$, $\frac{1}{12}$, and $\frac{1}{12}$, respectively.

6–24. The probabilities that a customer selects 1, 2, 3, 4, or 5 items at a convenience store are 0.32, 0.12, 0.23, 0.18, and 0.15, respectively.

6–25. The probabilities that a tutor sees 1, 2, 3, 4, or 5 students in any one day are 0.10, 0.25, 0.25, 0.20, and 0.20, respectively.

6–26. Three patients are given a headache relief tablet. The probabilities for 0, 1, 2, or 3 successes are 0.18, 0.52, 0.21, and 0.09, respectively.

6–27. A box contains three $1 bills, two $5 bills, one $10 bill, and one $20 bill. Construct a probability distribution for the data.

6–28. Construct a probability distribution for a family of three children. Let X represent the number of boys.

6–29. Construct a probability distribution for drawing a card from a deck of 40 cards consisting of 10 cards numbered 1, 10 cards numbered 2, 15 cards numbered 3, and 5 cards numbered 4.

6–30. Using the sample space for tossing two dice, construct a probability distribution for the sums 2 through 12.

A probability distribution can be written in formula notation as $P(X) = 1/X$, where $X = 2, 3, 6$. The distribution is shown as follows:

X	2	3	6
$P(X)$	$\frac{1}{2}$	$\frac{1}{3}$	$\frac{1}{6}$

For Exercises 6–31 through 6–36, write the distribution for the formula, and determine whether it is a probability distribution.

*6–31. $P(X) = X/6$ for $X = 1, 2, 3$

*6–32. $P(X) = X$ for $X = 0.2, 0.3, 0.5$

*6–33. $P(X) = X/6$ for $X = 3, 4, 7$

*6–34. $P(X) = X + 0.1$ for $X = 0.1, 0.02, 0.04$

*6–35. $P(X) = X/7$ for $X = 1, 2, 4$

*6–36. $P(X) = X/(X + 2)$ for $X = 0, 1, 2$

6–3 MEAN, VARIANCE, AND EXPECTATION

The mean, variance, and standard deviation for a probability distribution are computed differently from the mean, variance, and standard deviation for samples. This section explains how these statistics—as well as a new statistic called the expectation—are calculated for probability distributions.

Mean   In Chapter 3, the means for a sample or population were computed by adding the values and dividing by the total number of values, as shown in the formulas:

$$\overline{X} = \frac{\Sigma X}{n} \qquad \mu = \frac{\Sigma X}{N}$$

But how would one compute the mean of the number of spots that show on top when a die is rolled? One could try rolling the die, say, 10 times, recording the number of spots, and finding the mean; however, this answer would only be an approximation to the true mean. What about 50 rolls or 100 rolls? Actually, the more times that the die is rolled, the better the approximation. One might ask then, "How many times must the die be rolled to get the exact answer?" *It must be rolled an infinite number of times.* Since this task is impossible, the above formulas cannot be used because the denominators would be infinity. Hence, a new method of computing the mean is necessary. This method gives the exact theoretical value of the mean as if it were possible to roll the die an infinite number of times.

Before the formula is stated, an example will be used to explain the concept. Suppose two coins are tossed repeatedly, and the number of heads that occurred is recorded. What will the mean of the number of heads be? The sample space is

HH, HT, TH, TT

and each outcome has a probability of $\frac{1}{4}$. Now, in the long run, one would *expect* two heads (HH) to occur approximately $\frac{1}{4}$ of the time, one head to occur approximately $\frac{1}{2}$ of the time (HT or TH), and no heads (TT) to occur approximately $\frac{1}{4}$ of the time. Hence, on average, one would expect the number of heads to be

$$\tfrac{1}{4} \cdot 2 + \tfrac{1}{2} \cdot 1 + \tfrac{1}{4} \cdot 0 = 1$$

That is, if it were possible to toss the coins many times or an infinite number of times, the *average* of the number of heads would be 1.

Hence, in order to find the mean for a probability distribution, one must multiply each possible outcome by its corresponding probability and find the sum of the products.

> ## Formula for the Mean of a Probability Distribution
>
> The mean of the random variable of a probability distribution is
>
> $$\mu = X_1 \cdot P(X_1) + X_2 \cdot P(X_2) + X_3 \cdot P(X_3) + \cdots + X_n \cdot P(X_n)$$
> $$= \Sigma X \cdot P(X)$$
>
> where $X_1, X_2, X_3, \ldots, X_n$ are the outcomes and $P(X_1), P(X_2), P(X_3)$, $\ldots, P(X_n)$ are the corresponding probabilities.

The next four examples illustrate the use of the formula.

EXAMPLE 6–6   Find the mean of the number of spots that appear when a die is tossed.

Solution   In the toss of a die, the mean can be computed as follows.

Outcome, X	1	2	3	4	5	6
Probability, $P(X)$	$\frac{1}{6}$	$\frac{1}{6}$	$\frac{1}{6}$	$\frac{1}{6}$	$\frac{1}{6}$	$\frac{1}{6}$

$$\mu = \Sigma X \cdot P(X) = 1 \cdot \tfrac{1}{6} + 2 \cdot \tfrac{1}{6} + 3 \cdot \tfrac{1}{6} + 4 \cdot \tfrac{1}{6} + 5 \cdot \tfrac{1}{6} + 6 \cdot \tfrac{1}{6}$$
$$= \tfrac{21}{6} = 3\tfrac{1}{2} \text{ or } 3.5$$

That is, when a die is tossed many times, the theoretical mean will be 3.5. Note that even though the die cannot show a 3.5, the theoretical average is 3.5.   ◄

The reason why this formula gives the theoretical mean is that in the long run, each outcome would occur approximately $\frac{1}{6}$ of the time. Hence, multiplying the outcome by its corresponding probability and finding the sum would yield the theoretical mean. In other words, the outcome 1 would occur $\frac{1}{6}$ of the time, the outcome 2 would occur approximately $\frac{1}{6}$ of the time, etc.

EXAMPLE 6–7   In a family with two children, find the mean of the number of children who will be girls.

Solution   Recall from Example 6–2 that the probability distribution is as follows:

Number of Girls, X	0	1	2
Probability, $P(X)$	$\frac{1}{4}$	$\frac{1}{2}$	$\frac{1}{4}$

Hence, the mean is

$$\mu = \Sigma X \cdot P(X) = 0 \cdot \tfrac{1}{4} + 1 \cdot \tfrac{1}{2} + 2 \cdot \tfrac{1}{4} = 1$$ ◄

EXAMPLE 6–8   If three coins are tossed, find the mean number of heads that will occur.

Solution   The probability distribution is as follows:

Number of Heads, X	0	1	2	3
Probability, $P(X)$	$\frac{1}{8}$	$\frac{3}{8}$	$\frac{3}{8}$	$\frac{1}{8}$

The mean is

$$\mu = \Sigma X \cdot P(X) = 0 \cdot \tfrac{1}{8} + 1 \cdot \tfrac{3}{8} + 2 \cdot \tfrac{3}{8} + 3 \cdot \tfrac{1}{8} = \tfrac{12}{8} = 1\tfrac{1}{2} = 1.5$$ ◄

EXAMPLE 6-9   In a restaurant, the following probability distribution was obtained for the number of items a person ordered for a large pizza.

Number of Items, X	0	1	2	3	4
Probability, $P(X)$	0.3	0.4	0.2	0.06	0.04

Find the mean for the distribution.

Solution

$\mu = \Sigma X \cdot P(X)$

$= (0)(0.3) + (1)(0.4) + (2)(0.2) + (3)(0.06) + (4)(0.04) = 1.14$ ◄

Variance

For a probability distribution, the mean of the random variable describes the measure of the so-called long-run or theoretical average, but it does not tell anything about the spread of the distribution. Recall from Chapter 3 that in order to measure this spread or variability, statisticians use the variance and standard deviation. The following formulas were used:

$$\sigma^2 = \frac{\Sigma (X - \mu)^2}{N} \qquad \text{or} \qquad \sigma = \sqrt{\frac{\Sigma (X - \mu)^2}{N}}$$

These formulas cannot be used for a random variable of a probability distribution since N is infinite, so the variance and standard deviation must be computed differently.

In order to find the variance for the random variable of a probability distribution, one must subtract the theoretical mean of the random variable from each outcome and square the difference. Then each difference is multiplied by its corresponding probability, and the products are added. The formula is

$$\sigma^2 = \Sigma [(X - \mu)^2 \cdot P(X)]$$

Finding the variance by using the preceding formula is somewhat tedious. So for simplified computations, a shortcut formula can be used. This formula is algebraically equivalent to the one shown above and will be used in the examples that follow.

Formula for the Variance of a Probability Distribution

The variance of a probability distribution is found by multiplying the square of each outcome by its corresponding probability, summing those products, and subtracting the square of the mean. The formula for the variance of a probability distribution is

$$\sigma^2 = \Sigma [X^2 \cdot P(X)] - \mu^2$$

The standard deviation of a probability distribution is

$$\sigma = \sqrt{\sigma^2}$$

Remember that the variance and standard deviation cannot be negative.

EXAMPLE 6–10   Compute the variance and standard deviation for the data in Example 6–6.

Solution   Recall that the mean is $\mu = 3.5$, as computed in Example 6–6. Square each outcome and multiply by the corresponding probability, sum those products, and then subtract the square of the mean.

$$\sigma^2 = [1^2 \cdot \tfrac{1}{6} + 2^2 \cdot \tfrac{1}{6} + 3^2 \cdot \tfrac{1}{6} + 4^2 \cdot \tfrac{1}{6} + 5^2 \cdot \tfrac{1}{6} + 6^2 \cdot \tfrac{1}{6}] - (3.5)^2 = 2.917$$

To get the standard deviation, find the square root of the variance.

$$\sigma = \sqrt{2.917} = 1.71 \qquad \blacktriangleleft$$

EXAMPLE 6–11   Five balls numbered 0, 2, 4, 6, and 8 are placed in a bag. After the balls are mixed, one is selected, its number is noted, and then it is replaced. If this experiment is repeated many times, find the variance and standard deviation of the numbers on the balls.

Solution   Let X be the number on each ball. The probability distribution is as follows:

Number on Ball, X	0	2	4	6	8
Probability, $P(X)$	$\tfrac{1}{5}$	$\tfrac{1}{5}$	$\tfrac{1}{5}$	$\tfrac{1}{5}$	$\tfrac{1}{5}$

The mean is

$$\mu = \Sigma X \cdot P(X) = 0 \cdot \tfrac{1}{5} + 2 \cdot \tfrac{1}{5} + 4 \cdot \tfrac{1}{5} + 6 \cdot \tfrac{1}{5} + 8 \cdot \tfrac{1}{5} = 4.00$$

The variance is

$$\begin{aligned}
\sigma^2 &= \Sigma [X^2 \cdot P(X)] - \mu^2 \\
&= [0^2 \cdot (\tfrac{1}{5}) + 2^2 \cdot (\tfrac{1}{5}) + 4^2 \cdot (\tfrac{1}{5}) + 6^2 \cdot (\tfrac{1}{5}) + 8^2 \cdot (\tfrac{1}{5})] - 4^2 \\
&= [0 + \tfrac{4}{5} + \tfrac{16}{5} + \tfrac{36}{5} + \tfrac{64}{5}] - 16 \\
&= \tfrac{120}{5} - 16 \\
&= 24 - 16 = 8
\end{aligned}$$

The standard deviation is $\sigma = \sqrt{8} = 2.83$. $\qquad \blacktriangleleft$

The mean, variance, and standard deviation can also be found by using vertical columns, as shown. [0.2 is used for $P(X)$ since $\tfrac{1}{5} = 0.2$.]

X	$P(X)$	$X \cdot P(X)$	$X^2 \cdot P(X)$
0	0.2	0	0
2	0.2	0.4	0.8
4	0.2	0.8	3.2
6	0.2	1.2	7.2
8	0.2	1.6	12.8
		$\Sigma X \cdot P(X) = 4$	$\Sigma X^2 \cdot P(X) = 24$

The mean is found by summing the $X \cdot P(X)$ column, and the variance is found by summing the $X^2 \cdot P(X)$ column and subtracting the square of the mean:

$$\sigma^2 = 24 - 4^2 = 8.$$

EXAMPLE 6–12 The probability that 0, 1, 2, 3, or 4 people will be placed on hold when they call a radio talk show is shown in the distribution. Find the variance and standard deviation for the data. The radio station has four phone lines.

X	0	1	2	3	4
$P(X)$	0.18	0.34	0.23	0.21	0.04

Solution The mean is

$$\mu = \Sigma\, X \cdot P(X)$$
$$= 0 \cdot (0.18) + 1 \cdot (0.34) + 2 \cdot (0.23) + 3 \cdot (0.21) + 4 \cdot (0.04)$$
$$= 1.59$$

The variance is

$$\sigma^2 = [\Sigma\, X^2 \cdot P(X)] - \mu^2$$
$$= [0^2 \cdot (0.18) + 1^2 \cdot (0.34) + 2^2 \cdot (0.23)$$
$$+ 3^2 \cdot (0.21) + 4^2 \cdot (0.04)] - 1.59^2$$
$$= [0 + 0.34 + 0.92 + 1.89 + 0.64] - 2.5281$$
$$= 3.79 - 2.5281$$
$$= 1.2619$$

The standard deviation is $\sigma = \sqrt{\sigma^2}$, or $\sigma = \sqrt{1.2619} = 1.12$. ◄

Expectation Related to the concept of the mean for a probability distribution is the concept of   or expectation. Expected value is used in various types of games of chance, in insurance, and in other areas, such as decision theory.

The expected value of a discrete random variable of a probability distribution is the theoretical average of the variable. The formula is

$$\mu = E(X) = \Sigma\, X \cdot P(X)$$

The symbol $E(X)$ is used for the expected value.

The formula for the expected value is the same as the formula for the theoretical mean. The expected value, then, is the theoretical mean of the probability distribution. That is, $E(X) = \mu$.

EXAMPLE 6–13 One thousand tickets are sold at $1 each for a color television valued at $350. What is the expected value if a person purchases one ticket?

Solution The problem can be set up as follows:

	Win	Lose
Gain, X	$349	−$1
Probability, $P(X)$	$\dfrac{1}{1000}$	$\dfrac{999}{1000}$

Two things should be noted. First, for a win, the net gain is $349, since the person does not get the cost of the ticket ($1) back. Second, for a loss, the gain is represented by a negative number, in this case −$1. The solution, then, is

$$E(X) = \$349 \cdot \frac{1}{1000} + (-\$1) \cdot \frac{999}{1000} = -\$0.65$$ ◄

Expected value problems of this type can also be solved by finding the overall gain and subtracting the cost of the tickets, as shown:

$$E(X) = \$350 \cdot \frac{1}{1000} - \$1 = -\$0.65$$

Hence, the overall gain ($350) must be used.

Note that the expectation is $-\$0.65$. This does not mean that a person loses $0.65, since the person can only win a television set valued at $350 or lose $1 on the ticket. What this expectation means is that the average of the losses is $0.65 for each of the 1000 ticket holders. Here is another way of looking at this situation: If a person purchased one ticket each week over a long period of time, the average loss would be $0.65 per ticket, since theoretically, on average, that person would win the set once for each 1000 tickets purchased.

EXAMPLE 6–14 A ski resort loses $70,000 per season when it does not snow very much and makes $250,000 profit when it does snow a lot. The probability of it snowing at least 75 inches (i.e., a good season) is 40%. Find the expectation for the profit.

Solution

Profit, X	$250,000	$-\$70,000$
Probability, $P(X)$	0.40	0.60

$$E(X) = (\$250,000)(0.40) + (-\$70,000)(0.60) = \$58,000 \qquad \blacktriangleleft$$

EXAMPLE 6–15 One thousand tickets are sold at $1 each for four prizes of $100, $50, $25, and $10. What is the expected value if a person purchases two tickets?

Gain, X	$98	$48	$23	$8	$-\$2$
Probability, $P(X)$	$\frac{2}{1000}$	$\frac{2}{1000}$	$\frac{2}{1000}$	$\frac{2}{1000}$	$\frac{992}{1000}$

Solution

$$E(X) = \$98 \cdot \frac{2}{1000} + \$48 \cdot \frac{2}{1000} + \$23 \cdot \frac{2}{1000} + \$8 \cdot \frac{2}{1000} + (-\$2) \cdot \frac{992}{1000}$$

$$= -\$1.63$$

An alternative solution is

$$E(X) = \$100 \cdot \frac{2}{1000} + \$50 \cdot \frac{2}{1000} + \$25 \cdot \frac{2}{1000} + \$10 \cdot \frac{2}{1000} - \$2$$

$$= -\$1.63 \qquad \blacktriangleleft$$

In gambling games, if the expected value is zero, the game is said to be fair. That is, both participants have an equal chance of winning or losing. If the expected value of a game is positive, then the game is in favor of the player. That is, the player has a better-than-even chance of winning. If the expected value is negative, then the game is said to be in favor of the house. That is, in the long run, the players will lose money.

Speaking of Statistics

Tanker trucks and accidents

Trucks carrying gasoline or fuel oil account for 40% of all hazardous material accidents and all 10 related deaths on U.S. highways last year. Here's a look at the top five materials transported, total accidents and damages:

Material	Accidents	Damage
Gasoline	58	$13,284,806
Fuel oil	46	$3,494,874
Crude oil	9	$196,646
Sodium Hydroxide[1]	8	$378,133
Denatured alcohol	6	$451,374
Total	263[2]	$23,901,303

1 – Solution
2 – Some accidents may involve more than one hazardous material

Source: Department of Transportation

By Marty Baumann, USA TODAY

Source: Based on data from the Department of Transportation. Copyright 1992, USA TODAY. Reprinted with permission.

In this study, using the damage amount as X and the accidents as $P(X)$, find the mean, variance, and standard deviation of the probability distribution.

EXERCISES

6–37. From past experience, a company has found that in cartons of transistors, 92% contain no defective transistors, 3% contain one defective transistor, 3% contain two defective transistors and 2% contain three defective transistors. Find the mean, variance, and standard deviation for the defective transistors.

6–38. The number of coats sold per day at a retail store is shown in the table, with the corresponding probabilities. Find the mean, variance, and standard deviation of the distribution.

Number of Coats Sold, X	8	9	10	11	12
Probability, $P(X)$	0.1	0.2	0.2	0.3	0.2

6–39. A shoe store manager computes the probabilities of selling shoes for each size, as shown below. Find the mean, variance, and standard deviation for the distribution.

Size, X	8	9	10	11	12	13
Probability, $P(X)$	0.13	0.39	0.23	0.14	0.08	0.03

6–40. The probability distribution for the number of passengers per trip on the crosstown bus is shown below. Find the mean, variance, and standard deviation of the distribution.

Number of Passengers, X	40	41	42	43	44
Probability, $P(X)$	0.21	0.12	0.47	0.10	0.10

6–41. A public speaker computes the probabilities for the number of speeches she gives each week. Compute the mean, variance, and standard deviation of the distribution shown.

Number of Speeches, X	0	1	2	3	4	5
Probability, P(X)	0.06	0.42	0.22	0.12	0.15	0.03

6–42. A recent survey by an insurance company showed the following probabilities for the number of automobiles each policyholder owned. Find the mean, variance, and standard deviation for the distribution.

Number of Automobiles, X	1	2	3	4
Probability, P(X)	0.4	0.3	0.2	0.1

6–43. A concerned parents group determined the number of commercials shown in each of five children's programs over a period of time. Find the mean, variance, and standard deviation for the distribution shown.

Number of Commercials, X	5	6	7	8	9
Probability, P(X)	0.2	0.25	0.38	0.10	0.07

6–44. A study conducted by a television station showed the number of televisions per household and the corresponding probabilities for each. Find the mean, variance, and standard deviation.

Number of Televisions, X	1	2	3	4
Probability, P(X)	0.32	0.51	0.12	0.05

6–45. The following distribution shows the number of students enrolled in CPR classes offered by the local fire department. Find the mean, variance, and standard deviation for the distribution.

Number of Students, X	12	13	14	15	16
Probability, P(X)	0.15	0.20	0.38	0.18	0.09

6–46. A florist determines the probabilities for the number of flower arrangements she delivers each day. Find the mean, variance, and standard deviation for the distribution shown.

Number of Arrangements, X	6	7	8	9	10
Probability, P(X)	0.2	0.2	0.3	0.2	0.1

6–47. A cash prize of $2500 is to be awarded by the Community Ambulance Association. If 3000 tickets are sold at $2 each, find the expectation.

6–48. A box contains ten $1 bills, five $2 bills, three $5 bills, one $10 bill, and one $100 bill. A person is charged $20 to select one bill. Find the expectation. Is the game fair?

6–49. If a person rolls doubles when he tosses two dice, he wins $5. If the game is to be fair, how much should the person pay to play the game?

6–50. If a player rolls two dice and gets a sum of 2 or 12, he wins $20. If the person gets a 7, he wins $5. The cost to play the game is $3. Find the expectation of the game.

6–51. A lottery offers one $1000 prize, one $500 prize, and five $100 prizes. One thousand tickets are sold at $3 each. Find the expectation if a person buys one ticket.

6–52. In Exercise 6–51, find the expectation if a person buys two tickets. Assume that the player's ticket is replaced after each draw and the same ticket can win more than one prize.

6–53. For a daily lottery, a person selects a three-digit number. If the person plays for $1, she can win $500. Find the expectation. In the same daily lottery, if a person boxes a number, she will win $80. Find the expectation if the number 123 were played for $1 and boxed. (When a number is "boxed," it can win when the digits occur in any order.)

6–54. If a person, aged 60, buys a $1000 life insurance policy at a cost of $60 and has a probability of 0.972 of living to age 61, find the expectation of the policy.

6–55. A roulette wheel has 38 numbers, 1 through 36, 0, and 00. Half of the numbers from 1 through 36 are red and half are black. A ball is rolled, and it falls into one of the 38 slots, giving a number and a color. Green is the color for 0 and 00. The payoffs for a $1 bet are as follows:

Red or black	$1	0	$35
Odd or even	$1	00	$35
1–18	$1	Any single number	$35
19–36	$1	0 or 00	$17

If a person bets $1, find the expected value for each.

a. Red
b. Even (exclude zero)
c. 0

d. Any single value
e. 0 or 00

*6–56. Construct a probability distribution for the sum shown on the faces when two dice are rolled. Find the mean, variance, and standard deviation of the distribution.

*6–57. When one die is rolled, the expected value of the number of spots is 3.5. In Exercise 6–56, the mean number of spots was found for rolling two dice. What is the mean number of spots if three dice are rolled?

*6-58. Another formula for finding the variance for a probability distribution is

$$\sigma^2 = \Sigma (X - \mu)^2 \cdot P(X)$$

Verify, algebraically, that this formula gives the same result as the formula shown in this section.

*6-59. Roll a die 100 times. Compute the mean and standard deviation. How does the result compare with the theoretical results of Example 6-6?

*6-60. Roll two dice 100 times and find the mean, variance, and standard deviation of the sum of the spots. Compare the result with the theoretical results obtained in Exercise 6-56.

*6-61. Conduct a survey of the number of books your classmates have brought to class today. Construct a probability distribution and find the mean, variance, and standard deviation.

*6-62. In a recent promotional campaign, a company offered the following prizes and the corresponding probabilities. Find the expected value of winning. The tickets are free.

Number of Prizes	Amount	Probability
1	$100,000	$\dfrac{1}{1,000,000}$
2	$10,000	$\dfrac{1}{50,000}$
5	$1,000	$\dfrac{1}{10,000}$
10	$100	$\dfrac{1}{1,000}$

If the winner had to mail in the winning ticket to claim the prize, what would be the expectation if the cost of the stamp is considered?

6-4 THE BINOMIAL DISTRIBUTION

Many types of probability problems have only two outcomes, or they can be reduced to two outcomes. For example, when a coin is tossed, it can land heads or tails. When a baby is born, it will be either male or female. In a basketball game, a team either wins or loses. A true-false item can be answered in only two ways, true or false. Other situations can be reduced to two outcomes. For example, a medical treatment can be classified as effective or ineffective, depending on the results. A person can be classified as having normal or abnormal blood pressure, depending on the measure of the sphygmomanometer. A multiple-choice question, even though there are four or five answer choices, can be classified as correct or incorrect. Situations such as these are called *binomial experiments*.

A binomial experiment is a probability experiment that satisfies the following four requirements:

1. **Each trial can have only two outcomes or outcomes that can be reduced to two outcomes.**
2. **There must be a fixed number of trials.**
3. **The outcomes of each trial must be independent.**
4. **The probability of a success must remain the same for each trial.**

A binomial experiment and its results give rise to a special probability distribution called the binomial distribution.

The outcomes of a binomial experiment and the corresponding probabilities of these outcomes are called a binomial distribution.

In binomial experiments, the outcomes are usually classified as successes or failures. For example, the correct answer to a multiple-choice item can be classified as a success, but any of the other choices would be incorrect and hence classified as a failure. The notation that is commonly used for binomial experiments and the binomial distribution is defined next.

Notation for the Binomial Distribution

$P(S)$ The symbol for the probability of success.
$P(F)$ The symbol for the probability of failure.
p The numerical probability of a success.
q The numerical probability of a failure.

$$P(S) = p \quad \text{and} \quad P(F) = 1 - p = q$$

n The number of trials.
X The number of successes.

Note that $q = 1 - p$ and $0 \le X \le n$.

The probability of a success in a binomial experiment can be computed with the following formula.

Binomial Probability Formula

In a binomial experiment, the probability of exactly X successes in n trials is

$$P(X) = \frac{n!}{(n - X)!X!} \cdot p^X \cdot q^{n-X}$$

An explanation of why the formula works will be given following Example 6–16.

EXAMPLE 6–16   A coin is tossed three times. Find the probability of getting exactly two heads.

Solution   This problem can be solved by looking at the sample space. There are three ways to get two heads.

$$\text{HHH, } \underline{\text{HHT, HTH, THH,}} \text{ TTH, THT, HTT, TTT}$$

The answer is $\frac{3}{8}$, or 0.375.   ◄

Looking at the problem in Example 6–16 from the standpoint of a binomial experiment, one can show that it meets the four requirements.

1. There are only two outcomes for each trial, heads or tails.
2. There are a fixed number of trials (three).

3. The outcomes are independent (the outcome of one toss in no way affects the outcome of another toss).
4. The probability of a success (heads) is $\frac{1}{2}$ in each case.

In this case, $n = 3$, $X = 2$, $p = \frac{1}{2}$, and $q = \frac{1}{2}$. Hence, substituting in the formula gives

$$P(2 \text{ heads}) = \frac{3!}{(3-2)!2!} \cdot \left(\frac{1}{2}\right)^2 \left(\frac{1}{2}\right)^1 = \frac{3}{8} = 0.375$$

which is the same answer obtained by using the sample space.

An explanation of the formula can be demonstrated by using the same example. First, note that there are three ways to get exactly two heads and one tail from a possible eight ways. They are HHT, HTH, and THH. In this case, then, the number of ways of obtaining two heads from three coin tosses is $_3C_2$, or 3. In general, the number of ways to get X successes from n trials without regard to order is

$$_nC_X = \frac{n!}{(n-X)!X!}$$

and this is the first part of the binomial formula.

Next, each success has a probability of $\frac{1}{2}$, and can occur twice. Likewise, each failure has a probability of $\frac{1}{2}$ and can occur once, giving the $(\frac{1}{2})^2(\frac{1}{2})^1$ part of the formula. Generalizing, then, each success has a probability of p and can occur X times, and each failure has a probability of q and can occur $(n - X)$ times. Putting it all together yields the binomial probability formula.

EXAMPLE 6-17 If a student randomly guesses at five multiple-choice questions, find the probability that the student gets exactly three correct. Each question has five possible choices.

Solution In this case $n = 5$, $X = 3$, and $p = \frac{1}{5}$, since there is one chance in five of guessing a correct answer. Then,

$$P(3) = \frac{5!}{(5-3)!3!} \cdot \left(\frac{1}{5}\right)^3 \left(\frac{4}{5}\right)^2 = 0.0512 \qquad \blacktriangleleft$$

EXAMPLE 6-18 A survey found that 30% of teenage consumers received their spending money from part-time jobs.[1] If five teenagers are selected at random, find the probability that at least three of them will have part-time jobs.

Solution To find the probability that at least three have a part-time job, it is necessary to find the individual probabilities for 3, 4, and 5, and then add them to get the total probability.

$$P(3) = \frac{5!}{(5-3)!(3!)} \cdot (0.3)^3(0.7)^2 = 0.1323$$

$$P(4) = \frac{5!}{(5-4)!(4!)} \cdot (0.3)^4(0.7)^1 = 0.02835$$

$$P(5) = \frac{5!}{(5-5)!(5!)} \cdot (0.3)^5(0.7)^0 = 0.00243$$

[1]February 4, 1992, based on a survey from Teenage Research Unlimited, Northbrook, Ill.

Hence,

$$P(\text{at least three teenagers have part-time jobs})$$
$$= 0.1323 + 0.02835 + 0.00243 = 0.16308 \qquad \blacktriangleleft$$

Computing probabilities using the binomial probability formula can be quite tedious at times, so tables have been developed. Table B in the Appendix gives the probabilities for individual events. The next example shows how to use Table B to compute probabilities for binomial experiments.

EXAMPLE 6–19 Solve the problem in Example 6–16 by using Table B.

Solution Since $n = 3$, $X = 2$, and $p = 0.5$, the value 0.375 is found as shown in Figure 6–3.

FIGURE 6–3
Using Table B for
Example 6–19

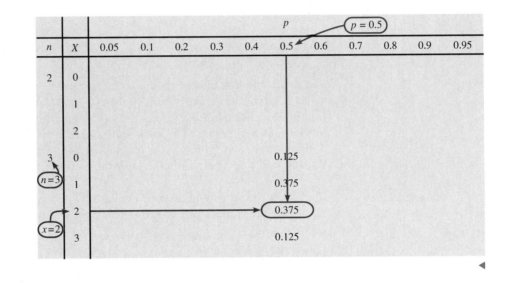

EXAMPLE 6–20 The probability of a defective telephone in a large office building is 0.05. If a sample of 20 telephones is selected, find these probabilities. Use the binomial table.

a. There are exactly 5 defective telephones.
b. There are at most 3 defective telephones.
c. There are at least 3 defective telephones.

Solution a. $n = 20$, $p = 0.05$, and $X = 5$. From the table, one gets 0.002.
b. $n = 20$ and $p = 0.05$. At most 3 defective telephones means 0, 1, 2, or 3. Hence, the solution is

$$P(0) + P(1) + P(2) + P(3) = 0.358 + 0.377 + 0.189 + 0.060$$
$$= 0.984$$

c. $n = 20$ and $p = 0.05$. At least 3 defective telephones means 3, 4, 5, . . . , 20. This problem can best be solved by finding $P(0) + P(1) + P(2)$ and subtracting from 1.

$$P(0) + P(1) + P(2) = 0.358 + 0.377 + 0.189 = 0.924$$
$$1 - 0.924 = 0.076 \qquad \blacktriangleleft$$

EXAMPLE 6—21   A manufacturer who produces VCR remote controls finds from past studies that 5% of them are defective. In a random sample of 15, find the probability that exactly 3 are defective.

Solution $n = 15$, $p = 0.05$, and $X = 3$. From Table B, $P(3) = 0.031$.   ◄

Remember that in the use of the binomial distribution, the outcomes must be independent. For example, in the selection of components from a batch to be tested, each component must be replaced before the next one is selected. Otherwise, the outcomes are not independent. However, a dilemma arises because there is a chance that the same component could be selected again. This situation can be avoided by not replacing the component and using the hypergeometric distribution to calculate the probabilities. The hypergeometric distribution is presented later in this chapter. Note that when the population is large and the sample is small, the binomial probabilities are nearly the same as the corresponding hypergeometric probabilities.

Mean, Variance, and Standard Deviation for the Binomial Distribution

The mean, variance, and standard deviation of a variable that has the binomial distribution can be found by using the following formulas.

$$
\begin{aligned}
\text{mean} \quad & \mu = n \cdot p \\
\text{variance} \quad & \sigma^2 = n \cdot p \cdot q \\
\text{standard deviation} \quad & \sigma = \sqrt{n \cdot p \cdot q}
\end{aligned}
$$

These formulas are algebraically equivalent to the formulas for the mean, variance, and standard deviation of the variables for probability distributions, but because they are for variables of the binomial distribution, they have been simplified using algebra. The algebraic derivation is omitted here, but their equivalence is shown in the next example.

EXAMPLE 6—22   A coin is tossed four times. Find the mean, variance, and standard deviation of the number of heads that will be obtained.

Solution   With the formulas for the binomial distribution and $n = 4$, $p = \frac{1}{2}$, and $q = \frac{1}{2}$, the results are

$$
\begin{aligned}
\mu &= n \cdot p = 4 \cdot \tfrac{1}{2} = 2 \\
\sigma^2 &= n \cdot p \cdot q = 4 \cdot \tfrac{1}{2} \cdot \tfrac{1}{2} = 1 \\
\sigma &= \sqrt{1} = 1
\end{aligned}
$$
    ◄

From Example 6—22, when four coins are tossed many, many times, the average of the number of heads that appear is two, and the standard deviation of the number of heads is one. Note that these are theoretical values.

As stated previously, this problem can be solved by using the expected value formulas. The distribution is shown as follows:

No. of Heads, X	0	1	2	3	4
Probability, $P(X)$	$\frac{1}{16}$	$\frac{4}{16}$	$\frac{6}{16}$	$\frac{4}{16}$	$\frac{1}{16}$

$$\mu = E(X) = \Sigma X \cdot P(X) = 0 \cdot \tfrac{1}{16} + 1 \cdot \tfrac{4}{16} + 2 \cdot \tfrac{6}{16} + 3 \cdot \tfrac{4}{16} + 4 \cdot \tfrac{1}{16} = \tfrac{32}{16} = 2$$

$$\sigma^2 = \Sigma X^2 \cdot P(X) - \mu^2$$

$$= 0^2 \cdot \tfrac{1}{16} + 1^2 \cdot \tfrac{4}{16} + 2^2 \cdot \tfrac{6}{16} + 3^2 \cdot \tfrac{4}{16} + 4^2 \cdot \tfrac{1}{16} - 2^2 = \tfrac{80}{16} - 4 = 1$$

$$\sigma = \sqrt{1} = 1$$

Hence, the simplified binomial formulas give the same results.

EXAMPLE 6–23   A die is rolled 240 times. Find the mean, variance, and standard deviation for the number of 3s that will be rolled.

Solution $n = 240$ and $p = \frac{1}{6}$.

$$\mu = n \cdot p = 240 \cdot \left(\tfrac{1}{6}\right) = 40$$

$$\sigma^2 = n \cdot p \cdot q = 240 \left(\tfrac{1}{6}\right) \left(\tfrac{5}{6}\right) = 33.33$$

$$\sigma = \sqrt{n \cdot p \cdot q} = \sqrt{33.33} = 5.77$$

On average, there will be forty 3s. The standard deviation is 5.77.   ◄

EXAMPLE 6–24   The probability of a defective wristwatch is 0.02. Given a shipment of 8000 watches, find the mean, variance, and standard deviation of the number of defective watches.

Solution

$$\mu = n \cdot p = (8000)(0.02) = 160$$

$$\sigma^2 = n \cdot p \cdot q = (8000)(0.02)(0.98) = 156.8$$

$$\sigma = \sqrt{n \cdot p \cdot q} = \sqrt{156.8} = 12.52$$

For each shipment, then, the average number of defective watches is 160, and the standard deviation of the number of defective watches is 12.52.   ◄

EXERCISES

6–63. Which of the following are binomial experiments or can be reduced to binomial experiments?

a. Surveying 100 people to determine if they like Sudsy Soap.
b. Tossing a coin 100 times to see how many heads occur.
c. Drawing a card from a deck and getting a heart.
d. Asking 1000 people which brand of cigarettes they smoked.
e. Testing four different brands of aspirin to see which brands are effective.
f. Testing one brand of aspirin using ten people to determine whether it is effective.
g. Asking 100 people if they smoke.
h. Checking 1000 applicants to see whether the applicants were admitted to White Oak College.

i. Surveying 300 prisoners to see how many different crimes they were convicted of.
j. Surveying 300 prisoners to see whether this is their first offense.

6–64. (ANS.) Compute the probability of X successes, using Table B in the Appendix.

a. $n = 2, p = 0.30, X = 1$
b. $n = 4, p = 0.60, X = 3$
c. $n = 5, p = 0.10, X = 0$
d. $n = 10, p = 0.40, X = 4$
e. $n = 12, p = 0.90, X = 2$
f. $n = 15, p = 0.80, X = 12$
g. $n = 17, p = 0.05, X = 0$
h. $n = 20, p = 0.50, X = 10$
i. $n = 16, p = 0.20, X = 3$

6-65. Compute the probability of X successes, using the binomial formula.

  a. $n = 6, X = 3, p = 0.03$
  b. $n = 4, X = 2, p = 0.18$
  c. $n = 5, X = 3, p = 0.63$
  d. $n = 9, X = 0, p = 0.42$
  e. $n = 10, X = 5, p = 0.37$

For Exercises 6-66 through 6-75, assume all variables are binomial.

6-66. A burglar alarm system has six fail-safe components. The probability of each failing is 0.05. Find these probabilities.

  a. Exactly three will fail.
  b. Fewer than two will fail.
  c. None will fail.
  d. Compare the answers for parts a, b, and c, and explain why the results are reasonable.

6-67. A student takes a ten-question, true-false exam and guesses on each question. Find the probability of passing if the lowest passing grade is six out of ten.

6-68. If the quiz in Exercise 6-67 was a multiple-choice quiz with five choices for each question, find the probability of guessing at least six out of ten correctly.

6-69. Suppose that 30% of all the automobiles in a shopping mall parking lot belong to employees. Find the probability that if nine cars are selected at random, three belong to employees.

6-70. In a recent survey, 80% of the citizens in a community favored the building of a post office. If 15 citizens are selected, find the probability that 9 favor the building of a post office.

6-71. If $\frac{3}{4}$ of the nursing majors are female, find the probability that 3 out of 7 will be female.

6-72. Suppose that 60% of all customers who buy used cars from a dealer return within one week for warrantied repairs. Find the probability that if 10 cars were bought, at most 3 customers will return for repairs.

6-73. A recent survey found that 40% of all people over the age of 62 wear hearing aids. If a random sample of five people over 62 is selected, find these probabilities.

  a. Exactly two people wear hearing aids.
  b. At most three people wear hearing aids.
  c. At least two people wear hearing aids.

6-74. If 60% of the children had German measles by the time they were 12 years old, find these probabilities for a sample of nine children.

  a. At least five have had German measles.
  b. Exactly seven have had German measles.
  c. More than three have had German measles.

6-75. If 20% of the people in a community use the emergency room at a hospital in one year, find these probabilities for a sample of ten people.

  a. At most three used the emergency room.
  b. Exactly three used the emergency room.
  c. At least five used the emergency room.

6-76. (ANS.) Find the mean, variance, and standard deviation for each of the values of n and p when the conditions for the binomial distribution are met.

  a. $n = 100, p = 0.75$
  b. $n = 300, p = 0.3$
  c. $n = 20, p = 0.5$
  d. $n = 10, p = 0.8$
  e. $n = 1000, p = 0.1$
  f. $n = 500, p = 0.25$
  g. $n = 50, p = \frac{2}{5}$
  h. $n = 36, p = \frac{1}{6}$

6-77. The failure rate for a driver's license test is 40%. If 80 people take the test, find the mean, variance, and standard deviation for the number of failures.

6-78. Find the mean, variance, and standard deviation for the number of heads when ten coins are tossed.

6-79. If 2% of automobile carburetors are defective, find the mean, variance, and standard deviation of a lot of 500 carburetors.

6-80. Tasteful Tomato Seeds have an 80% germination rate. Find the mean, variance, and standard deviation of a package of 200 seeds.

6-81. If the number of deaths due to accidents for people under 20 years of age is 34%, find the mean, variance, and standard deviation for the number of deaths due to accidents for 1000 people in this age range.

6-82. In a restaurant, a study found that 42% of all patrons smoked. If the seating capacity of the restaurant is 80 people, find the mean, variance, and standard deviation of the number of smokers.

*6–83. The graph shown in Figure 6–4 represents the probability distribution for the number of girls in a family of three children. From this graph, construct a probability distribution.

*6–84. Construct a binomial distribution graph for the number of defective computer chips in a lot of five if $p = 0.2$.

FIGURE 6–4
Binomial Distribution Graph for Exercise 6–83

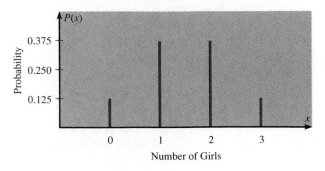

6–5 OTHER TYPES OF DISTRIBUTIONS (OPTIONAL)

In addition to the binomial distribution, there are other types of distributions that are used in statistics. Three of the most commonly used distributions are the multinomial distribution, the Poisson distribution, and the hypergeometric distribution. These distributions are described in the following subsections.

The Multinomial Distribution

Recall that in order for an experiment to be binomial, two outcomes are required for each trial. But if each trial in an experiment has more than two outcomes, a distribution called the **multinomial distribution** must be used. For example, a survey might require the responses of "approve," "disapprove," or "no opinion." In another situation, a person may have a choice of five activities for Friday night, such as a movie, dinner, baseball game, play, or party. Since these situations have more than two possible outcomes for each trial, the binomial distribution cannot be used to compute probabilities.

The multinomial distribution can be used for the situations above if the probabilities for each trial remain constant and the outcomes are independent for a fixed number of trials.

Formula for the Multinomial Distribution

If M consists of events $E_1, E_2, E_3, \ldots, E_k$, which have corresponding probabilities $p_1, p_2, p_3, \ldots, p_k$ of occurring, and X_1 is the number of times E_1 will occur, X_2 is the number of times E_2 will occur, X_3 is the number of times E_3 will occur, etc., then the probability that M will occur is

$$P(M) = \frac{n!}{X_1! \cdot X_2! \cdot X_3! \cdot \ldots \cdot X_k!} \cdot p_1^{X_1} \cdot p_2^{X_2} \cdot \ldots \cdot p_k^{X_k}$$

where $X_1 + X_2 + X_3 + \cdots + X_k = n$.

EXAMPLE 6-25   In a large city, 50% of the people choose a movie, 30% choose dinner and a play, and 20% choose shopping as a leisure activity. If a sample of five people is randomly selected, find the probability that three are planning to go to a movie, one to a play, and one to a shopping mall.

Solution $n = 5$, $X_1 = 3$, $X_2 = 1$, $X_3 = 1$, $p_1 = 0.50$, $p_2 = 0.30$, and $p_3 = 0.20$. Substituting in the formula gives

$$P(M) = \frac{5!}{3! \cdot 1! \cdot 1!} \cdot (0.50)^3 (0.30)^1 (0.20)^1 = 0.15 \qquad \blacktriangleleft$$

Again, note that the multinomial distribution can be used even though replacement is not done, provided that the sample is small in comparison with the population.

EXAMPLE 6-26   In a music store, a manager found that the probabilities that a person buys 0, 1, or 2 or more CDs are 0.3, 0.6, and 0.1, respectively. If six customers enter the store, find the probability that one won't buy any CDs, three will buy one CD, and two will buy two or more CDs.

Solution $n = 6$, $X_1 = 1$, $X_2 = 3$, $X_3 = 2$, $p_1 = 0.3$, $p_2 = 0.6$, and $p_3 = 0.1$. Then,

$$P(M) = \frac{6!}{1! \, 3! \, 2!} (0.3)^1 (0.6)^3 (0.1)^2$$
$$= 60(0.3)(0.216)(0.01) = 0.03888 \qquad \blacktriangleleft$$

EXAMPLE 6-27   In a sample of four families with three children, find the probability that one family has two girls, two families have three girls, and one family has three boys.

Solution $n = 4$, $X_1 = 1$, $X_2 = 2$, $X_3 = 1$, $p_1 = \frac{3}{8}$, $p_2 = \frac{1}{8}$, and $p_3 = \frac{1}{8}$. Then,

$$P(M) = \frac{4!}{1! \cdot 2! \cdot 1!} \cdot \left(\frac{3}{8}\right)^1 \left(\frac{1}{8}\right)^2 \left(\frac{1}{8}\right)^1 = \frac{9}{1024} = 0.00879 \qquad \blacktriangleleft$$

As one can see, the multinomial distribution is similar to the binomial distribution but has the advantage of allowing one to compute probabilities when there are more than two outcomes for each trial in the experiment. That is, the multinomial distribution is a general distribution, and the binomial distribution is a special case of the multinomial distribution.

The Poisson Distribution   A discrete probability distribution that is useful when n is large and p is small and when the independent variables occur over a period of time is called the **Poisson distribution,** named for Simeon D. Poisson (1781–1840). Poisson was a French mathematician and physicist who formulated the distribution. In addition to being used for the stated conditions (i.e., n is large, p is small, and the variables occur over a period of time), the Poisson distribution can be used when a density of items is distributed over a given area or volume, such as the number of plants growing per acre of woods or the number of defects in a given length of videotape.

Formula for the Poisson Distribution

The probability of X occurrences in an interval of time, volume, area, etc., for a variable where λ (Greek letter lambda) is the mean number of occurrences is

$$P(X; \lambda) = \frac{e^{-\lambda}\lambda^X}{X!} \qquad \text{where} \qquad X = 0, 1, 2, \ldots$$

The letter e is a constant approximately equal to 2.7183.

EXAMPLE 6–28 If there are 200 typographical errors randomly distributed in a 500-page manuscript, find the probability that a given page contains exactly 3 errors.

Solution First, find the mean number (λ) of errors. Since there are 200 errors distributed over 500 pages, each page has an average of

$$\lambda = \frac{200}{500} = \frac{2}{5} = 0.4$$

or 0.4 error per page. Since $X = 3$, substituting into the formula yields

$$P(X; \lambda) = \frac{e^{-\lambda}\lambda^X}{X!} = \frac{(2.7183)^{-0.4}(0.4)^3}{3!} = 0.0071501$$

Thus, there is less than a 1% probability that any given page will contain 3 errors. ◄

Since the mathematics involved in computing Poisson probabilities is somewhat complicated, tables have been compiled for these probabilities. Table C in the Appendix gives various values for λ and X.

In Example 6–28, where X is 3 and λ is 0.4, the table gives the value 0.0072 for the probability. See Figure 6–5.

FIGURE 6–5
Using Table C

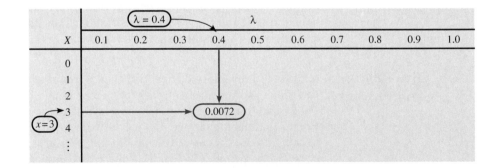

EXAMPLE 6–29 A sales firm receives, on the average, 3 calls per hour on its toll-free number. For any given hour, find the probability that it will receive the following.

a. At most 3 calls   b. At least 3 calls   c. Five or more calls

Solution   a. At most 3 calls means 0, 1, 2, or 3 calls. Hence,

$$P(0; 3) + P(1; 3) + P(2; 3) + P(3; 3)$$
$$= 0.0498 + 0.1494 + 0.2240 + 0.2240$$
$$= 0.6472$$

b. At least 3 calls means 3 or more calls. It is easier to find the probability of 0, 1, and 2 calls and then subtract this answer from 1 to get the probability of at least 3 calls.

$$P(0; 3) + P(1; 3) + P(2; 3) = 0.0498 + 0.1494 + 0.2240 = 0.4232$$

and $$1 - 0.4232 = 0.5768$$

c. For the probability of 5 or more calls, it is easier to find the probability of getting 0, 1, 2, 3, or 4 calls and subtract this answer from 1. Hence,

$$P(0; 3) + P(1; 3) + P(2; 3) + P(3; 3) + P(4; 3)$$
$$= 0.0498 + 0.1494 + 0.2240 + 0.2240 + 0.1680$$
$$= 0.8152$$

and $$1 - 0.8152 = 0.1848$$

Thus, for the events described, the part a event is most likely to occur and the part c event is least likely to occur. ◄

The Poisson distribution can also be used to approximate the binomial distribution when the expected value, $\lambda = n \cdot p$, is less than 5, as shown in the next example. (The same is true when $n \cdot q < 5$.)

EXAMPLE 6–30   If approximately 2% of the people are left-handed, find the probability that in a room of 200 people, there are exactly 5 people who are left-handed.

Solution   Since $\lambda = n \cdot p$, then $\lambda = (200)(0.02) = 4$. Hence,

$$P(X; \lambda) = \frac{(2.7183)^{-4}(4)^5}{5!} = 0.1563$$

which is verified by the table value of $P(5; 4) = 0.1563$. ◄

The Hypergeometric Distribution

When sampling is done *without* replacement, the binomial distribution does not give exact probabilities, since the trials are not independent. The smaller the size of the population, the less accurate the binomial probabilities will be.

For example, suppose a committee of 4 people were to be selected from 7 women and 5 men. What is the probability that the committee will consist of 3 women and 1 man?

In order to solve this problem, one must find the number of ways a committee of 3 women and 1 man can be selected from 7 women and 5 men. This answer can be found by using combinations, and it is

$$_7C_3 \cdot {}_5C_1 = 35 \cdot 5 = 175$$

Next, one must find the total number of ways a committee of 4 people can be selected from 12 people. Again, by the use of combinations, the answer is

$$_{12}C_4 = 495$$

Finally, the probability of getting a committee of 3 women and 1 man from 7 women and 5 men is

$$P(A) = \frac{175}{495} = \frac{35}{99}$$

The results of the problem can be generalized by using a special probability distribution called the hypergeometric distribution. The **hypergeometric distribution** is a distribution of a variable that has two outcomes when sampling is done without replacement.

The probabilities for the hypergeometric distribution can be calculated by using the formula given next.

Formula for the Hypergeometric Distribution

Given a population with only two types of objects (females and males, defective and nondefective, successes or failures, etc.), such that there are a items of one kind and b items of another kind and $a + b$ equals the total population, the probability $P(A)$ of selecting a sample of size n with X items of type a and $n - X$ items of type b is

$$P(A) = \frac{{}_aC_X \cdot {}_bC_{n-X}}{{}_{(a+b)}C_n}$$

The basis of the formula is that there are ${}_aC_X$ ways of selecting the first type of items, ${}_bC_{n-X}$ ways of selecting the second type of items, and ${}_{(a+b)}C_n$ ways of selecting n items from the entire population.

EXAMPLE 6–31   Ten people apply for a job as assistant manager of a restaurant. Five have completed college and five have not. If the manager selects three applicants at random, find the probability that all three are college graduates.

Solution   Assigning the values to the variables gives

$$a = 5 \text{ college graduates} \qquad n = 3$$
$$b = 5 \text{ nongraduates} \qquad X = 3$$

and $n - X = 0$. Substituting in the formula gives

$$P(A) = \frac{{}_5C_3 \cdot {}_5C_0}{{}_{10}C_3} = \frac{10}{120} = \frac{1}{12} \qquad \blacktriangleleft$$

EXAMPLE 6–32   A recent study found that 4 out of 9 houses were underinsured. If 5 houses are selected from the 9 houses, find the probability that 2 are underinsured.

Solution   In this problem

$$a = 4 \qquad b = 5 \qquad n = 5 \qquad X = 2 \qquad n - X = 3$$

Then,
$$P(A) = \frac{{}_4C_2 \cdot {}_5C_3}{{}_9C_5} = \frac{60}{126} = \frac{10}{21} \qquad \blacktriangleleft$$

In many situations where objects are manufactured and shipped to a company, the company selects a few items and tests them to see whether they are satisfactory or defective. If a certain percentage is defective, the company then can refuse the whole shipment. This procedure saves the time and cost of testing every single item. In order to make the judgment about whether to accept or reject the whole shipment based on a small sample of tests, the company must know the probability of getting a specific number of defective items. To calculate the probability, the company uses the hypergeometric distribution.

EXAMPLE 6–33   A lot of 12 compressor tanks is checked to see whether there are any defective tanks. Three tanks are checked for leaks. If 1 or more of the 3 is defective, the lot is rejected. Find the probability that the lot will be rejected if there are actually 3 defective tanks in the lot.

Solution   Since the lot is rejected if at least 1 tank is found to be defective, it is necessary to find the probability that none are defective and subtract this probability from 1.

Here, $a = 3$, $b = 9$, $n = 3$, $X = 0$; so

$$P(A) = \frac{{}_3C_0 \cdot {}_9C_3}{{}_{12}C_3} = \frac{1 \cdot 84}{220} = 0.382$$

Hence,

$$P(\text{at least one defective}) = 1 - P(\text{no defectives}) = 1 - 0.382 = 0.618$$

There is a 0.618, or 61.8%, probability that the lot will be rejected when 3 of the 12 tanks are defective.   ◄

EXERCISES

6–85. What is the relationship between the multinomial distribution and the binomial distribution?

6–86. (ANS.) Use the multinomial formula and find the probabilities for each.

a. $n = 6$, $X_1 = 3$, $X_2 = 2$, $X_3 = 1$, $p_1 = 0.5$, $p_2 = 0.3$, $p_3 = 0.2$
b. $n = 5$, $X_1 = 1$, $X_2 = 2$, $X_3 = 2$, $p_1 = 0.3$, $p_2 = 0.6$, $p_3 = 0.1$
c. $n = 4$, $X_1 = 1$, $X_2 = 1$, $X_3 = 2$, $p_1 = 0.8$, $p_2 = 0.1$, $p_3 = 0.1$
d. $n = 3$, $X_1 = 1$, $X_2 = 1$, $X_3 = 1$, $p_1 = 0.5$, $p_2 = 0.3$, $p_3 = 0.2$
e. $n = 5$, $X_1 = 1$, $X_2 = 3$, $X_3 = 1$, $p_1 = 0.7$, $p_2 = 0.2$, $p_3 = 0.1$

6–87. The probabilities that a textbook page will have 0, 1, 2, or 3 typographical errors are 0.79, 0.12, 0.07, and 0.02 respectively. If 8 pages are randomly selected, find the probability that 4 will contain no errors, 2 will contain 1 error, 1 will contain 2 errors, and 1 will contain 3 errors.

6–88. The probabilities are 0.35, 0.45, and 0.20 that an airliner will have no violations, 1 violation, or 2 or more violations when it is given a safety inspection. If 6 airliners are inspected, find the probability that 3 will have no violations, 2 will have 1 violation, and 1 will have 2 or more violations.

6–89. When a customer enters a pharmacy, the probability that he or she will have 0, 1, 2, or 3 prescriptions filled are 0.60, 0.25, 0.10, and 0.05, respectively. For a sample of 6 people who enter the pharmacy, find the probability that 2 will have no prescriptions, 2 will have 1 prescription, 1 will have 2 prescriptions, and 1 will have 3 prescriptions.

6–90. If a family has 3 children, the probabilities of 0, 1, 2, or 3 girls are $\frac{1}{8}$, $\frac{3}{8}$, $\frac{3}{8}$, and $\frac{1}{8}$, respectively. If 4 families with 3 children are selected, find the probability that 1 family will have no girls, 1 family will have 1 girl, 1 family will have 2 girls, and 1 family will have 3 girls.

6-91. According to Mendel's theory, if tall and colorful plants are crossed with short and colorless plants, the corresponding probabilities are $\frac{9}{16}, \frac{3}{16}, \frac{3}{16}$, and $\frac{1}{16}$ for tall and colorful, tall and colorless, short and colorful, and short and colorless, respectively. If 8 plants are selected, find the probability that 1 will be tall and colorful, 3 will be tall and colorless, 3 will be short and colorful, and 1 will be short and colorless.

6-92. Find each probability, $P(X; \lambda)$, using Table C in the Appendix.

a. $P(5; 4)$ b. $P(2; 4)$ c. $P(6; 3)$
d. $P(10; 7)$ e. $P(9; 8)$

6-93. If 2% of the containers manufactured by a company are defective, find the probability that in a case of 150 containers, there are 3 defective ones.

6-94. A recent study of robberies for a certain geographical region showed an average of 1 robbery per 20,000 people. In a city of 80,000 people, find the probability of the following.

a. 0 robberies   b. 1 robbery   c. 2 robberies
d. 3 or more robberies

6-95. In a 400-page manuscript, there are 200 randomly distributed misprints. If a page is selected, find the probability that it has 1 misprint.

6-96. A telephone-soliciting company obtains an average of 5 orders per 1000 solicitations. If the company reaches 250 potential customers, find the probability of obtaining at least 2 orders.

6-97. A mail-order company receives an average of 5 orders per 500 solicitations. If it sends out 100 advertisements, find the probability of receiving at least 2 orders.

6-98. A videotape has an average of one defect for every 1000 feet. Find the probability of at least one defect in 3000 feet.

6-99. If 3% of all cars fail the emissions inspection, find the probability that in a sample of 90 cars, 3 will fail.

6-100. The average number of phone inquiries per day at the poison control center is four. Find the probability it will receive five calls on a given day.

6-101. In a batch of 1000 fuses, there are, on average, 5 defective ones. If a random sample of 100 is selected, find the probability of 4 defective fuses.

6-102. In a camping club of 18 members, 9 preferred hoods and 9 preferred hats and earmuffs. On a recent winter outing attended by 6 members, find the probability that exactly 3 members wore earmuffs and hats.

6-103. A bookstore owner examines 5 books from each lot of 25 to check for missing pages. If he finds at least 2 books with missing pages, the entire lot is returned. If, indeed, there are 5 books with missing pages, find the probability that the lot will be returned.

6-104. Shirts are packed at random in two sizes, regular and extra large. Four shirts are selected from a box of 24 and checked for size. If there are 15 regular shirts in the box, find the probability that all 4 will be regular size.

6-105. A shipment of 20 pieces of furniture is rejected if 5 are checked for scratches and at least 1 piece is found to be scratched. Find the probability that the shipment will be returned if there are actually 8 pieces that are scratched.

6-106. A shipment of 24 electric typewriters is rejected if 3 are checked for defects and at least 1 is found to be defective. Find the probability that the shipment will be returned if there are actually 6 typewriters that are defective.

6-6 SUMMARY

Many variables have special probability distributions. In this chapter, several of the most common probability distributions were presented, including the binomial distribution, the multinomial distribution, the Poisson distribution, and the hypergeometric distribution.

The binomial distribution is used when there are only two independent outcomes for an experiment, there is a fixed number of trials, and the probability is the same for each trial. The multinomial distribution is an extension of the binomial distribution and is used when there are three or more outcomes for an experiment. The hypergeometric distribution is used when sampling is done without replacement. Finally, the Poisson distribution is used in special cases when independent events occur over a period of time, area, or volume.

A probability distribution can be graphed, and the mean, variance, and standard deviation can be found. The mathematical expectation can also be calculated for a probability distribution. Expectation can be used in insurance and games of chance.

Important Terms

Binomial distribution
Binomial experiment
Expected value
Hypergeometric distribution
Multinomial distribution
Poisson distribution
Probability distribution
Random variable

Important Formulas

Formula for the mean of a probability distribution:

$$\mu = \Sigma \, X \cdot P(X)$$

Formula for the variance of a probability distribution:

$$\sigma^2 = \Sigma \, [X^2 \cdot P(X)] - \mu^2$$

Formula for expected value:

$$E(X) = \Sigma \, X \cdot P(X)$$

Binomial probability formula:

$$P(X) = \frac{n!}{(n-X)!X!} \cdot p^X \cdot q^{n-X}$$

Formula for the mean of the binomial distribution:

$$\mu = n \cdot p$$

Formulas for the variance and standard deviation of the binomial distribution:

$$\sigma^2 = n \cdot p \cdot q \qquad \sigma = \sqrt{n \cdot p \cdot q}$$

Formula for the multinomial distribution:

$$P(M) = \frac{n!}{X_1! \cdot X_2! \cdot X_3! \cdot \ldots \cdot X_k!} \cdot p_1^{X_1} \cdot p_2^{X_2} \cdot \ldots \cdot p_k^{X_k}$$

Formula for the Poisson distribution:

$$P(X; \lambda) = \frac{e^{-\lambda}\lambda^X}{X!} \qquad \text{where} \qquad X = 0, 1, 2, \ldots$$

Formula for the hypergeometric distribution:

$$P(A) = \frac{{}_aC_X \cdot {}_bC_{n-X}}{{}_{(a+b)}C_n}$$

Review Exercises

For Exercises 6–107 through 6–110, determine whether the distribution represents a probability distribution. If it does not, state why.

6–107.

X	1	2	3	4	5
$P(X)$	$\frac{3}{10}$	$\frac{1}{10}$	$\frac{2}{10}$	$\frac{3}{10}$	$\frac{1}{10}$

6–108.

X	5	10	15	20
$P(X)$	0.5	0.3	0.1	0.4

6–109.

X	100	200	300
$P(X)$	0.3	0.3	0.1

6–110.

X	8	12	16	20
$P(X)$	$\frac{5}{6}$	$\frac{1}{12}$	$\frac{1}{12}$	$\frac{1}{12}$

6–111. The number of rescue calls a helicopter ambulance service receives per 24-hour period is distributed as shown below. Construct a graph for the data.

Number of Calls, X	6	7	8	9	10
Probability, $P(X)$	0.12	0.31	0.40	0.15	0.02

6–112. A study was conducted to determine the number of radios each household has. The data is shown below. Construct a probability distribution and draw a graph for the data.

Number of Radios	Probability
0	0.05
1	0.30
2	0.45
3	0.12
4	0.08

6–113. A box contains 5 pennies, 3 dimes, 1 quarter, and 1 half-dollar. Construct a probability distribution and draw a graph for the data.

6–114. At Tyler's Tie Shop, Tyler found the probabilities that a customer will buy 0, 1, 2, 3, or 4 ties, as shown. Construct a graph for the distribution.

Number of Ties, X	0	1	2	3	4
Probability, $P(X)$	0.30	0.50	0.10	0.08	0.02

6–115. A bank has a drive-through service. The number of customers arriving during a 15-minute period is distributed as shown. Find the mean, variance, and standard deviation for the distribution.

Number of Customers, X	0	1	2	3	4
Probability, $P(X)$	0.12	0.20	0.31	0.25	0.12

6–116. At a small community library, the number of visitors per hour during the day has the distribution shown. Find the mean, variance, and standard deviation for the data.

Number of Visitors, X	8	9	10	11	12
Probability, $P(X)$	0.15	0.25	0.29	0.19	0.12

6–117. During a recent paint sale at Corner Hardware, the number of cans of paint a customer purchased was distributed as shown. Find the mean, variance, and standard deviation of the distribution.

Number of Cans, X	1	2	3	4	5
Probability, $P(X)$	0.42	0.27	0.15	0.10	0.06

6–118. The number of inquiries received per day for a college catalog is distributed as shown. Find the mean, variance, and standard deviation for the data.

Number of Inquiries, X	22	23	24	25	26	27
Probability, $P(X)$	0.08	0.19	0.36	0.25	0.07	0.05

6–119. There are five envelopes in a box. The envelopes contain a penny, a nickel, a dime, a quarter, and a half-dollar. A person selects an envelope. Find the expected value of the draw.

6–120. A person selects a card from a deck. If it is a red card, he wins $1. If it is a black card between or including 2 and 10, he wins $5. If it is a black face card, he wins $10, and if it is a black ace, he wins $100. Find the expectation of the game.

6–121. If 30% of all commuters ride the train to work, find the probability that if ten workers are selected, five will ride the train.

6–122. If 90% of all people between the ages of 30 and 50 drive a car, find these probabilities for a sample of twenty 30–50-year-olds.

a. Exactly 20 drive a car.
b. At least 15 drive a car.
c. At most 15 drive a car.

6–123. If 10% of the people who are given a certain drug experience dizziness, find these probabilities for a sample of 15 people who take the drug.

a. At least 2 people will become dizzy.
b. Exactly 3 people will become dizzy.
c. At most 4 people will become dizzy.

6–124. If 75% of the nursing students are able to pass a drug calculation test, find the mean, variance, and standard deviation of the number of students who pass the test in a sample of 180 nursing students.

6–125. A club has 50 members. If there is a 10% absentee rate per meeting, find the mean, variance, and standard deviation of the number of people who will be absent from each meeting.

6–126. (OPT.) The probabilities that a person will make 0, 1, 2, or 3 errors on an insurance claim are 0.70, 0.20, 0.08, and 0.02, respectively. If 20 claims are selected, find the probability that 12 will contain no errors, 4 will contain 1 error, 3 will contain 2 errors, and 1 will contain 3 errors.

6–127. (OPT.) Before a VCR leaves the factory, it is given a quality control check. The probabilities that a VCR contains 0, 1, or 2 defects are 0.90, 0.06, and 0.04, respectively. In a sample of 12 recorders, find the probability that 8 have no defects, 3 have 1 defect, and 1 has 2 defects.

6–128. (OPT.) In a Christmas display, the probability that all the lights are the same color is 0.50; that 2 colors are used is 0.40; and that 3 or more colors are used is 0.10. If a sample of 10 displays is selected, find the probability that 5 have only 1 color of light, 3 have 2 colors, and 2 have 3 colors.

6–129. (OPT.) If 4% of the population carries a certain genetic trait, find the probability that in a sample of 100 people, there are exactly 8 people who have the trait. Assume the distribution is approximately Poisson.

6–130. (OPT.) Computer Help Hot Line receives, on the average, 6 calls per hour asking for assistance. The distribution is Poisson. For any randomly selected hour, find the probability that the company will receive the following.

a. At least 6 calls
b. Four or more calls
c. At most 5 calls

6–131. (OPT.) The number of boating accidents on Lake Emilie follows a Poisson distribution. The probability of an accident is 0.003. If there are 1000 boats on the lake during a summer month, find the probability that there will be 6 accidents.

6–132. (OPT.) If five cards are drawn from a deck, find the probability that two will be hearts.

6–133. (OPT.) There are 50 automobiles in a used-car lot. Ten are white. If 5 automobiles are selected to be sold at an auction, find the probability that exactly 2 will be white.

6–134. (OPT.) A board of directors consists of seven men and five women. If a slate of three officers is selected, find these probabilities.

a. Exactly two are men.
b. Exactly three are women.
c. Exactly two are women.

COMPUTER APPLICATIONS

1. Write a computer program to compute the probability for a binomial variable. Input n, X, and p.
2. Write a computer program to compute the probability for a Poisson variable. Input X and λ, and let $e = 2.7183$.
3. MINITAB can be used to calculate binomial probabilities. For example, suppose a student randomly guesses at each of the ten multiple-choice questions on an exam. Find the probability that the student gets exactly six correct. Each question has five answer choices, so the probability of guessing a correct response is $\frac{1}{5}$, or 0.2. This calculation can be accomplished by typing the following commands:

```
MTB > PDF 6;
SUBC > BINOMIAL n = 10 p = .2.
```

The computer will print the following:

```
   K        P(X = K)
6.00        0.0055
```

Hence, the answer is 0.0055.

To find the probability the student passes the test by guessing six or more correct, type the following commands:

```
MTB > CDF 5;
SUBC > BINOMIAL 10, .2.
```

The computer will print the following:

```
K          P(X less or = K)
.500            .9936
```

Since 0.9936 is the cumulative frequency up to six, the answer must be subtracted from 1.000 to get

$$1.000 - 0.9936 = 0.0064$$

Use MINITAB to find the solution to Exercise 6–66.

4. MINITAB can be used to simulate binomial experiments with two outcomes. Suppose two people play a tennis tournament consisting of five games, and the winner is the person who wins the most games. If they are evenly matched ($P = 0.5$), simulate five tournaments. This calculation can be accomplished by typing the following commands:

```
MTB > RANDOM 5 C1-C5;
SUBC > BERNOULLI .5.
MTB > PRINT C1-C5
```

The computer will print the following:

Row	C1	C2	C3	C4	C5
1	1	1	1	0	0
2	0	0	0	1	1
3	0	1	0	0	1
4	0	1	0	1	1
5	1	0	0	1	0

The digit 0 means player A won the game, and the digit 1 means player B won the game. Tabulate the results.

Suppose player B is the better of the two and the probability that player B wins is 0.7; repeat the experiment and tabulate the results.

5. MINITAB can be used to calculate Poisson probabilities. For example, suppose that there are an average of 5 defects on each computer disk manufactured. To find the Poisson probability that a disk has 0, 3, 5, or 10 defects, type the following commands:

```
MTB > SET C1
DATA > 0 3 5 10
DATA > END
MTB > PDF C1;
SUBC > POISSON 5.
```

The computer will print the following:

K	P(X = K)
0.00	0.0067
3.00	0.1404
5.00	0.1755
10.00	0.0181

The k values are printed in the left column, and the corresponding probabilities in the right column.

Find the probability of 2, 4, or 8 defects per disk.

"Can you settle a bet, Wellman? Herb says 53% of all statistics are meaningless and I say it's 56%."

Source: Reprinted with permission of *DollarSense*, © 1989.

✎ TEST

Directions: Determine whether each statement is true or false.

1. Random variables are variables whose values are determined by chance.
2. The sum of the probabilities of a probability distribution can be less than 1.
3. The expected value of a random variable can be thought of as the long-run average.
4. The number of courses a student is taking this semester is an example of a continuous random variable.
5. In a binomial experiment, there must be either 1, 2, or 3 outcomes.
6. The mean for a binomial variable can be found by using the formula $\mu = np$.
7. When the multinomial distribution is used, the trials of the experiment must be dependent.
8. The number of plants growing in a specific area can be approximated using the hypergeometric distribution.
9. The binomial distribution can be used to represent continuous random variables.
10. In a probability experiment, p must always equal q.

CHAPTER

7

The Normal Distribution

7–1 INTRODUCTION

Random variables can be either discrete or continuous. Discrete variables and their distributions were explained in Chapter 6. Recall that a discrete variable cannot assume all values between any two given values of the variables. On the other hand, a continuous variable can assume all values between any two given values of the variables. Examples of continuous variables are the heights of adult men, IQ scores, body temperatures of rats, and cholesterol levels of adults. Many continuous variables, such as the examples just mentioned, have distributions that are bell-shaped and are called *approximately normally distributed variables*. For example, if a researcher selects a random sample of 100 adult women, measures their heights, and constructs a histogram, the researcher gets a graph similar to the one shown in Figure 7–1a. Now, if the researcher increases the sample size and decreases the width of the classes, the histograms will look like the ones shown in Figures 7–1b and 7–1c. Finally, if it were possible to measure exactly the heights of all adult females in the United States and plot them, the histogram would approach what is called the *normal distribution,* shown in Figure 7–1d. This distribution is also known as the *bell curve* or the *Gaussian distribution,* named for the German mathematician Carl Friedrich Gauss (1777–1855) who derived its equation.

FIGURE 7–1
Histograms for the Distribution of Heights of Adult Women

Heights of Adult Women

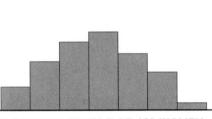

(a) *RANDOM SAMPLE OF 100 WOMEN*

Heights of Adult Women

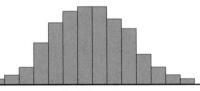

(b) *SAMPLE SIZE INCREASED AND CLASS WIDTH DECREASED*

Heights of Adult Women

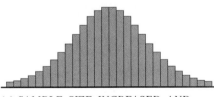

(c) *SAMPLE SIZE INCREASED AND CLASS WIDTH DECREASED FURTHER*

Heights of Adult Women

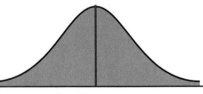

(d) *NORMAL DISTRIBUTION FOR THE POPULATION*

No variable fits the normal distribution perfectly, since the normal distribution is a theoretical distribution. However, the normal distribution can be used to describe many variables, because the deviations from the normal distribution are very small. This concept will be explained further in the next section.

The discovery of the equation for the normal distribution can be traced to three mathematicians. In 1733, the French mathematician Abraham DeMoivre derived an equation for the normal distribution based on the random variation of the number of heads appearing when a large number of coins were tossed. Not realizing any connection with other naturally occurring variables, he showed this formula to only a few friends. About a hundred years later, two mathematicians, Laplace in France and Gauss in Germany, derived the equation of the normal curve independently and without any knowledge of DeMoivre's work. In 1924 Karl Pearson found that DeMoivre had discovered the formula before Laplace or Gauss, although the normal curve has retained Gauss's name.

In this chapter, the properties of the normal distribution will be presented, and then its applications will be discussed. Next, a very important theorem called the *central limit theorem* will be explained. Finally, the chapter will explain how the normal curve distribution can be used as an approximation to other distributions, such as the binomial distribution. Since the binomial distribution is a discrete distribution, a correction for continuity may be employed when the normal distribution is used for its approximation.

7–2 PROPERTIES OF THE NORMAL DISTRIBUTION

In mathematics, curves can be represented by equations. For example, the equation of the circle shown in Figure 7–2 is $x^2 + y^2 = r^2$, where r is the radius. The circle can be used to represent many physical objects, such as a wheel or a gear. Even though it is not possible to manufacture a wheel that is perfectly round, the equation and the properties of the circle can be used to study the many aspects of the wheel, such as area, velocity, and acceleration. In a similar manner, the theoretical curve, called the normal distribution curve, can be used to study many variables that are not perfectly normally distributed but are nevertheless approximately normal.

FIGURE 7–2
Graph of a Circle and an Application

Circle

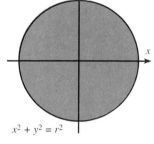

$x^2 + y^2 = r^2$

Wheel

The mathematical equation for the normal distribution is

$$y = \frac{e^{-(X-\mu)^2/2\sigma^2}}{\sigma\sqrt{2\pi}}$$

where

$e \approx 2.718$ ($\approx$ means "approximately equal to")

$\pi = 3.14$

$\mu =$ population mean

$\sigma =$ population standard deviation

This equation may look quite formidable, but in applied statistics, tables are used for specific problems instead of the equation.

Another important aspect in applied statistics is that *the area under the normal distribution curve is more important than the frequencies*. Therefore, when the normal distribution is pictured, the y axis, which indicates the frequencies, is sometimes omitted.

The shape and position of the normal distribution curve depends on two parameters, the mean and the standard deviation. Each normally distributed variable will have its own normal distribution curve, which will depend on the values

FIGURE 7–3
Shapes of Normal
Distributions

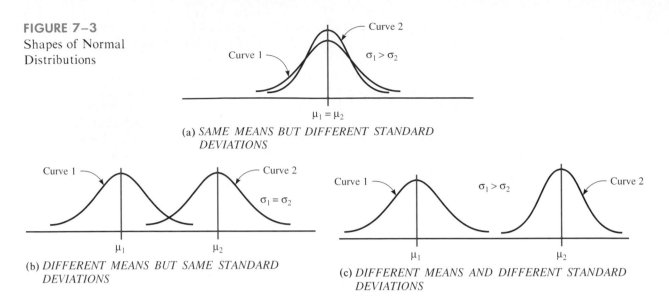

(a) *SAME MEANS BUT DIFFERENT STANDARD DEVIATIONS*

(b) *DIFFERENT MEANS BUT SAME STANDARD DEVIATIONS*

(c) *DIFFERENT MEANS AND DIFFERENT STANDARD DEVIATIONS*

of the variable's mean and standard deviation. Figure 7–3a shows two normal distributions with the same mean values but different standard deviations. The larger the standard deviation, the more dispersed or spread out the distribution is. Figure 7–3b shows two normal distributions with the same standard deviation but with different means. These curves have the same shapes but are located at different positions on the x axis. Figure 7–3c shows two normal distributions with different means and different standard deviations.

The normal distribution is defined formally as follows.

The normal distribution is a continuous, symmetric, bell-shaped distribution of a variable.

The properties of the normal distribution, including those mentioned in the definition, are explained next.

Summary of the Properties of the Theoretical Normal Distribution

1. The normal distribution curve is bell-shaped.
2. The mean, median, and mode are equal and located at the center of the distribution.
3. The normal distribution curve is unimodal (i.e., it has only one mode).
4. The curve is symmetrical about the mean, which is equivalent to saying that its shape is the same on both sides of a vertical line passing through the center.
5. The curve is continuous—i.e., there are no gaps or holes. For each value of X, there is a corresponding value of y.
6. The curve never touches the x axis. Theoretically, no matter how far in either direction the curve extends, it never meets the x axis but gets increasingly closer.

7. The total area under the normal distribution curve is equal to 1.00, or 100%. This feature may seem unusual, since the curve never touches the *x* axis, but this fact can be proven mathematically by using calculus. (The proof is beyond the scope of this textbook.)

8. The area under the normal curve that lies within one standard deviation of the mean is approximately 0.68, or 68%; within two standard deviations, about 0.95, or 95%; and within three standard deviations, about 0.997, or 99.7%. See Figure 7–4, which also shows the area in each region.

FIGURE 7–4
Areas Under the Normal
Distribution Curve

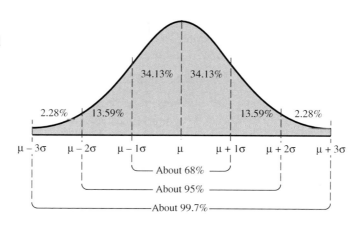

One must know these properties in order to solve problems using applications of the normal distribution.

If the data values are evenly distributed about the mean of the distribution, it is said to be a **symmetrical distribution.** When more of the data values fall to the left or to the right of the mean, the distribution is said to be *skewed*. Figure 7–5a shows a **negatively skewed distribution;** the majority of the data values fall

FIGURE 7–5
Skewed and Normal
Distributions

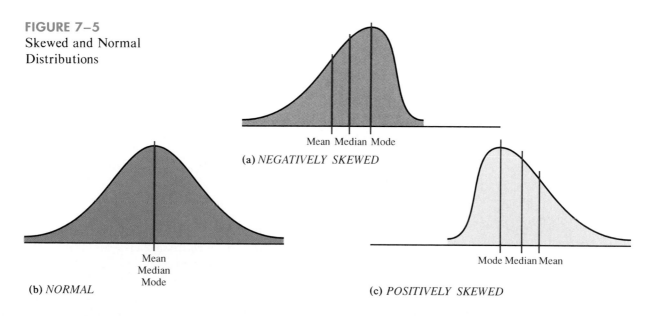

(a) *NEGATIVELY SKEWED*

(b) *NORMAL*

(c) *POSITIVELY SKEWED*

to the right of the mean. Note the positioning of the mean and mode: The mean is to the left of the median, and the mode is to the right of the median. Figure 7–5b shows a normally distributed variable. Figure 7–5c shows a **positively skewed distribution;** the majority of the data values fall to the left of the mean. The mean is to the right of the median, and the mode is to the left of the median. The "tail" of the curve indicates the direction of skewness (right: positive; left: negative). These distributions can be compared with the ones in Figure 3–2, Chapter 3.

7–3 THE STANDARD NORMAL DISTRIBUTION

Since each normally distributed variable has its own mean and standard deviation, as stated earlier, the shape and location of these curves will vary. In practical applications, then, one would have to have a table of areas under the curve for each variable. In order to simplify this situation, statisticians use what is called the standard normal distribution.

The standard normal distribution is a normal distribution with a mean of 0 and a standard deviation of 1.

The standard normal distribution is shown in Figure 7–6.

The values under the curve indicate the proportion of area in each section. For example, the area between the mean and one standard deviation is about 0.3413, or 34.13%.

FIGURE 7–6
Standard Normal
Distribution

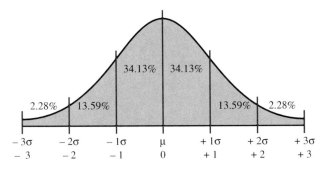

All normally distributed variables can be transformed into the standard normally distributed variable by using the formula for the standard score:

$$z = \frac{\text{value} - \text{mean}}{\text{standard deviation}} \qquad \text{or} \qquad z = \frac{X - \mu}{\sigma}$$

This is the same formula used in Section 3–4. The use of this formula will be explained in the next section.

As was stated earlier, the area under the normal distribution curve is used to solve practical application problems, such as finding the percentage of adult women whose height is between 5 feet 4 inches and 5 feet 7 inches, or finding the probability that a new battery will last longer than four years. Hence, the major emphasis of this section will be to show the procedure for finding the area under the normal distribution curve for any z value. The applications will be shown in the next section. Once the X values are transformed by using the above formula, they are called z values. The **z value** is actually the number of standard deviations that a particular X value is away from the mean. Table E in the Appendix gives the area (to four decimal places) under the standard normal curve for any z value from 0 to 3.09.

Finding Areas Under the Normal Distribution Curve

For the solution of problems using the normal distribution, a four-step procedure is recommended with the use of Procedure Table 6.

STEP 1 Draw a picture.

STEP 2 Shade the area desired.

STEP 3 Find the correct figure in Procedure Table 6; i.e., find the figure in Procedure Table 6 that is similar to the one you've drawn.

STEP 4 Follow the directions given in the appropriate block of Procedure Table 6 to get the desired area.

There are seven basic types of problems and all seven are summarized in Procedure Table 6. Note that this table is presented here as an aid in understanding how to use the normal distribution table and in visualizing the problems. After learning the procedures, one should *not* find it necessary to refer to the procedure table for every problem.

Procedure Table 6

Finding the Area Under the Normal Distribution Curve

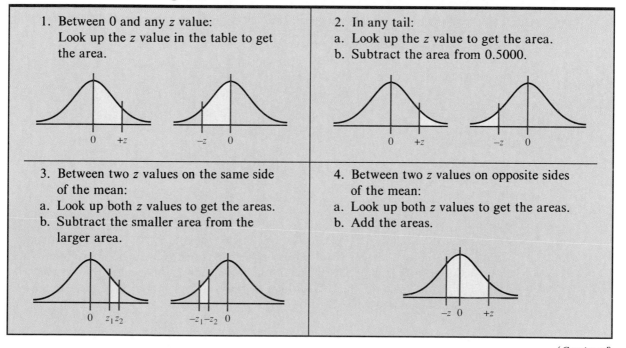

1. Between 0 and any z value:
   Look up the z value in the table to get the area.

2. In any tail:
   a. Look up the z value to get the area.
   b. Subtract the area from 0.5000.

3. Between two z values on the same side of the mean:
   a. Look up both z values to get the areas.
   b. Subtract the smaller area from the larger area.

4. Between two z values on opposite sides of the mean:
   a. Look up both z values to get the areas.
   b. Add the areas.

(Continued)

Procedure Table 6 (continued)

Finding the Area Under the Normal Distribution Curve

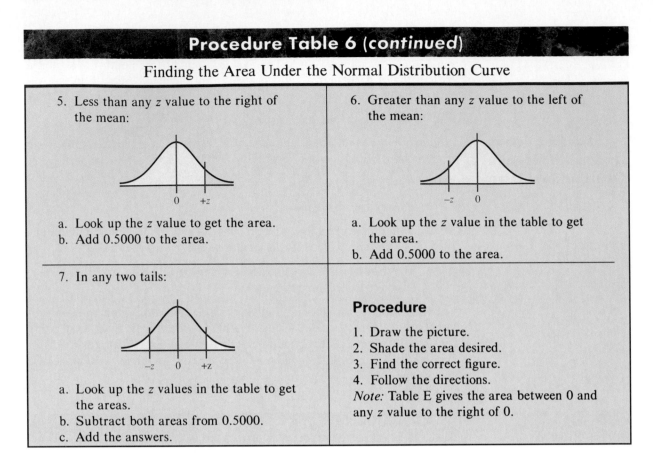

5. Less than any *z* value to the right of the mean:

a. Look up the *z* value to get the area.
b. Add 0.5000 to the area.

6. Greater than any *z* value to the left of the mean:

a. Look up the *z* value in the table to get the area.
b. Add 0.5000 to the area.

7. In any two tails:

a. Look up the *z* values in the table to get the areas.
b. Subtract both areas from 0.5000.
c. Add the answers.

Procedure

1. Draw the picture.
2. Shade the area desired.
3. Find the correct figure.
4. Follow the directions.
Note: Table E gives the area between 0 and any *z* value to the right of 0.

EXAMPLE 7–1   Find the area under the normal distribution curve between $z = 0$ and $z = 2.34$.

Solution   Draw the figure and represent the area as shown in Figure 7–7.

FIGURE 7–7
Area Under the
Standard Normal Curve
for Example 7–1

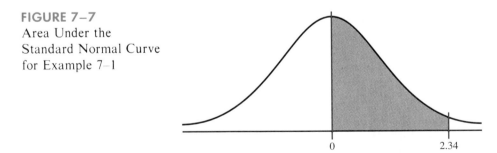

Since Table E gives the area between 0 and any *z* value to the right of 0, one need only look up the *z* value in the table. Find 2.3 in the left column and 0.04 in the top row. The value where the column and row meet in the table is the answer, 0.4904. See Figure 7–8. Hence, the area is 0.4904, or 49.04%.

FIGURE 7–8
Using Table E in the
Appendix for Example
7–1

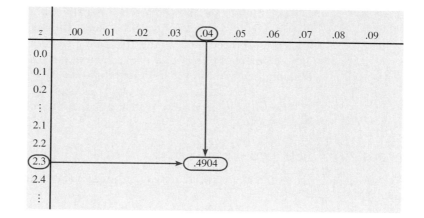

EXAMPLE 7–2   Find the area between $z = 0$ and $z = 1.5$.

Solution   Draw the figure and represent the area as shown in Figure 7–9.

FIGURE 7–9
Area Under the
Standard Normal Curve
for Example 7–2

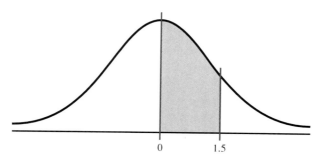

Find the area in Table E by finding 1.5 in the left column and .00 in the top row. The value is 0.4332, or 43.32%.   ◀

Next, one must be able to find the area for values that are not in Table E. This solution is accomplished by using the properties of the normal distribution described in Section 7–2.

EXAMPLE 7–3   Find the area between $z = 0$ and $z = -1.75$.

Solution   Represent the area as shown in Figure 7–10.

FIGURE 7–10
Area Under the
Standard Normal Curve
for Example 7–3

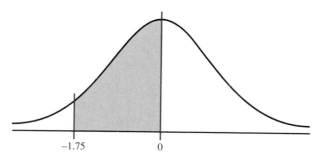

Table E does not give the area for negative values of *z*. But since the normal distribution is symmetric about the mean, the area to the left of the mean (in this case, 0) is the same as the area to the right of the mean. Hence, one needs only to look up the area for $z = +1.75$, which is 0.4599, or 45.99%. This solution is summarized in block 1 in Procedure Table 6. ◄

Remember that area is always a positive number, even though the z value may be negative.

EXAMPLE 7–4 Find the area to the right of $z = 1.11$.

Solution Draw the figure and represent the area as shown in Figure 7–11.

FIGURE 7–11
Area Under the
Standard Normal Curve
for Example 7–4

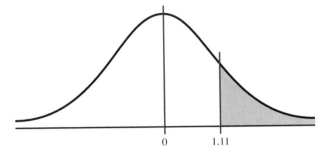

The required area is in the tail of the curve. Since Table E gives the area between $z = 0$ and $z = 1.11$, first find that area. Then subtract this value from 0.5000, since half of the area under the curve is to the right of $z = 0$. See Figure 7–12.

FIGURE 7–12
Finding the Area in the
Tail of the Curve
(Example 7–4)

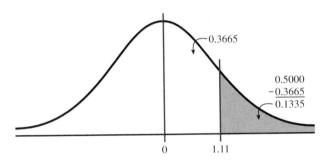

The area between $z = 0$ and $z = 1.11$ is 0.3665, and the area to the right of $z = 1.11$ is 0.1335, or 13.35%, obtained by subtracting 0.3665 from 0.5000. ◄

EXAMPLE 7–5   Find the area to the left of $z = -1.93$.

Solution   The desired area is shown in Figure 7–13.

FIGURE 7–13
Area Under the
Standard Normal Curve
for Example 7–5

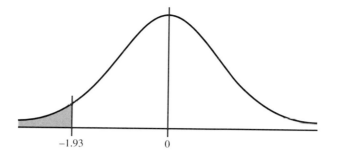

Again, Table E gives the area for positive z values. But from the symmetric property of the normal distribution, the area to the left of -1.93 is the same as the area to the right of $z = +1.93$, as shown in Figure 7–14.

FIGURE 7–14
Comparison of Areas to
the Right of $+1.93$ and
to the Left of -1.93
(Example 7–5)

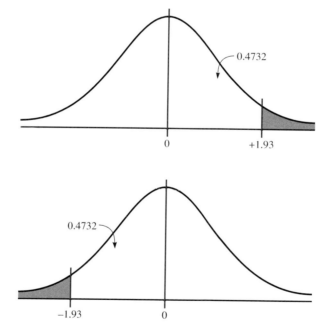

Now, one need only find the area between 0 and $+1.93$ and subtract it from 0.5000, as shown:

$$\begin{array}{r} 0.5000 \\ -\underline{0.4732} \\ 0.0268, \text{ or } 2.68\% \end{array}$$

This procedure is summarized in block 2 of Procedure Table 6.   ◄

EXAMPLE 7–6   Find the area between $z = 2.00$ and $z = 2.47$.

Solution   The desired area is shown in Figure 7–15.

FIGURE 7–15
Area Under the Curve
for Example 7–6

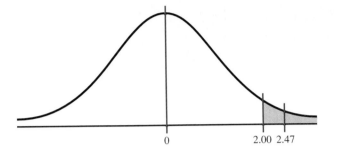

For this situation, look up the area from $z = 0$ to $z = 2.47$ and the area from $z = 0$ to $z = 2.00$. Then subtract the two areas, as shown in Figure 7–16.

FIGURE 7–16
Finding the Area Under
the Curve for Example
7–6

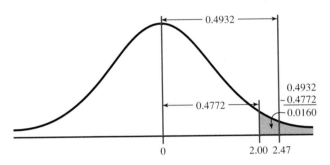

The area between $z = 0$ and $z = 2.47$ is 0.4932. The area between $z = 0$ and $z = 2.00$ is 0.4772. Hence, the desired area is $0.4932 - 0.4772 = 0.0160$, or 1.60%. This procedure is summarized in block 3 of Procedure Table 6.   ◄

Two things should be noted here. First, the areas, not the z values, are subtracted. Subtracting the z values will yield an incorrect answer. Second, the procedure in Example 7–6 is used when both z values are on the same side of the mean.

EXAMPLE 7–7   Find the area between $z = -2.48$ and $z = -0.83$.

Solution   The desired area is shown in Figure 7–17.

FIGURE 7–17
Area Under the Curve
for Example 7–7

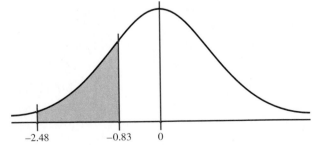

The area between $z = 0$ and $z = -2.48$ is 0.4934. The area between $z = 0$ and $z = -0.83$ is 0.2967. Subtracting yields $0.4934 - 0.2967 = 0.1967$, or 19.67%. This solution is summarized in block 3 of Procedure Table 6. ◀

EXAMPLE 7–8   Find the area between $z = +1.68$ and $z = -1.37$.

Solution   The desired area is shown in Figure 7–18.

FIGURE 7–18
Area Under the Curve
for Example 7–8

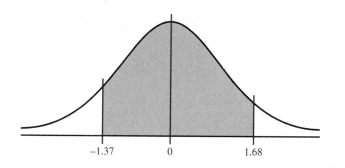

Now, since the two areas are on opposite sides of $z = 0$, one must find both areas and add them. The area between $z = 0$ and $z = 1.68$ is 0.4535. The area between $z = 0$ and $z = -1.37$ is 0.4147. Hence, the total area between $z = -1.37$ and $z = +1.68$ is $0.4535 + 0.4147 = 0.8682$, or 86.82%.

This type of problem is summarized in block 4 of Procedure Table 6. ◀

EXAMPLE 7–9   Find the area to the left of $z = 1.99$.

Solution   The desired area is shown in Figure 7–19.

FIGURE 7–19
Area Under the Curve
for Example 7–9

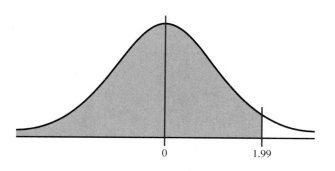

Since Table E only gives the area between $z = 0$ and $z = 1.99$, one must add 0.5000 to the table area, since 0.5000 of the total area lies to the left of $z = 0$. The area between $z = 0$ and $z = 1.99$ is 0.4767, and the total area is $0.4767 + 0.5000 = 0.9767$, or 97.67%.

This solution is summarized in block 5 of Procedure Table 6. ◀

The same procedure is used when the z value is to the left of the mean, as shown in the next example.

EXAMPLE 7–10   Find the area to the right of $z = -1.16$.

Solution   The desired area is shown in Figure 7–20.

FIGURE 7–20
Area Under the Curve
for Example 7–10

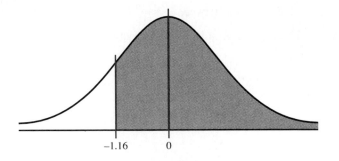

-1.16 0

The area between $z = 0$ and $z = -1.16$ is 0.3770. Hence, the total area is
$0.3770 + 0.5000 = 0.8770$, or 87.70%.
This type of problem is summarized in block 6 of Procedure Table 6.   ◄

The final type of problem is that of finding the area in two tails. This problem
is solved by finding the area in each tail and adding them, as shown in the next
example.

EXAMPLE 7–11   Find the area to the right of $z = +2.43$ and to the left of $z = -3.01$.

Solution   The desired area is shown in Figure 7–21.

FIGURE 7–21
Area Under the Curve
for Example 7–11

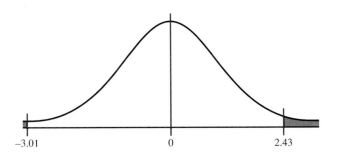

-3.01 0 2.43

The area to the right of 2.43 is $0.5000 - 0.4925 = 0.0075$. The area to
the left of $z = -3.01$ is $0.5000 - 0.4987 = 0.0013$. The total area, then, is
$0.0075 + 0.0013 = 0.0088$, or 0.88%.
This solution is summarized in block 7 of Procedure Table 6.   ◄

**The Normal
Distribution Curve
as a Probability
Distribution Curve**

The normal distribution curve can be used as a probability distribution curve for
normally distributed variables. Recall that the normal distribution is a *continuous distribution,* as opposed to a discrete probability distribution, as explained
in Chapter 6. The fact that it is continuous means that there are no gaps in the
curve. In other words, for every z value on the x axis, there is a corresponding
height, or frequency value.

However, as was stated earlier, the area under the curve is more important
than the frequencies. This area corresponds to a *probability.* That is, if it were
possible to select any z value at random, the probability of choosing one, say

between 0 and 2.00, would be the same as the area under the curve between 0 and 2.00. In this case, the area is 0.4772. Therefore, the probability of selecting any z value between 0 and 2.00 is 0.4772. The problems involving probability are solved in the same manner as the previous examples involving areas in this section. For example, if the problem is to find the probability of selecting a z value between 2.25 and 2.94, it is solved by using the method shown in block 3 of Procedure Table 6.

For probabilities, a special notation is used. For example, if the problem is to find the probability of any z value between 0 and 2.32, this probability is written as $P(0 < z < 2.32)$.

EXAMPLE 7–12 Find the probability for each.

a. $P(0 < z < 2.32)$
b. $P(z < 1.65)$
c. $P(z > 1.91)$

Solution a. $P(0 < z < 2.32)$ means to find the area under the normal distribution curve between 0 and 2.32. This area is found by looking up the area in Table E corresponding to $z = 2.32$. It is 0.4898, or 48.98%. The area is shown in Figure 7–22.

FIGURE 7–22
Area Under the Curve for Part a of Example 7–12

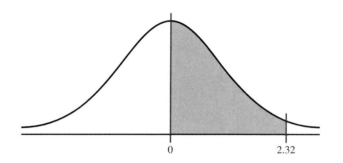

b. $P(z < 1.65)$ is represented in Figure 7–23.

FIGURE 7–23
Area Under the Curve for Part b of Example 7–12

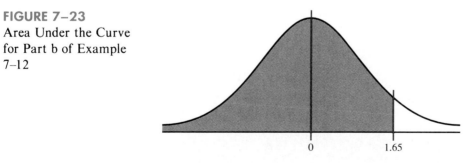

First, find the area between 0 and 1.65 in Table E. Then add it to 0.5000 to get $0.4505 + 0.5000 = 0.9505$, or 95.05%.

c. $P(z > 1.91)$ is shown in Figure 7–24.

FIGURE 7–24
Area Under the Curve
for Part c of Example
7–12

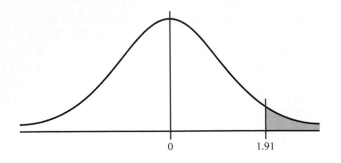

Since this area is a tail area, find the area between 0 and 1.91 and subtract it from 0.5000. Hence, $0.5000 - 0.4719 = 0.0281$. ◄

Sometimes, one must find a specific z value for a given area under the normal distribution. The procedure is to work backward, using Table E.

EXAMPLE 7–13   Find the z value such that the area under the normal distribution curve between 0 and the z value is 0.2123.

Solution   Draw the figure. The area is shown in Figure 7–25.

FIGURE 7–25
Area Under the Curve
for Example 7–13

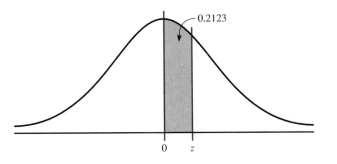

Next, find the area in Table E, as shown in Figure 7–26. Then read the correct z value in the left column as 0.5 and in the top row as 0.06 and add these two values to get 0.56.

FIGURE 7–26
Finding the z Value from
Table E (Example 7–13)

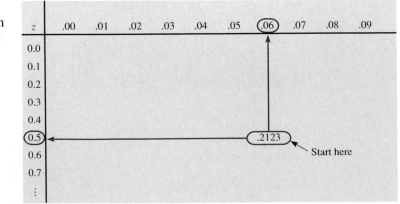

Finding the area under the standard normal distribution curve is the first step in solving a wide variety of practical applications in which the variables are normally distributed. Some of these applications will be presented in Section 7–4.

EXERCISES

7–1. What are the characteristics of the normal distribution?

7–2. Why is the normal distribution important in statistical analysis?

7–3. What is the total area under the normal distribution curve?

7–4. What percentage of the area falls below the mean? Above the mean?

7–5. What percentage of the area under the normal distribution curve falls within one standard deviation above and below the mean? Two standard deviations? Three standard deviations?

For Exercises 7–6 through 7–25, find the area under the normal distribution curve.

7–6. Between $z = 0$ and $z = 1.97$.

7–7. Between $z = 0$ and $z = 0.56$.

7–8. Between $z = 0$ and $z = -0.48$.

7–9. Between $z = 0$ and $z = -2.07$.

7–10. To the right of $z = 1.02$.

7–11. To the right of $z = 0.23$.

7–12. To the left of $z = -0.42$.

7–13. To the left of $z = -1.43$.

7–14. Between $z = 1.23$ and $z = 1.90$.

7–15. Between $z = 0.79$ and $z = 1.28$.

7–16. Between $z = -0.87$ and $z = -0.21$.

7–17. Between $z = -1.56$ and $z = -1.83$.

7–18. Between $z = 0.24$ and $z = -1.12$.

7–19. Between $z = 2.47$ and $z = -1.03$.

7–20. To the left of $z = +1.22$.

7–21. To the left of $z = 2.16$.

7–22. To the right of $z = -1.92$.

7–23. To the right of $z = -0.18$.

7–24. To the left of $z = -2.15$ and to the right of $z = 1.62$.

7–25. To the right of $z = 1.98$ and to the left of $z = -0.59$.

In Exercises 7–26 through 7–39, find probabilities for each, using the standard normal distribution.

7–26. $P(0 < z < 1.69)$

7–27. $P(0 < z < 0.67)$

7–28. $P(-1.23 < z < 0)$

7–29. $P(-1.57 < z < 0)$

7–30. $P(z > 2.59)$

7–31. $P(z > 2.83)$

7–32. $P(z < -1.77)$

7–33. $P(z < -1.51)$

7–34. $P(-0.05 < z < 1.10)$

7–35. $P(-2.46 < z < 1.74)$

7–36. $P(1.32 < z < 1.51)$

7–37. $P(1.46 < z < 2.97)$

7–38. $P(z > -1.39)$

7–39. $P(z < 1.42)$

For Exercises 7–40 through 7–45, find the z value that corresponds to the given area.

7–40.

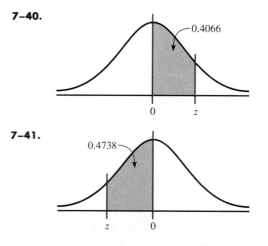

7–41.

7–42.

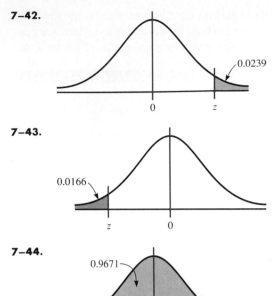

7–43.

7–44.

7–45.

*7–46. Find a z value to the right of the mean so that 53.98% of the distribution lies to the left of it.

*7–47. Find a z value to the left of the mean so that 96.86% of the area lies to the right of it.

*7–48. Find two z values so that 40% of the middle area is bounded by them.

*7–49. Find two z values, one positive and one negative, so that the areas in the two tails total the following values.

   a. 5%
   b. 10%
   c. 1%

*7–50. Find the z values that correspond to the 90th percentile, 80th percentile, 50th percentile, and 5th percentile.

*7–51. Draw a normal distribution with a mean of 100 and a standard deviation of 15.

*7–52. Find the equation for the standard normal distribution by substituting 0 for μ and 1 for σ in the equation

$$y = \frac{e^{-(X-\mu)^2/2\sigma^2}}{\sigma\sqrt{2\pi}}$$

*7–53. Graph the standard normal distribution by using the formula derived in Exercise 7–52. Let $\pi \approx 3.14$ and $e \approx 2.718$. Use X values of $-2, -1.5, -1, -0.5, 0, 0.5, 1, 1.5, 2$.

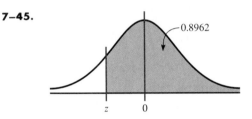

7–4 APPLICATIONS OF THE NORMAL DISTRIBUTION

The standard normal distribution curve can be used to solve a wide variety of practical problems. The only requirement is that the variable be normally distributed or approximately normally distributed. There are several mathematical tests that can be used to determine whether a variable is normally distributed. However, those tests are not included here; and for all the problems presented in this chapter, one can assume that the variable is normally or approximately normally distributed.

In order to solve problems by using the standard normal distribution, one must transform the original variable into a standard normal distribution variable by using the formula

$$z = \frac{\text{value} - \text{mean}}{\text{standard deviation}} \qquad \text{or} \qquad z = \frac{X - \mu}{\sigma}$$

This is the same formula presented in Section 3–4. This formula transforms the values of the variable into standard units or z values. Once the variable is transformed, then Procedure Table 6 and Table E in the Appendix can be used to solve problems.

For example, suppose that the scores for an IQ test are normally distributed, have a mean of 100, and have a standard deviation of 15. When the IQ scores are transformed into z values, the two distributions coincide, as shown in Figure 7-27. (Recall that the z distribution has a mean of 0 and a standard deviation of 1.)

FIGURE 7-27
IQ Values and Their
Corresponding
z Values

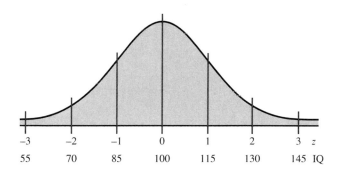

In order to solve the application problems in this section, one need only transform the values of the variable into z values and then use Procedure Table 6 and Table E in the Appendix, as shown in the next examples.

EXAMPLE 7-14   If the scores for an IQ test have a mean of 100 and a standard deviation of 15, find the percentage of IQ scores that will fall below 112.

Solution   **STEP 1**   Draw the figure and represent the area, as shown in Figure 7-28.

FIGURE 7-28
Area Under the Curve
for Example 7-14

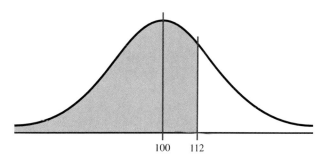

STEP 2   Find the z value corresponding to an IQ score of 112.

$$z = \frac{X - \mu}{\sigma} = \frac{112 - 100}{15} = \frac{12}{15} = 0.8$$

Hence, 112 is 0.8 standard deviation above the mean of 100, as shown for the z distribution in Figure 7-29.

FIGURE 7-29
Area and z Values for
Example 7-14

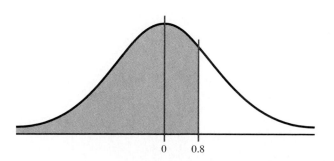

STEP 3 Find the area using Table E in the Appendix. The area between $z = 0$ and $z = 0.8$ is 0.2881. Since the area under the curve to the left of $z = 0.8$ is desired, it is necessary to add 0.5000 to 0.2881, to get 0.7881 (i.e., $0.5000 + 0.2881 = 0.7881$). Therefore, 78.81% of the IQ scores fall below 112. ◄

EXAMPLE 7–15 A study on recycling shows that in a certain city, each household accumulates an average 14 pounds of newspapers each month to be recycled. The standard deviation is 2 pounds. If a household is selected at random, find the probability it will accumulate the following.

a. Between 13 and 17 pounds of newspapers for one month.
b. More than 16.2 pounds of newspapers for one month.

Assume that the distribution is approximately normally distributed.

Solution a. **STEP 1** Draw the figure and represent the area. See Figure 7–30.

FIGURE 7–30
Area Under the Curve for Part a of Example 7–15

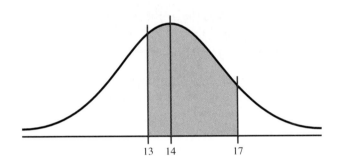

STEP 2 Find the two z values.

$$z_1 = \frac{X - \mu}{\sigma} = \frac{13 - 14}{2} = -\frac{1}{2} = -0.5$$

$$z_2 = \frac{X - \mu}{\sigma} = \frac{17 - 14}{2} = \frac{3}{2} = 1.5$$

STEP 3 Find the appropriate area, using Table E in the Appendix. The area between $z = 0$ and $z = -0.5$ is 0.1915. The area between $z = 0$ and $z = 1.5$ is 0.4332. The total area is then found by adding 0.1915 and 0.4332, to get 0.6247 (i.e., $0.1915 + 0.4332 = 0.6247$). See Figure 7–31.

Hence, the probability that a household accumulates between 13 and 17 pounds of newspapers is 0.6247, or 62.47%.

FIGURE 7–31
Area and z Values for Part a of Example 7–15

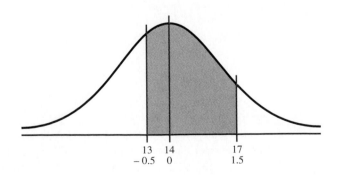

b. **STEP 1** Draw the figure and represent the area, as shown in Figure 7–32.

FIGURE 7–32
Area Under the Curve for Part b of Example 7–15

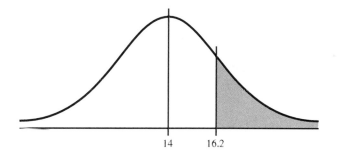

14   16.2

STEP 2 Find the z value for 16.2.

$$z = \frac{X - \mu}{\sigma} = \frac{16.2 - 14}{2} = \frac{2.2}{2} = 1.1$$

STEP 3 Find the appropriate area. The area between $z = 0$ and $z = 1.1$ obtained from Table E is 0.3643. Since the desired area is in the right tail, 0.3643 is subtracted from 0.5000.

$$0.5000 - 0.3643 = 0.1357$$

Hence, the probability that a household will accumulate more than 16.2 pounds of newspapers is 0.1357, or 13.57%.   ◄

The normal distribution can also be used to answer questions of "How many?" This application is shown in the next example.

EXAMPLE 7–16 A standardized achievement test has a mean of 50 and a standard deviation of 10. The scores are normally distributed. If the test is administered to 800 selected people, approximately how many will score between 48 and 62?

Solution First, find the area under the normal distribution curve between 48 and 62. See Figure 7–33.

FIGURE 7–33
Area Under the Curve for Example 7–16

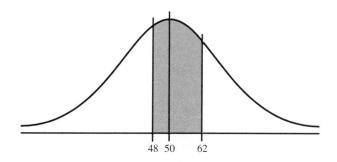

48 50   62

The z values are

$$z = \frac{48 - 50}{10} = -0.2 \quad \text{and} \quad z = \frac{62 - 50}{10} = 1.2$$

The area between 0 and -0.2 is 0.0793, and the area between 0 and 1.2 is 0.3849. When the z values are on opposite sides of the mean, the desired area is found by adding the two values. Therefore, $0.0793 + 0.3849 = 0.4642$, or 46.42%. If 800 people take the exam, approximately 46.42% will score between 48 and 62.

In order to find how many individuals will score between 48 and 62, multiply 800 by 46.42%. Hence, 800 · 0.4642 = 371.36, or approximately 371 people. ◄

Note: For problems like the one shown in Example 7–16, be sure to change the percentage to a decimal before multiplying. Also, round the answer to the nearest whole number, since it is not possible to have 371.36 people.

The normal distribution can also be used to find specific data values for given percentages. This application is shown in the next example.

EXAMPLE 7–17   An exclusive college desires to accept only the top 10% of all graduating seniors on the basis of the results of a national placement test. This test has a mean of 500 and a standard deviation of 100. Find the cutoff score for the exam.

Solution   Since the test scores are normally distributed, the test value (*X*) that cuts off the upper 10% of the area under the normal distribution curve is desired. This area is shown in Figure 7–34.

FIGURE 7–34
Area Under the Curve
for Example 7–17

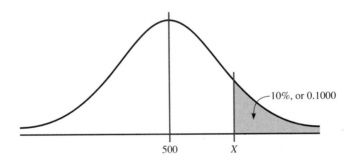

This problem can be solved by working backward. The steps are shown next.

STEP 1   Subtract 0.1000 from 0.5000 to get the area under the normal distribution between 500 and *X:* 0.5000 − 0.1000 = 0.4000.

STEP 2   Find the *z* value that corresponds to an area of 0.4000. This value is found by looking up 0.4000 in the area portion of Table E. If the specific value cannot be found, use the closest value—in this case, 0.3997, as shown in Figure 7–35. The corresponding *z* value is 1.28.

FIGURE 7–35
Finding the *z* Value from
Table E (Example 7–17)

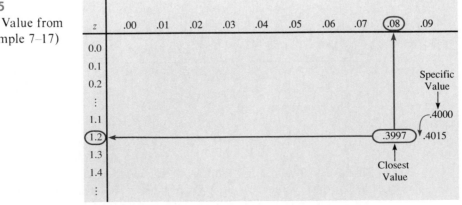

STEP 3 Substitute in the formula $z = (X - \mu)/\sigma$ and solve for X.

$$1.28 = \frac{X - 500}{100}$$

$$X = (1.28)(100) + 500 = 628$$

The score of 628 should be used as a cutoff. Anybody scoring below 628 should not be admitted. ◄

When one must solve for X, the following formula can be used:

$$X = z \cdot \sigma + \mu$$

EXAMPLE 7–18 For a medical study, a researcher wishes to select people in the middle 60% of the population based on blood pressure. If the mean systolic blood pressure is 120 and the standard deviation is 8, find the upper and lower readings that would qualify people to participate in the study.

Solution Assuming that blood pressure readings are normally distributed, the cutoff points are as shown in Figure 7–36.

FIGURE 7–36
Area Under the Curve
for Example 7–18

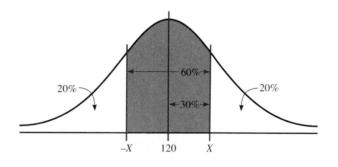

Note that two values are needed, one positive and one negative. The value to the right of the mean will be found first. The closest z value for an area of 0.3000 is 0.84. Substituting in the formula $X = z\sigma + \mu$, one gets

$$X = z\sigma + \mu = (0.84)(8) + 120 = 126.72$$

On the other side, $z = -0.84$; hence,

$$X = (-0.84)(8) + 120 = 113.28$$

Therefore, the middle 60% will have blood pressure readings of $113.28 < X < 126.72$. ◄

As shown in this section, the normal distribution is a useful tool in answering many questions about variables that are normally or approximately normally distributed.

Speaking of **Statistics**

A Low-fat Walk Away from Risk of Heart Disease

By Robert Woessner
USA TODAY

Eating less fat and taking a brisk, 45-minute daily walk can—in less than a month—virtually eliminate what researchers have called a "deadly quartet" of heart-disease risk factors, new research shows.

Millions of Americans—most of them men with normal cholesterol—carry these four danger factors: high blood pressure, obesity, elevated insulin and high triglycerides.

"Most doctors check some of those, but most don't look at triglycerides and no one looks at insulin," says UCLA's R. James Barnard, whose study is in the current *American Journal of Cardiology.* Sponsor was the Pritikin Longevity Center.

Barnard followed 40 men and 32 women, ages 21–78, 26% of whom had all four risk factors. All ate diets with about 10% fat—much lower than the average American's 37%—and cut out red meat, alcohol, tobacco, caffeine. They also took daily walks and exercise classes five days a week.

Findings after 26 days:

- Significant improvements in all risk factors; only 5% still had risk factors.
- Diabetics cut all risk factors most; some were able to go off medication. Others with hypertension did the same.
- Subjects lost about 10 pounds each.

While risk factors faded fast, "You probably get back into risk as quickly as you got out of it," Barnard says.

Test subjects were in supervised exercise classes, but Barnard says, "A good, brisk 45-minute daily walk (plus cutting fat) is probably enough."

Is This Diet Too Lean and Mean?

Is a diet with as little as 10% of its calories from fat realistic for most people?

Study sponsor Robert Pritikin thinks so. "It is the natural diet of man," he says.

But most health groups recommend about 30%.

"A 10% fat diet means a ton of fruits and vegetables, cereals and breads; you'll feel bloated and have a lot of gas," says Kim Galeaz Gioe, American Dietetic Association. "It means lean fish, poultry with no skin, very little pork or beef, all nonfat dairy products."

New fat-free products may make it easier, but with diets below 25% fat, "compliance falls way off," says John Foreyt, Nutrition Research Clinic, Baylor College of Medicine, Houston.

Source: Copyright 1992, USA TODAY. Reprinted with permission.

This study contains many variables, such as cholesterol level and blood pressure. List the variables and decide which ones you feel would be approximately normally distributed and which ones you consider would not be normal.

EXERCISES

For all exercises, assume that the variables are normally distributed.

7-54. If the average weight of a bag of pretzels is 16 ounces, with a standard deviation of 0.3 ounce, find the probability that a bag of pretzels will contain between 16 and 16.2 ounces.

7-55. If the mean salary of computer technicians in the United States is $32,550, and the standard deviation is $2000, find these probabilities for a randomly selected computer technician.

a. The technician earns more than $35,000.
b. The technician earns less than $28,000.

7-56. The Speedmaster IV automobile gets an average 22.0 miles per gallon in the city. The standard deviation is 3 miles per gallon. Find the probability that on any given day, the automobile will get more than 26 miles per gallon when driven in the city.

7-57. If the mean salary of high school teachers in the United States is $29,835, and the standard deviation is $3000, find these probabilities for a randomly selected teacher.

a. The teacher earns more than $35,000.
b. The teacher earns less than $25,000.

7-58. A recent survey conducted by a television magazine found that the average person watched 18.6 hours of television per week. The standard deviation was 2.3 hours. If a person is selected at random, find these probabilities.

a. The person watched less than 15 hours per week.
b. The person watched more than 22 hours per week.

7-59. A survey found that people keep their television sets an average of 4.8 years. The standard deviation is 0.89 year. If a person decides to buy a new television set, find the probability that the person has owned the old set for the following amount of time.

a. Less than 2.5 years
b. Between 3 and 4 years
c. More than 4.2 years

7-60. In a recent survey, the average age of CEOs was found to be 56 years. The standard deviation was 4 years. If a CEO is selected at random, find the probability that the individual is in the following age range.

a. Between 53 and 59 years old
b. Between 58 and 63 years old
c. Between 50 and 55 years old

7-61. The average life of a brand of automobile tires is 30,000 miles, with a standard deviation of 2000 miles. If a tire is selected and tested, find the probability that it will have the following lifetime.

a. Between 25,000 and 28,000 miles
b. Between 27,000 and 32,000 miles
c. Between 31,500 and 33,500 miles

7-62. The average time a person spends at the West Newton Zoo is 62 minutes. The standard deviation is 12 minutes. If a visitor is selected at random, find the probability that he or she will spend the following time at the zoo.

a. At least 82 minutes
b. At most 50 minutes

7-63. The average time for a courier to travel from Pittsburgh to Harrisburg is 200 minutes, and the standard deviation is 10 minutes. If one of these trips is selected at random, find the probability that the courier will have the following travel time.

a. At least 180 minutes
b. At most 205 minutes

7-64. The average amount of snow per season in Trafford is 44 inches. The standard deviation is 6 inches. Find the probability that next year Trafford will receive the following amount of snowfall.

a. At most 50 inches of snow
b. At least 53 inches of snow

7-65. The average waiting time for a drive-in window at a local bank is 9.2 minutes, with a standard deviation of 2.6 minutes. When a customer arrives at the bank, find the probability that the customer will have to wait the following time.

a. Between 5 and 10 minutes
b. Less than 6 minutes or more than 9 minutes

7-66. The average time it takes college freshmen to complete the Mason Basic Reasoning Test is 24.6 minutes. The standard deviation is 5.8 minutes. Find these probabilities.

a. It will take a student between 15 and 30 minutes to complete the test.
b. It will take a student less than 18 minutes or more than 28 minutes to complete the test.

7–67. A brisk walk at 4 miles per hour burns an average of 300 calories per hour. If the standard deviation of the distribution is 8 calories, find the probability that a person who walks one hour at the rate of 4 miles per hour will burn the following calories.

a. More than 280 calories
b. Less than 293 calories
c. Between 285 and 320 calories

7–68. During September, the average temperature of Laurel Lake is 64.2°, and the standard deviation is 3.2°. For a randomly selected day, find the probability that the temperature will be as follows:

a. Above 62°   c. Between 65° and 68°
b. Below 67°

7–69. If the systolic blood pressure for a certain group of obese people has a mean of 132 and a standard deviation of 8, find the probability that a randomly selected person will have the following blood pressure.

a. Above 130   c. Between 131 and 136
b. Below 140

7–70. In order to qualify for letter sorting, applicants are given a speed-reading test. The scores are normally distributed, with a mean of 80 and a standard deviation of 8. If only the top 15% of the applicants are selected, find the cutoff score.

7–71. The scores on an IQ test have a mean of 100 and a standard deviation of 15. If a personnel manager wishes to select from the top 75% of applicants who take the test, find the cutoff score.

7–72. For an educational study, a volunteer must place in the middle 50% in IQ. If the mean for the population is 100 and the standard deviation is 15, find the two limits (upper and lower) for IQ levels that would enable a volunteer to participate in the study.

7–73. A contractor decided to build homes that will include the middle 80% of the market. If the average size (in square feet) of homes built is 1810 (Congressional Research Service, in M. D. Shook and R. L. Shook, *The Book of Odds,* Penguin Books, 1991, p. 15), find the maximum and minimum sizes of the homes the contractor should build. Assume that the standard deviation is 92 square feet.

7–74. If the average price of a new home is $145,500 (Congressional Research Service, in M. D. Shook and R. L. Shook, *The Book of Odds,* Penguin Books, 1991, p. 15), find the maximum and minimum prices of the houses a contractor will build if the contractor wants to include the middle 80% of the market. Assume that the standard deviation of prices is $1500.

7–75. An athletic association wants to sponsor a bicycle race. The average time it takes to ride the course is 62.5 minutes, with a standard deviation of 5.8 minutes. If the association decides to include only the top 25% of the riders, what should this cutoff time be in the tryout run?

7–76. In order to help students improve their reading, a school district decides to implement a reading program. It is to be administered to the bottom 5% of the students in the district, based on the scores on a reading achievement exam. If the average score for the students in the district is 122.6, find the cutoff score that will make a student eligible for the program. The standard deviation is 18.

7–77. An automobile dealer finds that the average price of a previously owned automobile is $8256. He decides to sell automobiles that will appeal to the middle 60% of the market in terms of price. Find the maximum and minimum prices of the automobiles that the dealer will sell. The standard deviation is $1150.

7–78. A small publisher wishes to publish self-improvement books. After a survey of the market, the publisher finds that the average cost of the type of book that she wishes to publish is $12.80. If she wants to price her books to sell in the middle 70% range, what should the maximum and minimum prices of the books be? The standard deviation is $0.83.

7–79. A special enrichment program in mathematics is to be offered to the top 12% of students in a school district. A standardized mathematics achievement test given to all students has a mean of 57.3 and a standard deviation of 16. Find the cutoff score.

7–80. A pet shop owner decides to sell tropical fish that will appeal to the middle 60% of customers. The owner reads a study that states that the mean price of tropical fish sold is $9.52, with a standard deviation of $1.02. Find the maximum and minimum prices of tropical fish the owner should sell.

7–81. An advertising company plans to market a product to low-income families. A study states that for a particular area, the average income per family is $24,596, and the standard deviation is $6256. If the company plans to target the bottom 18% of the families, based on income, find the cutoff income.

7–82. If a one-person household spends an average of $40 per week on groceries (Food Marketing Institute, in M. D. Shook and R. L. Shook, *The Book of Odds,* Penguin Books, 1991, p. 192), find the maximum and minimum dollar amount spent per week for the middle 50% of one-person households. Assume that the standard deviation is $5.

7–83. The mean lifetime of a wristwatch is 25 months, with a standard deviation of 5 months. If the distribution is normal, for how many months should a guarantee be if the manufacturer does not want to exchange more than 10% of the watches?

7–84. In order to qualify for police academy training, recruits are given a test of stress tolerance. The scores are normally distributed, with a mean of 60 and a standard deviation of 10. If only the top 20% of recruits are selected, find the cutoff score.

7–85. In the distributions shown, find the mean and standard deviation for each.

a.

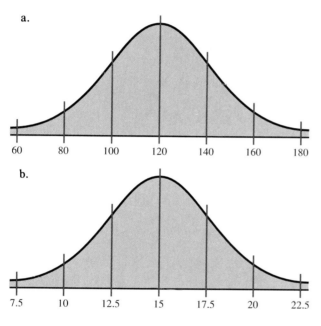

b.

c.

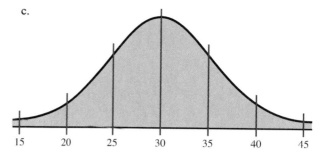

*7–86. Suppose that the mathematics SAT scores for high school seniors for a specific year have a mean of 456, a standard deviation of 100, and are approximately normally distributed. If a subgroup of these high school seniors, those that are in the National Honor Society, is selected, would you expect the distribution of scores to have the same mean and standard deviation?

*7–87. Given a data set, how could you decide if the distribution of the data was approximately normal?

*7–88. If a distribution of raw scores were plotted and then the scores were transformed into z scores, would the shape of the distribution change? Explain your answer.

*7–89. In a normal distribution, find σ when $\mu = 100$ and 2.68% of the area lies to the right of 105.

*7–90. In a normal distribution, find μ when σ is 6 and 3.75% of the area lies to the left of 85.

*7–91. In a certain normal distribution, 1.25% of the area lies to the left of 42 and 1.25% of the area lies to the right of 48. Find μ and σ.

*7–92. An instructor gives a 100-point examination in which the grades are normally distributed. The mean is 60 and the standard deviation is 10. If there are 5% A's and 5% F's, 15% B's and 15% D's, and 60% C's, find the scores that divide the distribution into those categories.

7–5 THE CENTRAL LIMIT THEOREM

In addition to knowing how individual data values vary about the mean for a population, statisticians are also interested in knowing about the distribution of the means of samples taken from a population. This topic is discussed in the subsections that follow.

Distribution of Sample Means

Suppose a researcher selects 100 samples of a specific size from a large population and computes the mean of the same variable for each of the 100 samples. These sample means, $\overline{X}_1, \overline{X}_2, \overline{X}_3, \ldots, \overline{X}_{100}$, constitute a sampling distribution of sample means.

A sampling distribution of sample means is a distribution obtained by using the means computed from random samples of a specific size taken from a population.

If the samples are randomly selected, the sample means, for the most part, will be somewhat different from the population mean μ. These differences are caused by sampling error.

Sampling error is the difference between the sample measure and the corresponding population measure due to the fact that the sample is not a perfect representation of the population.

When all possible samples of a specific size are selected from a population, the distribution of the sample means for a variable has two important properties, which are explained next.

Properties of the Distribution of Sample Means
1. The mean of the sample means will be the same as the population mean.
2. The standard deviation of the sample means will be smaller than the standard deviation of the population, and it will be equal to the population standard deviation, divided by the square root of the sample size.

The following example illustrates these two properties. Suppose a professor gave an 8-point quiz to a small class of 4 students. The results of the quiz were 2, 6, 4, and 8. For the sake of discussion, assume that the 4 students constitute the population. The mean of the population is

$$\mu = \frac{2 + 6 + 4 + 8}{4} = 5$$

The standard deviation of the population is

$$\sigma = \sqrt{\frac{(2-5)^2 + (6-5)^2 + (4-5)^2 + (8-5)^2}{4}} = 2.236$$

The graph of the original distribution is shown in Figure 7–37.

FIGURE 7–37
Distribution of
Quiz Scores

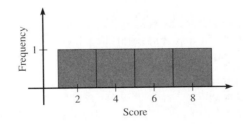

Now, if all samples of size 2 are taken with replacement, and the mean of each sample is found, the distribution is as shown next.

Sample	Mean	Sample	Mean
2, 2	2	6, 2	4
2, 4	3	6, 4	5
2, 6	4	6, 6	6
2, 8	5	6, 8	7
4, 2	3	8, 2	5
4, 4	4	8, 4	6
4, 6	5	8, 6	7
4, 8	6	8, 8	8

A frequency distribution of sample means is as follows.

$\overline{X}$	f
2	1
3	2
4	3
5	4
6	3
7	2
8	1

The mean of the sample means, denoted by $\mu_{\overline{X}}$, is

$$\mu_{\overline{X}} = \frac{2 + 3 + \cdots + 8}{16} = \frac{80}{16} = 5$$

which is the same as the population mean. Hence,

$$\mu_{\overline{X}} = \mu$$

The standard deviation of sample means, denoted by $\sigma_{\overline{X}}$, is

$$\sigma_{\overline{X}} = \sqrt{\frac{(2 - 5)^2 + (3 - 5)^2 + \cdots + (8 - 5)^2}{16}} = 1.581$$

which is the same as the population standard deviation divided by $\sqrt{2}$:

$$\sigma_{\overline{X}} = \frac{2.236}{\sqrt{2}} = 1.581$$

In summary, if all possible samples of size n are taken from the same population, the mean of the sample means, denoted by $\mu_{\overline{X}}$, equals the population mean μ; and the standard deviation of the sample means, denoted by $\sigma_{\overline{X}}$, equals $\sigma/\sqrt{n}$. The standard deviation of the sample means is called the **standard error of the mean.** Hence,

$$\sigma_{\overline{X}} = \frac{\sigma}{\sqrt{n}}$$

A third property of the sampling distribution of sample means pertains to the shape of the distribution and is explained by the **central limit theorem.**

The Central Limit Theorem

As the sample size n increases, the shape of the distribution of the sample means taken from a population with mean μ and standard deviation of σ will approach a normal distribution. As previously shown, this distribution will have a mean μ and a standard deviation $\sigma/\sqrt{n}$.

For the data from the example discussed above, Figure 7–38 shows the graph of the sample means. The graph appears to be somewhat normal, even though it is a histogram.

FIGURE 7–38
Distribution of Sample Means

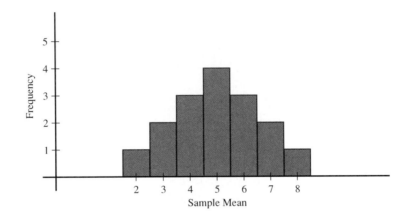

The central limit theorem can be used to answer questions about sample means in the same manner that the normal distribution can be used to answer questions about individual values. The only difference is that a new formula must be used for the z values. It is

$$z = \frac{\overline{X} - \mu}{\sigma/\sqrt{n}}$$

Notice that $\overline{X}$ is the sample mean, and the denominator is the standard error of the mean.

If a large number of means were selected from a large population, the distribution would look like the one shown in Figure 7–39. The percentages indicate the areas of each of the regions.

FIGURE 7–39
Distribution of Sample Means for Large Number of Samples

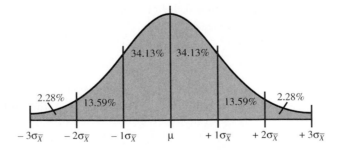

When using the central limit theorem, one must remember two things:

1. When the original variable is normally distributed, the distribution of the sample means will be normally distributed, for any sample size n.
2. When the distribution of the original variable departs from normality, a sample size of 30 or more is needed to use the normal distribution as an approximation for the distribution of the sample means. The larger the sample, the better the approximation will be.

The next several examples show how the standard normal distribution can be used to answer questions about sample means.

EXAMPLE 7–19   The average amount of a purchase at a newsstand is \$2.12. The standard deviation is \$0.45. If 25 customers purchase items at the stand, find the probability that the mean of the purchases is greater than \$2.30. Assume that the distribution is normal.

Solution   Since the distribution is normal, the sample means are normally distributed with a mean of \$2.12. The standard deviation of the sample means is

$$\sigma_{\overline{X}} = \frac{\sigma}{\sqrt{n}} = \frac{0.45}{\sqrt{25}} = 0.09$$

The distribution of the means is shown in Figure 7–40.

FIGURE 7–40
Distribution of
the Means for
Example 7–19

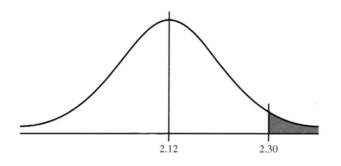

2.12   2.30

The z value is

$$z = \frac{2.30 - 2.12}{0.45/\sqrt{25}} = 2.00$$

The area between 0 and 2.00 is 0.4772. Since the desired area is in the tail, one must subtract from 0.5000. Hence, $0.5000 - 0.4772 = 0.0228$, or 2.28%.   ◄

EXAMPLE 7–20   The average lifetime of smoke alarms manufactured by a company is 60 months, with a standard deviation of 12 months. If a sample of 36 smoke alarms is selected, find the probability that the mean of the sample will be between 57 and 62 months.

FIGURE 7–41
Area Under the Curve
for Example 7–20

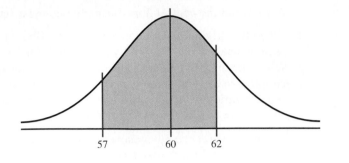

Solution   The desired area is shown in Figure 7–41.
The two z values are

$$z_1 = \frac{62 - 60}{12/\sqrt{36}} = 1.00 \qquad z_2 = \frac{57 - 60}{12/\sqrt{36}} = -1.5$$

The two areas are 0.3413 and 0.4332. Since the z values are on opposite sides of the mean, the probability is found by adding the areas: $0.3413 + 0.4332 = 0.7745$, or 77.45%.   ◀

When the sample size is 30 or larger, the normality assumption is not necessary, as shown in Example 7–20.
Students sometimes have difficulty in deciding whether to use

$$z = \frac{\overline{X} - \mu}{\sigma/\sqrt{n}} \qquad \text{or} \qquad z = \frac{X - \mu}{\sigma}$$

The formula

$$z = \frac{\overline{X} - \mu}{\sigma/\sqrt{n}}$$

should be used when one is attempting to gain information about a sample mean, as shown in this section. The formula

$$z = \frac{X - \mu}{\sigma}$$

is used when one is attempting to gain information about an individual data value obtained from the population. Notice that the first formula contains $\overline{X}$, the symbol for the sample mean, while the second formula contains X, the symbol for an individual data value. The next example illustrates the uses of the two formulas.

EXAMPLE 7–21   The average number of pounds of meat a person consumes a year is 218.4 pounds (American Dietetic Association, in M. D. Shook and R. L. Shook, *The Book of Odds*, Penguin Books, 1991, p. 164). Assume that the standard deviation is 25 pounds and the distribution is approximately normal.

a. If a person is selected at random, find the probability that a person consumes less than 224 pounds per year.
b. If a sample of 40 individuals is selected, find the probability that the mean of the sample will be less than 224 pounds per year.

Solution a. Since the question asks about an individual person, the formula $z = (X - \mu)/\sigma$ is used.

The distribution is shown in Figure 7–42.

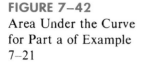

FIGURE 7–42
Area Under the Curve for Part a of Example 7–21

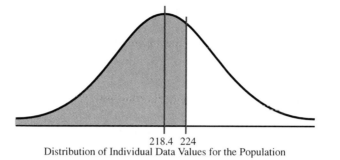

218.4 224
Distribution of Individual Data Values for the Population

The z value is

$$z = \frac{X - \mu}{\sigma} = \frac{224 - 218.4}{25} = 0.22$$

The area between 0 and 0.22 is 0.0871, and this area must be added to 0.5000 to get the total area less than $z = 0.22$.

$$0.0871 + 0.5000 = 0.5871$$

Hence, the probability of selecting an individual who consumes less than 224 pounds of meat per year is 0.5871, or 58.71%.

b. Since the question concerns the mean of a sample with a size of 40, the formula $z = (\overline{X} - \mu)/(\sigma/\sqrt{n})$ is used.

The area is shown in Figure 7–43.

FIGURE 7–43
Area Under the Curve for Part b of Example 7–21

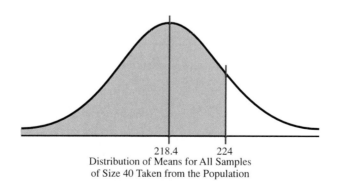

218.4   224
Distribution of Means for All Samples of Size 40 Taken from the Population

The z value is

$$z = \frac{\overline{X} - \mu}{\sigma/\sqrt{n}} = \frac{224 - 218.4}{25/\sqrt{40}} = 1.42$$

The area between $z = 0$ and $z = 1.42$ is 0.4222, and this value must be added to 5.000 to get the total area.

$$0.4222 + 0.5000 = 0.9222$$

Hence, the probability that the mean of a sample of 40 individuals is less than 224 pounds per year is 0.9222, or 92.22%.

Comparing the two probabilities, one can see that the probability of selecting an individual who consumes less than 224 pounds of meat per year is 58.71%, whereas the probability of selecting a sample of 40 people with a mean consumption of meat that is less than 224 pounds per year is 92.22%. This is a rather large difference, and it is due to the fact that the distribution of sample means is much less variable than the distribution of individual data values.　◀

Finite Population Correction Factor

The formula for the standard error of the mean, $\sigma/\sqrt{n}$, is accurate when the samples are drawn with replacement or are drawn without replacement from a very large or infinite population. Since sampling with replacement is for the most part unrealistic, a *correction factor* is necessary when one is computing the standard error of the mean for samples drawn without replacement from a finite population. The correction factor is computed by using the following formula:

$$\sqrt{\frac{N - n}{N - 1}}$$

where N is the population size and n is the sample size.

This correction factor is necessary if relatively large samples are taken from a small population, because the sample mean will then more accurately estimate the population mean and there will be less error in the estimation. Therefore, the standard error of the mean must be multiplied by the correction factor to adjust it for large samples taken from a small population. That is,

$$\sigma_{\bar{X}} = \frac{\sigma}{\sqrt{n}} \cdot \sqrt{\frac{N - n}{N - 1}}$$

When the population is large and the sample is small, the correction factor is generally not used, since it will be very close to 1.00. The questions are, "How large is large?" and "How small is small?" Statisticians generally agree that *the correction factor should be used when the sample is greater than 5% of the population.*

When the sample size n is greater than 5% of the population size N, the finite population correction factor is used to correct the standard error of the mean.

The formula for the correction factor is

$$\sqrt{\frac{N - n}{N - 1}}$$

The formula for the standard error of the mean in this situation is

$$\sigma_{\bar{X}} = \frac{\sigma}{\sqrt{n}} \cdot \sqrt{\frac{N - n}{N - 1}}$$

and the formula for computing the z value is

$$z = \frac{\bar{X} - \mu}{\dfrac{\sigma}{\sqrt{n}} \cdot \sqrt{\dfrac{N - n}{N - 1}}}$$

EXAMPLE 7–22 A sample of 50 students is taken from 500 students. The standard deviation of the variable (IQ) is 15. Determine whether the finite population correction is necessary. If so, compute it. Finally, find the adjusted standard error of the mean.

Solution Since $n = 50$, $N = 500$, and 50 is greater than 5% of 500, which is 25, the correction factor should be used. It is

$$\sqrt{\frac{N - n}{N - 1}} = \sqrt{\frac{500 - 50}{500 - 1}} = 0.9496$$

The adjusted standard error of the mean is

$$\sigma_{\overline{X}} = \frac{\sigma}{\sqrt{n}} \cdot \sqrt{\frac{N - n}{N - 1}} = \frac{15}{\sqrt{50}} (0.9496) = 2.014 \qquad \blacktriangleleft$$

The correction factor should be used when one is comparing a sample mean with a given population mean, as shown in the next example.

EXAMPLE 7–23 The average weight of a group of young adult males is 160 pounds. The standard deviation is 10 pounds. If a sample of 30 people is selected from the population of 300, find the probability that the mean will be less than 156 pounds.

Solution Since the sample size 30 is larger than 5% of the population, which is 15 (i.e., $0.05 \cdot 300 = 15$), the correction factor must be used. The desired area is shown in Figure 7–44.

FIGURE 7–44
Area Under the Curve
for Example 7–23

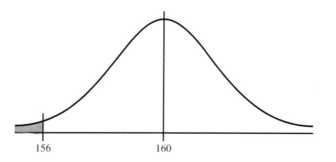

156   160

The formula for the z value must include the correction factor. Thus, the z value is

$$z = \frac{\overline{X} - \mu}{\dfrac{\sigma}{\sqrt{n}} \sqrt{\dfrac{N - n}{N - 1}}}$$

$$= \frac{156 - 160}{\dfrac{10}{\sqrt{30}} \sqrt{\dfrac{300 - 30}{300 - 1}}} = -2.31$$

The area for this z value is 0.4896, and it must be subtracted from 0.5000 to get the desired area, which is 0.0104, or 1.04%. Hence, the probability that a sample mean will be less than 156 pounds is 1.04%. $\qquad \blacktriangleleft$

As this section has shown, the central limit theorem enables the researcher to determine probabilities associated with sample means of sampling distributions.

Speaking of **Statistics**

WASHINGTON (AP)—A state-by-state list showing the average number of days people had to wait for answers to initial claims for benefits under Disability Insurance or Supplemental Security Income. The averages are based on experience through the first nine months of 1992. The figures were compiled by Families USA Foundation.

State	Days
Alabama	121
Alaska	114
Arizona	155
Arkansas	91
California	160
Colorado	121
Connecticut	91
Delaware	111
D.C.	126
Florida	81
Georgia	95
Hawaii	167
Idaho	75
Illinois	103
Indiana	97
Iowa	79
Kansas	113
Kentucky	107
Louisiana	138
Maine	77
Maryland	110
Massachusetts	94
Michigan	115
Minnesota	88
Mississippi	93
Missouri	74
Montana	77
Nebraska	72
Nevada	156
New Hampshire	102
New Jersey	161
New Mexico	113
New York	119
North Carolina	69
North Dakota	77
Ohio	134
Oklahoma	120
Oregon	126
Pennsylvania	107
Rhode Island	110
South Carolina	106
South Dakota	87
Tennessee	92
Texas	123
Utah	155
Vermont	107
Virginia	79
Washington	109
West Virginia	106

Source: Associated Press. Used with permission.

The listing shows a distribution of sample means. Construct a frequency distribution, and comment on the shape of the distribution (approximately normal, skewed, etc.). Suggest some reasons why the distribution is shaped as it is. What information is missing from the article?

EXERCISES

7–93. If samples of a specific size are selected from a population and the means are computed, what is this distribution of means called?

7–94. Why do most of the sample means differ somewhat from the population mean? What is this difference called?

7–95. What is the mean of the sample means?

7–96. What is the standard deviation of the sample means called? What is the formula for this standard deviation?

7–97. What does the central limit theorem say about the shape of the distribution of sample means?

7–98. When is the finite population correction factor necessary?

7–99. For each of the following, determine whether the finite population correction factor is necessary, and if so, calculate it.

a. $N = 500$, $n = 50$
b. $N = 100$, $n = 20$
c. $N = 2000$, $n = 50$
d. $N = 5000$, $n = 200$
e. $N = 981$, $n = 50$
f. $N = 2032$, $n = 100$
g. $N = 562$, $n = 30$
h. $N = 3215$, $n = 150$
i. $N = 8000$, $n = 500$
j. $N = 10,000$, $n = 400$

For Exercises 7–100 through 7–117, assume that the sample is taken from a large population and the correction factor can be ignored. If the sample size is less than 30, assume that the variable is normally distributed.

7–100. In a certain city it was found that the average monthly amount of glass that a family of four uses is 12.4 pounds. The standard deviation is 2.5 pounds. Find the probability that the mean of a sample of 55 families will be between 12 and 13 pounds.

7–101. The mean serum cholesterol of a large population of overweight adults is 220 milligrams per deciliter (mg/dl), and the standard deviation is 16.3 mg/dl. If a sample of 30 adults is selected, find the probability that the mean will be between 220 and 222 mg/dl.

7–102. For a certain large group of individuals, the mean hemoglobin level in the blood is 21.0 grams per milliliter (g/ml). The standard deviation is 2 g/ml. If a sample of 25 individuals is selected, find the probability that the mean will be greater than 21.3 g/ml.

7–103. The mean weight of 18-year-old females is 126 pounds, and the standard deviation is 15.7. If a sample of 25 females is selected, find the probability that the mean of the sample will be greater than 128.3 pounds.

7–104. The mean grade point average of the engineering majors at a large university is 3.23, with a standard deviation of 0.72. In a class of 48 students, find the probability that the mean grade point average of the students is less than 3.15.

7–105. The mean of the number of ounces of coffee a vending machine dispenses is 10. The standard deviation is 0.1. If a sample of 10 cups is selected, find the probability that the mean of the sample will be less than 9.93 ounces.

7–106. The average hourly wage of fast-food workers employed by a nationwide chain is $5.55. The standard deviation is $1.15. If a sample of 50 workers is selected, find the probability that the mean of the sample will be between $5.90 and $5.25.

7–107. The mean score on a dexterity test for 12-year-olds is 30. The standard deviation is 5. If a psychologist administers the test to a class of 22 students, find the probability that the mean of the sample will be between 27 and 31.

7–108. A recent study of the life span of portable radios found the average to be 3.1 years, with a standard deviation of 0.9 year. If the number of radios owned by the students in one dormitory is 47, find the probability that the mean lifetime of these radios will be less than 2.7 years.

7–109. The average age of accountants is 43 years, with a standard deviation of 5 years. If an accounting firm employs 30 accountants, find the probability that the average age of the group is greater than 44.2 years old.

7–110. The average annual precipitation for Des Moines is 30.83 inches, with a standard deviation of 5 inches. If a random sample of 10 years is selected, find the probability that the mean will be between 32 and 33 inches.

7–111. The average national monthly welfare payment is $360, with a standard deviation of $50. If 50 families are randomly selected, find the probability that the mean will be between $344 and $350.

7-112. An article written for the Associated Press (December 23, 1992) stated that the average annual salary in Pennsylvania was $24,393. Assume that salaries are normally distributed for a certain group of wage earners, and the standard deviation of this group is $4362.

a. Find the probability that a randomly selected individual will earn less than $26,000.
b. Find the probability that for a randomly selected sample of 25 individuals, the mean salary will be less than $26,000.
c. Why is the probability for part a higher than the probability for part b?

7-113. The average time it takes a group of adults to complete a certain achievement test is 46.2 minutes. The standard deviation is 8 minutes.

a. Find the probability that a randomly selected adult will complete the test in less than 43 minutes.
b. Find the probability that if 50 randomly selected adults take the test, the mean time it takes the group to complete the test will be less than 43 minutes.
c. Does it seem reasonable that an adult would finish the test in less than 43 minutes? Explain.
d. Does it seem reasonable that the mean of the 50 adults could be less than 43 minutes?

7-114. Assume that the mean systolic blood pressure of normal adults is 120 millimeters of mercury (mmHg), and the standard deviation is 5.6.

a. If an individual is selected, find the probability that the individual's pressure will be between 120 and 121.8 mmHg.
b. If a sample of 30 adults is randomly selected, find the probability that the sample mean will be between 120 and 121.8 mmHg.
c. Why is the answer to part a so much smaller than the answer to part b?

7-115. The average cholesterol content of a certain brand of eggs is 210 milligrams, and the standard deviation is 15 milligrams.

a. If a single egg is selected, find the probability that the cholesterol content will be more than 215 milligrams.
b. If a sample of 25 eggs is selected, find the probability that the mean of the sample will be larger than 215 milligrams.

7-116. At a large university, the mean age of graduate students who are majoring in psychology is 32.6 years, and the standard deviation is 3 years.

a. If an individual from the department is randomly selected, find the probability that the individual's age will be between 31.0 and 33.2 years.

b. If a random sample of 15 individuals is selected, find the probability that the mean age of the students in the sample will be between 31.0 years and 33.2 years.

7-117. The average labor cost for automobile repairs for a large chain of automobile repair shops is $48.25. The standard deviation is $4.20.

a. If a store is selected at random, find the probability that the labor cost will range between $46 and $48.
b. If 20 stores are selected at random, find the probability that the mean of the sample will be between $46 and $48.
c. Which answer is larger? Explain why.

For Exercises 7-118 through 7-123, check to see whether the correction factor should be used. If so, be sure to include it in the calculations.

7-118. In the study of the life span of 500 runners, the mean age at death was 72.0 years, and the standard deviation was 5.3 years. If a sample of 50 runners is selected, find the probability that the mean age of the sample will be less than 70 years.

7-119. A study of 800 homeowners in a certain area showed that the average value of the homes was $82,000, and the standard deviation was $5000. If 50 homes are for sale, find the probability that the mean of the values of these homes is greater than $83,500.

7-120. The Central Coal Company employs 150 miners. The average number of miles the employees travel to work each day is 8 miles, and the standard deviation is 2 miles. Find the probability that the average number of miles the workers on the day shift travel to work is greater than 7.5 miles. There are 50 people who work on the day shift.

7-121. The average price of ties sold by an exclusive men's shop is $40. The standard deviation is $5. If 8 ties are sold from a lot of 20, find the probability that the average price will be greater than $43.

7-122. The average weight of 100 five-year-old boys is 48 pounds. The standard deviation is 7 pounds. If 15% of the boys are weighed, find the probability that the average weight of the sample will be between 48 and 50 pounds.

7-123. If the average sodium content of 150 microwave dinners is 795 milligrams, find the probability that a sample of 10 dinners will have an average sodium content less than 725 milligrams. The standard deviation of the population is 75 milligrams.

*7-124. The average breaking strength of a certain brand of steel cable is 2000 pounds, with a standard deviation of 100 pounds. A sample of 20 cables is selected and tested. Find the sample mean that will cut off the upper 95% of all of the samples of size 20 taken from the population.

*7–125. The standard deviation of a variable (IQ) is 15. If a sample of 100 individuals is selected, compute the standard error of the mean. What size sample is necessary to double the standard error of the mean?

*7–126. In Exercise 7–125, what size sample is needed to cut the standard error of the mean in half?

*7–127. A professor gives a quiz to three students. The scores are 6, 8, and 10. Find all samples of size 2 with replacement. Compute the means of these samples. Compare this result with the population mean. Compute the standard deviation of the means. Compare this result with the population standard deviation. Show that

$$\sigma_{\overline{X}} = \frac{\sigma}{\sqrt{n}}$$

7–6 THE NORMAL APPROXIMATION TO THE BINOMIAL DISTRIBUTION

The normal distribution can be used to solve problems that involve the binomial distribution. Recall from Chapter 6 that a binomial distribution has the following characteristics:

1. There must be a fixed number of trials.
2. The outcomes of each trial must be independent.
3. Each experiment can have only two outcomes or be reduced to two outcomes.
4. The probability of a success must remain the same for each trial.

Also, recall that a binomial distribution is determined by n (the number of trials) and p (the probability of a success). When p is approximately 0.5, and as n increases, the shape of the binomial distribution becomes similar to the normal distribution. It should be emphasized that the larger n is and the closer p is to 0.5, the more similar the binomial distribution is to the normal distribution.

But when p is close to 0 or 1, and n is relatively small, the normal approximation is inaccurate. As a rule of thumb, statisticians generally agree that the normal approximation should only be used when $n \cdot p$ and $n \cdot q$ are both greater than or equal to 5. (*Note:* $q = 1 - p$.) For example, if p is 0.3 and n is 10, then $np = (10)(0.3) = 3$; and the normal distribution should not be used as an approximation. On the other hand, if $p = 0.5$ and $n = 10$, then $np = (10)(0.5) = 5$ and $nq = (10)(0.5) = 5$; and the normal distribution can be used as an approximation. See Figure 7–45.

In addition to the previous condition of $np \geq 5$ and $nq \geq 5$, a correction for continuity may be used in the normal approximation.

A correction for continuity is a correction employed when a continuous distribution is used to approximate a discrete distribution.

The continuity correction means that for any specific value of X, say 8, the boundaries of X in the binomial distribution, 7.5 to 8.5, must be used. (See Chapter 1, Section 1–3.) Hence, when one employs the normal distribution to approximate the binomial, the boundaries of any specific value X must be used as they are shown in the binomial distribution. For example, for $P(X = 8)$, the correction is $P(7.5 < X < 8.5)$. For $P(X \leq 7)$, the correction is $P(X < 7.5)$. For $P(X \geq 3)$, the correction is $P(X > 2.5)$.

FIGURE 7–45

Comparison of the
Binomial Distribution
and the Normal
Distribution

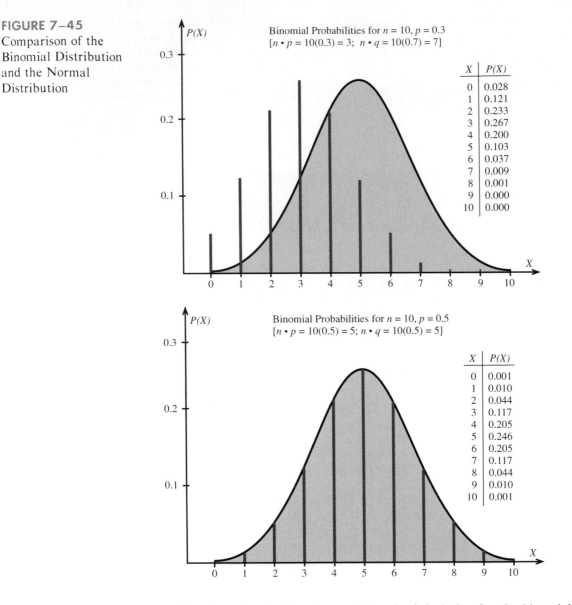

The formulas for the mean and standard deviation for the binomial distribution are necessary for calculations. They are

$$\mu = n \cdot p \qquad \sigma = \sqrt{n \cdot p \cdot q}$$

Procedure for the Normal Approximation to the Binomial Distribution

Step 1: Check to see whether the normal approximation can be used.

Step 2: Find the mean μ and the standard deviation σ.

Step 3: Write the problem in probability notation, using X.

Step 4: Rewrite the problem by using the continuity correction factor, and show the corresponding area under the normal distribution.

Step 5: Find the corresponding z values.

Step 6: Find the solution.

EXAMPLE 7–24   A construction company found that 6% of its telephone solicitations resulted in positive responses –a customer requested that a representative of the company visit his or her home. If 300 calls are made, find the probability of getting exactly 25 positive responses.

Solution   Here, $p = 0.06$, $q = 0.94$, and $n = 300$.

STEP 1   Check to see whether the normal approximation can be used.

$$np = (300)(0.06) = 18 \qquad nq = (300)(0.94) = 282$$

Since $np \geq 5$ and $nq \geq 5$, the normal distribution can be used.

STEP 2   Find the mean and standard deviation.

$$\mu = np = (300)(0.06) = 18$$
$$\sigma = \sqrt{npq} = \sqrt{(300)(0.06)(0.94)} = \sqrt{16.92} = 4.11$$

STEP 3   Write the problem in probability notation: $P(X = 25)$.

STEP 4   Rewrite the problem by using the continuity correction factor: $P(24.5 < X < 25.5)$. Show the corresponding area under the normal distribution curve (see Figure 7–46).

FIGURE 7–46
Area Under the Curve
and the X Values for
Example 7–24

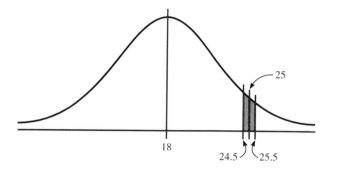

STEP 5   Find the corresponding z values. Since 25 represents any value between 24.5 and 25.5, find both z values.

$$z_1 = \frac{25.5 - 18}{4.11} = 1.82 \qquad z_2 = \frac{24.5 - 18}{4.11} = 1.58$$

STEP 6   Find the solution. Find the corresponding areas in the table: The area for $z = 1.82$ is 0.4656, and the area for $z = 1.58$ is 0.4429. Subtract the areas to get the approximate value: $0.4656 - 0.4429 = 0.0227$, or 2.27%.   ◄

EXAMPLE 7–25   If another company found that the positive response rate to its telephone solicitations was 10%, find the probability of getting 10 or more positive responses if 200 calls were made.

Solution   Here, $p = 0.10$, $q = 0.90$, and $n = 200$.

STEP 1   Since np is $(200)(0.10) = 20$ and nq is $(200)(0.90) = 180$, the normal approximation can be used.

STEP 2 $\mu = np = (200)(0.10) = 20$
$$\sigma = \sqrt{npq} = \sqrt{(200)(0.10)(0.90)} = \sqrt{18} = 4.24$$

STEP 3 $P(X \geq 10)$.

STEP 4 $P(X > 9.5)$. The desired area is shown in Figure 7–47.

FIGURE 7–47
Area Under the Curve
and X Value for
Example 7–25

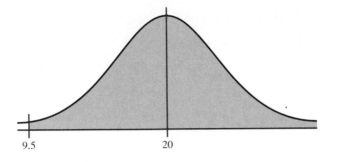

9.5   20

STEP 5 Since the problem is to find the probability of 10 or more positive responses, the normal distribution graph is as shown in Figure 7–47. Hence, the area between 9.5 and 20 must be added to 0.5000 to get the correct approximation.

The z value is

$$z = \frac{9.5 - 20}{4.24} = -2.48$$

STEP 6 The area between 20 and 9.5 is 0.4934. Thus, the probability of getting 10 or more responses is $0.5000 + 0.4934 = 0.9934$, or 99.34%. ◄

EXAMPLE 7–26 If a baseball player's batting average is 0.320 (32%), find the probability that the player will get at most 26 hits at 100 times at bat.

Solution Here, $p = 0.32$, $q = 0.68$, and $n = 100$.

STEP 1 Since $np = (100)(0.320) = 32$ and $nq = (100)(0.680) = 68$, the normal distribution can be used to approximate the binomial distribution.

STEP 2 $\mu = np = (100)(0.32) = 32$
$\sigma = \sqrt{npq} = \sqrt{(100)(0.32)(0.68)} = \sqrt{21.76} = 4.66$

STEP 3 $P(X \leq 26)$.

STEP 4 $P(X < 26.5)$. The desired area is shown in Figure 7–48.

FIGURE 7–48
Area Under the Curve
for Example 7–26

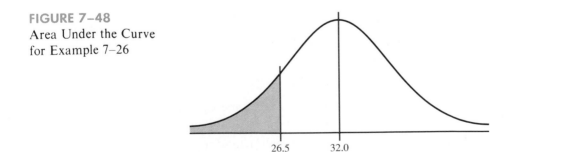

26.5   32.0

STEP 5   The z value is

$$z = \frac{26.5 - 32}{4.66} = -1.18$$

STEP 6   The area between the mean and 26.5 is 0.3810. Since the area in the left tail is desired, 0.3810 must be subtracted from 0.5000. So the probability is $0.5000 - 0.3810 = 0.1190$, or 11.9%.   ◄

The closeness of the normal approximation is shown in the next example.

EXAMPLE 7–27   When $n = 10$ and $p = 0.5$, use the binomial distribution table to find the probability that $X = 6$. Then, use the normal approximation to find the probability that $X = 6$.

Solution   From Table B, for $n = 10$, $p = 0.5$, and $X = 6$, the probability is 0.205.
For the normal approximation,

$$\mu = np = (10)(0.5) = 5$$
$$\sigma = \sqrt{npq} = \sqrt{(10)(0.5)(0.5)} = 1.58$$

Now, $X = 6$ is represented by the boundaries 5.5 and 6.5. So the z values are

$$z_1 = \frac{6.5 - 5}{1.58} = 0.95 \qquad z_2 = \frac{5.5 - 5}{1.58} = 0.32$$

The corresponding area for 0.95 is 0.3289, and the corresponding area for 0.32 is 0.1255.

The solution is $0.3289 - 0.1255 = 0.2034$, which is very close to the binomial table value of 0.205. The desired area is shown in Figure 7–49.

FIGURE 7–49
Area Under the Curve
for Example 7–27

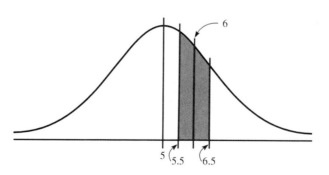

The normal approximation also can be used to approximate other distributions, such as the Poisson distribution.

Students sometimes have difficulty deciding whether to add 0.5 to or to subtract 0.5 from the data value for the correction factor. Table 7–1 summarizes the different situations.

Table 7–1

Summary of the Normal Approximation to the Binomial Distribution	
Binomial	**Normal**
When Finding	Use
1. $P(X = a)$	$P(a - 0.5 < X < a + 0.5)$
2. $P(X \geq a)$	$P(X > a - 0.5)$
3. $P(X > a)$	$P(X > a + 0.5)$
4. $P(X \leq a)$	$P(X < a + 0.5)$
5. $P(X < a)$	$P(X < a - 0.5)$

For all cases, $\mu = n \cdot p$, $\sigma = \sqrt{n \cdot p \cdot q}$, and $n \cdot p \geq 5$ and $n \cdot q \geq 5$.

EXERCISES

7–128. (ANS.) Use the normal approximation to the binomial to find the probabilities for the specific value(s) of X.

a. $n = 30$, $p = 0.5$, $X = 18$
b. $n = 50$, $p = 0.8$, $X = 44$
c. $n = 100$, $p = 0.1$, $X = 12$
d. $n = 10$, $p = 0.5$, $X \geq 7$
e. $n = 20$, $p = 0.7$, $X \leq 12$
f. $n = 50$, $p = 0.6$, $X \leq 40$

7–129. Check each binomial distribution to see whether it can be approximated by the normal distribution (i.e., are $np \geq 5$ and $nq \geq 5$?).

a. $n = 20$, $p = 0.5$
b. $n = 10$, $p = 0.6$
c. $n = 40$, $p = 0.9$
d. $n = 50$, $p = 0.2$
e. $n = 30$, $p = 0.8$
f. $n = 20$, $p = 0.85$

7–130. A telephone-soliciting company has a 12% success rate for sales. If the callers contact 500 people, find the probability of getting at least 55 sales.

7–131. A bank found that 1 person in 20 will be overdrawn on his or her checking account each month. If a bank has 400 customers, find the probability that 15 or less people will be overdrawn.

7–132. An airline company has found that 5% of its passengers will not show up for their scheduled flights. If the plane has 100 seats, find the probability that 6 or more people will not show up for the fully booked flight.

7–133. The Lake Inn has 80 rooms, and the occupancy rate for March is 60%. Find these probabilities.

a. At least 60 rooms will be rented.
b. Fewer than 50 rooms will be rented.

7–134. If the probability that a newborn child will be female is 50%, find the probability that in 100 births, 55 or more will be females.

7–135. If 20% of all families in a country have at least 2 automobiles, find the probability that in a random sample of 100 families, exactly 20 will have 2 or more automobiles.

7–136. An automobile dealer states that 90% of all automobiles sold have air-conditioning. If the dealer sells 250 automobiles, find the probability that fewer than 5 automobiles sold will not have air-conditioning.

7–137. If 3% of hair dryer motors are defective, find the probability that in a shipment of 180, 6 or fewer will be defective.

7–138. If the failure rate in a certain mathematics course is 20%, find the probability that in a class of 25, exactly 5 students will fail.

7–139. A political candidate estimates that 30% of the voters in his party favor his proposed tax reform bill. If there are 400 people at a rally, find the probability that at least 100 voters will favor his tax bill.

*7–140. Recall that for use of the normal distribution as an approximation to the binomial distribution, the condition $np \geq 5$ and $nq \geq 5$ must be met. For each given probability, compute the minimum sample size needed for use of the normal approximation.

a. $p = 0.1$
b. $p = 0.3$
c. $p = 0.5$
d. $p = 0.8$
e. $p = 0.9$

7–7 SUMMARY

The normal distribution can be used to describe a variety of variables, such as heights, weights, temperature, and IQ. The normal distribution is bell-shaped, unimodal, symmetric, and continuous; and its mean, median, and mode are equal. Since each variable has its own distribution with mean μ and standard deviation σ, mathematicians use the standard normal distribution, which has a mean of 0 and a standard deviation of 1. Other approximately normally distributed variables can be transformed into the standard normal distribution by using the formula $z = (X - \mu)/\sigma$.

The normal distribution can also be used to describe a sampling distribution of sample means. These samples must be of the same size and randomly selected from the population. The means of the samples will differ somewhat from the population mean. These differences are due to sampling errors, since the samples are not perfect representations of the populations. The mean of the sample means will be equal to the population mean, and the standard deviation of the sample means will be equal to the population standard deviation divided by the square root of the sample size. The central limit theorem states that as the size of the samples increases, the distribution of sample means will be approximately normal.

The normal distribution can be used to approximate other distributions such as the binomial distribution. For the use of the normal distribution as an approximation, the condition $np \geq 5$ and $nq \geq 5$ must be met. Also, a correction for continuity may be used for more accurate results.

Important Terms

Central limit theorem
Correction for continuity
Finite population correction
    factor
Negatively skewed distribution
Normal distribution
Positively skewed distribution
Sampling distribution of sample
    means
Sampling error
Standard error of the mean
Standard normal distribution
Symmetrical distribution
z value

Important Formulas

Formula for the z value (or standard score):

$$z = \frac{X - \mu}{\sigma}$$

Formula for finding a specific data value:

$$X = z \cdot \sigma + \mu$$

Formula for the mean of the sample means:

$$\mu_{\overline{X}} = \mu$$

Formula for the standard error of the mean:

$$\sigma_{\overline{X}} = \frac{\sigma}{\sqrt{n}}$$

Formula for the z value for the central limit theorem:

$$z = \frac{\overline{X} - \mu}{\sigma/\sqrt{n}}$$

Formula for the z value when the finite correction factor is needed $(n > 0.05N)$:

$$z = \frac{\overline{X} - \mu}{\frac{\sigma}{\sqrt{n}} \sqrt{\frac{N - n}{N - 1}}}$$

Formula for the mean and standard deviation for the binomial distribution:

$$\mu = n \cdot p \qquad \sigma = \sqrt{n \cdot p \cdot q}$$

Review Exercises

7–141. Find the area under the standard normal distribution curve for each.

a. Between $z = 0$ and $z = 1.95$
b. Between $z = 0$ and $z = 0.37$
c. Between $z = 1.32$ and $z = 1.82$
d. Between $z = -1.05$ and $z = 2.05$
e. Between $z = -0.03$ and $z = 0.53$
f. Between $z = +1.10$ and $z = -1.80$
g. To the right of $z = 1.99$
h. To the right of $z = -1.36$
i. To the left of $z = -2.09$
j. To the left of $z = 1.68$

7–142. Using the standard normal distribution, find each probability.

a. $P(0 < z < 2.07)$
b. $P(-1.83 < z < 0)$
c. $P(-1.59 < z < +2.01)$
d. $P(1.33 < z < 1.88)$
e. $P(-2.56 < z < 0.37)$
f. $P(z > 1.66)$
g. $P(z < -2.03)$
h. $P(z > -1.19)$
i. $P(z < 1.93)$
j. $P(z > -1.77)$

7–143. The reaction time for a laboratory animal to respond to a stimulus is 1.5 seconds, with a standard deviation of 0.3 second. Find the probability that it will take the animal the following time to react.

a. Between 1.0 and 1.3 seconds
b. More than 2.2 seconds
c. Less than 1.1 seconds

7–144. The average diastolic blood pressure of a certain age group of people is 85 mmHg. The standard deviation is 6. If an individual is selected, find the probability that the individual's pressure will be the following.

a. Greater than 90   c. Between 85 and 95
b. Below 80   d. Between 88 and 92

7–145. The average number of miles a mail carrier walks on a typical route in a certain city is 5.6. The standard deviation is 0.8 mile. If a carrier is chosen at random, find the probability that the carrier walks the following miles.

a. Between 5 and 6 miles
b. Less than 4 miles
c. More than 6.3 miles

7–146. The average number of years a person lives after being diagnosed as having a certain disease is 4 years. The standard deviation is 6 months. If an individual is diagnosed as having the disease, find the probability that the individual lives for the following number of years.

a. More than 5 years
b. Less than 2 years
c. Between 3.5 and 5.2 years
d. Between 4.2 and 4.8 years

7–147. For a certain group of taxpayers who get refunds, the average amount of the refund is $918. The standard deviation is $45. If a taxpayer is selected at random, find the probability that the refund is the following amount.

a. Between $918 and $950
b. Less than $900
c. More than $925
d. Between $875 and $910

7–148. The average weight of a suitcase for an airline passenger is 45 pounds. The standard deviation is 2 pounds. If 15% of the suitcases are overweight, find the maximum weight allowed by the airline.

7–149. An educational study to be conducted requires an IQ in the middle 40% range. If $\mu = 100$ and $\sigma = 15$, find the highest and lowest acceptable IQ values that would enable a candidate to participate in the study.

7–150. The average repair cost for automatic washers is $73, with a standard deviation of $8. The costs are normally distributed. If 9 washers are repaired, find the probability that the mean of the repair bills will be less than $70.

7–151. The average heating bill for a residential area is $123 for the month of November. The standard deviation is $8. If the amounts of the heating bills are normally distributed, find the probability that the mean of the bill for 12 residents will be more than $128.

7–152. In a recent study of appliances, a manufacturer found that for 100 toasters, the average lifetime was 14 months. The standard deviation was 4 months. If 20 toasters are selected, find the probability that the mean lifetime of the sample will be less than 1 year. Assume that the variable is normally distributed.

7–153. The average of the purchases of 500 customers at the Train Station Hobby Shop was $12.50. The standard deviation was $2.50. If a week was selected at random, find the probability that the average of the purchases of 100 customers was more than $12.00.

7–154. The probability of winning on a slot machine is 5%. If a person plays the machine 500 times, find the probability of winning 30 times. Use the normal approximation to the binomial distribution.

7–155. If 12% of all people in a certain geographic region suffer from allergies, find the probability that in a sample of 200, there are more than 15 people who suffer from allergies.

7–156. In a corporation, 30% of the people elect to enroll in the financial investment program offered by the company. Find the probability that if 800 people are selected, at least 260 have enrolled in the program.

7–157. A company that installs decks for residential customers finds that it will sell a deck package to 20% of the customers its representatives visit. If the sales representatives visit 200 customers, find the probability of making at least 50 sales.

COMPUTER APPLICATIONS

1. MINITAB can be used to find the areas under the normal distribution curve for specific values of data. Suppose that a distribution has a mean of 50 and a standard deviation of 10. To find the probability that an individual value will be less than or equal to 48, type the following commands:

```
MTB > CDF 48;
SUBC > NORMAL 50, 10.
```

The computer will respond as follows:

```
48.0000    .4207
```

This means that the probability of $X < 48$ is 0.4207. MINITAB will only print the cumulative probability for values less than or equal to a specific value. If the parameters 50 and 10 are omitted, the values for the standard normal distribution, $\mu = 0$ and $\sigma = 1$, will be assumed.

Using the same distribution, find the probability that an individual value will be less than or equal to 56.

TEST

Directions: Determine whether each statement is true or false.

1. The total area under the normal distribution is infinite.
2. The mean of the standard normal distribution is equal to 0.
3. The standard normal distribution is a continuous distribution.
4. All variables that are normally distributed can be transformed into a variable that follows the standard normal distribution.
5. The z value of a data value that is less than the mean of the data will always be negative.
6. Approximately 5% of the area under the normal distribution falls within one standard deviation of the mean.
7. The area to the left of $z = 0$ under the standard normal distribution will be negative.
8. The normal distribution is also a probability distribution.
9. $P(z \leq 0) = -0.5$.
10. The sample means will usually differ from the population mean due to sampling error.
11. When the samples are small, the distribution of sample means will be normally distributed about the population mean.
12. The mean of the sample means will be equal to the population mean.
13. The standard deviation of the sample means will be equal to the population standard deviation.
14. The normal distribution can be used to approximate the binomial distribution when p is approximately equal to 0.
15. Because the normal distribution is continuous and the binomial distribution is discrete, a correction for continuity should be used when one employs the normal approximation to the binomial distribution.
16. The central limit theorem applies to the means of samples selected from different populations.
17. When the sample size is greater than or equal to 30, approximately 68% of the sample means will fall within one standard error of the mean.

CHAPTER

8

Confidence Intervals and Sample Size

8-1 INTRODUCTION

One aspect of inferential statistics is **estimation,** which is the process of estimating the value of a parameter from information obtained from a sample. For example, in a recent book entitled *The Book of Odds,* by Michael D. Shook and Robert L. Shook (Penguin Books, 1991), one will find the following statements:

"One out of 4 Americans is currently dieting."[1]

"Seventy-two percent of Americans have flown in commercial airlines."[2]

"The average kindergarten student has seen more than 5,000 hours of television."[3]

"The average school nurse makes $25,300 a year."[4]

"The average amount of life insurance is $108,000 per household with life insurance."[5]

Since the populations from which these values were obtained are large, these values are only *estimates* of the true parameters and are derived from data collected from samples.

The statistical procedures for estimating the population mean and the population proportion will be explained in this chapter. The technique for estimating the population variance is explained in Chapter 12.

An important question in estimation is that of sample size—i.e., "How large should the sample be in order to make an accurate estimate?" This question is not easy to answer since the size of the sample depends on several factors, such as the accuracy desired and the probability of making a correct estimate. The question of sample size will be explained in this chapter also.

8-2 CONFIDENCE INTERVALS FOR THE MEAN (σ KNOWN OR $n >$ 30) AND SAMPLE SIZE

Suppose a college president wishes to estimate the average age of the students attending classes this semester. She could select a random sample of 100 students and find the average age of these students, say 27.3 years. From the sample mean, she could infer that the average age of all the students is 27.3 years. This type of estimate is called a point estimate.

A point estimate is a specific numerical value estimate of a parameter. The best point estimate of the population mean μ is the sample mean $\overline{X}$.

One might ask why other measures of central tendency, such as the median and mode, are not used to estimate the population mean. The reason is that the means of samples vary less than other statistics (such as medians and modes) when many samples are selected from the same population. Therefore, the sample mean is the best estimate of the population mean.

[1]Calorie Control Council. Used with permission.
[2]"The Bristol Meyers Report: Medicine in the Next Century." Used with permission.
[3]U.S. Department of Education.
[4]National Association of School Nurses. Used with permission.
[5]American Council of Life Insurance. Used with permission.

Sample measures (i.e., statistics) are used to estimate population measures (i.e., parameters). These statistics are called **estimators.** As previously stated, the sample mean rather than the sample median or sample mode is the best estimator of the population mean.

A good estimator must satisfy the three properties described next.

Three Properties of a Good Estimator

1. The estimator must be an **unbiased estimator**—i.e., the expected value or the mean of the estimator must be equal to the mean of the parameter being estimated.
2. The estimator must be consistent. For a **consistent estimator,** as sample size increases, the value of the estimator approaches the value of the parameter estimated.
3. The estimator must be a **relatively efficient estimator**—i.e., of all the statistics that can be used to estimate a parameter, the relatively efficient estimator has the smallest variance.

Confidence Intervals

As stated in Chapter 7, the sample mean will be, for the most part, somewhat different from the population mean due to sampling error. Therefore, one might ask a second question, "How good is a point estimate?" The answer is that there is no way of knowing how close the point estimate is to the population mean.

This answer places some doubt on the accuracy of point estimates. For this reason, statisticians prefer another type of estimate called an interval estimate.

An interval estimate of a parameter is a range of values used to estimate the parameter.

In an interval estimate, the parameter is specified as falling between two values. For example, an interval estimate for the average age of all students might be $26.9 < \mu < 27.7$, or 27.3 ± 0.4 years.

When an interval estimate is made, a probability of being correct can be assigned. For instance, one may wish to be 95% sure that the interval contains the true population mean. Another question then arises, "Why 95%? Why not 99% or 99.5%?"

If one desires to be more confident, such as 99% or 99.5% confident, then the interval must be larger. For example, a 99% confidence interval for the mean age of college students might be $26.7 < \mu < 27.9$, or 27.3 ± 0.6. Hence, a trade-off occurs. To be more confident that the interval contains the true population mean, one must make the interval larger.

A confidence interval is a specific interval estimate of a parameter determined by using data obtained from a sample and the specific confidence level of the estimate.
The confidence level of an interval estimate of a parameter is the probability that the interval estimate will contain the parameter.

Intervals constructed in this way are called confidence intervals. Two common confidence intervals are used: the 95% and the 99% confidence intervals.

The algebraic derivation of the formula for determining a confidence interval for a mean will be omitted. Instead, a brief intuitive explanation will be given.

The central limit theorem states that when the sample size is large, approximately 95% of the sample means will fall within 1.96 standard errors of the population mean. That is,

$$\mu \pm 1.96 \left(\frac{\sigma}{\sqrt{n}} \right)$$

Now, if a specific sample mean is selected, say $\overline{X}$, there is a 95% probability that it falls within the range of $\mu \pm 1.96 \, (\sigma/\sqrt{n})$. Likewise, there is a 95% probability that the interval specified by

$$\overline{X} \pm 1.96 \left(\frac{\sigma}{\sqrt{n}} \right)$$

will contain μ, as will be shown later. Stated another way,

$$\overline{X} - 1.96 \left(\frac{\sigma}{\sqrt{n}} \right) < \mu < \overline{X} + 1.96 \left(\frac{\sigma}{\sqrt{n}} \right)$$

Hence, one can be 95% confident that the population mean is contained within that interval when the values of the variable are normally distributed in the population.

The value used for the 95% confidence interval, 1.96, is obtained from Table E. For a 99% confidence interval, the value 2.58 is used instead of 1.96 in the formula. This value is also obtained from Table E and is based on the standard normal distribution. Since other confidence intervals are used in statistics, the symbol $z_{\alpha/2}$ (read "zee sub alpha over two") is used in the general formula for confidence intervals. The Greek letter α (alpha) represents the total area in both of the tails of the standard normal distribution curve. $\alpha/2$ represents the area in each one of the tails. More will be said after Examples 8–1 and 8–2 about finding other values for $z_{\alpha/2}$.

The relationship between α and the confidence level is that the stated confidence level is the percentage equivalent to the decimal value of $1 - \alpha$, and vice versa. When the 95% confidence interval is to be found, $\alpha = 0.05$, since $1 - 0.05 = 0.95$, or 95%. When $\alpha = 0.01$, then $1 - \alpha = 1 - 0.01 = 0.99$. Hence, the 99% confidence interval is being calculated.

Formula for the Confidence Interval of the Mean for a Specific α

$$\overline{X} - z_{\alpha/2}\left(\frac{\sigma}{\sqrt{n}} \right) < \mu < \overline{X} + z_{\alpha/2}\left(\frac{\sigma}{\sqrt{n}} \right)$$

For a 95% confidence interval, $z_{\alpha/2} = 1.96$; and for a 99% confidence interval, $z_{\alpha/2} = 2.58$.

The term $z_{\alpha/2}(\sigma/\sqrt{n})$ is called the maximum error of estimate. For a specific value, say $\alpha = 0.05$, 95% of the sample means will fall within this error value on either side of the population mean, as previously explained. See Figure 8–1.

FIGURE 8–1
95% Confidence Interval

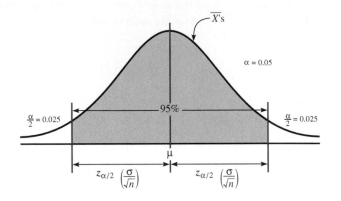

The maximum error of estimate is the maximum difference between the point estimate of a parameter and the actual value of the parameter.

A more detailed explanation of the error of estimate follows the examples illustrating the computation of confidence intervals.

EXAMPLE 8–1

The president of a large university wishes to estimate the average age of the students presently enrolled. From past studies, the standard deviation is known to be 2 years. A sample of 50 students is selected, and the mean is found to be 23.2 years. Find the 95% confidence interval of the population mean.

Solution

Since the 95% confidence interval is desired, $z_{\alpha/2} = 1.96$. Hence, substituting in the formula

$$\overline{X} - z_{\alpha/2}\left(\frac{\sigma}{\sqrt{n}}\right) < \mu < \overline{X} + z_{\alpha/2}\left(\frac{\sigma}{\sqrt{n}}\right)$$

one gets

$$23.2 - (1.96)\left(\frac{2}{\sqrt{50}}\right) < \mu < 23.2 + (1.96)\left(\frac{2}{\sqrt{50}}\right)$$

$$23.2 - 0.55 < \mu < 23.2 + 0.55$$

$$22.65 < \mu < 23.75$$

or 23.2 ± 0.55 years. Hence, the president can say, with 95% confidence, that the average age of the students is between 22.65 and 23.75 years. ◄

EXAMPLE 8–2 A certain medication is known to increase the pulse rate of its users. The standard deviation of the pulse rate is known to be 5 beats per minute. A sample of 30 users had an average pulse rate of 104 beats per minute. Find the 99% confidence interval of the true mean.

Solution Since the 99% confidence interval is desired, $z_{\alpha/2} = 2.58$. Hence, substituting in the formula

$$\overline{X} - z_{\alpha/2}\left(\frac{\sigma}{\sqrt{n}}\right) < \mu < \overline{X} + z_{\alpha/2}\left(\frac{\sigma}{\sqrt{n}}\right)$$

one gets
$$104 - (2.58)\left(\frac{5}{\sqrt{30}}\right) < \mu < 104 + (2.58)\left(\frac{5}{\sqrt{30}}\right)$$
$$104 - 2.36 < \mu < 104 + 2.36$$
$$101.64 < \mu < 106.36 \text{ or } 104 \pm 2.36. \qquad \blacktriangleleft$$

Another way of looking at a confidence interval is shown in Figure 8–2. According to the central limit theorem, approximately 95% of the sample means fall within 1.96 standard deviations of the population mean if the sample size is 30 or more. If it were possible to build a confidence interval about each sample mean, as was done in the previous examples, 95% of these intervals would contain the population mean, as shown in Figure 8–3. Hence, there is a 95% probability that a confidence interval built around a specific sample mean would contain the population mean.

FIGURE 8–2
95% Confidence Interval for Sample Means

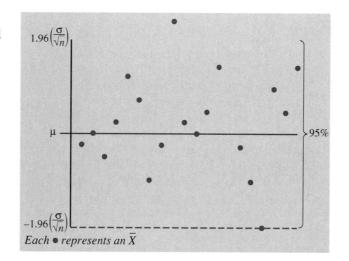

Each ● represents an $\overline{X}$

FIGURE 8–3
95% Confidence
Intervals for Each
Sample Mean

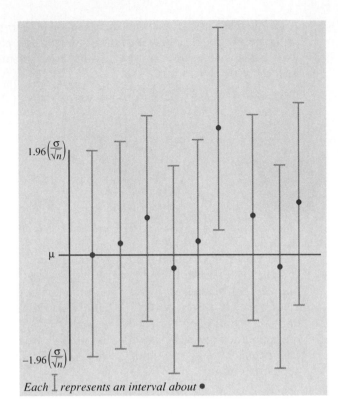

If one desires to be 99% confident, the confidence intervals must be enlarged so that 99 out of every 100 intervals contain the population mean.

Since other confidence intervals (besides 95% and 99%) are sometimes used in statistics, an explanation of how to find the values for $z_{\alpha/2}$ is necessary. As stated previously, the Greek letter α represents the total of the areas in both tails of the normal distribution. The value for α is found by subtracting the decimal equivalent for the desired confidence level from 1. For example, if one wanted to find the 98% confidence interval, one would change 98% to 0.98 and find $\alpha = 1 - 0.98$, or 0.02. Then $\alpha/2$ is obtained by dividing α by 2. So $\alpha/2$ is 0.02/2, or 0.01. Finally, $z_{0.01}$ is the z value that will give an area of 0.01 in the right tail of the standard normal distribution curve. See Figure 8–4.

FIGURE 8–4
Finding $\alpha/2$ for a 98%
Confidence Interval

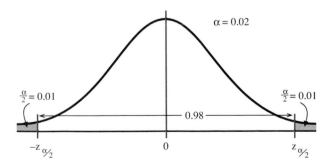

Once $\alpha/2$ is determined, the corresponding $z_{\alpha/2}$ value can be found by using the procedure shown in Chapter 7 (see Example 7–17), which is reviewed here. To get the $z_{\alpha/2}$ value for a 98% confidence interval, subtract 0.01 from 0.5000 to get 0.49. Next, locate the area that is closest to 0.49 (in this case, 0.4901) in Table E, and then find the corresponding z value. In this example, it is 2.33. See Figure 8–5.

FIGURE 8–5
Finding $z_{\alpha/2}$ for a 98% Confidence Interval

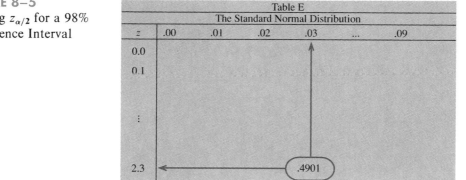

For confidence intervals, only the positive value is used in the formula.

When the original variable is normally distributed and σ is known, the standard normal distribution can be used to find confidence intervals regardless of the size of the sample. When $n \geq 30$, the distribution of means will be approximately normal even if the original distribution of the variable departs from normality. Also, if $n \geq 30$ (some authors use $n > 30$), s can be substituted for σ in the formula for confidence intervals; and the standard normal distribution can be used to find confidence intervals for means, as shown in the next example.

EXAMPLE 8–3   A sample of 50 days showed that a fast-food restaurant served an average of 182 customers during lunch time (11:00 A.M. to 2:00 P.M.). The standard deviation of the sample was 8. Find the 90% confidence interval for the mean.

Solution   **STEP 1**   Find $\alpha/2$. Since the 90% confidence interval is to be used, $\alpha = 1 - 0.90 = 0.10$, and

$$\frac{\alpha}{2} = \frac{0.10}{2} = 0.05$$

STEP 2   Find $z_{\alpha/2}$. Subtract 0.05 from 0.5000 to get 0.4500. The corresponding z value obtained from Table E is 1.65. (*Note:* This value is found by using the z value for an area between 0.4495 and 0.4505. A more precise z value obtained mathematically is 1.645 and is sometimes used; however, 1.65 will be used in this textbook.)

STEP 3   Substitute in the formula

$$\overline{X} - z_{\alpha/2}\left(\frac{s}{\sqrt{n}}\right) < \mu < \overline{X} + z_{\alpha/2}\left(\frac{s}{\sqrt{n}}\right)$$

(s is used in place of σ when σ is unknown, since $n \geq 30$.)

$$182 - 1.65\left(\frac{8}{\sqrt{50}}\right) < \mu < 182 + 1.65\left(\frac{8}{\sqrt{50}}\right)$$
$$182 - 1.65\,(1.1314) < \mu < 182 + 1.65\,(1.1314)$$
$$180.13 < \mu < 183.87$$

Hence, one can be 90% confident that the true population mean is between 180.13 and 183.87. ◀

Sample Size

Closely related to statistical estimating is determining a sample size. Quite often, one asks, "How large a sample is necessary to make an accurate estimate?" The answer is not simple, since it depends upon three things: the maximum error of estimate, the population standard deviation, and the degree of confidence. For example, how close to the true mean does one want to be (2 units, 5 units, etc.), and how confident (90%, 95%, 99%, etc.) does one wish to be? For the purpose of this chapter, it will be assumed that the population standard deviation of the variable is known or has been estimated from a previous study.

The formula for sample size is derived from the maximum error of estimate formula,

$$E = z_{\alpha/2}\left(\frac{\sigma}{\sqrt{n}}\right)$$

and this formula is solved for n as follows:

$$E\sqrt{n} = z_{\alpha/2}(\sigma)$$
$$\sqrt{n} = \frac{z_{\alpha/2}\sigma}{E}$$

Hence,
$$n = \left(\frac{z_{\alpha/2} \cdot \sigma}{E}\right)^2$$

Formula for the Minimum Sample Size Needed for an Interval Estimate of the Population Mean

$$n = \left(\frac{z_{\alpha/2} \cdot \sigma}{E}\right)^2$$

where E is the maximum error of estimate. If necessary, round the answer up to obtain a whole number. That is, if there is any fraction or decimal portion in the answer, the next whole number must be used for sample size.

TV Survey Eyes On-Off Habits

NEW YORK (AP)—You're watching TV and the show offends you. Do you sit there and take it or do you get up and go?

A survey released yesterday on the public's attitudes toward television said 12 percent of the respondents do nothing, 45 percent change channels and 15 percent turn off their sets.

National Association of Broadcasters and the Network Television Association, an alliance of the Big Three broadcast networks, co-sponsored the survey, which has been taken every two years since 1959.

The Roper Organization polled 4,000 adults nationwide in personal interviews between November and December last year. The results had a margin of sampling error of plus or minus 3 percentage points.

The study, "America's Watching," said that even in an age of channel "grazing" about two in three viewers regularly make a special effort to watch shows they particularly like.

And despite increasing cable alternatives, about three-quarters of these "appointment" viewers with cable said that ABC, CBS and NBC still offer most of their favorites.

Viewers find TV news highly credible, the survey also said. In the event of conflicting news reports, 56 percent of respondents said they would believe television's account above others.

A hefty 69 percent said they get most of their news from TV. Newspapers were second, with 43 percent. Radio, word-of-mouth and magazines fell far behind.

A majority of the respondents reported having seen something on television that they found "either personally offensive or morally objectionable" in the past few weeks. But only 42 percent could specify whether it was profanity, sex or violence.

Source: Associated Press. Reprinted with permission.

This news article reports on a survey conducted by the Roper Organization concerning the habits of people who watch television. From the survey results, answer the following questions.

How large was the sample?

What was the maximum error of estimate?

How would you define the population?

What would be the confidence interval for the proportion of viewers who get most of their news from television?

What information is missing from the article?

EXAMPLE 8–4 The college president asks the statistics teacher to estimate the average age of the students at their college. How large a sample is necessary? The statistics teacher decides the estimate should be accurate within 1 year and be 99% confident. From a previous study, the standard deviation of the ages was found to be 2 years.

Solution Since $\alpha = 0.01$ (or $1 - 0.99$), $z_{\alpha/2} = 2.58$, and $E = 1$, substituting in the formula, one gets

$$n = \left(\frac{z_{\alpha/2} \cdot \sigma}{E}\right)^2 = \left[\frac{(2.58)(2)}{1}\right]^2 = 26.6256 \approx 27$$

Therefore, in order to be 99% sure that the estimate is within 1 year of the true mean age, the teacher needs a sample size of at least 27. ◀

Notice that when one is finding the sample size, the size of the population is irrelevant when the population is large or infinite or when sampling is done with replacement. When these conditions cannot be met, an adjustment is made in the formula for computing sample size. This adjustment is beyond the scope of this book.

Sometimes, interval estimates rather than point estimates are reported. For instance, one may read a statement such as "On the basis of a sample of 200 families, the survey estimates that the American family of two spends an average of $84 per week for groceries. There is a 95% probability that this estimate is accurate within $3 of the true mean." This statement means that the 95% confidence interval of the true mean is

$$\$81 < \mu < \$87$$

EXERCISES

8–1. What is the difference between a point estimate and an interval estimate of a parameter? Which is better? Why?

8–2. What information is necessary to calculate a confidence interval?

8–3. What is the maximum error of estimate?

8–4. What is meant by the 95% confidence interval of the mean?

8–5. What are three properties of a good estimator?

8–6. What statistic best estimates μ?

8–7. What is necessary to determine sample size?

8–8. When one is determining the sample size for a confidence interval, is the size of the population relevant?

8–9. Find the critical values for each.

a. $z_{\alpha/2}$ for the 99% confidence interval
b. $z_{\alpha/2}$ for the 98% confidence interval
c. $z_{\alpha/2}$ for the 95% confidence interval
d. $z_{\alpha/2}$ for the 90% confidence interval
e. $z_{\alpha/2}$ for the 94% confidence interval

8–10. A study of 36 camels showed that they could walk at an average rate of 2.6 miles per hour. The sample standard deviation is 0.4. Find the 95% confidence interval of the mean for all camels.

8–11. A sample of the reading scores of 35 fifth-graders had a mean of 82. The standard deviation of the sample is 15.

a. Find the 95% confidence interval of the mean reading scores of all fifth-graders.
b. Find the 99% confidence interval of the mean reading scores of all fifth-graders.
c. Which interval is larger? Explain why.

8–12. In a recent study of 35 ninth-grade students, the mean number of hours per week that they watched television was 22.6. The standard deviation of the sample was 2.8. Find the 98% confidence interval of the mean.

8–13. A study of 40 English composition professors showed that they spent, on average, 12.6 minutes correcting a student's term paper.

a. Find the 90% confidence interval of the mean time for all composition papers when $\sigma = 2.5$ minutes.
b. If a professor stated that he spent, on average, 30 minutes correcting a term paper, what would be your reaction?

8–14. A study of 40 bowlers showed that their average score was 186. The standard deviation of the population is 6.

a. Find the 95% confidence of the mean score for all bowlers.
b. Find the 95% confidence interval of the mean score if a sample of 100 bowlers is used instead of a sample of 40.
c. Which interval is smaller? Explain why.

8–15. A sociologist found that in a sample of 49 retired men, the average number of jobs they had during their lifetimes was 7.2. The standard deviation of the sample was 2.1. Find the 90% confidence interval of the mean for the number of jobs a man had during his lifetime.

8–16. Researchers found that for a certain ethnic group, the females lose an average of 65 hairs per week from their heads. The standard deviation of the sample of 36 females was 4. Find the 95% confidence interval of the population mean.

8–17. A random sample of 48 days taken at a large hospital shows that an average of 38 patients were treated in the emergency room per day. The standard deviation of the population is 4.

a. Find the 99% confidence interval of the mean number of emergency room patients treated each day at the hospital.
b. Find the 99% confidence interval of the mean number of emergency room patients treated each day if the standard deviation was 8 instead of 4.
c. Why is the confidence interval for part b larger than the one for part a?

8–18. A random sample of 49 female shoppers showed that they spent an average of $23.45 per visit at a grocery store. The standard deviation of the sample was $2.80. Find the 90% confidence interval of the true mean.

8–19. In Exercise 8–18, the researcher found that the female shoppers spent an average of 18.6 minutes per visit, with a sample standard deviation of 5 minutes. Find the 90% confidence interval of the mean time a female spends grocery shopping.

8–20. A researcher is interested in estimating the average salary of police officers in a large city. She wants to be 95% sure that her estimate is correct. If the standard deviation is $1050, how large a sample is necessary to get the desired information and to be accurate within $200?

8–21. A university dean wishes to estimate the average number of hours his part-time instructors teach per week. The standard deviation from a previous study is 2.6 hours. How large a sample must be selected if he wants to be 99% confident of finding whether the true mean differs from the sample mean by 1 hour?

8–22. In order to determine rates, a motel manager wishes to find the 90% confidence interval of the mean rate that competing motels charge. He wishes to be accurate within $2.00. If the standard deviation of the rates is $10.00, how large a sample must be selected to get the desired information?

8–23. An insurance company is trying to estimate the average number of sick days that full-time food service workers use per year. A pilot study found the standard deviation to be 2.5 days. How large of a sample must be selected if the company wants to be 95% sure of getting an interval that contains the true mean with a maximum error of 1 day?

8–24. A restaurant owner wishes to find the 99% confidence interval of the true mean cost of a dry martini. How large should the sample be if she wishes to be accurate within $0.10? A previous study showed that the standard deviation of the price was $0.12.

8–25. A health care professional wishes to estimate the birth weights of infants. How large a sample must she select if she desires to be 90% confident that the true mean is within 6 ounces of the sample mean? The standard deviation of the birth weights is known to be 8 ounces.

8-3 CONFIDENCE INTERVALS FOR THE MEAN (σ UNKNOWN AND $n < 30$)

When σ is known and the variable is normally distributed, or when σ is unknown and $n \geq 30$, the standard normal distribution is used to find confidence intervals for the mean. However, in many situations, the population standard deviation is not known and the sample size is less than 30. In such situations, the standard deviation from the sample can be used in place of the population standard deviation for confidence intervals. But a somewhat different distribution, called the *t* **distribution,** must be used when the sample size is less than 30 and the variable is normally or approximately normally distributed.

The *t* distribution was formulated in 1908 by an Irish brewery employee named W. S. Gosset. Gosset was involved in researching new methods of manufacturing ale. Because brewery employees were not allowed to publish results, Gosset published his findings using the pseudonym Student; hence, the *t* distribution is sometimes called the Student's *t* distribution.

Some important characteristics of the *t* distribution are described next.

Characteristics of the *t* Distribution

The *t* distribution shares some characteristics of the normal distribution and differs from it in others. The *t* distribution is similar to the standard normal distribution in the following ways.

1. It is bell-shaped.
2. It is symmetrical about the mean.
3. The mean, median, and mode are equal to 0 and are located at the center of the distribution.
4. The curve never touches the *x* axis.

The *t* distribution differs from the standard normal distribution in the following ways.

1. The variance is greater than 1.
2. The *t* distribution is actually a family of curves based on the concept of *degrees of freedom,* which is related to sample size.
3. As the sample size increases, the *t* distribution approaches the standard normal distribution. See Figure 8–6.

FIGURE 8–6
The t Family of Curves

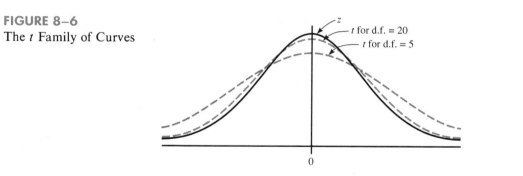

Many statistical distributions use the concept of degrees of freedom, and the formulas for finding the degrees of freedom vary for different statistical tests. The **degrees of freedom** are the number of values that are free to vary after a sample statistic has been computed, and they tell the researcher which specific curve to use when a distribution consists of a family of curves.

For example, if the mean of 5 values is 10, then 4 of the 5 values are free to vary. But once 4 values are selected, the fifth value must be a specific number to get a sum of 50, since $50 \div 5 = 10$. Hence, the degrees of freedom are $5 - 1 = 4$, and this value tells the researcher which t curve to use.

The symbol d.f. will be used for degrees of freedom. The degrees of freedom for a confidence interval for the mean are found by subtracting 1 from the sample size. That is, d.f. $= n - 1$.

The formula for finding a confidence interval about the mean using the t distribution is given next.

Formula for a Specific Confidence Interval for the Mean When σ Is Unknown and $n < 30$

$$\overline{X} - t_{\alpha/2}\left(\frac{s}{\sqrt{n}}\right) < \mu < \overline{X} + t_{\alpha/2}\left(\frac{s}{\sqrt{n}}\right)$$

The degrees of freedom are $n - 1$.

The values for $t_{\alpha/2}$ are found in Table F in the Appendix. The top row of Table F, labeled "Confidence Intervals," is used to get these values. The other two rows, labeled "One Tail" and "Two Tails," will be explained in the next chapter and should not be used here.

The next two examples show how to find the value in Table F for $t_{\alpha/2}$.

EXAMPLE 8–5   Find the $t_{\alpha/2}$ value for a 95% confidence interval when the sample size is 22.

Solution   d.f. = 22 − 1, or 21. Find 21 in the left column and 95% in the row labeled "Confidence Intervals." The intersection where the two meet gives the value for $t_{\alpha/2}$, which is 2.080. See Figure 8–7.

FIGURE 8–7

Finding $t_{\alpha/2}$ for Example 8–5

	Table F						
		The *t* Distribution					
	Confidence Intervals	50%	80%	90%	95%	98%	99%
d.f.	One Tail α	0.25	0.10	0.05	0.025	0.01	0.005
	Two Tails α	0.50	0.20	0.10	0.05	0.02	0.01
1							
2							
3							
⋮							
21						2.080	2.518	2.831
⋮							
z ∞		.674	1.282[a]	1.645[b]	1.960	2.326[c]	2.576[d]

Source: Beyer, W.H., *Handbook of Tables for Probability and Statistics, 2nd Edition*, CRC Press, Boca Raton, Florida, 1986. With permission.

[a] This value has been rounded to 1.28 in the textbook.
[b] This value has been rounded to 1.65 in the textbook.
[c] This value has been rounded to 2.33 in the textbook.
[d] This value has been rounded to 2.58 in the textbook.

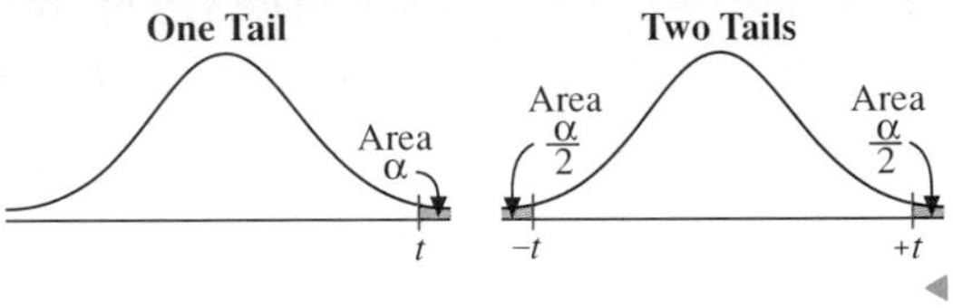

One Tail   Area α

Two Tails   Area $\frac{\alpha}{2}$ Area $\frac{\alpha}{2}$

EXAMPLE 8–6   Find the $t_{\alpha/2}$ value for a 99% confidence interval when the sample size is 14.

Solution   d.f. = n − 1 = 14 − 1 = 13. Look up 13 in the column labeled "d.f." and look up 99% in the row labeled "Confidence Intervals." The value where the two meet is 3.012.

Note: At the bottom of Table F where d.f. = ∞, the $z_{\alpha/2}$ values can be found for specific confidence intervals. The reason is that as the degrees of freedom increase, the *t* distribution approaches the standard normal distribution.

The next example shows how to find the confidence interval when one is using the *t* distribution.

EXAMPLE 8–7   Ten randomly selected automobiles were stopped, and the tread depth of the right front tire was measured. The mean was 0.32 inch, and the standard deviation was 0.08 inch. Find the 95% confidence interval of the mean depth. Assume that the variable is approximately normally distributed.

Solution   Since σ is unknown and s must replace it, the t distribution (Table F) must be used. Hence, with 9 degrees of freedom, $t_{\alpha/2} = 2.262$.

The 95% confidence interval of the population mean is found by substituting in the formula

$$\overline{X} - t_{\alpha/2}\left(\frac{s}{\sqrt{n}}\right) < \mu < \overline{X} + t_{\alpha/2}\left(\frac{s}{\sqrt{n}}\right)$$

Hence, $0.32 - (2.262)\left(\dfrac{0.08}{\sqrt{10}}\right) < \mu < 0.32 + (2.262)\left(\dfrac{0.08}{\sqrt{10}}\right)$

$$0.32 - 0.057 < \mu < 0.32 + 0.057$$
$$0.263 < \mu < 0.377$$

Therefore, one can be 95% confident that the population mean is between 0.263 and 0.377.   ◀

Students sometimes have difficulty deciding whether to use $z_{\alpha/2}$ or $t_{\alpha/2}$ values when finding confidence intervals for the mean. As stated previously, when σ is known, $z_{\alpha/2}$ values can be used no matter what the sample size is, as long as the variable is normally distributed or $n \geq 30$. When σ is unknown and $n \geq 30$, s can be used in the formula and $z_{\alpha/2}$ values can be used. Finally, when σ is unknown and $n < 30$, s is used in the formula and $t_{\alpha/2}$ values are used, as long as the variable is approximately normally distributed. These rules are summarized in Figure 8–8.

FIGURE 8–8
When to use the z or t
Distribution

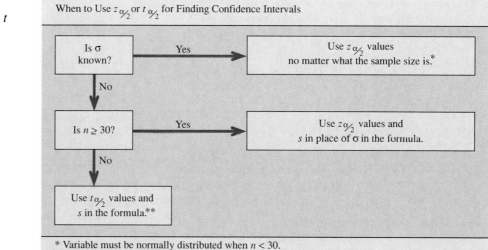

When to Use $z_{\alpha/2}$ or $t_{\alpha/2}$ for Finding Confidence Intervals

Is σ known? — Yes → Use $z_{\alpha/2}$ values no matter what the sample size is.*

No ↓

Is $n \geq 30$? — Yes → Use $z_{\alpha/2}$ values and s in place of σ in the formula.

No ↓

Use $t_{\alpha/2}$ values and s in the formula.**

* Variable must be normally distributed when $n < 30$.
** Variable must be approximately normally distributed.

It should be pointed out that some statisticians have a different point of view. They use $z_{\alpha/2}$ values when σ is known and $t_{\alpha/2}$ values when σ is unknown. In these circumstances, a different t table, one that contains t values for sample sizes greater than or equal to 30, would be needed. The procedure shown in Figure 8–8 is the one used throughout this textbook.

EXERCISES

8–26. What are the properties of the t distribution?

8–27. Who developed the t distribution?

8–28. What is meant by degrees of freedom?

8–29. When should the t distribution be used to find a confidence interval for the mean?

8–30. (ANS.) Find the values for each.

a. $t_{\alpha/2}$ and $n = 18$ for the 99% confidence interval
b. $t_{\alpha/2}$ and $n = 23$ for the 95% confidence interval
c. $t_{\alpha/2}$ and $n = 15$ for the 98% confidence interval
d. $t_{\alpha/2}$ and $n = 10$ for the 90% confidence interval
e. $t_{\alpha/2}$ and $n = 20$ for the 95% confidence interval

For Exercises 8–31 through 8–46, assume that all variables are approximately normally distributed.

8–31. The average hemoglobin reading for a sample of 20 teachers was 16 grams per 100 milliliters, with a sample standard deviation of 2 grams. Find the 99% confidence interval of the true mean.

8–32. A meteorologist who sampled 15 cold weather fronts found that the average speed at which they traveled across a certain state was 18 miles per hour. The standard deviation of the sample was 2 miles per hour. Find the 95% confidence interval of the mean.

8–33. A sample of 25 two-year-old chickens shows that they lay an average of 21 eggs per month. The standard deviation of the sample was 2 eggs. Find the 99% confidence interval of the true mean.

8–34. A sample of 20 tuna fish showed that they swim an average of 8.6 miles per hour. The standard deviation for the sample was 1.6. Find the 90% confidence interval of the true mean.

8–35. A sample of 6 adult elephants had an average weight of 12,200 pounds, with a sample standard deviation of 200 pounds. Find the 95% confidence interval of the true mean.

8–36. A sample of 15 private-duty nurses showed an average weekly wage of $480.75. The standard deviation of the sample was $56. Find the 98% confidence interval of the true mean.

8–37. A recent study of 28 city residents showed that the mean of the number of years they lived at their present address was 9.3 years. The standard deviation of the sample was 2 years. Find the 90% confidence interval of the true mean.

8–38. An automobile shop manager timed 6 employees and found that the average time it took them to change a water pump was 18 minutes. The standard deviation of the sample was 3 minutes. Find the 99% confidence interval of the true mean.

8–39. A recent study of 25 students showed that they spent an average of $18.53 for gasoline per week. The standard deviation of the sample was $3.00. Find the 95% confidence interval of the true mean.

8–40. For a group of 10 men subjected to a stress situation, the mean of the number of heartbeats per minute was 126, and the standard deviation was 4. Find the 95% confidence interval of the true mean.

8–41. For the stress test described in Exercise 8–40, 6 females had an average heartbeat rate of 115 beats per minute. The standard deviation of the sample was 6 beats. Find the 95% confidence interval of the true mean for the females.

8–42. A random sample of 29 male wrestlers showed that their average weight was 256 pounds. The standard deviation of the sample was 8 pounds. Find the 99% confidence interval of the true mean.

8–43. A sample of 15 waitresses showed an average weekly income of $320.20. The standard deviation of the sample was $12. Find the 98% confidence interval of the true mean.

8–44. For a group of 20 students taking a final exam, the mean heart rate was 96 beats per minute, and the standard deviation was 5. Find the 95% confidence interval of the true mean.

8–45. The average yearly income for 28 married couples living in city C is $58,219. The standard deviation of the sample is $56. Find the 95% confidence interval of the true mean.

*8–46. A random sample of 20 parking meters in a large municipality showed the following incomes for a day.

$2.60	$1.05	$2.45	$2.90
$1.30	$3.10	$2.35	$2.00
$2.40	$2.35	$2.40	$1.95
$2.80	$2.50	$2.10	$1.75
$1.00	$2.75	$1.80	$1.95

Find the 95% confidence interval of the true mean.

THE MAN IN THE STREET
(SUBJECT TO A SAMPLING ERROR OF PLUS OR MINUS THREE PERCENTAGE POINTS)

Source: Drawing by Nurit: © 1988 *The New Yorker Magazine, Inc.*

8–4 CONFIDENCE INTERVALS AND SAMPLE SIZE FOR PROPORTIONS

A recent USA TODAY "Snapshots" feature (July 2, 1993) stated that 12% of the pleasure boats in the United States were named *Serenity*. The parameter 12% is called a **proportion.** It means that of all the pleasure boats in the United States, 12 out of every 100 are named *Serenity*. A proportion represents a part of a whole. It can be expressed as a fraction, decimal, or percentage. In this case, $12\% = 0.12 = \frac{12}{100}$ or $\frac{3}{25}$. Proportions can also represent probabilities. In this case, if a pleasure boat is selected at random, the probability that it is called *Serenity* is 0.12.

Proportions can be obtained from samples or populations. The following symbols will be used.

Symbols Used in Proportion Notation

p = Symbol for the population proportion
$\hat{p}$ (read "p hat") = Symbol for the sample proportion

For a sample proportion,

$$\hat{p} = \frac{X}{n} \quad \text{and} \quad \hat{q} = \frac{n - X}{n} \quad \text{or} \quad 1 - \hat{p}$$

where X = number of sample units that possess the characteristics of interest and n = sample size.

For example, in a study, 200 people were asked if they were satisfied with their job or profession; 162 said that they were. In this case, $n = 200$, $X = 162$, and $\hat{p} = X/n = 162/200 = 0.81$. It can be said that for this sample, 0.81 or 81% of those surveyed were satisfied with their job or profession. The sample proportion is $\hat{p} = 0.81$.

The proportion of people who did not respond favorably when asked if they were satisfied with their job or profession constitutes $\hat{q}$, where $\hat{q} = (n - X)/n$. For the survey, $\hat{q} = (200 - 162)/200 = 38/200$, or 0.19, or 19%.

When $\hat{p}$ and $\hat{q}$ are given in decimals or fractions, $\hat{p} + \hat{q} = 1$; and when $\hat{p}$ and $\hat{q}$ are given in percentages, $\hat{p} + \hat{q} = 100\%$. It follows, then, that $\hat{q} = 1 - \hat{p}$, or $\hat{p} = 1 - \hat{q}$, when $\hat{p}$ and $\hat{q}$ are in decimal or fraction form. For the sample survey on job satisfaction, $\hat{q}$ can also be found by using $\hat{q} = 1 - \hat{p}$, or $1 - 0.81 = 0.19$.

Similar reasoning applies to population proportions; i.e., $p = 1 - q$, $q = 1 - p$, and $p + q = 1$, when p and q are expressed in decimal or fraction form. When p and q are expressed as percentages, $p + q = 100\%$, $p = 100\% - q$, and $q = 100\% - p$.

EXAMPLE 8–8   In a recent survey of 150 households, 54 had central air-conditioning. Find $\hat{p}$ and $\hat{q}$.

Solution   Since $X = 54$ and $n = 150$,

$$\hat{p} = \frac{X}{n} = \frac{54}{150} = 0.36 = 36\%$$

$$\hat{q} = \frac{n - X}{n} = \frac{150 - 54}{150} = \frac{96}{150} = 0.64 = 64\%$$

$\hat{q}$ can also be found by using the formula $\hat{q} = 1 - \hat{p}$. In this case, $\hat{q} = 1 - 0.36 = 0.64$.   ◄

As with means, the statistician, given the sample population, tries to estimate the population proportion. Point and interval estimates for a population proportion can be made by using the sample proportion. For a point estimate of p (the population proportion), $\hat{p}$ (the sample proportion) is used. On the basis of the three properties of a good estimator, $\hat{p}$ is unbiased, consistent, and relatively efficient. But as with means, one is not able to decide how good the point estimate of p is. Therefore, statisticians also use an interval estimate for a proportion, and they can assign a probability that the interval will contain the population proportion.

Confidence Intervals   To construct a confidence interval about a proportion, one must use the maximum error of estimate, which is

$$E = z_{\alpha/2} \sqrt{\frac{\hat{p}\,\hat{q}}{n}}$$

Confidence intervals about proportions must meet the criteria that $np \geq 5$ and $nq \geq 5$.

Speaking of **Statistics**

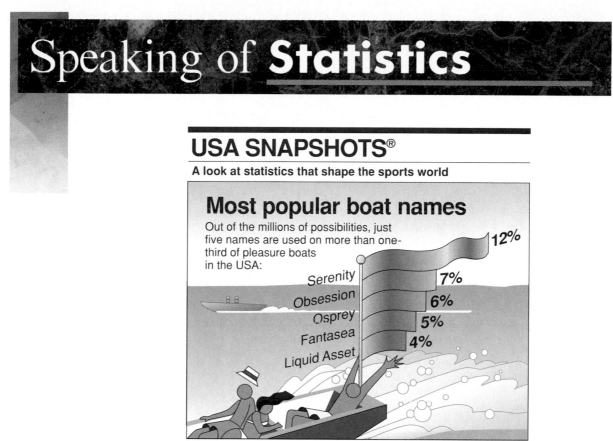

USA SNAPSHOTS®

A look at statistics that shape the sports world

Most popular boat names

Out of the millions of possibilities, just five names are used on more than one-third of pleasure boats in the USA:

Serenity — 12%
Obsession — 7%
Osprey — 6%
Fantasea — 5%
Liquid Asset — 4%

Source: Boat/U.S. (sample of 200,000 boat names)   By Sam Ward, USA TODAY

Source: Based on a survey by Boat/U.S. Copyright 1993, USA TODAY. Reprinted with permission.

This study uses proportions. The sample size was 200,000. Find the actual number of boats in each group.

Formula for a Specific Confidence Interval for a Proportion

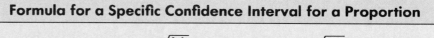

$$\hat{p} - (z_{\alpha/2}) \sqrt{\frac{\hat{p}\,\hat{q}}{n}} < p < \hat{p} + (z_{\alpha/2}) \sqrt{\frac{\hat{p}\,\hat{q}}{n}}$$

when np and nq are greater than or equal to 5.

EXAMPLE 8–9   A sample of 500 nursing applications included 60 from males. Find the 90% confidence interval of the true proportion of males who applied to the nursing program.

Solution   Since $\alpha = 1 - 0.90 = 0.10$, and $z_{\alpha/2} = 1.65$, substituting in the formula

$$\hat{p} - (z_{\alpha/2}) \sqrt{\frac{\hat{p}\,\hat{q}}{n}} < p < \hat{p} + (z_{\alpha/2}) \sqrt{\frac{\hat{p}\,\hat{q}}{n}}$$

when $\hat{p} = 60/500 = 0.12$ and $\hat{q} = 1 - 0.12 = 0.88$, one gets

$$0.12 - (1.65) \sqrt{\frac{(0.12)(0.88)}{500}} < p < 0.12 + (1.65) \sqrt{\frac{(0.12)(0.88)}{500}}$$

$$0.12 - 0.024 < p < 0.12 + 0.024$$

$$0.096 < p < 0.144$$

or $$9.6\% < p < 14.4\%$$

Hence, one can be 90% sure that the proportion of males who applied is between 9.6% and 14.4%. ◀

When a specific percentage is given, the percentage becomes $\hat{p}$ when it is changed to a decimal. For example, if the problem states that 12% of the applicants were males, then $\hat{p} = 0.12$.

EXAMPLE 8–10   In a sample of 100 teenage girls, 30% used hair coloring. Find the 95% confidence interval of the true proportion of teenage girls who use hair coloring.

Solution   Since $\hat{p} = 0.30$ and $z_{\alpha/2} = 1.96$, substituting in the formula

$$\hat{p} - z_{\alpha/2} \sqrt{\frac{\hat{p}\,\hat{q}}{n}} < p < \hat{p} + z_{\alpha/2} \sqrt{\frac{\hat{p}\,\hat{q}}{n}}$$

one gets

$$0.30 - (1.96) \sqrt{\frac{(0.30)(0.70)}{100}} < p < 0.30 + (1.96) \sqrt{\frac{(0.30)(0.70)}{100}}$$

$$0.30 - 0.0898 < p < 0.30 + 0.0898$$

$$0.2102 < p < 0.3898$$

$$21.02\% < p < 38.98\%$$

Hence, one can say with 95% confidence that the true proportion of teenage girls who use hair coloring is between 21.02% and 38.98%. ◀

Sample Size for Proportions

In order to find the sample size necessary for determining a confidence interval about a proportion, one uses the following formula.

Formula for the Minimum Sample Size Needed for an Interval Estimate of a Population Proportion

$$n = \hat{p}\,\hat{q} \left(\frac{z_{\alpha/2}}{E}\right)^2$$

If necessary, round up to obtain a whole number.

This formula can be found by solving the maximum error of estimate value for *n:*

$$E = z_{\alpha/2} \sqrt{\frac{\hat{p}\hat{q}}{n}}$$

There are two situations to consider. First, if some approximation of $\hat{p}$ is known (e.g., from a previous study), that value can be used in the formula. Second, if no approximation of $\hat{p}$ is known, one should use $\hat{p} = 0.5$. This value will give a sample size sufficiently large to guarantee an accurate prediction, given the confidence interval and the error of estimate.

EXAMPLE 8–11   A researcher wishes to estimate, with 95% confidence, the number of people who own a home computer. A previous study shows that 40% of those interviewed had a computer at home. The researcher wishes to be accurate within 2% of the true proportion. Find the minimum sample size necessary.

Solution   Since $\alpha = 0.05$, $z_{\alpha/2} = 1.96$, $E = 0.02$, $\hat{p} = 0.40$, and $\hat{q} = 0.60$, then

$$n = \hat{p}\hat{q}\left(\frac{z_{\alpha/2}}{E}\right)^2 = (0.40)(0.60)\left(\frac{1.96}{0.02}\right)^2 = 2304.96$$

which, when rounded up, is 2305.   ◄

EXAMPLE 8–12   The same researcher wishes to estimate the proportion of executives who own an automobile telephone. She wants to be 90% confident and be accurate within 5% of the true proportion. Find the minimum sample size necessary.

Solution   Since no prior knowledge of *p* is known, statisticians assign the values $\hat{p} = 0.5$ and $\hat{q} = 0.5$. The sample size obtained by using these values will be large enough to ensure the specified degree of confidence. Hence,

$$n = \hat{p}\hat{q}\left(\frac{z_{\alpha/2}}{E}\right)^2 = (0.5)(0.5)\left(\frac{1.65}{0.05}\right)^2 = 272.25$$

which, when rounded up, is 273.   ◄

In determination of the sample size, the size of the population is irrelevant. Only the degree of confidence and the maximum error are necessary to make the determination.

EXERCISES

8–47. In each case, find $\hat{p}$ and $\hat{q}$.

a. $n = 80$ and $X = 40$
b. $n = 200$ and $X = 90$
c. $n = 130$ and $X = 60$
d. $n = 60$ and $X = 35$
c. $n = 95$ and $X = 43$

8–48. (ANS.) Find $\hat{p}$ and $\hat{q}$ for each percentage. (Use each percentage for $\hat{p}$.)

a. 12%
b. 29%
c. 65%
d. 53%
e. 67%

8–49. In a sample of 200 people, 168 said that they watched educational television. Find the 95% confidence interval of the true proportion who watch educational television.

8–50. In a recent study of 100 people, 85 said that they were satisfied with their present home. Find the 90% confidence interval of the true proportion of individuals who are satisfied with their present home.

8–51. An employment counselor found that in a sample of 100 unemployed workers, 65% were not interested in returning to work. Find the 95% confidence interval of the true proportion of workers who do not wish to return to work.

8–52. A survey of 100 new-home buyers showed that 56% preferred whole-house air-conditioning. Find the 95% confidence interval of the true proportion of buyers who prefer whole-house air-conditioning.

8–53. A nutritionist found that in a sample of 80 families, 22% indicated that they ate apples at least once a week. Find the 90% confidence interval of the true proportion of families who eat apples at least once a week.

8–54. A survey of 120 female freshmen shows that 18 did not wish to work after marriage. Find the 95% confidence interval of the true proportion of females who do not wish to work after marriage.

8–55. In an article on education, a researcher stated that only 10% of a sample of 200 people feel that teachers should pass competency examinations before teaching. Find the 99% confidence interval of the true proportion.

8–56. A survey of 80 recent fatal traffic accidents showed that 46 were alcohol-related. Find the 95% confidence interval of the true proportion of people who die from alcohol-related accidents.

8–57. A survey of 90 families showed that 40 owned at least one gun. Find the 95% confidence interval of the true proportion of families who own at least one gun.

8–58. In a certain state, a survey of 500 workers showed that 45% belonged to a union. Find the 90% confidence interval of the true proportion of workers who belong to a union.

8–59. For a certain age group, a study of 100 people showed that 25% died of cancer. Find the true proportion of individuals who die of cancer for that age group. Use the 98% confidence interval.

8–60. A researcher wishes to be 95% confident that her estimate of the true proportion of individuals who travel overseas is within 0.04 of the true proportion. Find the sample size necessary. In a prior study, a sample of 200 people showed that 80 traveled overseas last year.

8–61. A medical researcher wishes to determine the percentage of females under 20 years of age who take vitamins. He wishes to be 99% confident that the estimate is within 2 percentage points of the true proportion. How large should the sample size be? A recent study of 180 females who took vitamins showed that 25% were under 20 years of age.

8–62. How large a sample must one take to be 90% confident that the estimate is within 0.05 of the true proportion of women over 55 who are widows? A recent study indicated that 29% of the 100 women over 55 in the study were widows.

8–63. A researcher wishes to estimate the proportion of adult males who are under 5 feet 5 inches tall. She wants to be 90% confident that her estimate is within 5% of the true proportion. How large a sample should be taken if in a sample of 300 males, 30 were under 5 feet 5 inches tall?

8–64. An educator desires to estimate, within 0.03, the true proportion of high school students who study at least 1 hour each school night. He wants to be 98% confident. How large a sample is necessary? Previously, he conducted a study and found that 60% of the 250 students surveyed spent at least 1 hour each school night studying.

*8–65. If a sample of 600 people is chosen and the researcher desires to have a maximum error of estimate of 4% on the specific proportion who favor gun control, find the degree of confidence. A recent study showed that 50% were in favor of some form of gun control.

*8–66. A sample of 800 people is selected and interviewed concerning their opinions on increasing the school year by 20 days. If 30% are in favor, find the maximum error of estimate if the confidence interval is 95%.

8–5 SUMMARY

An important aspect of inferential statistics is estimation. Estimations of parameters of populations are accomplished by selecting a random sample from that population and choosing and computing a statistic that is the best estimator of the parameter. A good estimator must be unbiased, consistent, and relatively efficient. The best estimators of μ and p are $\overline{X}$ and $\hat{p}$, respectively.

There are two types of estimators of a parameter: point estimates and interval estimates. A point estimate is a specific value. For example, if a researcher wishes to estimate the average length of a certain adult fish, a sample of the fish is selected and measured. The mean of this sample is computed—e.g., 3.2 centimeters. From this sample mean, the researcher estimates the population mean to be 3.2 centimeters.

The problem with point estimates is that the accuracy of the estimate cannot be determined. For this reason, statisticians prefer to use the interval estimate. By computing an interval about the sample value, statisticians can be 95% or 99% (or some other percentage) confident that their estimate contains the true parameter. The confidence level is determined by the researcher. The higher the confidence level, the wider the interval of the estimate must be. For example, a 95% confidence interval of the true mean length of a certain species of fish might be

$$3.17 < \mu < 3.23$$

whereas the 99% confidence interval might be

$$3.15 < \mu < 3.25$$

When the confidence interval of the mean is computed, the z or t values are used, depending on whether or not the population standard deviation is known and depending on the size of the sample. If σ is known, the z values can be used. If σ is not known, the t values must be used when the sample size is less than 30.

Closely related to computing confidence intervals is determining the sample size to make an estimate of the mean. In order to determine the minimum sample size necessary, one must have the following information:

1. The degree of confidence must be stated.
2. The population standard deviation must be known or be able to be estimated.
3. The maximum error of estimate must be stated.

Confidence intervals and sample sizes can also be computed for proportions.

Important Terms

Confidence interval
Confidence level
Consistent estimator
Degrees of freedom
Estimation
Estimator
Interval estimate
Maximum error of estimate
Point estimate
Proportion
Relatively efficient estimator
t distribution
Unbiased estimator

Important Formulas

Formula for the confidence interval of the mean when σ is known (when $n \geq 30$, s can be used if σ is unknown):

$$\overline{X} - z_{\alpha/2}\left(\frac{\sigma}{\sqrt{n}}\right) < \mu < \overline{X} + z_{\alpha/2}\left(\frac{\sigma}{\sqrt{n}}\right)$$

Formula for the sample size for means:

$$n = \left(\frac{z_{\alpha/2} \cdot \sigma}{E}\right)^2$$

where E is the maximum error.
Formula for the confidence interval of the mean when σ is unknown and $n < 30$:

$$\overline{X} - t_{\alpha/2}\left(\frac{s}{\sqrt{n}}\right) < \mu < \overline{X} + t_{\alpha/2}\left(\frac{s}{\sqrt{n}}\right)$$

Formula for the confidence interval for a proportion:

$$\hat{p} - (z_{\alpha/2})\sqrt{\frac{\hat{p}\hat{q}}{n}} < p < \hat{p} + (z_{\alpha/2})\sqrt{\frac{\hat{p}\hat{q}}{n}}$$

where $\hat{p} = X/n$ and $\hat{q} = 1 - \hat{p}$
Formula for the sample size for proportions:

$$n = \hat{p}\hat{q}\left(\frac{z_{\alpha/2}}{E}\right)^2$$

Review Exercises

8–67. A study of 50 adults from a certain population showed the mean diastolic blood pressure to be 72 millimeters of mercury (mmHg) and the population standard deviation to be 11.6. Find the 90% confidence interval of the true mean of the population.

8–68. The owner of a small business complained that the rent was too high. He randomly surveyed 32 other businesses in his locality and found that the mean rent on their small stores was $1250 per month. The standard deviation of the sample was $33. Find the 90% confidence interval of the population mean.

8–69. The average weight of 60 randomly selected compact automobiles was 2627 pounds. The sample standard deviation was 400 pounds. Find the 99% confidence interval of the true mean weight of the automobiles.

8–70. In a study of 12 nurses from a large city hospital, the average age was 26.8 years, and the standard deviation of the sample was 4.8 years. Find the 95% confidence interval of the population mean of all nurses in the hospital.

8–71. In a hospital, a sample of 10 weeks was selected, and the researcher found that an average of 12 babies were born each week. The standard deviation of the sample was 2. Find the 99% confidence interval of the true mean.

8–72. For a certain urban area, in a sample of five months, an average of 28 mail carriers were bitten by dogs each month. The standard deviation of the sample was 3. Find the 90% confidence interval of the true mean number of mail carriers who are bitten by dogs each month.

8–73. How large a sample is necessary to be 95% sure that the estimate of the mean income of draftsmen is within $200 of the true mean? The standard deviation of income is $800.

8–74. A researcher wishes to estimate, within $25, the true average amount of postage a community college spends each year. If she wishes to be 90% confident, how large a sample is necessary? The standard deviation is known to be $80.

8–75. A recent study of 150 students found that 86 resided in off-campus housing. Find the 95% confidence interval of the proportion of all students who reside in off-campus housing.

8–76. In a study of 200 accidents that required treatment in an emergency room, 40% occurred at home. Find the 90% confidence interval of the true proportion of accidents that occur at home.

8–77. A political analyst found that 60% of 300 Republican voters feel that the federal government has too much power. Find the 95% confidence interval of the population proportion of Republican voters who feel this way.

8–78. A nutritionist wishes to determine, within 2%, the true proportion of adults who snack before bedtime. If she wishes to be 95% confident that her estimate contains the population proportion, how large a sample will she need? A previous study found that 18% of the 100 people surveyed said they did snack before bedtime.

8–79. A survey of 200 adults showed that 15% played basketball for regular exercise. If a researcher desires to find the 99% confidence interval of the true proportion of adults who play basketball and be within 1% of the true population, how large a sample should be selected?

COMPUTER APPLICATIONS

1. Write a computer program to generate a 90%, 95%, or 99% confidence interval for the mean of a population, given the population standard deviation. The program should use the raw data for samples of 30 or more.
2. Write a computer program to compute the minimum sample size needed to estimate the mean for a variable. Input the 90%, 95%, or 99% confidence level desired, the standard deviation, and the maximum error.
3. MINITAB can be used to compute a confidence interval about a mean using the z or t values. For example, a researcher wishes to find the 95% confidence interval of the average age of the managers of a nationwide chain store. A sample of 30 managers is selected and the following data are obtained:

43   52   18   20   25   20   45   43   21   42   32   24   32   19   25
26   44   42   41   41   53   22   25   23   21   27   33   36   47   19

The standard deviation (from past records) is 11 years. Using MINITAB, enter the data by using the SET C1 command. To find the 95% confidence interval of the mean using the z values, type ZINT 95 SIGMA = 11 C1, as shown below. This command tells MINITAB to use the z values and find the 95% confidence interval when $\sigma = 11$. The program follows.

```
MTB  > SET C1
DATA > 43 52 18 20 25 20 45 43 21 42 32 24 32 19 25
DATA > 26 44 42 41 41 53 22 25 23 21 27 33 36 47 19
DATA > END
MTB  > ZINT  95  SIGMA = 11 C1
```

The computer will print the following:

```
THE ASSUMED SIGMA = 11.0
      N   MEAN   STDEV SE MEAN   95.0 PERCENT C.I.
C1   30  32.03  11.01    2.01 (   28.09,    35.98)
```

The sample size is 30. The sample mean is 32.03, and the standard deviation of the sample is 11.01. The standard error of the mean is 2.01. The 95% confidence interval is

$$28.09 < \mu < 35.98$$

For a 99% confidence interval, substitute 99 for 95 in the ZINT line and find the interval.

4. When the population standard deviation is unknown and the sample is small, MINITAB will compute the confidence interval for the population mean using the t values. For example, 10 statistics books were selected and the number of pages in each were 625, 675, 535, 406, 512, 680, 483, 522, 619, and 575. Find the 90% confidence interval of the population mean of the number of pages in all statistics books.

The data is entered using the SET C1 command. To find the 90% confidence interval using the t values, type TINT 90 C1, as shown. Thus, using MINITAB, enter the following:

```
MTB  > SET C1
DATA > 625 675 535 406 512
DATA > 680 483 522 619 575
DATA > END
MTB  > TINT 90 C1
```

The computer will print the following:

```
      N   MEAN   STDEV  SE MEAN  90.0 PERCENT C.I.
C1   10  563.2   87.9     27.8 (   512.3,    614.1)
```

The sample size is 10. The sample mean is 563.2. The sample standard deviation is 87.9. The standard error of the mean is 27.8. The 90% confidence interval is

$$512.3 < \mu < 614.1$$

Find the 95% and 99% confidence intervals for the data given in this problem.

√ DATA ANALYSIS Applying the Concepts

The Data Bank is located in the Appendix.

1. From the Data Bank, choose a variable, find the mean, and construct the 95% and 99% confidence intervals of the population mean. Use a sample of at least 30 subjects. Find the mean of the population, and determine whether it falls within the confidence interval.

2. Repeat Exercise 1 using a different variable and a sample of 15.

3. Repeat Exercise 1 using a proportion. For example, construct a confidence interval for the proportion of individuals who did not complete high school.

✎ TEST

Directions: Determine whether the statement is true or false.

1. The best point estimate of the population mean is the sample median.
2. The two most often used confidence intervals are 90% and 95%.
3. For a specific confidence interval, the larger the sample size, the smaller the maximum error will be.
4. When the population standard deviation is not known and the sample size is less than 30, the t values must be used for computing confidence intervals.
5. "The average height of an adult male is 5 feet 10 inches" is an example of a point estimate.
6. Interval estimates are preferred over point estimates since a confidence level can be determined.
7. When a 99% confidence interval is calculated instead of a 95% confidence interval, with the same sample size, the maximum error of estimate will be smaller.
8. In order to determine the sample size needed to estimate a parameter, one must know the maximum error of estimate.
9. A good estimator must be unbiased, consistent, and relatively efficient.
10. An estimator is consistent if, as the sample size decreases, the value of the estimator approaches the value of the parameter estimated.

CHAPTER

9

Hypothesis Testing

9–1 INTRODUCTION

Researchers are interested in answering many types of questions. For example, a scientist may want to know whether the earth is warming up. A physician might want to know whether a new medication will lower a person's blood pressure. An educator may wish to see whether a new teaching technique is better than a traditional one. A retail merchant might want to know whether the public prefers a certain color in a new line of fashion. Automobile manufacturers are interested in determining whether automobile air bags will reduce the severity of injuries caused by accidents. These types of questions can be addressed through statistical **hypothesis testing,** which is a decision-making process for evaluating claims about a population. In hypothesis testing, the researcher must define the population under study, state the particular hypotheses that will be investigated, give the significance level, select a sample from the population, collect the data, perform the calculations required for the statistical test, and reach a conclusion.

Hypotheses concerning parameters such as means and proportions can be investigated. There are two specific statistical tests used for hypotheses concerning means: the *z test* and the *t test*. This chapter will explain in detail the hypothesis-testing procedure along with the *z* test and the *t* test.

9–2 STEPS IN HYPOTHESIS TESTING

Every hypothesis-testing situation begins with the statement of a hypothesis.

A statistical hypothesis is a conjecture about a population parameter. This conjecture may or may not be true.

There are two types of statistical hypotheses for each situation: the null hypothesis and the alternative hypothesis.

The null hypothesis, symbolized by H_0, is a statistical hypothesis that states that there is no difference between a parameter and a specific value or that there is no difference between two parameters.
The alternative hypothesis, symbolized by H_1, is a statistical hypothesis that states a specific difference between a parameter and a specific value or states that there is a difference between two parameters.

As an illustration of how hypotheses should be stated, three different statistical studies will be used as examples.

Situation A. A medical researcher is interested in finding out whether a new medication will have any undesirable side effects. She is particularly concerned with the pulse rate of the patients who take the medication. Will the pulse rate increase, decrease, or remain unchanged after a patient takes the medication?

Since the researcher knows that the mean pulse rate for the population under study is 82 beats per minute, the hypotheses for this situation are

$$H_0: \mu = 82$$
$$H_1: \mu \neq 82$$

The null hypothesis specifies that the mean will remain unchanged, and the alternative hypothesis states that it will be different. This test is called a *two-tailed test* (a term that will be formally defined later in this section), since the possible side effects of the medicine could be to raise or lower the pulse rate.

Situation B. A chemist invents an additive to increase the life of an automobile battery. If the mean lifetime of the automobile battery is 36 months, then his hypotheses are

$$H_0: \mu \leq 36$$
$$H_1: \mu > 36$$

In this situation, the chemist is only interested in increasing the lifetime of the batteries, so his alternative hypothesis is that the mean is greater than 36 months. The null hypothesis is that the mean is less than or equal to 36 months. This test is called a *one-tailed test*, since the interest is in an increase only.

Situation C. A contractor wishes to lower heating bills by using a special type of insulation in houses. If the average of the monthly heating bills is $78, his hypotheses about heating costs with the use of insulation are

$$H_0: \mu \geq \$78$$
$$H_1: \mu < \$78$$

This test is also a one-tailed test, since the contractor is only interested in lowering heating costs.

In order to state hypotheses correctly, researchers must translate the *conjecture or claim* from words into mathematical symbols. The basic symbols used are as follows:

Equal	=	Less than	<
Not equal	≠	Greater than or equal to	≥
Greater than	>	Less than or equal to	≤

The null and alternative hypotheses are stated together, and the null hypothesis contains the equal sign, as shown.

Two-Tailed Test	**One-Tailed Right Test**	**One-Tailed Left Test**
$H_0: =$	$H_0: \leq$	$H_0: \geq$
$H_1: \neq$	$H_1: >$	$H_1: <$

The formal definitions of the different types of tests are given later in this section.

Table 9-1 shows some common phrases that are used in hypotheses conjectures and the corresponding symbols. This table should be helpful when one is translating verbal conjectures into mathematical symbols.

Table 9–1

Hypothesis-Testing Common Phrases	
>	**<**
Is greater than	Is less than
Is more than	Is below
Is larger than	Is lower than
Is longer than	Is shorter than
Is bigger than	Is smaller than
Is better than	Is reduced from
≥	**≤**
Is greater than or equal to	Is less than or equal to
Is at least	Is not more than
Is not less than	Is at most
=	**≠**
Is equal to	Is not equal to
Is exactly the same as	Is different from
Has not changed from	Has changed from
Is the same as	Is not the same as

EXAMPLE 9–1   State the null and alternative hypotheses for each conjecture.

a. A researcher thinks that if expectant mothers use vitamin pills, the birth weight of the babies will increase. The average of the birth weights of the population is 8.6 pounds.

b. An engineer hypothesizes that she can decrease the mean number of defects in a manufacturing process of compact discs by using robots instead of humans for certain tasks. The mean number of defective discs per 1000 is 18.

c. A psychologist feels that if he plays soft music during a test, the results of the test will be changed. He is not sure whether the grades will be higher or lower. In the past, the mean of the scores was 73.

Solution   a. $H_0: \mu \leq 8.6$ and $H_1: \mu > 8.6$.   c. $H_0: \mu = 73$ and $H_1: \mu \neq 73$.
b. $H_0: \mu \geq 18$ and $H_1: \mu < 18$.   ◄

After stating the hypothesis, the researcher's next step is to design the study. In designing the study, the researcher selects the correct *statistical test,* chooses an appropriate *level of significance,* and formulates a plan for conducting the study. In Situation A, for instance, the researcher will select a sample of patients who will be given the drug. After allowing a suitable period of time for the drug to be absorbed, the researcher will measure each person's pulse rate. Recall that the sample means vary about the population mean. So even if the null hypothesis is true, the mean of the sample will not, in most cases, be exactly equal to the population mean of 82.

Now a question arises. If the mean of the sample is not exactly equal to the population mean, how does one know whether the medication affects the pulse rate? That is, is the difference due to chance, or is it due to the effects of the medication?

If the mean pulse rate of the sample were, say, 83, the researcher would probably conclude that this difference was due to chance and would not reject the null hypothesis. But if the sample mean were, say, 90, then in all likelihood the researcher would conclude that the medication increased the pulse rate of the users and would reject the null hypothesis. The question is, "Where does the researcher draw the line?" This decision is not made on feelings or intuition; it is made statistically. That is, the difference must be significant and in all likelihood not due to chance. Here is where the concepts of statistical test and level of significance are used.

A statistical test uses the data obtained from a sample to make a decision about whether or not the null hypothesis should be rejected.
The numerical value obtained from a statistical test is called the test value

---

In a statistical test, the mean is computed for the data obtained from the sample and is compared with the population mean. Then a decision is made to reject or not reject the null hypothesis on the basis of the value obtained from the statistical test. If the difference is significant, the null hypothesis is rejected. If it is not, then the null hypothesis is not rejected.

In the hypothesis-testing situation, there are four possible outcomes. The null hypothesis may or may not be true, and a decision is made to reject or not reject it on the basis of the data obtained from a sample. The four possible outcomes are shown in Figure 9–1. Notice that there are two possibilities for a correct decision and two possibilities for an incorrect decision.

FIGURE 9–1
Possible Outcomes of a Hypothesis Test

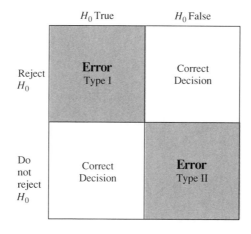

If the null hypothesis is true and it is rejected, then a *type I error* is made. In Situation A, for instance, the medication might not significantly change the pulse rate of all the users in the population; but it might change the rate, by chance, of the subjects in the sample. In this case, the researcher will reject the null hypothesis when it is really true, thus committing a type I error.

On the other hand, the medication might not change the pulse rate of the subjects in the sample; but when it is given to the general population, it might cause a significant increase or decrease in the pulse rate of the users. The researcher, on the basis of the data obtained from the sample, will not reject the null hypothesis, thus committing a *type II error.*

A type I error occurs if one rejects the null hypothesis when it is true.
A type II error occurs if one does not reject the null hypothesis when it is false.

The decision to reject or not reject the null hypothesis does not *prove* anything. *The only way to prove anything statistically is to use the entire population,* which, in most cases, is not possible. The decision, then, is made on the basis of probabilities. That is, when there is a large difference between the mean obtained from the sample and the hypothesized mean, the null hypothesis is probably not true. The question is, "How large a difference is necessary to reject the null hypothesis?" Here is where the level of significance is used.

The level of significance is the maximum probability of committing a type I error. This probability is symbolized by α (Greek letter alpha).

The probability of a type II error is symbolized by β (Greek letter **beta**). In most hypothesis-testing situations, β cannot be computed.

Statisticians generally agree on using two arbitrary significance levels: the 0.05 level and the 0.01 level. That is, if the null hypothesis is rejected, the probability of a type I error will be 5% or 1%, and the probability of a correct decision will be 95% or 99%, depending on which level of significance is used. Here is another way of putting it: When $\alpha = 0.05$, there is a 5% chance of rejecting a true null hypothesis; and when $\alpha = 0.01$, there is a 1% chance of rejecting a true null hypothesis.

In a hypothesis-testing situation, the researcher decides what level of significance to use. It does not have to be the 0.05 or 0.01 level but can be any level, depending on the seriousness of the type I error. After a significance level is chosen, a *critical value* is selected from a table for the appropriate test. If a *z* test is used, for example, the *z* table (Table E) is consulted to find the critical value. The critical value determines the critical and noncritical regions.

The critical value(s) separates the critical region from the noncritical region. The symbol for critical value is C.V.
The critical or rejection region is the range of values of the test value that indicates that there is a significant difference and that the null hypothesis should be rejected.

The noncritical or nonrejection region is the range of values of the test value that indicates that the difference was probably due to chance and that the null hypothesis should not be rejected.

The critical value can be on the right side of the mean or on the left side of the mean for a one-tailed test. Its location depends on the inequality sign of the alternative hypothesis. For example, in Situation B, where the chemist is interested in increasing the average lifetime of automobile batteries, the alternative hypothesis is $H_1: \mu > 36$. Since the inequality sign is $>$, the null hypothesis will be rejected only when the sample mean is significantly greater than 36. Hence, the critical value must be on the right side of the mean. Therefore, this test is called one-tailed right test.

A one-tailed test indicates that the null hypothesis should be rejected when the test value is in the critical region on one side of the mean. A one-tailed test is either *right* or *left*, depending on the direction of the inequality of the alternative hypothesis.

To obtain the critical value, the researcher must choose an alpha level. In Situation B, suppose the researcher chose $\alpha = 0.01$. Then, the researcher must find a z value such that 1% of the area falls to the right of the z value and 99% falls to the left of the z value, as shown in Figure 9-2a.

Next, the researcher must find the value in Table E closest to 0.4900. Note that because the table gives the area between 0 and the z, 0.5000 must be subtracted from 0.9900 to get 0.4900. The critical z value is 2.33, since that value gives the area closest to 0.4900, as shown in Figure 9-2b.

FIGURE 9-2
Finding the Critical Value for $\alpha = 0.01$ (One-Tailed Test)

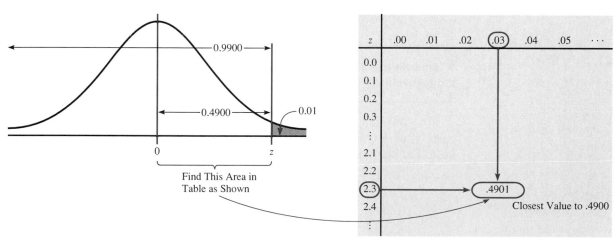

(a) *THE CRITICAL REGION*   (b) *THE CRITICAL VALUE FROM TABLE E*

The critical and noncritical regions and the critical value are shown in Figure 9–3.

FIGURE 9–3
Critical and Noncritical
Regions for $\alpha = 0.01$
(One-Tailed Right Test)

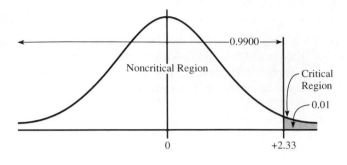

Now, move on to Situation C, where the contractor is interested in lowering the heating bills. The alternative hypothesis is $H_1: \mu < \$78$. Hence, the critical value falls to the left of the mean. This test is thus a one-tailed left test. At $\alpha = 0.01$, the critical value is -2.33, as shown in Figure 9–4.

FIGURE 9–4
Critical and Noncritical
Regions for $\alpha = 0.01$
(One-Tailed Left Test)

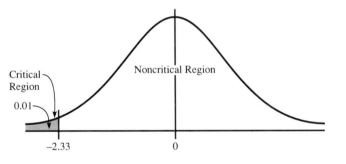

When a researcher conducts a two-tailed test, as in Situation A, the null hypothesis can be rejected when there is a significant difference in either direction, above or below the mean.

A two-tailed test indicates that the null hypothesis should be rejected when the test value is in either of the two critical regions.

For a two-tailed test, then, the critical region must be split into two equal parts. If $\alpha = 0.01$, then half of the area, or 0.005, must be to the right of the mean and half must be to the left of the mean, as shown in Figure 9–5.

FIGURE 9–5
Finding the Critical
Values for $\alpha = 0.01$
(Two-Tailed Test)

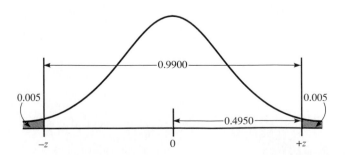

In this case, the area to be found in Table E is 0.4950. The critical values are $+2.58$ and -2.58, as shown in Figure 9–6.

FIGURE 9–6
Critical and Noncritical
Regions for $\alpha = 0.01$
(Two-Tailed Test)

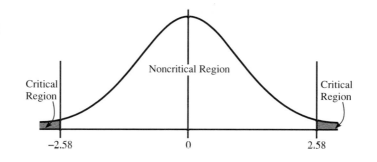

A similar procedure is used for other values of α.

Figure 9–7 shows the critical values (C.V.) for the three situations discussed in this section for α values of $\alpha = 0.05$ and $\alpha = 0.01$. The procedure for finding critical values is outlined on page 308.

FIGURE 9–7
Summary of Hypothesis
Testing and Critical
Values

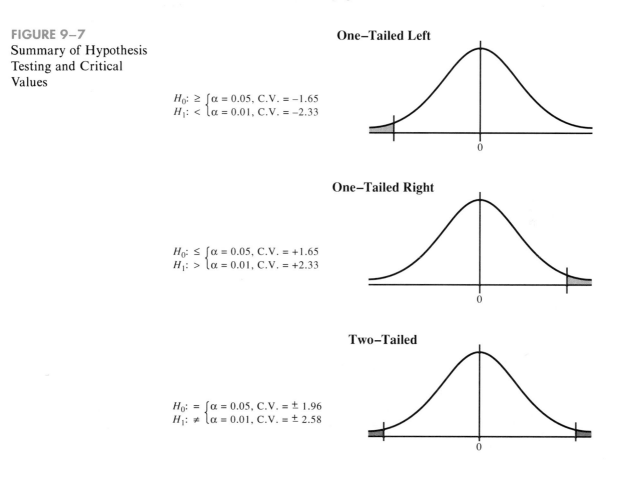

One–Tailed Left

$H_0: \geq$ $\begin{cases} \alpha = 0.05, \text{ C.V.} = -1.65 \\ \alpha = 0.01, \text{ C.V.} = -2.33 \end{cases}$
$H_1: <$

One–Tailed Right

$H_0: \leq$ $\begin{cases} \alpha = 0.05, \text{ C.V.} = +1.65 \\ \alpha = 0.01, \text{ C.V.} = +2.33 \end{cases}$
$H_1: >$

Two–Tailed

$H_0: =$ $\begin{cases} \alpha = 0.05, \text{ C.V.} = \pm 1.96 \\ \alpha = 0.01, \text{ C.V.} = \pm 2.58 \end{cases}$
$H_1: \neq$

Procedure for Finding the Critical Values for Specific α Values, Using Table E

1. Draw the figure and indicate the appropriate area.
   a. If the test is one-tailed left, the critical region, with an area equal to α, will be on the left side of the mean.
   b. If the test is one-tailed right, the critical region, with an area equal to α, will be on the right side of the mean.
   c. If the test is two-tailed, α must be divided by 2; half of the area will be to the right of the mean, and the other half will be to the left of the mean.
2. For a one-tailed test, subtract the area (equivalent to α) in the critical region from 0.5000, since Table E gives the area under the standard normal distribution curve between 0 and any z to the right of 0. For a two-tailed test, subtract the area (equivalent to $\alpha/2$) from 0.5000.
3. Find the area in Table E corresponding to the value obtained in step 2. If the exact value cannot be found in the table, use the closest value.
4. Find the z value that corresponds to the area. This will be the critical value.
5. Determine the sign of the critical value for a one-tailed test.
   a. If the test is one-tailed left, the critical value will be negative.
   b. If the test is one-tailed right, the critical value will be positive.
   For a two-tailed test, one value will be positive and the other negative.

EXAMPLE 9–2   Using Table E in the Appendix, find the critical value(s) for each situation and draw the appropriate figure, showing the critical region.

a. A one-tailed left test with $\alpha = 0.10$.
b. A two-tailed test with $\alpha = 0.02$.
c. A one-tailed right test with $\alpha = 0.005$.

Solution   a.

STEP 1   Draw the figure and indicate the appropriate area. Since this test is a one-tailed left test, the area of 0.10 is located in the left tail, as shown in Figure 9–8.

STEP 2   Subtract 0.10 from 0.5000 to get 0.4000.

STEP 3   In Table E, find an area that is closest to 0.4000; in this case, it is 0.3997.

STEP 4   Find the z value that corresponds to this area. It is 1.28.

STEP 5   Determine the sign of the critical value (i.e., the z value). Since this test is a one-tailed left test, the sign of the critical value is negative. Hence, the critical value is -1.28. See Figure 9–8.

FIGURE 9–8
Critical Value
and Critical Region
for Part a of
Example 9–2

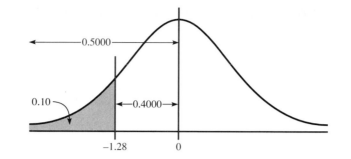

b.

STEP 1 Draw the figure and indicate the appropriate area. In this case, there are two areas equivalent to $\alpha/2$, or $0.02/2 = 0.01$.

STEP 2 Subtract 0.01 from 0.5000 to get 0.4900.

STEP 3 Find the area in Table E equal to or closest to 0.4900. In this case, it is 0.4901.

STEP 4 Find the z value that corresponds to this area. It is 2.33.

STEP 5 Determine the sign of the critical value. Since this test is a two-tailed test, there are two critical values; one is positive and the other is negative. They are $+2.33$ and -2.33. See Figure 9–9.

FIGURE 9–9
Critical Values
and Critical Regions
for Part b of
Example 9–2

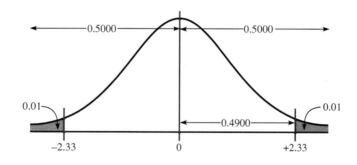

c.

STEP 1 Draw the figure and indicate the appropriate area. Since this is a one-tailed right test, the area 0.005 is located in the right tail, as shown in Figure 9–10.

FIGURE 9–10
Critical Value
and Critical Region
for Part c of
Example 9–2

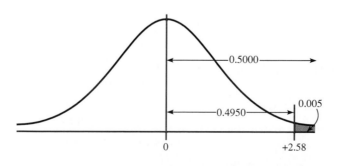

STEP 2 Subtract 0.005 from 0.5000 to get 0.4950.

STEP 3 Find the area in Table E equal to 0.4950.

STEP 4 Find the z value that corresponds to this area. It is 2.58.

STEP 5 Determine the sign of the critical value. Since this test is a one-tailed right test, the sign is positive; hence, the critical value is $+2.58$. ◄

In hypothesis testing, the following steps are recommended.

1. State the hypotheses. Be sure to state both the null and the alternative hypotheses.
2. Design the study. This step includes selecting the correct statistical test, choosing a level of significance, and formulating a plan to carry out the study. In the plan, information such as the definition of the population, the way that the sample will be selected, and the methods that will be used to collect the data should be included.
3. Conduct the study and collect the data.
4. Evaluate the data. The data should be tabulated in this step, and the statistical test should be conducted. Finally, a decision should be made about whether to reject or not reject the null hypothesis.
5. Summarize the results.

For the purposes of this chapter, a simplified version of the hypothesis-testing procedure will be used, since designing the study and collecting the data will be omitted. This procedure is summarized in Procedure Table 7.

Procedure Table 7

Procedure for Solving Hypothesis-Testing Problems

Step 1: State the hypotheses, and identify the claim.
Step 2: Find the critical value(s) from Table E in the Appendix.
Step 3: Compute the test value.
Step 4: Make the decision to reject or not reject the null hypothesis.
Step 5: Summarize the results.

EXERCISES

9–1. Define *null* and *alternative hypothesis,* and give an example of each.

9–2. What is meant by a type I error? A type II error? How are they related?

9–3. What is meant by a statistical test?

9–4. Explain the difference between a one-tailed and a two-tailed test.

9–5. What is meant by the critical region? The noncritical region?

9–6. What symbols are used to represent the null hypothesis and the alternative hypothesis?

9–7. What symbols are used to represent the probabilities of type I and type II errors?

9–8. Explain what is meant by a significant difference.

9-9. When should a one-tailed test be used? A two-tailed test?

9-10. List the steps in hypothesis testing.

9-11. In hypothesis testing, why can't the hypothesis be proved true?

9-12. (ANS.) Using the z table (Table E), find the critical value (or values) for each.

a. $\alpha = 0.01$, two-tailed test
b. $\alpha = 0.05$, one-tailed right test
c. $\alpha = 0.005$, one-tailed left test
d. $\alpha = 0.10$, one-tailed left test
e. $\alpha = 0.05$, two-tailed test
f. $\alpha = 0.04$, one-tailed right test
g. $\alpha = 0.01$, one-tailed left test
h. $\alpha = 0.10$, two-tailed test
i. $\alpha = 0.02$, one-tailed right test
j. $\alpha = 0.02$, two-tailed test

9-13. For each conjecture, state the null and alternative hypotheses.

a. The average age of taxi drivers in New York City is 36.3 years.
b. The average income of nurses is $36,250.
c. The average IQ of disc jockeys is greater than 117.3.
d. The average pulse rate of female joggers is less than 72 beats per minute.
e. The average bowling score of people who enrolled in a basic bowling class is less than 100.
f. The average cost of a VCR is $297.75.
g. The average electric bill for residents of White Pine Estates exceeds $52.98 per month.
h. The average number of calories of brand A's low-calorie meals is at most 300.
i. The average weight loss of people who use brand A's low-calorie meals for six weeks is at least 3.6 pounds.

9-3 THE z TEST

In this chapter, two statistical tests will be explained: the z test and the t test. This section explains the z test, and Section 9-4 explains the t test.

Many hypotheses are tested by using a statistical test based on the following general formula:

$$\text{test value} = \frac{(\text{observed value}) - (\text{expected value})}{\text{standard error}}$$

The observed value is the statistic (such as the mean) that is computed from the sample data. The expected value is the parameter (such as the mean) that one would expect to obtain if the null hypothesis were true—in other words, the hypothesized value. The denominator is the standard error of the statistic being tested (in this case, the standard error of the mean).

The z test is defined formally as follows.

The z test is a statistical test for the mean of a population. It can be used when $n \geq 30$, or when the population is normally distributed and σ is known. The formula for the z test is

$$z = \frac{\overline{X} - \mu}{\sigma/\sqrt{n}}$$

where
$\overline{X}$ = **sample mean**
μ = **hypothesized population mean**
σ = **population standard deviation**
n = **sample size**

For the z test, the observed value is the value of the sample mean. The expected value is the value of the population mean, assuming that the null hypothesis is true. The denominator $\sigma/\sqrt{n}$ is the standard error of the mean.

The formula for the *z* test is the same formula shown in Chapter 7 for the situation where one is using a distribution of sample means. Recall that the central limit theorem allows one to use the standard normal distribution to approximate the distribution of sample means when $n \geq 30$ or when the variable is normally distributed and σ is known.

Speaking of Statistics

Stick Margarine May Boost Risk of Heart Attacks

By Nanci Hellmich
USA TODAY

The bad news about margarine continues to mount.

Eating stick margarine increases the risk of heart attack, according to new findings from the Harvard Nurses' Health Study, an ongoing analysis of the diets of 90,000 nurses.

This new report adds to growing evidence that trans fatty acids—the fats that form when liquid vegetable oils are processed or hydrogenated—raise blood cholesterol in much the same way that saturated fat does.

Other culprits besides margarine: any solid vegetable shortening, some fried foods at chains such as McDonald's and Burger King and processed foods made with partially hydrogenated vegetable oils (check the ingredient list).

The new study reported in Saturday's *Lancet* shows:

• Women who frequently use margarine have more than a 50% higher risk of heart disease than those who infrequently use margarine.

• Those who eat a couple of cookies a day (cookies often contain partially hydrogenated oils) are at a 50% higher risk.

• Women who eat lots of foods high in trans fatty acids have a 70% higher risk of heart disease than those who don't.

In recent years, many people have switched to margarine from butter, which is high in saturated fat.

But don't go back to butter just because margarine is unhealthy, says Dr. Walter Willett, study author. Instead, consider switching to liquid oils instead of solid margarine or shortening.

He's a strong supporter of olive oil because "there's a strong indication that olive oil is at least safe." And it's a mono-unsaturated fat that lowers bad cholesterol (LDL) without lowering good cholesterol (HDL).

Also, some tub margarines contain more water and air and therefore less fat and fewer trans fatty acids than stick margarines.

Source: Copyright 1993, USA TODAY.
Reprinted with permission.

In this study on margarine consumption, many conclusions are stated. State hypotheses that may have been used to test these conclusions. Define the population and the sample used. Do you feel that the sample would be representative of all adults? Explain your answer. Comment on the sample size.

Note: The student's first encounter with hypothesis testing can be somewhat challenging and confusing, since there are many new concepts being introduced at the same time. *In order to understand all the concepts, the student must carefully follow each step in the examples and try each exercise that is assigned.* Only after careful study and patience will these concepts become clear.

As stated in the previous section, there are five steps for solving *hypothesis-testing* problems:

STEP 1 State the hypotheses, and identify the claim.

STEP 2 Find the critical value(s).

STEP 3 Compute the test value.

STEP 4 Make the decision to reject or not reject the null hypothesis.

STEP 5 Summarize the results.

The next example illustrates these steps.

EXAMPLE 9–3 A researcher reports that the average salary of veterinarians is more than $42,000. A sample of 30 veterinarians has a mean salary of $43,260. At $\alpha = 0.05$, test the claim that veterinarians earn more than $42,000 a year. The standard deviation of the population is $5230.

Solution **STEP 1** State the hypotheses.

$$H_0: \mu \leq \$42,000 \qquad H_1: \mu > \$42,000 \text{ (claim)}$$

STEP 2 Find the critical value. Since $\alpha = 0.05$ and the test is a one-tailed right test, the critical value is $z = +1.65$.

STEP 3 Compute the test value.

$$z = \frac{\overline{X} - \mu}{\sigma/\sqrt{n}} = \frac{\$43,260 - 42,000}{5230/\sqrt{30}} = 1.320$$

STEP 4 Make the decision. Since the test value $+1.320$ is less than the critical value $+1.65$, the decision is, "Do not reject the null hypothesis." This test is summarized in Figure 9–11.

FIGURE 9–11
Summary of the *z* Test
of Example 9–3

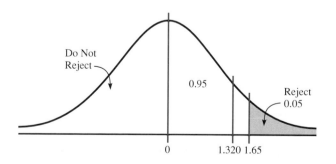

STEP 5 Summarize the results. There is not enough evidence to support the claim that veterinarians earn more than $42,000 a year. ◄

Comment: Even though in Example 9–3, the sample mean, $43,260, is higher than the hypothesized population mean of $42,000, it is not *significantly* higher. Hence, the difference may be due to chance. When the null hypothesis is not rejected, there is still a probability of a type II error—i.e., of not rejecting the null hypothesis when it is false.

The probability of a type II error is not easily ascertained. Further explanation about the type II error is given in Section 9–6. For now, it is only necessary to realize that the type II error exists.

It should also be noted that when the null hypothesis is not rejected, it cannot be accepted as true. There is merely not enough evidence to say that it is false. This guideline may sound a little confusing, but the situation is analogous to a jury trial. The verdict is either guilty or not guilty and is based on the evidence presented. If a person is judged not guilty, it does not mean that the person is proved innocent; it only means that there was not enough evidence to reach the guilty verdict.

EXAMPLE 9–4   A national magazine claims that the average college student watches less television than the general public. The national average is 29.4 hours per week, with a standard deviation of 2 hours. A sample of 25 college students has a mean of 27 hours. Test the claim at $\alpha = 0.01$.

Solution   **STEP 1**   State the hypotheses, and identify the claim.

$$H_0: \mu \geq 29.4 \qquad \text{and} \qquad H_1: \mu < 29.4 \text{ (claim)}$$

STEP 2   Find the critical value. Since $\alpha = 0.01$ and the test is a one-tailed left test, the critical value is -2.33.

STEP 3   Compute the test value.

$$z = \frac{\overline{X} - \mu}{\sigma/\sqrt{n}} = \frac{27 - 29.4}{2/\sqrt{25}} = -6$$

STEP 4   Make the decision. Since the test value -6 falls in the critical region, the decision is to reject the null hypothesis. The test is summarized in Figure 9–12.

FIGURE 9–12
Summary of the z Test
of Example 9–4

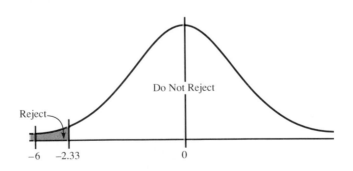

STEP 5   Summarize the results. There is enough evidence to support the claim that college students watch less television than the general public. ◄

Comment: In Example 9–4, the difference is said to be significant. However, when the null hypothesis is rejected, there is always a chance of a type I error. In this case, the probability of a type I error is at most 0.01, or 1%.

EXAMPLE 9–5   A merchant believes that the average age of customers who purchase a certain brand of jeans is 15 years of age. A sample of 35 customers had an average age of 15.6 years. At $\alpha = 0.01$, should his conjecture be rejected? The standard deviation of the ages is 1 year.

Solution   **STEP 1**   State the hypotheses, and identify the claim.

$$H_0: \mu = 15 \text{ (claim)} \qquad \text{and} \qquad H_1: \mu \neq 15$$

STEP 2   Find the critical values. Since $\alpha = 0.01$ and the test is a two-tailed test, the critical values are $+2.58$ and -2.58.

STEP 3   Compute the test value.

$$z = \frac{\overline{X} - \mu}{\sigma/\sqrt{n}} = \frac{15.6 - 15}{1/\sqrt{35}} = 3.55$$

STEP 4   Make the decision. Reject the null hypothesis, since the test value falls in the critical region, as shown in Figure 9–13.

FIGURE 9–13
Critical and
Test Values for
Example 9–5

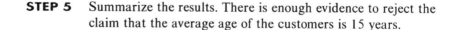

-2.58   0   2.58 3.55

STEP 5   Summarize the results. There is enough evidence to reject the claim that the average age of the customers is 15 years. ◄

As with confidence intervals, the central limit theorem states that when the population standard deviation σ is unknown, the sample standard deviation s can be used in the formula as long as the sample size is 30 or more. The formula for the z test in this case is

$$z = \frac{\overline{X} - \mu}{s/\sqrt{n}}$$

When n is less than 30 and σ is unknown, the t test must be used. The t test will be explained in the next section.

Students sometimes have difficulty summarizing the results of a hypothesis test. Figure 9–14 shows the four possible outcomes and the summary statements for each situation.

First of all, the claim can be either the null or alternative hypothesis, and one should identify which it is. Second, after the study is completed, the null hypothesis is either rejected or not rejected. From these two facts, the decision can be identified in the appropriate block of Figure 9–14.

FIGURE 9–14
Outcomes of a Hypothesis-Testing Situation

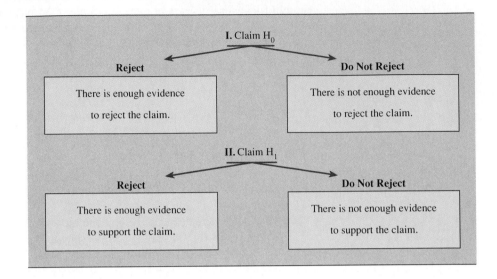

For example, suppose that a researcher claims that the mean weight of a specific adult animal is 42 pounds. In this case, the claim would be the null hypothesis, H_0: $\mu = 42$. If the null hypothesis is rejected, the conclusion would be that there is enough evidence to reject the claim that the mean weight of the adult animal is 42 pounds. See Figure 9–15a.

FIGURE 9–15
Outcomes of a Hypothesis-Testing Situation for Two Specific Cases

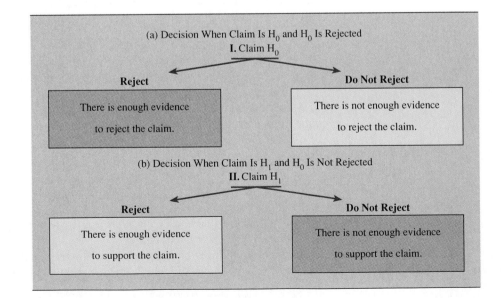

On the other hand, suppose that the researcher claims that the mean weight of the adult animals is not 42 pounds. The claim would be the alternative hypothesis, H_1: $\mu \neq 42$. Furthermore, suppose that the null hypothesis is not rejected. The conclusion, then, would be that there is not enough evidence to support the claim that the mean weight of the adult animals is not 42 pounds. See Figure 9-15b.

Again, remember that nothing is being proven true or false. The statistician is only stating that there is or is not enough evidence to say that a claim is *probably* true or false. As noted previously, the only way to prove something would be to use the entire population under study, and usually this cannot be done, especially when the population is large.

Speaking of **Statistics**

Why You Hate Snakes

They hiss and they slither, but they pose less of a threat to our lives than cars or ovens. So why do snakes set off more clinical fears and phobias than knives and guns?

Here's one scientific theory: Humans and other primates are predisposed to acquire fears of critters that once threatened our ancestors' lives.

Psychologist Susan Mineka of Northwestern University contends that we have a predisposition to such "evolutionary memories" because our ancestors once had to face snakes, certainly more so than, say, ovens. Because they survived, those who rapidly acquired the fear were most favored in natural selection.

Mineka, along with University of Wisconsin psychologist Michael Cook, put the theory to a test in six rhesus monkeys. Reared in the lab, the animals had no prior exposure to snakes. The psychologists showed them a videotape of wild-reared monkeys reacting with horror to snakes. Within 24 minutes, the lab monkeys acquired a fear of snakes.

The psychologists then edited fake flowers, a toy snake, a toy rabbit, and a toy crocodile into the video. Tests later showed that after 40 to 60 seconds of exposure to each object, the monkeys feared only the toy snakes and crocodiles. Of the four objects, only snakes and crocodiles preyed on our ancestors. Coincidence?

Meanwhile, the search for evidence continues. The next time snakes inhabit your nightmares, ask whether it's that viper horror you watched, or are you just connecting with the fears of your forefathers.

Source: Psychology Today (Vol. 25, No. 2, March/April 1992). Reprinted with permission.

Psychologists use statistics to differentiate facts and relationships from coincidence. In this study on the fear of snakes, can you cite other factors besides coincidence and predisposition that might affect the results?

EXERCISES

For Exercises 9–14 through 9–29, perform each of the following steps.

a. State the hypotheses, and identify the claim.
b. Find the critical value(s).
c. Compute the test value.
d. Make the decision.
e. Summarize the results.

Use diagrams to show the critical region (or regions). For samples smaller than 30, assume that the population is normally distributed.

9–14. A shoe store manager claims that the average cost of a pair of tennis shoes is $89.95. A sample of 68 pairs of tennis shoes has an average cost of $90.26. The standard deviation of the sample is $3.00. At $\alpha = 0.05$, test the manager's claim.

9–15. A doctor claims that the IQs of nursing students are higher than average. If a sample of 25 students produces a mean of 111.3, test the doctor's claim at $\alpha = 0.05$. The population mean is 100, and the population standard deviation is 15.

9–16. The manager of a large factory believes that the average hourly wage of the employees is below $9.78 per hour. A sample of 18 employees has a mean hourly wage of $9.60. Test the claim at $\alpha = 0.10$. The standard deviation of all salaries is $1.42.

9–17. The average SAT score in mathematics is 483, with a standard deviation of 100. A special preparation course states that it can increase scores. A sample of 32 students completed the course, and the average of their scores was 494. At $\alpha = 0.05$, does the course do what it claims?

9–18. A maker of frozen meals claims that the average caloric content of its meals is 800, and the standard deviation is 25. A researcher tested 12 meals and found that the average number of calories was 873. Is there enough evidence to reject the claim at $\alpha = 0.02$?

9–19. Little Slope Ski Resort advertises that it has an average of at least 80 guests per week during ski season. The standard deviation of the population is 8. For the past 10 weeks there was an average of 71 guests per week. Is the advertisement correct? Use $\alpha = 0.01$.

9–20. A diet clinic states that there is an average loss of 24 pounds for patients who stay on the program for 20 weeks. The standard deviation is 5 pounds. The clinic tries a new diet, reducing the salt intake to see whether that strategy will produce a greater weight loss. A group of 40 volunteers loses an average of 16.3 pounds each over 20 weeks. Should the clinic change to the new diet? Use $\alpha = 0.05$.

9–21. A travel agent claims that the average cost of a three-day trip to Atlantic City is $915. The standard deviation is $35. Sixty people who scheduled the trip paid an average of $927 for the trip. At $\alpha = 0.05$, should the agent's claim be rejected?

9–22. A bank manager claims that the average loan to the bank's customers is $4800. The standard deviation of the population is $800. A sample of 25 customers had an average loan of $4235. At $\alpha = 0.10$, is there enough evidence to reject the manager's claim?

9–23. A manufacturer states that the average lifetime of its lightbulbs is 3 years, or 36 months. The standard deviation is 8 months. Fifty bulbs are selected, and the average lifetime is found to be 32 months. Should the manufacturer's statement be rejected at $\alpha = 0.01$?

9–24. A real estate agent claims that the average price of a condominium in Naples, Florida, is at most $56,900. The standard deviation is $2500. A sample of 36 condominiums has an average selling price of $57,374. Does the evidence support the claim at $\alpha = 0.05$?

9–25. The average serum cholesterol level in a certain group of patients is 240 milligrams. The standard deviation is 18 milligrams. A new medication is designed to lower the cholesterol level if taken for one month. A sample of 40 people used the medication for 30 days, and their average cholesterol level was 229 milligrams. At $\alpha = 0.01$, does the medication lower the cholesterol level of the patients?

9–26. The state's education secretary claims that the average cost of one year's tuition for all private high schools in the state is $2350. A sample of 28 private high schools is selected, and the average tuition is $2315. The standard deviation for the population is $38. Test the hypothesis, at $\alpha = 0.05$, that the average cost of tuition is equal to $2350.

9–27. An admissions counselor states that the average cost for one year's tuition for all colleges in the state of Pennsylvania is $2126. A sample of 32 colleges is selected, and the average tuition is $2159, with a standard deviation of $123. Test the hypothesis, at $\alpha = 0.02$, that the average cost for tuition is equal to $2126.

•9–28. A motorist claims that the South Boro Police issue an average of 60 speeding tickets per day. The following data show the number of speeding tickets issued each day for a period of one month. Assume σ is 13.42, and test the motorist's claim at $\alpha = 0.05$.

72	45	36	68	69	71	57	60
83	26	60	72	58	87	48	59
60	56	64	68	42	57	57
58	63	49	73	75	42	63

*9-29. A manager states that in his factory, the average number of days per year missed by the employees due to illness is less than the national average of 10. The following data show the number of days missed by 40 employees last year. Is there sufficient evidence to believe the manager's statement, at $\alpha = 0.05$? (Use s to estimate σ.)

0	6	12	3	3	5	4	1
3	9	6	0	7	6	3	4
7	4	7	1	0	8	12	3
2	5	10	5	15	3	2	5
3	11	8	2	2	4	1	9

*9-30. Suppose a statistician chose to test a hypothesis at $\alpha = 0.01$. The critical value for a one-tailed right test is $+2.33$. If the test value was 1.97, what would the decision be? What would happen if, after seeing the test value, he decided to choose $\alpha = 0.05$? What would the decision be? Explain the contradiction, if there is one.

*9-31. The president of a company states that the average hourly wage of her employees is \$8.65. A sample of 50 employees has the distribution shown. At $\alpha = 0.05$, is the president's statement believable? (Use s to approximate σ.)

Class	Frequency
8.35–8.43	2
8.44–8.52	6
8.53–8.61	12
8.62–8.70	18
8.71–8.79	10
8.80–8.88	2

9-4 THE t TEST

When the population standard deviation is unknown and the sample size is less than 30, the z test is inappropriate for testing hypotheses involving means. A different test, called the t test, is used. The t test is used when σ is unknown and $n < 30$. (Some authors use $n \leq 30$.)

As stated in Chapter 8, the t distribution was developed by W. S. Gosset and is similar to the standard normal distribution in the following ways.

1. It is bell-shaped.
2. It is symmetrical about the mean.
3. The mean, median, and mode are equal to 0 and are located at the center of the distribution.
4. The curve never touches the x axis.

The t distribution differs from the standard normal distribution in the following ways.

1. The variance is greater than 1.
2. The t distribution is a family of curves based on the **degree of freedom,** which is a number related to sample size. (Recall that the symbol for degrees of freedom is d.f. See Section 8-3 for an explanation of degrees of freedom.)
3. As the sample size increases, the t distribution approaches the normal distribution.

The t test is defined formally next.

The t test is a statistical test for the mean of a population and is used when the population is normally or approximately normally distributed, σ is unknown, and $n < 30$.

The formula for the t test is

$$t = \frac{\overline{X} - \mu}{s/\sqrt{n}}$$

The degrees of freedom are d.f. $= n - 1$.

The formula for the *t* test is similar to the formula for the *z* test. But since the population standard deviation σ is unknown, the sample standard deviation *s* is used instead.

The critical values for the *t* test are given in Table F in the Appendix. For a one-tailed test, the α level is found by looking at the top row of the table and finding the appropriate column. The degrees of freedom are found by looking down the left-hand column. Notice that degrees of freedom are given for values from 1 through 28. When the degrees of freedom are 29 or more, the row with ∞ (infinity) is used. Note that the values in this row are the same as the values for the *z* distribution, since as the sample size increases, the *t* distribution approaches the *z* distribution. When the sample size is 30 or more, statisticians generally agree that the two distributions can be considered identical, since the difference between the values of each is relatively small.

EXAMPLE 9–6 Find the critical *t* value for $\alpha = 0.05$ with d.f. = 16 for a one-tailed right *t* test.

Solution Find the 0.05 column in the top row and 16 in the left-hand column. Where the row and column meet, the appropriate critical value is found, which is +1.746. See Figure 9–16. ◄

FIGURE 9–16
Finding the
Critical Value for the
t Test in Table F
(Example 9–6)

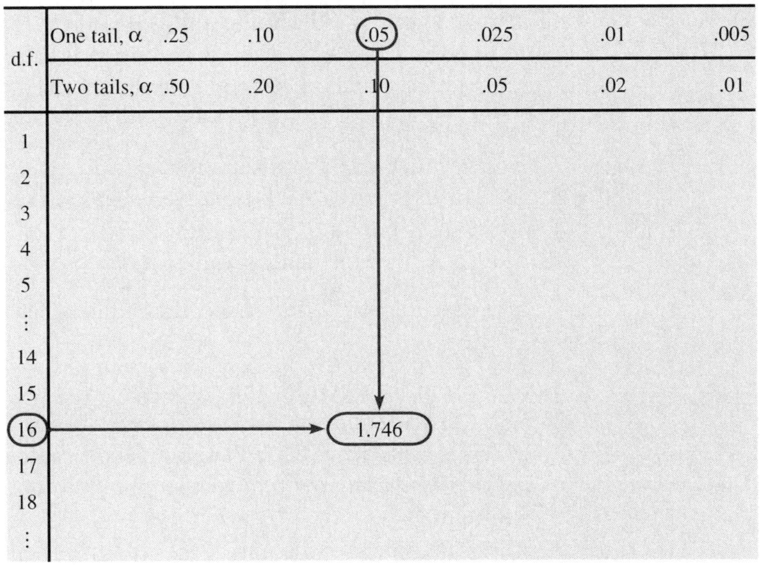

One tail, α	.25	.10	.05	.025	.01	.005
d.f.						
Two tails, α	.50	.20	.10	.05	.02	.01
1						
2						
3						
4						
5						
⋮						
14						
15						
16			1.746			
17						
18						
⋮						

EXAMPLE 9–7 Find the critical *t* value for $\alpha = 0.01$ with d.f. = 22 for a one-tailed left test.

Solution Find the 0.01 column in the row labeled "One Tail" and find 22 in the left column. The critical value is −2.508 since the test is a one-tailed left test. ◄

EXAMPLE 9–8 Find the critical values for $\alpha = 0.10$ with d.f. = 18 for a two-tailed *t* test.

Solution Find the 0.10 column in the row labeled "Two Tails" and find 18 in the column labeled "d.f." The critical values are +1.734 and −1.734. ◄

EXAMPLE 9–9   Find the critical value for $\alpha = 0.05$ with d.f. = 28 for a one-tailed right t test.

Solution   Find the 0.05 column in the "One Tail" row and 28 in the left column. The critical value is $+1.701$. ◄

When one is testing hypotheses by using the t test, the same procedure as for the z test is followed.

STEP 1   State the hypotheses, and identify the claim.

STEP 2   Find the critical value(s).

STEP 3   Compute the test value.

STEP 4   Make the decision to reject or not reject the null hypothesis.

STEP 5   Summarize the results.

Remember that the t test should be used when the population is approximately normally distributed, the population standard deviation is unknown, and the sample size is less than 30.

The next three examples illustrate the application of the t test.

EXAMPLE 9–10   A job placement director claims that the average starting salary for nurses is $24,000. A sample of 10 nurses has a mean of $23,450 and a standard deviation of $400. Test the director's claim at $\alpha = 0.05$.

Solution   **STEP 1** H_0: $\mu = \$24,000$ (claim) and H_1: $\mu \neq \$24,000$.

STEP 2   The critical values are $+2.262$ and -2.262 for $\alpha = 0.05$ and d.f. = 9.

STEP 3   The test value is

$$t = \frac{\overline{X} - \mu}{s/\sqrt{n}} = \frac{23{,}450 - 24{,}000}{400/\sqrt{10}} = -4.35$$

STEP 4   Reject the null hypothesis, since $-4.35 < -2.262$, as shown in Figure 9–17.

FIGURE 9–17
Summary of the
t Test of
Example 9–10

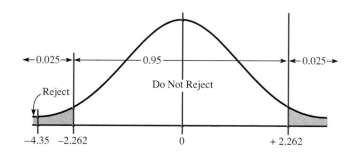

STEP 5   There is enough evidence to reject the claim that the starting salary of nurses is $24,000. ◄

EXAMPLE 9–11   A machine is designed to fill jars with 16 ounces of coffee. A consumer suspects that the machine is not filling the jars completely. A sample of 8 jars has a mean of 15.6 ounces and a standard deviation of 0.3 ounces. At $\alpha = 0.10$, test the consumer's claim.

Solution   **STEP 1** H_0: $\mu \geq 16$ and H_1: $\mu < 16$ (claim).

STEP 2   At $\alpha = 0.10$ and d.f. $= 7$, the critical value is -1.415.

STEP 3   The test value is

$$t = \frac{\overline{X} - \mu}{s/\sqrt{n}} = \frac{15.6 - 16}{0.3/\sqrt{8}} = -3.77$$

STEP 4   Reject the null hypothesis, since the test value falls in the critical region, as shown in Figure 9–18.

FIGURE 9–18
Summary of the
t Test of
Example 9–11

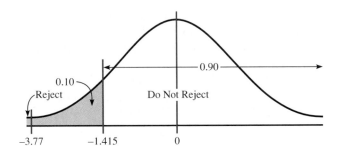

STEP 5   There is enough evidence to support the claim that the machine is not filling the jars with 16 ounces of coffee.   ◄

EXAMPLE 9–12   A physician claims that joggers' maximal volume oxygen uptake is greater than the average of all adults. A sample of 15 joggers has a mean of 43.6 milliliters per kilogram (ml/kg) and a standard deviation of 6 ml/kg. If the average of all adults is 36.7 ml/kg, test the physician's claim at $\alpha = 0.01$.

Solution   **STEP 1** H_0: $\mu \leq 36.7$ and H_1: $\mu > 36.7$ (claim).

STEP 2   The critical value is $+2.624$, since $\alpha = 0.01$ and d.f. $= 14$.

STEP 3   The test value is

$$t = \frac{\overline{X} - \mu}{s/\sqrt{n}} = \frac{43.6 - 36.7}{6/\sqrt{15}} = 4.45$$

STEP 4   Reject the null hypothesis, since 4.45 falls in the critical region, as shown in Figure 9–19.

FIGURE 9–19
Summary of the
t Test of
Example 9–12

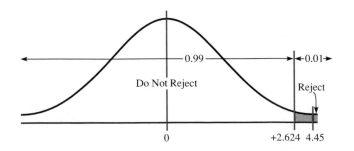

STEP 5   There is enough evidence to support the claim that the joggers' maximal volume oxygen uptake is greater than 36.7 ml/kg.   ◄

As for confidence intervals, students sometimes have difficulty deciding whether to use the *z* test or *t* test. The rules are the same as those pertaining to confidence intervals. These rules follow.

1. If σ is known, always use the *z* test regardless of the sample size. (If $n < 30$, the population must be approximately normally distributed.)
2. If σ is unknown but $n \geq 30$, use the *z* test and use *s* in place of σ in the formula.
3. If σ is unknown and $n < 30$, use the *t* test. (The population must be approximately normally distributed.)

These rules are summarized in Figure 9–20.

FIGURE 9–20
Using the *z* or *t* Test

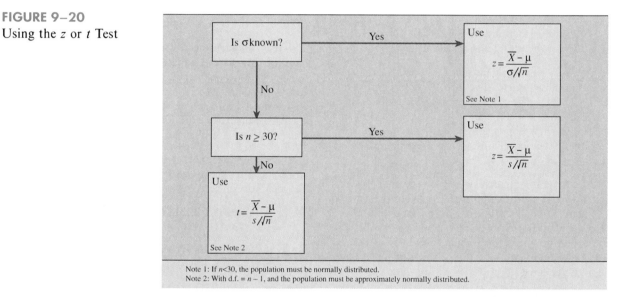

Note 1: If *n*<30, the population must be normally distributed.
Note 2: With d.f. = *n* − 1, and the population must be approximately normally distributed.

EXERCISES

9–32. In what ways is the *t* distribution similar to the standard normal distribution?

9–33. In what ways is the *t* distribution different from the standard normal distribution?

9–34. What are the degrees of freedom for the *t* test?

9–35. How does the formula for the *t* test differ from the formula for the *z* test?

9–36. (ANS.) Find the critical value (or values) for the *t* test for each.

a. $n = 10$, $\alpha = 0.05$, one-tailed right
b. $n = 18$, $\alpha = 0.10$, two-tailed

c. $n = 6$, $\alpha = 0.01$, one-tailed left
d. $n = 9$, $\alpha = 0.025$, one-tailed right
e. $n = 15$, $\alpha = 0.05$, two-tailed
f. $n = 23$, $\alpha = 0.005$, one-tailed left
g. $n = 28$, $\alpha = 0.01$, two-tailed
h. $n = 17$, $\alpha = 0.02$, two-tailed

For Exercises 9–37 through 9–50, perform each of the following steps.

a. State the hypotheses, and identify the claim.
b. Find the critical value(s).
c. Find the test value.
d. Make the decision.
e. Summarize the results.

Assume that the population is approximately normally distributed.

9–37. A sports store manager claims that the shop sells $850 in baseball equipment each day during the months of April, May, and June. A seven-day period is selected at random, and the average of the sales was found to be $920, with a sample standard deviation of $85. Test the claim at $\alpha = 0.05$.

9–38. The manager of a car rental agency claims that the average mileage of the car rented is less than 8000. A sample of 5 automobiles has an average mileage of 7723, with a standard deviation of 500 miles. At $\alpha = 0.01$, test the manager's claim.

9–39. A rental agency claims that the average rent that small-business establishments pay in Eagle City is $800. A sample of 10 establishments shows an average rental rate of $863. The standard deviation of the sample is $20. At $\alpha = 0.05$, is there evidence to reject the claim?

9–40. A large hospital instituted a fitness program in order to reduce absenteeism. The director reported that the average number of working hours lost due to illness per employee is 48 hours per year. After one year, a sample of 18 employees showed an average of 41 hours work loss, with a standard deviation of 5. Did the program reduce absenteeism? Use $\alpha = 0.10$.

9–41. In order to increase customer service, a muffler repair shop says that its mechanics can replace a muffler in less than 12 minutes. A time management specialist selected 6 repair jobs and found the mean time of the 6 jobs to be 11.6 minutes. The standard deviation of the sample was 2.1 minutes. At $\alpha = 0.025$, is there enough evidence to support the claim?

9–42. A travel agency claims that it books an average of 45 people per trip to Washington, D.C. A sample of 15 trips showed a mean of 41 people booked and a standard deviation of 5. At $\alpha = 0.05$, test the claim.

9–43. A rental agent states that the average rent for a studio apartment in Shadyside is $750. A sample of 12 renters shows that the mean is $732 and the standard deviation is $17. Is there enough evidence to reject the agent's claim? Use $\alpha = 0.01$.

9–44. A manufacturer claims that a new snowmaking machine can save money for ski resort owners. He states that in a sample of 10 tons, the cost of making a ton of snow was $5.75. The average of the old machine was $6.62. The standard deviation of the sample was $1.05. At $\alpha = 0.10$, does the new machine save money?

9–45. An officer states that the average fine levied by the Office of Safety against companies is at most $350. A company owner suspects it is higher. She samples 12 companies and finds that the average is $358. The standard deviation of the sample is $16. Test the claim that the average is higher than $350, at $\alpha = 0.05$.

9–46. West Newton advertises that it is the cleanest town in New Jersey by stating that the average price of a car wash is $3.00. A sample of 5 car washes had an average price of $3.70 and a standard deviation of $0.30. At $\alpha = 0.05$, is the average price of a car wash really $3.00?

9–47. A recent survey stated that the average single-person household receives at least 37 telephone calls per month. To test the claim, a researcher surveyed 29 single-person households and found that the average number of calls was 34.9. The standard deviation of the sample was 6. At $\alpha = 0.05$, can the claim be substantiated?

9–48. A student believes that the average cost of a Saturday night date is no longer $30.00. In order to test her hypothesis, she randomly selected 16 men from the dormitory and asked them how much they spent on a date last Saturday. She found that the average cost was $31.17. The standard deviation of the sample was $5.51. At $\alpha = 0.05$, is there enough evidence to support her claim?

*9–49. From past experience, a teacher believes that the average score on a real estate exam is 75. A sample of 20 students' exam scores is as follows:

80, 68, 72, 73, 76, 81, 71, 71, 65, 50,
63, 71, 70, 70, 76, 75, 69, 70, 72, 74

Test the hypothesis that the students' average score is still 75 this year. Use $\alpha = 0.01$.

*9–50. A new laboratory technician read a report that the average number of students using the computer laboratory per hour was 16. In order to test this hypothesis, he selects a day at random and keeps track of the number of students who use the lab for an 8-hour period. The results are as follows:

20, 24, 18, 16, 16, 19, 21, 23

At $\alpha = 0.05$, can the technician conclude that the average is actually 16?

9–5 TEST FOR PROPORTIONS

Many hypothesis-testing situations involve proportions. Recall from Chapter 8 that a *proportion* is the same as a percentage of the population. For example, the *Harper's Index Book* (Henry Holt, 1987) lists the following proportions:

Percentage of Americans who think they look younger than they are: 57.
Percentage of Americans who believe heaven exists: 84.
Percentage of American families headed by a single parent: 26.
Percentage of motor vehicles with vanity plates in Illinois: 1.1.

A hypothesis test involving a population proportion can be considered as a binomial experiment when there are only two outcomes and the probability of a success does not change from trial to trial. Recall from Section 6–4 in Chapter 6 that the mean is $\mu = np$ and the standard deviation is $\sigma = \sqrt{npq}$ for the binomial distribution.

Since the normal distribution can be used as an approximation of the binomial distribution when $np \geq 5$ and $nq \geq 5$, the standard normal distribution can be used to test hypotheses for proportions.

Formula for the z Test for Proportions

$$z = \frac{X - \mu}{\sigma} \quad \text{or} \quad z = \frac{X - np}{\sqrt{npq}}$$

where $\mu = np$ and $\sigma = \sqrt{npq}$.

The following examples illustrate the hypothesis-testing procedure for proportions. *Note that there is an additional step (Step 2) in this procedure for finding the mean and standard deviation.* (A detailed explanation of proportions can be found at the beginning of Section 8–4.)

EXAMPLE 9–13 An educator estimates that the dropout rate for seniors at high schools in Ohio is 15%. Last year, 38 seniors from a random sample of 200 Ohio seniors withdrew. At $\alpha = 0.05$, test the educator's claim.

Solution **STEP 1** State the hypotheses, and identify the claim.

$$H_0: p = 0.15 \text{ (claim)} \quad \text{and} \quad H_1: p \neq 0.15$$

STEP 2 Find the mean and the standard deviation.

$$\mu = np = (200)(0.15) = 30$$
$$\sigma = \sqrt{npq} = \sqrt{(200)(0.15)(0.85)} = 5.05$$

STEP 3 Find the critical value(s). Since $\alpha = 0.05$ and the test is a two-tailed test, the critical values are ± 1.96.

STEP 4 Compute the test value.

$$z = \frac{X - \mu}{\sigma} = \frac{38 - 30}{5.05} = 1.58$$

STEP 5 Make the decision. Do not reject the null hypothesis, since the test value falls outside the critical region, as shown in Figure 9–21.

FIGURE 9–21
Critical and
Test Values for
Example 9–13

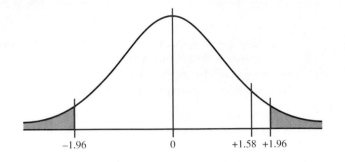

−1.96   0   +1.58 +1.96

STEP 6 Summarize the results. There is not enough evidence to reject the claim that the dropout rate for seniors in high schools in Ohio is 15%. ◄

EXAMPLE 9–14 A telephone company representative estimates that 40% of its customers want call-waiting service. To test this hypothesis, she selected a sample of 100 customers and found that 37% had the call-waiting service. At $\alpha = 0.01$, is her estimate appropriate?

Solution **STEP 1** State the hypotheses, and identify the claim.

$$H_0: p = 0.40 \text{ (claim)} \qquad \text{and} \qquad H_1: p \neq 0.40$$

STEP 2 Find the mean and the standard deviation.

$$\mu = np = (100)(0.40) = 40$$
$$\sigma = \sqrt{npq} = \sqrt{(100)(0.40)(0.60)} = 4.90$$

Note: $X = (100)(0.37) = 37$.

STEP 3 Find the critical value(s). Since $\alpha = 0.01$ and this test is a two-tailed test, the critical values are ± 2.58.

STEP 4 Compute the test value.

$$z = \frac{X - \mu}{\sigma} = \frac{37 - 40}{4.90} = -0.612$$

STEP 5 Make the decision. Do not reject the null hypothesis, since the test value falls in the noncritical region, as shown in Figure 9–22.

FIGURE 9–22
Critical and
Test Values for
Example 9–14

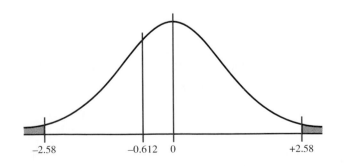

−2.58   −0.612 0   +2.58

STEP 6 Summarize the results. There is not enough evidence to reject the claim that 40% of the telephone company's customers have call-waiting service. ◄

EXAMPLE 9-15 A physician claims that she spends at least 15 minutes per visit with more than 60% of her patients. A researcher found that in a study of 80 patients, 56 patients' visits lasted 15 minutes or longer. Does the evidence support the physician's claim? Use $\alpha = 0.01$.

Solution **STEP 1** $H_0: p \leq 0.60$ and $H_1: p > 0.60$ (claim).

STEP 2
$$\mu = np = (80)(0.60) = 48$$
$$\sigma = \sqrt{npq} = \sqrt{(80)(0.60)(0.40)} = 4.38$$

STEP 3 The critical value is $+2.33$.

STEP 4
$$z = \frac{X - \mu}{\sigma} = \frac{56 - 48}{4.38} = 1.83$$

STEP 5 Do not reject the null hypothesis, since the test value falls in the noncritical region, as shown in Figure 9-23.

FIGURE 9-23
Critical and
Test Values for
Example 9-15

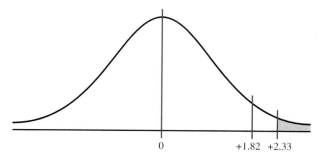

0   +1.82 +2.33

STEP 6 There is not enough evidence to support the physician's claim that she spends at least 15 minutes per visit with more than 60% of her patients. ◄

EXAMPLE 9-16 An attorney claims that more than 25% of all lawyers advertise. A sample of 200 lawyers in a certain city showed that 63 had used some form of advertising. At $\alpha = 0.05$, test the attorney's claim.

Solution **STEP 1** $H_0: p \leq 0.25$ and $H_1: p > 0.25$ (claim).

STEP 2
$$\mu = np = (200)(0.25) = 50$$
$$\sigma = \sqrt{npq} = \sqrt{(200)(0.25)(0.75)} = 6.12$$

STEP 3 The critical value is $+1.65$.

STEP 4
$$z = \frac{X - \mu}{\sigma} = \frac{63 - 50}{6.12} = 2.12$$

STEP 5 Reject the null hypothesis, since the test value falls in the critical region, as shown in Figure 9–24.

FIGURE 9–24
Critical and
Test Values for
Example 9–16

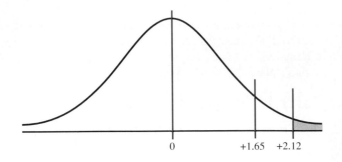

$$0 \qquad +1.65 \quad +2.12$$

STEP 6 There is enough evidence to support the attorney's claim that more than 25% of the lawyers use some form of advertising. ◄

Tests for proportions can also be conducted by using a formula equivalent to the one shown in the preceding examples. This formula is

$$z = \frac{\hat{p} - p}{\sqrt{pq/n}}$$

where

$\hat{p} = X/n$, the sample proportion

p = hypothesized population proportion

n = sample size

The denominator $\sqrt{pq/n}$ is the standard error of the proportions.

EXERCISES

9–51. Give three examples of proportions.

9–52. Why is a proportion considered a binomial variable?

9–53. When one is testing hypotheses using proportions, what are the necessary requirements?

9–54. What are the mean and the standard deviation of a proportion?

For Exercises 9–55 through 9–65, perform each of the following steps.

a. State the hypotheses, and identify the claim.
b. Find the mean and the standard deviation.
c. Find the critical value(s).
d. Compute the test value.
e. Make the decision.
f. Summarize the results.

9–55. A toy manufacturer claims that at least 23% of the fourteen-year-old residents of a certain city own a skateboard. A sample of 40 fourteen-year-olds shows that 7 own a skateboard. Test the claim. Use $\alpha = 0.05$.

9–56. From past records, a hospital found that 37% of all full-term babies born in the hospital weighed more than 7 pounds 2 ounces. This year a sample of 100 babies showed that 23 weighed over 7 pounds 2 ounces. At $\alpha = 0.01$, is there enough evidence to say the percentage has changed?

9–57. A study on crime suggested that at least 40% of all arsonists were under 21 years of age. Checking local crime statistics, a researcher found that 30 out of 80 arson suspects were under 21. At $\alpha = 0.10$, should the crime study's statement be believed?

9–58. A recent study found that, at most, 32% of people who have been in a plane crash have died. In a sample of 100 people who were in a plane crash, 38 died. Is the study's statement believable? Use $\alpha = 0.05$.

9–59. An engineering school dean claims that 21% of the school's graduates are women. In a graduating class of 143 students, there were 38 females. Does this result suggest that the dean's claim is believable? Use $\alpha = 0.05$.

9–60. Experts claim that 29% of all robberies are committed by people less than 18 years of age. Test this claim at $\alpha = 0.05$ if in a sample of 83 robberies, 17 were committed by people under the age of 18.

9–61. A recent study claimed that at least 15% of all eighth-grade students are overweight. In a sample of 80 students, 9 were found to be overweight. At $\alpha = 0.05$, test the claim.

9–62. At a large university, a study found that no more than 25% of the students who commute travel more than 14 miles to campus. At $\alpha = 0.10$, are the findings supported if in a sample of 100 students, 30 drove more than 14 miles?

9–63. A telephone company wants to advertise that more than 30% of all its customers have at least two telephones. To support this advertisement, the company selects a sample of 200 customers and finds that 72 have more than two telephones. Does the evidence support the advertisement? Use $\alpha = 0.05$.

9–64. Experts claim that 10% of murders are committed by women. Test this claim at $\alpha = 0.01$ if in a sample of 67 murders, 10 were committed by women.

9–65. Researchers suspect that 18% of all high school students smoke at least one pack of cigarettes a day. At Wilson High School, with an enrollment of 300 students, a study found that 50 students smoked at least one pack of cigarettes a day. At $\alpha = 0.05$, can the study conclude that 18% of all high school students smoke at least one pack of cigarettes a day?

When *np* or *nq* is not 5 or more, the binomial table (Table B) must be used to find critical values in hypothesis tests involving proportions.

*9–66. A coin is tossed 9 times and 3 heads appear. Can one conclude that the coin is not balanced? Use $\alpha = 0.10$. [*Hint:* Use the binomial table, Table B in the Appendix, and find $P(X = 3)$ with $p = 0.5$ and $n = 9$.]

*9–67. Fashion buyers have suggested that 30% of all men purchase blue suits. If in a sample of 10 men, 6 who purchase suits select blue suits, should the null hypothesis be rejected at $\alpha = 0.05$?

*9–68. In the past, 20% of all airline passengers flew first class. In a sample of 15 passengers, 5 flew first class. At $\alpha = 0.10$, can one conclude that the proportions have changed?

9–6 ADDITIONAL TOPICS REGARDING HYPOTHESIS TESTING (OPTIONAL)

In hypothesis testing, there are several other concepts that might be of interest to students in elementary statistics. These topics include *P*-values, the relationship between hypothesis testing and confidence intervals, and some additional information about the type II error.

P-Values

Statisticians usually test hypotheses at the common α levels of 0.05 or 0.01 and sometimes at 0.10. Recall that the choice of the level depends on the seriousness of the type I error. Besides listing an α value, many computer statistical packages give a *P*-value for hypothesis tests. The **P-value** is the actual probability of getting the sample mean value or a more extreme sample mean value in the direction of the alternative hypothesis ($<$ or $>$) if the null hypothesis is true. In other words, the *P*-value is the actual area under the standard normal distribution curve, representing the probability of the particular sample mean or a more extreme sample mean occurring if the null hypothesis is true.

For example, suppose that a null hypothesis is H_0: $\mu \le 50$ and the mean of a sample is $\overline{X} = 52$. If the computer printed a *P*-value of 0.0356 for a statistical test, then the probability of getting a sample mean of 52 is 0.0356 if the true population mean is 50. The relationship between the *P*-value and the α value can be explained in this manner. For $P = 0.0356$ the null hypothesis would be rejected at $\alpha = 0.01$ but not at $\alpha = 0.05$. See Figure 9–25.

FIGURE 9–25
Comparison of α Values and P-Values

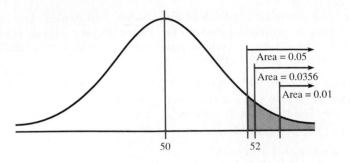

When the hypothesis test is two-tailed, the P-value must be doubled. For a two-tailed test, if α is 0.05 and the P-value given is 0.0356, the actual P-value would be $2(0.0356) = 0.0712$. That is, the null hypothesis should not be rejected at $\alpha = 0.05$, since 0.0712 is greater than 0.05.

The next example shows how to find a P-value for use of the z test.

EXAMPLE 9–17   A researcher wishes to test the claim that the average age of lifeguards in Ocean City is greater than 24 years. She selects a sample of 36 guards and finds the mean of the sample to be 24.7 years, with a standard deviation of 2 years. Is there evidence to support the claim at $\alpha = 0.05$? Find the P-value.

Solution   **STEP 1**   State the hypotheses, and identify the claim.

$$H_0\text{: } \mu \leq 24 \qquad \text{and} \qquad H_1\text{: } \mu > 24 \text{ (claim)}$$

STEP 2   Compute the test value.

$$z = \frac{24.7 - 24}{2/\sqrt{36}} = 2.10$$

STEP 3   Using Table E in the Appendix, find the corresponding area under the normal distribution for $z = 2.10$. It is 0.4821.

STEP 4   Subtract this value for the area from 0.5000 to find the area in the right tail.

$$0.5000 - 0.4821 = 0.0179$$

Hence, the P-value is 0.0179.

STEP 5   Make the decision. Since the P-value is less than 0.05, the decision is to reject the null hypothesis.

STEP 6   Summarize the results. There is enough evidence to support the claim that the average age of lifeguards in Ocean City is greater than 24 years.   ◀

Note: Had the researcher chosen $\alpha = 0.01$, the null hypothesis would not have been rejected, since the P-value (0.0179) is greater than 0.01.

A clear distinction between the α value and the P-value should be made. The α value is chosen by the researcher before the statistical test is conducted. The P-value is computed after the sample mean has been found.

There are two schools of thought on P-values. Some researchers do not choose an α value but report the P-value and allow the reader to decide whether the null hypothesis should be rejected. Others decide on the α value in advance and use the P-value to make the decision, as shown in the previous example. A note of caution is needed here: If a researcher selects $\alpha = 0.01$ and the P-value is 0.03, the researcher may decide to change the α value from 0.01 to 0.05 so that the null hypothesis would be rejected. This, of course, should not be done. Hence, if the α level is selected in advance, it should be used in making the decision. If the researcher is using the P-value, then no α value should be selected and the reader should make the decision about whether the null hypothesis should be rejected.

One additional note on hypothesis testing is that the researcher should distinguish between *statistical significance* and *practical significance*. When the null hypothesis is rejected at a specific significance level, it can be concluded that the difference is probably not due to chance and thus is said to be statistically significant. However, the results may not have any practical significance. For example, suppose that a new fuel additive increases the miles per gallon a car can get by $\frac{1}{4}$ of a mile for a sample of 1000 automobiles. The results may be statistically significant at the 0.05 level, but it would hardly be worthwhile to market the product for such a small increase. Hence, there is no practical significance to the results. It is up to the researcher to use common sense when interpreting the results of a statistical test.

Confidence Intervals and Hypothesis Testing

There is a relationship between confidence intervals and hypothesis testing. When the null hypothesis is rejected in a hypothesis-testing situation, the confidence interval for the mean using the same level of significance *will not* contain the hypothesized mean. Likewise, when the null hypothesis is not rejected, the confidence interval computed using the same level of significance *will* contain the hypothesized mean. The next two examples show this concept for two-tailed tests.

EXAMPLE 9-18   Sugar is packed in 5-pound bags. An inspector feels that the bags may not contain 5 pounds, as stated. A sample of 50 bags produces a mean of 4.6 pounds and a standard deviation of 0.7 pound. Test the claim at $\alpha = 0.05$ that the bags do not contain 5 pounds, as stated. Also, find the 95% confidence interval of the true mean.

Solution H_0: $\mu = 5$ and H_1: $\mu \neq 5$ (claim). The critical values are $+1.96$ and -1.96. The test value is

$$z = \frac{\overline{X} - \mu}{s/\sqrt{n}} = \frac{4.6 - 5.00}{0.7/\sqrt{50}} = \frac{-0.4}{0.099} = -4.04$$

Since $-4.04 < -1.96$, the null hypothesis is rejected. There is enough evidence to support the claim that the bags do not weigh 5 pounds.

The 95% confidence for the mean is given by

$$\overline{X} - z_{\alpha/2} \cdot \frac{s}{\sqrt{n}} < \mu < \overline{X} + z_{\alpha/2} \cdot \frac{s}{\sqrt{n}}$$

$$4.6 - (1.96) \left(\frac{0.7}{\sqrt{50}} \right) < \mu < 4.6 + (1.96) \left(\frac{0.7}{\sqrt{50}} \right)$$

$$4.406 < \mu < 4.794$$

Notice that the 95% confidence interval of μ does *not* contain the hypothesized value $\mu = 5$. Hence, there is agreement between the hypothesis test and the confidence interval.   ◄

EXAMPLE 9–19   A researcher claims that adult hogs fed a special diet will have an average weight of 200 pounds. A sample of 10 hogs has an average weight of 198.2 pounds and a standard deviation of 3.3 pounds. At $\alpha = 0.05$, is the claim justified? Also, find the 95% confidence interval of the true mean.

Solution H_0: $\mu = 200$ pounds (claim) and H_1: $\mu \neq 200$ pounds. The t test must be used since σ is unknown and $n < 30$. The critical values at $\alpha = 0.05$ with 9 degrees of freedom are $+2.262$ and -2.262. The test value is

$$t = \frac{\overline{X} - \mu}{s/\sqrt{n}} = \frac{198.2 - 200}{3.3/\sqrt{10}} = \frac{-1.8}{1.0436} = -1.72$$

Thus, the null hypothesis is not rejected. There is not enough evidence to reject the claim that the weight of the adult hogs is 200 pounds.

The 95% confidence interval of the mean is

$$\overline{X} - t_{\alpha/2} \cdot \frac{s}{\sqrt{n}} < \mu < \overline{X} + t_{\alpha/2} \cdot \frac{s}{\sqrt{n}}$$

$$198.2 - (2.262) \left(\frac{3.3}{\sqrt{10}} \right) < \mu < 198.2 + (2.262) \left(\frac{3.3}{\sqrt{10}} \right)$$

$$198.2 - 2.361 < \mu < 198.2 + 2.361$$

$$195.84 < \mu < 200.56$$

The 95% confidence interval does contain the hypothesized mean $\mu = 200$.   ◄

In summary, then, when the null hypothesis is rejected, the confidence interval computed at the same significance level will not contain the value of the mean that is stated in the null hypothesis. On the other hand, when the null hypothesis is not rejected, the confidence interval computed at the same significance level will contain the value of the mean stated in the null hypothesis. These results are true for other hypothesis-testing situations and are not limited to means tests.

The relationship between confidence intervals and hypothesis testing presented here is valid for two-tailed tests. The relationship between one-tailed hypothesis tests and one-sided or one-tailed confidence intervals is also valid; however, this technique is beyond the scope of the textbook.

IMPORTANT FORMULAS

CHAPTER 2

Range = highest value − lowest value

Midpoint $(X_m) = \dfrac{\text{lower boundary} + \text{higher boundary}}{2}$

Class width = upper boundary − lower boundary

Degrees for each section of the pie graph:

$$\text{degrees} = \frac{f}{n} \cdot 360°$$

f = frequency n = total number of subjects

Percent for each section of the pie graph: $\% = \dfrac{f}{n} \cdot 100$

CHAPTER 3

Mean for individual data: $\overline{X} = \dfrac{\Sigma X}{n}$ or $\mu = \dfrac{\Sigma X}{N}$

Mean for grouped data: $\overline{X} = \dfrac{\Sigma f \cdot X_m}{n}$

Median for grouped data: $\text{MD} = \dfrac{(n/2) - \text{cf}}{f}\,(w) + L_m$

cf = cumulative frequency

w = upper boundary − lower boundary

L_m = lower boundary of the median class

Weighted mean: $\overline{X} = \dfrac{\Sigma w \cdot X}{\Sigma w}$

Midrange: $\text{MR} = \dfrac{\text{lower value} + \text{highest value}}{2}$

Variance for a population: $\sigma^2 = \dfrac{\Sigma (X - \mu)^2}{N}$

Variance for a sample: $s^2 = \dfrac{\Sigma X^2 - [(\Sigma X)^2/n]}{n - 1}$

Variance for grouped data:

$$s^2 = \frac{\Sigma f \cdot X_m^2 - [(\Sigma f \cdot X_m)^2/n]}{n - 1}$$

Standard deviation for a population: $\sigma = \sqrt{\dfrac{\Sigma (X - \mu)^2}{N}}$

Standard deviation for a sample:

$$s = \sqrt{\frac{\Sigma X^2 - [(\Sigma X)^2/n]}{n - 1}}$$

Standard deviation for grouped data:

$$s = \sqrt{\frac{\Sigma f \cdot X_m^2 - [(\Sigma f \cdot X_m)^2/n]}{n - 1}}$$

Coefficient of variation: $\text{CVar} = \dfrac{s}{\overline{X}} \cdot 100\%$ or

$$\frac{\sigma}{\mu} \cdot 100\%$$

z score: $z = \dfrac{X - \mu}{\sigma}$ or $z = \dfrac{X - \overline{X}}{s}$

Percentile rank of a value x:

$$\text{Percentile} = \frac{\text{number of values below} + 0.5}{\text{total number of values}} \cdot 100\%$$

Value corresponding to a given percentile: $c = \dfrac{n \cdot p}{100}$

Interquartile range $= Q_3 - Q_1$

CHAPTER 4

Multiplication rule 1: Total number of outcomes of a sequence when each event has k possibilities $= k^n$

Multiplication rule 2: Total number of outcomes of a sequence when each event has a different number of possibilities $k_1 \cdot k_2 \cdot k_3 \cdot \ldots \cdot k_n$

Permutation rule 1: Number of permutations of n objects is $n!$

Permutation rule 2: Number of permutations of n objects taken r at a time is $_nP_r = \dfrac{n!}{(n - r)!}$

Permutation rule 3: Number of permutations of n objects in which k_1 are alike, k_2 are alike, etc., is

$$\frac{n!}{k_1! \cdot k_2! \cdot \ldots \cdot k_n!}$$

Combination rule: Number of combinations of r objects selected from n objects is $_nC_r = \dfrac{n!}{(n - r)!r!}$

CHAPTER 5

Classical probability:

$$P(E) = \frac{\text{number of ways } E \text{ can occur}}{\text{total number of outcomes in the sample space}}$$

$$= \frac{n(E)}{n(S)}$$

Empirical probability:

$$P(E) = \frac{\text{frequency of class}}{\text{total frequencies in the distribution}} = \frac{f}{n}$$

Addition rule 1 (mutually exclusive events):

$$P(A \text{ or } B) = P(A) + P(B)$$

Addition rule 2 (events not mutually exclusive):

$$P(A \text{ or } B) = P(A) + P(B) - P(A \text{ and } B)$$

Multiplication rule 1 (independent events):

$$P(A \text{ and } B) = P(A) \cdot P(B)$$

Multiplication rule 2 (dependent events):

$$P(A \text{ and } B) = P(A) \cdot P(B \mid A)$$

Conditional probability: $P(B \mid A) = \dfrac{P(A \text{ and } B)}{P(A)}$

Complementary events: $P(\overline{E}) = 1 - P(E)$

$$P(E) = 1 - P(\overline{E})$$

Bayes's theorem: $P(A_1 \mid B_1) =$

$$\frac{P(A_1) \cdot P(B_1 \mid A_1)}{P(A_1) \cdot P(B_1 \mid A_1) + P(A_2) \cdot P(B_1 \mid A_2) + \cdots + P(A_n) \cdot P(B_1 \mid A_n)}$$

CHAPTER 6

Mean for a probability distribution: $\mu = \Sigma\, X \cdot P(X)$

Variance for a probability distribution:

$$\sigma^2 = \Sigma[X^2 \cdot P(X)] - \mu^2$$

Expectation: $E(X) = \Sigma\, X \cdot P(X)$

Binomial probability: $P(X) = \dfrac{n!}{(n - X)!X!} \cdot p^X \cdot q^{n-X}$

Mean for binomial distribution: $\mu = n \cdot p$

Variance and standard deviation for the binomial
distribution: $\sigma^2 = n \cdot p \cdot q \qquad \sigma = \sqrt{n \cdot p \cdot q}$

Multinomial probability:

$$P(M) = \frac{n!}{X_1! X_2! X_3! \ldots X_k!} \cdot p_1{}^{X_1} \cdot p_2{}^{X_2} \cdot p_3{}^{X_3} \cdots p_k{}^{X_k}$$

Poisson probability: $P(X;\lambda) = \dfrac{e^{-\lambda} \lambda^X}{X!}$ where

$$X = 0,1,2,\ldots$$

Hypergeometric probability: $P(A) = \dfrac{{}_aC_X \cdot {}_bC_{n-X}}{{}_{(a+b)}C_n}$

CHAPTER 7

Mean of sample means: $\mu_{\bar{x}} = \mu$

Standard error of the mean: $\sigma_{\bar{x}} = \dfrac{\sigma}{\sqrt{n}}$

Central limit theorem formula: $z = \dfrac{\bar{X} - \mu}{\sigma / \sqrt{n}}$

Central limit theorem formula when $n > 0.05N$:

$$z = \frac{\bar{X} - \mu}{(\sigma / \sqrt{n}) \, \sqrt{(N - n)/(N - 1)}}$$

Mean for binomial variable:

$$\mu = n \cdot p$$

Standard deviation for a binomial variable:

$$\sigma = \sqrt{n \cdot p \cdot q}$$

Finding a specific data value: $X = z \cdot \sigma + \mu$

CHAPTER 8

z confidence interval for means:

$$\bar{X} - z_{\alpha/2}\left(\frac{\sigma}{\sqrt{n}}\right) < \mu < \bar{X} + z_{\alpha/2}\left(\frac{\sigma}{\sqrt{n}}\right)$$

t confidence interval for means:

$$\bar{X} - t_{\alpha/2}\left(\frac{s}{\sqrt{n}}\right) < \mu < \bar{X} + t_{\alpha/2}\left(\frac{s}{\sqrt{n}}\right)$$

Sample size for means: $n = \left(\dfrac{z_{\alpha/2} \cdot \sigma}{E}\right)^2$

where E is the maximum error of estimate

Confidence interval for a proportion:

$$\hat{p} - z_{\alpha/2} \cdot \sqrt{\frac{\hat{p}\hat{q}}{n}} < p < \hat{p} + (z_{\alpha/2}) \sqrt{\frac{\hat{p}\hat{q}}{n}}$$

Sample size for a proportion: $n = \hat{p}\hat{q}\left(\dfrac{z_{\alpha/2}}{E}\right)^2$

where $\hat{p} = \dfrac{X}{n}$ and $\hat{q} = 1 - \hat{p}$

CHAPTER 9

z test: $z = \dfrac{\bar{X} - \mu}{\sigma / \sqrt{n}}$

$z = \dfrac{\bar{X} - \mu}{s / \sqrt{n}}$ for σ unknown and $n \geq 30$

t test: $t = \dfrac{\bar{X} - \mu}{s / \sqrt{n}}$ for $n < 30$

z test for proportions: $z = \dfrac{X - np}{\sqrt{npq}}$

IMPORTANT FORMULAS

CHAPTER 10

z test for comparing two means:

$$z = \frac{(\bar{X}_1 - \bar{X}_2) - (\mu_1 - \mu_2)}{\sqrt{\dfrac{\sigma_1^2}{n_1} + \dfrac{\sigma_2^2}{n_2}}}$$

t test for comparing two means (independent samples, variances not equal):

$$t = \frac{(\bar{X}_1 - \bar{X}_2) - (\mu_1 - \mu_2)}{\sqrt{\dfrac{s_1^2}{n_1} + \dfrac{s_2^2}{n_2}}}$$

d.f. = the smaller of $n_1 - 1$ or $n_2 - 1$

t test for comparing two means (independent samples, variances equal):

$$t = \frac{(\bar{X}_1 - \bar{X}_2) - (\mu_1 - \mu_2)}{\sqrt{\dfrac{(n_1 - 1)s_1^2 + (n_2 - 1)s_2^2}{(n_1 + n_2 - 2)}}\sqrt{\dfrac{1}{n_1} + \dfrac{1}{n_2}}}$$

t test for comparing two means for dependent samples:

$$t = \frac{\bar{D} - \mu_D}{s_D/\sqrt{n}} \quad \text{where} \quad \bar{D} = \frac{\sum D}{n} \quad \text{and}$$

$$s_D = \sqrt{\frac{\sum D^2 - [(\sum D)^2/n]}{n - 1}}$$

z test for comparing two proportions:

$$z = \frac{(\hat{p}_1 - \hat{p}_2) - (p_1 - p_2)}{\sqrt{\bar{p}\,\bar{q}\left(\dfrac{1}{n_1} + \dfrac{1}{n_2}\right)}}$$

$$\text{where} \quad \bar{p} = \frac{X_1 + X_2}{n_1 + n_2} \qquad \hat{p}_1 = \frac{X_1}{n_1}$$

$$\bar{q} = 1 - \bar{p} \qquad \hat{p}_2 = \frac{X_2}{n_2}$$

CHAPTER 11

Correlation coefficient:

$$r = \frac{n(\sum xy) - (\sum x)(\sum y)}{\sqrt{[n(\sum x^2) - (\sum x)^2][n(\sum y^2) - (\sum y)^2]}}$$

t test for correlation coefficient: $t = r\sqrt{\dfrac{n - 2}{1 - r^2}}$

The regression line equation: $y' = a + bx$

$$\text{where} \quad a = \frac{(\sum y)(\sum x^2) - (\sum x)(\sum xy)}{n(\sum x^2) - (\sum x)^2}$$

$$b = \frac{n(\sum xy) - (\sum x)(\sum y)}{n(\sum x^2) - (\sum x)^2}$$

Coefficient of determination: $r^2 = \dfrac{\text{explained variation}}{\text{total variation}}$

Standard error of estimate:

$$s_{est} = \sqrt{\frac{\sum y^2 - a\sum y - b\sum xy}{n - 2}}$$

Prediction interval when $n \geq 100$:

$$y' - z_{\alpha/2}s_{est} < y < y' + z_{\alpha/2}s_{est}$$

CHAPTER 12

Chi-square test for a single variance: $\chi^2 = \dfrac{(n - 1)s^2}{\sigma^2}$

Chi-square test for goodness-of-fit and independence test:

$$\chi^2 = \sum \frac{(O - E)^2}{E}$$

Confidence interval for variance:

$$\frac{(n - 1)s^2}{\chi^2_{larger}} < \sigma^2 < \frac{(n - 1)s^2}{\chi^2_{smaller}}$$

Confidence interval for standard deviation:

$$\sqrt{\frac{(n - 1)s^2}{\chi^2_{larger}}} < \sigma < \sqrt{\frac{(n - 1)s^2}{\chi^2_{smaller}}}$$

CHAPTER 13

F test for comparing two variances: $F = \dfrac{s_1^2}{s_2^2}$

where s_1^2 is the larger variance and d.f.N. $= n_1 - 1$
d.f.D. $= n_2 - 1$

ANOVA test: $F = \dfrac{s_B^2}{s_W^2}$ where $\bar{X}_{GM} = \dfrac{\sum X}{N}$

$$s_B^2 = \frac{\sum n_i(\bar{X}_i - \bar{X}_{GM})^2}{k - 1}$$

$$s_W^2 = \frac{\sum (n_i - 1)s_i^2}{\sum (n_i - 1)}$$

d.f.N. $= k - 1$ where $N = n_1 + n_2 + \ldots + n_k$
d.f.D. $= N - k$ where k = number of groups

Scheffé test: $F_S = \dfrac{(\bar{X}_i - \bar{X}_j)^2}{s_W^2\,(1/n_i + 1/n_j)}$ and

$$F' = (k - 1)(\text{C.V.})$$

Tukey test: $q = \dfrac{\bar{X}_i - \bar{X}_j}{\sqrt{s_W/n}}$

Formulas for two-way ANOVA:

$$MS_A = \frac{SS_A}{a-1} \qquad F_A = \frac{MS_A}{MS_W}$$

$$MS_B = \frac{SS_B}{b-1} \qquad F_B = \frac{MS_B}{MS_W}$$

$$MS_{A \times B} = \frac{SS_{A \times B}}{(a-1)(b-1)} \qquad F_{A \times B} = \frac{MS_{A \times B}}{MS_W}$$

$$MS_W = \frac{SS_W}{ab(n-1)}$$

CHAPTER 14

z test value in the sign test: $z = \dfrac{(X + 0.5) - (n/2)}{\sqrt{n}/2}$

where n = sample size (greater than or equal to 26)
  X = smaller number of $+$ or $-$ signs

Wilcoxon rank sum test: $z = \dfrac{R - \mu_R}{\sigma_R}$

where

$$\mu_R = \frac{n_1(n_1 + n_2 + 1)}{2}$$

$$\sigma_R = \sqrt{\frac{n_1 n_2(n_1 + n_2 + 1)}{12}}$$

R = sum of the ranks for the smaller sample size (n_1)
n_1 = smaller of the sample sizes
n_2 = larger of the sample sizes

Wilcoxon signed-rank test: $z = \dfrac{w_s - \dfrac{n(n+1)}{4}}{\sqrt{\dfrac{n(n+1)(2n+1)}{24}}}$

where

  n = number of pairs where the difference is not 0
  w_s = smaller sum in absolute value of the signed ranks

Kruskal-Wallis test:

$$H = \frac{12}{N(N+1)}\left(\frac{R_1^2}{n_1} + \frac{R_2^2}{n_2} + \cdots + \frac{R_k^2}{n_k}\right) - 3(N+1)$$

where R_1 = sum of the ranks of sample 1
  n_1 = size of sample 1
  R_2 = sum of the ranks of sample 2
  n_2 = size of sample 2
  .
  .
  .
  R_k = sum of the ranks of sample k
  n_k = size of sample k
  N = $n_1 + n_2 + \cdots + n_k$
  k = number of samples

Spearman rank correlation coefficient:

$$r_S = 1 - \frac{6 \Sigma d^2}{n(n^2 - 1)}$$

where d = difference in the ranks
  n = number of data pairs

CHAPTER 16

Formulas for the $\overline{X}$ chart:

$$\overline{X}_{GM} = \frac{\Sigma \overline{X}}{k} \qquad UCL_{\overline{X}} = \overline{X}_{GM} + A_2\overline{R}$$

$$\overline{R} = \frac{\Sigma R}{k} \qquad LCL_{\overline{X}} = \overline{X}_{GM} - A_2\overline{R}$$

where k = number of samples and A_2 is obtained from Table O

Formulas for the R chart:

$$UCL_R = D_4\overline{R} \qquad LCL_R = D_3\overline{R}$$

where D_3 and D_4 are obtained from Table O

Formulas for the $\overline{p}$ chart:

$$\overline{p} = \frac{\Sigma p}{k} \qquad UCL_p = \overline{p} + 3\sqrt{\frac{\overline{p}(1 - \overline{p})}{n}}$$

$$LCL_p = \overline{p} - 3\sqrt{\frac{\overline{p}(1 - \overline{p})}{n}}$$

where k = number of samples n = sample size

Formulas for the $\overline{c}$ chart:

$$\overline{c} = \frac{\Sigma c}{n} \qquad UCL_c = \overline{c} + 3\sqrt{\overline{c}} \qquad LCL_c = \overline{c} - 3\sqrt{\overline{c}}$$

where c = number of defects per item
  n = number of items

Type II Error and the Power of a Test

Recall that in hypothesis testing, there are two possibilities: Either the null hypothesis (H_0) is true, or it is false. Furthermore, on the basis of the statistical test, the null hypothesis is either rejected or not rejected. These results give rise to four possibilities, as shown in Figure 9–26. This figure is similar to the one shown in Section 9–2.

FIGURE 9–26
Possibilities in Hypothesis Testing

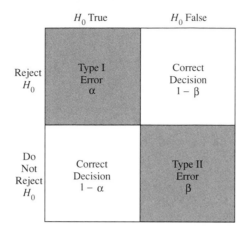

As stated previously, there are two types of errors: type I and type II. A type I error can only occur when the null hypothesis is rejected. By choosing a level of significance, say 0.05 or 0.01, the researcher can then determine the probability of committing a type I error. For example, suppose that the null hypothesis was H_0: $\mu \le 50$, and it was rejected. At the 0.05 level, the researcher has only a 5% chance of being wrong, i.e., of rejecting a true null hypothesis.

On the other hand, if the null hypothesis is not rejected, then either it is true or a type II error has been committed. A type II error occurs when the null hypothesis is indeed false, but it is not rejected. The probability of committing a type II error is denoted as β.

The value of β is not easy to compute. It depends on several things, including the value of α, the size of the sample, the population standard deviation, and the actual difference between the hypothesized value of the parameter being tested and the true parameter. The researcher has control over two of these factors, namely, the selection of α and the size of the sample. The standard deviation of the population is sometimes known, or it can be estimated. The major problem, then, is knowing the actual difference between the hypothesized parameter and the true parameter. If this difference were known, then the value of the parameter would be known; and if the parameter were known, then there would be no need to do any hypothesis testing. Hence, the value of β cannot be computed. But this does not mean that it should be ignored. What the researcher usually does is try to minimize the size of β or maximize what is called the *power of a test*.

The **power of a test** is defined as the probability of rejecting the null hypothesis when it is false. The power of a test is equal to $1 - \beta$ (shown in Figure 9–26). If somehow it is known that $\beta = 0.04$, then the power of the statistical test would be $1 - \beta$, or $1 - 0.04$, or 0.96, meaning that the probability of rejecting the null hypothesis when it is false is 96%.

The relationship between α, β, and the power of a test is shown in Figure 9–27. In the figure, the researcher has hypothesized that H_0: $\mu \leq 50$ when, in fact, $\mu = 55$. Of course, the researcher would not know this fact, but it is presented here for the sake of discussion. The shaded area under the hypothesized distribution curve to the right of the critical value represents the probability of

FIGURE 9–27
Relationship Between α and β

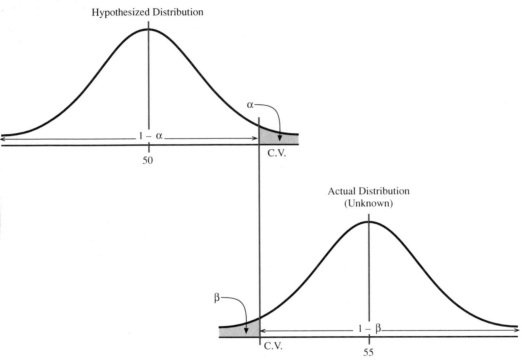

the type I error or α. The shaded area under the actual distribution curve to the left of the critical value is the probability of the type II error or β. The area to the right of the critical value under the actual distribution curve is the power of the statistical test. As β is decreased, the power of the test $(1 - \beta)$ is increased.

Even though the value of β cannot be computed, there are several ways to minimize it and thereby increase the power of the test. One way is to increase the value of α. For example, select $\alpha = 0.10$ instead of 0.05.

Figure 9–28 shows that when α is increased, β is decreased and $1 - \beta$ is increased. On the right, $\alpha = 0.05$; and on the left, $\alpha = 0.10$. The area $1 - \beta$ becomes larger as α becomes larger. Likewise, if α is decreased, β becomes larger and $1 - \beta$ decreases.

Another way to increase the power of a test is to select a larger sample size. A larger sample size would make the standard error of the mean smaller and consequently reduce β. (The derivation is omitted.)

FIGURE 9–28
Relationship Between α and β When α Is Increased

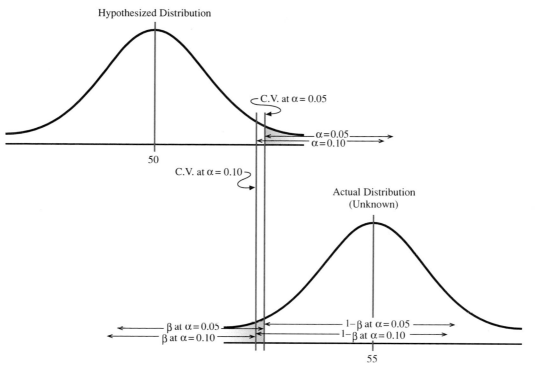

These two methods should not be used at the whim of the researcher. Before α can be increased, the researcher must consider the consequences of committing a type I error. If these consequences are more serious than the consequences of committing a type II error, then α should not be increased.

Likewise, there are consequences to increasing sample size. These consequences might include an increase in the amount of money required to do the study and an increase in the time needed to tabulate the data. When these consequences result, increasing the sample size is impractical.

There are several other methods a researcher can use to increase the power of a statistical test, but these methods are beyond the scope of this book.

The relationship between α, β, and the power of a test can be analyzed in more detail than the explanation given here. However, it is hoped that this explanation will help the student become aware that there is no magic formula or statistical test that can guarantee foolproof results when a decision is made about the validity of H_0. Whether the decision is to reject H_0 or not to reject H_0, there is in either case a chance of being wrong. The goal, then, is to try to keep the probabilities of type I and type II errors as small as possible.

EXERCISES

9–69. What is meant by a *P*-value?

9–70. (ANS.) State whether or not the null hypothesis should be rejected on the basis of the given *P*-value.

  a. *P*-value = 0.258, α = 0.05, one-tailed test
  b. *P*-value = 0.6841, α = 0.10, two-tailed test
  c. *P*-value = 0.0153, α = 0.01, one-tailed test
  d. *P*-value = 0.0232, α = 0.05, two-tailed test
  e. *P*-value = 0.002, α = 0.01, one-tailed test

9–71. A college professor claims that the average cost of a paperback textbook is greater than $27.50. A sample of 50 books has an average cost of $29.30. The standard deviation of the sample is $5.00. Test the claim and find the *P*-value for the test. On the basis of the *P*-value, should the null hypothesis be rejected at α = 0.05?

9–72. A study found that the average stopping distance of a school bus traveling 50 miles per hour was 264 feet (USA TODAY, "USA Snapshots," March 12, 1992). A group of automotive engineers decided to conduct a study of its school buses and found that for 20 buses, the average stopping distance of buses traveling 50 miles per hour was 262.3 feet. The standard deviation of the population was 3 feet. Test the claim that the average stopping distance of the company's buses is actually less than 264 feet. Find the *P*-value. On the basis of the *P*-value, should the null hypothesis be rejected at α = 0.01? Assume that the variable is normally distributed.

9–73. For a certain group of individuals, the average cost of a trip to the Super Bowl was $875. The standard deviation of the population was $50. This year, 49 fans who scheduled the trip paid an average of $890 for the three-day trip. Test the claim that the average cost is greater than last year's cost. Find the *P*-value. On the basis of the *P*-value, should the null hypothesis be rejected at α = 0.01?

9–74. A manufacturer states that the average lifetime of its television sets is more than 84 months. The standard deviation of the population is 10 months. One hundred sets are randomly selected and tested. The average lifetime of the sample is 85.1 months. Test the claim that the average lifetime of the sets is more than 84 months, and find the *P*-value. On the basis of the *P*-value, should the null hypothesis be rejected at α = 0.01?

9–75. A special cable has a breaking strength of 800 pounds. The standard deviation of the population is 12 pounds. A researcher selects a sample of 20 cables and finds that the average breaking strength is 793 pounds. Test the claim that the breaking strength is 800 pounds, and find the *P*-value. Should the null hypothesis be rejected at α = 0.01? Assume that the variable is normally distributed.

9–76. The average hourly wage last year for members of the hospital clerical staff in a large city was $6.32. The standard deviation of the population was $0.54. This year a sample of 50 workers had an average hourly wage of $6.51. Test the claim, at α = 0.05, that the average has not changed by finding the *P*-value for the test.

9–77. Ten years ago, the average acreage of farms in a certain geographic region was 65 acres. The standard deviation of the population was 7 acres. A recent study consisting of 22 farms showed that the average was 63.2 acres per farm. Test the claim, at α = 0.10, that the average has not changed by finding the *P*-value for the test. Assume that σ has not changed and the variable is normally distributed.

9–78. An automobile dealer recommends that the transmission of a car should be serviced at 30,000 miles. In order to see whether her customers are adhering to this recommendation, the dealer selects a sample of 40 customers and finds that the average mileage of the automobiles serviced is 30,456. The standard deviation of the sample is 1684 miles. By finding the *P*-value, determine whether the owners are having their transmissions serviced at 30,000 miles, for α = 0.10. Do you think the α value of 0.10 is an appropriate significance level?

9–79. Explain how confidence intervals are related to hypothesis testing.

9–80. A ski shop manager claims that the average of the sales for her shop is $1800 a day during the winter months. Ten winter days are selected at random, and the mean of the sales is $1830. The standard deviation of the population is $200. Test the claim at α = 0.05. Find the 95% confidence interval of the mean. Does the confidence interval interpretation agree with the hypothesis test results? Explain. Assume that the variable is normally distributed.

9–81. Charter bus records show that in past years, the buses carried an average of 42 people per trip to Niagara Falls. The standard deviation of the population in the past was found to be 8. This year, the average of 10 trips showed a mean of 48 people booked. Test the claim, at α = 0.10, that the average is still the same. Find the 90% confidence interval of the mean. Does the confidence interval interpretation agree with the hypothesis-testing results? Explain. Assume that the variable is normally distributed.

9–82. The sales manager of a rental agency claims that the monthly maintenance fee for a condominium in the Lakewood region is $86. Past surveys showed that the population standard deviation of the population is $6. A sample of 15 owners shows that they pay an average of $84. Test the manager's claim at $\alpha = 0.01$. Find the 99% confidence interval of the mean. Does the confidence interval interpretation agree with the results of the hypothesis test? Explain. Assume that the variable is normally distributed.

9–83. The average time it takes a person in a one-person canoe to complete a certain river course is 47 minutes. Because of rapid currents in the spring, a group of ten people traverse the course in 42 minutes. The standard deviation, known from previous trips, is 7 minutes. Test the claim that this group's time was different because of the strong currents. Use $\alpha = 0.10$. Find the 90% confidence level of the true mean. Does the confidence interval interpretation agree with the results of the hypothesis test? Explain. Assume that the variable is normally distributed.

9–84. From past studies, the average time college freshmen spend studying is 22 hours per week. The standard deviation is 4 hours. This year, 60 students were surveyed, and the average time that they spent studying was 20.8 hours. Test the claim that the time students spend studying has changed. Use $\alpha = 0.01$. It is believed that the standard deviation is unchanged. Find the 99% confidence interval of the mean. Do the results agree? Explain.

9–85. A survey taken several years ago found that the average time a person spent reading the local daily newspaper was 10.8 minutes. The standard deviation of the population was 3 minutes. In order to see whether the average time had changed since the newspaper's format was revised, the newspaper editor surveyed 36 individuals. The average time that the 36 people spent reading the paper was 12.2 minutes. At $\alpha = 0.02$, is there a change in the average time an individual spends reading the newspaper? Find the 98% confidence interval of the mean. Do the results agree? Explain.

9–86. What is meant by the power of a test?

9–87. How is the power of a test related to the type II error?

9–88. How can the power of a test be increased?

9–7 SUMMARY

This chapter introduces the basic concepts of hypothesis testing. A statistical hypothesis is a conjecture about a population. There are two types of statistical hypotheses: the null and the alternative hypotheses. The null hypothesis states that there is no difference, and the alternative hypothesis specifies the difference. In order to test the null hypothesis, researchers use a statistical test. Many test values are computed by using

$$\text{test value} = \frac{(\text{observed value}) - (\text{expected value})}{\text{standard error}}$$

Two common statistical tests are the z test and the t test. A test value is computed from the sample data in order to make a decision about whether the null hypothesis should or should not be rejected. Statistical tests can be one-tailed or two-tailed, depending on the hypotheses.

The null hypothesis is rejected when the difference between the population parameter and the sample statistic is said to be significant. The difference is significant when the test value falls in the critical region of the distribution. The critical region is determined by α, the level of significance of the test. The level is the probability of committing a type I error. This error occurs when the null hypothesis is rejected when it is true. Two generally agreed-upon significance levels are 0.05 and 0.01.

A second kind of error, the type II error, can occur when the null hypothesis is not rejected when it is false.

The z test is used when the population standard deviation is known or when it is not known but the sample size is greater than or equal to 30. The z test is also used to test proportions when $np \geq 5$ and $nq \geq 5$.

When the population standard deviation is not known, the sample standard deviation can be used, but a *t* test should be conducted in this situation when the sample size is less than 30.

All hypothesis-testing situations should include the following steps:

1. State the null and alternative hypotheses, and identify the claim.
2. State an alpha level, and find the critical value(s).
3. Compute the test value.
4. Make the decision to reject or not reject the null hypothesis.
5. Summarize the results.

Important Terms

α (alpha)
Alternative hypothesis
β (beta)
Critical or rejection region
Critical value
Degree of freedom
Hypothesis testing
Level of significance
Noncritical or nonrejection region
Null hypothesis
One-tailed test
Power of a test
P-value
Statistical hypothesis
Statistical test
Test value
t test
Two-tailed test
Type I error
Type II error
z test

Important Formulas

Formula for the *z* test for means:

$$z = \frac{\overline{X} - \mu}{\sigma / \sqrt{n}} \qquad \text{for any value } n$$

$$z = \frac{\overline{X} - \mu}{s / \sqrt{n}} \qquad \text{for} \quad n \geq 30$$

Formula for the *t* test for means:

$$t = \frac{\overline{X} - \mu}{s / \sqrt{n}} \qquad \text{for} \quad n < 30$$

Formula for the *z* test for proportions:

$$z = \frac{X - \mu}{\sigma} \qquad \text{or} \qquad z = \frac{X - np}{\sqrt{npq}}$$

Review Exercises

For Exercises 9–89 through 9–104, perform each of the following steps.

a. State the hypotheses, and identify the claim.
b. Find the critical value(s).
c. Compute the test value.
d. Make the decision.
e. Summarize the results.

9–89. A sociologist wishes to see whether for a certain group of professional women, the average age at which they deliver their first baby is 31.2 years. The sociologist selects a sample of 25 women and finds that the mean age at which they had their first baby was 29.7 years. The standard deviation of the population was 1.1 years. At $\alpha = 0.05$, does the evidence support the hypothesis? Assume that the variable is approximately normally distributed.

9–90. An automobile dealer believes that the average cost of accessories in new automobiles is $3000 over the base sticker price. He selects 50 new automobiles at random and finds that the average cost of the accessories is $3256. The standard deviation of the sample is $2300. Test his belief at $\alpha = 0.05$.

9–91. A recent study stated that if a person smoked, the average of the number of cigarettes he or she smoked was 14 per day. To test the claim, a researcher selected a random sample of 40 smokers and found that the mean number of cigarettes smoked per day was 18. The standard deviation of the sample was 6. At $\alpha = 0.05$, is the number of cigarettes a person smokes per day actually equal to 14?

9–92. A high school counselor wishes to test the theory that the average age of the dropouts in his school district is 16.3 years. He samples 32 recent dropouts and finds that their mean age is 16.9 years. At $\alpha = 0.01$, is the theory correct? The standard deviation of the population is 0.3.

9–93. In a certain city, a researcher wishes to determine whether the average age of its citizens is really 61.2 years, as the mayor claims. A sample of 22 residents has an average age of 59.8 years. The standard deviation of the sample is 1.5 years. At $\alpha = 0.01$, is the average age of the residents really 61.2 years? Assume that the variable is approximately normally distributed.

9–94. An airline claims that the average number of years of experience of its pilots is at least 18.3. In a study of 20 pilots, the average number of years of experience was 17.4. The standard deviation of the sample was 3. At $\alpha = 0.10$, is the number of years experience of the pilots really what the airline claimed? Assume that the variable is approximately normally distributed.

9–95. A recent study claims that the average age of murder victims in a small city was less than or equal to 23.2 years. A sample of 18 recent victims had a mean of 22.6 years and a standard deviation of 2 years. At $\alpha = 0.05$, is the average age higher than originally believed? Assume that the variable is approximately normally distributed.

9–96. A magazine article stated that the average age of men who were getting divorced for the first time was less than 40 years. A researcher decided to test this theory at $\alpha = 0.025$. She selected a sample of 20 men who were recently divorced and found that the average age was 38.6 years. The standard deviation of the sample was 4 years. Should the null hypothesis be rejected on the basis of the sample? Assume that the variable is approximately normally distributed.

9–97. The financial aid director of a college believes that at least 30% of the students are receiving some sort of financial aid. To see whether his belief is correct, the director selects a sample of 60 students and finds that 15 are receiving financial aid. At $\alpha = 0.05$, can the director conclude that at least 30% of the students are receiving financial aid?

9–98. A dietitian read a survey that states that at least 60% of adults eat eggs for breakfast at least four times a week. To test this claim, she selected a random sample of 100 adults and asked them how many days a week they ate eggs. In her sample, 54% responded that they ate eggs at least four times a week. At $\alpha = 0.10$, do her results support the survey?

9–99. A contractor desires to build new homes with fireplaces. He read the results of a survey that stated that 80% of all home buyers wanted a fireplace. In order to test this figure, he selected a sample of 30 home buyers and found that 20 wanted a fireplace. At $\alpha = 0.02$, should he arrive at the same conclusion as the survey?

9–100. A radio manufacturer claims that 65% of teenagers (13–16 years old) have their own portable radios. A researcher wishes to test the claim and selects a random sample of 80 teenagers. She finds that 57 have their own portable radios. At $\alpha = 0.05$, should the claim be rejected?

9–101. (OPT.) A football coach claims that the average weight of all the opposing teams' members is 225 pounds. For a test of the claim, a sample of 50 players is taken from all the opposing teams. The mean is found to be 230 pounds and a standard deviation 15 pounds. At $\alpha = 0.01$, test the coach's claim. Find the *P*-value, and make the decision.

9–102. (OPT.) An advertisement claims that Fasto Stomach Calm will provide relief from indigestion in less than 10 minutes. For a test of the claim, 25 individuals were given the product, and the average time was 9.25 minutes. From past studies, the standard deviation is known to be 2 minutes. Can one conclude that the claim is justified? Find the *P*-value, and let $\alpha = 0.05$. Assume that the variable is approximately normally distributed.

9–103. (OPT.) In order to see whether people are keeping their automobile tires inflated to the correct level (35 pounds per square inch, psi), a tire company manager selects a sample of 36 tires and checks the pressure. The mean of the sample is 33.5 psi, and the standard deviation is 3. Are the tires properly inflated? Use $\alpha = 0.10$. Find the 90% confidence interval of the mean. Do the results agree? Explain.

9–104. (OPT.) A biologist knows that the average length of a leaf of a certain full-grown plant is 4 inches. The standard deviation of the population is 0.6 inch. A sample of 20 leaves of that type of plant given a new type of plant food had an average length of 4.2 inches. Is there reason to believe that the new food is responsible for a change in the growth of the leaves? Use $\alpha = 0.01$. Find the 99% confidence interval of the mean. Do the results concur? Explain. Assume that the variable is approximately normally distributed.

🖥 COMPUTER APPLICATIONS

1. Write a computer program that will compare a sample mean to a specific population mean. Use the z test. The program should compute the mean and standard deviation of the sample and conduct a one-tailed or two-tailed test at $\alpha = 0.01$ or 0.05. The sample size should be at least 30.

2. Write a computer program similar to the one written in Exercise 1 but using the t test. This program will only compute the mean, standard deviation, and t test value. Do not include the critical values in this program.

3. MINITAB can be used to perform the z and t hypothesis tests. For example, a researcher hypothesizes that during rush hour, a bus has an average of 32 passengers per run. A sample of 30 rush-hour buses shows the following number of passengers per bus:

32	28	24	18	35	41	27	28	28	31
25	31	35	40	36	32	29	28	19	21
26	24	20	19	27	26	23	21	30	20

The population standard deviation is 3. Test the claim that the average number of passengers is 32 at $\alpha = 0.05$.

Using MINITAB for the z test, enter the following:

```
MTB > SET C1
DATA> 32 28 24 18 35 41 27 28 28 31
DATA> 25 31 35 40 36 32 29 28 19 21
DATA> 26 24 20 19 27 26 23 21 30 20
DATA> END
MTB > ZTEST MU=32, SIGMA 3 C1;
SUBC> ALTERNATIVE 0.
```

The computer will print the following:

```
TEST OF MU = 32.000 VS MU N.E. 32.000
THE ASSUMED SIGMA = 3.00
```

	N	MEAN	STDEV	SE MEAN	Z	P VALUE
C1	30	27.467	6.118	0.548	−8.28	0.0000

The test value is -8.28, which is significant beyond the 0.01 level, as indicated by the P-value. Note that if a one-tailed test is to be conducted, type -1 or 1 when the computer requests ALTERNATIVE.

Try Exercise 9–29 by using MINITAB.

4. MINITAB can be used to conduct a t test for a small sample when σ is unknown. For example, a medical researcher wishes to test the hypothesis that the average number of cavities for eighth-grade students is 2. A sample of 12 students yields the following data:

$$6, 8, 3, 2, 0, 0, 1, 5, 4, 3, 3, 2$$

Test the hypothesis at $\alpha = 0.05$.

Using MINITAB, enter the following:

```
MTB  > SET C1
DATA > 6 8 3 2 0 0 1 5 4 3 3 2
DATA > END
MTB  > TTEST MU= 2 C1;
SUBC > ALTERNATIVE 0.
```

The computer will print the following:

```
TEST OF MU = 2.000 VS MU N.E. 2.000
             N      MEAN     STDEV    SE MEAN         T     P VALUE
C1          12     3.083     2.392      0.690      1.57      0.14
```

From the t table, the null hypothesis should not be rejected, since the test value is 1.57 and the critical value is 2.201 at $\alpha = 0.05$.

Try Exercise 9–49 by using MINITAB.

√ DATA ANALYSIS Applying the Concepts

The Data Bank is located in the Appendix.

1. From the Data Bank, select a random sample of at least 30 individuals, and test one or more of the following hypotheses by using the z test.
   a. For serum cholesterol, H_0: $\mu = 220$ milligram percent (mg%).
   b. For systolic pressure, H_0: $\mu = 120$ millimeters of mercury (mmHg).
   c. For IQ, H_0: $\mu = 100$.
   d. For sodium level, H_0: $\mu = 140$ milliequivalents per liter (mEq/l).

2. Select a random sample of 15 individuals and test one or more of the hypotheses in Exercise 1 by using the t test.

3. Select a random sample of at least 30 individuals, and using the z test for proportions, test one or more of the following hypotheses.
   a. For educational level, H_0: $p = 0.50$ for level 2.
   b. For smoking status, H_0: $p = 0.20$ for level 1.
   c. For exercise level, H_0: $p = 0.10$ for level 1.
   d. For gender, H_0: $p = 0.50$ for males.

✎ TEST

Directions: Determine whether the statement is true or false.

1. Rejecting the null hypothesis when it is true is an example of a type II error.
2. The probability of a type II error is referred to as α.
3. No error is committed when the null hypothesis is rejected when it is false.
4. When the value of α is increased, the probability of committing a type I error is also increased.
5. If one wishes to test the claim that the mean of the population is 100, the appropriate null hypothesis is $\mu = 100$.

6. When one is conducting the z test, the population must be approximately normally distributed.
7. A conjecture about a population parameter is called a statistical hypothesis.
8. A test value separates the critical region from the noncritical region.
9. To test the hypothesis H_0: $\mu \leq 87$, one could use either a one-tailed test or a two-tailed test.
10. If the critical value was ± 1.96 and the test value was 2.05 for a two-tailed test, the null hypothesis would be rejected.

11. The test value is computed from data obtained from a sample.
12. When the null hypothesis is rejected, it has been proven to be false.
13. The degrees of freedom for the z test are equal to $n - 1$.
14. The t test must be used when $n \geq 30$ and σ is unknown.
15. The t distribution is a family of curves based on sample size.
16. As the sample size decreases, the t distribution approaches the normal distribution.

17. When σ is unknown and $n \geq 30$, the t test and the z test are approximately the same.
18. The t test uses the sample standard deviation rather than the population standard deviation.
19. When the sample size is less than 30 but the population standard deviation is known, the z test can be used to test hypotheses concerning a mean.
20. Hypotheses about proportions can be tested by using the normal distribution when $np \geq 5$ and $nq \geq 5$.

Testing the Difference Between Means and Proportions

10-1 INTRODUCTION

The basic concepts of hypothesis testing were explained in Chapter 9. With the z and t tests, a sample mean or proportion can be compared with a specific population mean or proportion in order to determine whether the null hypothesis should be rejected.

There are, however, many instances when researchers wish to compare two sample means. For example, the average lifetimes of two different brands of bus tires might be compared to see whether there is any difference in tread wear. Two different brands of fertilizer might be tested to see whether one is better than the other for growing plants. Or two brands of cough syrup might be tested to see whether one brand is more effective than the other.

In the comparison of two means, the same basic steps for hypothesis testing shown in Chapter 9 are used, and the z and t tests are also used. However, the specific formulas used in these tests are somewhat different. The methods and formulas for comparing two means will be shown in Sections 10–2 through 10–4. Finally, the z test can be used to compare two proportions, as shown in Section 10–5.

10-2 TESTING THE DIFFERENCE BETWEEN TWO MEANS: z TEST

Suppose a researcher wishes to determine whether there is a difference in the average age of nursing students who enroll in a nursing program at a community college and those who enroll in a nursing program at a university. In this case, the researcher is not interested in the average age of all beginning nursing students; instead, he is interested in *comparing* the means of the two groups. His research question is, Does the mean age of students who enroll at a community college differ from the mean age of nursing students who enroll at a university? Here, the hypotheses are

$$H_0: \mu_1 = \mu_2$$
$$H_1: \mu_1 \neq \mu_2$$

where μ_1 = mean age of all beginning nursing students at the community college

μ_2 = mean age of all beginning nursing students at the university

Another way of stating the hypotheses for this situation is

$$H_0: \mu_1 - \mu_2 = 0$$
$$H_1: \mu_1 - \mu_2 \neq 0$$

If there is no difference in population means, subtracting them will give a difference of zero. Conversely, if they are different, the result of subtracting will be a number other than zero. Both methods of stating hypotheses are correct; however, the first method will be used in this book.

> ## Assumptions for the Test to Determine the Difference Between Two Means
>
> 1. The samples must be independent of each other. That is, there can be no relationship between the subjects in each sample.
> 2. The populations from which the samples were obtained must be normally distributed, and the standard deviations of the variable must be known, or the samples must be greater than or equal to 30.

The theory behind testing the difference between two means is based on selecting pairs of samples and comparing the means of the pairs. The population means need not be known.

All possible pairs of samples are taken from the populations. The means for each of the pairs of samples are computed and then subtracted, and the differences are plotted. If both populations have the same mean, then most of the differences will be zero or close to zero. Occasionally, there will be a few large differences due to chance alone, some positive and some negative. If the differences are plotted, the curve will be shaped like the normal distribution and have a mean of zero, as shown in Figure 10–1.

FIGURE 10–1

Differences of Means of Pairs of Samples

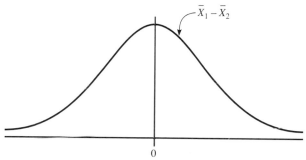

$\bar{X}_1 - \bar{X}_2$

0

The variance of the difference $\bar{X}_1 - \bar{X}_2$ is equal to the sum of the individual variances of $\bar{X}_1$ and $\bar{X}_2$. That is,

$$\sigma^2_{\bar{X}_1 - \bar{X}_2} = \sigma^2_{\bar{X}_1} + \sigma^2_{\bar{X}_2}$$

where

$$\sigma^2_{\bar{X}_1} = \frac{\sigma^2_1}{n_1} \qquad \text{and} \qquad \sigma^2_{\bar{X}_2} = \frac{\sigma^2_2}{n_2}$$

So the standard deviation of $\bar{X}_1 - \bar{X}_2$ is

$$\sqrt{\frac{\sigma^2_1}{n_1} + \frac{\sigma^2_2}{n_2}}$$

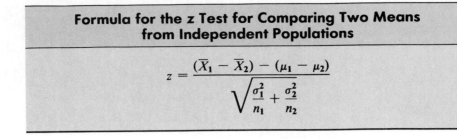

Formula for the z Test for Comparing Two Means from Independent Populations

$$z = \frac{(\overline{X}_1 - \overline{X}_2) - (\mu_1 - \mu_2)}{\sqrt{\frac{\sigma_1^2}{n_1} + \frac{\sigma_2^2}{n_2}}}$$

This formula is based on the general format of

$$\text{test value} = \frac{(\text{observed value}) - (\text{expected value})}{\text{standard error}}$$

where $\overline{X}_1 - \overline{X}_2$ is the observed difference, and the expected difference $\mu_1 - \mu_2$ is zero when the null hypothesis is $\mu_1 = \mu_2$, since that is equivalent to $\mu_1 - \mu_2 = 0$. Finally, the standard error of difference is

$$\sqrt{\frac{\sigma_1^2}{n_1} + \frac{\sigma_2^2}{n_2}}$$

In the comparison of two sample means, the difference may be due to chance, in which case the null hypothesis would not be rejected, and the researcher can assume that the means of the populations are basically the same. The difference in this case is not significant. See Figure 10–2a. On the other hand, if the difference is significant, the null hypothesis is rejected; and the researcher can conclude that the population means are different. See Figure 10–2b.

FIGURE 10–2
Hypothesis-Testing Situations in the Comparison of Means

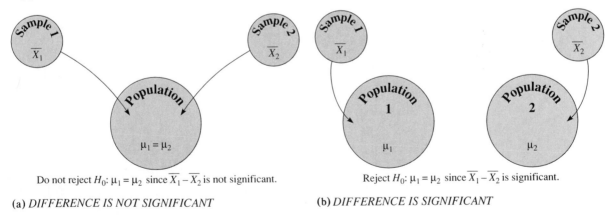

Do not reject H_0: $\mu_1 = \mu_2$ since $\overline{X}_1 - \overline{X}_2$ is not significant.

(a) *DIFFERENCE IS NOT SIGNIFICANT*

Reject H_0: $\mu_1 = \mu_2$ since $\overline{X}_1 - \overline{X}_2$ is significant.

(b) *DIFFERENCE IS SIGNIFICANT*

These tests can also be one-tailed, using the following hypotheses.

	One-Tailed Right			**One-Tailed Left**	
H_0: $\mu_1 \leq \mu_2$		H_0: $\mu_1 - \mu_2 \leq 0$	H_0: $\mu_1 \geq \mu_2$		H_0: $\mu_1 - \mu_2 \geq 0$
	or			or	
H_1: $\mu_1 > \mu_2$		H_1: $\mu_1 - \mu_2 > 0$	H_1: $\mu_1 < \mu_2$		H_1: $\mu_1 - \mu_2 < 0$

The same critical values used in Section 9–3 are used here. They can be obtained from Table E in the Appendix.

If σ_1^2 and σ_2^2 are not known, the researcher can use the variances obtained from each sample, s_1^2 and s_2^2, but both sample sizes must be 30 or more. The formula then is

$$z = \frac{(\overline{X}_1 - \overline{X}_2) - (\mu_1 - \mu_2)}{\sqrt{\dfrac{s_1^2}{n_1} + \dfrac{s_2^2}{n_2}}}$$

provided that $n_1 \geq 30$ and $n_2 \geq 30$. See Example 10–2.

When one or both sample sizes are less than 30 and σ_1 and σ_2 are unknown, the t test must be used, as shown in the next section.

The basic format for hypothesis testing is reviewed here.

STEP 1 State the hypotheses, and identify the claim.

STEP 2 Find the critical value(s).

STEP 3 Compute the test value.

STEP 4 Make the decision.

STEP 5 Summarize the results.

EXAMPLE 10–1 A private research firm tested two different makes of automobiles to determine whether there was a difference in the amount of damage each would sustain in a crash test at 10 miles per hour. The average dollar amounts of the damage to each is shown below. Assume that the populations are normally distributed.

Make 1	Make 2
$\overline{X}_1 = \$1560$	$\overline{X}_2 = \$1432$
$\sigma_1 = \$252$	$\sigma_2 = \$280$
$n_1 = 20$	$n_2 = 15$

At $\alpha = 0.05$, test the claim that there is no difference in the dollar amount of damage sustained by each automobile.

Solution **STEP 1** State the hypotheses, and identify the claim.

$$H_0\colon \mu_1 = \mu_2 \text{ (claim)} \qquad \text{and} \qquad H_1\colon \mu_1 \neq \mu_2$$

STEP 2 Find the critical values. Since $\alpha = 0.05$, the critical values are $+1.96$ and -1.96.

STEP 3 Compute the test value.

$$z = \frac{(\overline{X}_1 - \overline{X}_2) - (\mu_1 - \mu_2)}{\sqrt{\dfrac{\sigma_1^2}{n_1} + \dfrac{\sigma_2^2}{n_2}}} = \frac{(1560 - 1432) - 0}{\sqrt{\dfrac{252^2}{20} + \dfrac{280^2}{15}}} = 1.40$$

STEP 4   Make the decision. Do not reject the null hypothesis at $\alpha = 0.05$, since $1.40 < 1.96$. See Figure 10–3.

FIGURE 10–3
Critical and
Test Values for
Example 10–1

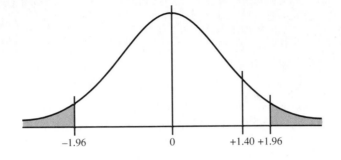

-1.96 0 $+1.40$ $+1.96$

STEP 5   Summarize the results. There is not enough evidence to reject the claim that the means are equal.   ◄

EXAMPLE 10–2   A college instructor hypothesizes that students miss fewer classes on average in required courses than in elective courses. After obtaining two random samples of 40 students each, the following data were obtained.

Days Missed, Required Course	**Days Missed, Elective Course**
$\overline{X}_1 = 6.1$	$\overline{X}_2 = 9.3$
$s_1 = 2$	$s_2 = 3$
$n_1 = 40$	$n_2 = 40$

At $\alpha = 0.05$, test the instructor's claim.

Solution   **STEP 1**   State the hypotheses, and identify the claim.

$$H_0\text{: } \mu_1 \geq \mu_2 \quad \text{and} \quad H_1\text{: } \mu_1 < \mu_2 \text{ (claim)}$$

STEP 2   Find the critical value. Since $\alpha = 0.05$, the critical value is -1.65.

STEP 3   Compute the test value.

$$z = \frac{(\overline{X}_1 - \overline{X}_2) - (\mu_1 - \mu_2)}{\sqrt{\dfrac{s_1^2}{n_1} + \dfrac{s_2^2}{n_2}}} = \frac{(6.1 - 9.3) - 0}{\sqrt{\dfrac{2^2}{40} + \dfrac{3^2}{40}}} = \frac{-3.2}{0.57} = -5.61$$

STEP 4   Make the decision. Reject the null hypothesis, since $-5.61 < -1.65$. See Figure 10–4.

FIGURE 10–4
Critical and
Test Values for
Example 10–2

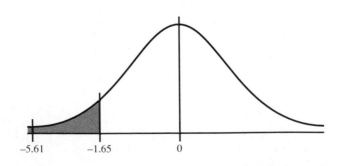

-5.61 -1.65 0

STEP 5 Summarize the results. There is enough evidence to support the claim that the students miss fewer classes in required courses than in elective courses. ◄

Sometimes, the researcher is interested in testing a specific difference in means other than zero. For example, he or she might hypothesize that the nursing students at a community college are, on the average, 3.2 years older than those at a university. In this case, the hypotheses are

$$H_0: \mu_1 - \mu_2 \le 3.2 \qquad \text{and} \qquad H_1: \mu_1 - \mu_2 > 3.2$$

The formula for the z test is then

$$z = \frac{(\overline{X}_1 - \overline{X}_2) - (\mu_1 - \mu_2)}{\sqrt{\dfrac{\sigma_1^2}{n_1} + \dfrac{\sigma_2^2}{n_2}}}$$

where $\mu_1 - \mu_2$ is the hypothesized difference or expected value. In this case, $\mu_1 - \mu_2 = 3.2$. See Exercise 10-17.

EXERCISES

10-1. Explain the difference between testing a single mean and testing the difference between two means.

10-2. When a researcher selects all possible pairs of samples from a population in order to find the difference between the means of each pair, what will be the shape of the distribution of the differences when the original distributions are normally distributed? What will be the mean of the distribution? What will be the standard deviation of the distribution?

10-3. What two assumptions must be met when one is using the z test to test differences between two sample means? When can the sample standard deviations, s_1 and s_2, be used in place of the population standard deviations σ_1 and σ_2?

10-4. Show two different ways to state that the means of two populations are equal.

For Exercises 10-5 through 10-17, perform each of the following steps.

a. State the hypotheses, and identify the claim.
b. Find the critical value(s).
c. Compute the test value.
d. Make the decision.
e. Summarize the results.

10-5. A publishing company wishes to test the claim that there is a difference between two overnight delivery companies in the speed with which their material is delivered. The average speed of material delivered over a 30-day period is shown below. Test the claim at $\alpha = 0.01$.

Company 1	Company 2
$\overline{X}_1 = 16$ hours	$\overline{X}_2 = 18$ hours
$\sigma_1 = 3.2$	$\sigma_2 = 3$
$n_1 = 30$	$n_2 = 30$

10-6. A manager has eight photocopy machines at two different locations. Each location is serviced by a different repair company. He wishes to see whether there is a difference in repair costs between the two companies. The means of the repair costs for a period of six months are given in the table. Is there a significant difference between repair costs of the two companies? Use $\alpha = 0.10$. Assume that the populations are normally distributed.

Company 1	Company 2
$\overline{X}_1 = \$105$	$\overline{X}_2 = \$96$
$\sigma_1 = \$15$	$\sigma_2 = \$10$
$n_1 = 8$	$n_2 = 8$

10-7. A medical researcher wishes to see whether the pulse rates of smokers are higher than the pulse rates of nonsmokers. Samples of 100 smokers and 100 nonsmokers are selected. The results are shown below. Can the researcher conclude, at $\alpha = 0.05$, that smokers have higher pulse rates than nonsmokers?

Smokers	Nonsmokers
$\overline{X}_1 = 90$	$\overline{X}_2 = 88$
$s_1 = 5$	$s_2 = 6$
$n_1 = 100$	$n_2 = 100$

10-8. A statistician claims that the average IQ of students who major in psychology is greater than that of students who major in mathematics. The results of an IQ test given to 50 students in each group are shown below. Test the claim at $\alpha = 0.01$.

Psychology	Mathematics
$\overline{X}_1 = 118$	$\overline{X}_2 = 115$
$\sigma_1 = 15$	$\sigma_2 = 15$
$n_1 = 50$	$n_2 = 50$

10-9. A researcher wishes to test the reaction times of taxi drivers against those of state police officers. The results of the study are shown below. Can the researcher conclude, at $\alpha = 0.02$, that taxi drivers have a slower reaction time than state police officers? Assume that the populations are normally distributed.

Taxi Drivers	State Police Officers
$\overline{X}_1 = 0.38$	$\overline{X}_2 = 0.37$
$\sigma_1 = 0.02$	$\sigma_2 = 0.01$
$n_1 = 10$	$n_2 = 9$

10-10. A real estate agent compares the selling prices of homes in two different suburbs of Seattle to see whether there is a difference in price. The results of the study are shown below. Is there evidence to reject the claim that the average cost of a home in both locations is the same? Use $\alpha = 0.01$.

Suburb 1	Suburb 2
$\overline{X}_1 = \$63,255$	$\overline{X}_2 = \$59,102$
$s_1 = \$5602$	$s_2 = \$4731$
$n_1 = 35$	$n_2 = 40$

10-11. A teacher wishes to see whether the evening students in her physics class score higher than the day students. The results of a final examination are shown below. Can the teacher conclude that the evening students have a higher score? Use $\alpha = 0.02$.

Day	Evening
$\overline{X}_1 = 73$	$\overline{X}_2 = 76$
$s_1 = 3$	$s_2 = 2.6$
$n_1 = 33$	$n_2 = 36$

10-12. A college admissions officer believes that students enrolling from Shawnee School District have higher SAT scores than those who enroll from West River School District. A sample of 30 students from each district is selected, and the results are as shown below. Is the belief of the admissions officer substantiated? Use $\alpha = 0.05$.

Shawnee	West River
$\overline{X}_1 = 656$	$\overline{X}_2 = 560$
$\sigma_1 = 100$	$\sigma_2 = 100$
$n_1 = 30$	$n_2 = 30$

10-13. Test to see whether there is a difference in cranking power at 32° between two brands of automobile batteries. Use $\alpha = 0.10$.

Brand 1	Brand 2
$\overline{X}_1 = 320$ amperes	$\overline{X}_2 = 383$ amperes
$s_1 = 15$ amperes	$s_2 = 10$ amperes
$n_1 = 36$	$n_2 = 36$

10-14. Is there a difference in average miles traveled for each of two taxi companies during a randomly selected week? The data are shown below. Use $\alpha = 0.05$. Assume that the populations are normally distributed.

Moonview Cab Company	Starlight Taxi Company
$\overline{X}_1 = 837$	$\overline{X}_2 = 753$
$\sigma_1 = 30$	$\sigma_2 = 40$
$n_1 = 15$	$n_2 = 20$

10-15. Jiffy Meals, a fast-food restaurant, boasts that it has the fastest service around. A competitor checks the customer waiting times of the fast-food restaurant and of his restaurant. The data are shown below. Test Jiffy Meals' claim at $\alpha = 0.02$.

Jiffy Meals	Competitor
$\overline{X}_1 = 3$ minutes	$\overline{X}_2 = 4.5$ minutes
$s_1 = 1$ minute	$s_2 = 0.8$ minute
$n_1 = 40$	$n_2 = 50$

*10-16. The dean of students wants to see whether there is a significant difference in age between resident students and commuting students. She selects a sample of 50 students from each group. Choose a significance level and test the hypothesis that there is no difference in age between the two groups. The ages are shown below. Use the standard deviations from the samples.

Resident Students

22	25	27	23	26	28	26	24
25	20	26	24	27	26	18	19
18	30	26	18	18	19	32	23
19	19	18	29	19	22	18	22
26	19	19	21	23	18	20	18
22	21	19	21	21	22	18	20
19	23						

Commuter Students

18	20	19	18	22	25	24	35
23	18	23	22	28	25	20	24
26	30	22	22	22	21	18	20
19	26	35	19	19	18	19	32
29	23	21	19	36	27	27	20
20	21	18	19	23	20	19	19
20	25						

*10–17. A researcher claims that students in a private school have an IQ that is 8 points higher than that of students in public schools. Random samples of 60 students from each type of school are selected and given an IQ exam. The results are shown below. At $\alpha = 0.05$, test the claim.

Private School	Public School
$\overline{X}_1 = 110$	$\overline{X}_2 = 104$
$s_1 = 15$	$s_2 = 15$
$n_1 = 60$	$n_2 = 60$

The true difference between two means can be estimated by using the following formula:

$$(\overline{X}_1 - \overline{X}_2) - (z_{\alpha/2})\sqrt{\frac{\sigma_1^2}{n_1} + \frac{\sigma_2^2}{n_2}} < \mu_1 - \mu_2$$

$$< (\overline{X}_1 - \overline{X}_2) + (z_{\alpha/2})\sqrt{\frac{\sigma_1^2}{n_1} + \frac{\sigma_2^2}{n_2}}$$

Note: When n_1 and n_2 are both 30 or more, s_1^2 and s_2^2 can be substituted for σ_1^2 and σ_2^2. Use this formula for Exercises 10–18 through 10–21.

*10–18. A study of teenagers found the following information on the length of time (in minutes) each talked on the telephone. Find the 95% confidence level of the true differences in means.

Boys	Girls
$\overline{X}_1 = 21$	$\overline{X}_2 = 18$
$\sigma_1 = 2.1$	$\sigma_2 = 3.2$
$n_1 = 50$	$n_2 = 50$

*10–19. Two brands of cigarettes are selected and the nicotine content of each is compared. The data are shown below. Find the 99% confidence interval of the true difference in the means of each.

Brand A	Brand B
$\overline{X}_1 = 28.6$ milligrams	$\overline{X}_2 = 32.9$ milligrams
$\sigma_1 = 5.1$ milligrams	$\sigma_2 = 4.4$ milligrams
$n_1 = 30$	$n_2 = 40$

*10–20. Two brands of batteries are tested, and the voltage of each is compared. The data follow. Find the 95% confidence interval of the true difference in the mean of each. Assume that both variables are normally distributed.

Brand X	Brand Y
$\overline{X}_1 = 9.2$ volts	$\overline{X}_2 = 8.8$ volts
$\sigma_1 = 0.3$ volt	$\sigma_2 = 0.1$ volt
$n_1 = 27$	$n_2 = 30$

*10–21. Two groups of students are given a problem-solving test, and the results are compared. Find the 90% confidence interval of the true difference in means.

Mathematics Majors	Computer Science Majors
$\overline{X}_1 = 83.6$	$\overline{X}_2 = 79.2$
$s_1 = 4.3$	$s_2 = 3.8$
$n_1 = 36$	$n_2 = 36$

10–3 TESTING THE DIFFERENCE BETWEEN TWO MEANS: *t* TEST, INDEPENDENT SAMPLES

In the previous section, the *z* test was used to test the difference between two means when the population standard deviations were known and the variables were normally or approximately normally distributed, or when both sample sizes were greater than or equal to 30. In many situations, however, these conditions cannot be met—i.e., the population standard deviations are not known, and one or both sample sizes are less than 30. In these cases, a *t* test is used to test the difference between means when the two samples are independent and when the samples are taken from two normally or approximately normally distributed populations. Samples are **independent** when they are not related.

There are actually two different options for the use of *t* tests. *One option is used when the variances of the samples are not equal, and the other option is used when the variances are equal.* To determine whether two sample variances are equal, the researcher can use an *F* test. A discussion of this test is given in

Speaking of **Statistics**

Mother's Milk: Food for Smarter Kids

By Mike Snider
USA TODAY

Children who got mother's milk as premature infants score higher on IQ tests than children who didn't, a new study finds.

Other studies have suggested breast feeding boosts development and IQ, but some researchers have attributed that to having a nurturing mother.

This new study of 300 preemies—fed by tube—offers "strong evidence . . . that receiving breast milk was associated with the IQ advantage, rather than the process of breast feeding," says Alan Lucas of the Medical Research Council, Cambridge, England.

He and others studied the preemies in two groups: those who got milk from their natural mothers and those who didn't. Infants in both groups may also have been fed infant formula and milk from a woman other than their mother.

Preemies who got milk from their mothers during the first four to five weeks of life averaged 8.3 points higher on IQ tests at age 7½–8, show findings in the British medical journal *Lancet,* out Saturday.

"We've produced strong evidence that breast milk may be beneficial for long-term development," Lucas says.

Researchers did not identify a specific amount of mother's milk needed for the IQ advantage, but the more mother's milk they got, the better the children seemed to do.

He says new formulas—with a better supplement of hormones, growth factors and certain fats found in mother's milk, coupled with at least some milk from the natural mother—could give preemies "a double advantage."

Source: Copyright 1992, USA TODAY.
Reprinted with permission.

In this study, what two groups were used? What was the difference in IQ scores? What level of significance do you feel was used? What was the sample size? What other factors could have resulted in the difference in IQs? Explain why these differences may have caused the difference in IQ.

Section 13–3 of Chapter 13. Note, however, that not all statisticians are in agreement about using the *F* test before using the *t* test. Some feel that conducting the *F* and *t* tests at the same level of significance will change the overall level of significance of the *t* test. Their reasons are beyond the scope of this textbook; therefore, each problem will specify whether the variances can be assumed to be equal or unequal.

Formulas for the Two t Tests

Variances are assumed to be unequal:

$$t = \frac{(\overline{X}_1 - \overline{X}_2) - (\mu_1 - \mu_2)}{\sqrt{\dfrac{s_1^2}{n_1} + \dfrac{s_2^2}{n_2}}}$$

where the degrees of freedom are equal to the smaller of $n_1 - 1$ or $n_2 - 1$.

Variances are assumed to be equal:

$$t = \frac{(\overline{X}_1 - \overline{X}_2) - (\mu_1 - \mu_2)}{\sqrt{\dfrac{(n_1 - 1)s_1^2 + (n_2 - 1)s_2^2}{n_1 + n_2 - 2}} \sqrt{\dfrac{1}{n_1} + \dfrac{1}{n_2}}}$$

where the degrees of freedom are equal to $n_1 + n_2 - 2$.

When the variances are unequal, the first formula

$$t = \frac{(\overline{X}_1 - \overline{X}_2) - (\mu_1 - \mu_2)}{\sqrt{\dfrac{s_1^2}{n_1} + \dfrac{s_2^2}{n_2}}}$$

follows the format of

$$\text{test value} = \frac{(\text{observed value}) - (\text{expected value})}{\text{standard error}}$$

where $\overline{X}_1 - \overline{X}_2$ is the observed difference between sample means and where the expected value $\mu_1 - \mu_2$ is equal to zero when no difference between population means is hypothesized. The denominator $\sqrt{(s_1^2/n_1) + (s_2^2/n_2)}$ is the standard error of the difference between two means. Since mathematical derivation of the standard error is somewhat complicated, it will be omitted here.

When the variances are assumed to be equal, the second formula

$$t = \frac{(\overline{X}_1 - \overline{X}_2) - (\mu_1 - \mu_2)}{\sqrt{\dfrac{(n_1 - 1)s_1^2 + (n_2 - 1)s_2^2}{n_1 + n_2 - 2}} \sqrt{\dfrac{1}{n_1} + \dfrac{1}{n_2}}}$$

also follows the format of

$$\text{test value} = \frac{(\text{observed value}) - (\text{expected value})}{\text{standard error}}$$

For the numerator, the terms are the same as in the first formula. However, a note of explanation is needed for the denominator of the second test statistic. Since both populations are assumed to have the same variance, the standard error is computed by using what is called a pooled estimate of the variance. **A pooled estimate of the variance** is a weighted average of the variance using the two sample variances and the *degrees of freedom* of each variance as the weights. Again, since the algebraic derivation of the standard error is somewhat complicated, it is omitted.

EXAMPLE 10–3 A coach wishes to see whether the weights of wrestlers differ from the weights of power lifters. Two samples are selected, and the information obtained is as given below. Can the coach conclude that the average weight of the wrestlers differs from the average weight of the power lifters? Use $\alpha = 0.05$. Assume that the variances are unequal, and the populations are approximately normally distributed.

Wrestlers	Power Lifters
$\overline{X}_1 = 262$ pounds	$\overline{X}_2 = 271$ pounds
$s_1 = 6$ pounds	$s_2 = 14$ pounds
$n_1 = 10$	$n_2 = 8$

Solution   **STEP 1**   State the hypotheses, and identify the claim.

$$H_0\colon \mu_1 = \mu_2 \quad \text{and} \quad H_1\colon \mu_1 \neq \mu_2 \text{ (claim)}$$

STEP 2   Find the critical values. Since the test is two-tailed, since $\alpha = 0.05$, and since the variances are unequal, the degrees of freedom are the smaller of $n_1 - 1$ or $n_2 - 1$. In this case, the degrees of freedom are $8 - 1 = 7$. Hence, from Table F, the critical values are $+2.365$ and -2.365.

STEP 3   Compute the test value. Since the variances are unequal, use the first formula.

$$t = \frac{(\overline{X}_1 - \overline{X}_2) - (\mu_1 - \mu_2)}{\sqrt{\dfrac{s_1^2}{n_1} + \dfrac{s_2^2}{n_2}}} = \frac{(262 - 271) - 0}{\sqrt{\dfrac{6^2}{10} + \dfrac{14^2}{8}}} = -1.70$$

STEP 4   Make the decision. Do not reject the null hypothesis, since $-1.70 > -2.365$. See Figure 10–5.

FIGURE 10–5
Critical and
Test Values for
Example 10–3

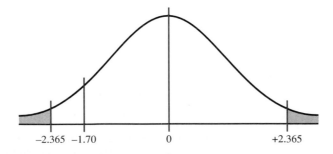

-2.365 -1.70 0 $+2.365$

STEP 5   Summarize the results. There is not enough evidence to support the claim that the weights of wrestlers differ from the weights of power lifters.   ◀

EXAMPLE 10–4 A researcher wishes to determine whether the salaries of professional nurses employed by private hospitals are higher than those of nurses employed by government-owned hospitals. She selects a sample of nurses from each type of hospital and calculates the means and standard deviations of their salaries. At $\alpha = 0.01$, can she conclude that the private hospitals pay more than the government hospitals? Assume that the variances are equal, and the populations are approximately normally distributed.

Private	Government
$\overline{X}_1 = \$26,800$	$\overline{X}_2 = \$25,400$
$s_1 = \$600$	$s_2 = \$450$
$n_1 = 10$	$n_2 = 8$

Solution **STEP 1** State the hypotheses, and identify the claim.

$$H_0: \mu_1 \leq \mu_2 \quad \text{and} \quad H_1: \mu_1 > \mu_2 \text{ (claim)}$$

STEP 2 Find the critical value. Since $\alpha = 0.01$ and the test is a one-tailed right test with equal variances, the degrees of freedom are $n_1 + n_2 - 2$, which is $10 + 8 - 2 = 16$. The critical value is $+2.583$.

STEP 3 Compute the test value. Use the second formula, since the variances are assumed to be equal.

$$t = \frac{(\overline{X}_1 - \overline{X}_2) - (\mu_1 - \mu_2)}{\sqrt{\dfrac{(n_1 - 1)s_1^2 + (n_2 - 1)s_2^2}{n_1 + n_2 - 2}}\sqrt{\dfrac{1}{n_1} + \dfrac{1}{n_2}}}$$

$$= \frac{(26{,}800 - 25{,}400) - 0}{\sqrt{\dfrac{(10 - 1)(600)^2 + (8 - 1)(450)^2}{10 + 8 - 2}}\sqrt{\dfrac{1}{10} + \dfrac{1}{8}}}$$

$$= 5.47$$

STEP 4 Make the decision. Reject the null hypothesis, since $5.47 > 2.583$. See Figure 10–6.

FIGURE 10–6
Critical and
Test Values for
Example 10–4

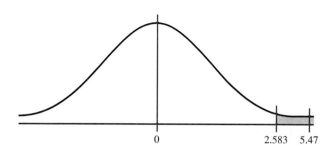

0   2.583   5.47

STEP 5 Summarize the results. There is enough evidence to support the claim that the salaries paid to nurses employed by private hospitals are higher than those paid to nurses employed by government-owned hospitals. ◄

Remember that when one is testing the difference between two independent means, two different statistical test formulas can be used. One formula is used when the variances are equal, and the other formula is used when the variances are not equal. As previously stated, some statisticians use an F test to determine whether two variances are equal; this test is explained in Chapter 13. In this chapter, the assumption will be stated.

EXERCISES

For Exercises 10–22 through 10–32, perform each of the following steps. Assume that all variables are normally or approximately normally distributed.

a. State the hypotheses, and identify the claim.
b. Find the critical value(s).
c. Compute the test value.
d. Make the decision.
e. Summarize the results.

10–22. A real estate agent wishes to determine whether tax assessors and real estate appraisers agree on the values of homes. A random sample of the two groups appraised 10 homes. The data are shown below. Is there a significant difference in the values of the homes for each group? Let $\alpha = 0.05$. Assume that the variances are equal.

Real Estate Appraisers	Tax Assessors
$\overline{X}_1 = \$83{,}256$	$\overline{X}_2 = \$88{,}354$
$s_1 = \$3256$	$s_2 = \$2341$
$n_1 = 10$	$n_2 = 10$

10–23. A researcher suggests that male nurses earn more than female nurses. A survey of 16 male nurses and 20 female nurses reports the following data. Test the claim at $\alpha = 0.05$. Assume that the variances are not equal.

Male	Female
$\overline{X}_1 = \$23{,}800$	$\overline{X}_2 = \$23{,}750$
$s_1 = \$300$	$s_2 = \$250$
$n_1 = 16$	$n_2 = 20$

10–24. An instructor thinks that math majors can write computer programs faster than business majors. A sample of 12 math majors took an average of 36 minutes to write a specific program and debug it; a sample of 18 business majors took an average of 39 minutes. The standard deviations of each group were 4 minutes and 9 minutes, respectively. At $\alpha = 0.10$, is there evidence to support the claim? Assume that the variances are unequal.

10–25. A researcher estimates that high school girls miss more days of school than high school boys. A sample of 16 girls showed that they missed an average of 3.9 days of school

per school year; a sample of 22 boys showed that they missed an average of 3.6 days of school per year. The standard deviations are 0.6 and 0.8, respectively. At $\alpha = 0.01$, test the claim. Assume that the variances are equal.

10–26. A leading manufacturer claims that his brand of television (brand A) will cost less to repair over a five-year period than any other brand. A researcher selects a sample of 10 sets of brand A and 10 sets of another leading brand (brand B) and finds the average repair costs over a five-year period. The data are shown below. At $\alpha = 0.01$, test the manufacturer's claim. Assume that the variances are equal.

Brand A	Brand B
$\overline{X}_1 = \$896$	$\overline{X}_2 = \$902$
$s_1 = \$14$	$s_2 = \$16$
$n_1 = 10$	$n_2 = 10$

10–27. An automotive engineer wishes to see whether a new automobile will get more miles per gallon with high-octane gasoline than with regular octane. She selects two samples of automobiles and uses regular gasoline for one and high-octane for the other. At $\alpha = 0.05$, do the automobiles using the high-octane gasoline get better mileage? Assume that the variances are unequal.

High-Octane	Regular
$\overline{X}_1 = 26.7$	$\overline{X}_2 = 23$
$s_1 = 2.1$	$s_2 = 1.6$
$n_1 = 5$	$n_2 = 12$

10–28. The local branch of the Internal Revenue Service spent an average of 21 minutes helping each of 10 people prepare their tax returns. The standard deviation was 5.6 minutes. A volunteer tax preparer spent an average of 27 minutes helping 14 people prepare their taxes. The standard deviation was 4.3 minutes. At $\alpha = 0.02$, is there a difference in the average time spent by the two services? Assume that the variances are equal.

10–29. The average price of seven ABC dishwashers was $815, and the average price of nine XYZ dishwashers was $845. The standard deviations were $19 and $9, respectively. At $\alpha = 0.05$, can one conclude that the XYZ dishwashers cost more? Assume that the variances are unequal.

10–30. The average monthly premium paid by 12 administrators for hospitalization insurance is $56. The standard deviation is $3. The average monthly premiums paid by 27 nurses is $63. The standard deviation is $5.75. At $\alpha = 0.05$, do the nurses pay more for hospitalization insurance? Assume that the variances are unequal.

10–31. Over a two-week period (14 days), the Coulterville police issued an average of 56 speeding tickets per day. The standard deviation was 6 tickets. For the same period, the Circleville police issued an average of 58 tickets. The standard deviation was 12. Can one conclude that the Circleville police issue more speeding tickets? Use $\alpha = 0.05$, and assume that the variances are unequal.

*10–32. The times (in minutes) it took six white mice to learn to run a simple maze and the times it took six brown mice to learn to run the same maze are given below. At $\alpha = 0.05$, does the color of the mice make a difference in the learning rate of the mice? Assume that the variances are equal.

White Mice	18	24	20	13	15	12
Brown Mice	25	16	19	14	16	10

Speaking of **Statistics**

Public School Teachers Show Bias for Boys

By Katy Kelly
USA TODAY

Public schools favor boys, says a report out today.

"There is an illusion that schools are treating boys and girls equally," says Alice McKee, American Association of University Women, which commissioned the report.

Based on a review of recent research, it suggests:

- Teachers ask academic questions of boys 80% more often than of girls.
- School curricula generally ignore or stereotype females.
- Most standardized tests are biased against girls.
- Preschool boys get more individual attention, even hugs, from teachers than girls do.

Bias is not intentional: "When teachers are made aware they're doing this . . . they are willing and eager to change," McKee says.

The report lists 40 recommendations, including requiring course work on gender issues for teacher certification.

New classroom materials are needed too, McKee says: "Boys and girls should be able to study women Nobel Prize winners in addition to Betsy Ross sewing the flag."

Source: Copyright 1992, USA TODAY.
Reprinted with permission.

On the basis of the conclusions in this study, state several hypotheses that may have been used to support these conclusions. Comment on how you think these hypotheses were tested. For example, how would one determine whether standardized tests are biased against girls?

10–4 TESTING THE DIFFERENCE BETWEEN TWO MEANS: *t* TEST, DEPENDENT SAMPLES

In the previous section, the *t* test was used to compare two sample means when the samples were independent. In this section, a different version of the *t* test is discussed. This version is used when the samples are dependent. Samples are considered to be **dependent samples** when the subjects are paired or matched in some way.

For example, suppose a medical researcher wants to see whether a drug will affect the reaction time of its users. In order to test this hypothesis, the researcher must pretest the subjects in the sample first. That is, they are given a test to ascertain their normal reaction times. Then after taking the drug, the subjects are tested again, using a posttest. Finally, the means of the two tests are compared to see whether there is a difference. Since the same subjects are used in both cases, the samples are *related;* i.e., subjects scoring high on the pretest will generally score high on the posttest, even after consuming the drug. Likewise, those scoring lower on the pretest will tend to score lower on the posttest. In order to take this effect into account, the researcher employs a *t* test using the differences between the pretest values and the posttest values. In this way, only the gain or loss in values is compared.

Here are some other examples of dependent samples. A researcher may want to design an SAT preparation course to help students raise their test scores the second time they take the SAT exam. Hence, the differences between the two exams are compared. A medical specialist may want to see whether a new counseling program will help subjects lose weight. Therefore, the preweights of the subjects will be compared with the postweights.

Besides samples in which the same subjects are used in a pre-post situation, there are other cases where the samples are considered dependent. For example, students might be matched or paired according to some variable that is pertinent to the study; then one student is assigned to one group and the other student is assigned to a second group. For instance, in a study involving learning, students can be selected and paired according to their IQs. That is, two students with the same IQ will be paired. Then one will be assigned to one sample group (which might receive instruction by computers), and the other student will be assigned to another sample group (which might receive instruction by the lecture-discussion method). These assignments will be done randomly. Since a student's IQ is important to learning, it is a variable that should be controlled. By matching subjects on IQ, the researcher can eliminate the variable's influence, for the most part. Matching, then, helps to reduce type II error by eliminating extraneous variables.

Two notes of caution should be mentioned. First, when subjects are matched according to one variable, the matching process does not eliminate the influence of other variables. In the previous example, matching students according to IQ does not account for their mathematical ability or their familiarity with computers. Since all variables influencing a study cannot be controlled, it is up to the researcher to determine which variables should be used in matching. Second, when the same subjects are used for a pre-post study, sometimes the knowledge that they are participating in a study can influence the results. For example, if people are placed in a special program, they may be more highly motivated to succeed owing to the fact that they have been selected to participate; the program itself may have little effect on the subject's success.

When the samples are dependent, a special *t* test for dependent means is used. This test employs the difference in the values of the matched pairs. The hypotheses are as follows, where μ_D is the symbol for the expected mean of the differences of the matched pairs.

Two-Tailed	One-Tailed Left	One-Tailed Right
$H_0\colon \mu_D = 0$	$H_0\colon \mu_D \geq 0$	$H_0\colon \mu_D \leq 0$
$H_1\colon \mu_D \neq 0$	$H_1\colon \mu_D < 0$	$H_1\colon \mu_D > 0$

The general procedure for finding the test value involves several steps. First, find the differences of the values of the pairs of data.

$$D = X_1 - X_2$$

Second, find the mean $(\overline{D})$ of the differences, using the formula

$$\overline{D} = \frac{\Sigma D}{n}$$

where n = the number of data pairs.

Third, find the standard deviation (s_D) of the differences, using the formula

$$s_D = \sqrt{\frac{\Sigma D^2 - \dfrac{(\Sigma D)^2}{n}}{n - 1}}$$

Fourth, find the estimated standard error $(s_{\overline{D}})$ of the differences, which is

$$s_{\overline{D}} = \frac{s_D}{\sqrt{n}}$$

Finally, find the test value, using the formula

$$t = \frac{\overline{D} - \mu_D}{\dfrac{s_D}{\sqrt{n}}} \quad \text{with d.f.} = n - 1$$

The formula in the final step follows the basic format of

$$\text{test value} = \frac{(\text{observed value}) - (\text{expected value})}{\text{standard error}}$$

where the observed value is the mean of the differences. The expected value μ_D is zero if the hypothesis is $\mu_D = 0$. The standard error of the difference is the standard deviation of the difference divided by the square root of the sample size. Both populations must be normally or approximately normally distributed.

Example 10–5 illustrates the hypothesis-testing procedure in detail.

EXAMPLE 10–5 A physical education director claims that by taking 800 international units (IU) of vitamin E, a weight lifter can increase his strength. Eight athletes are selected and given a test of strength, using the standard bench press. After two weeks of regular training, supplemented with vitamin E, they are tested again. Test the effectiveness of the vitamin E regimen at $\alpha = 0.05$. Each value in the data that follow represents the maximum number of pounds the athlete can bench-press.

Assume that the variable is approximately normally distributed.

Athlete	1	2	3	4	5	6	7	8
Before (X_1)	210	230	182	205	262	253	219	216
After (X_2)	219	236	179	204	270	250	222	216

Solution **STEP 1** State the hypotheses, and identify the claim. In order for the vitamin to be effective, the "before" weights must be significantly less than the "after" weights; hence, the mean of the differences must be less than zero.

$$H_0: \mu_D \geq 0 \quad \text{and} \quad H_1: \mu_D < 0 \text{ (claim)}$$

STEP 2 Find the critical value. The degrees of freedom are $n - 1$. In this case, d.f. $= 8 - 1 = 7$. The critical value for a one-tailed left test with $\alpha = 0.05$ is -1.895.

STEP 3 Compute the test value.

a. Make a table.

Before (X_1)	After (X_2)	A $D = (X_1 - X_2)$	B $D^2 = (X_1 - X_2)^2$
210	219		
230	236		
182	179		
205	204		
262	270		
253	250		
219	222		
216	216		

b. Find the differences and place the results in column A.

$$210 - 219 = -9$$
$$230 - 236 = -6$$
$$182 - 179 = +3$$
$$205 - 204 = +1$$
$$262 - 270 = -8$$
$$253 - 250 = +3$$
$$219 - 222 = -3$$
$$216 - 216 = \underline{\quad 0\quad}$$
$$\Sigma D = -19$$

c. Find the mean of the differences.

$$\overline{D} = \frac{\Sigma D}{n} = \frac{-19}{8} = -2.375$$

d. Square the differences and place the results in column B.

$$(-9)^2 = 81$$
$$(-6)^2 = 36$$
$$(+3)^2 = 9$$
$$(+1)^2 = 1$$
$$(-8)^2 = 64$$
$$(+3)^2 = 9$$
$$(-3)^2 = 9$$
$$0^2 = \underline{0}$$
$$\Sigma D^2 = 209$$

The completed table is shown next.

Before (X_1)	After (X_2)	**A** $D = (X_1 - X_2)$	**B** $D^2 = (X_1 - X_2)^2$
210	219	-9	81
230	236	-6	36
182	179	$+3$	9
205	204	$+1$	1
262	270	-8	64
253	250	$+3$	9
219	222	-3	9
216	216	$\underline{0}$	$\underline{0}$
		$\Sigma D = -19$	$\Sigma D^2 = 209$

e. Find the standard deviation of the differences.

$$s_D = \sqrt{\dfrac{\Sigma D^2 - \dfrac{(\Sigma D)^2}{n}}{n-1}} = \sqrt{\dfrac{209 - \dfrac{(-19)^2}{8}}{8-1}} = 4.84$$

f. Find the test value.

$$t - \dfrac{\overline{D} - \mu_D}{\dfrac{s_D}{\sqrt{n}}} = \dfrac{-2.375 - 0}{\dfrac{4.84}{\sqrt{8}}} = -1.39$$

STEP 4   Make the decision. The decision is, Do not reject the null hypothesis at $\alpha = 0.05$, since $-1.39 > -1.895$, as shown in Figure 10-7.

FIGURE 10-7
Critical and
Test Values for
Example 10-5

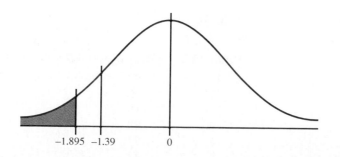

-1.895 -1.39 0

STEP 5 Summarize the results. There is not enough evidence to support the claim that vitamin E increases the strength of the weight lifter. ◄

The formulas for this t test are summarized next.

Formulas for the t Test for Dependent Samples

$$t = \frac{\overline{D} - \mu_D}{\frac{s_D}{\sqrt{n}}}$$

with d.f. $= n - 1$ and where

$$\overline{D} = \frac{\Sigma D}{n} \quad \text{and} \quad s_D = \sqrt{\frac{\Sigma D^2 - \frac{(\Sigma D)^2}{n}}{n - 1}}$$

EXAMPLE 10–6 A dietitian hopes to lower a person's cholesterol level by using a special diet supplemented with a combination of vitamin pills. Six subjects were pretested and then placed on the diet for two weeks. Their cholesterol levels were checked after the two-week period. The results are shown in the table. (The cholesterol levels are measured in milligrams per deciliter.) Can the dietitian conclude that the diet and the vitamin pill supplement have significantly lowered the cholesterol levels of the subjects involved? Use $\alpha = 0.10$. Assume that the variable is approximately normally distributed.

Subject	1	2	3	4	5	6
Before (X_1)	210	235	208	190	172	244
After (X_2)	190	170	210	188	173	228

Solution **STEP 1** State the hypotheses, and identify the claim. If the diet is effective, the "before" cholesterol level should be greater than the "after" level; hence, the mean of the differences should be greater than zero.

$$H_0: \mu_D \le 0 \quad \text{and} \quad H_1: \mu_D > 0 \text{ (claim)}$$

STEP 2 Find the critical value. The degrees of freedom are 5. At $\alpha = 0.10$, the critical value is $+1.476$.

STEP 3 Compute the test value.

a. Make a table.

Before (X_1)	After (X_2)	**A** $D = (X_1 - X_2)$	**B** $D^2 = (X_1 - X_2)^2$
210	190		
235	170		
208	210		
190	188		
172	173		
244	228		

b. Find the differences and place the results in column A.

$$210 - 190 = 20$$
$$235 - 170 = 65$$
$$208 - 210 = -2$$
$$190 - 188 = 2$$
$$172 - 173 = -1$$
$$244 - 228 = \underline{16}$$
$$\Sigma D = 100$$

c. Find the mean of the difference.

$$\overline{D} = \frac{\Sigma D}{n} = \frac{100}{6} = 16.7$$

d. Square the differences and place the results in column B.

$$(20)^2 = 400$$
$$(65)^2 = 4225$$
$$(-2)^2 = 4$$
$$(2)^2 = 4$$
$$(-1)^2 = 1$$
$$(16)^2 = \underline{256}$$
$$\Sigma D^2 = 4890$$

Then, complete the table as shown.

Before (X_1)	After (X_2)	A $D = (X_1 - X_2)$	B $D^2 = (X_1 - X_2)^2$
210	190	20	400
235	170	65	4225
208	210	−2	4
190	188	2	4
172	173	−1	1
244	228	16	256
		$\Sigma D = 100$	$\Sigma D^2 = 4890$

e. Find the standard deviation of the differences.

$$s_D = \sqrt{\frac{\Sigma D^2 - \dfrac{(\Sigma D)^2}{n}}{n - 1}} = \sqrt{\frac{4890 - \dfrac{(100)^2}{6}}{5}} = 25.39$$

f. Find the test value.

$$t = \frac{\overline{D} - \mu_D}{\dfrac{s_D}{\sqrt{n}}} = \frac{16.7 - 0}{\dfrac{25.39}{\sqrt{6}}} = 1.61$$

STEP 4   Make the decision. The decision is to reject the null hypothesis, since the test value 1.61 is greater than 1.476, as shown in Figure 10–8.

FIGURE 10–8
Critical and
Test Values for
Example 10–6

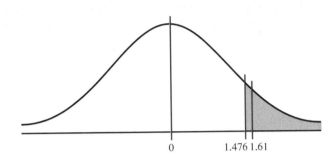

0   1.476 1.61

STEP 5   Summarize the results. There is enough evidence to support the claim that the diet and the vitamin pill supplement seem to lower the cholesterol level. ◄

The steps for this *t* test are summarized in Procedure Table 8.

Procedure Table 8

Testing the Difference Between Means for Dependent Samples

Step 1: State the hypotheses, and identify the claim.
Step 2: Find the critical value(s).
Step 3: Compute the test value.

a. Make a table, as shown.

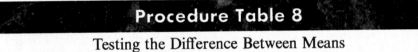

		A	**B**
X_1	X_2	$D = (X_1 - X_2)$	$D^2 = (X_1 - X_2)^2$
⋮	⋮		
		$\Sigma D =$ _____	$\Sigma D^2 =$ _____

b. Find the differences and place the results in column A.

$$D = (X_1 - X_2)$$

c. Find the mean of the differences.

$$\overline{D} = \frac{\Sigma D}{n}$$

d. Square the differences and place the results in column B. Complete the table.

$$D^2 = (X_1 - X_2)^2$$

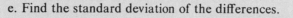

e. Find the standard deviation of the differences.

$$s_D = \sqrt{\frac{\Sigma D^2 - \dfrac{(\Sigma D)^2}{n}}{n - 1}}$$

f. Find the test value.

$$t = \frac{\overline{D} - \mu_D}{\dfrac{s_D}{\sqrt{n}}}$$

Step 4: Make the decision.

Step 5: Summarize the results.

If a specific difference is hypothesized, the following formula should be used:

$$t = \frac{\overline{D} - \mu_D}{s_D / \sqrt{n}}$$

where μ_D is the hypothesized difference.

For example, if a dietitian claims that people on a specific diet will lose an average of 3 pounds in a week, the hypotheses are

$$H_0: \mu_D = 3 \quad \text{and} \quad H_1: \mu_D \neq 3$$

The value 3 will be substituted in the test statistic formula for μ_D.

EXERCISES

10-33. Explain the difference between independent and dependent samples.

10-34. (ANS.) Classify each as independent or dependent samples.

a. Heights of identical twins.
b. Test scores of the same students in English and psychology.
c. The effectiveness of two different brands of aspirin.
d. Effects of a drug on reaction time, measured by a "before" and an "after" test.
e. The effectiveness of two different diets on two different groups of individuals.

For Exercises 10-35 through 10-43, perform each of the following steps. Assume that all variables are normally or approximately normally distributed.

a. State the hypotheses, and identify the claim.
b. Find the critical value(s).
c. Compute the test value.
d. Make the decision.
e. Summarize the results.

10-35. A program for improving the hygiene habits of kindergarten students was conducted. The school administrator hypothesized that after the program, the students would miss fewer days of school because of illness. The table shows the number of days missed per month before the program and after completion of the program by ten students. Test the claim, at $\alpha = 0.05$, that the students missed fewer days after the program.

Before	3	5	2	0	4	3	1	1	2	6
After	2	4	3	1	5	2	0	0	2	3

10–36. As an aid for improving students' study habits, nine students were randomly selected to attend a seminar on the importance of education in life. The table shows the number of hours each student studied per week before the seminar and after attending the seminar. At $\alpha = 0.10$, did attending the seminar increase the number of hours the students studied per week?

Before	9	12	6	15	3	18	10	13	7
After	9	17	9	20	2	21	15	22	6

10–37. A sociologist is interested in determining whether a film about stress management will change the attitudes of 12 people on the subject. The results of her questionnaires are shown below. A higher numerical value shows a more favorable attitude toward stress management. Test the claim, at $\alpha = 0.05$, that there will be a change in attitude.

Before	12	15	6	20	2	5	9	16	14	17	8	5
After	11	13	3	21	5	7	6	10	9	12	4	1

10–38. Suppose a researcher wishes to test the effects of a new diet designed to reduce the blood sodium level of ten patients. The patients' sodium levels are checked before being placed on the diet and after two weeks on the diet. Test the claim, at $\alpha = 0.05$, that the diet is effective in lowering the sodium content of the patients' blood.

Patient	Before	After
1	146	135
2	138	133
3	152	147
4	163	156
5	136	138
6	147	141
7	148	139
8	141	132
9	143	138
10	142	131

10–39. An office manager wishes to see whether the typing speed of ten secretaries can be increased by changing over to computers. The number of words typed per minute is given below. Test the claim, at $\alpha = 0.10$, that by using computers, the secretaries can type more words per minute.

Secretary	Typewriter	Computer
1	63	68
2	72	80
3	85	95
4	97	93
5	82	80
6	101	106
7	73	82
8	62	78
9	58	65
10	75	83

10–40. A composition teacher wishes to see whether a new grammar program will reduce the number of grammatical errors her students make when writing a two-page essay. The data are shown below. At $\alpha = 0.025$, test the claim that the number of errors has been reduced.

Student	1	2	3	4	5	6
Errors Before	12	9	0	5	4	3
Errors After	9	6	1	3	2	3

10–41. A sports shoe manufacturer claims that joggers who wear its brand of shoe will jog faster than those who don't use its product. A sample of eight joggers is taken, and the joggers agree to test the claim on a 1-mile track. The rates (in minutes) of the joggers while wearing the manufacturer's shoe and while wearing any other brand of shoe are shown here. Test the claim at $\sigma = 0.025$.

Runner	1	2	3	4	5	6	7	8
Manufacturer's Brand	8.2	6.3	9.2	8.6	6.8	8.7	8.0	6.9
Other Brand	7.1	6.8	9.8	8.0	5.8	8.0	7.4	8.0

10–42. A researcher wanted to compare the pulse rates of identical twins to see whether there was any difference in their average pulse rates. Eight sets of twins were selected. The rates are given in the table as number of beats per minute. At $\alpha = 0.01$, is there a significant difference in the pulse rates of twins?

Twin A	87	92	78	83	88	90	84	93
Twin B	83	95	79	83	86	93	80	86

10–43. A physician claims that a person's diastolic blood pressure can be lowered if the person listens to a special relaxation tape each evening. Ten patients are pretested and then given the tape, which is to be played each evening for one week. Then their blood pressure readings are taken again. The data are shown in the table (in millimeters of mercury). At $\alpha = 0.025$, can the physician conclude that using the tape may lower a person's blood pressure?

Patient	1	2	3	4	5	6	7	8	9	10
Before	86	92	95	84	80	78	98	95	94	96
After	84	83	81	78	82	74	86	83	80	82

*10–44. Instead of finding the mean of the differences between X_1 and X_2 by subtracting $X_1 - X_2$, one can find this mean by finding the means of X_1 and X_2 and then subtracting the means. Show that these two procedures will yield the same results.

10–5 TESTING THE DIFFERENCE BETWEEN PROPORTIONS

The z test with some modifications can be used to test the equality of two proportions. For example, a researcher might ask the question, Is the proportion of men who exercise regularly less than the proportion of women who exercise regularly? Is there a difference in the percentage of students who own a personal computer and the percentage of nonstudents who own a personal computer? Is there a difference in the proportion of college graduates who pay cash for purchases and the proportion of non-college graduates who pay cash for a purchase?

Recall from Chapter 8 that the symbol $\hat{p}$ ("p hat") is the sample proportion used to estimate the population proportion, denoted by p. For example, if in a sample of 30 college students, 9 were on probation, then the sample proportion is $\hat{p} = \frac{9}{30}$, or 0.3. The population proportion p is the number of all students who were on probation divided by the number of students who attended the college. The formula for $\hat{p}$ is

$$\hat{p} = \frac{X}{n}$$

where X = number of units that possess the characteristic of interest
          n = sample size

When one is testing the difference between two population proportions, p_1 and p_2, the hypotheses can be stated as follows, if no difference between the proportions is hypothesized.

$$H_0: p_1 = p_2 \qquad H_0: p_1 - p_2 = 0$$
$$\qquad\qquad \text{or}$$
$$H_1: p_1 \neq p_2 \qquad H_1: p_1 - p_2 \neq 0$$

Similar statements using $\geq$ and $<$ or $\leq$ and $>$ can be formed for one-tailed tests.

For two proportions, $\hat{p}_1 = X_1/n_1$ is used to estimate p_1, and $\hat{p}_2 = X_2/n_2$ is used to estimate p_2. The standard error of difference is

$$\sigma_{(\hat{p}_1 - \hat{p}_2)} = \sqrt{\sigma_{p_1}^2 + \sigma_{p_2}^2} = \sqrt{\frac{p_1 q_1}{n_1} + \frac{p_2 q_2}{n_2}}$$

where $\sigma_{p_1}^2$ and $\sigma_{p_2}^2$ are the variances of the proportions, $q_1 = 1 - p_1$, $q_2 = 1 - p_2$, and n_1 and n_2 are the respective sample sizes.

Since p_1 and p_2 are unknown, a weighted estimate of p can be computed by using the formula

$$\overline{p} = \frac{n_1 \hat{p}_1 + n_2 \hat{p}_2}{n_1 + n_2}$$

and $\overline{q} = 1 - \overline{p}$. This weighted estimate is based on the hypothesis that $p_1 = p_2$. Hence, $\overline{p}$ is a better estimate than either $\hat{p}_1$ or $\hat{p}_2$ since it is a combined average using both $\hat{p}_1$ and $\hat{p}_2$.

Since $\hat{p}_1 = X_1/n_1$ and $\hat{p}_2 = X_2/n_2$, $\overline{p}$ can be simplified to

$$\overline{p} = \frac{X_1 + X_2}{n_1 + n_2}$$

Finally, the standard error of difference in terms of the weighted estimate is

$$\sigma_{(\hat{p}_1 - \hat{p}_2)} = \sqrt{\overline{p}\,\overline{q}\left(\frac{1}{n_1} + \frac{1}{n_2}\right)}$$

The formula for the test value is as shown next.

Formula for the z Test for Comparing Two Proportions

$$z = \frac{(\hat{p}_1 - \hat{p}_2) - (p_1 - p_2)}{\sqrt{\overline{p}\,\overline{q}\left(\frac{1}{n_1} + \frac{1}{n_2}\right)}}$$

where

$$\overline{p} = \frac{X_1 + X_2}{n_1 + n_2} \qquad \hat{p}_1 = \frac{X_1}{n_1}$$

$$\overline{q} = 1 - \overline{p} \qquad \hat{p}_2 = \frac{X_2}{n_2}$$

This formula follows the format

$$\text{test value} = \frac{(\text{observed value}) - (\text{expected value})}{\text{standard error}}$$

There are two requirements necessary for use of the z test: (1) The samples must be independent of each other, and (2) $n_1 p_1$ and $n_1 q_1$ must be 5 or more and $n_2 p_2$ and $n_2 q_2$ must be 5 or more.

EXAMPLE 10–7   A sample of 100 students at a university showed that 43 had taken one or more remedial college courses. A sample of 200 students at a junior college showed that 90 had taken one or more remedial college courses. At $\alpha = 0.05$, test the hypothesis that there is no difference in the proportion of students who complete remedial courses at a university or a junior college.

Solution   Let $\hat{p}_1$ be the proportion of the university students who complete a remedial course, and let $\hat{p}_2$ be the proportion of the junior college students who complete a remedial course. Then,

$$\hat{p}_1 = \frac{X_1}{n_1} = \frac{43}{100} = 0.43 \qquad \text{and} \qquad \hat{p}_2 = \frac{X_2}{n_2} = \frac{90}{200} = 0.45$$

$$\overline{p} = \frac{X_1 + X_2}{n_1 + n_2} = \frac{43 + 90}{100 + 200} = \frac{133}{300} = 0.443$$

$$\overline{q} = 1 - \overline{p} = 1 - 0.443 = 0.557$$

Now, follow the steps in hypothesis testing.

STEP 1   State the hypotheses, and identify the claim.

$$H_0: p_1 = p_2 \text{ (claim)} \qquad \text{and} \qquad H_1: p_1 \neq p_2$$

STEP 2   Find the critical values. Since $\alpha = 0.05$, the critical values are $+1.96$ and -1.96.

STEP 3 Compute the test value.

$$z = \frac{(\hat{p}_1 - \hat{p}_2) - (p_1 - p_2)}{\sqrt{\bar{p}\,\bar{q}\left(\dfrac{1}{n_1} + \dfrac{1}{n_2}\right)}}$$

$$= \frac{(0.43 - 0.45) - 0}{\sqrt{(0.443)(0.557)\left(\dfrac{1}{100} + \dfrac{1}{200}\right)}} = -0.33$$

STEP 4 Make the decision. Do not reject the null hypothesis, since $-0.33 > -1.96$. See Figure 10–9.

FIGURE 10–9
Critical and
Test Values for
Example 10–7

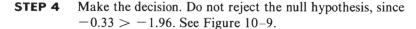

$-1.96 \qquad -0.33\ 0 \qquad +1.96$

STEP 5 Summarize the results. There is not enough evidence to reject the claim that there is no difference in the proportions. ◀

EXAMPLE 10–8 A sample of 50 randomly selected men with high triglyceride levels consumed 2 tablespoons of oat bran daily for six weeks. After six weeks, 60% of the men had lowered their triglyceride level. A sample of 80 men consumed 2 tablespoons of wheat bran for six weeks. After six weeks, 25% had lower triglyceride levels. Is there a significant difference in the two proportions, at the 0.01 significance level?

Solution Since the statistics are given in percentages, $\hat{p}_1 = 60\%$, or 0.60, and $\hat{p}_2 = 25\%$, or 0.25. In order to compute $\hat{p}$, one must find X_1 and X_2.

$$X_1 = (0.60)(50) = 30 \qquad X_2 = (0.25)(80) = 20$$

$$\bar{p} = \frac{X_1 + X_2}{n_1 + n_2} = \frac{30 + 20}{50 + 80} = \frac{50}{130} = 0.385$$

$$\bar{q} = 1 - \bar{p} = 1 - 0.385 = 0.615$$

STEP 1 State the hypotheses, and identify the claim.

$$H_0: p_1 = p_2 \qquad \text{and} \qquad H_1: p_1 \neq p_2 \text{ (claim)}$$

STEP 2 Find the critical values. Since $\alpha = 0.01$, the critical values are $+2.58$ and -2.58.

STEP 3 Compute the test value.

$$z = \frac{(\hat{p}_1 - \hat{p}_2) - (p_1 - p_2)}{\sqrt{\overline{p}\overline{q}\left(\dfrac{1}{n_1} + \dfrac{1}{n_2}\right)}}$$

$$= \frac{(0.60 - 0.25) - 0}{\sqrt{(0.385)(0.615)\left(\dfrac{1}{50} + \dfrac{1}{80}\right)}} = 3.99$$

STEP 4 Make the decision. Reject the null hypothesis, since $3.99 > 2.58$. See Figure 10–10.

FIGURE 10–10
Critical and
Test Values for
Example 10–8

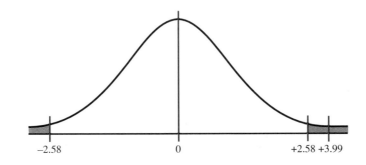

STEP 5 Summarize the results. There is enough evidence to support the claim that there is a difference in proportions. ◀

EXERCISES

10–45. Find the proportions $\hat{p}$ and $\hat{q}$ for each.

a. $n = 48, X = 34$
b. $n = 75, X = 28$
c. $n = 100, X = 50$
d. $n = 24, X = 6$
e. $n = 144, X = 12$

10–46. (ANS.) Find each X given $\hat{p}$.

a. $\hat{p} = 0.16, n = 100$
b. $\hat{p} = 0.08, n = 50$
c. $\hat{p} = 6\%, n = 80$
d. $\hat{p} = 52\%, n = 200$
e. $\hat{p} = 20\%, n = 150$

10–47. Find $\overline{p}$ and $\overline{q}$ for each.

a. $X_1 = 60, n_1 = 100, X_2 = 40, n_2 = 100$
b. $X_1 = 22, n_1 = 50, X_2 = 18, n_2 = 30$
c. $X_1 = 18, n_1 = 60, X_2 = 20, n_2 = 80$
d. $X_1 = 5, n_1 = 32, X_2 = 12, n_2 = 48$
e. $X_1 = 12, n_1 = 75, X_2 = 15, n_2 = 50$

For Exercises 10–48 through 10–59, perform the following steps.

a. State the hypotheses, and identify the claim.
b. Find the critical value(s).
c. Compute the test value.
d. Make the decision.
e. Summarize the results.

10–48. A sample of 150 people from a certain industrial community showed that 80 people suffered from a lung disease. A sample of 100 people from a rural community showed that 30 suffered from the same lung disease. At $\alpha = 0.05$, is there a difference between the proportion of people who suffer from the disease in the two communities?

10–49. A recent survey showed that in a sample of 50 general practitioners, 5 smoked; in a sample of 80 surgeons, 15 smoked. At $\alpha = 0.05$, is the proportion of surgeons who smoke higher than the proportion of general practitioners who smoke?

Speaking of **Statistics**

Less Exotic but More Relaxing Trips

By Cathy Lynn Grossman
USA TODAY

A happy vacation depends more on who you go with than where you go.

So says the travel research firm D. K. Shifflet & Associates, that asked 500 people nationwide whether they enjoyed their vacations last year more or less than before the recession. Among findings:

- 35% said they enjoyed their '91 vacations more than two or three years ago; only 19% had a worse trip. The other 46% reported no change.
- 53% who traveled with children enjoyed their '91 trips more, and 43% who stayed with family or friends did, too.

"The '90s emphasis on personal ties seems to be generating a 'happiness dividend' for travelers," Shifflet says.

"When times are good, people are more adventurous, they go farther. In difficult times, they need a more secure, relaxing vacation."

The findings match the annual lifestyles survey by Yankelovich Clancy Shulman, which concluded that vacations in the '90s are more for R&R than for bragging rights.

In 1988, 60% described traveling for pleasure as a sign of success. In '91 the number had dropped to 52%. What topped it? Being satisfied with life rose to 63%.

Source: Copyright 1992, USA TODAY.
Reprinted with permission.

In this study, there is a change in the percentage from 1988 to 1991. Do you feel that the change is significant? State a hypothesis that would test this claim, and using 1988 as the population proportion, test the claim at a significance level of your choice. Comment on the results.

10–50. In a sample of 100 customers, 43 used a MasterCard. In another sample of 100, 58 used a Visa card. At $\alpha = 0.05$, is there a difference in the proportion of people who use each type of credit card?

10–51. In Cleveland, a sample of 73 mail carriers showed that 10 had received an animal bite during one week. In Philadelphia, in a sample of 80 mail carriers, 16 had received animal bites. Is there a significant difference in the proportions? Use $\alpha = 0.05$.

10–52. A survey found that 83% of the men questioned preferred computer-assisted instruction to lecture, and 75% of the women preferred computer-assisted instruction to lecture. There were 100 individuals in each sample. At $\alpha = 0.05$, test the claim that there is no difference in the proportion of men and the proportion of women who favor computer-assisted instruction over lecture.

10–53. In a sample of 200 surgeons, 15% felt that the government should control health care. In a sample of 200 general practitioners, 21% felt this way. At $\alpha = 0.10$, is there a difference in the proportions?

10–54. In a sample of 80 Americans, 55% wished that they were rich. In a sample of 90 Europeans, 45% wished that they were rich. At $\alpha = 0.01$, is there a difference in the proportions?

10–55. In a sample of 200 men, 130 said they used seat belts; and in a sample of 300 women, 63 indicated that they used seat belts. Test the claim that men are more safety-conscious than women, at $\alpha = 0.01$.

10–56. A survey of 80 homes in a Washington, D.C., suburb showed that 45 homes were air-conditioned. A sample of 120 homes in a Pittsburgh suburb showed that 63 had air-conditioning. At $\alpha = 0.05$, is there a difference in the two proportions?

10–57. A recent study showed that in a sample of 100 people, 30% had visited Disneyland; and in another sample of 100 people, 24% had visited Disneyworld. Are the proportions of people who visited each park different? Use $\alpha = 0.02$.

10–58. A sample of 200 teenagers shows that 50 believe that war is inevitable, and a sample of 300 people over 60 shows that 93 believe war is inevitable. Is the proportion of teenagers who believe war is inevitable different from the proportion of people over 60? Use $\alpha = 0.01$.

10–59. In a sample of 50 high school seniors, 8 had their own cars. In a sample of 75 college freshmen, 20 had their own cars. At $\alpha = 0.05$, can it be concluded that there is a higher proportion of college freshmen who have their own cars?

*10–60. If there is a significant difference between p_1 and p_2 and between p_2 and p_3, can one conclude that there is a significant difference between p_1 and p_3?

The formula for finding the confidence interval for this difference between two population proportions p_1 and p_2 is

$$(\hat{p}_1 - \hat{p}_2) - z_{\alpha/2} \sqrt{\frac{\hat{p}_1 \hat{q}_1}{n_1} + \frac{\hat{p}_2 \hat{q}_2}{n_2}}$$

$$< (p_1 - p_2) < (\hat{p}_1 - \hat{p}_2) + z_{\alpha/2} \sqrt{\frac{\hat{p}_1 \hat{q}_1}{n_1} + \frac{\hat{p}_2 \hat{q}_2}{n_2}}$$

Use the formula for Exercises 10–61 and 10–62.

*10–61. Find the 95% confidence interval for the true difference in proportion for the data of a study in which 40% of the 200 males surveyed opposed the death penalty and 56% of the 100 females surveyed opposed the death penalty.

*10–62. Find the 99% confidence interval for the difference in the population proportion for the data of a study in which 80% of the 150 Republicans surveyed favored the bill for a salary increase and 60% of the 200 Democrats surveyed favored the bill for a salary increase.

10–6 SUMMARY

Many times, researchers are interested in comparing two population parameters, such as means or proportions. This comparison can be accomplished by using special z and t tests. If the samples are independent and the variances are known, the z test is used. The z test is also used when the variances are unknown but both sample sizes are 30 or more. If the variances are not known and one or both sample sizes are less than 30, the t test must be used. For independent samples, a further requirement is that one must determine whether the variances of the populations are equal. Different formulas are used in each case. If the samples are dependent, the t test for dependent samples is used. Finally, a z test is used to compare two proportions.

The goal of inferential statistics is to enable the student to do two things: (1) identify which test to use in a hypothesis-testing situation and (2) realize that the test can only be used when the assumptions for that test are met. For this reason, hypothesis test summaries are provided at the end of this chapter, Chapter 13, and Chapter 14.

Important Terms

Dependent samples
Independent samples
Pooled estimate of the variance

Important Formulas

Formula for the z test for comparing two means from independent populations:

$$z = \frac{(\overline{X}_1 - \overline{X}_2) - (\mu_1 - \mu_2)}{\sqrt{\dfrac{\sigma_1^2}{n_1} + \dfrac{\sigma_2^2}{n_2}}}$$

Formula for the t test for comparing two means (independent samples, variances not equal):

$$t = \frac{(\overline{X}_1 - \overline{X}_2) - (\mu_1 - \mu_2)}{\sqrt{\dfrac{s_1^2}{n_1} + \dfrac{s_2^2}{n_2}}}$$

and d.f. = the smaller of $n_1 - 1$ or $n_2 - 1$.

Formula for the t test for comparing two means (independent samples, variances equal):

$$t = \frac{(\overline{X}_1 - \overline{X}_2) - (\mu_1 - \mu_2)}{\sqrt{\dfrac{(n_1 - 1)s_1^2 + (n_2 - 1)s_2^2}{(n_1 + n_2 - 2)}} \sqrt{\dfrac{1}{n_1} + \dfrac{1}{n_2}}}$$

and d.f. = $n_1 + n_2 - 2$.

Formula for the t test for comparing two means from dependent samples:

$$t = \frac{\overline{D} - \mu_D}{\dfrac{s_D}{\sqrt{n}}}$$

where $\overline{D}$ is the mean of the differences,

$$\overline{D} = \frac{\Sigma D}{n}$$

and s_D is the standard deviation of the differences,

$$s_D = \sqrt{\frac{\Sigma D^2 - \dfrac{(\Sigma D)^2}{n}}{n - 1}}$$

Formula for the z test for comparing two proportions:

$$z = \frac{(\hat{p}_1 - \hat{p}_2) - (p_1 - p_2)}{\sqrt{\overline{p}\,\overline{q}\left(\dfrac{1}{n_1} + \dfrac{1}{n_2}\right)}}$$

where

$$\overline{p} = \frac{X_1 + X_2}{n_1 + n_2} \qquad \hat{p}_1 = \frac{X_1}{n_1}$$

$$\overline{q} = 1 - \overline{p} \qquad \hat{p}_2 = \frac{X_2}{n_2}$$

Review Exercises

For each problem, perform the following steps. Assume that all variables are normally or approximately normally distributed.

a. State the hypotheses, and identify the claim.
b. Find the critical value(s).
c. Compute the test value.
d. Make the decision.
e. Summarize the results.

10–63. A researcher wishes to see whether there is a difference in the triglyceride levels between two groups of women. A random sample of 25 women between the ages of 20 and 35 is selected and tested. The average level is 112 milligrams per deciliter (mg/dl). A second sample of 25 women between the ages of 36 and 50 is selected and tested. The average of this group is 114 mg/dl. The population standard deviation for both groups is 4 mg/dl. At $\alpha = 0.01$, is there a difference in the triglyceride levels between the two age groups?

10–64. Two groups of people are surveyed. In a sample of 50 drivers who are single people, the drivers average 106 miles per week for pleasure trips. In a sample of 65 married people, the drivers average 68 miles per week for pleasure trips. The sample standard deviations are 15 and 9 miles, respectively. At $\alpha = 0.01$, test the claim that single people do more driving for pleasure trips than married people.

10–65. The average price of 15 cans of tomato soup taken from different stores is $0.73, and the standard deviation is $0.05. The average price of 24 cans of chicken noodle soup is $0.91, and the standard deviation is $0.03. At $\alpha = 0.01$, is there a significant difference in price? Assume that the variances are equal.

10–66. The average temperature over a 25-day period in June for Birmingham, Alabama, is 78°, and the standard deviation is 3°. The average temperature over the same 25 days for Chicago, Illinois, is 75°, with a standard deviation of 5°. At $\alpha = 0.01$, can one conclude that it is warmer in Birmingham? Assume that the variances are equal.

10–67. A sample of 15 teachers from Rhode Island has an average salary of $35,270, with a standard deviation of $3256. A sample of 30 teachers from New York has an average salary of $29,512, with a standard deviation of $1432. Is there a significant difference in teachers' salaries between the two states? Use $\alpha = 0.02$. Assume that the variances are not equal.

10–68. The average income of 16 families who reside in a large metropolitan city is $54,356, and the standard deviation is $8256. The average income of 12 families who reside in a suburb of the same city is $46,512, with a standard deviation of $1311. At $\alpha = 0.05$, can one conclude that the income of the families who reside within the city is greater than that of those who reside in the suburb? Assume that the variances are not equal.

10–69. In an effort to improve the vocabulary of ten students, a teacher provides a weekly 1-hour tutoring session for the students. A pretest is given before the sessions and a posttest is given afterward. The results are shown in the table. At $\alpha = 0.01$, can the teacher conclude that the tutoring sessions helped to improve the students' vocabulary?

Student	1	2	3	4	5	6	7	8	9	10
Pretest	83	76	92	64	82	68	70	71	72	63
Posttest	88	82	100	72	81	75	79	68	81	70

10–70. In an effort to increase production of an automobile part, the factory manager decides to play music in the manufacturing area. Eight workers are selected, and the number of items each produced for a specific day is recorded. After one week of music, the same workers are monitored again. The data are given in the following table. At $\alpha = 0.05$, can the manager conclude that the music has increased production?

Worker	1	2	3	4	5	6	7	8
Before	6	8	10	9	5	12	9	7
After	10	12	9	12	8	13	8	10

10–71. In a sample of 75 workers from a factory in city A, 8% did not graduate from high school. In a sample of 60 workers from a factory in city B, 9% did not graduate from high school. Can one conclude that there is a difference in the proportions of workers who did not graduate from high school in the two cities? Use $\alpha = 0.05$.

10–72. In a recent survey of 50 apartment residents, 32 had microwave ovens. In another survey of 60 homeowners, 24 had microwave ovens. At $\alpha = 0.05$, can one conclude that the proportions are equal?

COMPUTER APPLICATIONS

1. Write a computer program to perform a *t* test comparing the means of two independent samples. Have the program compute the test statistic when the variances are assumed equal and when the variances are not equal.
2. Write a computer program to compute the *t* test value comparing two means when the samples are dependent.
3. MINITAB can be used to compare two sample means when they are independent. For example, suppose a researcher wishes to compare the average lifetimes (in hours) of two samples of 9-volt batteries. The data are as follows:

Type 1	12	15	18	13	10	9	6	3	11	5
Type 2	18	20	13	7	9	4				

Test the hypothesis that the type 2 battery is better.

   Data are entered by using the SET command. In order to conduct the *t* test, one uses the command TWOSAMPLE T C1 C2 ;. For a one-tailed left test, the command ALTERNATIVE −1 . is used. Type in the following:

```
MTB  > SET C1
DATA > 12    15   18   13   10   9   6   3   11   5
DATA > END
MTB  > SET C2
DATA > 18   20   13   7   9   4
DATA > END
MTB  > TWOSAMPLE T C1 C2;
SUBC > ALTERNATIVE    -1.
```

The computer will print the following:

```
TWOSAMPLE T FOR C1 VS C2
        N       MEAN      STDEV    SE MEAN
C1  10         10.20      4.64       1.5
C2   6         11.83      6.31       2.6

95 PCT CI FOR MU C1 - MU C2: (-8.5, 5.2)

TTEST MU C1 = MU C2 (VS LT): T= -0.55 P=0.30 DF= 8
```

The table gives the sample sizes, the means, the standard deviations, and the standard error for the means. The *t* value is −0.55. The *P*-value is 0.30 with 8 degrees of freedom. Since the *P*-value is greater than 0.05, the decision is to not reject the null hypothesis.

   Try Exercise 10–32 by using MINITAB.
4. MINITAB can also be used to compare two means when the samples are dependent. For example, suppose a researcher wishes to compare the lifetimes of two brands of tires used on the same cars. The data (in thousands of miles) are as follows:

Brand A	33	35	28	29	32	34	30	34
Brand B	27	29	36	34	30	29	28	24

Test the claim that there is no difference in the average lifetimes of both brands.

The data are entered by using the SET commands. After the data have been entered, it is necessary to instruct the computer to subtract the values by using the LET C3 = C1 − C2 command. Finally, the TTEST DIFFERENCE 0 C3 command is used to get the computer to conduct a dependent t test with a difference of zero. The ALTERNATIVE = 0 statement instructs the computer to conduct a two-tailed test.

Type in the following:

```
MTB  > SET C1
DATA > 33 35 28 29 32 34 30 34
DATA > END
MTB  > SET C2
DATA > 27 29 36 34 30 29 28 24
DATA > END
MTB  > LET C3 = C1 - C2
MTB  > TTEST DIFFERENCE 0 C3;
SUBC > ALTERNATIVE = 0.
```

The computer will print the following:

```
TEST OF MU = 0.0000 VS MU N.E. 0.0000
     N   MEAN   STDEV   SE MEAN    T    P-VALUE
C3   8  2.250   6.028   2.128    1.06    0.33
```

In this case, the null hypothesis is not rejected, since $t = 1.06$ and the P-value is 0.33.

Try Exercise 10–35 by using MINITAB.

√ DATA ANALYSIS Applying the Concepts

The Data Bank is located in the Appendix.

1. From the Data Bank, select a variable and compare the mean of the variable for a random sample of at least 30 men with the mean of the variable for a random sample of at least 30 women. Use a z test.

2. Repeat the experiment in Exercise 1 using a different variable and two samples of size 15. Compare the means by using a t test. Assume that the variances are equal.

3. Compare the proportion of men who are smokers with the proportion of women who are smokers. Use the data in the Data Bank. Choose random samples of size 30 or more. Use the z test for proportions.

✎ TEST

Directions: Determine whether each statement is true or false.

1. If one hypothesizes that there is no difference between two means, then $\mu_1 = \mu_2$.

2. If one selected all possible pairs of samples from a population, found the difference between the means of each pair, and plotted these differences, the shape of the distribution would be normal if the sample sizes were larger than or equal to 30.

3. When one is testing the difference between means of two samples and using the t test, it is not important to distinguish whether or not the samples are independent° of each other.

4. When one is testing the difference between two means using the t test, a pooled estimate of the variances is used if the variances are unequal.

5. If a researcher wishes to see whether the mean number of defective items produced by a machine has changed after it has been repaired, the researcher can use the t test for dependent samples.

6. If the same diet is given to two groups of randomly selected individuals, the samples are considered to be dependent.

7. The z test with some modification can be used to test the equality of two proportions.
8. When one is testing proportions, the sample proportion is usually a good estimate of the population proportion.

9. When one is testing two proportions for equality, the standard error of difference is computed by using a pooled estimate of the variance.

HYPOTHESIS-TESTING SUMMARY 1

1. Comparison of a sample mean with a specific population mean.

    example: $H_0: \mu = 100$

    a. Use the z test when σ is known:

    $$z = \frac{\overline{X} - \mu}{\dfrac{\sigma}{\sqrt{n}}}$$

    b. Use the t test when σ is unknown:

    $$t = \frac{\overline{X} - \mu}{\dfrac{s}{\sqrt{n}}} \quad \text{with} \quad \text{d.f.} = n - 1$$

2. Comparison of two sample means.

    example: $H_0: \mu_1 = \mu_2$

    a. Use the z test when the population variances are known:

    $$z = \frac{(\overline{X}_1 - \overline{X}_2) - (\mu_1 - \mu_2)}{\sqrt{\dfrac{\sigma_1^2}{n_1} + \dfrac{\sigma_2^2}{n_2}}}$$

    b. Use the t test for independent samples when the population variances are unknown and the sample variances are unequal:

    $$t = \frac{(\overline{X}_1 - \overline{X}_2) - (\mu_1 - \mu_2)}{\sqrt{\dfrac{s_1^2}{n_1} + \dfrac{s_2^2}{n_2}}}$$

    with

    d.f. = the smaller of $n_1 - 1$ or $n_2 - 1$

    c. Use the t test for independent samples when the population variances are unknown and the sample variances are equal:

    $$t = \frac{(\overline{X}_1 - \overline{X}_2) - (\mu_1 - \mu_2)}{\sqrt{\dfrac{(n_1 - 1)s_1^2 + (n_2 - 1)s_2^2}{n_1 + n_2 - 2}} \sqrt{\dfrac{1}{n_1} + \dfrac{1}{n_2}}}$$

    with   d.f. $= n_1 + n_2 - 2$

    d. Use the t test for dependent samples when the means are related:

    example: $H_0: \mu_D = 0$

    $$t = \frac{\overline{D} - \mu_D}{\dfrac{s_D}{\sqrt{n}}} \quad \text{with} \quad \text{d.f.} = n - 1$$

    where n = number of pairs.

3. Comparison of a sample proportion with a specific population proportion.

    example: $H_0: p = 0.32$

    Use the z test:

    $$z = \frac{X - np}{\sqrt{npq}} \quad \text{or} \quad z = \frac{X - \mu}{\sigma}$$

4. Comparison of two sample proportions.

    example: $H_0: p_1 = p_2$

    Use the z test:

    $$z = \frac{(\hat{p}_1 - \hat{p}_2) - (p_1 - p_2)}{\sqrt{\overline{p}\,\overline{q}\left(\dfrac{1}{n_1} + \dfrac{1}{n_2}\right)}}$$

    where $\overline{p} = \dfrac{X_1 + X_2}{n_1 + n_2} \qquad \hat{p}_1 = \dfrac{X_1}{n_1}$

    $\overline{q} = 1 - \overline{p} \qquad \hat{p}_2 = \dfrac{X_2}{n_2}$

CHAPTER

11

Correlation and Regression

11–1 INTRODUCTION

In the previous chapters, two areas of inferential statistics, hypothesis testing and confidence intervals, were explained. Another area of inferential statistics involves determining whether a relationship between two or more variables exists. For example, a businessperson may want to know whether the volume of sales for a given month is related to the amount of advertising the firm does that month. Educators are interested in determining whether the number of hours a student studies is related to the student's score on a particular exam. Medical researchers are interested in questions such as "Is caffeine related to heart damage?" or "Is there a relationship between a person's age and his or her blood pressure?" A zoologist may want to know whether the birth weight of a certain animal is related to the life span of the animal. These are only a few of the many questions that can be answered by using the techniques of correlation and regression analysis. **Correlation** is a statistical method used to determine whether a relationship between variables exists. **Regression** is a statistical method used to describe the nature of the relationship between variables—i.e., a positive or negative, linear or nonlinear relationship.

The purpose of this chapter is to answer the following questions statistically:

1. Are two or more variables related?
2. If so, what is the strength of the relationship?
3. What type of relationship exists?
4. What kind of predictions can be made from the relationship?

In order to answer the first two questions, statisticians use a measure to determine whether two or more variables are related and also to determine the strength of the relationship between or among the variables. This measure is called a *correlation coefficient*. For example, there are many variables that contribute to heart disease, such as lack of exercise, smoking, heredity, age, stress, and diet. Of these variables, some are more important than others; therefore, if a physician is interested in helping a patient, the physician must know which factors are most important.

Another consideration in determining relationships is to ascertain what type of relationship exists. There are two types of relationships: simple and multiple. In a **simple relationship,** there are only two variables under study. For example, a manager may wish to see whether the number of years the salespeople have been working for the company has anything to do with the amount of sales of the representatives. This type of study involves a simple relationship, since there are only two variables, years of experience and amount of sales.

In **multiple relationships,** many variables are under study. For example, an educator may wish to investigate the relationship between a student's success in college and factors such as the number of hours devoted to studying, the student's IQ, and the student's high school background. This type of study involves several variables.

Simple relationships can also be positive or negative. A **positive relationship** exists when both variables increase or decrease at the same time. For instance, a person's height and weight are related; and the relationship is positive, since the taller a person is, generally, the more the person weighs. In a **negative relationship,** as one variable increases, the other variable decreases, and vice versa. For example, if one compares the strength of people over 60 years of age, one will find that as age increases, strength generally decreases. The word *generally* is used here since there are exceptions.

Finally, the fourth question asks what type of predictions can be made. Predictions are made in all areas and on a daily basis. Examples include weather forecasting, stock market analyses, sales predictions, crop predictions, gasoline predictions, and sports predictions. Some predictions are more accurate than others, due to the strength of the relationship. That is, the stronger the relationship is between variables, the more accurate the prediction is.

11-2 SCATTER PLOTS

In simple correlation and regression studies, the researcher collects data on two variables to see whether a relationship exists between the variables. For example, if a researcher wishes to see whether there is a relationship between number of hours studied by students and their test scores on an exam, he must select a random sample of students, determine the hours each studied, and obtain their grades on the exam. A table can be made for the data, as shown here.

Student	Hours Studied, x	Grade, y (%)
A	6	82
B	2	63
C	1	57
D	5	88
E	2	68
F	3	75

The two variables for this study are called the independent variable and the dependent variable. The **independent variable** is the variable in regression that can be controlled or manipulated. In this case, the variable "number of hours studied" is the independent variable and is designated as the x variable. The **dependent variable** is the variable in regression that cannot be controlled or manipulated. The grade the student received on the exam is the dependent variable, designated as the y variable. The reason for this distinction between the variables is that one assumes that the grade the student earns *depends* on the number of hours the student studied. Also, one assumes that, to some extent, the student can regulate or *control* the number of hours he or she studies for the exam.

The determination of the x and y variables is not always clear-cut and sometimes is an arbitrary decision. For example, if a researcher studies the effects of age on a person's blood pressure, the researcher can generally assume that a person's age affects the person's blood pressure. Hence, the variable "age" can be called the independent variable and the variable "blood pressure" can be called the dependent variable. On the other hand, if a researcher is studying the attitudes of husbands on a certain issue and the attitudes of their wives on the same issue, it is difficult for the researcher to say which variable is the independent variable and which is the dependent variable. In this study, the researcher can arbitrarily designate the variables as independent and dependent.

The independent and dependent variables can be plotted on a graph called a scatter plot. A scatter plot is an aid for understanding the correlation and regression techniques.

A scatter plot is a graph of the independent and dependent variables in regression and correlation analysis.

The procedure for drawing a scatter plot is shown in the next three examples.

EXAMPLE 11-1 Construct a scatter plot for the data obtained in a study of age and systolic blood pressure of six randomly selected subjects. The data are shown in the following table.

Subject	Age, x	Pressure, y
A	43	128
B	48	120
C	56	135
D	61	143
E	67	141
F	70	152

Solution   **STEP 1**   Draw and label the x and y axes, as shown in Figure 11-1.

STEP 2   Plot each point on the graph, as shown in Figure 11-2.

FIGURE 11-1
Labeling the Axes for the Scatter Plot for Example 11-1

FIGURE 11-2
Scatter Plot for Example 11-1

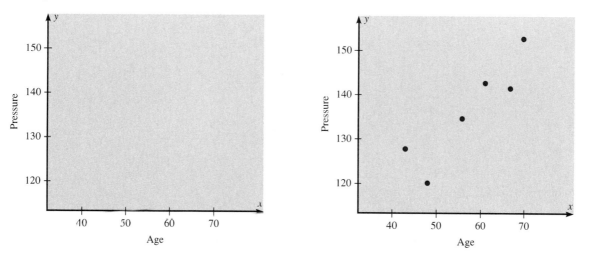

EXAMPLE 11-2 Construct a scatter plot for the data obtained in a study on the number of absences and the final grades of seven randomly selected students from a statistics class. The data are shown below.

Student	Number of Absences, x	Final Grade, y (%)
A	6	82
B	2	86
C	15	43
D	9	74
E	12	58
F	5	90
G	8	78

Solution **STEP 1** Draw and label the *x* and *y* axes.

STEP 2 Plot each point on the graph, as shown in Figure 11–3.

FIGURE 11–3
Scatter Plot for Example 11–2

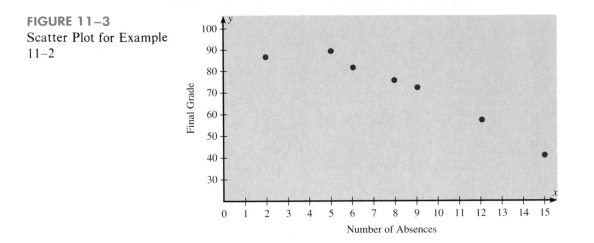

EXAMPLE 11–3 Construct a scatter plot for the data obtained in a study on the number of hours a person exercises each week and the amount of milk (in ounces) each person consumes per week. The data follow.

Subject	Hours, *x*	Amount, *y*
A	3	48
B	0	8
C	2	32
D	5	64
E	8	10
F	5	32
G	10	56
H	2	72
I	1	48

Solution **STEP 1** Draw and label the *x* and *y* axes.

STEP 2 Plot each point on the graph, as shown in Figure 11–4.

FIGURE 11–4
Scatter Plot for Example 11–3

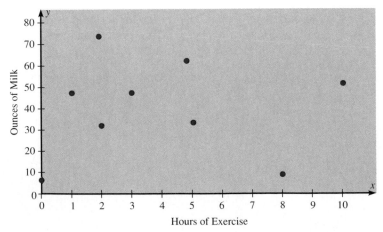

After the plot is drawn, it should be analyzed to determine which type of relationship, if any, exists. For example, the plot shown in Figure 11–2 suggests a positive relationship, since as a person's age increases, the blood pressure tends to increase also. The plot of the data shown in Figure 11–3 suggests a negative relationship, since as the number of absences increases, the final grade decreases. Finally, the plot of the data shown in Figure 11–4 shows no specific type of relationship, since no pattern is discernible.

Note that the data shown in Figures 11–2 and 11–3 also suggest a linear relationship, since the points seem to fit a straight line, although not perfectly. Sometimes a scatter plot, such as the one in Figure 11–5, will show a curvilinear relationship between the data. In this situation, the methods shown in this section and the next cannot be used; instead, methods for curvilinear relationships can be used. These methods are beyond the scope of this book.

FIGURE 11–5
Scatter Plot Suggesting
a Curvilinear
Relationship

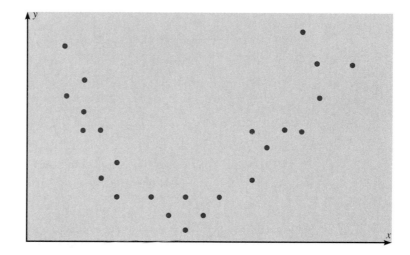

11–3 CORRELATION

Correlation Coefficient

As indicated in Section 11–1, statisticians use a measure called the correlation coefficient to determine the strength of the relationship between two variables. There are several different types of correlation coefficients. The one explained in this section is called the **Pearson product moment correlation coefficient** (PPMC), named after statistician Karl Pearson, who pioneered the research in this area.

The **correlation coefficient** computed from the sample data measures the strength and direction of a relationship between two variables. The symbol for the sample correlation coefficient is r. The symbol for the population correlation coefficient is ρ (Greek letter rho).

The range of the correlation coefficient is from −1 *to* +1. If there is a *strong, positive linear relationship* between the variables, the value of *r* will be close to +1. If there is a *strong, negative linear relationship* between the variables, the value of r will be close to -1. When there is no linear relationship between the variables or only a weak relationship, the value of r will be close to 0. See Figure 11–6.

FIGURE 11–6

Range of Values for the Correlation Coefficient

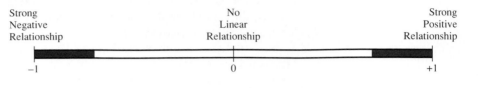

There are several ways to compute the value of the correlation coefficient. One simple method is to use the formula shown next.

> ## Formula for the Correlation Coefficient r
>
> $$r = \frac{n(\Sigma\, xy) - (\Sigma\, x)(\Sigma\, y)}{\sqrt{[n(\Sigma\, x^2) - (\Sigma\, x)^2][n(\Sigma\, y^2) - (\Sigma\, y)^2]}}$$
>
> where n is the number of data pairs.

The formula looks somewhat complicated, but using a table to compute the values, as shown in the next example, makes it somewhat easier to determine the value of r.

EXAMPLE 11–4 Compute the value of the correlation coefficient for the data obtained in the study of age and blood pressure given in Example 11–1.

Solution **STEP 1** Make a table, as shown here.

Subject	Age, x	Pressure, y	xy	x^2	y^2
A	43	128			
B	48	120			
C	56	135			
D	61	143			
E	67	141			
F	70	152			

STEP 2 Find the products of the x and y values and place the products in the column labeled xy.

$$43 \cdot 128 = 5504, \quad \text{etc.}$$

STEP 3 Square the x values and place them in the column labeled x^2.

$$43^2 = 1849, \quad \text{etc.}$$

STEP 4 Square the y values and place them in the column labeled y^2.

$$128^2 = 16{,}384, \quad \text{etc.}$$

STEP 5 Find the sum of each column. The completed table is shown next.

Subject	Age, x	Pressure, y	xy	x^2	y^2
A	43	128	5,504	1,849	16,384
B	48	120	5,760	2,304	14,400
C	56	135	7,560	3,136	18,225
D	61	143	8,723	3,721	20,449
E	67	141	9,447	4,489	19,881
F	70	152	10,640	4,900	23,104
	$\Sigma x = 345$	$\Sigma y = 819$	$\Sigma xy = 47,634$	$\Sigma x^2 = 20,399$	$\Sigma y^2 = 112,443$

STEP 6 Substitute in the formula and solve for r.

$$r = \frac{n(\Sigma xy) - (\Sigma x)(\Sigma y)}{\sqrt{[n(\Sigma x^2) - (\Sigma x)^2][n(\Sigma y^2) - (\Sigma y)^2]}}$$

$$= \frac{(6)(47,634) - (345)(819)}{\sqrt{[(6)(20,399) - (345)^2][(6)(112,443) - (819)^2]}} = 0.897$$

The correlation coefficient suggests a strong positive relationship. ◄

EXAMPLE 11–5 Compute the value of the correlation coefficient for the data obtained in the study of the number of student absences and the final grade of the seven students in the statistics class given in Example 11–2.

Solution **STEP 1** Make a table, which is shown completed in Step 5.

STEP 2 Find the products of the x and y values and place the products in the column labeled xy.

$$6 \cdot 82 = 492, \quad \text{etc.}$$

STEP 3 Square the x values and place them in the column labeled x^2.

$$6^2 = 36, \quad \text{etc.}$$

STEP 4 Square the y values and place them in the column labeled y^2.

$$82^2 = 6724, \quad \text{etc.}$$

STEP 5 Find the sum of each column. The completed table follows.

Student	Number of Absences, x	Final Grade, y (%)	xy	x^2	y^2
A	6	82	492	36	6,724
B	2	86	172	4	7,396
C	15	43	645	225	1,849
D	9	74	666	81	5,476
E	12	58	696	144	3,364
F	5	90	450	25	8,100
G	8	78	624	64	6,084
	$\Sigma x = 57$	$\Sigma y = 511$	$\Sigma xy = 3745$	$\Sigma x^2 = 579$	$\Sigma y^2 = 38,993$

STEP 6 Substitute in the formula and solve for r.

$$r = \frac{n(\Sigma\, xy) - (\Sigma\, x)(\Sigma\, y)}{\sqrt{[n(\Sigma\, x^2) - (\Sigma\, x)^2][n(\Sigma\, y^2) - (\Sigma\, y)^2]}}$$

$$= \frac{(7)(3745) - (57)(511)}{\sqrt{[(7)(579) - (57)^2][(7)(38{,}993) - (511)^2]}} = -0.944$$

The value of r suggests a strong negative relationship between a student's final grade and the number of absences a student has. ◄

EXAMPLE 11–6 Compute the value of the correlation coefficient for the data given in Example 11–3 for the number of hours a person exercises and the amount of milk a person consumes per week.

Solution **STEP 1** Make a table, which is shown completed in Step 5.

STEP 2 Find the products of the x and y values and place the products in the column labeled xy.

$$3 \cdot 48 = 144, \qquad \text{etc.}$$

STEP 3 Square the x values and place them in the column labeled x^2.

$$3^2 = 9, \qquad \text{etc.}$$

STEP 4 Square the y values and place them in the column labeled y^2.

$$48^2 = 2304, \qquad \text{etc.}$$

STEP 5 Find the sum of each column. The completed table follows.

Subject	Hours, x	Amount, y	xy	x^2	y^2
A	3	48	144	9	2,304
B	0	8	0	0	64
C	2	32	64	4	1,024
D	5	64	320	25	4,096
E	8	10	80	64	100
F	5	32	160	25	1,024
G	10	56	560	100	3,136
H	2	72	144	4	5,184
I	1	48	48	1	2,304
	$\Sigma x = 36$	$\Sigma y = 370$	$\Sigma xy = 1520$	$\Sigma x^2 = 232$	$\Sigma y^2 = 19{,}236$

STEP 6 Substitute in the formula and solve for r.

$$r = \frac{n(\Sigma\, xy) - (\Sigma\, x)(\Sigma\, y)}{\sqrt{[n(\Sigma\, x^2) - (\Sigma\, x)^2][n(\Sigma\, y^2) - (\Sigma\, y)^2]}}$$

$$= \frac{(9)(1520) - (36)(370)}{\sqrt{[(9)(232) - (36)^2][9(19{,}236) - (370)^2]}} = 0.067$$

The value of r indicates a very weak positive relationship between the variables. ◄

In Example 11–4, the value of r was high (close to 1.00); the value of r in Example 11–6 was much lower (close to 0). This question then arises: "When is the value of r due to chance and when does it suggest a significant relationship between the variables?" This question will be answered next.

The Significance of the Correlation Coefficient

As stated before, the range of the correlation coefficient is between -1 and $+1$. When the value of r is near $+1$ or -1, there is a strong relationship. When the value of r is near 0, the relationship is weak or nonexistent. Since the value of r is computed from data obtained from samples, there are two possibilities—either the value of r is high enough to conclude that there is a significant relationship between the variables, or the value of r is due to chance.

In order to make this determination, one uses a hypothesis-testing procedure. This procedure, which follows, is similar to the one used in the previous chapters.

STEP 1 State the hypotheses.

STEP 2 Find the critical values.

STEP 3 Compute the test value.

STEP 4 Make the decision.

STEP 5 Summarize the results.

There are several methods that can be used to test the significance of the correlation coefficient. Two methods will be shown in this section. The first method uses the t test.

Formula for the *t* test for the Correlation Coefficient

$$t = r\sqrt{\frac{n-2}{1-r^2}}$$

with degrees of freedom equal to $n - 2$.

Although hypothesis tests can be one-tailed, most hypotheses involving the correlation coefficient are two-tailed. Recall that ρ represents the population correlation coefficient. Also, if there is no relationship, the value of the correlation coefficient will be 0. Hence, the hypotheses will be

$$H_0: \rho = 0 \qquad H_1: \rho \neq 0$$

One does not have to identify the claim here, since the question will always be whether or not there is a significant relationship between the variables.

The two-tailed critical values are used. These values are found in Table F in the Appendix. Also, when one is testing the significance of a correlation coefficient, both variables, x and y, must come from normally distributed populations.

EXAMPLE 11–7 Test the significance of the correlation coefficient found in Example 11–4. Use $\alpha = 0.05$ and $r = 0.897$.

Solution **STEP 1** State the hypotheses.

$$H_0: \rho = 0 \qquad \text{and} \qquad H_1: \rho \neq 0$$

STEP 2 Find the critical values. Since $\alpha = 0.05$ and there are $6 - 2 = 4$ degrees of freedom, the critical values obtained from Table F are ± 2.776, as shown in Figure 11–7.

FIGURE 11–7
Critical Values for
Example 11–7

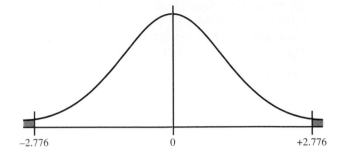

-2.776   0   +2.776

STEP 3 Compute the test value.

$$t = r \sqrt{\frac{n-2}{1-r^2}} = 0.897 \sqrt{\frac{6-2}{1-(0.897)^2}} = 4.059$$

STEP 4 Make the decision. Reject the null hypothesis, since the test value falls in the critical region, as shown in Figure 11–8.

FIGURE 11–8
Test Value for Example
11–7

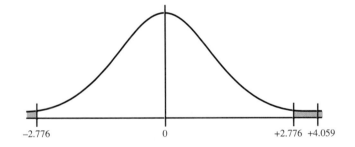

-2.776   0   +2.776   +4.059

STEP 5 Summarize the results. There is a significant relationship between the variables. ◀

The second method of testing the significance of r is to use Table I in the Appendix. This table shows the values of the correlation coefficient that are significant for a specific α level and a specific number of degrees of freedom. For example, for 7 degrees of freedom and $\alpha = 0.05$, the table gives a critical value of 0.666. Any value of r greater than $+0.666$ or less than -0.666 will be significant, and the null hypothesis will be rejected. See Figure 11–9. When Table I is used, one need not compute the t test value.

FIGURE 11–9
Finding the Critical
Value from Table I

d.f.	$\alpha = 0.05$	$\alpha = 0.01$
1		
2		
3		
4		
5		
6		
7	.666	

EXAMPLE 11–8   Using Table I, test the significance of the correlation coefficient, $r = 0.067$, obtained in Example 11–6, at $\alpha = 0.01$.

Solution   Since the sample size is 9, there are 7 degrees of freedom. When $\alpha = 0.01$ and with 7 degrees of freedom, the value obtained from Table I is 0.798. For a significant relationship, a value of r greater than $+0.798$ or less than -0.798 is needed. Since $r = 0.067$, the null hypothesis is not rejected. Hence, there is not enough evidence to say that there is a significant relationship between the variables. See Figure 11–10.

FIGURE 11–10
Rejection and
Nonrejection Regions for
Example 11–8

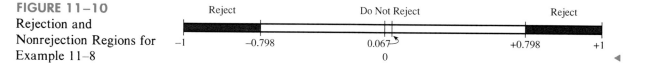

| Reject | Do Not Reject | Reject |

-1 -0.798 0.067 $+0.798$ $+1$
0

Correlation and Causation

Researchers must understand the nature of the relationship between the independent variable x and the dependent variable y. When a hypothesis test indicates that a significant relationship exists between the variables, researchers must consider the possibilities outlined below.

Possible Relationships Between Variables

When the null hypothesis has been rejected for a specific α value, any of the following five possibilities can exist.

1. *There is a direct cause-and-effect relationship between the variables.* That is, x causes y. For example, water causes plants to grow, poison causes death, and heat causes ice to melt.
2. *There is a reverse cause-and-effect relationship between the variables.* For example, suppose a researcher feels that excessive coffee consumption causes nervousness in subjects consuming the coffee, but the researcher fails to consider that the reverse situation may occur. That is, it may be that an extremely nervous person craves coffee to calm his or her nerves.   *(continued)*

3. *The relationship between the variables may be caused by a third variable.* For example, if a statistician correlated the number of deaths due to drowning and the number of cans of soft drink consumed during the summer, he or she would probably find a significant relationship. However, the soft drink is not necessarily responsible for the deaths, since both variables may be related to heat and humidity.

4. *There may be a complexity of interrelationships among many variables.* For example, a researcher may find a significant relationship between students' high school grades and college grades. But there probably are many other variables involved, such as IQ, hours of study, influence of parents, motivation, age, and instructors.

5. *The relationship may be coincidental.* For example, a researcher may be able to find a significant relationship between the increase in the number of people who are exercising and the increase in the number of people who are committing crimes. But common sense dictates that any relationship between these two variables must be due to coincidence.

Thus, when the null hypothesis is rejected, the researcher must consider all possibilities and select the appropriate one as determined by the study.

EXERCISES

11–1. What is meant by the statement that two variables are related?

11–2. How is a relationship between two variables measured in statistics? Explain.

11–3. What is the symbol for the sample correlation coefficient? The population correlation coefficient?

11–4. What is the range of values for the correlation coefficient?

11–5. What is meant when the relationship between the two variables is positive? Negative?

11–6. Give an example of two variables that are positively correlated. That are negatively correlated.

11–7. Give an example of a correlation study, and identify the independent and dependent variables.

11–8. What is the diagram of the independent and dependent variables called? Why is drawing this diagram important?

11–9. What is the name of the correlation coefficient used in this section?

11–10. What statistical test is used to test the significance of the correlation coefficient?

11–11. When two variables are correlated, can the researcher be sure that one variable causes the other variable? Why or why not?

For Exercises 11–12 through 11–27, perform the following steps.

a. Draw the scatter plot for the variables.
b. Compute the value of the correlation coefficient.
c. State the hypotheses.
d. Test the significance of the correlation coefficient at $\alpha = 0.05$, using Table I.
e. Give a brief explanation of the type of relationship.

11–12. A manager wishes to find out whether there is a relationship between the number of radio ads aired per week and the amount of sales (in thousands of dollars) of a product. The data for the sample follow. (The information in this exercise will be used for Exercises 11–40 and 11–64.)

No. of Ads, x	2	5	8	8	10	12
Sales, y	$2	$4	$7	$6	$9	$10

Speaking of **Statistics**

Coffee Not Disease Culprit, Study Says

NEW YORK (AP)—Two new studies suggest that coffee drinking, even up to 5½ cups per day, does not increase the risk of heart disease, and other studies that claim to have found increased risks might have missed the true culprits, a researcher says.

"It might not be the coffee cup in one hand, it might be the cigarette or coffee roll in the other," said Dr. Peter W. F. Wilson, the author of one of the new studies.

He noted in a telephone interview Thursday that many coffee drinkers, particularly heavy coffee drinkers, are smokers. And one of the new studies found that coffee drinkers had excess fat in their diets.

The findings of the new studies conflict sharply with a study reported in November 1985 by Johns Hopkins University scientists in Baltimore.

The Hopkins scientists found that coffee drinkers who consumed five or more cups of coffee per day had three times the heart-disease risk of non–coffee drinkers.

The reason for the discrepancy appears to be that many of the coffee drinkers in the Hopkins study also smoked—and it was the smoking that increased their heart-disease risk, said Wilson.

Wilson, director of laboratories for the Framingham Heart Study in Framingham, Mass., said Thursday at a conference sponsored by the American Heart Association in Charleston, S.C., that he had examined the coffee intake of 3,937 participants in the Framingham study during 1956–66 and an additional 2,277 during the years 1972–1982.

In contrast to the subjects in the Hopkins study, most of these coffee drinkers consumed two or three cups per day, Wilson said. Only 10 percent drank six or more cups per day.

He then looked at blood cholesterol levels, heart and blood vessel disease in the two groups. "We ran these analyses for coronary heart disease, heart attack, sudden death and stroke and in absolutely every analysis, we found no link with coffee," Wilson said.

He found that coffee consumption was linked to a significant decrease in total blood cholesterol in men, and to a moderate increase in total cholesterol in women.

Source: Associated Press. Reprinted with permission

In correlation and regression studies, it is difficult to control all variables. This study shows some of the consequences when researchers overlook certain aspects in studies. Suggest ways that the extraneous variables might be controlled in future studies.

11–13. A researcher wishes to determine whether a person's age is related to the number of hours he or she watches television daily. The data for the sample follow. (The information in this exercise will be used for Exercises 11–41, 11–64, and 11–81.)

Age, x	18	24	36	40	58
Hours, y	3.9	2.6	2	2.3	1.2

11–14. A study was conducted to determine the relationship between a person's monthly income and the number of meals that person eats away from home per month. The data from the sample are shown here. (The information in this exercise will be used for Exercises 11–42, 11–64, and 11–82.)

Income, x	$500	$1200	$1500	$945	$850	$400	$540
Meals, y	8	12	16	10	9	3	7

11–15. The director of an alumni association for a small college wants to determine whether there is any type of relationship between the amount of an alumnus's contribution and the years the alumnus has been out of school. The data follow. (The information in this exercise will be used for Exercises 11–43, 11–65, and 11–83.)

Years, x	1	5	3	10	7	6
Contribution, y	$500	$100	$300	$50	$75	$80

11–16. A store manager wishes to find out whether there is a relationship between the age of his employees and the number of sick days they take each year. The data for the sample follow. (The information in this exercise will be used for Exercises 11–44, 11–65, and 11–84.)

Age	18	26	39	48	53	58
Days	16	12	9	5	6	2

11–17. An educator wants to see how strong the relationship is between a student's IQ and his or her grade point average. The data obtained from the sample follow. (The information in this exercise will be used for Exercises 11–45 and 11–65.)

IQ, x	98	105	100	100	106	95	116	112
GPA, y	2.1	2.4	3.2	2.7	2.2	2.3	3.8	3.4

11–18. An English instructor is interested in finding the strength of a relationship between the final exam grades of students enrolled in Composition I and Composition II classes. The data are given below in percentages. (The information in this exercise will be used for Exercises 11–46 and 11–66.)

Comp I	83	97	80	95	73	78	91	86
Comp II	78	95	83	97	78	72	90	80

11–19. An insurance company wants to determine the strength of the relationship between the number of hours a person works per week and the number of injuries or accidents that person has. The data follow. (The information in this exercise will be used for Exercises 11–47 and 11–66.)

Hours Worked, x	40	32	36	44	41
No. of Accidents, y	1	0	3	8	5

11–20. An emergency service wishes to see whether a relationship exists between the outside temperature and the number of emergency calls it receives. The data follow. (The information in this exercise will be used for Exercises 11–48 and 11–66.)

Temperature, x	68	74	82	88	93	99	101
No. of Calls, y	7	4	8	10	11	9	13

11–21. An educator wants to see how the number of hours of tutoring a student receives affects the final grade of the student. The data obtained from a sample follow. (The information in this exercise will be used for Exercise 11–49.)

No. of Hours, x	10	8	12	4	8	5
Final Grade, y	83	91	92	71	82	75

11–22. A medical researcher wishes to determine how the dosage (in milligrams) of a drug affects the heart rate of the patient. The data for seven patients are given below. (The information in this exercise will be used for Exercise 11–50.)

Drug Dosage, x	0.125	0.20	0.25	0.30	0.35	0.40	0.50
Heart Rate, y	95	90	93	92	88	80	82

11–23. A physician wishes to know whether there is a relationship between the number of pounds a mother gained during her pregnancy and the birth weight of her baby. The data are shown below. (The information in this exercise will be used for Exercise 11–51.)

Mother's Weight Gain, x	22	30	38	15	27	35	19	23
Baby's Weight, y	6.8	7.2	8.1	7.6	8.3	8.9	9.2	8.6

11–24. A sales manager wishes to determine the relationship between the number of miles his sales representatives travel and the amount of their monthly sales (in hundreds of dollars). The data are shown below. (The information in this exercise will be used in Exercise 11–52.)

Miles Traveled, x	50	250	120	300	200
Sales, y	$25	$175	$100	$210	$150

11-25. A psychologist selects six families with two children each, a boy and a girl. She is interested in comparing the IQs of each gender to determine whether there is a relationship between them. The data are given below. (The information in this exercise will be used for Exercise 11-53.)

IQ Females, x	107	95	116	109	101	98
IQ Males, y	107	102	112	104	105	103

11-26. A researcher wishes to determine whether there is a relationship between the age of a copy machine and the monthly maintenance cost. The data follow. (The information in this exercise will be used for Exercise 11-54.)

Age, x	3	5	2	1	2	4	3
Cost, y	$80	$100	$75	$60	$80	$93	$84

11-27. A study was conducted at a large hospital to determine whether there was a relationship between the number of years nurses were employed and the number of nurses who voluntarily resign. The results are shown below. (The information in this exercise will be used for Exercise 11-55.)

Years Employed, x	5	6	3	7	8	2
Resignations, y	7	4	7	1	2	8

*11-28. There are several formulas for computing r; one is

$$r = \frac{\Sigma\,(x - \bar{x})(y - \bar{y})}{(n - 1)(s_x)(s_y)}$$

Using the data in Exercise 11-27, compute r by using this formula. Compare the results.

*11-29. Compute r for the following data set. What is the value of r? Explain the reason for this value of r. Now, interchange the values of x and y and compute r again. Compare this value with the previous one. Explain the results of the comparison.

x	1	2	3	4	5
y	3	5	7	9	11

*11-30. Compute r for the following data and test the hypothesis H_0: $\rho = 0$. Draw the scatter plot; then explain the results.

x	-3	-2	-1	0	1	2	3
y	9	4	1	0	1	4	9

11-4 REGRESSION

In studying relationships between two variables, the researcher, after collecting the data, constructs a scatter plot. The purpose of the scatter plot, as indicated previously, is to determine the nature of the relationship. The possibilities include a positive linear relationship, a negative linear relationship, a curvilinear relationship, or no discernible relationship. After the scatter plot is drawn, the next step is to compute the value of the correlation coefficient and to test the significance of the relationship. If the value of the correlation coefficient is significant, the next step is to determine the equation of the **regression line,** which is the line of best fit of the data. (*Note:* Determining the regression line when r is not significant and then making predictions using the regression line is meaningless.) The purpose of the regression line is to enable the researcher to see the trend and make predictions on the basis of the data.

Line of Best Fit   Figure 11–11 shows a scatter plot for the data of two variables. It also shows that several lines can be drawn on the graph near the points. Given a scatter plot, one must be able to draw the *line of best fit. Best fit* means that the sum of the squares of the vertical distances from each point to the line are at a minimum. The reason one needs a line of best fit is that the values of *y* will be predicted from the values of *x;* hence, the closer the points are to the line, the better the fit and the prediction will be. See Figure 11–12.

FIGURE 11–11
Scatter Plot with Three
Lines Fit to the Data

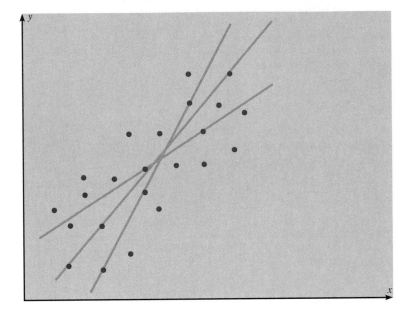

FIGURE 11–12
Line of Best Fit for a Set
of Data Points

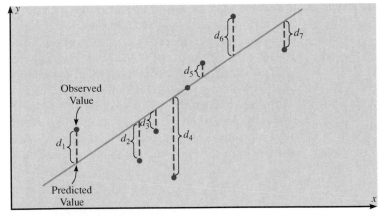

Figure 11–13 shows the relationship between the values of the correlation coefficient and the variability of the scores about the regression line. The closer the points fit the regression line, the higher the absolute value of *r* is and the closer it will be to +1 or −1. Figure 11–13a shows *r* as positive and approximately equal to 0.50; Figure 11–13b shows *r* approximately equal to 0.90; and Figure 11–13c shows *r* equal to +1. For positive values of *r,* the line slopes upward from left to right.

Speaking of **Statistics**

Study Links Analgesic with Kidney Damage

BOSTON—Long-time daily users of acetaminophen, one of the nation's most popular over-the-counter painkillers, face three times the usual risk of kidney damage, a study concludes.

Acetaminophen is sold under many brand names, including Tylenol, Anacin-3 and Datril. The medicine appears to be safe when used occasionally, but was associated with kidney failure and less severe forms of the disease when taken every day for years.

While the findings, published in today's *New England Journal of Medicine,* raise concern about overuse of painkillers, experts cautioned that they should be considered tentative until confirmed by other studies.

Acetaminophen abuse could be responsible for "a small but not inconsequential amount" of kidney disease among Americans, according to the study's director, Dr. Dale P. Sandler of the National Institute of Environmental Health Science.

However, she added, "This study does not demonstrate any effect for acetaminophen when taken as directed, which is when needed for symptoms, but when symptoms persist, you should seek medical advice."

Source: Associated Press. Reprinted with permission.

This study suggests a relationship between the overuse of painkillers and possible kidney damage. Describe the independent variable and the dependent variable. Is the relationship positive or negative?

FIGURE 11–13
Relationship Between the Correlation Coefficient and the Line of Best Fit

(a) r = 0.50

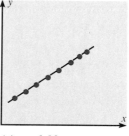

(b) r = 0.90

(c) r = 1.00

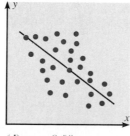

(d) r = –0.50

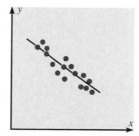

(e) r = –0.90

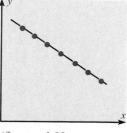

(f) r = –1.00

Figure 11–13d shows r as negative and approximately equal to -0.50; Figure 11–13e shows r approximately equal to -0.90; and Figure 11–13f shows r equal to -1. For negative values of r, the line slopes downward from left to right.

Determination of the Regression Line Equation

In algebra, the equation of a line is usually given as $y = mx + b$, where m is the slope of the line and b is the y intercept. (Students who need an algebraic review of the properties of a line should refer to Appendix A–3 before studying this section.) In statistics, the equation of the regression line is written as $y' = a + bx$, where a is the y intercept and b is the slope of the line. See Figure 11–14.

FIGURE 11–14
A Line as Represented in Algebra and in Statistics

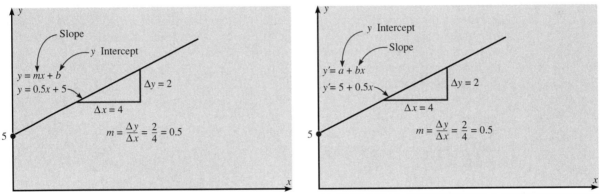

(a) *ALGEBRA OF A LINE*   (b) *STATISTICAL NOTATION FOR A REGRESSION LINE*

There are several methods for finding the equation of the regression line. In this section, two formulas will be given. *These formulas use the same values that are used in computing the value of the correlation coefficient.* The mathematical development of these formulas requires calculus and is beyond the scope of this textbook.

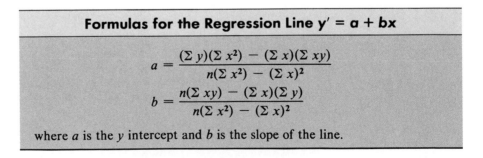

Formulas for the Regression Line $y' = a + bx$

$$a = \frac{(\Sigma y)(\Sigma x^2) - (\Sigma x)(\Sigma xy)}{n(\Sigma x^2) - (\Sigma x)^2}$$

$$b = \frac{n(\Sigma xy) - (\Sigma x)(\Sigma y)}{n(\Sigma x^2) - (\Sigma x)^2}$$

where a is the y intercept and b is the slope of the line.

EXAMPLE 11-9   Find the equation of the regression line for the data in Example 11-4, and graph the line on the scatter plot of the data.

Solution   The values needed for the equation are $n = 6$, $\Sigma\ x = 345$, $\Sigma\ y = 819$, $\Sigma\ xy = 47{,}634$, and $\Sigma\ x^2 = 20{,}399$. Substituting in the formulas, one gets

$$a = \frac{(\Sigma\ y)(\Sigma\ x^2) - (\Sigma\ x)(\Sigma\ xy)}{n(\Sigma\ x^2) - (\Sigma\ x)^2} = \frac{(819)(20{,}399) - (345)(47{,}634)}{(6)(20{,}399) - (345)^2} = 81.048$$

$$b = \frac{n(\Sigma\ xy) - (\Sigma\ x)(\Sigma\ y)}{n(\Sigma\ x^2) - (\Sigma\ x)^2} = \frac{(6)(47{,}634) - (345)(819)}{(6)(20{,}399) - (345)^2} = 0.964$$

Hence, the equation of the regression line, $y' = a + bx$, is

$$y' = 81.048 + 0.964x$$

The graph of the line is shown in Figure 11-15.

FIGURE 11-15
Regression Line for
Example 11-9

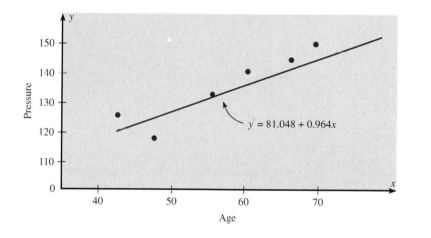

$y' = 81.048 + 0.964x$

EXAMPLE 11-10   Find the equation of the regression line for the data in Example 11-5, and graph the line on the scatter plot.

Solution   The values needed for the equation are $n = 7$, $\Sigma\ x = 57$, $\Sigma\ y = 511$, $\Sigma\ xy = 3745$, and $\Sigma\ x^2 = 579$. Substituting in the formulas, one gets

$$a = \frac{(\Sigma\ y)(\Sigma\ x^2) - (\Sigma\ x)(\Sigma\ xy)}{n(\Sigma\ x^2) - (\Sigma\ x)^2} = \frac{(511)(579) - (57)(3745)}{(7)(579) - (57)^2} = 102.493$$

$$b = \frac{n(\Sigma\ xy) - (\Sigma\ x)(\Sigma\ y)}{n(\Sigma\ x^2) - (\Sigma\ x)^2} = \frac{(7)(3745) - (57)(511)}{(7)(579) - (57)^2} = -3.622$$

Hence, the equation of the regression line, $y' = a + bx$, is

$$y' = 102.493 - 3.622x$$

The graph of the line is shown in Figure 11–16.

FIGURE 11–16
Regression Line for
Example 11–10

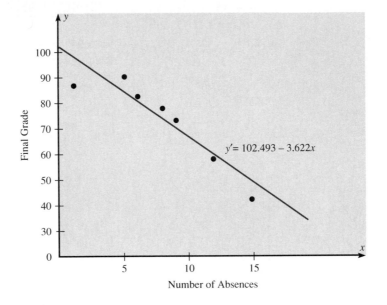

$y' = 102.493 - 3.622x$

Number of Absences

The sign of the correlation coefficient and of the slope of the regression line will always be the same. That is, if r is positive, then b will be positive; and if r is negative, then b will be negative.

The regression line can be used to make predictions for the dependent variable. The method for making predictions is shown in the next example.

EXAMPLE 11–11 Using the equation of the regression line found in Example 11–9, predict the blood pressure for a person who is 50 years old.

Solution Substituting 50 for x in the regression line $y' = 81.048 + 0.964x$ gives

$$y' = 81.048 + (0.964)(50) = 129.248$$

The value obtained in Example 11–11 is a point prediction, and with point predictions, no degree of accuracy or confidence can be determined. More information on prediction is given in Section 11–6, including prediction intervals.

For valid predictions, the value of the correlation coefficient must be significant. Also, two other assumptions must be met.

Assumptions for Valid Predictions in Regression

1. For any specific value of the independent variable x, the value of the dependent variable y must be normally distributed about the regression line. See Figure 11–17a.
2. The standard deviation of each of the dependent variables must be the same for each value of the independent variable. See Figure 11–17b.

FIGURE 11–17
Assumptions for Predictions

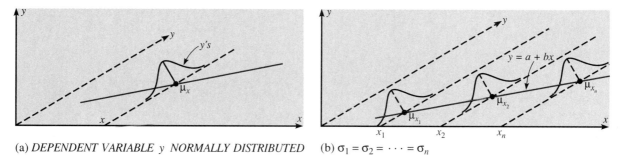

(a) *DEPENDENT VARIABLE y NORMALLY DISTRIBUTED*   (b) $\sigma_1 = \sigma_2 = \cdots = \sigma_n$

When predictions are made beyond the bounds of the data, caution should be used in the interpretation of the results. For example, in 1979, some experts predicted that the United States would run out of oil by the year 2003. This prediction was based on the current consumption and on known oil reserves at that time. However, since then, the automobile industry has produced many new fuel-efficient automobiles. Also, there are many as-yet undiscovered oil fields. Finally, science may someday discover a way to run an automobile on something as unlikely but as common as peanut oil. *Remember that when predictions are made, they are based on present conditions or on the premise that present trends will continue.* This assumption may or may not prove true in the future.

The procedures for finding the value of the correlation coefficient and the regression line equation are summarized in Procedure Table 9.

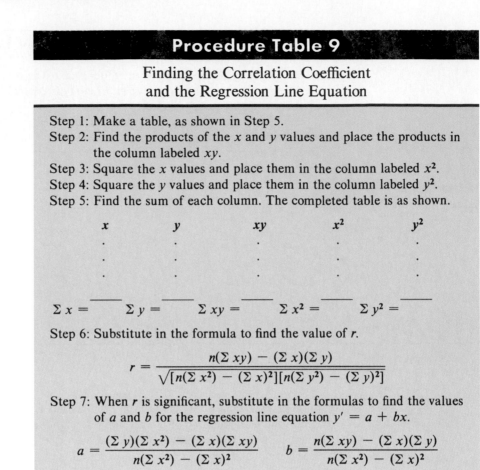

Procedure Table 9

Finding the Correlation Coefficient and the Regression Line Equation

Step 1: Make a table, as shown in Step 5.

Step 2: Find the products of the x and y values and place the products in the column labeled xy.

Step 3: Square the x values and place them in the column labeled x^2.

Step 4: Square the y values and place them in the column labeled y^2.

Step 5: Find the sum of each column. The completed table is as shown.

x	y	xy	x^2	y^2
.	.	.	.	.
.	.	.	.	.
.	.	.	.	.

$$\Sigma x = \underline{\quad}\quad \Sigma y = \underline{\quad}\quad \Sigma xy = \underline{\quad}\quad \Sigma x^2 = \underline{\quad}\quad \Sigma y^2 = \underline{\quad}$$

Step 6: Substitute in the formula to find the value of r.

$$r = \frac{n(\Sigma xy) - (\Sigma x)(\Sigma y)}{\sqrt{[n(\Sigma x^2) - (\Sigma x)^2][n(\Sigma y^2) - (\Sigma y)^2]}}$$

Step 7: When r is significant, substitute in the formulas to find the values of a and b for the regression line equation $y' = a + bx$.

$$a = \frac{(\Sigma y)(\Sigma x^2) - (\Sigma x)(\Sigma xy)}{n(\Sigma x^2) - (\Sigma x)^2} \qquad b = \frac{n(\Sigma xy) - (\Sigma x)(\Sigma y)}{n(\Sigma x^2) - (\Sigma x)^2}$$

EXERCISES

11–31. What two things should be done before one performs a regression analysis?

11–32. What are the assumptions for regression analysis?

11–33. What is the general form for the regression line used in statistics?

11–34. What is the symbol for the slope? For the y intercept?

11–35. What is meant by the *line of best fit?*

11–36. When all the points fall on the regression line, what is the value of the correlation coefficient?

11–37. What is the relationship of the sign of the correlation coefficient and the sign of the slope of the regression line?

11–38. As the value of the correlation coefficient increases from 0 to 1, or decreases from 0 to -1, how do the points of the scatter plot fit the regression line?

11–39. How is the value of the correlation coefficient related to the accuracy of the predicted value for a specific value of x?

11–40. For the data in Exercise 11–12, find the equation of the regression line, and predict y' when $x = 7$ ads.

11–41. For the data in Exercise 11–13, find the equation of the regression line, and predict y' when $x = 35$ years.

11–42. For the data in Exercise 11–14, find the equation of the regression line, and predict y' when $x = \$1100$.

11–43. For the data in Exercise 11–15, find the equation of the regression line, and predict y' when $x = 4$ years.

11–44. For the data in Exercise 11–16, find the equation of the regression line, and predict y' when $x = 47$ years.

11–45. For the data in Exercise 11–17, find the equation of the regression line, and predict y' when $x = 104$.

11–46. For the data in Exercise 11–18, find the equation of the regression line, and predict y' when $x = 88$.

11–47. For the data in Exercise 11–19, find the equation of the regression line, and predict y' when $x = 35$ hours.

11–48. For the data in Exercise 11–20, find the equation of the regression line, and predict y' when $x = 80$ degrees.

11–49. For the data in Exercise 11–21, find the equation of the regression line, and predict y' when $x = 9$ hours.

11–50. For the data in Exercise 11–22, find the equation of the regression line, and predict y' when $x = 0.27$ milligram.

11–51. For the data in Exercise 11–23, find the equation of the regression line, and predict y' when $x = 20$ pounds.

11–52. For the data in Exercise 11–24, find the equation of the regression line, and predict y' when $x = 100$ miles.

11–53. For the data in Exercise 11–25, find the equation of the regression line, and predict y' when $x = 104$.

11–54. For the data in Exercise 11–26, find the equation of the regression line, and predict y' when $x = 4$ years.

11–55. For the data in Exercise 11–27, find the equation of the regression line, and predict y' when $x = 4$ years.

LAFF - A - DAY

"Explain that to me."

Source: Reprinted with special permission of King Features Syndicate, Inc.

For Exercises 11–56 and 11–63, do a complete regression analysis by performing the following steps.

a. Draw a scatter plot.
b. Compute the correlation coefficient.
c. State the hypotheses.
d. Test the hypotheses at $\alpha = 0.05$.
e. Determine the regression line equation.
f. Plot the regression line on the scatter plot.
g. Summarize the results.

11–56. The following data were obtained for the amount of rainfall (in inches) and the yield of bushels of oats per acre for a selected sample. Predict the yield when there are 5 inches of rain.

Rainfall, x	2.5	3	5.7	8	9.3	9.7	10.2
Bushels, y	38	42	41	45	46	44	48

11–57. The following data were obtained from a survey of the number of years a person smoked and the percentage of lung damage. Predict the percentage of lung damage for a person who has smoked for 30 years.

Years, x	22	14	31	36	9	41	19
Damage, y	20	14	54	63	17	71	23

11–58. A researcher wishes to determine whether a person's age is related to the number of hours he or she jogs per week. The data for the sample follow.

Age, x	34	22	48	56	62
Hours, y	5.5	7	3.5	3	1

11–59. A study was conducted to determine the relationship between a person's monthly income and the amount that person spends per month for recreation. The data from the sample follow.

Income, x	$800	$1200	$1000	$900	$850	$907	$1100
Amount, y	$60	$200	$160	$135	$45	$90	$150

11–60. A statistics instructor is interested in finding the strength of a relationship between the final exam grades of students enrolled in Statistics I and in Statistics II. The data are given below in percentages.

Statistics I, x	87	92	68	72	95	78	83	98
Statistics II, y	83	88	70	74	90	74	83	99

11–61. An educator wants to see how the number of absences a student in her class has affects the final grade of the student. The data obtained from a sample follow.

No. of Absences, x	10	12	2	0	8	5
Final Grade, y	70	65	96	94	75	82

11–62. A physician wishes to know whether there is a relationship between a father's weight (in pounds) and his newborn son's weight (in pounds). The data are given below.

Father's Weight, x	176	160	187	210	196	142	205	215
Son's Weight, y	6.6	8.2	9.2	7.1	8.8	9.3	7.4	8.6

11–63. A sales manager wishes to determine the relationship between the number of years of experience a salesperson has and the amount of his or her monthly sales. The data are given below.

Years of Experience, x	2	10	4	6	2
Sales, y	$1000	$5000	$3000	$5000	$3000

*11–64. For Exercises 11–12, 11–13, and 11–14, find the mean of the x and y variables, then substitute the mean of the x variable into the corresponding regression line equation, and find y'. Compare the value of y' with $\bar{y}$ for each exercise. Generalize the results.

*11–65. The y intercept value, a, can also be found by using the following equation:

$$a = \bar{y} - b\bar{x}$$

Verify this result by using the data in Exercises 11–15, 11–16, and 11–17.

*11–66. The value of the correlation coefficient can also be found by using the formula

$$r = \frac{b s_x}{s_y}$$

Verify this result for Exercises 11–18, 11–19, and 11–20.

11–5 MULTIPLE REGRESSION (OPTIONAL)

In Sections 11–3 and 11–4, the concepts of simple linear regression and correlation were explained. In this section, multiple regression is explained.

In multiple regression, there are several independent variables and one dependent variable. For example, suppose a nursing instructor wishes to see whether there is a relationship between a student's grade point average, the student's age, and the student's score on the state board nursing examination. The two independent variables are the grade point average, denoted by x_1, and the age, denoted by x_2. The instructor will collect the data for all three variables from a sample of nursing students. Rather than conduct two separate regression studies, the instructor will conduct a multiple regression analysis using two independent variables instead of one.

In multiple regression, the equation is of the form

$$y' = a + b_1 x_1 + b_2 x_2 + \cdots + b_n x_n$$

where $x_1, x_2, \ldots, x_n$ are the independent variables. Also, a multiple correlation coefficient can be computed from the data.

Multiple regression analysis is used when there are several independent variables contributing to the variation of the dependent variable and one cannot control all of them. This analysis can be used to make more accurate predictions for the dependent variable.

Here are some other examples of the use of multiple regression analysis: A store manager may wish to see whether the amount spent on advertising and the amount of floor space used for a display affect the amount of sales of a product. A sociologist may wish to see whether the amount of time children spend watching television and playing video games is related to the individual's weight.

Since the computations for multiple regression analysis are quite complicated, they will be omitted. Only illustrative examples will be shown.

A nursing instructor wishes to see whether the student's grade point average and age are related to the student's score on the state board nursing examination. She selects five students and obtains the following data.

Student	Grade Point Average, x_1	Age, x_2	State Board Score, y
A	3.2	22	550
B	2.7	27	570
C	2.5	24	525
D	3.4	28	670
E	2.2	23	490

The multiple regression equation for the data is

$$y' = -44.81 + 87.64x_1 + 14.53x_2$$

If a student has a grade point average of 3.0 and is 25 years old, her predicted state board score can be computed by substituting these values in the equation for x_1 and x_2, respectively:

$$y' = -44.81 + 87.64(3.0) + 14.53(25) = 581.36 \text{ or about } 581$$

In addition, a multiple correlation coefficient R for the variables can be computed. In this case, it is 0.989. The multiple correlation coefficient will always be higher than the individual correlation coefficients. For this specific example, the multiple correlation coefficient is higher than the two individual correlation coefficients computed by using grade point average and state board scores ($r = 0.845$) and age and state board scores ($r = 0.791$).

The theory of multiple regression is quite complicated. It is introduced here only to make the student aware that other types of correlation and regression analysis can be performed.

11–6 COEFFICIENT OF DETERMINATION AND STANDARD ERROR OF ESTIMATE (OPTIONAL)

In the previous sections, it was stated that if the correlation coefficient is significant, the equation of the regression line can be determined. Also, for various values of the independent variable x, the corresponding values of the dependent variable y can be predicted. In addition, several other measures are associated with the correlation and regression techniques. They include the coefficient of determination, the standard error of estimate, and the prediction interval. But before these concepts can be explained, definitions of the different types of variation associated with the regression model must be given.

Types of Variation for the Regression Model

Consider the following hypothetical regression model.

x	1	2	3	4	5
y	10	8	12	16	20

The equation of the regression line is $y' = 4.8 + 2.8x$, and $r = 0.919$. The actual y values are 10, 8, 12, 16, and 20. The predicted values, designated by y', for each x can be found by substituting each x value in the regression equation and finding y'. For example, when $x = 1$,

$$y' = 4.8 + 2.8x = 4.8 + (2.8)(1) = 7.6$$

Now, for each *x*, there is an observed *y* value and a predicted *y′* value. For example, when $x = 1$, $y = 10$ and $y′ = 7.6$. Recall that the closer the observed values are to the predicted values, the better the fit is and the closer *r* is to $+1$ or -1. Hence, the *total variation* is defined by $\Sigma\,(y - \bar{y})^2$. The variation obtained from the predicted *y′* values is $\Sigma\,(y′ - \bar{y})^2$ and is called the *explained variation*. The variation of the observed values and the predicted values is $\Sigma\,(y - y′)^2$ and is called the *unexplained variation*. Finally, total variation = explained variation + unexplained variation, or

$$\Sigma\,(y - \bar{y})^2 = \Sigma\,(y′ - \bar{y})^2 + \Sigma\,(y - y′)^2$$

These values are shown in Figure 11–18. For a single point, the differences are called *deviations*. For the previous example, for $x = 1$ and $y = 10$, one gets $y′ = 7.6$ and $\bar{y} = 13.2$.

FIGURE 11–18
Deviations for the
Regression Equation

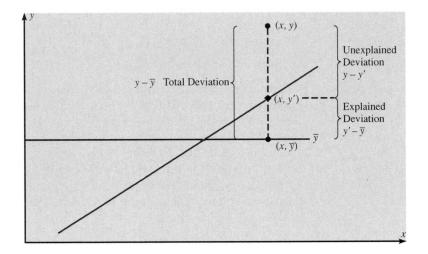

The procedure for finding the three types of variation is illustrated next.

STEP 1 Find the predicted *y′* values.

for $x = 1$ $y′ = 4.8 + 2.8x = 4.8 + (2.8)(1) = 7.6$
for $x = 2$ $y′ = 4.8 + (2.8)(2) = 10.4$
for $x = 3$ $y′ = 4.8 + (2.8)(3) = 13.2$
for $x = 4$ $y′ = 4.8 + (2.8)(4) = 16$
for $x = 5$ $y′ = 4.8 + (2.8)(5) = 18.8$

Hence, the values for this example are as follows:

x	*y*	*y′*
1	10	7.6
2	8	10.4
3	12	13.2
4	16	16
5	20	18.8

STEP 2 Find the mean of the *y* values.

$$\bar{y} = \frac{10 + 8 + 12 + 16 + 20}{5} = 13.2$$

STEP 3   Find the total variation $\Sigma (y - \bar{y})^2$.

$$(10 - 13.2)^2 = 10.24$$
$$(8 - 13.2)^2 = 27.04$$
$$(12 - 13.2)^2 = 1.44$$
$$(16 - 13.2)^2 = 7.84$$
$$(20 - 13.2)^2 = \underline{46.24}$$
$$\Sigma (y - \bar{y})^2 = 92.8$$

STEP 4   Find the explained variation $\Sigma (y' - \bar{y})^2$.

$$(7.6 - 13.2)^2 = 31.36$$
$$(10.4 - 13.2)^2 = 7.84$$
$$(13.2 - 13.2)^2 = 0.00$$
$$(16 - 13.2)^2 = 7.84$$
$$(18.8 - 13.2)^2 = \underline{31.36}$$
$$\Sigma (y' - \bar{y})^2 = 78.4$$

STEP 5   Find the unexplained variation $\Sigma (y - y')^2$.

$$(10 - 7.6)^2 = 5.76$$
$$(8 - 10.4)^2 = 5.76$$
$$(12 - 13.2)^2 = 1.44$$
$$(16 - 16)^2 = 0.00$$
$$(20 - 18.8)^2 = \underline{1.44}$$
$$\Sigma (y - y')^2 = 14.4$$

Notice that

total variation = explained variation + unexplained variation
92.8   =   78.4   +   14.4

Coefficient of Determination

The *coefficient of determination* is the ratio of the explained variation to the total variation and is denoted by r^2. That is,

$$r^2 = \frac{\text{explained variation}}{\text{total variation}}$$

For the example, $r^2 = 78.4/92.8 = 0.845$. The term r^2 is usually expressed as a percentage. So in this case, 84.5% of the total variation is explained by the regression line using the independent variable.

Another way to arrive at the value for r^2 is to square the correlation coefficient. In this case, $r = 0.919$, and $r^2 = 0.845$, which is the same value found by using the variation ratio.

The coefficient of determination is a measure of the variation of the dependent variable that is explained by the regression line and the independent variable. The symbol for the coefficient of determination is r^2.

Of course, it is usually easier to find the coefficient of determination by squaring r and converting it to a percentage. Therefore, if $r = 0.90$, then $r^2 = 0.81$, which is equivalent to 81%. This result means that 81% of the variation in the dependent variable is accounted for by the variations in the independent variable. The rest of the variation, 0.19 or 19%, is unexplained. This value is called the *coefficient of nondetermination*. The coefficient of nondetermination is found by subtracting the coefficient of determination from 1. As the value of r approaches 0, r^2 decreases more rapidly. For example, if $r = 0.6$, then $r^2 = 0.36$, which means that only 36% of the variation in the dependent variable can be attributed to the variation in the independent variable.

Formula for the Coefficient of Nondetermination
$1.00 - r^2$

Standard Error of Estimate

When a y' value is predicted for a specific x value, the prediction is a point prediction. However, a prediction interval about the y' value can be constructed, just as a confidence interval was constructed for an estimate of the population mean. The prediction interval uses a statistic called the standard error of estimate.

The standard error of estimate, denoted by s_{est} is the standard deviation of the observed y values about the predicted y' values. The formula for the standard error of estimate is

$$s_{est} = \sqrt{\frac{\Sigma (y - y')^2}{n - 2}}$$

The standard error of estimate is similar to the standard deviation, but the mean is not used. As can be seen from the formula, the standard error of estimate is the square root of the unexplained variation—i.e., the variation due to the difference of the observed values and the expected values—divided by $n - 2$. So the closer the observed values are to the predicted values, the smaller the standard error of estimate will be.

The next example shows how to compute the standard error of estimate.

EXAMPLE 11–12   A researcher collects the following data and determines that there is a significant relationship between the age of a copy machine and the monthly maintenance cost of the machine. The regression equation is $y' = 55.57 + 8.13x$. Find the standard error of estimate.

Machine	Age, x (Years)	Monthly Cost, y
A	1	$ 62
B	2	78
C	3	70
D	4	90
E	4	93
F	6	103

Solution   **STEP 1**   Make a table, as shown.

x	y	y'	$y - y'$	$(y - y')^2$
1	62			
2	78			
3	70			
4	90			
4	93			
6	103			

STEP 2   Using the regression line equation, $y' = 55.57 + 8.13x$, compute the predicted values y' for each x and place the results in the column labeled y'.

$$x = 1 \qquad y' = 55.57 + (8.13)(1) = 63.7$$
$$x = 2 \qquad y' = 55.57 + (8.13)(2) = 71.83$$
$$x = 3 \qquad y' = 55.57 + (8.13)(3) = 79.96$$
$$x = 4 \qquad y' = 55.57 + (8.13)(4) = 88.09$$
$$x = 6 \qquad y' = 55.57 + (8.13)(6) = 104.35$$

STEP 3   For each y, subtract y' and place the answer in the column labeled $y - y'$.

$$62 - 63.7 = -1.7 \qquad\qquad 90 - (88.09) = 1.91$$
$$78 - 71.83 = 6.17 \qquad\qquad 93 - (88.09) = 4.91$$
$$70 - 79.96 = -9.96 \qquad 103 - (104.35) = -1.35$$

STEP 4   Square the numbers found in Step 3 and place the squares in the column labeled $(y - y')^2$.

STEP 5   Find the sum of the numbers in the last column. The completed table follows.

x	y	y'	$y - y'$	$(y - y')^2$
1	62	63.7	-1.7	2.89
2	78	71.83	6.17	38.0689
3	70	79.96	-9.96	99.2016
4	90	88.09	1.91	3.6481
4	93	88.09	4.91	24.1081
6	103	104.35	-1.35	1.8225
				169.7392

STEP 6   Substitute in the formula and find s_{est}.

$$s_{est} = \sqrt{\frac{\Sigma (y - y')^2}{n - 2}} = \sqrt{\frac{169.7392}{6 - 2}} = 6.51$$

In this case, the standard deviation of observed values about the predicted values is 6.51.   ◀

The standard error of estimate can also be found by using the formula

$$s_{est} = \sqrt{\frac{\Sigma y^2 - a \Sigma y - b \Sigma xy}{n - 2}}$$

EXAMPLE 11–13 Find the standard error of estimate for the data of Example 11–12 by using the preceding formula. The equation of the regression line is $y' = 55.57 + 8.13x$.

Solution **STEP 1** Make a table, which is shown completed in Step 4.

STEP 2 Find the product of x and y values and place the results in the third column.

STEP 3 Square the y values and place the results in the fourth column.

STEP 4 Find the sums of the second, third, and fourth columns. The completed table is shown here.

x	y	xy	y^2
1	62	62	3,844
2	78	156	6,084
3	70	210	4,900
4	90	360	8,100
4	93	372	8,649
6	103	618	10,609
	$\Sigma y = 496$	$\Sigma xy = 1778$	$\Sigma y^2 = 42,186$

STEP 5 From the regression equation, $y' = 55.57 + 8.13x$, $a = 55.57$ and $b = 8.13$.

STEP 6 Substitute in the formula and solve for s_{est}:

$$s_{est} = \sqrt{\frac{\Sigma y^2 - a\,\Sigma y - b\,\Sigma xy}{n - 2}}$$

$$= \sqrt{\frac{42,186 - (55.57)(496) - (8.13)(1778)}{6 - 2}} = 6.48$$

This value is close to the value found in Example 11–12. The difference is due to rounding error. ◄

Prediction Interval The standard error of estimate can be used for constructing a **prediction interval** (similar to a confidence interval) about a value y' only when the sample size is large (usually 100 or more), since there are two other possible sources of prediction errors. The other sources of error are the errors made in estimating the slope and the y intercept of the regression line. If the sample size is large, these errors are relatively small compared with s_{est}. Hence, when the sample size is large, the 95% prediction interval about a value y' is given by

$$y' - 1.96 s_{est} < y < y' + 1.96 s_{est}$$

A 99% prediction interval is given by

$$y' - 2.58 s_{est} < y < y' + 2.58 s_{est}$$

When the sample size is less than 100, an adjustment to s_{est} must be made. This adjustment is beyond the scope of this book.

Formula for the Prediction Interval about a Value y' When the Sample Size Is Large

$$y' - z_{\alpha/2} s_{est} < y < y' + z_{\alpha/2} s_{est}$$

EXAMPLE 11–14   For a sample of 100, the regression line equation for age and blood pressures is $y' = 100 + 0.96x$, and the standard error of estimate is 5. Find the 95% prediction interval of the blood pressure of a person who is 43 years old.

Solution   **STEP 1**   Using the regression equation, find the predicted value of y'.

$$y' = 100 + 0.96x = 100 + (0.96)(43) = 141.28$$

STEP 2   Since $\alpha = 0.05$ and $s_{est} = 5$, substituting in the formula for the prediction interval yields

$$y' - 1.96s_{est} < y < y' + 1.96s_{est}$$
$$141.28 - (1.96)(5) < y < 141.28 + (1.96)(5)$$
$$131.48 < y < 151.08$$ ◄

Similar prediction intervals can be computed for smaller samples, but the technique is beyond the scope of this textbook.

Speaking of Statistics

Smoker's Sex Affects Risk

By Anita Manning
USA TODAY

Female smokers may be more than twice as likely as men to contract lung cancer, a new study suggests.

Among heavy smokers—those who smoke two packs a day for 30 years—the risk for women is far greater than for men, say University of Toronto researchers.

They found cancer risk increases with the number of cigarettes and length of time a person smokes.

But "it increased faster for females than males," says Harvey Risch, Yale University School of Medicine. Researchers found:

- At 40 "pack-years"—smoking a pack a day for 40 years—women face 28 times greater risk of lung cancer than non-smoking women; among men, the risk is 10 times greater.

- At 60 pack-years, women showed 82 times the risk; for men it was 23-fold.
- At 29 pack-years or fewer, the risk was seven times greater for women, five times for men.

Studied were 442 women with cancer, 410 without, 403 men with, 362 without.

Says Risch: "We don't have a good biological reason" for the disparity, but one possibility is that women's smaller lungs receive a more concentrated exposure to the smoke.

Source: Copyright 1993, USA TODAY.
Reprinted with permission.

In this study, the researchers found several relationships. Identify them, and state some hypotheses that may have been used to test the significance of these relationships.

EXERCISES

11–67. Define *multiple regression*.

11–68. Why can't a scatter plot be drawn for a multiple regression problem?

11–69. What is the general form of the multiple regression equation?

11–70. Define *coefficient of determination*.

11–71. How is the coefficient of determination found?

11–72. Define *coefficient of nondetermination*.

11–73. How is the coefficient of nondetermination found?

11–74. Define *standard error of estimate* for regression.

11–75. When can the standard error of estimate be used to construct a prediction interval about a value y'?

11–76. A researcher has determined that a significant relationship exists among an employee's age x_1, grade point average x_2, and income y. The multiple regression equation is

$$y' = -34,127 + 132x_1 + 20,805x_2$$

Predict the income of a person who is 32 years old and who had a grade point average of 3.4.

11–77. A manufacturer found that a significant relationship exists among the number of hours x_1 an assembly line employee works per shift, the total number of items produced x_2, and the number of defective items produced y. The multiple regression equation is $y' = 9.6 + 2.2x_1 - 1.08x_2$. Predict the number of defective items produced by an employee who has worked 9 hours and produced 24 items.

11–78. A real estate agent found that there is a significant relationship among the number of acres x_1 on a farm, the number of rooms x_2 in the farmhouse, and the selling price y (in thousands of dollars) of farms in a specific area. The regression equation is

$$y' = 44.9 - 0.0266x_1 + 7.56x_2$$

Predict the selling price of a farm that has 371 acres and a farmhouse with 6 rooms.

11–79. An educator has found a significant relationship among a high school student's IQ, x_1, grade point average x_2, and the score y on the mathematics section of the SAT exam. The regression equation is

$$y' = 230 + 0.96x_1 + 77.1x_2$$

Predict the SAT score of a student whose IQ is 109 and whose grade point average is 3.12.

11–80. A medical researcher found a significant relationship among a person's age x_1, cholesterol level x_2, sodium level of the blood x_3, and the systolic blood pressure y. The regression equation is

$$y' = 97.7 + 0.691x_1 + 219x_2 - 299x_3$$

Predict a person's blood pressure if the person is 35 years old, has a cholesterol level of 194 milligrams per deciliter (mg/dl), and a sodium blood level of 142 milliequivalents per liter (mEq/l).

11–81. Compute the standard error of estimate for Exercise 11–13.

11–82. Compute the standard error of estimate for Exercise 11–14.

11–83. Compute the standard error of estimate for Exercise 11–15.

11–84. Compute the standard error of estimate for Exercise 11–16.

11–85. A researcher found a significant relationship between a student's IQ and grade point average, using a sample of 200 students. The regression equation is $y' = -5 + 0.07x$, and the standard error of estimate is 0.50. Find the 95% prediction interval of the grade point average of a student who has an IQ of 105.

11–86. An instructor found a significant relationship between a student's final grades in Algebra I and in Algebra II in high school. A sample of 150 students was used. The regression equation is $y' = 10.250 + 0.80x$, and the standard error of estimate is 3.3. Find the 99% prediction interval of the student's grade in Algebra II if the student scored 78 in Algebra I.

11–87. A researcher found a significant relationship between the number of hours a person works per week and the number of accidents the person has per year. A sample of 300 workers was used. The regression equation is $y' = -18 + 0.50x$, and the standard error of estimate is 1.4. Find the 95% prediction interval of the number of accidents per year for a person who works 46 hours per week.

11–88. The owner of an emergency care center found a significant relationship between the outside temperature and the number of emergency cases treated. A sample of 150 days was used. The regression equation is $y' = 7.5 + 0.20x$. The standard error of estimate is 1.80. Find the 90% prediction interval for the number of emergency cases treated when the outside temperature is 80 degrees.

11–7 OTHER TYPES OF CORRELATION COEFFICIENTS (OPTIONAL)

The Pearson product moment correlation coefficient is the most commonly used measure of correlation; however, there are several other types of correlation coefficients that are used in statistics. Each type depends upon the nature of the data.

The **Spearman's rank correlation coefficient** is used when the researcher wishes to compare the rankings of two individuals. For example, suppose a freshman and a senior were given a list of ten problematic issues facing their university and asked to rank each on a scale of 1 to 10, 1 being the most pressing and 10 being the least pressing. The rankings of the two students can then be compared to see whether there is a relationship between them. A detailed explanation of the Spearman rank correlation coefficient is given in Chapter 14.

The **point-biserial correlation coefficient** is used when one of the variables is a true dichotomous variable and the other variable is a continuous, normally distributed variable. A *true dichotomous variable* is one in which there are only two naturally occurring values of the variable. For example, the gender of a subject can only have two values, male or female. Other examples of true dichotomous variables are owning an automobile versus not owning an automobile, or being a smoker versus being a nonsmoker.

Using the point-biserial correlation coefficient, one can then determine whether there is a relationship between a continuous variable and a dichotomous variable. For example, a researcher can determine whether there is a relationship between smoking versus not smoking and blood pressure. For the computation of the point-biserial correlation coefficient, two values, 0 or 1, are assigned to the dichotomous variable. In this example, 0 could be assigned to a nonsmoker and 1 to a smoker.

The **biserial correlation coefficient** is used when the variables are continuous but one variable is reduced to two categories. Thus, one variable is an artificial dichotomy. For example, suppose that students complete a final examination in a course, but then the students are divided into two groups—those who were successful and those who were not successful, on the basis of whether they scored above or below a certain score on the exam. A biserial correlation coefficient would then be used to see whether there was a relationship between successful or unsuccessful completion of a course and the ages of the students who completed the course. Caution must be used with this coefficient: The underlying distribution of the variables must be approximately normally distributed; otherwise, values of the biserial correlation may exceed $+1.00$ or be less than -1.00.

The **phi correlation coefficient** is used when the variables are both truly dichotomous. For example, a researcher may wish to see whether there is a relationship between being married or not being married and owning a home or not owning a home. Phi is computed much like the Pearson product moment correlation coefficient.

When two continuous variables have been artificially reduced to two categories, statisticians use the **tetrachoric correlation coefficient** to see whether the variables are related. For instance, a researcher may wish to divide students into two groups according to their IQs—a group of high-IQ students and a group of low-IQ students—to see whether IQ is related to, say, a reflex response that has been divided into two categories, fast and slow.

In addition to the correlation coefficients just described, there are several others, such as Kendall's tau and the contingency correlation coefficients. These coefficients are beyond the scope of this book.

Speaking of **Statistics**

Obesity: There Are Many Causes

Obesity is one of the USA's most common diet-related problems.

It affects 34 million of us—about a quarter of the population, according to the nutrition report. And those most likely to be overweight are the poor and minorities.

People who are severely overweight have an increased risk for high blood pressure, high cholesterol levels and diabetes—all risk factors for heart disease. Obesity also increases the risk for gall bladder disease and some cancers—tumors of the kidneys, cervix and thyroid gland.

Effective treatment for obesity is still a problem because the causes aren't completely understood. Overeating is only one factor. Obesity is also associated with:

- Heredity.
- Altered or slow metabolism of fatty tissue.
- A condition in which the body doesn't convert calories into heat properly.
- Too little exercise.
- Certain prescription drugs.

The most sensible treatment: a lifelong combination of diet and exercise.

Extreme low-calorie diets—300 to 400 calories a day—are very dangerous and have resulted in deaths. Closely supervised 800-calorie diets now being tried in some clinics may be more successful.

So far, there's no effective long-term success with drug therapies.

Source: Copyright 1988, *USA Today.*
Reprinted with permission.

If you were doing an obesity study based on the findings of this article, describe some of the variables that might be used in a multiple regression equation.

11–8 SUMMARY

Many relationships between variables exist in the real world. One way to determine whether a relationship exists between variables is to use the statistical techniques known as correlation and regression. The strength and direction of the relationship is measured by the value of the correlation coefficient. It can assume values between and including -1 and $+1$. The closer the value of the correlation coefficient is to $+1$ or -1, the stronger the relationship is between the variables. A value of $+1$ or -1 indicates a perfect relationship. A positive relationship between two variables means that for small values of the independent variable, the values of the dependent variable will be small; and that for large values of the independent variable, the values of the dependent variable will be large. A negative relationship between two variables means that for small values of the independent variable, the values of the dependent variable will be large, and that for large values of the independent variable, the values of the dependent variable will be small.

Relationships can be linear or curvilinear. In order to determine the shape, one draws a scatter plot of the variables. If the relationship is linear, the data can be approximated by a straight line, called the regression line or the line of best fit. The closer the value of r is to $+1$ or -1, the closer the points will fit the line.

In addition, relationships can be multiple—i.e., there can be two or more independent variables and one dependent variable. A coefficient of correlation and a regression equation can be found for multiple relationships just as they can be found for simple relationships.

The coefficient of determination is a better indicator of the strength of a relationship than the correlation coefficient. It is a better indicator because it identifies the percentage of variation of the dependent variable that is directly attributable to the variation of the independent variable. The coefficient of determination is obtained by squaring the correlation coefficient and then converting the result to a percentage.

Another statistic used in correlation and regression is the standard error of estimate, which is the standard deviation of the y values about the predicted y' values. When the sample size is 100 or more, the standard error of estimate can be used to construct a prediction interval about a specific value y'.

Finally, remember that a significant relationship between two variables does not necessarily mean that one variable is a direct cause of the other variable—although in some cases this is true. Other possibilities that should be considered include a complex relationship involving other (perhaps unknown) variables, a third variable interacting with both variables, or a relationship due solely to chance.

Important Terms

Biserial correlation coefficient
Coefficient of determination
Correlation
Correlation coefficient
Dependent variable
Independent variable
Multiple relationship
Negative relationship
Pearson product moment
  correlation coefficient
Phi correlation coefficient
Point-biserial correlation
  coefficient
Positive relationship
Prediction interval
Regression
Regression line
Scatter plot
Simple relationship
Spearman's rank correlation
  coefficient
Standard error of estimate
Tetrachoric correlation
  coefficient

Important Formulas

Formula for the correlation coefficient:

$$r = \frac{n(\Sigma\, xy) - (\Sigma\, x)(\Sigma\, y)}{\sqrt{[n(\Sigma\, x^2) - (\Sigma\, x)^2][n(\Sigma\, y^2) - (\Sigma\, y)^2]}}$$

Formula for the t test for the correlation coefficient:

$$t = r\sqrt{\frac{n-2}{1-r^2}}$$

The regression line equation: $y' = a + bx$, where

$$a = \frac{(\Sigma\, y)(\Sigma\, x^2) - (\Sigma\, x)(\Sigma\, xy)}{n(\Sigma\, x^2) - (\Sigma\, x)^2}$$

$$b = \frac{n(\Sigma\, xy) - (\Sigma\, x)(\Sigma\, y)}{n(\Sigma\, x^2) - (\Sigma\, x)^2}$$

Formula for the standard error of estimate:

$$s_{\text{est}} = \sqrt{\frac{\Sigma\,(y - y')^2}{n-2}}$$

or

$$s_{\text{est}} = \sqrt{\frac{\Sigma\, y^2 - a\,\Sigma\, y - b\,\Sigma\, xy}{n-2}}$$

Formula for the prediction interval for a value y' when $n \geq 100$:

$$y' - z_{\alpha/2}s_{\text{est}} < y < y' + z_{\alpha/2}s_{\text{est}}$$

Review Exercises

For Exercises 11–89 through 11–95, do a complete regression analysis by performing the following steps.

a. Draw the scatter plot.
b. Compute the value of the correlation coefficient.
c. Test the significance of the correlation coefficient at $\alpha = 0.01$, using Table I.
d. Determine the regression line equation.
e. Plot the regression line on the scatter plot.
f. Predict y' for a specific value of x.

11–89. A study is done to see whether there is a relationship between a student's grade point average and the number of hours the student studies per week. The data are shown below. If there is a significant relationship, predict the grade point average of a student who studies 10 hours per week.

Hours, x	3	12	9	15	5	7	16
GPA, y	2.1	3.5	3.0	4.0	1.7	3.2	3.7

11–90. A study is conducted to determine the relationship between the number of calls a salesman makes and the number of sales he has over a one-week period. The data are shown below. If there is a significant relationship, predict the number of sales when the salesman makes 19 calls.

Calls, x	23	16	9	32	6	12
Sales, y	15	9	7	22	3	8

11–91. A study is done to see whether there is a relationship between a mother's age and the number of children she has. The data are shown below. If there is a significant relationship, predict the number of children of a mother whose age is 34.

Mother's Age, x	18	22	29	20	27	32	33	36
No. of Children, y	2	1	3	1	2	4	3	5

11–92. A study is conducted to determine the relationship between a driver's age and the number of accidents a person has over a one-year period. The data are shown below. If there is a significant relationship, predict the number of accidents of a driver who is 28.

Driver's Age, x	16	24	18	17	23	27	32
No. of Accidents, y	3	2	5	2	0	1	1

11–93. A researcher desires to know whether the typing speed of a secretary (in words per minute) is related to the time (in hours) that it takes the secretary to learn to use a word-processing program on a new computer. The data follows.

Speed, x	Time, y
48	7
74	4
52	8
79	3.5
83	2
56	6
85	2.3
63	5
88	2.1
74	4.5
90	1.9
92	1.5

If there is a significant relationship, predict the time it will take the average secretary to learn the word-processing program if he or she has a typing speed of 72 words per minute. (This information will be used for Exercises 11–99 and 11–101.)

11–94. A study was conducted with vegetarians to see whether the number of grams of protein each ate per day was related to diastolic blood pressure. The data are given below. (This information will be used for Exercises 11–100 and 11–102.)

Grams, x	4	6.5	5	5.5	8	10	9	8.2	10.5
Pressure, y	73	79	83	82	84	92	88	86	95

If there is a significant relationship, predict the diastolic pressure of a vegetarian who consumed 8 grams of protein per day.

11–95. A study was conducted to determine whether there is a relationship between strength and speed. A sample of 20-year-old men was selected. Each was asked to do push-ups and to run a specific course. The number of push-ups and time it took to run the course (in seconds) are given in the table.

Push-ups, x	5	8	10	10	11	13	15	18	23
Time, y	61	65	43	56	62	73	48	49	50

If there is a significant relationship, predict the running time for a person who can do 18 push-ups.

11–96. (OPT.) A study was conducted and a significant relationship was found among a student's grade point average x_1, age x_2, and score y on the state teacher's examinations. The regression equation is

$$y' = 440 - 19.1x_1 + 6.3x_2$$

Predict the score for a student who is 25 years old and has a grade point average of 3.11.

11–97. (OPT.) A study was conducted and a significant relationship was found among the number of hours a teenager watches television per day x_1, the number of hours the teenager plays video games per day x_2, and the teenager's weight y. The regression equation is

$$y' = 99.8 + 4.71x_1 + 7.11x_2$$

Predict a teenager's weight if he watches television an average of 2 hours and plays video games for 1 hour per day.

11–98. (OPT.) A study found a significant relationship among a person's years of experience on a particular job x_1, the number of years of higher education x_2, and the person's income y. The regression equation is

$$y' = 11.2 + 1.88x_1 + 0.547x_2$$

Predict a person's income if he or she has been employed for 3 years and completed 4 years of college.

11–99. (OPT.) For Exercise 11–93, find the standard error of estimate. This information will be used in Exercise 11–101.

11–100. (OPT.) For Exercise 11–94, find the standard error of estimate. This information will be used in Exercise 11–102.

11–101. (OPT.) Assuming that the regression equation for Exercise 11–93 was found for a sample of 100 or more, find the 90% prediction interval for the time it takes a secretary to learn the word-processing program if he or she can type 72 words per minute. Use the value for the standard error of estimate found in Exercise 11–99.

11–102. (OPT.) Assuming that the regression equation for Exercise 11–94 was found for a sample of 100 or more, find the 95% prediction interval of the blood pressure of a person who eats 8 grams of protein. Use the value for the standard error of estimate found in Exercise 11–100.

COMPUTER APPLICATIONS

1. Write a program that will find the correlation coefficient, the regression line equation, and compute a t test statistic for a sample of two related variables.
2. MINITAB can be used to do a complete correlation and regression analysis. For example, a researcher collected data on the temperature and the number of ice cream cones sold at a local dairy store. The following data were obtained.

Temperature, x	65	68	73	76	80	84	88	92	95	98
No. of Cones, y	15	22	20	24	27	27	32	38	42	40

The data are entered by using the READ command, where C1 is the x variable and C2 is the y variable. To get a scatter plot, use the PLOT C2 VS C1 command. The y variable must be entered first in the PLOT command. Using MINITAB, enter the following:

```
MTB  > READ C1 C2
DATA > 65 15
DATA > 68 22
DATA > 73 20
DATA > 76 24
DATA > 80 27
DATA > 84 27
DATA > 88 32
DATA > 92 38
DATA > 95 42
DATA > 98 40
DATA > END
      10 ROWS READ
MTB  > PLOT C2 VS C1
```

The computer will print the following:

To find the correlation coefficient, use the CORRELATION C1 C2 command, as follows:

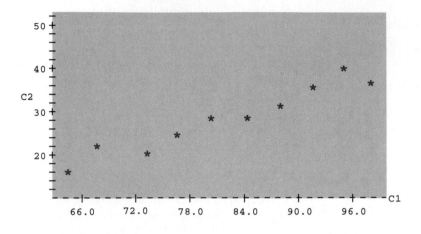

```
MTB > CORRELATION C1, C2
```

The computer will print the following:

```
Correlation of C1 and C2 = 0.968
```

To find the equation of the regression line, use the REGRESS C2 1 C1 command. The *y* variable, C2, is entered first in the command, then a 1, followed by the *x* variable, C1. The 1 tells MINITAB that there is only one independent variable designated as C1.

```
MTB > REGRESS C2 1 C1
```

The computer will print the following:

```
The regression equation is
C2 = - 34.2 + 0.768 C1
```

Hence, the value of the correlation coefficient is 0.968, and the equation of the regression line is $y = -34.2 + 0.768x$.

Do a complete regression analysis for the data given in Exercise 11–89.

√ DATA ANALYSIS Applying the Concepts

The Data Bank is located in the Appendix.

From the Data Bank, choose two variables that might be related—e.g., IQ and educational level; age and cholesterol level; exercise and weight; or weight and systolic pressure. Do a complete correlation and regression analysis by performing the following steps. Select a random sample of at least ten subjects.

1. Draw a scatter plot.
2. Compute the correlation coefficient.
3. Test the hypothesis H_0: $\rho = 0$.
4. Find the regression line equation.
5. Summarize the results.

✎ TEST

Directions: Determine whether the statement is true or false.

1. A statistical graph of two variables is called a box plot.
2. The variable designated as x is called the dependent variable.
3. The strength of the relationship between two variables is measured by the correlation coefficient.
4. The range of the correlation coefficient is from 0 to 1.
5. A negative relationship between two variables means that on average, if the x values are large, the y values will be small; and that if the x values are small, the y values will be large.
6. A correlation coefficient of -1 implies a perfect linear relationship between the variables.
7. Even though the value of the correlation coefficient is high, it may not be significant.
8. In order to test the significance of r, one uses the t test.
9. The t test for testing the significance of the correlation coefficient has $n - 1$ degrees of freedom.
10. When the correlation coefficient is significant, one can assume that there is a direct cause-and-effect relationship between the variables.
11. It is not possible to have a significant correlation by chance alone.
12. The sign (positive or negative) of r will determine the nature of the relationship (positive or negative).
13. The regression line is also called the line of best fit.
14. The equation for the regression line that is used in statistics is $x = a + by$.
15. All the points on the scatter plot will always fall on the regression line if $r = +1$.
16. For valid predictions, r must be significant.
17. Valid predictions can be made even though the present conditions do not continue.
18. In multiple regression, there are several dependent variables and one independent variable.
19. The standard error of estimate can be used to construct a prediction interval about a value y' when the sample size is usually 100 or more.
20. The coefficient of determination is $1 - r^2$.

CHAPTER

12

Chi-Square

12–1 INTRODUCTION

So far, two statistical distributions, the z and the t distributions, have been explained. Another frequently used distribution is the chi-square (χ^2) distribution. This distribution can be used to test hypotheses concerning *variances,* such as "Is the variance of IQ scores obtained from a sample equal to a specific value, say 225?" It can also be used for tests concerning *frequency distributions,* such as "If a sample of buyers is given a choice of automobile colors, will each color be selected with the same frequency?" Finally, the chi-square distribution can be used to test the *independence* of two variables. For example, "Is the opinion of the senators on gun control independent of party affiliation?" That is, do the Republicans feel one way and the Democrats feel differently, or do they have the same opinion?

This chapter explains the chi-square distribution and its applications. In addition to the applications mentioned above, the chi-square has many other uses in statistics.

12–2 THE CHI-SQUARE DISTRIBUTION

The chi-square variable is similar to the t variable in that its distribution is a family of curves based on the number of degrees of freedom. The symbol for chi-square is χ^2 (Greek letter chi, pronounced "kī"). Several of the distributions are shown in Figure 12–1, along with the corresponding degrees of freedom. The **chi-square distribution** is obtained from the values of $(n-1)s^2/\sigma^2$ when random samples are selected from a normally distributed population whose variance is σ^2.

FIGURE 12–1
The Chi-Square Family of Curves

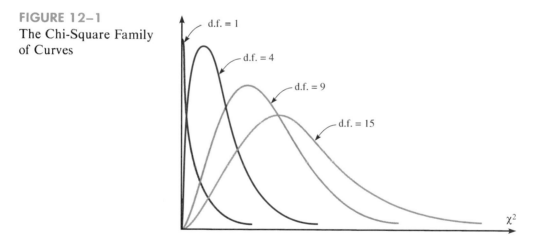

A chi-square variable cannot be negative, and the distributions are positively skewed. At about 100 degrees of freedom, the chi-square distribution becomes somewhat symmetrical. The area under each chi-square distribution is equal to 1.00, or 100%.

In order to find the area under the chi-square distribution, one uses Table G in the Appendix. There are three cases to consider:

1. Finding the chi-square critical value for a specific α when the hypothesis test is one-tailed right.

2. Finding the chi-square critical value for a specific α when the hypothesis test is one-tailed left.
3. Finding the chi-square critical values for a specific α when the hypothesis test is two-tailed.

The next three examples explain how to use Table G.

EXAMPLE 12–1 Find the critical chi-square value for 15 degrees of freedom when $\alpha = 0.05$ and the test is one-tailed right.

Solution The distribution is shown in Figure 12–2.

Find the α value at the top of Table G and find the corresponding degrees of freedom in the left column. The critical value is located where the two columns meet—in this case, 24.996. See Figure 12–3.

FIGURE 12–2
Chi-Square Distribution
for Example 12–1

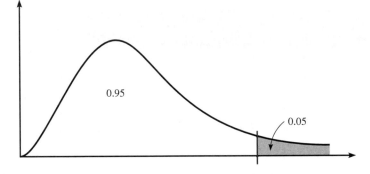

FIGURE 12–3
Locating the Critical
Value in Table G for
Example 12–1

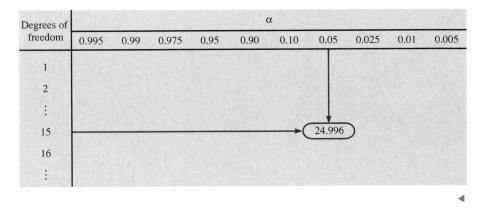

Degrees of freedom	α									
	0.995	0.99	0.975	0.95	0.90	0.10	0.05	0.025	0.01	0.005
1										
2										
⋮										
15								24.996			
16										
⋮										

EXAMPLE 12–2 Find the critical chi-square value for 10 degrees of freedom when $\alpha = 0.05$ and the test is one-tailed left.

Solution This distribution is shown in Figure 12–4.

When the test is one-tailed left, the α value must be subtracted from 1: $1 - 0.05 = 0.95$. Then, the left side of the table is used, because the chi-square table gives the area to the right of the critical value, and the chi-square statistic cannot be negative. The table is set up so that it gives the values for the area to the right of the critical value. In this case, 95% of the area will be to the right of the value.

FIGURE 12–4
Chi-Square Distribution
for Example 12–2

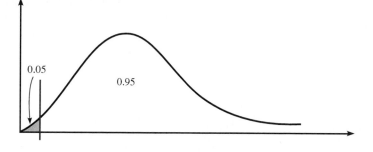

For 0.95 and 10 degrees of freedom, the critical value is 3.940. See Figure 12–5.

FIGURE 12–5
Locating the Critical
Value in Table G for
Example 12–2

Degrees of freedom	α									
	0.995	0.99	0.975	0.95	0.90	0.10	0.05	0.025	0.01	0.005
1										
2										
⋮										
10				3.940						
⋮										

◄

EXAMPLE 12–3   Find the critical chi-square values for 22 degrees of freedom when $\alpha = 0.05$ and a two-tailed test is conducted.

Solution   When a two-tailed test is conducted, the area must be split, as shown in Figure 12–6. Note that the area to the right of the larger value is 0.025 (0.05/2), and the area to the right of the smaller value is 0.975 (1.00 − 0.05/2).

Remember that chi-square values cannot be negative. Hence, one must use α values in the table of 0.025 and 0.975. With 22 degrees of freedom, the critical values are 36.781 and 10.982, respectively.

FIGURE 12–6
Chi-Square Distribution
for Example 12–3

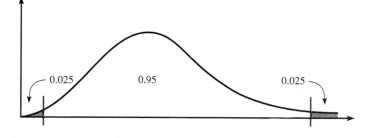

◄

After the degrees of freedom reach 30, Table G only gives values for multiples of 10 (40, 50, 60, etc.). When the exact degrees of freedom one is seeking are not specified in the table, the closest smaller value should be used. For example, if the given degrees of freedom are 36, the table value for 30 degrees of freedom should be used. This guideline is used to keep the type I error equal to or below the α value.

12–3 TEST FOR A SINGLE VARIANCE

Chi-Square Test

When one is testing a claim about a single variance, there are three possible test situations: one-tailed right test, one-tailed left test, and two-tailed test.

If a researcher believes the variance of a population to be greater than some specific value, say 225, then the researcher states the hypotheses as

$$H_0: \sigma^2 \leq 225$$
$$H_1: \sigma^2 > 225$$

and conducts a one-tailed right test.

If the researcher believes the variance of a population to be less than 225, then the researcher states the hypotheses as

$$H_0: \sigma^2 \geq 225$$
$$H_1: \sigma^2 < 225$$

and conducts a one-tailed left test.

Finally, if a researcher does not wish to specify a direction, he or she states the hypotheses as

$$H_0: \sigma^2 = 225$$
$$H_1: \sigma^2 \neq 225$$

and conducts a two-tailed test.

Formula for the Chi-Square Test for a Single Variance

$$\chi^2 = \frac{(n-1)s^2}{\sigma^2}$$

with degrees of freedom equal to $n - 1$ and where

n = sample size
s^2 = sample variance
σ^2 = population variance

One might ask, "Why is it important to test variances?" There are several reasons. First, in any situation where consistency is required, such as in manufacturing, one would like to have the smallest variation possible in the products. For example, when bolts are manufactured, the variation in diameters due to the process must be kept to a minimum, or the nuts will not fit them properly. In education, consistency is required on a test. That is, if the same students take the same test several times, they should get approximately the same grades, and the variance of each of the students' grades should be small. On the other hand, if the test is to be used to judge learning, the overall standard deviation of all of the grades should be large so that one can differentiate those who have learned the subject from those who have not learned it.

Three assumptions are made for the chi-square test, as outlined next.

Assumptions for the Chi-Square Test for a Single Variance

1. The sample must be randomly selected from the population.
2. The population must be normally distributed for the variable under study.
3. The observations must be independent of each other.

The hypothesis-testing procedure follows the same five steps listed for hypothesis tests in the preceding chapters.

STEP 1 State the hypotheses, and identify the claim.

STEP 2 Find the critical value(s).

STEP 3 Compute the test value.

STEP 4 Make the decision.

STEP 5 Summarize the results.

The next three examples illustrate the hypothesis-testing procedure for variances.

EXAMPLE 12–4 An instructor wishes to see whether the variation in IQ of the 23 students in her class is less than the variance of the population. The variance of the class is 198. Test the claim that the variation of the students is less than the population variance ($\sigma^2 = 225$) at $\alpha = 0.05$. Assume that IQ is normally distributed.

Solution **STEP 1** State the hypotheses, and identify the claim.

$$H_0: \sigma^2 \geq 225 \qquad \text{and} \qquad H_1: \sigma^2 < 225 \text{ (claim)}$$

STEP 2 Find the critical value. Since this test is a one-tailed left test and $\alpha = 0.05$, the value $1 - 0.05 = 0.95$ should be used. The degrees of freedom are $n - 1 = 23 - 1 = 22$. Hence, the critical value is 12.338. Note that the critical region is on the left, as shown in Figure 12–7.

FIGURE 12–7
Critical Value for
Example 12–4

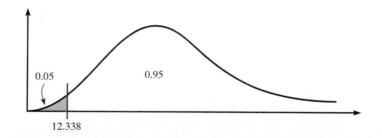

STEP 3 Compute the test value.

$$\chi^2 = \frac{(n-1)s^2}{\sigma^2} = \frac{(23-1)(198)}{225} = 19.36$$

STEP 4 Make the decision. Since the test value 19.36 falls in the noncritical region, as shown in Figure 12–8, the decision is to not reject the null hypothesis.

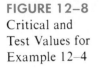
FIGURE 12–8
Critical and
Test Values for
Example 12–4

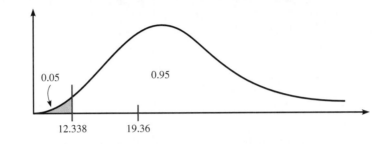

0.05

0.95

12.338   19.36

STEP 5 Summarize the results. There is not enough evidence to support the claim that the variation in IQ scores of the instructor's students is less than the variation in IQ scores of the population. ◄

EXAMPLE 12–5 A medical researcher believes that the standard deviation of the temperatures of newborn infants is greater than 0.6°. A sample of 15 infants was found to have a standard deviation of 0.8°. At $\alpha = 0.10$, does the evidence support the researcher's belief? Assume that the variable is normally distributed.

Solution **STEP 1** State the hypotheses, and identify the claim.

$$H_0: \sigma^2 \le 0.36 \qquad \text{and} \qquad H_1: \sigma^2 > 0.36 \text{ (claim)}$$

Note that since standard deviations are given in the problem, they must be squared to get variances.

STEP 2 Find the critical value. Since this test is a one-tailed right test with degrees of freedom of $15 - 1 = 14$ and $\alpha = 0.10$, the critical value is 21.064.

STEP 3 Compute the test value.

$$\chi^2 = \frac{(n-1)s^2}{\sigma^2} = \frac{(15-1)(0.64)}{0.36} = 24.89$$

STEP 4 Make the decision. The decision is to reject the null hypothesis, since the test value 24.89 is greater than the critical value 21.064 and falls in the critical region, as shown in Figure 12–9.

FIGURE 12-9
Critical and
Test Values for
Example 12-5

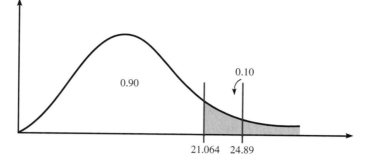

STEP 5 Summarize the results. There is enough evidence to support the claim that the standard deviation of the temperatures is greater than 0.6°. ◄

EXAMPLE 12-6 A cigarette manufacturer wishes to test the claim that the variance of the nicotine content of the cigarette his company manufactures is equal to 0.644. Nicotine content is measured in milligrams, and assume that it is normally distributed. A sample of 20 cigarettes has a standard deviation of 1.00 milligram. At $\alpha = 0.05$, test the manufacturer's claim.

Solution **STEP 1** State the hypotheses, and identify the claim.

$$H_0\text{: } \sigma^2 = 0.644 \text{ (claim)} \qquad \text{and} \qquad H_1\text{: } \sigma^2 \neq 0.644$$

STEP 2 Find the critical values. Since this test is a two-tailed test at $\alpha = 0.05$, the critical values for 0.025 and 0.975 must be found. The degrees of freedom are 19; hence, the critical values are 32.852 and 8.907. The critical or rejection regions are shown in Figure 12-10.

FIGURE 12-10
Critical Values for
Example 12-6

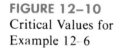
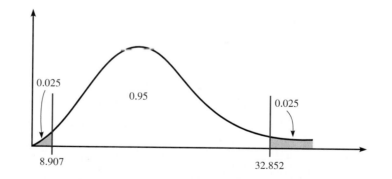

STEP 3 Compute the test value.

$$\chi^2 = \frac{(n-1)s^2}{\sigma^2} = \frac{(20-1)(1.0)^2}{0.644} = 29.5$$

Since the standard deviation s is given in the problem, it must be squared for the formula.

STEP 4 Make the decision. Do not reject the null hypothesis, since the test value falls between the critical values ($8.907 < 29.5 < 32.852$) and in the noncritical region, as shown in Figure 12–11.

FIGURE 12–11
Critical and
Test Values for
Example 12–6

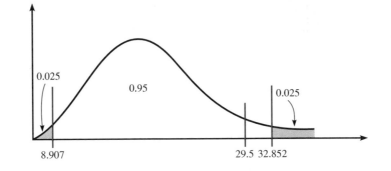

STEP 5 Summarize the results. There is not enough evidence to reject the manufacturer's claim that the variance of the nicotine content of the cigarettes is equal to 0.644. ◄

Confidence Intervals for Variances

The chi-square distribution can also be used to find a specific confidence interval for the variance or standard deviation of a variable.

Formulas for the Confidence Interval of a Variance and of a Standard Deviation

The formula for the confidence interval of a variance is

$$\frac{(n-1)s^2}{\chi^2_{\text{larger}}} < \sigma^2 < \frac{(n-1)s^2}{\chi^2_{\text{smaller}}}$$

where

$n =$ sample size
$s^2 =$ sample variance
$\chi^2_{\text{larger}} =$ larger two-tailed table value for a specific α
$\chi^2_{\text{smaller}} =$ smaller two-tailed table value for a specific α

The formula for the confidence interval of a standard deviation is

$$\sqrt{\frac{(n-1)s^2}{\chi^2_{\text{larger}}}} < \sigma < \sqrt{\frac{(n-1)s^2}{\chi^2_{\text{smaller}}}}$$

The degrees of freedom for both formulas are $n-1$.

Remember that in confidence intervals, two-tailed values are used. Hence, if a 95% confidence interval is desired, the critical values obtained from Table G must correspond to the 0.025 and 0.975 levels. Also, place the larger value (corresponding to 0.025) in the denominator of the left side of the formula, and place the smaller value (corresponding to 0.975) in the denominator of the right side.

EXAMPLE 12-7   Find the 95% confidence interval of the variance and standard deviation for the nicotine content of Example 12-6.

Solution   Since $\alpha = 0.05$, the two critical values for the 0.025 and 0.975 levels for 19 degrees of freedom are 32.852 and 8.907. The 95% confidence interval for the variance is found by substituting in the formula:

$$\frac{(n-1)s^2}{\chi^2_{larger}} < \sigma^2 < \frac{(n-1)s^2}{\chi^2_{smaller}}$$

$$\frac{(20-1)(1^2)}{32.852} < \sigma^2 < \frac{(20-1)(1^2)}{8.907}$$

$$0.578 < \sigma^2 < 2.133$$

Hence, one can be 95% confident that the true variance for the nicotine content is between 0.578 and 2.133. (*Note:* This interval does include the manufacturer's claimed variance of 0.644.)

For the standard deviation, the confidence interval is

$$\sqrt{0.578} < \sigma < \sqrt{2.133}$$

$$0.760 < \sigma < 1.46$$

Hence, one can be 95% confident that the true standard deviation is between 0.760 and 1.46. ◄

EXERCISES

12-1. List three properties of the chi-square distribution.

12-2. Can the variance test be one-tailed? Two-tailed?

12-3. What are the assumptions for the chi-square variance test?

12-4. (ANS.) Using Table G, find the critical value(s) for each, show the critical and noncritical regions, and state the appropriate null and alternate hypotheses for $\sigma^2 = 225$.

a. $\alpha = 0.05$, $n = 18$, one-tailed right
b. $\alpha = 0.10$, $n = 23$, one-tailed left
c. $\alpha = 0.05$, $n = 15$, two-tailed
d. $\alpha = 0.10$, $n = 8$, two-tailed
e. $\alpha = 0.01$, $n = 17$, one-tailed right
f. $\alpha = 0.025$, $n = 20$, one-tailed left
g. $\alpha = 0.01$, $n = 13$, two-tailed
h. $\alpha = 0.025$, $n = 29$, one-tailed left

For Exercises 12-5 through 12-18, assume that the variables are normally or approximately normally distributed.

12-5. A manufacturer of pacemakers desires that the batteries used in the devices have a standard deviation less than or equal to 1.5 months. A sample of 25 batteries has a standard deviation of 1.8 months. Should the manufacturer use the batteries? Select $\alpha = 0.05$.

12-6. A state trooper knows from past studies that the standard deviation of the time it takes to inspect a truck for safety violations is 18.6 minutes. This year, the standard deviation of the 22 trucks was 15.3 minutes. At $\alpha = 0.10$, can the trooper conclude that the standard deviation has changed?

12-7. Twelve 1-gallon containers of ice cream are selected, and each is tested to determine its weight in ounces. The variance of the sample is 3.0. Test the claim that the population variance is greater than 2.8, at $\alpha = 0.01$.

12-8. A company claims that the variance of the sugar content of its yogurt is less than or equal to 25. (The sugar content is measured in milligrams per ounce.) A sample of 20 servings is selected, and the sugar content is measured. The variance of the sample was found to be 36. At $\alpha = 0.10$, is there enough evidence to reject the claim?

12-9. A researcher suggests that the variance of the ages of the seniors at White Oak College is greater than 1.8, as past studies have shown. A sample of 60 seniors is selected, and the variance is found to be 2.1 years. At $\alpha = 0.05$, is the variation of the seniors' ages greater than 1.8?

12–10. The manager of a large company claims that the standard deviation of the time (in minutes) that it takes a telephone call to be transferred to the correct office in her company is less than or equal to 1.2 minutes. A sample of 15 calls is selected and the calls timed. The standard deviation of the sample is 1.8 minutes. At $\alpha = 0.01$, is the standard deviation less than or equal to 1.2 minutes?

12–11. The standard deviation of the diameter of 28 oranges was 0.34 inch. Find the 99% confidence interval of the true standard deviation of the diameters of the oranges.

12–12. A random sample of 22 power lawn mowers was selected, and the motors were tested to see how many miles per gallon of gasoline each one obtained. The variance of the measurements was 2.6. Find the 95% confidence interval of the true variance.

12–13. A flu vaccine is tested to see how long (in months) it prevents patients from contracting the flu. The variance obtained from five tests was 1.6. Find the 98% confidence interval of the true variance.

12–14. The heights of 28 police officers from a large-city police force were measured. The standard deviation of the sample was 1.83 inches. Find the 95% confidence interval of the standard deviation of the heights of the officers.

12–15. A random sample of 15 television sets was selected, and the lifetimes (in months) of the picture tubes were measured. The variance of the sample was 8.6. Find the 90% confidence interval of the true variance.

12–16. A service station advertises that customers will have to wait no more than 30 minutes for an oil change. A sample of 28 oil changes has a standard deviation of 5.2 minutes. Find the 95% confidence interval of the population standard deviation of the time spent waiting for an oil change.

*12–17. A machine fills 12-ounce bottles with soda. In order for the machine to function properly, the standard deviation of the sample must be less than or equal to 0.03 ounce. A sample of 8 bottles is selected, and the number of ounces of soda in each bottle is as given below. At $\alpha = 0.05$, is the machine functioning properly?

    12.03, 12.1, 12.02, 11.98, 12.00, 12.05, 11.97, 11.99

*12–18. A construction company purchases steel cables. A sample of 12 cables is selected, and the breaking strength (in pounds) of each is found. The data are shown below. The lot is rejected if the standard deviation of the sample is greater than 2 pounds. At $\alpha = 0.01$, should the lot be rejected?

| 2001 | 1998 | 2002 | 2000 | 1998 | 1999 |
| 1997 | 2005 | 2003 | 2001 | 1999 | 2006 |

12–4 TEST FOR GOODNESS OF FIT

In addition to being used to test a single variance, the chi-square statistic can be used to see whether a frequency distribution fits a specific pattern. For example, in order to meet customer demands, a manufacturer of running shoes may wish to see whether the buyers show a preference for a specific style. A traffic engineer may wish to see whether accidents occur more often on some days than on others, so that he can increase police patrols accordingly. An emergency service may want to see whether it receives more calls at certain times of the day than others, so that it can provide adequate staffing.

When one is testing to see whether a frequency distribution fits a specific pattern, the chi-square **goodness-of-fit test** is used. For example, suppose a market analyst wishes to see whether consumers have any preference among five flavors of a new fruit soda. A sample of 100 people provided the following data:

Cherry	**Strawberry**	**Orange**	**Lime**	**Grape**
32	28	16	14	10

If there were no preference, one would expect that each flavor would be selected with equal frequency. In this case, the equal frequency is $100/5 = 20$. That is, *approximately* 20 people would select each flavor.

Since the frequencies for each flavor were obtained from a sample, these actual frequencies are called the **observed frequencies.** The frequencies obtained by calculation (as if there were no preference) are called the **expected frequencies.** A completed table for the test is shown as follows:

Frequency	Cherry	Strawberry	Orange	Lime	Grape
Observed	32	28	16	14	10
Expected	20	20	20	20	20

The observed frequencies will almost always differ from the expected frequencies due to sampling error; i.e., the values differ from sample to sample. But the question is: "Are these differences significant (a preference exists), or are they due to chance?" The chi-square goodness-of-fit test will enable the researcher to determine the answer.

Before computing the test value, one must state the hypotheses. The null hypothesis should be a statement indicating that there is no difference or no change. For this example, the hypotheses are as follows:

H_0: Consumers show no preference for flavor of the fruit soda.
H_1: Consumers show a preference.

In the goodness-of-fit test, the degrees of freedom are equal to the number of categories minus 1. For this example, there are five categories, namely, cherry, strawberry, orange, lime, and grape; hence, the degrees of freedom are $5 - 1 = 4$. The degrees of freedom are equal to the number of categories minus 1 because the number of subjects in each of the first four categories is free to vary. But in order for the sum to be 100—which is the total number of subjects—the number of subjects in the last category is fixed.

Formula for the Chi-Square Goodness-of-Fit Test

$$\chi^2 = \Sigma \frac{(O - E)^2}{E}$$

with degrees of freedom equal to the number of categories minus 1, and where

$$O = \text{observed frequency}$$
$$E = \text{expected frequency}$$

Two assumptions are needed for the goodness-of-fit test. These assumptions are given next.

Assumptions for the Chi-Square Goodness-of-Fit Test

1. The data are obtained from a random sample.
2. The expected frequency for each category must be 5 or more.

This test is always a one-tailed right test, since when the $(O - E)$ values are squared, the answer will always be positive. This formula is explained in the next example.

EXAMPLE 12–8   Test the claim that there is no preference in the selection of fruit soda flavors, using the data shown previously. Let $\alpha = 0.05$.

Solution   STEP 1   State the hypotheses, and identify the claim.

H_0: Consumers show no preference for the flavors (claim).
H_1: Consumers show a preference.

STEP 2   Find the critical value. The degrees of freedom are $5 - 1 = 4$, and $\alpha = 0.05$. Hence, the critical value from Table G is 9.488.

STEP 3   Compute the test value. It is helpful to make a table with the following columns:

$$O \qquad E \qquad (O - E) \qquad (O - E)^2 \qquad \frac{(O - E)^2}{E}$$

a. Place the observed values in the first column.
b. Find the expected values (as if the null hypothesis were true). In this case, each expected value is $100/5 = 20$, and place these expected values in the second column.
c. Subtract the expected values from the observed values, and place the answers in the third column. That is,

$$32 - 20 = 12$$
$$28 - 20 = 8$$
$$.$$
$$.$$
$$.$$
$$10 - 20 = -10$$

d. Square the numbers in the third column, and place the results in the fourth column. That is,

$$12^2 = 144$$
$$8^2 = 64$$
$$.$$
$$.$$
$$.$$
$$(-10)^2 = 100$$

e. Divide each square by the corresponding expected value for that class, and place the answers in the last column. That is,

$$\frac{144}{20} = 7.2$$

$$\frac{64}{20} = 3.2$$

.

.

.

$$\frac{100}{20} = 5$$

The completed table follows.

O	E	(O − E)	(O − E)²	$\dfrac{(O - E)^2}{E}$
32	20	12	144	7.2
28	20	8	64	3.2
16	20	−4	16	0.8
14	20	−6	36	1.8
10	20	−10	100	5
				18.0

f. Find the sum of the values in the last column. This sum is the test value. For this example, the test value is 18.0.

STEP 4 Make the decision. The decision is to reject the null hypothesis, since 18.0 > 9.488, as shown in Figure 12–12.

FIGURE 12–12
Critical and
Test Values for
Example 12–8

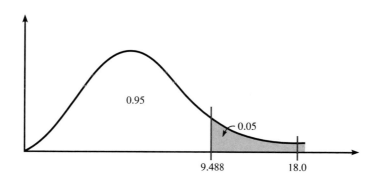

STEP 5   Summarize the results. There is enough evidence to reject the claim that consumers show no preference for the flavors.   ◄

One need not always make a table, since the test value can be computed directly from the formula, as shown:

$$\chi^2 = \Sigma \frac{(O - E)^2}{E}$$
$$= \frac{(32 - 20)^2}{20} + \frac{(28 - 20)^2}{20} + \frac{(16 - 20)^2}{20} + \frac{(14 - 20)^2}{20} + \frac{(10 - 20)^2}{20}$$
$$= 18.0$$

In order to get some idea of why this test is called the goodness-of-fit test, one can examine graphs of the observed values and expected values. See Figure 12–13. From the graphs, one can see whether the observed values and expected values are close together or far apart.

FIGURE 12–13
Graphs of the Observed and Expected Values for the Soda Flavors

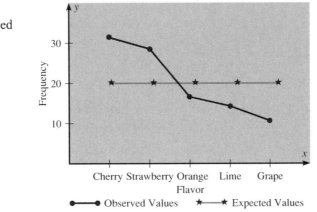

When the observed values and expected values are close together, the chi-square test value will be small. Then, the decision will be to not reject the null hypothesis—hence, there is a "good fit." See Figure 12–14a. When the observed values and the expected values are far apart, the chi-square test value will be large. Then, the null hypothesis will be rejected—hence, there is "not a good fit." See Figure 12–14b.

FIGURE 12–14
Results of the Goodness-of-Fit Test

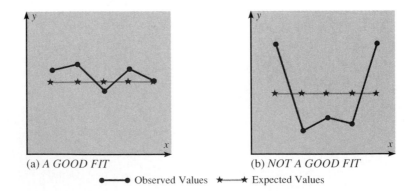

(a) *A GOOD FIT*   (b) *NOT A GOOD FIT*

●——● Observed Values   ★——★ Expected Values

The procedure for the chi-square goodness-of-fit test is summarized in Procedure Table 10.

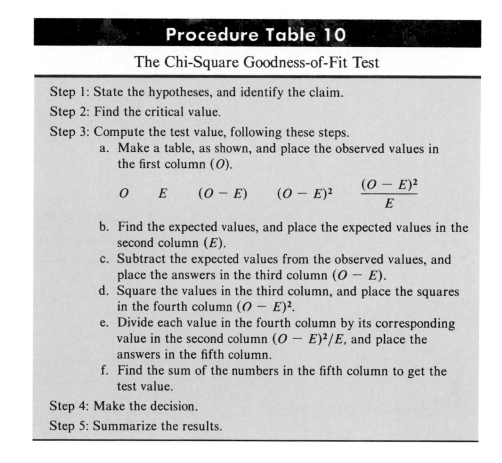

Procedure Table 10

The Chi-Square Goodness-of-Fit Test

Step 1: State the hypotheses, and identify the claim.

Step 2: Find the critical value.

Step 3: Compute the test value, following these steps.

    a. Make a table, as shown, and place the observed values in the first column (O).

$$O \quad E \quad (O - E) \quad (O - E)^2 \quad \frac{(O - E)^2}{E}$$

    b. Find the expected values, and place the expected values in the second column (E).

    c. Subtract the expected values from the observed values, and place the answers in the third column ($O - E$).

    d. Square the values in the third column, and place the squares in the fourth column ($O - E)^2$.

    e. Divide each value in the fourth column by its corresponding value in the second column ($O - E)^2/E$, and place the answers in the fifth column.

    f. Find the sum of the numbers in the fifth column to get the test value.

Step 4: Make the decision.

Step 5: Summarize the results.

EXAMPLE 12-9 The advisor of an ecology club at a large college believes that the group consists of 10% freshmen, 20% sophomores, 40% juniors, and 30% seniors. The membership for the club this year consisted of 11 freshmen, 19 sophomores, 47 juniors, and 23 seniors. At $\alpha = 0.10$, test the advisor's conjecture.

Solution **STEP 1** State the hypotheses, and identify the claim.

H_0: The club consists of 10% freshmen, 20% sophomores, 40% juniors, and 30% seniors (claim).

H_1: The distribution is not the same as stated in the null hypothesis.

STEP 2 Find the critical value. Since $\alpha = 0.10$ and the degrees of freedom are $4 - 1 = 3$, the critical value is 6.251.

STEP 3 Compute the test value, following these steps.

    a. Make a table, as shown completed in part f, and place the observed values in the first column (O).

b. Compute and place the expected values in the second column (E). Since the sample consists of 100 students and the percentage of each group is given, the expected values can be found as follows:

freshmen	10% of 100 = 10
sophomores	20% of 100 = 20
juniors	40% of 100 = 40
seniors	30% of 100 = 30

c. Subtract the expected values from the observed values, and place the answers in the third column ($O - E$).

$$11 - 10 = 1$$
$$19 - 20 = -1$$
$$47 - 40 = 7$$
$$23 - 30 = -7$$

d. Square the values obtained in the third column, and place the squares in the fourth column ($O - E)^2$.

$$1^2 = 1$$
$$(-1)^2 = 1$$
$$7^2 = 49$$
$$(-7)^2 = 49$$

e. Divide each value in the fourth column by its corresponding value in the second column, and place the answers in the fifth column ($O - E)^2/E$.

$$\frac{1}{10} = 0.1$$
$$\frac{1}{20} = 0.05$$
$$\frac{49}{40} = 1.225$$
$$\frac{49}{30} = 1.63$$

f. Find the sum of the values in the fifth column to get the test value. In this case, it is 3.005. The completed table follows.

O	E	$(O - E)$	$(O - E)^2$	$\dfrac{(O - E)^2}{E}$
11	10	1	1	0.1
19	20	−1	1	0.05
47	40	7	49	1.225
23	30	−7	49	1.63
				3.005

STEP 4 Make the decision. Since $3.005 < 6.251$, the decision is to not reject the null hypothesis. See Figure 12–15.

FIGURE 12–15
Critical and Test Values
for Example 12–9

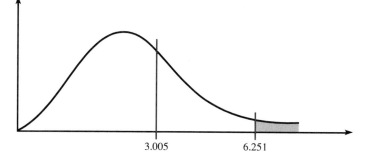

3.005   6.251

STEP 5   Summarize the results. There is not enough evidence to reject the advisor's claim.   ◄

The test value can be computed by using the formula as follows:

$$\chi^2 = \Sigma \frac{(O - E)^2}{E}$$

$$= \frac{(11 - 10)^2}{10} + \frac{(19 - 20)^2}{20} + \frac{(47 - 40)^2}{40} + \frac{(23 - 30)^2}{30} = 3.005$$

For use of the chi-square goodness-of-fit test, statisticians have determined that the expected frequencies should be at least 5, as stated in the assumptions. The reasoning is as follows: The chi-square distribution is continuous, whereas the goodness-of-fit test is discrete. However, the continuous distribution is a good approximation and can be used when the expected value for each class is at least 5. If an expected frequency of a class is less than 5, then that class can be combined with another class so that the expected frequency is 5 or more.

EXERCISES

12–19. How does the goodness-of-fit test differ from the chi-square variance test?

12–20. How are the degrees of freedom computed for the goodness-of-fit test?

12–21. How are the expected values computed for the goodness-of-fit test?

12–22. When the expected frequencies are less than 5 for a specific class, what should be done so that one can use the goodness-of-fit test?

For Exercises 12–23 through 12–35, perform the following steps.

a. State the hypotheses, and identify the claim.
b. Find the critical value.
c. Compute the test value.
d. Make the decision.
e. Summarize the results.

12–23. A staff member of an emergency medical service wishes to determine whether the number of accidents is equally distributed during the week. A week was selected at random, and the following data were obtained. Test the hypothesis that the number of accidents is equally distributed throughout the week, at $\alpha = 0.05$.

Day	Mon.	Tues.	Wed.	Thurs.	Fri.	Sat.	Sun.
No. of Accidents	28	32	15	14	38	43	19

12–24. A clothes manufacturer wants to know whether customers prefer any specific color over other colors in shirts. She selects a random sample of 100 shirts sold and notes the color. The data are shown here. At $\alpha = 0.10$, is there a color preference for the shirts?

Color	White	Blue	Black	Red	Yellow	Green
No. Sold	43	22	16	10	5	4

12–25. The concessions manager at Twin Rivers Stadium wishes to see whether there is any preference in the flavors of popcorn that are sold during sporting events. A random sample of sales is selected, and the data are as shown here. At $\alpha = 0.01$, are the flavors selected with equal frequency?

Flavor	Plain	Barbecue	Butter	Cheddar Cheese
No. Sold	25	18	32	45

12–26. A bank manager wishes to see whether there is any preference in the times that customers use the bank. Six hours are selected, and the numbers of customers visiting the bank during each hour are as shown here. At $\alpha = 0.05$, do the customers show a preference for specific times?

Time	10:00	11:00	12:00	1:00	2:00	3:00
No. of Customers	26	33	42	36	24	19

12–27. A county medical examiner hypothesizes that the blood types of the residents are distributed as 28% type A, 32% type O, 25% type B, and 15% type AB. A random sample of 200 residents is selected, and the following data are tallied. At $\alpha = 0.10$, can the examiner conclude that his hypothesis is correct?

Type	A	O	B	AB
Frequency	58	65	55	22

12–28. The chairperson of the history department of a college hypothesizes that the final grades are distributed as 40% A's, 30% B's, 20% C's, 5% D's, and 5% F's. At the end of the semester, the following numbers of grades were earned. For $\alpha = 0.05$, is the grade distribution for the department different than expected?

Grade	A	B	C	D	F
Number	45	52	39	8	6

12–29. During the summer months, a medical emergency services director claims that his center receives 30% of its weekly calls on Saturday, 30% on Sunday, and 8% on each weekday. A week is selected at random, and the number of calls is found to be distributed as shown below. At $\alpha = 0.05$, test the director's claim.

Day	Sun.	Mon.	Tues.	Wed.	Thurs.	Fri.	Sat.
No. of Calls	32	7	10	5	9	6	21

12–30. In a check to see whether a die is loaded, each number on the face should appear approximately $\frac{1}{6}$ of the time. A suspected die is selected and rolled 180 times, and the results are tallied as shown. At $\alpha = 0.10$, can one conclude that the die is loaded?

Number	1	2	3	4	5	6
Frequency	34	38	25	21	30	32

12–31. A fast-food restaurant is offering three new menu items, chili dogs (CD), bacon cheeseburgers (BC), and taco salads (TS). A day is selected at random, and the number of each new item ordered is tallied. The data follow. At $\alpha = 0.01$, can the restaurant conclude that the customers show no preference in the menu selection?

Item	CD	BC	TS
Frequency	57	68	43

12–32. The owner of a sporting goods store wishes to see whether his customers show any preference for the month in which they purchase hunting rifles. The sales of rifles for the end of last year are shown here. At $\alpha = 0.05$, test the claim that there is no preference for the month in which the customers purchase guns.

Month	Sept.	Oct.	Nov.	Dec.
No. Sold	18	23	28	15

12–33. The dean of students of a college wishes to test the claim that the distribution of students is as follows: 40% business (BU), 25% computer science (CS), 15% science (SC), 10% social science (SS), 5% liberal arts (LA), and 5% general studies (GS). Last semester, the program enrollment was distributed as shown below. At $\alpha = 0.10$, is the distribution of students the same as hypothesized?

Major	BU	CS	SC	SS	LA	GS
Number	72	53	32	20	16	7

12–34. A quality control engineer for a manufacturing plant wishes to determine whether the number of defective items manufactured during the week is approximately the same on each day. A week is selected at random, and the number of defective items produced each day is as shown. At $\alpha = 0.05$, can the engineer conclude that the defective items are produced with the same frequency each day?

Day	Mon.	Tues.	Wed.	Thurs.	Fri.
Number	32	16	23	19	40

12–35. A computer store owner feels that 50% of her customers purchase word-processing programs, 25% purchase spreadsheet programs, and 25% purchase data base programming. A sample of purchases shows the following distribution. At $\alpha = 0.05$, is her assumption correct?

Program	Word-processing	Spreadsheet	Data base
No. of Purchases	38	23	19

*12–36. Three coins are tossed 72 times and the number of heads is as shown. At $\alpha = 0.05$, test the null hypothesis that the coins are balanced and randomly tossed. (*Hint:* Use the binomial distribution.)

No. of Heads	0	1	2	3
Frequency	3	10	17	42

*12–37. Select a three-digit state lottery number over a period of 50 days. Count the number of times each digit, 0 through 9, occurs. Test the claim, at $\alpha = 0.05$, that the digits occur at random.

12–5 TEST FOR INDEPENDENCE

When data are tabulated in table form in terms of frequencies, a chi-square **independence test** can be used to test the independence of two variables. For example, suppose a new postoperative procedure is administered to a number of patients in a large hospital. One can ask the question, "Do the doctors feel differently about this procedure than the nurses do, or do they feel basically the same way?" Note that the question is not whether or not they prefer the procedure but whether there is a difference of opinion between the two groups.

In order to answer this question, a researcher selects a sample of nurses and doctors and tabulates the data in table form, as shown.

Group	Prefer New Procedure	Prefer Old Procedure	No Preference
Nurses	100	80	20
Doctors	50	120	30

As the survey indicates, 100 nurses prefer the new procedure, 80 prefer the old procedure and 20 have no preference; whereas 50 doctors prefer the new procedure, 120 like the old procedure, and 30 have no preference. Since the main question is whether there is a difference in opinion, the null hypothesis is stated as follows:

H_0: The opinion about the procedure is *independent* of the profession.

The alternative hypothesis is stated as follows:

H_1: The opinion about the procedure is *dependent* on the profession.

If the null hypothesis is not rejected, the test means that both professions feel basically the same way about the procedure, and the differences are due to chance. If the null hypothesis is rejected, the test means that one group feels differently about the procedure than the other. Remember that rejection does *not* mean that one group favors the procedure and the other does not. Perhaps both groups favor it or both dislike it, but in different proportions.

In order to test the null hypothesis using the chi-square independence test, one must compute the expected frequencies, assuming that the null hypothesis is true. These frequencies are computed by using the observed frequencies given in the table.

The Contingency Table

When data are arranged in table form for the chi-square independence test, the table is called a **contingency table.** The table is made up of R rows and C columns. The table shown previously has two rows and three columns.

Group	Prefer New Procedure	Prefer Old Procedure	No Preference
Nurses	100	80	20
Doctors	50	120	30

Note that row and column headings do not count in determining the number of rows and columns.

A contingency table is designated as an $R \times C$ (rows times columns) table. In this case, $R = 2$ and $C = 3$; hence, this table is a 2×3 contingency table. Each block in the table is called a *cell* and is designated by its row and column position. For example, the cell with a frequency of 8 is designated as $C_{1,2}$, i.e., row 1, column 2. The cells are shown below.

	Column 1	Column 2	Column 3
Row 1	$C_{1,1}$	$C_{1,2,}$	$C_{1,3}$
Row 2	$C_{2,1}$	$C_{2,2}$	$C_{2,3}$

The degrees of freedom for any contingency table are (rows $-$ 1) times (columns $-$ 1); that is, d.f. $= (R - 1)(C - 1)$. In this case, $(2 - 1)(3 - 1) = (1)(2) = 2$. The reason for this formula for d.f. is that all the expected values except one are free to vary in each row and in each column.

Computation of Expected Frequencies

Using the previous table, one can compute the expected frequencies for each block (or cell) as shown next.

a. Find the sum of each row and each column, and find the grand total, as shown.

Group	Prefer New Procedure	Prefer Old Procedure	No Preference	
Nurses	100	80	20	Row 1 sum 200
Doctors	+ 50	+ 120	+ 30	Row 2 sum 200
	150	200	50	(400)
	Column 1 sum	Column 2 sum	Column 3 sum	Grand total

b. For each cell, multiply the corresponding row sum by the column sum and divide by the grand total, to get the expected value:

$$\text{expected value} = \frac{\text{row sum} \times \text{column sum}}{\text{grand total}}$$

For example, for $C_{1,2}$, the expected value denoted by $E_{1,2}$ is (refer to the previous tables)

$$E_{1,2} = \frac{(200)(200)}{400} = 100$$

For each cell, the expected values are computed as follows:

$$E_{1,1} = \frac{(200)(150)}{400} = 75 \qquad E_{2,1} = \frac{(200)(150)}{400} = 75$$

$$E_{1,2} = \frac{(200)(200)}{400} = 100 \qquad E_{2,2} = \frac{(200)(200)}{400} = 100$$

$$E_{1,3} = \frac{(200)(50)}{400} = 25 \qquad E_{2,3} = \frac{(200)(50)}{400} - 25$$

The expected values can now be placed in the corresponding cells along with the observed values, as shown.

Group	Prefer New Procedure	Prefer Old Procedure	No Preference	
Nurses	100 (75)	80 (100)	20 (25)	200
Doctors	50 (75)	120 (100)	30 (25)	200
	150	200	50	400

The rationale for the computation of the expected frequencies for a contingency table uses proportions. For $C_{1,1}$ a total of 150 out of 400 people prefer the new procedure. And since there are 200 nurses, one would expect, if the null hypothesis were true, (150/400)(200), or 75, of the nurses to be in favor of the new procedure.

Statistical Test

The formula for the test value for the independence test is the same as the one used for the goodness-of-fit test. It is

$$\chi^2 = \sum \frac{(O - E)^2}{E}$$

For the previous example, compute the $(O - E)^2/E$ values for each cell, and then find the sum.

$$\chi^2 = \sum \frac{(O - E)^2}{E}$$

$$= \frac{(100 - 75)^2}{75} + \frac{(80 - 100)^2}{100} + \frac{(20 - 25)^2}{25} + \frac{(50 - 75)^2}{75}$$

$$+ \frac{(120 - 100)^2}{100} + \frac{(30 - 25)^2}{25}$$

$$= 26.67$$

The final steps are to make the decision and summarize the results. This test is always a one-tailed right test, and the degrees of freedom are $(R - 1)(C - 1) = (2 - 1)(3 - 1) = 2$. If $\alpha = 0.05$, the critical value is 5.991. Hence, the decision is to reject the null hypothesis, since $26.67 > 5.991$. See Figure 12–16.

FIGURE 12–16
Critical and Test Values for the Postoperative Procedures Example

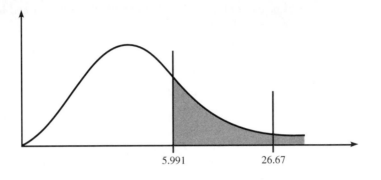

5.991   26.67

The conclusion is that there is enough evidence to support the claim that opinion is related to (dependent on) profession—i.e., that the doctors and nurses differ in their opinions about the procedure.

The next two examples illustrate the procedure for the chi-square test of independence.

EXAMPLE 12–10   A sociologist wishes to see whether the number of years of college a person has completed is related to his or her place of residence. A sample of 70 is selected and classified as shown.

Location	No College	B.S. Degree	Advanced Degree	Total
Urban	15	12	8	35
Suburban	8	15	9	32
Rural	6	8	7	21
Total	29	35	24	88

At $\alpha = 0.05$, can the sociologist conclude that the years of college education are dependent upon the residence location?

Solution   **STEP 1**   State the hypotheses, and identify the claim.

H_0: A person's place of residence is independent of the number of years of college completed.

H_1: A person's place of residence is dependent on the number of years of college completed (claim).

STEP 2   Find the critical value. The critical value is 9.488, since the degrees of freedom are $(3 - 1)(3 - 1) = (2)(2) = 4$.

STEP 3 Compute the test value. To compute the test value, one must first compute the expected values.

$$E_{1,1} = \frac{(35)(29)}{88} = 11.53 \qquad E_{2,3} = \frac{(32)(24)}{88} = 8.73$$

$$E_{1,2} = \frac{(35)(35)}{88} = 13.92 \qquad E_{3,1} = \frac{(21)(29)}{88} = 6.92$$

$$E_{1,3} = \frac{(35)(24)}{88} = 9.55 \qquad E_{3,2} = \frac{(21)(35)}{88} = 8.35$$

$$E_{2,1} = \frac{(32)(29)}{88} = 10.55 \qquad E_{3,3} = \frac{(21)(24)}{88} = 5.73$$

$$E_{2,2} = \frac{(32)(35)}{88} = 12.73$$

The completed table is as shown.

Location	No College	B.S. Degree	Advanced Degree	Total
Urban	15 (11.53)	12 (13.92)	8 (9.55)	35
Suburban	8 (10.55)	15 (12.73)	9 (8.73)	32
Rural	6 (6.92)	8 (8.35)	7 (5.73)	21
	29	35	24	88

Then, the chi-square test value is

$$\chi^2 = \sum \frac{(O - E)^2}{E}$$

$$= \frac{(15 - 11.53)^2}{11.53} + \frac{(12 - 13.92)^2}{13.92} + \frac{(8 - 9.55)^2}{9.55}$$

$$+ \frac{(8 - 10.55)^2}{10.55} + \frac{(15 - 12.73)^2}{12.73} + \frac{(9 - 8.73)^2}{8.73}$$

$$+ \frac{(6 - 6.92)^2}{6.92} + \frac{(8 - 8.35)^2}{8.35} + \frac{(7 - 5.73)^2}{5.73}$$

$$= 3.03$$

STEP 4 Make the decision. The decision is to not reject the null hypothesis, since $3.03 < 9.488$. See Figure 12–17.

FIGURE 12–17
Critical and Test Values
for Example 12–10

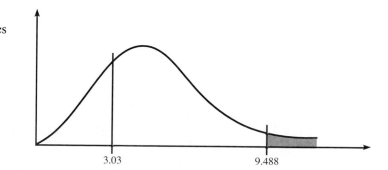

3.03   9.488

STEP 5 Summarize the results. There is not enough evidence to support the claim that a person's place of residence is dependent on the number of years of college completed. ◀

EXAMPLE 12–11 A researcher wishes to determine whether there is a relationship between the gender of an individual and the amount of alcohol consumed. A sample of 68 people was selected, and the following data were obtained.

Alcohol Consumption

Gender	Low	Moderate	High	Total
Male	10	9	8	27
Female	13	16	12	41
Total	23	25	20	68

At $\alpha = 0.10$, can the researcher conclude that alcohol consumption is related to the gender of the individual?

Solution **STEP 1** State the hypotheses, and identify the claim.

H_0: The amount of alcohol that a person consumes is independent of the gender of the individual.

H_1: The amount of alcohol that a person consumes is dependent on the gender of the individual (claim).

STEP 2 Find the critical value. The critical value is 4.605, since the degrees of freedom are $(2 - 1)(3 - 1) = 2$.

STEP 3 Compute the test value. First, compute the expected values.

$$E_{1,1} = \frac{(27)(23)}{68} = 9.13 \qquad E_{2,1} = \frac{(41)(23)}{68} = 13.87$$

$$E_{1,2} = \frac{(27)(25)}{68} = 9.93 \qquad E_{2,2} = \frac{(41)(25)}{68} = 15.07$$

$$E_{1,3} = \frac{(27)(20)}{68} = 7.94 \qquad E_{2,3} = \frac{(41)(20)}{68} = 12.06$$

The completed table is as shown next.

Alcohol Consumption

Gender	Low	Moderate	High	Total
Male	10 (9.13)	9 (9.93)	8 (7.94)	27
Female	13 (13.87)	16 (15.07)	12 (12.06)	41
	23	25	20	68

Then, the test value is

$$\chi^2 = \sum \frac{(O - E)^2}{E}$$
$$= \frac{(10 - 9.13)^2}{9.13} + \frac{(9 - 9.93)^2}{9.93} + \frac{(8 - 7.94)^2}{7.94}$$
$$+ \frac{(13 - 13.87)^2}{13.87} + \frac{(16 - 15.07)^2}{15.07} + \frac{(12 - 12.06)^2}{12.06}$$
$$= 0.283$$

STEP 4 Make the decision. The decision is to not reject the null hypothesis, since $0.283 < 4.605$. See Figure 12–18.

FIGURE 12–18
Critical and Test Values
for Example 12–11

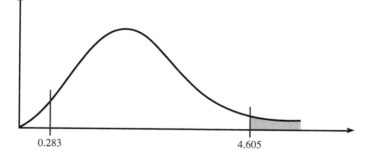

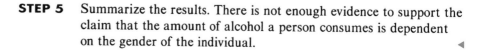

0.283   4.605

STEP 5   Summarize the results. There is not enough evidence to support the claim that the amount of alcohol a person consumes is dependent on the gender of the individual.   ◄

When the degrees of freedom for a contingency table are equal to 1—i.e., the table is a 2 × 2 table—some statisticians suggest using the *Yates correction for continuity*. The formula for the test is then

$$\chi^2 = \sum \frac{(|O - E| - 0.5)^2}{E}$$

Since the chi-square test is already conservative, most statisticians agree that the Yates correction is not necessary. (See Exercise 12–55.)

The procedure for the chi-square independence test is summarized in Procedure Table 11.

Procedure Table 11

The Chi-Square Independence Test

Step 1: State the hypotheses, and identify the claim.

Step 2: Find the critical value.

Step 3: Compute the test value. To compute the test value, first find the expected values. For each cell of the contingency table, use the formula

$$E = \frac{(\text{row sum})(\text{column sum})}{\text{grand total}}$$

to get the expected value. To find the test value, use the formula

$$\chi^2 = \sum \frac{(O - E)^2}{E}$$

Step 4: Make the decision.

Step 5: Summarize the results.

Speaking of **Statistics**

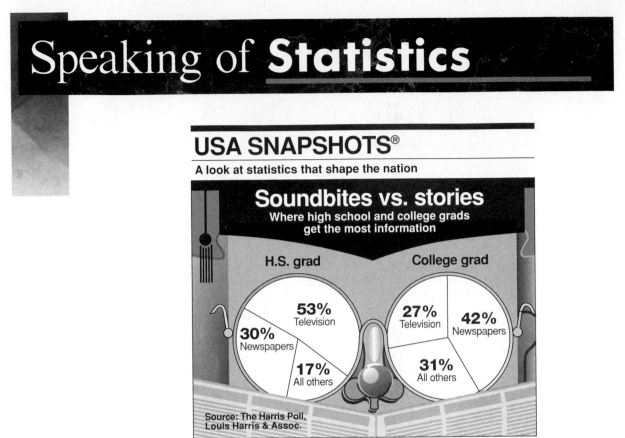

USA SNAPSHOTS®

A look at statistics that shape the nation

Soundbites vs. stories
Where high school and college grads get the most information

H.S. grad

College grad

53% Television

30% Newspapers

17% All others

27% Television

42% Newspapers

31% All others

Source: The Harris Poll, Louis Harris & Assoc.

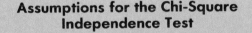

By Patti Stang and Marcia Staimer, USA TODAY

Source: Based on the Harris Poll, Louis Harris and Associates. Copyright 1993, USA TODAY. Reprinted with permission.

This illustration shows how high school and college graduates receive information. How could one use a chi-square independence test to see whether the two variables are related?

Assumptions for the Chi-Square Independence Test

1. The data are obtained from a random sample.
2. The expected value in each cell must be 5 or more.

EXERCISES

12–38. How is the chi-square independence test similar to the goodness-of-fit test? How is it different?

12–39. How are the degrees of freedom computed for the independence test?

12–40. When the observed frequencies are close to the expected frequencies, what is the value of chi-square?

12–41. Generally, how would the null and alternative hypotheses be stated for the chi-square independence test?

12–42. What is the name of the table used in the independence test?

12–43. How are the expected values computed for each cell in the table?

For Exercises 12–44 through 12–53, perform the following steps.

  a. State the hypotheses, and identify the claim.
  b. Find the critical value.
  c. Compute the test value.
  d. Make the decision.
  e. Summarize the results.

12–44. A study is being conducted to determine whether there is a relationship between jogging and blood pressure. A random sample of 210 subjects is selected, and they are classified as shown in the table. At $\alpha = 0.05$, test the claim that jogging and blood pressure are not related.

	Blood Pressure		
Jogging Status	**Low**	**Moderate**	**High**
Joggers	34	57	21
Nonjoggers	15	63	20

12–45. A researcher wishes to see whether the age of an individual is related to milk consumption. A sample of 152 individuals is selected, and they are classified as shown in the table. At $\alpha = 0.10$, is there a relationship between milk consumption and age?

	Milk Consumption		
Age	**Low**	**Moderate**	**High**
21–30	12	16	18
31–40	15	27	9
41–50	12	10	5
51 and over	6	9	13

12–46. A survey of the 164 state representatives is conducted to see whether their opinions on a bill are related to their party affiliation. The following data were obtained. At $\alpha = 0.01$, can the researcher conclude that opinions are related to party affiliations?

	Opinion		
Party	**Approve**	**Disapprove**	**No Opinion**
Republican	27	15	13
Democrat	43	18	12
Independent	9	15	12

12–47. An automobile insurance company wishes to determine whether the age of the insured is related to the amount of liability the driver carries. A sample of 222 drivers shows the following data. At $\alpha = 0.05$, is the amount of coverage independent of the age of the driver?

	Amount		
Age	**Under $50,000**	**$50,000– $100,000**	**Over $100,000**
21–30	16	25	3
31–40	23	44	15
41–50	15	31	18
51 and over	9	12	11

12–48. In order to test the effectiveness of a new drug, a researcher gives one group of individuals the new drug and another group a placebo. The results of the study are as shown in the table. At $\alpha = 0.10$, can the researcher conclude that the drug is effective?

Medication	**Effective**	**Not Effective**
Drug	32	9
Placebo	12	18

12–49. A greenhouse owner wishes to determine whether the number of plants she sells is related to the location of her portable plant stands. A sample of locations is selected, and the number of sales at each area is tallied over a three-day period. The results are shown in the table. At $\alpha = 0.01$, is the number of plants sold related to the location of the stand?

	Number of Plants Sold		
Location of Plant Stands	**Mon.**	**Tues.**	**Wed.**
Mall	17	12	4
Grocery	12	23	47
Roadside	23	15	52

12–50. A study was conducted to determine whether the preference for a two-wheel drive or a four-wheel drive automobile is related to the gender of the purchaser. A sample of 90 buyers was selected, and the data are as shown in the table. At $\alpha = 0.05$, can the researcher conclude that the automobile preference is independent of the gender of the buyer?

Gender	Two-Wheel Drive	Four-Wheel Drive
Male	23	43
Female	18	6

12–51. A university official wishes to determine whether the degree of the instructor is related to the students' opinion of the quality of instruction received. A sample of students' evaluations of various instructors is selected, and the data below are obtained. At $\alpha = 0.10$, can the official conclude that the degree of the instructor is related to the opinions of the students about that instructor's effectiveness in the classroom?

	Degree		
Rating	B.S.	M.S.	Ph.D.
Excellent	14	9	4
Average	16	5	7
Poor	3	12	16

12–52. A researcher wishes to determine whether the marital status of a student is related to his or her grade in a statistics course. The data below were obtained from a random sample of 142 students. At $\alpha = 0.05$, is the marital status independent of the grade received in the course?

Marital	Grade			
Status	A	B	C	D or F
Single	27	32	16	10
Married	14	19	16	8

12–53. A study is being conducted to determine whether the age of the customer is related to the type of video movie he or she rents. A sample of renters gives the data shown in the table. At $\alpha = 0.10$, is the type of movie selected related to the age of the customer?

	Type of Movie		
Age	Documentary	Comedy	Mystery
12–20	14	9	8
21–29	15	14	9
30–38	9	21	39
39–47	7	22	17
48 and over	6	38	12

*12–54. For a 2×2 table, a, b, c, and d are the observed values for each cell, as shown.

a	b
c	d

The chi-square test value can be computed as

$$\chi^2 = \frac{n(ad - bc)^2}{(a + b)(a + c)(c + d)(b + d)}$$

where $n = a + b + c + d$. Compute the χ^2 test value by using the above formula and by using the formula $\Sigma (O - E)^2/E$, and compare the results for the following table.

12	15
9	23

*12–55. For the contingency table shown in Exercise 12–54, compute the chi-square test value by using Yates's correction for continuity.

*12–56. When the chi-square test value is significant, and there is a relationship between the variables, the strength of this relationship can be measured by using the *contingency coefficient*. The formula for the contingency coefficient is

$$C = \sqrt{\frac{\chi^2}{\chi^2 + n}}$$

where χ^2 is the test value and n is the sum of frequencies of the cells. The contingency coefficient will always be less than 1. Compute the contingency coefficient for Exercises 12–44 and 12–48.

12–6 SUMMARY

Three uses of the chi-square distribution were explained in this chapter. The variance test is used to test the hypothesis that a variance or standard deviation of a population is equal to a specific value. For example, a researcher may wish to determine whether the variance of IQ scores for a population is equal to 225. This test can be one-tailed (either right or left) or two-tailed.

Speaking of **Statistics**

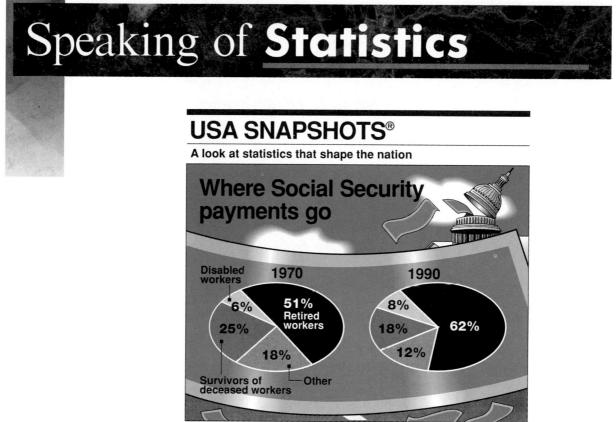

Source: Based on data from the U.S. Social Security Administration. Copyright 1993, USA TODAY. Reprinted with permission.

This "USA SNAPSHOTS" feature shows how Social Security money was spent in 1970 and in 1990. Do you feel that there is a difference in the two distributions? If so, suggest several reasons for the differences in the two pie graphs.

The second chi-square test is the goodness-of-fit test, and it is used to determine whether the frequencies of a distribution are the same as the hypothesized frequencies. For example, is the number of defective parts produced by a factory the same each day? This test is always a one-tailed right test.

The third chi-square test is the test of independence, and it is used to determine whether two variables are related or are independent. This test uses a contingency table and is always a one-tailed right test. An example of the use of this test is a test to determine whether the attitudes of urban residents about the recycling of trash differ from the attitudes of rural residents.

The chi-square distribution is also used for other types of statistical hypothesis tests, such as the Kruskal-Wallis test, which is explained in Chapter 14.

Important Terms

Chi-square distribution
Contingency table
Expected frequency
Goodness-of-fit test
Independence test
Observed frequency

Important Formulas

Formula for the chi-square test for a single variance:

$$\chi^2 = \frac{(n-1)s^2}{\sigma^2}$$

with degrees of freedom equal to $n - 1$ and where

n = sample size
s^2 = sample variance
σ^2 = population variance

Formula for the confidence interval of a variance:

$$\frac{(n-1)s^2}{\chi^2_{larger}} < \sigma^2 < \frac{(n-1)s^2}{\chi^2_{smaller}}$$

Formula for the chi-square test for goodness of fit:

$$\chi^2 = \sum \frac{(O-E)^2}{E}$$

with degrees of freedom equal to the number of categories minus 1 and where

O = observed frequency
E = expected frequency

Formula for the chi-square independence test:

$$\chi^2 = \sum \frac{(O-E)^2}{E}$$

with degrees of freedom equal to (rows − 1)(columns − 1).
Formula for the expected value for each cell:

$$E = \frac{(\text{row sum})(\text{column sum})}{\text{grand total}}$$

Review Exercises

For Exercises 12–57 through 12–64, assume that the variables are normally distributed. For Exercises 12–57, 12–59, 12–61, 12–63, and 12–65 through 12–72, perform the following steps.

a. State the hypotheses, and identify the claim.
b. Find the critical value(s).
c. Compute the test value.
d. Make the decision.
e. Summarize the results.

12–57. A film editor feels that the standard deviation for the number of minutes in a music video is 3.4 minutes. A sample of 24 videos has a standard deviation of 4.2 minutes. At $\alpha = 0.05$, is the sample standard deviation different from what the editor hypothesized?

12–58. Find the 95% confidence interval of the true standard deviation in Exercise 12–57.

12–59. A researcher hypothesizes that the standard deviation of the gestation time of births of males is 2 weeks. A sample of 50 male births had a standard deviation of 18 days. At $\alpha = 0.01$, can the researcher conclude that the hypothesis is correct?

12–60. Construct the 99% confidence interval of the true standard deviation of the gestation time in Exercise 12–59.

12–61. The standard deviation of the fuel consumption of a certain automobile is hypothesized to be greater than or equal to 4.3 miles per gallon. A sample of 20 automobiles produced a standard deviation of 2.6 miles per gallon. Is the standard deviation really less than previously thought? Use $\alpha = 0.05$.

12–62. A sample of the ages of nine contestants on a television game show had a standard deviation of 6.2 years. Find the 90% confidence interval of the population standard deviation of all the contestants that appeared on the show.

12–63. A manufacturer claims that the standard deviation of the drying time of a certain type of paint is 18 minutes. A sample of five test panels produced a standard deviation of 21 minutes. Can one conclude, at $\alpha = 0.05$, that the claim is correct?

12–64. Find the 95% confidence interval of the standard deviation of the drying time of paint in Exercise 12–63.

12–65. A company owner wishes to determine whether the number of sales of a product is equally distributed over five regions. A month is selected at random, and the number of sales is recorded. The data are as shown in the table. At $\alpha = 0.05$, can the owner conclude that the number of items sold in each region is the same?

Region	NE	SE	MW	NW	SW
Sales	236	324	182	221	365

12–66. An ad is placed in newspapers in four different counties asking for volunteers to test a new medication for reducing a person's blood pressure. The number of inquiries received in each area is as shown in the table. At $\alpha = 0.01$, can one conclude that all ads produced the same number of responses?

County	1	2	3	4
No. of Inquiries	87	62	56	93

12–67. A bartender decides to test his theory that 50% of the customers order beer, 35% order a cocktail, and 15% order a nonalcoholic drink. An hour is selected at random, and the number of drinks sold is recorded as shown in the table. At $\alpha = 0.05$, is the bartender's claim valid?

Drink	Beer	Cocktail	Nonalcoholic
Number	93	65	42

12–68. A recent survey shows that the number of each type of video game purchased for a randomly selected week of the year is as given in the table. At $\alpha = 0.01$, can one conclude that each type of video game was purchased with equal frequency?

Game	Haunted House	Chess-Mate	Crazy Cars
No. Purchased	324	312	225

12–69. An educator wishes to determine whether there is a difference in the favorite subjects selected by males and females in college. A random sample of males and females provides the data given in the table. At $\alpha = 0.05$, can the educator conclude that subject selection is independent of the gender of the individual?

Gender	Subject		
	Math	**History**	**English**
Male	324	223	191
Female	152	104	312

12–70. A pet store owner wishes to determine whether the type of pet a person selects is related to the individual's gender. The data obtained from a sample are as shown in the table. At $\alpha = 0.10$, is the gender of the purchaser related to the type of pet purchased?

Gender of Purchaser	Pet Purchased		
	Dog	**Cat**	**Bird**
Male	32	27	16
Female	23	4	8

12–71. A survey at a ballpark shows the following selection of condiments made for the hot dogs purchased. At $\alpha = 0.10$, is the condiment chosen independent of the gender of the individual?

Gender	Condiment		
	Catsup	**Mustard**	**Relish**
Men	15	18	10
Women	25	14	8

12–72. A researcher suggests that there is a relationship between the type of television service viewers receive and the way the viewers get the programming information. A sample of viewers shows the data given in the table. At $\alpha = 0.05$, is there a relationship between the type of service and the way a person gets program information?

Type of Service	Program Information	
	Newspaper	**TV Magazine**
Cable	26	54
Satellite	43	52
Antenna	63	27

Speaking of Statistics

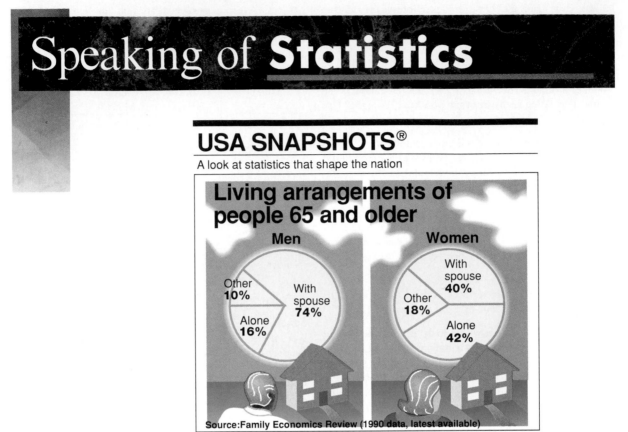

USA SNAPSHOTS®

A look at statistics that shape the nation

Living arrangements of people 65 and older

Men

Other **10%**
With spouse **74%**
Alone **16%**

Women

With spouse **40%**
Other **18%**
Alone **42%**

Source:Family Economics Review (1990 data, latest available)

By Elys A. McLean, USA TODAY

Source: Based on data from the Family Economics Review (1990 data, latest available). Copyright 1993, USA TODAY. Reprinted with permission.

How would one use the chi-square independence test to see whether the living arrangements of people 65 and older are related to the gender of the individual?

🖥 COMPUTER APPLICATIONS

1. Write a computer program for the chi-square goodness-of-fit test. Enter the number of classes, the observed values, and the expected values. Compute the chi-square test value.
2. Write a computer program for the chi-square independence test. Enter the number of rows and columns and the observed values. Have the computer calculate the expected values, print a table with the observed and expected values, and compute the chi-square test value.

3. MINITAB can be used for the chi-square goodness-of-fit test and the independence test. For example, suppose that a researcher wishes to test the hypothesis that the number of want ads placed in a local newspaper is approximately the same each day of the week. A week is selected at random, and the number of ads is recorded as follows:

Mon.	Tues.	Wed.	Thurs.	Fri.	Sat.
87	62	51	65	92	105

Test the hypothesis at $\alpha = 0.05$.

In this case, the expected values must be computed. For this problem, the expected value for each day is $462/6 = 77$. The observed values are entered by using the SET C1 command, and the expected values are entered by using the SET C2 command. To get MINITAB to compute the chi-square test value, use the command LET K1 = SUM((C1 − C2) ** 2/C2). To have MINITAB print the test value, use the PRINT K1 command.

Type the following information:

```
MTB  > SET C1
DATA > 87 62 51 65 92 105
DATA > END
MTB  > SET C2
DATA > 77 77 77 77 77 77
DATA > END
MTB  > LET K1=SUM((C1-C2)**2/C2)
MTB  > PRINT K1
```

The computer will print the following:

```
K1  27.9740
```

Find the critical value in the table by using d.f. $= 6 - 1 = 5$. It is 11.0705 (see below). The decision, then, is to reject the null hypothesis.

To get the critical value, type the following:

```
MTB > INVCDF 0.95
SUBC > CHISQUARE DF 5.
```

The computer will print

```
0.9500  11.0705
```

Try Exercise 12–65 by using this program.

4. The independence test is shown in this exercise. A study showed how three newspapers reported stories about the president over a one-month period.

Newspaper	Favorable	Unfavorable
Sun Daily	75	25
Tribune Times	54	46
Evening Star	60	40

Test the claim that the newspapers do not view the president differently. Use $\alpha = 0.05$.

To get MINITAB to perform an independence test, enter the data using the READ C1 C2 command. After the data have been entered, type the CHISQUARE C1-C2 command as shown.

```
MTB  > READ C1 C2
DATA > 75 25
DATA > 54 46
DATA > 60 40
DATA > END
        3 ROWS READ
MTB  > CHISQUARE C1-C2
```

The computer will print the following (as noted, expected counts are printed below observed counts):

```
             C1        C2       Total
    1        75        25        100
           63.00     37.00
    2        54        46        100
           63.00     37.00
    3        60        40        100
           63.00     37.00
Total      189       111        300

ChiSq = 2.286 + 3.892 +
        1.286 + 2.189 +
        0.143 + 0.243 = 10.039
df = 2
```

To get the critical value, type the following:

```
MTB > INVCDF    0.95;
SUBC > CHISQUARE DF   2.
0.95    5.9915
```

The decision is to reject the null hypothesis, since $10.039 > 5.9915$.

Using this program, try Exercise 12–71.

√ DATA ANALYSIS Applying the Concepts

The Data Bank is located in the Appendix.

1. From the Data Bank, select a sample of 20 individuals and test the hypothesis H_0: $\sigma^2 = 225$ for IQ level.
2. Test the hypothesis that the marital status of the individuals is equally distributed among the four groups. Use the chi-square goodness-of-fit test. Choose a sample of at least 50 individuals.
3. Use the chi-square test of independence to test the hypothesis that smoking is independent of the gender of the individual. Use a sample size of at least 50 individuals.
4. From the Data Bank, choose a variable and sample 30 individuals. Find the 95% and 99% confidence intervals of the true variance.

✎ TEST

Directions: Determine whether the statement is true or false.

1. The values of the chi-square variable cannot be negative.
2. The degrees of freedom for a 4×3 contingency table are 12.
3. When the expected values are nearly the same as the observed values, the chi-square test value will be large.
4. The chi-square goodness-of-fit test is always one-tailed.
5. The chi-square test of independence is always two-tailed.
6. The null hypothesis for the chi-square test of independence is that the variables are dependent or related.
7. The degrees of freedom for the goodness-of-fit test are the number of categories minus 1.
8. An important assumption for the chi-square tests is that the observations must be independent of each other.
9. The test value for the chi-square goodness-of-fit test and for the chi-square independence test is computed by using the same formula.
10. When the null hypothesis is rejected in the chi-square variance test, there is a high likelihood that the population variance is not equal to the value hypothesized in H_0.

CHAPTER

13

The F Test and Analysis of Variance

13–1 INTRODUCTION

In Chapter 10, the z and t tests were used when the means of two populations were compared; however, many statistical studies involve comparing two variances or standard deviations. For example, is the variance of IQ scores for college freshmen greater than the variance of IQ scores of college juniors? In another situation, a researcher may be interested in comparing the variance of the cholesterol of men with the variance of the cholesterol of women. For the comparison of two variances or standard deviations, an *F* **test** is used. The *F* test should not be confused with the chi-square test, which compares a single sample variance to a specific population variance, as shown in Chapter 12.

Another use for the *F* test is in a statistical technique called *analysis of variance* (ANOVA). This technique is used to test claims involving three or more means. (*Note:* The *F* test can also be used to test the equality of two means. But since it is equivalent to the *t* test in this case, the *t* test is usually used instead of the *F* test when there are only two means.) For example, suppose a researcher wishes to see whether the means of the time it takes three groups of students to solve a computer problem using FORTRAN, BASIC, and PASCAL are different. The researcher will use the ANOVA technique for this test. The *z* and *t* tests should not be used when three or more means are compared, for reasons given later in this chapter.

For three groups, the *F* test can only indicate whether or not a difference exists among the three means. It cannot indicate where the difference lies—i.e., between $\overline{X}_1$ and $\overline{X}_2$, or $\overline{X}_1$ and $\overline{X}_3$, or $\overline{X}_2$ and $\overline{X}_3$. If the *F* test indicates that there is a difference among the means, other statistical tests are used to find where the difference exists. The most commonly used tests are the Scheffé test and the Tukey test. These tests are also explained in this chapter.

The analysis of variance that is used to compare three or more means is called a *one-way analysis of variance* since it contains only one variable. In the example above, the variable is the type of computer language used. The analysis of variance can be extended to studies involving two variables, such as type of computer language used and mathematical background of the students. These studies involve a *two-way analysis of variance*. Section 13–6 explains the two-way analysis of variance.

13–2 THE *F* TEST

If two independent samples are selected from two normally distributed populations in which the variances are equal ($\sigma_1^2 = \sigma_2^2$) and if the variances s_1^2 and s_2^2 are compared as $\dfrac{s_1^2}{s_2^2}$, the sampling distribution of the variances is called the *F* **distribution.**

Characteristics of the F Distribution

1. The values of F cannot be negative, because variances are always positive or zero.
2. The distribution is positively skewed.
3. The mean value of F is approximately equal to 1.
4. The F distribution is a family of curves based upon the degrees of freedom of the variance of the numerator and the degrees of freedom of the variance of the denominator.

Figure 13–1 shows the shapes of several curves for the F distribution.

FIGURE 13–1
The F Family of Curves

$$F$$

0

| **Formula for the F Test** |

$$F = \frac{s_1^2}{s_2^2}$$

where s_1^2 is the larger of the two variances.

The F test has two terms for the degrees of freedoms: that of the numerator, $n_1 - 1$, and that of the denominator, $n_2 - 1$, where n_1 is the sample size from which the larger variance was obtained.

When one is finding the F test value, *the larger of the variances is placed in the numerator of the F formula;* this is not necessarily the variance of the larger of the two sample sizes.

Table H in the Appendix gives the F critical values for $\alpha = 0.005, 0.01, 0.025, 0.05,$ and 0.10 (each α value involves a separate table in Table H). These values are one-tailed values; if a two-tailed test is being conducted, then the $\alpha/2$ value must be used. For example, if a two-tailed test with $\alpha = 0.05$ is being conducted, then the $0.05/2 = 0.025$ table of Table H should be used.

EXAMPLE 13–1   Find the critical value for a one-tailed right F test when $\alpha = 0.05$ and the degrees of freedom for the numerator (abbreviated d.f.N.) are 15 and the degrees of freedom for the denominator (abbreviated d.f.D.) are 21.

Solution   Since this test is a one-tailed test with $\alpha = 0.05$, the 0.05 table should be used. The d.f.N. is listed across the top, and the d.f.D. is listed in the left column. The critical value is found where the row and column intersect in the table. In this case, it is 2.18. See Figure 13–2.

FIGURE 13–2
Finding the Critical
Value in Table H for
Example 13–1

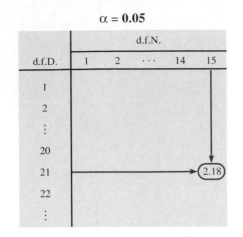

As noted previously, when the F test is used, the larger variance is always placed in the numerator of the formula. Hence, when one is conducting a two-tailed test, α is split; and even though there are two values, only the right one is used. The reason is that the F test value is always greater than or equal to 1.

EXAMPLE 13–2   Find the critical value for a two-tailed F test with $\alpha = 0.05$ when the sample size from which the variance for the numerator was obtained was 21 and the sample size from which the variance for the denominator was obtained was 12.

Solution   Since this test is a two-tailed test with $\alpha = 0.05$, the $0.05/2 = 0.025$ table must be used. Here, d.f.N. $= 21 - 1 = 20$, and d.f.D. $= 12 - 1 = 11$; hence, the critical value is 3.23. See Figure 13–3.

FIGURE 13–3
Finding the Critical
Value in Table H for
Example 13–2

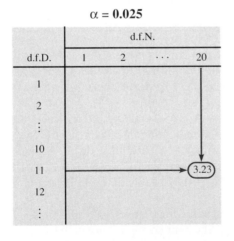

When the degrees of freedom values cannot be found in the table, the closest value on the smaller side should be used. For example, if d.f.N. $= 14$, this value is between the given table values of 12 and 15; therefore, 12 should be used to be on the safe side.

13–3 TESTING THE DIFFERENCE BETWEEN TWO VARIANCES

When one is testing the equality of two variances, the following hypotheses are used.

One-Tailed Right	One-Tailed Left	Two-Tailed
$H_0: \sigma_1^2 \leq \sigma_2^2$	$H_0: \sigma_1^2 \geq \sigma_2^2$	$H_0: \sigma_1^2 = \sigma_2^2$
$H_1: \sigma_1^2 > \sigma_2^2$	$H_1: \sigma_1^2 < \sigma_2^2$	$H_1: \sigma_1^2 \neq \sigma_2^2$

There are four key points to keep in mind when one is using the F test.

Notes for the Use of the F Test

1. The larger variance should always be designated as s_1^2 and be placed in the numerator of the formula.

$$F = \frac{s_1^2}{s_2^2}$$

2. For a two-tailed test, the α value must be divided by 2 and the critical value be placed on the right side of the F curve.
3. If the standard deviations instead of the variances are given in the problem, they must be squared for the formula for the F test.
4. When the degrees of freedom cannot be found in Table H, the closest value on the smaller side should be used.

Assumptions for Testing the Difference Between Two Variances

1. The populations from which the samples were obtained must be normally distributed. (*Note:* The test should not be used when the distributions depart from normality.)
2. The samples must be independent of each other.

Remember, also, that in tests of hypotheses, the following five steps should be used.

STEP 1 State the hypotheses, and identify the claim.

STEP 2 Find the critical value.

STEP 3 Compute the test value.

STEP 4 Make the decision.

STEP 5 Summarize the results.

EXAMPLE 13–3 A medical researcher wishes to see whether the variances of the heart rates (in beats per minute) of smokers are different from the variances of heart rates of those who do not smoke. Two samples are selected, and the data are as shown. Use $\alpha = 0.05$.

Smokers	**Nonsmokers**
$n_1 = 26$	$n_2 = 18$
$s_1^2 = 36$	$s_2^2 = 10$

Solution   **STEP 1**   State the hypotheses, and identify the claim.

$$H_0: \sigma_1^2 = \sigma_2^2 \quad \text{and} \quad H_1: \sigma_1^2 \neq \sigma_2^2 \text{ (claim)}$$

STEP 2   Find the critical value. Use the 0.025 table in Table H since $\alpha = 0.05$ and this test is a two-tailed test. Here, d.f.N. $= 26 - 1 = 25$, and d.f.D. $= 18 - 1 = 17$. The critical value is 2.56 (d.f.N. $= 24$ was used). See Figure 13–4.

FIGURE 13–4
Critical Value for
Example 13–3

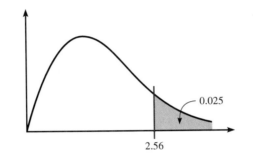

0.025

2.56

STEP 3   Compute the test value.

$$F = \frac{s_1^2}{s_2^2} = \frac{36}{10} = 3.6$$

STEP 4   Make the decision. Reject the null hypothesis, since $3.6 > 2.56$.

STEP 5   Summarize the results. There is enough evidence to support the claim that the variances are different.   ◀

EXAMPLE 13–4 An instructor hypothesizes that the standard deviation of the final exam grades in her statistics class is larger for the male students than it is for the female students. The data from the final exam for the last semester are as shown. Test her claim, using $\alpha = 0.01$.

Males	**Females**
$n_1 = 16$	$n_2 = 18$
$s_1 = 4.2$	$s_2 = 2.3$

Solution   **STEP 1**   State the hypotheses, and identify the claim.

$$H_0: \sigma_1^2 \leq \sigma_2^2 \quad \text{and} \quad H_1: \sigma_1^2 > \sigma_2^2 \text{ (claim)}$$

STEP 2   Find the critical value. Here, d.f.N. $= 16 - 1 = 15$, and d.f.D. $= 18 - 1 = 17$. From the 0.01 table, the critical value is 3.31.

STEP 3 Compute the test value.

$$F = \frac{s_1^2}{s_2^2} = \frac{(4.2)^2}{(2.3)^2} = 3.33$$

STEP 4 Make the decision. Reject the null hypothesis, since $3.33 > 3.31$.

STEP 5 Summarize the results. There is enough evidence to support the claim that the standard deviation of the final exam grades for the male students is larger than the standard deviation of the final exam grades for the female students.   ◄

EXERCISES

13–1. What is the formula for the F test?

13–2. When one is computing the F test value, what condition is placed on the variance that is in the numerator?

13–3. Why is the critical region always on the right side in the use of the F test?

13–4. What are two applications of the F test?

13–5. What are the two different degrees of freedom associated with the F distribution?

13–6. What are the characteristics of the F distribution?

13–7. Using Table H, find the critical value for each.

a. sample 1: $\sigma_1^2 = 128$, $n_1 = 23$
   sample 2: $\sigma_2^2 = 162$, $n_2 = 16$
   two-tailed, $\alpha = 0.01$
b. sample 1: $\sigma_1^2 = 37$, $n_1 = 14$
   sample 2: $\sigma_2^2 = 89$, $n_2 = 25$
   one-tailed, $\alpha = 0.01$
c. sample 1: $\sigma_1^2 = 232$, $n_1 = 30$
   sample 2: $\sigma_2^2 = 387$, $n_2 = 46$
   two-tailed, $\alpha = 0.05$
d. sample 1: $\sigma_1^2 = 164$, $n_1 = 21$
   sample 2: $\sigma_2^2 = 53$, $n_2 = 17$
   two-tailed, $\alpha = 0.10$
e. sample 1: $\sigma_1^2 = 92.8$, $n_1 = 11$
   sample 2: $\sigma_2^2 = 43.6$, $n_2 = 11$
   one-tailed, $\alpha = 0.05$

For Exercises 13–8 through 13–21, perform the following steps. Assume that all variables are normally distributed.

a. State the hypotheses, and identify the claim.
b. Find the critical value.
c. Compute the test value.
d. Make the decision.
e. Summarize the results.

13–8. An instructor feels that when a composition course is taught in conjunction with a word-processing course, the variance in the final grades will be larger than when the composition course is taught without the word-processing component. Two groups are randomly selected. The variance of the exams of the group that also had word-processing instruction is 103, and the variance of the exams of the students who did not have the word-processing component is 73. Each sample consists of 20 students. At $\alpha = 0.05$, can the instructor's claim be supported?

13–9. A consumer advocate hypothesizes that there is no difference in the variance of the number of hours that each of two companies' lightbulbs will last. A sample of 12 lightbulbs is selected from company A, and the variance of hours is 36. A sample of 12 lightbulbs selected from company B has a variance of 50. At $\alpha = 0.10$, can the consumer advocate conclude that there is no difference in the variance of the life of the lightbulbs?

13–10. A researcher hypothesizes that the variance of the IQ scores of women who major in psychology is larger than the variance of the IQ scores of men who major in psychology. A sample of IQ scores of 22 women had a variance of 192, and a sample of IQ scores of 18 men had a variance of 84. At $\alpha = 0.05$, can the researcher conclude that the hypothesis is correct?

13–11. A quality control inspector wishes to test the variations in the breaking strength of two types of rope her company manufactures. A sample of 31 nylon ropes is selected, and the standard deviation of the breaking strength is 8.2 pounds. A sample of 24 cotton ropes has a standard deviation of 4.6 pounds. At $\alpha = 0.05$, is there a difference in the standard deviations of the breaking strength of the two types of rope?

13–12. A nurse hypothesizes that the variations of the lengths of newborn males is different from the variations of the lengths of newborn females. A sample of 15 newborn males is selected, and the standard deviation is 1.3 inches. The standard deviation of a sample of 15 newborn females is 0.9 inch. At $\alpha = 0.10$, can the nurse conclude that the variation of the lengths is different?

13–13. Two different brands of refrigerators are selected and tested to determine whether there is a difference in the variation in temperatures. Each is set at 36°. After 3 hours, the standard deviation of the temperature of a sample of five brand A refrigerators was 1.6°, and the standard deviation of the temperatures of nine brand B refrigerators was 0.7°. At $\alpha = 0.05$, can one conclude that the standard deviations of the temperatures of both brands are equal?

13–14. A researcher feels that the variation of blood pressure of overweight individuals is greater than the variation of blood pressure of normal-weight individuals. The standard deviation of the pressures of 28 overweight people was found to be 6.2 mmHg, and the standard deviation of the pressures of 25 normal-weight people was 2.7 mmHg. At $\alpha = 0.01$, can the researcher conclude that the blood pressures of overweight individuals are more variable than those of individuals who are of normal weight?

13–15. An educator hypothesizes that the variation in the number of years of teaching experience of senior high school teachers is greater than the variation in the number of years of teaching experience of elementary school teachers. Two groups are randomly selected. The variance of the number of years of teaching experience of 18 elementary teachers was 1.9, and the variance of the number of years of teaching experience of 26 senior high school teachers was 2.8. At $\alpha = 0.10$, can the educator conclude that the variation in the number of years of teaching experience of senior high school teachers is greater than the variation of the elementary school teachers?

13–16. A researcher wishes to test the variation in the number of pounds lost by women who follow two popular liquid diets. Ten women followed diet A for four months, and the standard deviation of the weight loss was 6.3 pounds. Twelve women followed diet B for four months, and the standard deviation of the weight loss was 4.8 pounds. At $\alpha = 0.05$, can the researcher conclude that the variation in pounds lost following diet A is greater than the variation in pounds lost following diet B?

13–17. The variations of stopping distances of two different brands of automobile tires are tested. A sample of 25 Eagle Claw tires produces a standard deviation of stopping distance of 10.6 feet. A sample of 18 Mega Tread tires has a standard deviation of 14.2 feet. At $\alpha = 0.10$, can one conclude that there is no difference in the standard deviations of the stopping distances of the two brands of tires?

13–18. Two fast-food restaurants are selected, and the variations in the time it takes to prepare meals are compared. A sample of 9 meals selected from the first restaurant had a standard deviation of 5.3 minutes, and a sample of 15 meals from the second restaurant had a standard deviation of 8.4 minutes. At $\alpha = 0.01$, can one conclude that there is a difference in the standard deviations?

13–19. A researcher hypothesizes that the variation in the salaries of elementary school teachers is greater than the variation in the salaries of secondary school teachers. A sample of the salaries of 30 elementary school teachers has a variance of $8324, and a sample of the salaries of 30 secondary school teachers has a variance of $2862. At $\alpha = 0.05$, can the researcher conclude that the variation in the salaries of the elementary school teachers is greater than the variation in the salaries of the secondary teachers?

*13–20. The sodium content (in milligrams) in one serving of two brands of cereal is shown below. Calculate the variance for each sample, and test the claim that the variances are equal, at $\alpha = 0.05$.

Brand A			Brand B		
170	158	162	160	154	153
163	154	171	142	180	144
182	163	165	183	191	142

*13–21. The number of calories per serving of two brands of yogurt is calculated and shown below. Test the claim that the variance of the calories in the two brands of yogurt are the same, at $\alpha = 0.05$.

Brand A			Brand B		
63	73	80	86	93	64
60	86	83	82	81	75
65	59	67	92	67	69
70	72	82	88	63	63

13–4 ONE-WAY ANALYSIS OF VARIANCE

When an *F* test is used to test a hypothesis concerning the means of three or more populations, the technique is called **analysis of variance** (commonly abbreviated as ANOVA). At first glance, one might think that when comparing the means of three or more samples, the *t* test can be used, comparing two means at a time. There are several reasons why the *t* test should not be done.

First, when one is comparing two means at a time, the rest of the means under study are ignored. With the F test, all the means are compared simultaneously. Second, when one is comparing two means at a time and making all pairwise comparisons, the probability of rejecting the null hypothesis when it is true is increased, since the more t tests that are conducted, the greater is the likelihood of getting significant differences by chance alone. Third, the greater the number of means there are to compare, the greater is the number of t tests that are needed. For example, for the comparison of 3 means two at a time, 3 t tests are required. For the comparison of 5 means two at a time, 10 tests are required. And for the comparison of 10 means two at a time, 45 tests are required.

Assumptions for the F Test for Comparing Three or More Means

1. The populations from which the samples were obtained must be normally or approximately normally distributed.
2. The samples must be independent of each other.
3. The variances of the populations must be equal.

Even though one is comparing three or more means in this use of the F test, *variances* are used in the test instead of means.

With the F test, two different estimates of the population variance are made. The first estimate is called the **between-group variance,** and it involves computing the variance by using the means of the groups or between the groups. The second estimate, the **within-group variance,** is made by computing the variance using all the data and is not affected by differences in the means. If there is no difference in the means, the between-group variance estimate will be approximately equal to the within-group variance estimate, and the F test value will be approximately equal to 1. However, when the means differ significantly, the between-group variance will be much larger than the within-group variance; the F test value will be significantly greater than 1; and the null hypothesis will be rejected. Since variances are compared, this procedure is called *analysis of variance* (ANOVA).

For a test of the difference among three or more means, the following hypotheses should be used:

H_0: $\mu_1 = \mu_2 = \cdots = \mu_n$.
H_1: At least one mean is different from the others.

As stated previously, a significant test value means that there is a high probability that this difference in means is not due to chance, but it does not indicate where the difference lies.

The degrees of freedom for this F test are

$$\text{d.f.N.} = k - 1$$

where k is the number of groups, and

$$\text{d.f.D.} = N - k$$

where N is the sum of the sample sizes of the groups, $N = n_1 + n_2 + \cdots + n_k$. The F test to compare means is always one-tailed right.

The next two examples illustrate the computational procedure for the ANOVA technique for comparing three or more means, and the steps are summarized in Procedure Table 12, shown after the examples.

EXAMPLE 13–5 A researcher wishes to try three different techniques to lower the blood pressure of individuals diagnosed with high blood pressure. The subjects are randomly assigned to three groups; the first group takes medication, the second group exercises, and the third group diets. After four weeks, the reduction in each person's blood pressure is recorded. At $\alpha = 0.05$, test the claim that there is no difference among the means. The data follow.

Medication	Exercise	Diet
10	6	5
12	8	9
9	3	12
15	0	8
13	2	4
$\overline{X}_1 = 11.8$	$\overline{X}_2 = 3.8$	$\overline{X}_3 = 7.6$
$s_1^2 = 5.7$	$s_2^2 = 10.2$	$s_3^2 = 10.3$

Solution   **STEP 1**   State the hypotheses, and identify the claim.

H_0: $\mu_1 = \mu_2 = \mu_3$ (claim).
H_1: At least one mean is different from the others.

STEP 2   Find the critical value. Since $k = 3$ and $N = 15$,

$$\text{d.f.N.} = k - 1 = 3 - 1 = 2$$
$$\text{d.f.D.} = N - k = 15 - 3 = 12$$

The critical value is 3.89, obtained from Table H with $\alpha = 0.05$.

STEP 3   Compute the test value, using the procedure outlined here.

a. Find the mean and variance of each sample (these values are shown below the data).

b. Find the grand mean. The *grand mean,* denoted by $\overline{X}_{\text{GM}}$, is the mean of all values in the samples.

$$\overline{X}_{\text{GM}} = \frac{\Sigma X}{N} = \frac{10 + 12 + 9 + \cdots + 4}{15} = \frac{116}{15} = 7.73$$

c. Find the between-group variance, denoted by s_B^2.

$$s_B^2 = \frac{\Sigma n_i(\overline{X}_i - \overline{X}_{\text{GM}})^2}{k - 1}$$

$$= \frac{5(11.8 - 7.73)^2 + 5(3.8 - 7.73)^2 + 5(7.6 - 7.73)^2}{3 - 1}$$

$$= \frac{160.13}{2} = 80.07$$

Note: This formula finds the variance between the means using the sample sizes as weights and considers the differences in the means.

d. Find the within-group variance, denoted by s_W^2.

$$s_W^2 = \frac{\Sigma (n_i - 1)s_i^2}{\Sigma (n_i - 1)}$$

$$= \frac{(5 - 1)(5.7) + (5 - 1)(10.2) + (5 - 1)(10.3)}{(5 - 1) + (5 - 1) + (5 - 1)}$$

$$= \frac{104.80}{12} = 8.73$$

Note: This formula finds the variance within the means, again using the sample sizes as weights, but it does not involve the differences in the means.

e. Find the F test value.

$$F = \frac{s_B^2}{s_W^2} = \frac{80.07}{8.73} = 9.17$$

STEP 4   Make the decision. The decision is to reject the null hypothesis, since $9.17 > 3.89$.

STEP 5   Summarize the results. There is enough evidence to reject the claim and conclude that all of the means are not equal.   ◄

The numerator of the fraction obtained in Step 3, part c, of the computational procedure is called the **sum of squares between groups,** denoted by SS_B. The numerator of the fraction obtained in Step 3, part d, of the computational procedure is called the **sum of squares within groups,** denoted by SS_W. This statistic is also called the sum of squares for the error. When SS_B is divided by d.f.N., the between-group variance is obtained. When SS_W is divided by d.f.D., the within-group variance is obtained. These two variances are sometimes called **mean squares,** denoted by MS_B and MS_W. These terms are used to summarize the analysis of variance and are placed in a summary table, as shown in Table 13-1.

Table 13-1

Analysis of Variance Summary Table				
Source	**Sum of Squares**	**d.f.**	**Mean Square**	**F**
Between	SS_B	$k - 1$	MS_B	
Within (error)	SS_W	$N - k$	MS_W	
Total				

In the table,

SS_B = sum of squares between groups

SS_W = sum of squares within groups

k = number of groups

$N = n_1 + n_2 + \cdots + n_k$ = sum of the sample sizes for the groups

$$MS_B = \frac{SS_B}{k - 1}$$

$$MS_W = \frac{SS_W}{N - k}$$

$$F = \frac{MS_B}{MS_W}$$

The totals are obtained by adding the corresponding columns.

For Example 13–5, the ANOVA summary table is as shown in Table 13–2.

Table 13–2

Analysis of Variance Summary Table for Example 13–5				
Source	**Sum of Squares**	**d.f.**	**Mean Square**	**_F_**
Between	160.13	2	80.07	9.17
Within (error)	104.80	12	8.73	
Total	264.93	14		

Most computer programs will print out an ANOVA summary table.

EXAMPLE 13–6

A marketing specialist wishes to see whether there is a difference in the average time (in minutes) a customer has to wait in a checkout line in three large self-service department stores. The times (in minutes) are shown next.

Store A	Store B	Store C
3	5	1
2	8	3
5	9	4
6	6	2
3	2	7
1	5	3
$\overline{X}_1 = 3.33$	$\overline{X}_2 = 5.83$	$\overline{X}_3 = 3.33$
$s_1^2 = 3.47$	$s_2^2 = 6.17$	$s_3^2 = 4.27$

At $\alpha = 0.05$, is there a significant difference in the mean waiting times of customers for each store?

Solution   **STEP 1**   State the hypotheses, and identify the claim.

H_0: $\mu_1 = \mu_2 = \mu_3$.

H_1: At least one mean is different from the others (claim).

STEP 2 Find the critical value. Since $k = 3$, $N = 18$, and $\alpha = 0.05$,

$$\text{d.f.N.} = k - 1 = 3 - 1 = 2$$
$$\text{d.f.D.} = N - k = 18 - 3 = 15$$

The critical value is 3.68.

STEP 3 Compute the test value.

a. Find the mean and variance of each sample (these values are shown below the data).

b. Find the grand mean.

$$\overline{X}_{\text{GM}} = \frac{\Sigma X}{N} = \frac{3 + 2 + \cdots + 3}{18} = \frac{75}{18} = 4.17$$

c. Find the between-group variance.

$$s_B^2 = \frac{\Sigma n_i(\overline{X}_i - \overline{X}_{\text{GM}})^2}{k - 1}$$
$$= \frac{6(3.33 - 4.17)^2 + 6(5.83 - 4.17)^2 + 6(3.33 - 4.17)^2}{3 - 1}$$
$$= \frac{25}{2} = 12.5$$

d. Find the within-group variance.

$$s_W^2 = \frac{\Sigma (n_i - 1)s_i^2}{\Sigma (n_i - 1)}$$
$$= \frac{(6 - 1)(3.47) + (6 - 1)(6.17) + (6 - 1)(4.27)}{(6 - 1) + (6 - 1) + (6 - 1)}$$
$$= \frac{69.6}{15} = 4.64$$

e. Find the F test value.

$$F = \frac{s_B^2}{s_W^2} = \frac{12.5}{4.64} = 2.69$$

STEP 4 Make the decision. Since $2.69 < 3.68$, the decision is to not reject the null hypothesis.

STEP 5 Summarize the results. There is not enough evidence to support the claim that there is a difference among the means. The analysis of variance summary table for this example is shown in Table 13–3.

Table 13–3

Analysis of Variance Summary Table for Example 13–6				
Source	**Sum of Squares**	**d.f.**	**Mean Square**	**F**
Between	25	2	12.5	2.69
Within	69.6	15	4.64	
Total	94.6	17		

The procedure for computing the F test value for the ANOVA is summarized in Procedure Table 12.

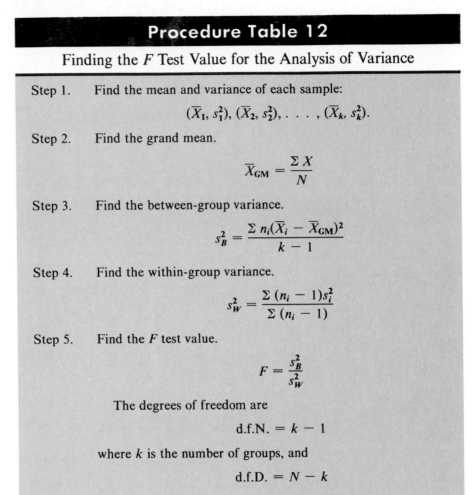

Procedure Table 12

Finding the F Test Value for the Analysis of Variance

Step 1.   Find the mean and variance of each sample:

$$(\overline{X}_1, s_1^2), (\overline{X}_2, s_2^2), \ldots, (\overline{X}_k, s_k^2).$$

Step 2.   Find the grand mean.

$$\overline{X}_{GM} = \frac{\Sigma X}{N}$$

Step 3.   Find the between-group variance.

$$s_B^2 = \frac{\Sigma n_i(\overline{X}_i - \overline{X}_{GM})^2}{k - 1}$$

Step 4.   Find the within-group variance.

$$s_W^2 = \frac{\Sigma (n_i - 1)s_i^2}{\Sigma (n_i - 1)}$$

Step 5.   Find the F test value.

$$F = \frac{s_B^2}{s_W^2}$$

The degrees of freedom are

$$d.f.N. = k - 1$$

where k is the number of groups, and

$$d.f.D. = N - k$$

where N is the sum of the sample sizes of the groups, $N = n_1 + n_2 + \cdots + n_k$.

13–5 THE SCHEFFÉ TEST AND THE TUKEY TEST

When the null hypothesis is rejected using the F test, the researcher may like to know where the difference among the means is. Several procedures have been developed to determine where the significant differences in the means lie after the ANOVA procedure has been performed. One of the most commonly used tests is the *Scheffé test.* Another frequently used test is the *Tukey test.* The Scheffé test will be explained first.

Scheffé Test   In order to conduct the **Scheffé test,** one must compare the means two at a time, using all possible combinations of means. For example, if there are three means, the following comparisons must be done:

$$\overline{X}_1 \text{ versus } \overline{X}_2 \qquad \overline{X}_1 \text{ versus } \overline{X}_3 \qquad \overline{X}_2 \text{ versus } \overline{X}_3$$

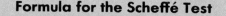

Formula for the Scheffé Test

$$F_S = \frac{(\overline{X}_i - \overline{X}_j)^2}{s_W^2[(1/n_i) + (1/n_j)]}$$

where $\overline{X}_i$ and $\overline{X}_j$ are the means of the samples being compared, n_i and n_j are the respective sample sizes, and s_W^2 is the within-group variance.

In order to find the critical value F' for the Scheffé test, one must multiply the critical value for the F test by $k - 1$:

$$F' = (k - 1)(\text{C.V.})$$

There is a significant difference between the two means being compared when F_S is greater than F'. Example 13–7 illustrates the use of the Scheffé test.

EXAMPLE 13–7   Using the Scheffé test, test each pair of means in Example 13–5 to see whether a specific difference exists, at $\alpha = 0.05$.

Solution   a. For $\overline{X}_1$ versus $\overline{X}_2$,

$$F_S = \frac{(\overline{X}_1 - \overline{X}_2)^2}{s_W^2[(1/n_1) + (1/n_2)]} = \frac{(11.8 - 3.8)^2}{8.73[(1/5) + (1/5)]} = 18.33$$

b. For $\overline{X}_2$ versus $\overline{X}_3$,

$$F_S = \frac{(\overline{X}_2 - \overline{X}_3)^2}{s_W^2[(1/n_2) + (1/n_3)]} = \frac{(3.8 - 7.6)^2}{8.73[(1/5) + (1/5)]} = 4.135$$

c. For $\overline{X}_1$ versus $\overline{X}_3$,

$$F_S = \frac{(\overline{X}_1 - \overline{X}_3)^2}{s_W^2[(1/n_1) + (1/n_3)]} = \frac{(11.8 - 7.6)^2}{8.73[(1/5) + (1/5)]} = 5.05$$

The critical value at $\alpha = 0.05$ is

$$F' = (k - 1)(\text{C.V.}) = (3 - 1)(3.89) = 7.78$$

with d.f.N. $= 2$ and d.f.D. $= 12$.

Since only the F test value for part a ($\overline{X}_1$ versus $\overline{X}_2$) is greater than the critical value 7.78, the only significant difference is between $\overline{X}_1$ and $\overline{X}_2$, i.e., between medication and exercise.   ◀

On occasion, when F is significant, the Scheffé test may not show any significant differences in the pairs of means. This result occurs because the difference may actually lie in the average of two or more means when compared with the other mean. The Scheffé test can be used to make these types of comparisons, but the technique is beyond the scope of this book.

Tukey Test   The **Tukey test** can also be used after the analysis of variance has been completed to make pairwise comparisons between means when the groups have the same sample size. The symbol for the test value in the Tukey test is q.

**Formula for the Test Value
for the Tukey Test**

$$q = \frac{\overline{X}_i - \overline{X}_j}{\sqrt{s_W^2/n}}$$

where $\overline{X}_i$ and $\overline{X}_j$ are the means of the samples being compared, n is the size of the samples, and s_W^2 is the within-group variance.

When the absolute value of q is greater than the critical value for the Tukey test, there is a significant difference between the two means being compared. The procedures for finding q and the critical value for the Tukey test are shown in the next example.

EXAMPLE 13–8 Using the Tukey test, test each pair of means in Example 13–5 to see whether a specific difference exists, at $\alpha = 0.05$.

Solution a. For $\overline{X}_1$ versus $\overline{X}_2$,

$$q = \frac{\overline{X}_1 - X_2}{\sqrt{s_W^2/n}} = \frac{11.8 - 3.8}{\sqrt{8.73/5}} = \frac{8}{1.32} = 6.06$$

b. For $\overline{X}_1$ versus $\overline{X}_3$,

$$q = \frac{\overline{X}_1 - \overline{X}_3}{\sqrt{s_W^2/n}} = \frac{11.8 - 7.6}{\sqrt{8.73/5}} = \frac{4.2}{1.32} = 3.18$$

c. For $\overline{X}_2$ versus $\overline{X}_3$,

$$q = \frac{\overline{X}_2 - \overline{X}_3}{\sqrt{s_W^2/n}} = \frac{3.8 - 7.6}{\sqrt{8.73/5}} = -\frac{3.8}{1.32} = -2.88$$

In order to find the critical value for the Tukey test, use Table N in the Appendix. The number of means k is found in the row at the top, and the number of degrees of freedom for s_W^2 is found in the left column (denoted by v). Since $k = 3$, d.f. $= 12$, and $\alpha = 0.05$, the critical value is 3.77. See Figure 13–5. Hence, the only q value that is greater in absolute value than the critical value is the one for the difference between $\overline{X}_1$ and $\overline{X}_2$. The conclusion, then, is that there is a significant difference in means for medication and exercise. These results agree with the Scheffé analysis.

FIGURE 13–5
Finding the Critical Value in Table N for the Tukey Test (Example 13–8)

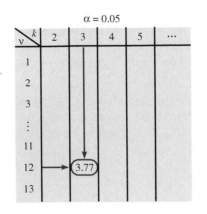

Speaking of Statistics

Coffee Doesn't Increase Risk of Heart Disease

By Tim Friend
USA TODAY

In the ongoing debate over coffee's effects on the heart, the latest research scores one in favor of coffee drinkers:

You can safely drink four six-ounce cups of coffee a day.

"Coffee drinkers appear to have no grounds for concern," says Dr. Roy Fried, Kaiser Permanente Medical Center, Kensington, Md.

Previous studies linked coffee to a higher risk of heart disease. The strongest link came from observations that boiled coffee raises bad cholesterol levels, thus raising heart disease risk. But studies also suggested filtered coffee had no effect on cholesterol.

To sort out the confusion, Fried and researchers at the Johns Hopkins Medical Institutions, Baltimore, studied 100 healthy men for eight weeks after the men were taken off coffee and all caffeine. Fried says the first surprise was cholesterol levels dropped.

Then the men were divided into groups: one got four cups of caffeinated coffee a day, another got two cups, a third got four cups of decaf and a fourth got two cups of decaf. A fifth group still got no coffee.

Results, out today in the *Journal of the American Medical Association:*

- Levels of LDL (bad) cholesterol increased by six points in the four-cup-a-day caffeinated group, raising the heart disease risk by 9%.
- Levels of HDL (good) cholesterol increased three points in the same group, lowering heart disease risk 7% to 9%.

- No other group had cholesterol changes except the two-cup-a-day caffeinated group, which had a slight increase in good cholesterol only.

Overall, coffee does raise cholesterol levels, but the effect of LDL was offset by the HDL, Fried says.

Fried says it's still unclear how coffee raises cholesterol.

The coffee used in the study was brewed in automatic-drip makers with paper filters. Fried says the only sweetener used was aspartame; the only lighteners, skim and non-fat powdered dry milk.

What's in the Daily Grind

The USA drinks more coffee than any other country—an average 1¾ cups per person per day.

Some coffee facts:

- It contains 393 chemicals, including caffeine, tannins, caramelized sugar and carbon dioxide.
- A 6-oz cup contains about 4 calories.
- Caffeine content is highest in drip-brewed, then percolated, then instant. Even decaf contains a small amount.
- Caffeine is both a stimulant and a diuretic; it's absorbed rapidly, appearing in all tissues and organs within about five minutes.

Source: *The Mount Sinai School of Medicine Complete Book of Nutrition.*

Source: Copyright 1992, USA TODAY. Reprinted with permission.

This article contains many facts about coffee and its effects. Which facts are based on descriptive studies and which are based on inferential studies?

The student might wonder why there are two different tests that can be used after the ANOVA. Actually, there are several other tests that can be used in addition to the Scheffé and Tukey tests. It is up to the researcher to select the most appropriate test. The Scheffé test is the most general of the tests, and it can be used when the samples are of different sizes. Furthermore, the Scheffé test can be used to make comparisons such as the average of $\overline{X}_1$ and $\overline{X}_2$ compared with $\overline{X}_3$. However, the Tukey test is more powerful than the Scheffé test when making pairwise comparisons between means. A rule of thumb for pairwise comparisons is to use the Tukey test when the samples are equal in size and to use the Scheffé test when the samples are unequal in size. This rule will be followed in this textbook.

EXERCISES

13–22. What test is used to compare three or more means?

13–23. State three reasons why multiple t tests cannot be used when comparing three or more means.

13–24. What are the assumptions for ANOVA?

13–25. Define *between-group variance* and *within-group variance*.

13–26. What is the formula for comparing three or more means?

13–27. State the hypotheses used in the ANOVA test.

13–28. What two tests are used to compare individual means if the null hypothesis is rejected when using the ANOVA technique?

If the null hypothesis is rejected in Exercises 13–29 through 13–40, use the Scheffé test when the sample sizes are unequal to test the differences between the means, and use the Tukey test when the sample sizes are equal. Assume that all variables are normally distributed, that the samples are independent, and that the population variances are equal. Also, for each exercise, perform the following steps.

a. State the hypotheses, and identify the claim.
b. Find the critical value.
c. Compute the test value.
d. Make the decision.
e. Summarize the results.

13–29. A researcher wishes to see whether there is a difference in the average age of nurses, doctors, and X-ray technicians at a local hospital. Employees are randomly selected, and their ages are recorded as shown in the table. At $\alpha = 0.05$, can the researcher conclude that there is a difference in the average ages of each group?

Nurses	Doctors	X-Ray Technicians
23	60	33
25	36	28
26	29	35
35	56	29
42	32	23
22	54	41
	58	

13–30. Three different types of computer disks are selected, and the number of defects in each is as recorded below. At $\alpha = 0.05$, can one conclude that there is a difference in the means of the number of defects for each group?

Type A	Type B	Type C
0	2	1
1	0	0
0	3	1
2	5	1
3	3	0
2	4	2
0	6	0
1	0	0
1	2	1
0	5	2

13-31. The grade point averages of students participating in college sports programs are to be compared. The data are shown in the table. At $\alpha = 0.10$, can one conclude that there is a difference in the mean grade point average of the three groups?

Football	Basketball	Hockey
3.2	3.8	2.6
2.6	3.1	1.9
2.4	2.6	1.7
2.4	3.9	2.5
1.8	3.3	1.9

13-32. Three different relaxation techniques are given to randomly selected patients in an effort to reduce their stress levels. A special instrument has been designed to measure the percentage of stress reduction in each person. The data are shown in the table. At $\alpha = 0.05$, can one conclude that there is a difference in the means of the percentages?

Technique I	Technique II	Technique III
3	12	15
10	12	14
5	17	18
1	13	14
13	18	20
3	9	22
4	14	16

13-33. A researcher wishes to see whether there is any difference in the weight gains of athletes following one of three special diets. Athletes are randomly assigned to three groups and placed on the diet for six weeks. The weight gains (in pounds) are shown below. At $\alpha = 0.05$, can the researcher conclude that there is a difference in the diets?

Diet A	Diet B	Diet C
3	10	8
6	12	3
7	11	2
4	14	5
	8	
	6	

13-34. Three different teaching methods are used to teach statistics. At the end of the semester, a final exam is given to all the students. The results are shown below. At $\alpha = 0.10$, is there a difference in the means of the students who received instruction in the different methods?

Programmed Instruction	Lecture	Computer-Assisted
87	99	98
64	86	67
93	74	88
82	82	72
78	63	86

13-35. Workers are randomly assigned to four different machines on an assembly line. The number of defective parts produced by each worker for one day is recorded. The data are as shown below. At $\alpha = 0.05$, can one conclude that the mean number of defective parts produced by the workers is the same?

Machine 1	Machine 2	Machine 3	Machine 4
3	8	10	9
2	6	9	15
0	2	8	3
6	0	11	0
4	1	12	2
3	9	15	0
5	7	17	1

13-36. Five different types of fertilizers are used on strawberry plants. The number of strawberries on each randomly selected plant is as recorded below. At $\alpha = 0.10$, can one conclude that the type of fertilizer makes a difference in the mean number of strawberries per plant?

A	B	C	D	E
3	2	3	3	4
3	0	7	6	3
6	1	7	2	4
8	1	3	2	4
8	2	8	1	3
9	1	7	1	2
3	2	4	2	0

13-37. A researcher tests the lifetimes (in hours) of three different cassette tapes. The data are shown below. At $\alpha = 0.10$, is there a difference in the means?

Tape 1	Tape 2	Tape 3
196	98	94
183	91	106
112	101	85
107	99	102
189	84	101

13-38. A sample of real estate agents located in three different cities is selected, and the weekly commissions of the agents are as recorded below. At $\alpha = 0.10$, is there a difference in the commissions of the agents?

A	B	C
$3500	$7,329	$8400
4000	8,641	7615
3271	12,312	3141
4268	9,610	2321
	9,100	1001
	10,300	

13–39. The time it takes (in minutes) to treat randomly selected patients in an emergency room for three shifts is as recorded below. At $\alpha = 0.05$, can one conclude that there is a significant difference in the mean time it takes to treat patients for each shift?

Morning (7–3)	Afternoon (3–11)	Night (11–7)
12	9	6
18	8	15
18	16	8
21	20	9
19	15	5

13–40. Four different types of pain relief medications are given to randomly selected patients. The time it takes to relieve the pain (in minutes) is shown below for each medication. At $\alpha = 0.01$, can one conclude that there is a difference in the pain relief ability of the medications?

A	B	C	D
3	8	7	14
2	12	4	16
5	15	9	8
4	9	2	15
3	6	1	12

13–6 TWO-WAY ANALYSIS OF VARIANCE

The analysis of variance technique shown previously is called a **one-way analysis of variance** since there is only *one independent variable*. The **two-way analysis of variance** is an extension of the one-way analysis of variance; it involves *two independent variables*. The independent variables are also called **factors.**

The two-way analysis of variance is quite complicated, and many aspects of the subject should be considered when one is using a research design involving a two-way analysis of variance. For the purposes of this textbook, only a brief introduction to the subject will be given.

In doing a study that involves a two-way analysis of variance, the researcher is able to test the effects of two independent variables or factors on one *dependent variable*. In addition, the interaction effect of the two variables can also be tested.

For example, suppose a researcher wishes to test the effects of two different types of plant food and two different types of soil on the growth of certain plants. The two independent variables are the type of plant food and the type of soil, while the dependent variable is the plant growth. Other factors, such as water, temperature, and sunlight, are held constant.

In order to conduct this experiment, the researcher sets up four groups of plants. See Figure 13–6.

FIGURE 13–6
Treatment Groups for the Plant Food–Soil Type Experiment

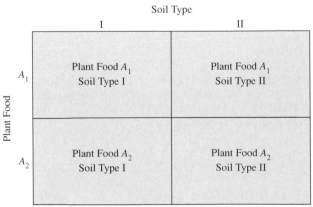

Assume that the plant food type is designated by the letters A_1 and A_2 and the soil type by the Roman numerals I and II. The groups for such a two-way ANOVA are sometimes called **treatment groups.** The four groups are as follows:

Group 1   Plant food A_1, soil type I
Group 2   Plant food A_1, soil type II
Group 3   Plant food A_2, soil type I
Group 4   Plant food A_2, soil type II

The plants are assigned to the groups at random. This design is called a 2 × 2 (read "two-by-two") design, since each variable consists of two **levels,** i.e., two different treatments.

The two-way ANOVA enables the researcher to test the effects of the plant food and the soil type in a single experiment rather than in separate experiments involving the plant food alone and the soil type alone. Furthermore, the researcher can test an additional hypothesis about the effect of the *interaction* of the two variables, plant food and soil type, on plant growth. For example, is there a difference in the growth of plants using plant food A_1 and soil type II and the plant growth using plant food A_2 and soil type I? When a difference of this type occurs, the experiment is said to have a significant **interaction effect.** That is, the types of plant food affect the plant growth differently in different soil types.

There are many different kinds of two-way ANOVA designs, depending on the number of levels of each variable. Figure 13–7 shows a few of these designs. As stated previously, the plant food–soil type experiment uses a 2 × 2 ANOVA.

FIGURE 13-7
Some Types of Two-Way ANOVA Designs

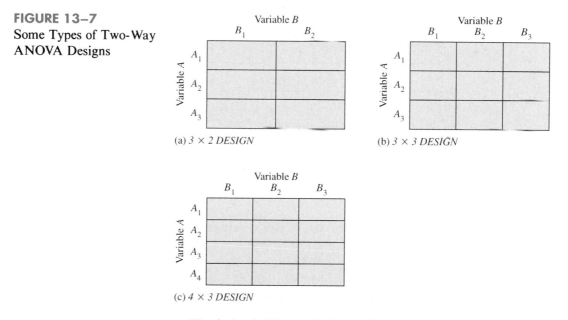

(a) *3 × 2 DESIGN*

(b) *3 × 3 DESIGN*

(c) *4 × 3 DESIGN*

The design in Figure 13–7a is called a 3 × 2 design, since one factor has three levels and the other factor has two levels. The design in Figure 13–7b is a 3 × 3 design, since each factor has three levels. The design in Figure 13–7c is a 4 × 3 design.

The two-way ANOVA design has several null hypotheses. There is one for each independent variable and one for the interaction. In the plant food–soil type problem, the hypotheses are as follows:

1. H_0: There is no interaction between the type of plant food used and the type of soil used on the plant growth.
   H_1: There is an interaction effect between the plant food type and the soil type on plant growth.
2. H_0: There is no difference in the means of the heights of the plants grown using different plant foods.
   H_1: There is a difference in the means of the heights of the plants grown using different plant foods.
3. H_0: There is no difference in the means of the heights of the plants grown in the different soil types.
   H_1: There is a difference in the means of the heights of the plants grown in the different soil types.

The first set of hypotheses concerns the interaction effect; the second and third sets test the effects of the independent variables. The effects of the independent variables are sometimes called the **main effects.**

As with the one-way ANOVA, a between-group variance estimate is calculated and a within-group variance estimate is calculated. An F test is then performed for each of the independent variables and the interaction. The results of the two-way ANOVA are summarized in a two-way table, as shown in Table 13–4 for the plant experiment.

Table 13–4

ANOVA Summary Table for Plant Food and Soil Type				
Source	**Sum of Squares**	**d.f.**	**Mean Square**	**F**
Plant food				
Soil type				
Interaction				
Within (error)				
Total				

In general, the two-way **ANOVA summary table** is set up as shown in Table 13–5.

Table 13–5

ANOVA Summary Table				
Source	**Sum of Squares**	**d.f.**	**Mean Square**	**F**
A	SS_A	$a - 1$	MS_A	F_A
B	SS_B	$b - 1$	MS_B	F_B
$A \times B$	$SS_{A \times B}$	$(a - 1)(b - 1)$	$MS_{A \times B}$	$F_{A \times B}$
Within (error)	SS_W	$ab(n - 1)$	MS_W	
Total				

In the table,

SS_A = sum of squares for factor A

SS_B = sum of squares for factor B

$SS_{A \times B}$ = sum of squares for the interaction

SS_W = sum of squares for the error term (within-group)

a = number of levels of factor A

b = number of levels of factor B

n = number of subjects in each group

$$MS_A = \frac{SS_A}{a - 1}$$

$$MS_B = \frac{SS_B}{b - 1}$$

$$MS_{A \times B} = \frac{SS_{A \times B}}{(a - 1)(b - 1)}$$

$$MS_W = \frac{SS_W}{ab(n - 1)}$$

$$F_A = \frac{MS_A}{MS_W}, \text{ with d.f.N.} = a - 1 \text{ and d.f.D.} = ab(n - 1)$$

$$F_B = \frac{MS_B}{MS_W}, \text{ with d.f.N.} = b - 1 \text{ and d.f.D.} = ab(n - 1)$$

$$F_{A \times B} = \frac{MS_{A \times B}}{MS_W}, \text{ with d.f.N.} = (a - 1)(b - 1) \text{ and d.f.D.} = ab(n - 1)$$

The assumptions for the two-way analysis of variance arc basically the same as those for the one-way analysis of variance, except for sample size.

Assumptions for the Two-Way ANOVA

1. The populations from which the samples were obtained must be normally or approximately normally distributed.
2. The samples must be independent.
3. The variances of the populations from which the samples were selected must be equal.
4. The groups must be equal in sample size.

The computational procedure for the two-way ANOVA is quite lengthy. For this reason, it will be omitted in Example 13–9, and only the two-way ANOVA summary table will be shown. The table used in the example is similar to the one generated by most computer programs. The student should be able to interpret the table and summarize the results.

The next example illustrates the two-way ANOVA.

EXAMPLE 13–9   A researcher wishes to see whether the type of gasoline a person uses and the type of automobile a person drives have any effect on gasoline consumption. Two types of gasoline, regular and high-octane, will be used, and two types of automobiles, two-wheel and four-wheel drive, will be used in each group. There will be two automobiles in each group, for a total of eight automobiles used. Using a two-way analysis of variance, the researcher will perform the following steps.

STEP 1   State the hypotheses.

STEP 2   Find the critical value for each F test, using $\alpha = 0.05$.

STEP 3   Complete the summary table to get the test value.

STEP 4   Make the decision.

STEP 5   Summarize the results.

The data (in miles per gallon) are shown below, and the summary table is given in Table 13–6.

	Type of Automobile	
Gas	**Two-Wheel**	**Four-Wheel**
Regular	26.7	28.6
	25.2	29.3
High-octane	32.3	26.1
	32.8	24.2

Table 13–6

ANOVA Summary Table for Example 13–9				
Source	**SS**	**d.f.**	**MS**	**F**
Gasoline, A	3.781			
Automobile, B	9.461			
Interaction, $(A \times B)$	53.561			
Within (error)	3.356			
Total	70.159			

Solution   **STEP 1**   State the hypotheses. The hypotheses for the interaction are as follows:

H_0: There is no interaction effect between the type of gasoline used and the type of automobile a person drives on gasoline consumption.

H_1: There is an interaction effect between the type of gasoline used and the type of automobile a person drives on gasoline consumption.

The hypotheses for the gasoline types are as follows:

H_0: There is no difference between the means of the gasoline consumption for the two types of gasoline.

H_1: There is a difference between the means of the gasoline consumption for the two types of gasoline.

The hypotheses for the types of automobile driven are as follows:

H_0: There is no difference between the means of the gasoline consumption for the two-wheel drive and the four-wheel drive automobiles.

H_1: There is a difference between the means of the gasoline consumption for the two-wheel drive and the four-wheel drive automobiles.

STEP 2   Find the critical values for each F test. In this case, each independent variable or factor has two levels. Hence, a 2×2 ANOVA table is used. Factor A is designated as the gasoline type. It has two levels, regular and high-octane; therefore, $a = 2$. Factor B is designated as the automobile type. It also has two levels; therefore, $b = 2$. The degrees of freedom for each factor are as follows:

$$\begin{aligned}
\text{factor } A\text{:} \quad &\text{d.f.N.} = a - 1 = 2 - 1 = 1 \\
\text{factor } B\text{:} \quad &\text{d.f.N.} = b - 1 = 2 - 1 = 1 \\
\text{interaction } (A \times B)\text{:} \quad &\text{d.f.N.} = (a - 1)(b - 1) \\
&\qquad\quad = (2 - 1)(2 - 1) = 1 \cdot 1 = 1 \\
\text{within (error):} \quad &\text{d.f.D.} = ab(n - 1) \\
&\qquad\quad = 2 \cdot 2(2 - 1) = 4
\end{aligned}$$

where n is the number of data values in each group. In this case, $n = 2$.

The critical value for the F_A test is found by using $\alpha = 0.05$, d.f.N. $= 1$, and d.f.D. $= 4$. In this case, $F_A = 7.71$. The critical value for the F_B test is found by using $\alpha = 0.05$, d.f.N. $= 1$, and d.f.D. $= 4$; F_B is also 7.71. Finally, the critical value for the $F_{A \times B}$ test is found by using d.f.N. $= 1$ and d.f.D. $= 4$; it is also 7.71.

Note: If there are different levels of the factors, the critical values will not all be the same. For example, if factor A has three levels and factor B has four levels, and if there are two subjects in each group, then the degrees of freedom are as follows:

d.f.N. $= a - 1 = 3 - 1 = 2$	for factor A
d.f.N. $= b - 1 = 4 - 1 = 3$	for factor B
d.f.N. $= (a - 1)(b - 1) = (3 - 1)(4 - 1) = 2 \cdot 3 = 6$	for factor $A \times B$
d.f.D. $= ab(n - 1) = 3 \cdot 4(2 - 1) = 12$	for the within (error) factor

STEP 3   Complete the ANOVA summary table to get the test values. The mean squares are computed first.

$$\text{MS}_A = \frac{\text{SS}_A}{a - 1} = \frac{3.781}{2 - 1} = 3.781$$

$$\text{MS}_B = \frac{\text{SS}_B}{b - 1} = \frac{9.461}{2 - 1} = 9.461$$

$$\text{MS}_{A \times B} = \frac{\text{SS}_{A \times B}}{(a - 1)(b - 1)} = \frac{53.561}{(2 - 1)(2 - 1)} = 53.561$$

$$\text{MS}_W = \frac{\text{SS}_W}{ab(n - 1)} = \frac{3.356}{4} = 0.839$$

The F values are computed next.

$$F_A = \frac{MS_A}{MS_W} = \frac{3.781}{0.839} = 4.507 \qquad \text{d.f.N.} = a - 1 = 1 \qquad \text{d.f.D.} = ab(n - 1) = 4$$

$$F_B = \frac{MS_B}{MS_W} = \frac{9.461}{0.839} = 11.277 \qquad \text{d.f.N.} = b - 1 = 1 \qquad \text{d.f.D.} = ab(n - 1) = 4$$

$$F_{A \times B} = \frac{MS_{A \times B}}{MS_W} = \frac{53.561}{0.839} = 63.839 \qquad \text{d.f.N.} = (a - 1)(b - 1) = 1 \qquad \text{d.f.D.} = ab(n - 1) = 4$$

The completed ANOVA table is shown in Table 13–7.

Table 13–7

Completed ANOVA Summary Table for Example 13–9				
Source	**SS**	**d.f.**	**MS**	**F**
Gasoline, A	3.781	1	3.781	4.507
Automobile, B	9.461	1	9.461	11.277
Interaction, $(A \times B)$	53.561	1	53.561	63.839
Within (error)	3.356	4	0.839	
Total	70.159	7		

STEP 4 Make the decision. Since $F_B = 11.277$ and $F_{A \times B} = 63.839$ are greater than the critical value 7.71, the null hypotheses concerning the type of automobile driven and the interaction effect should be rejected.

STEP 5 Summarize the results. Since the null hypothesis for the interaction effect was rejected, it can be concluded that the combination of type of gasoline and type of automobile does affect the gasoline consumption.   ◄

In the preceding analysis, the effect of the type of gasoline used and the effect of the type of automobile driven are called the *main effects*. If there is no significant interaction effect, the main effects can be interpreted independently. However, if there is a significant interaction effect, caution must be used in the interpretation of the main effects.

In order to interpret the results of a two-way analysis of variance, researchers suggest drawing a graph, plotting the means of each of the groups, analyzing the graph, and interpreting the results. In Example 13–9, the means for each group or cell can be found by adding the data values in each cell and dividing by n. The means for each cell are shown in the chart below.

Gas	Type of Automobile	
	Two-Wheel	**Four-Wheel**
Regular	$\overline{X} = \dfrac{26.7 + 25.2}{2} = 25.95$	$\overline{X} = \dfrac{28.6 + 29.3}{2} = 28.95$
High-octane	$\overline{X} = \dfrac{32.3 + 32.8}{2} = 32.55$	$\overline{X} = \dfrac{26.1 + 24.2}{2} = 25.15$

The graph of the means for each of the variables is shown in Figure 13–8. In this graph, the lines cross each other. When such an intersection occurs and the interaction is significant, the interaction is said to be **disordinal.** When there is a disordinal interaction, the main effects should not be interpreted without considering the interaction effect, as was done in Example 13–9.

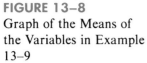

FIGURE 13–8
Graph of the Means of the Variables in Example 13–9

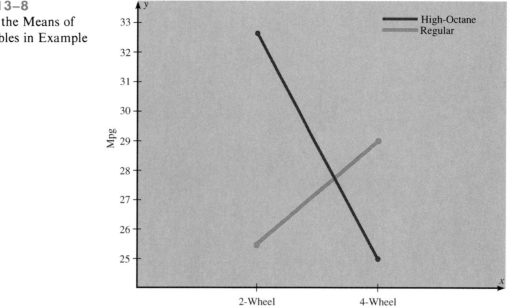

The other type of interaction that can occur is an *ordinal interaction.* Figure 13–9 shows a graph of means in which an ordinal interaction occurs between two variables. The lines do not cross each other, and they are not parallel. If the *F*

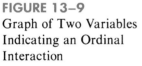

FIGURE 13–9
Graph of Two Variables Indicating an Ordinal Interaction

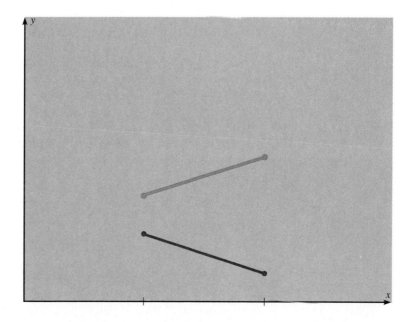

test value for the interaction is significant and the lines do not cross each other, then the interaction is said to be **ordinal.** The main effects in this case can be interpreted independently of each other.

Finally, when there is no significant interaction effect, the lines in the graph will be parallel or approximately parallel. When this situation occurs, the main effects can be interpreted independently of each other because there is no significant interaction. Figure 13–10 shows the graph of two variables when the interaction effect is not significant; the lines are parallel.

FIGURE 13–10

Graph of Two Variables Indicating No Interaction

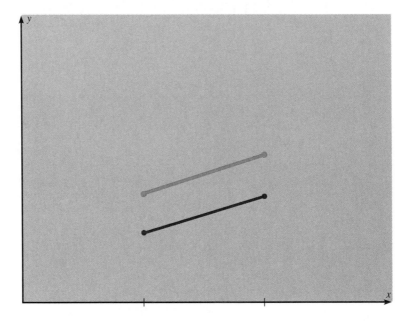

Example 13–9 was an example of a 2 × 2 two-way analysis of variance since each independent variable had two levels. For other types of variance problems, such as a 3 × 2 or a 4 × 3 ANOVA, the interpretation of the results can be quite complicated. Procedures using tests like the Tukey and Scheffé tests for analyzing the cell means exist and are similar to the tests shown for the one-way ANOVA, but they are beyond the scope of this textbook and will not be presented here. Also, many other designs for analysis of variance are available to the researcher, such as three-factor designs and repeated-measure designs; these designs are also beyond the scope of this book.

In summary, the two-way ANOVA is an extension of the one-way ANOVA. It can be used to test the effects of two independent variables and a possible interaction effect on a dependent variable.

EXERCISES

13–41. How does the two-way ANOVA differ from the one-way ANOVA?

13–42. Explain what is meant by the *main effects* and the *interaction effect*.

13–43. What is another name for the independent variable?

13–44. How are the values for the mean squares computed?

13–45. How are the F test values computed?

13–46. In the following two-way ANOVA, variable A has three levels and variable B has two levels. There are five data values in each cell. Find each degrees of freedom value.

a. d.f.N. for factor A
b. d.f.N. for factor B
c. d.f.N. for factor $A \times B$
d. d.f.D. for the within (error) factor

13–47. In the following two-way ANOVA, variable A has five levels and variable B has four levels. There are six data values in each cell. Find each degrees of freedom value.

a. d.f.N. for factor A
b. d.f.N. for factor B
c. d.f.N. for factor $A \times B$
d. d.f.D. for the within (error) factor

13–48. What are the two types of interactions that can occur in the two-way ANOVA?

13–49. When can the main effects for the two-way ANOVA be interpreted independently?

13–50. Describe what the graph of the variables would look like for each situation in a two-way ANOVA experiment.

a. No interaction effect occurs.
b. An ordinal interaction effect occurs.
c. A disordinal interaction effect occurs.

For Exercises 13–51 through 13–56, perform the following steps. Assume that all variables are normally or approximately normally distributed, that the samples are independent, and that the population variances are equal.

a. State the hypotheses.
b. Find the critical value for each F test.
c. Complete the summary table and find the test value.
d. Make the decision.
e. Summarize the results. (Draw a graph of the cell means if necessary.)

13–51. A company wishes to test the effectiveness of its advertising. A product is selected, and two types of ads are written; one is serious and one is humorous. Also, one ad is used on television, and one is used on radio. Sixteen potential customers are selected and assigned randomly to one of four groups. After seeing or listening to the ad, each customer is asked to rate the effectiveness of the ad on a scale of 1 to 20. Various points are assigned for clarity, conciseness, etc. The data are shown below. At $\alpha = 0.01$, analyze the data using a two-way ANOVA.

Type of Ad	Medium	
	Radio	Television
Humorous	6, 10, 11, 9	15, 18, 14, 16
Serious	8, 13, 12, 10	19, 20, 13, 17

ANOVA Summary Table, Exercise 13–51

Source	SS	d.f.	MS	F
Type	10.563			
Medium	175.563			
Interaction	0.063			
Within	66.250			
Total	252.439			

13–52. A medical researcher wishes to test the effects of two diets and the time of day on the sodium level in a person's blood. Eight people are randomly selected and two are randomly assigned to each of the four groups. Analyze the data shown in the tables below using a two-way ANOVA at $\alpha = 0.05$. The sodium content is measured in milliequivalents per liter.

Time	Diet Type			
	I		II	
8:00 A.M.	135	145	138	141
8:00 P.M.	155	162	171	191

ANOVA Summary Table, Exercise 13–52

Source	SS	d.f.	MS	F
Time	1800			
Diet	242			
Interaction	264.5			
Within	279			
Total	2585.5			

13–53. A contractor wishes to see whether there is a difference in the time (in days) it takes two subcontractors to build three different types of homes. At $\alpha = 0.05$, analyze the data shown below using a two-way ANOVA.

Home Type

Subcontractor	I	II	III
A	25, 28, 26, 30, 31	30, 32, 35, 29, 31	43, 40, 42, 49, 48
B	15, 18, 22, 21, 17	21, 27, 18, 15, 19	23, 25, 24, 17, 13

ANOVA Summary Table, Exercise 13–53

Source	SS	d.f.	MS	F
Subcontractor	1672.553			
Home type	444.867			
Interaction	313.267			
Within	328.8			
Total	2759.487			

13–54. Two special training programs in outdoor survival are available for army recruits. One lasts one week, and the other lasts two weeks. The officer wishes to test the effectiveness of the programs and see whether there are any gender differences. Six subjects are randomly assigned to each of the programs according to gender. After completing the program, each is given a written test on his or her knowledge of survival skills. The test consists of 100 questions. The scores of the groups are shown at the right. Use $\alpha = 0.10$ and analyze the data using a two-way ANOVA.

Duration

Gender	One Week	Two Weeks
Female	86, 92, 87, 88, 78, 95	78, 62, 56, 54, 65, 63
Male	52, 67, 53, 42, 68, 71	85, 94, 82, 84, 78, 91

ANOVA Summary Table, Exercise 13–54

Source	SS	d.f.	MS	F
Gender	57.042			
Duration	7.042			
Interaction	3978.375			
Within	1365.5			
Total	5407.959			

13–55. Two different types of outdoor paint, enamel and latex, were tested to see how long (in months) each lasted before it began to crack, flake, and peel. In addition, they were tested in four geographic locations in the United States to study the effects of climate on the paint. At $\alpha = 0.01$, analyze the data shown below using a two-way ANOVA. Each group contained five test panels.

Geographic Location

Type of Paint	North	East	South	West
Enamel	60, 53, 58, 62, 57	54, 63, 62, 71, 76	80, 82, 62, 88, 71	62, 76, 55, 48, 61
Latex	36, 41, 54, 65, 53	62, 61, 77, 53, 64	68, 72, 71, 82, 86	63, 65, 72, 71, 63

ANOVA Summary Table, Exercise 13–55

Source	SS	d.f.	MS	F
Paint type	12.1			
Location	2501			
Interaction	268.1			
Within	2326.8			
Total	5108			

13–56. A company sells three different items: swimming pools, spas, and saunas. The owner decides to see whether the age of the sales representative and the type of item affect the monthly sales. At $\alpha = 0.05$, analyze the data shown below using a two-way ANOVA. The amount of sales is given in hundreds of dollars for a randomly selected month, and five salespeople were selected for each group.

Age of Salesperson	Product		
	Pool	**Spa**	**Sauna**
Over 30	56, 23, 52, 28, 35	43, 25, 16, 27, 32	47, 43, 52, 61, 74
30 and under	16, 14, 18, 27, 31	58, 62, 68, 72, 83	15, 14, 22, 16, 27

ANOVA Summary Table, Exercise 13–56

Source	SS	d.f.	MS	F
Age	168.033			
Product	1762.067			
Interaction	7955.267			
Within	2574			
Total	12,459.367			

13–7 SUMMARY

The F test can be used to compare two sample variances in order to determine whether they are equal. It also can be used to compare three or more means. When three or more means are compared, the technique is called analysis of variance (ANOVA). The ANOVA technique uses two estimates of the population variance. One is called the between-group variance, which is the variance of the sample means; and the other is called the within-group variance, which is the overall variance of all the values. When there is no significant difference among the means, the two estimates will be approximately equal, and the F test value will be close to 1. If there is a significant difference among the means, the between-group variance estimate will be larger than the within-group variance estimate, and a significant test value will result.

If there is a significant difference among means, the researcher may wish to see where this difference lies. There are several statistical tests that can be used to compare the sample means after the ANOVA technique has been done. The most commonly used tests are the Scheffé test and the Tukey test. When the sample sizes are the same, the Tukey test can be used. The Scheffé test is more general and can be used when the sample sizes are equal or not equal.

When there is one independent variable, the analysis of variance is called a one-way ANOVA. When there are two independent variables, the analysis of variance is called a two-way ANOVA. The two-way ANOVA enables the researcher to test the effects of two independent variables and a possible interaction effect on one dependent variable.

Speaking of **Statistics**

Therapies' Results May Be in the Mind

By Tim Friend
USA TODAY

Meditation, biofeedback and cognitive therapies touted for lowering blood pressure may owe their success to the placebo effect, research suggests.

With the placebo effect, patients improve because they expect their treatment to make them better, even though the treatment is ineffective.

Findings, in today's *Annals of Internal Medicine,* show that the mental techniques were:

• Slightly better at reducing high blood pressure than doing nothing at all.
• No better than sham meditation and sham biofeedback. For sham meditation, false instructions are given on how to meditate. For fake biofeedback, bogus blood pressure readings were provided.

"Frankly, this will be a terribly controversial paper because there are a lot of people making good money and drawing on the patient's expectations about biofeedback, meditation and relaxation techniques," says Dr. Thomas Delbanco, Beth Israel Hospital, Boston. "Just a patient taking their own blood pressure may have the same impact."

In the research, led by Beth Israel's Dr. David Eisenberg, more than 1,000 studies of cognitive therapy for high blood pressure were found. Only 26 met standards of having a control group for comparison.

But even though no benefit beyond placebo was found when data from the 26 studies were pooled and analyzed, if one of the techniques works, should you stop?

Says Eisenberg: "Expecting improvement appears to play a critical role in shaping the response to cognitive therapies. But the patient and physician can review the study and come to their own conclusions."

Source: Copyright 1993, USA TODAY.
Reprinted with permission.

In this study, several different methods were used in an attempt to lower a person's blood pressure. The results are a little surprising. Suggest several hypotheses that may have been used in this study. What statistical tests might have been used to test these hypotheses?

Important Terms

Analysis of variance (ANOVA)
ANOVA summary table
Between-group variance
Disordinal interaction
Factors
F distribution
F test
Interaction effect
Level
Main effect
Mean square
One-way ANOVA
Ordinal interaction
Scheffé test
Sum of squares between groups
Sum of squares within groups
Treatment groups
Tukey test
Two-way ANOVA
Within-group variance

Important Formulas

Formula for the F test for comparing two variances:

$$F = \frac{s_1^2}{s_2^2}$$

where s_1^2 is the larger variance and

$$\text{d.f.N.} = n_1 - 1 \quad \text{and} \quad \text{d.f.D.} = n_2 - 1$$

Formulas for the ANOVA test:

$$\overline{X}_{\text{GM}} = \frac{\Sigma X}{N}$$

$$F = \frac{s_B^2}{s_W^2}$$

where

$$s_B^2 = \frac{\Sigma n_i(\overline{X}_i - \overline{X}_{\text{GM}})^2}{k - 1} \qquad s_W^2 = \frac{\Sigma (n_i - 1)s_i^2}{\Sigma (n_i - 1)}$$

$$\text{d.f.N.} = k - 1 \qquad\qquad N = n_1 + n_2 + \cdots + n_k$$
$$\text{d.f.D.} = N - k \qquad\qquad k = \text{number of groups}$$

Formulas for the Scheffé test:

$$F_S = \frac{(\overline{X}_i - \overline{X}_j)^2}{s_W^2[(1/n_i) + (1/n_j)]} \quad \text{and} \quad F' = (k - 1)(\text{C.V.})$$

Formula for the Tukey test:

$$q = \frac{\overline{X}_i - \overline{X}_j}{\sqrt{s_W^2/n}}$$

$$\text{d.f.N.} = k \quad \text{and} \quad \text{d.f.D.} = \text{degrees of freedom for } s_W^2$$

Formulas for the two-way ANOVA:

$$\text{MS}_A = \frac{\text{SS}_A}{a - 1} \qquad F_A = \frac{\text{MS}_A}{\text{MS}_W} \qquad \begin{array}{l} \text{d.f.N.} = a - 1 \\ \text{d.f.D.} = ab(n - 1) \end{array}$$

$$\text{MS}_B = \frac{\text{SS}_B}{b - 1} \qquad F_B = \frac{\text{MS}_B}{\text{MS}_W} \qquad \begin{array}{l} \text{d.f.N.} = b - 1 \\ \text{d.f.D.} = ab(n - 1) \end{array}$$

$$\text{MS}_{A \times B} = \frac{\text{SS}_{A \times B}}{(a - 1)(b - 1)} \qquad F_{A \times B} = \frac{\text{MS}_{A \times B}}{\text{MS}_W} \qquad \begin{array}{l} \text{d.f.N.} = (a - 1)(b - 1) \\ \text{d.f.D.} = ab(n - 1) \end{array}$$

$$\text{MS}_W = \frac{\text{SS}_W}{ab(n - 1)}$$

Review Exercises

For Exercises 13–57 through 13–70, perform the following steps.

a. State the hypotheses, and identify the claim.
b. Find the critical value(s).
c. Compute the test value.
d. Make the decision.
e. Summarize the results.

13–57. An educator wishes to compare the variances of the amount of money spent per pupil for two different states. The data are given below. At $\alpha = 0.05$, is there a significant difference in the variances of the amounts each state spends per pupil?

State 1	State 2
$s_1^2 = \$585$	$s_2^2 = \$261$
$n_1 = 18$	$n_2 = 16$

13–58. A researcher wants to compare the variances of the heights (in inches) of major league baseball players with those of players in the minor leagues. A sample of 25 players from each league is selected, and the variances of the heights for each league are 2.25 and 4.85, respectively. At $\alpha = 0.10$, is there a significant difference between the variances of the heights for the two leagues?

13–59. A traffic safety commissioner feels that the variation in the number of speeding tickets given on Route 19 is larger than the variation in the number of speeding tickets given on Route 22. Ten weeks are randomly selected; the standard deviation of the number of tickets issued for Route 19 is 6.3, and the standard deviation of the number of tickets issued for Route 22 is 2.8. At $\alpha = 0.05$, can the commissioner conclude that the variance of speeding tickets issued on Route 19 is greater than the variance of speeding tickets issued on Route 22?

13–60. The variances of the cholesterol amounts of two different types of eggs are to be compared. A sample of 10 eggs from type A has a variance of 19.6, and a sample of 14 eggs from type B has a variance of 7.3. At $\alpha = 0.01$, can one conclude that there is a difference in the variances of the two types of eggs? (Cholesterol is measured in milligrams.)

13–61. A researcher claims that the variation in the number of days factory workers miss per year due to illness is greater than the variation in the number of days hospital workers miss per year. A sample of 42 workers from a large hospital has a standard deviation of 2.1 days, and a sample of 65 workers from a large factory has a standard deviation of 3.2 days. Test the claim, at $\alpha = 0.10$.

13–62. The variations in the number of absentees per day in two schools are being compared. A sample of 30 days is selected; the standard deviation of the number of absentees in school A is 4.9, and for school B it is 2.5. At $\alpha = 0.01$, can one conclude that there is a difference in the two standard deviations?

If the null hypothesis is rejected in Exercises 13–63 through 13–68, use the Scheffé test when the sample sizes are unequal to test the differences between the means, and use the Tukey test when the sample sizes are equal.

13–63. The number of cars that park in three different city-owned lots is being compared. A week is selected at random, and the number of cars parked each day is as shown in column 2. Test the claim, at $\alpha = 0.01$, that there is no difference in the number of cars parked in each lot.

Lot A	Lot B	Lot C
203	319	89
162	321	126
190	271	115
219	194	92
188	342	106
209	423	100
212	199	94

13–64. Three composition instructors recorded the number of grammatical errors their students made on a term paper. The data are as shown below. At $\alpha = 0.01$, is there a significant difference in the average number of errors in each of the instructors' classes?

Instructor A	Instructor B	Instructor C
2	6	1
3	7	4
5	12	0
4	4	1
8	9	2
	1	2
	0	

13–65. A plant owner wants to see whether the average time (in minutes) it takes his employees to commute to work is different for three groups. The data are as shown below. At $\alpha = 0.05$, can the owner conclude that there is a significant difference among the means?

Managers	Salespeople	Stock Clerks
35	9	15
18	3	6
27	12	27
24	6	22
	14	
	8	
	21	

13–66. The coliform levels (in parts per million) of three different lakes were checked for a period of five days. The data are as shown below. At $\alpha = 0.05$, is there a difference in the means of the coliform levels of the lakes?

Sunset Lake	South Lake	Indian Lake
62	97	33
53	82	35
41	99	31
38	84	28
55	79	26

13–67. Students are randomly assigned to three different reading classes. Each class is taught by a different method. At the end of the course, a comprehensive reading examination is given, and the results are as shown on page 487. At $\alpha = 0.05$, is there a significant difference in the means of the examination results?

Class A	Class B	Class C
87	82	97
92	78	90
61	41	83
83	65	92
47	63	91

13–68. Four hospitals are being compared to see whether there is any significant difference in the mean number of operations performed in each. A sample of six days provided the following number of operations performed each day. Can one conclude, at $\alpha = 0.05$, that there is no difference in the means?

Hospital A	Hospital B	Hospital C	Hospital D
8	4	5	10
5	9	6	12
6	3	3	13
3	1	7	9
2	0	7	0
7	1	3	1

13–69. A teacher wishes to test the math anxiety level of her students in two classes at the beginning of the semester. The classes are Calculus I and Statistics. Furthermore, she wishes to see whether there is a difference owing to the ages of the students. Math anxiety is measured by the score on a 100-point anxiety test. Use $\alpha = 0.10$ and a two-way analysis of variance to see whether there is a difference. Five students are randomly assigned to each group. The data are shown below.

	Class	
Age	Calculus I	Statistics
Under 20	43, 52, 61, 57, 55	19, 20, 31, 36, 24
20 and over	56, 55, 42, 48, 61	63, 78, 67, 71, 75

ANOVA Summary Table, Exercise 13–69

Source	SS	d.f.	MS	F
Age	2376.2			
Class	105.8			
Interaction	2645			
Within	763.2			
Total	5890.2			

13–70. A medical researcher wishes to test the effects of two different diets and two different exercise programs on the glucose level in a person's blood. The glucose level is measured in milligrams per deciliter (mg/dl). Three subjects are randomly assigned to each group. Analyze the data shown below using a two-way ANOVA with $\alpha = 0.05$.

	Diet	
Exercise Program	A	B
I	62, 64, 66	58, 62, 53
II	65, 68, 72	83, 85, 91

ANOVA Summary Table, Exercise 13–70

Source	SS	d.f.	MS	F
Exercise	816.750			
Diet	102.083			
Interaction	444.083			
Within	108			
Total	1470.916			

COMPUTER APPLICATIONS

1. Write a computer program that will compute the ANOVA test value. Enter the number of groups and the number of values in each group.
2. MINITAB can be used to perform the ANOVA F test. For example, suppose that three bank tellers record the number of customers each received over a 5-hour period, as follows:

A	B	C
9	8	12
6	7	15
15	12	18
4	3	9
3	5	10

At $\alpha = 0.05$, can one conclude that the averages are equal?

The data for each level are entered by using the SET commands. The ANOVA will be performed by the AOVONEWAY C1 C2 C3 command. Type in the following:

```
MTB  > SET C1
DATA > 9   6   15   4   3
DATA > END
MTB  > SET C2
DATA > 8   7   12   3   5
DATA > END
MTB  > SET C3
DATA > 12   15   18   9   10
DATA > END
MTB  > AOVONEWAY  C1 C2 C3
```

The computer will print the ANOVA table as shown.

```
ANALYSIS OF VARIANCE
SOURCE DF     SS    MS    F     P
FACTOR   2  104.9  52.5  3.25  0.075
ERROR   12  194.0  16.2
TOTAL   14  298.9
```

The *F* test value is 3.25. Since the *P*-value, 0.075, is greater than 0.05, the null hypothesis is not rejected.

To find the *F* critical value with 2 and 12 degrees of freedom, type INVCDF 0.95 and F 2 12.

```
MTB  >  INVCDF  0.95;
SUBC >  F  2  12.
```

The computer will print the following:

```
0.9500    3.89
```

The critical value is 3.89. Since 3.25 is less than 3.89, the null hypothesis is not rejected at $\alpha = 0.05$, as verified previously by the *P*-value.

Try Exercise 13–35 by using MINITAB.

3. MINITAB can be used to conduct a two-way ANOVA. For example, suppose a researcher wishes to test the reading abilities of students (as measured by scores on a 100-point examination) who have learned to read by using three different textbooks. Furthermore, one group also receives instruction via a computer. Using a two-way ANOVA, with $\alpha = 0.05$, analyze the data shown.

Used	Textbook		
Computer	1	2	3
Yes	80, 72, 81	90, 98, 91	60, 68, 57
No	72, 71, 76	51, 54, 53	78, 82, 76

The data are entered by rows, using a SET C1 command. Next, it is necessary to show MINITAB how the two-way ANOVA table is set up, which is done by coding the data. The data values in row 1 are coded by using a 0 for each data value, and the data values in row 2 are coded by using a 1. (Any number can be used as a code.) The coded data are entered by using the SET C2 command. It shows two rows with nine data values in each row.

Next, the data values must be coded for the columns. A 1 is used for a data value in column 1, a 2 represents a data value in column 2, and a 3 represents a data value in column 3. The SET C3 command is used to enter the coded values. To check the data to make sure that they have been placed in the right rows or columns, use a TABLE C2 C3 command followed by a DATA C1 command.

Type in the following:

```
MTB  >  SET  C1
DATA >  80 72 81 90 98 91 60 68 57
DATA >  72 71 76 51 54 53 78 82 76
DATA >  END
MTB  >  SET  C2
DATA >  0 0 0  0 0 0  0 0 0
DATA >  1 1 1  1 1 1  1 1 1
DATA >  END
MTB  >  SET  C3
DATA >  1 1 1  2 2 2  3 3 3
DATA >  1 1 1  2 2 2  3 3 3
DATA >  END
MTB  >  TABLE  C2  C3;
SUBC >  DATA  C1.
```

The computer will print the following:

```
ROWS: C2      COLUMNS: C3

         1        2        3

0    80.000   90.000   60.000
     72.000   98.000   68.000
     81.000   91.000   57.000

1    72.000   51.000   78.000
     71.000   54.000   82.000
     76.000   53.000   76.000
```

```
CELL CONTENTS --
          C1:DATA
```

Next, the cell means can be computed and printed out using the TABLE C2 C3 command followed by a MEAN C1 command.

```
MTB  >  TABLE  C2  C3;
SUBC >  MEAN  C1.
```

The computer will print the following:

```
ROWS: C2      COLUMNS: C3
          1       2       3      ALL

  0   77.667  93.000  61.667  77.444
  1   73.000  52.667  78.667  68.111
ALL   75.333  72.833  70.167  72.778
```

```
CELL CONTENTS --
          C1:MEAN
```

Finally, the two-way ANOVA summary table will be generated if a TWOWAY C1 C2 C3 command is typed in, as shown.

```
MTB > TWOWAY C1 C2 C3
```

The computer will print the following:

```
ANALYSIS OF VARIANCE C1
SOURCE        DF      SS       MS
C2            1      392.0    392.0
C3            2       80.1     40.1
INTERACTION   2     2514.3   1257.2
ERROR        12      188.7     15.7
TOTAL        17     3175.1
```

MINITAB will not print the F test values. They must be calculated by the user. For example, the F test value for C2 is

$$F = \frac{MS_{C2}}{MS_{error}} = \frac{392}{15.7} = 24.968$$

Finish this example by computing the F test values and summarizing the results.

Try Exercise 13–69 by using MINITAB.

√ DATA ANALYSIS Applying the Concepts

The Data Bank is located in the Appendix.

1. From the Data Bank, choose a variable and test the equality of the variances for the males and females.
2. From the Data Bank, select a random sample of subjects, and test the hypothesis that the mean cholesterol levels of the nonsmokers, less-than-one-pack-a-day smokers, and one-pack-plus smokers are equal. Use an ANOVA test. If the null hypothesis is rejected, conduct the Scheffé test to find where the difference is. Summarize the results.
3. Repeat Exercise 2 for the mean IQs of the various educational levels of the subjects.
4. Using the Data Bank, randomly select 12 subjects and randomly assign them to one of the four groups in the following classifications.

	Smoker	**Nonsmoker**
Male		
Female		

Use one of the following variables—weight, cholesterol, or systolic pressure—as the dependent variable and perform a two-way ANOVA on the data. A computer program should be used to generate the ANOVA table.

✎ TEST

Directions: Determine whether the statement is true or false.

1. The analysis of variance uses an F test.
2. When three or more means are to be compared, the t test can be used.
3. The null hypothesis for the analysis of variance is that all means are unequal.
4. In analysis of variance, the null hypothesis will be rejected only when there is a significant difference between all pairs of means.
5. The F test does not use the concept of degrees of freedom.
6. When one conducts the F test for analysis of variance, two estimates of the population variance are compared.
7. When the F test value is close to 1, the null hypothesis will be rejected.
8. Two sample variances can be compared by using an F test.
9. When one calculates the F test value, the larger variance is always placed in the numerator.
10. The Scheffé test should be used even though the null hypothesis has not been rejected.
11. The Tukey test is generally more powerful than the Scheffé test for pairwise comparisons after the rejection of the null hypothesis in the one-way ANOVA.
12. The two-way ANOVA can test two main hypotheses and one interaction hypothesis.

HYPOTHESIS-TESTING SUMMARY 2*

5. Test of the significance of the correlation coefficient.

Example: $H_0: \rho = 0$

Use a t test:

$$t = r \sqrt{\frac{n-2}{1-r^2}} \quad \text{with} \quad \text{d.f.} = n - 2$$

6. Comparison of a sample variance or standard deviation with a specific population variance or standard deviation.

Example: $H_0: \sigma^2 = 225$

Use the chi-square test:

$$\chi^2 = \frac{(n-1)s^2}{\sigma^2} \quad \text{with} \quad \text{d.f.} = n - 1$$

7. Comparison of a sample distribution with a specific population.

Example: H_0: There is no difference between the two distributions.

Use the chi-square goodness-of-fit test:

$$\chi^2 = \sum \frac{(O-E)^2}{E} \quad \text{with d.f.} = \text{no. of categories} - 1$$

8. Comparison of the independence of two variables.

Example: H_0: Variable A is independent of variable B.

Use the chi-square independence test:

$$\chi^2 = \sum \frac{(O-E)^2}{E} \quad \text{with} \quad \text{d.f.} = (R-1)(C-1)$$

9. Comparison of two sample variances or standard deviations.

Example: $H_0: \sigma_1^2 = \sigma_2^2$

Use the F test:

$$F = \frac{s_1^2}{s_2^2}$$

where

s_1^2 = larger variance   d.f.N. $= n_{\text{numerator}} - 1$
s_2^2 = smaller variance   d.f.D. $= n_{\text{denominator}} - 1$

10. Comparison of three or more sample means.

Example: $H_0: \mu_1 = \mu_2 = \mu_3$

Use the analysis of variance test:

$$F = \frac{s_B^2}{s_W^2}$$

where

$$s_B^2 = \frac{\sum n_i(\overline{X}_i - \overline{X}_{\text{GM}})^2}{k-1}$$

$$s_W^2 = \frac{\sum (n_i - 1)s_i^2}{\sum (n_i - 1)}$$

d.f.N. $= k - 1$ $N = n_1 + n_2 + \cdots + n_k$
d.f.D. $= N - k$ k = number of groups

11. Test when the F value for the ANOVA is significant. Use the Scheffé test to find what pairs of means are significantly different:

$$F_S = \frac{(\overline{X}_i - \overline{X}_j)^2}{s_W^2[(1/n_i) + (1/n_j)]} \quad \text{and} \quad F' = (k-1)(\text{C.V.})$$

Use the Tukey test to find which pairs of means are significantly different:

$$q = \frac{\overline{X}_i - \overline{X}_j}{\sqrt{s_W^2/n}} \quad \text{with} \quad \begin{array}{l} \text{d.f.N.} = K \\ \text{d.f.D.} = \text{degrees of freedom for } s_W^2 \end{array}$$

12. Test for the two-way ANOVA.

Example: H_0: There is no significant difference for the main effects.
H_0: There is no significant difference for the interaction effect.

$$\text{MS}_A = \frac{\text{SS}_A}{a-1}$$

$$\text{MS}_B = \frac{\text{SS}_B}{b-1}$$

$$\text{MS}_{A \times B} = \frac{\text{SS}_{A \times B}}{(a-1)(b-1)}$$

$$\text{MS}_W = \frac{\text{SS}_W}{ab(n-1)}$$

$$F_A = \frac{\text{MS}_A}{\text{MS}_W} \qquad \begin{array}{l} \text{d.f.N.} = a - 1 \\ \text{d.f.D.} = ab(n-1) \end{array}$$

$$F_B = \frac{\text{MS}_B}{\text{MS}_W} \qquad \begin{array}{l} \text{d.f.N.} = b - 1 \\ \text{d.f.D.} = ab(n-1) \end{array}$$

$$F_{A \times B} = \frac{\text{MS}_{A \times B}}{\text{MS}_W} \qquad \begin{array}{l} \text{d.f.N.} = (a-1)(b-1) \\ \text{d.f.D.} = ab(n-1) \end{array}$$

*This summary is a continuation of Hypothesis-Testing Summary 1.

CHAPTER

14

Nonparametric Statistics

14–1 INTRODUCTION

Statistical tests, such as the z, t, and F tests, are called parametric tests. **Parametric tests** are statistical tests for population parameters such as means, variances, and proportions that involve assumptions about the populations from which the samples were selected. One assumption is that these populations are normally distributed. But what if the population in a particular hypothesis-testing situation is *not* normally distributed? Statisticians have developed a branch of statistics known as **nonparametric statistics** or **distribution-free statistics** to use when the population from which the samples are selected is not normally distributed. In addition, nonparametric statistics can be used to test hypotheses that do not involve specific population parameters, such as μ, σ, or p.

For example, a sportswriter may wish to know whether there is a relationship between the rankings of two judges on the diving abilities of ten Olympic swimmers. In another situation, a sociologist may wish to determine whether men and women enroll at random for a specific drug rehabilitation program. The statistical tests used in these situations are nonparametric or distribution-free tests. The term *nonparametric* is used for both situations.

The nonparametric tests explained in this chapter are the sign test, the Wilcoxon rank sum test, the Wilcoxon signed-rank test, the Kruskal-Wallis test, and the runs test. In addition, the Spearman rank correlation coefficient, a statistic for determining the relationship between ranks, will be explained.

14–2 ADVANTAGES AND DISADVANTAGES OF NONPARAMETRIC METHODS

As stated previously, nonparametric tests and statistics can be used in place of their parametric counterparts (z, t, and F) when the assumption of normality cannot be met. However, one should not assume that these statistics are a better alternative than the parametric statistics. There are both advantages and disadvantages in the use of nonparametric methods.

Advantages   There are five advantages in the use of nonparametric methods.

1. They can be used to test population parameters when the variable is not normally distributed.
2. They can be used when the data is nominal or ordinal.
3. They can be used to test hypotheses that do not involve population parameters.
4. In most cases, the computations are easier than those for the parametric counterparts.
5. They are easier to understand.

Disadvantages   There are three disadvantages in the use of nonparametric methods.

1. They are *less sensitive* than their parametric counterparts when the assumptions of the parametric methods are met. Therefore, larger differences are needed before the null hypothesis can be rejected.

2. They tend to use less information than the parametric tests. For example, the sign test requires the researcher to determine only whether the data values are above or below the median, not how much above or below the median each value is.

3. They are *less efficient* than their parametric counterparts when the assumptions of the parametric methods are met. That is, larger sample sizes are needed to overcome the loss of information. For example, the nonparametric sign test is about 60% as efficient as its parametric counterpart, the z test. Thus, a sample size of 100 is needed for use of the sign test, compared with a sample size of 60 for use of the z test to obtain the same results.

Since there are both advantages and disadvantages to the nonparametric methods, the researcher should use caution in selection of these methods. If the assumptions can be met, the parametric methods are preferred. However, when parametric assumptions cannot be met, the researcher has another valuable tool, the nonparametric methods, for analyzing the data.

Ranking

Many nonparametric tests involve the **ranking** of data, i.e., the positioning of a data value in a data array according to some rating scale. For example, suppose a judge decides to rate five speakers on a scale of 1 to 10, using a certain number of points for categories such as voice, gestures, logical presentation, and platform personality. The ratings are as shown in the chart.

Speaker	A	B	C	D	E
Rating	8	6	10	3	1

Now, the rankings are as shown in the next chart.

Speaker	C	A	B	D	E
Rating	10	8	6	3	1
Ranking	1	2	3	4	5

Since speaker C received the highest score, 10 points, he or she is ranked first; speaker A received the next highest total, 8 points, so he or she is ranked second; etc.

Now what happens if two or more speakers receive the same number of points? Suppose the judge awards points as follows:

Speaker	A	B	C	D	E
Rating	8	6	10	6	3

The speakers are then ranked as follows:

Speaker	C	A	B	D	E
Rating	10	8	6	6	3
Ranking	1	2	Tie for 3rd and 4th		5

When there is a tie for two or more places, the average of the ranks must be used. In this case, each would be ranked as

$$\frac{3+4}{2} = \frac{7}{2} = 3.5$$

Speaking of **Statistics**

Health Disparities Widen by Region

By Anita Manning
USA TODAY

The gap between the healthiest and the unhealthiest people in the USA is widening, says a report released today.

Northwestern National Life Insurance Co.'s 1993 state health rankings show the disparity between the health of people in states at the top of the list and those at the bottom has grown since 1990.

"The states with the healthiest populations generally have more resources, better education and more affluent populations," says research director Candyce Wisner. "The least healthy are states that have been hit hardest by the recession. A lot of it is resources at the state level."

Improvements in health in 27 states were recorded: Top gainers were Oregon and New Hampshire, up 9%, and Vermont and Connecticut, up 8%.

Another 19 states dropped in health measures, with Wyoming down 11% since 1990, primarily due to an increase in cancer and infectious disease. South Carolina was down 10% due to a rise in violent crime.

Most states where health gains were noted showed a drop in smoking and greater public support for health care, Wisner says. "The most effective thing . . . is for people to take greater responsibility for their own health in terms of lifestyle choices. To the extent that health care reform can encourage people to do that, and provide better access (to care), that's where the real differences will come."

Source: Based on data from Northwestern National Life Insurance Company. Copyright 1993, USA TODAY. Reprinted with permission.

Many statistical studies involve ranking. In this study, the Northwestern National Life Insurance Company ranked the states according to the health of its citizens. Rank each state according to the criteria given in this chapter. For example, Hawaii and Connecticut are tied; hence, they would be ranked as 3.5.

Hence, the rankings are as follows:

Speaker	C	A	B	D	E
Rating	10	8	6	6	3
Ranking	1	2	3.5	3.5	5

Many times, the data are already ranked, and therefore, no additional computations must be done. For example, if the judge does not have to award points but can simply select the best, second-best, third-best speaker, etc., then these ranks can be used directly.

How the States Rank

Rank/State	Rank/State
1. Minnesota	26. Montana
2. New Hampshire	*Oregon
3. Hawaii	*Arizona
*Connecticut	29. Missouri
5. Utah	*Oklahoma
6. Vermont	31. Texas
*Kansas	*Idaho
8. Massachusetts	33. Illinois
9. Iowa	*South Dakota
*Colorado	35. Wyoming
11. Wisconsin	36. New York
*Nebraska	37. North Carolina
13. Virginia	38. Georgia
14. Maine	*Kentucky
15. New Jersey	40. Tennessee
16. Maryland	*Florida
*Ohio	42. Nevada
18. North Dakota	43. Alabama
*Rhode Island	*Alaska
*Pennsylvania	*Arkansas
21. Washington	46. South Carolina
22. Indiana	*New Mexico
23. Delaware	48. West Virginia
24. California	49. Louisiana
25. Michigan	50. Mississippi

*Tie.

EXERCISES

14–1. What is meant by *nonparametric statistics?*

14–2. When should nonparametric statistics be used?

14–3. List the advantages and disadvantages of nonparametric statistics.

For Exercises 14–4 through 14–10, rank each set of data.

14–4. 7, 5, 9, 8, 4, 2, 1

14–5. 21, 65, 31, 41, 61, 41, 72, 34

14–6. 73, 320, 432, 186, 241

14–7. 8, 6, 3, 8, 7, 5, 5, 9, 12, 15, 17, 14

14–8. 22, 25, 28, 28, 18, 32, 37, 41, 41, 43

14–9. 190, 236, 187, 190, 321, 532, 673

14–10. 3.8, 7.9, 3.6, 4.1, 2.5, 7.9, 4.12, 3.21, 4.1

14–3 THE SIGN TEST

Single-Sample Sign Test

The simplest nonparametric test is the **sign test** for single samples, which is used to test the value of a median for a specific sample. When using the sign test, the researcher hypothesizes the specific value for the median of a population; then he or she selects a sample of data and compares each value with the median. If the data value is above the median, it is assigned a $+$ sign. If it is below the median, it is assigned a $-$ sign. And if it is exactly the same as the median, it is assigned a 0. Then, the number of $+$ and $-$ signs are compared. If the null hypothesis is true, the number of $+$ signs should be approximately equal to the number of $-$ signs. If the null hypothesis is not true, there will be a disproportionate number of $+$ or $-$ signs.

Test Value for the Sign Test

The test value is the smaller number of $+$ or $-$ signs.

For example, if there are 8 positive signs and 3 negative signs, the test value is 3. When the sample size is 25 or less, Table J in the Appendix is used to determine the critical value. For a specific α, if the test value is less than or equal to the critical value obtained from the table, the null hypothesis should be rejected. The values in Table J are obtained from the binomial distribution. The derivation is omitted here.

EXAMPLE 14–1 A convenience store owner hypothesizes that the median number of snow cones he sells per day is 40. A random sample of 20 days yields the following data for the number of snow cones sold each day.

18	43	40	16	22
30	29	32	37	36
39	34	39	45	28
36	40	34	39	52

At $\alpha = 0.05$, is the owner's claim correct?

Solution

STEP 1 State the hypotheses, and identify the claim.

H_0: median = 40 (claim) and H_1: median $\neq$ 40

STEP 2 Find the critical value. Compare each value of the data with the median. If the value is greater than the median, replace the value with a $+$ sign. If it is less than the median, replace it with a $-$ sign. And if it is equal to the median, replace it with a 0. The completed table follows.

$-$	$+$	0	$-$	$-$
$-$	$-$	$-$	$-$	$-$
$-$	$-$	$-$	$+$	$-$
$-$	0	$-$	$-$	$+$

Refer to Table J in the Appendix, using $n = 18$ (the total number of $+$ and $-$ signs; omit the zeros) and $\alpha = 0.05$ for a two-tailed test; the critical value is 4. See Figure 14–1.

FIGURE 14–1
Finding the Critical Value in Table J for Example 14–1

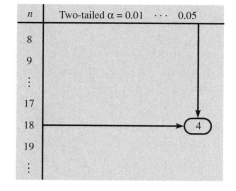

n	Two-tailed $\alpha = 0.01$ $\cdots$ 0.05
8	
9	
$\vdots$	
17	
18	4
19	
$\vdots$	

STEP 3 Compute the test value. Count the number of $+$ and $-$ signs obtained in Step 2, and use the smaller value as the test value. Since there are $3 +$ signs and $15 -$ signs, 3 is the test value.

STEP 4 Make the decision. Compare the test value 3 with the critical value 4. If the test value is less than or equal to the critical value, the null hypothesis is rejected. In this case, the null hypothesis is rejected since $3 < 4$.

STEP 5 Summarize the results. There is enough evidence to reject the claim that the median number of snow cones sold per day is 40. ◄

When the sample size is 26 or more, the normal approximation can be used to find the test value. The formula is given below. The critical value is found in Table E in the Appendix.

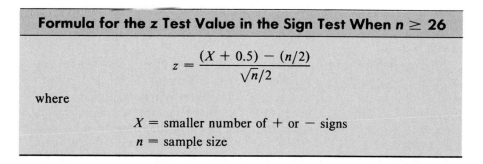

Formula for the z Test Value in the Sign Test When $n \geq 26$

$$z = \frac{(X + 0.5) - (n/2)}{\sqrt{n}/2}$$

where

$X =$ smaller number of $+$ or $-$ signs
$n =$ sample size

EXAMPLE 14–2 A manufacturer claims that the median lifetime of a rubber washer is at least 8 years. A sample of 50 washers showed that 21 lasted more than 8 years. At $\alpha = 0.05$, test the manufacturer's claim.

Solution **STEP 1** State the hypotheses, and identify the claim.

$$H_0: \text{median} \geq 8 \text{ (claim)} \qquad \text{and} \qquad H_1: \text{median} < 8$$

STEP 2 Find the critical value. Since $\alpha = 0.05$, and since this is a one-tailed left test, the critical value is -1.65, obtained from Table E.

STEP 3 Compute the test value.

$$z = \frac{(X + 0.5) - (n/2)}{\sqrt{n}/2} = \frac{(21 + 0.5) - (50/2)}{\sqrt{50}/2}$$

$$= \frac{-3.5}{3.5355} = -0.99$$

STEP 4 Make the decision. Since the test value of -0.99 is greater than -1.65, the decision is to not reject the null hypothesis.

STEP 5 Summarize the results. There is not enough evidence to reject the claim that the median lifetime of the washers is at least 8 years. ◀

In Example 14–2, the sample size was 50 and 21 washers lasted more than 8 years; so $50 - 21$, or 29, washers did not last 8 years. The value of X corresponds to the smaller of the two numbers 21 and 29. In this case, $X = 21$ is used in the formula. The reason is that there would be 21 positive signs, since subtracting 8 years from the value in years of a washer that lasted longer than 8 years would result in a positive answer. When 8 is subtracted from the value in years of a washer that did not last 8 years, the answer would be negative. Assuming that no washer lasted exactly 8 years would result in 21 positive answers and 29 negative answers. Since 21 is the smaller of the two numbers, the value of X is 21.

Suppose that a researcher hypothesizes that the median age of houses in a certain municipality is 40 years. In a random sample of 100 houses, 68 were older than 40 years. Then the value that would be used for X in the formula would be $100 - 68$, or 32, since it is the smaller of the two numbers, 68 and 32. When 40 is subtracted from the age of a house older than 40 years, the resultant answer would be positive. When 40 is subtracted from the age of a house that is less than 40 years old, the resultant answer would be negative. There would be 68 positive signs and 32 negative signs (assuming that no house would be exactly 40 years old). Hence, 32 would be used for X, since it is the smaller of the two values.

Paired-Sample Sign Test The sign test can also be used to test sample means in a comparison of two dependent samples, such as a "before" and "after" test. Recall that when dependent samples are taken from normally distributed populations, the t test is used (Section 10–4). When the condition of normality cannot be met, the nonparametric sign test can be used, as shown in the next example.

Speaking of Statistics

A look at doctors' income

Specialists, like radiologists, lead in income among doctors. Here's a look at median physician net income before taxes:

Radiology	$223,000
Anesthesiology	$210,000
Surgery	$200,000
Obstetrics/gynecology	$200,000
Pathology	$153,000
Internal medicine	$125,000
Psychiatry	$110,000
Pediatrics	$105,000
General/ family practice	$98,000

Source: USA TODAY research   By Marcia Staimer, USA TODAY

Source: Copyright 1993, USA TODAY. Reprinted with permission.

Here is a study that uses the median. Speculate why the median rather than the mean was used.

EXAMPLE 14–3   A medical researcher feels that the number of ear infections in swimmers can be reduced if the swimmers use earplugs. A sample of 10 people is selected, and the number of infections for a four-month period is recorded. During the first two months, the swimmers did not use the earplugs; during the second two months, they did. At the beginning of the second two-month period, each swimmer was examined to make sure that no infections were present. The data are as shown below. At $\alpha = 0.05$, can the researcher conclude that using earplugs reduced the number of ear infections?

Number of Ear Infections

Swimmer	Before, X_B	After, X_A
A	3	2
B	0	1
C	5	4
D	4	0
E	2	1
F	4	3
G	3	1
H	5	3
I	2	2
J	1	3

Solution   **STEP 1**   State the hypotheses, and identify the claim.

H_0: The number of ear infections will not be reduced.
H_1: The number of ear infections will be reduced (claim).

STEP 2   Find the critical value. Subtract the "after" values X_A from the "before" values X_B, and indicate the difference by a $+$ or $-$ sign, according to the value, as shown in the following table.

Swimmer	Before, X_B	After, X_A	Sign of Difference
A	3	2	$+$
B	0	1	$-$
C	5	4	$+$
D	4	0	$+$
E	2	1	$+$
F	4	3	$+$
G	3	1	$+$
H	5	3	$+$
I	2	2	0
J	1	3	$-$

From Table J, with $n = 9$ (the total number of $+$ and $-$ signs; the 0 is not counted) and $\alpha = 0.05$ (one-tailed), at most $1 - $ sign is needed to reject the null hypothesis.

STEP 3   Compute the test value. Count the number of $+$ and $-$ signs found in Step 2, and use the smaller value as the test value. There are $2 - $ signs, so the test value is 2.

STEP 4   Make the decision. Since there are $2 - $ signs, the decision is to not reject the null hypothesis.

STEP 5   Summarize the results. There is not enough evidence to support the claim that the number of ear infections was reduced by the use of earplugs.   ◀

When the sample size is greater than 25, the normal approximation can be used in the same manner as in Example 14–2. The procedure for conducting the sign test for single or paired samples is given in Procedure Table 13.

Procedure Table 13

Sign Test for Single and Paired Samples

Step 1: State the hypotheses, and identify the claim.

Step 2: Find the critical value(s). For the single-sample test, compare each value with the median. If the value is larger than the median, replace it with a $+$ sign. If it is smaller than the median, replace it with a $-$ sign.

For the paired-sample sign test, subtract the "before" and "after" values, and indicate the difference with a $+$ or $-$ sign, according to the value. Use Table J and n = total number of $+$ and $-$ signs.

Step 3: Compute the test value. Count the number of $+$ and $-$ signs found in Step 2, and use the smaller value as the test value.

Step 4: Make the decision. Compare the test value with the critical value in Table J. If the test value is less than or equal to the critical value, reject the null hypothesis.

Step 5: Summarize the results.

Note: If the sample size n is 26 or more, use Table E and the following formula for the test value:

$$z = \frac{(X + 0.5) - (n/2)}{\sqrt{n}/2}$$

where X = smaller number of $+$ or $-$ signs
n = sample size

EXERCISES

14–11. Why is the sign test the simplest nonparametric test to use?

14–12. What population parameter can be tested by using the sign test?

14–13. In the sign test, what is used as the test value when $n < 26$?

14–14. When $n \geq 26$, what is used in place of Table J for the sign test?

For Exercises 14–15 through 14–30, perform the following steps.

a. State the hypotheses, and identify the claim.
b. Find the critical value(s).
c. Compute the test value.
d. Make the decision.
e. Summarize the results.

14–15. An oceanographer believes that the average height of the waves at Ocean City is 2.8 feet. The wave heights are measured for a random sample of 20 days. The data are as shown. At $\alpha = 0.05$, test the oceanographer's claim.

3.6	2.1	2.3	2.1	2.7
3.2	3.9	3.4	3.0	2.9
2.0	1.9	3.2	3.5	2.8
1.8	2.3	3.7	3.9	4.2

14–16. A meteorologist suggests that the median temperature for the month of July in Jacksonville, Florida, is 81°. The sample shown below shows the temperatures taken at noon in Jacksonville during 20 days in July. Test the claim at $\alpha = 0.01$.

81	83	87	92	91
78	73	81	93	96
79	80	84	86	82
85	77	72	73	80

14–17. A real estate agent suggests that the median rent for a one-bedroom apartment in Blue View is $325 per month. A sample of 12 one-bedroom apartments shows the following monthly rent costs for a one-bedroom apartment. At $\alpha = 0.05$, test the agent's claim.

$420	$460	$514	$405
320	435	531	450
560	309	312	350

14–18. A government economist estimates that the median cost per pound of beef is $5.00. A sample of 22 livestock buyers shows the following costs per pound of beef. Test the economist's hypothesis at $\alpha = 0.10$.

$5.35	$5.16	$4.97	$4.83	$5.05	$5.19
4.78	4.93	4.86	5.00	4.63	5.06
5.19	5.00	5.05	5.10	5.16	5.25
5.16	5.42	5.13	5.27		

14–19. Sixteen out of 35 high school seniors ran a distance of 1 mile in less than 6.25 minutes. At $\alpha = 0.05$, test the claim that the median time to complete the mile is 6.25 minutes.

14–20. One hundred people are placed on a special diet. After three weeks, 62 lost weight, 29 gained weight, and 9 weighed the same as before. Test the hypothesis that the diet is effective, at $\alpha = 0.10$. (*Note:* It will be effective if less than 50% of the people did not lose weight.)

14–21. Fifty students were surveyed, and 29 favored single-room dormitories. At $\alpha = 0.02$, test the hypothesis that more than 50% of the students favor single-room dormitories.

14–22. One hundred fifty psychology majors were tested, and 86 had an IQ greater than or equal to 103. Test the claim, at $\alpha = 0.05$, that half of the psychology majors have an IQ of at least 103.

14–23. One hundred students are asked if they favor increasing the school year by 20 days. The responses were 62 "no," 36 "yes," and 2 "undecided." At $\alpha = 0.10$, test the hypothesis that 50% of the students are against extending the school year.

14–24. Test the hypothesis that the median age of television newscasters is 31, if, out of 32 television newscasters, 8 are older than 31 years. Use $\alpha = 0.05$.

14–25. A study was conducted to see whether a certain diet medication had an effect on the weights (in pounds) of eight women. Their weights were taken before and six weeks after daily administration of the medication. The data are as shown below. At $\alpha = 0.05$, can one conclude that the medication had an effect (increase or decrease) on the weights of the women?

Subject	A	B	C	D	E	F	G	H
Weight Before	187	163	201	158	139	143	198	154
Weight After	178	162	188	156	133	150	175	150

14–26. Two different laboratory machines measure the sodium content of the same 10 blood samples. The data are as shown in the table. At $\alpha = 0.01$, can one conclude that both machines gave the same reading?

Sample	1	2	3	4	5	6	7	8	9	10
Machine 1	138	136	142	151	154	141	140	138	132	136
Machine 2	140	136	141	150	153	144	143	136	131	138

14–27. An educator has designed a reasoning skills course. Nine students are selected and given a pretest to determine their reasoning abilities. After completing the course, the same students are given an equivalent form of the test to see whether their reasoning skills have improved. The data are as shown in the table. At $\alpha = 0.05$, did the course improve their reasoning skills?

Student	1	2	3	4	5	6	7	8	9
Pretest	80	76	74	83	92	78	91	74	88
Posttest	82	78	73	85	95	79	93	78	90

14–28. A researcher wishes to test the effects of a pill on a person's appetite. Twelve subjects are allowed to eat a meal of their choice, and their caloric intake is measured. The next day, the same subjects take the pill and eat a meal of their choice. The caloric intake of the second meal is measured. The data are as shown in the table. At $\alpha = 0.02$, can the researcher conclude that the pill had an effect on a person's appetite?

Subject	1	2	3	4	5	6	7	8
Meal 1	856	732	900	1321	843	642	738	1005
Meal 2	843	721	872	1341	805	531	740	900

Subject	9	10	11	12
Meal 1	888	756	911	998
Meal 2	805	695	878	914

14-29. In order to test a theory that alcohol consumption can have an effect on a person's IQ, a researcher conducts a study on ten adults. Each is given an IQ test. Then for one week, each subject is required to consume a certain amount of alcohol, and his or her IQ is retested. The results are as shown in the table. At $\alpha = 0.10$, can the researcher conclude that alcohol does not affect a person's IQ?

Subject	IQ Before	IQ After
1	105	106
2	109	105
3	98	94
4	112	109
5	109	105
6	117	115
7	123	125
8	114	114
9	95	98
10	101	100

14-30. A manufacturer feels that if routine maintenance (cleaning and oiling of machines) is increased to once a day rather than once a week, the number of defective parts produced by the machines will decrease. Nine machines are selected, and the number of defective parts produced over a 24-hour operating period is counted. Maintenance is then increased to once a day for a week, and the number of defective parts each machine produces is again counted over a 24-hour operating period. The data are as shown in the table. At $\alpha = 0.01$, can the manufacturer conclude that increased maintenance reduces the number of defective parts manufactured by the machines?

Machine	1	2	3	4	5	6	7	8	9
Before	6	18	5	4	16	13	20	9	3
After	5	16	7	4	18	12	14	7	1

The confidence interval for the median of a set of values less than or equal to 25 in number can be found by ordering the data from smallest to largest, finding the median, and using Table J. For example, to find the 95% confidence interval of the true median for 17, 19, 3, 8, 10, 15, 1, 23, 2, 12, order the data:

1, 2, 3, 8, 10, 12, 15, 17, 19, 23

From Table J, select $n = 10$ and $\alpha = 0.05$, and find the critical value. Use the two-tailed row. In this case, the critical value is 1. Add 1 to this value to get 2. In the ordered list, count from the left two numbers and from the right two numbers, and use these numbers to get the confidence interval, as shown:

1, 2, 3, 8, 10, 12, 15, 17, 19, 23
          2 ≤ MD ≤ 19

Always add 1 to the number obtained from the table before counting. For example, if the critical value is 3, then count 4 values from the left and right.

For Exercises 14-31 through 14-35, find the confidence interval of the median, indicated in parentheses, for each set of data.

*14-31. 3, 12, 15, 18, 16, 15, 22, 30, 25, 4, 6, 9 (95%)

*14-32. 101, 115, 143, 106, 100, 142, 157, 163, 155, 141, 145, 153, 152, 147, 143, 115, 164, 160, 147, 150 (90%)

*14-33. 8.2, 7.1, 6.3, 5.2, 4.8, 9.3, 7.2, 9.3, 4.5, 9.6, 7.8, 5.6, 4.7, 4.2, 9.5, 5.1 (98%)

*14-34. 1, 8, 2, 6, 10, 15, 24, 33, 56, 41, 58, 54, 5, 3, 42, 31, 15, 65, 21 (99%)

*14-35. 12, 15, 18, 14, 17, 19, 25, 32, 16, 47, 14, 23, 27, 42, 33, 35, 39, 41, 21, 19 (95%)

14-4 THE WILCOXON RANK SUM AND SIGNED-RANK TESTS

The sign test does not consider the magnitude of the data. For example, if a value is 1 point or 100 points below the median, it will still receive a − sign. And when one compares values in the pretest-posttest situation, the magnitude of the differences is not considered. The Wilcoxon tests consider differences in magnitudes by using ranks.

The two tests considered in this section are the **Wilcoxon rank sum test,** which is used for independent samples, and the **Wilcoxon signed-rank test,** which is used for dependent samples. Both tests are used to compare distributions. The parametric equivalents are the z and t tests for independent and dependent samples (Sections 10–2 and 10–3) and the t test for dependent samples (Section 10–4). The parametric tests, as stated previously, require the samples to be selected from approximately normally distributed populations, but no requirements about the population are required for the Wilcoxon tests.

In the Wilcoxon tests, the values of the data for both samples are combined and then ranked. If the null hypothesis is true—meaning that there is no difference in the population distributions—then the values in each sample should be ranked approximately the same. Therefore, when the ranks are summed for each sample, the sums should be approximately equal, and the null hypothesis will not be rejected. If there is a large difference in the sums of the ranks, then the distributions are not identical, and the null hypothesis will be rejected.

The first test to be considered is the Wilcoxon rank sum test for independent samples. The formulas needed for the test are given next.

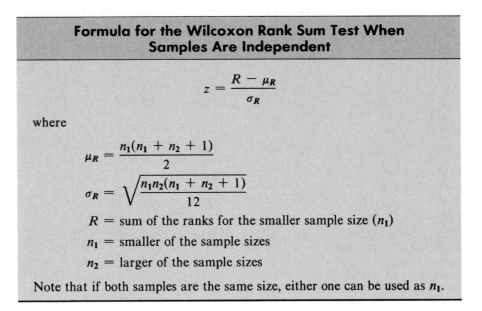

Formula for the Wilcoxon Rank Sum Test When Samples Are Independent

$$z = \frac{R - \mu_R}{\sigma_R}$$

where

$$\mu_R = \frac{n_1(n_1 + n_2 + 1)}{2}$$

$$\sigma_R = \sqrt{\frac{n_1 n_2(n_1 + n_2 + 1)}{12}}$$

R = sum of the ranks for the smaller sample size (n_1)

n_1 = smaller of the sample sizes

n_2 = larger of the sample sizes

Note that if both samples are the same size, either one can be used as n_1.

The next example illustrates the Wilcoxon rank sum test for independent samples.

EXAMPLE 14–4 Two independent samples of army and marine recruits are selected, and the time it takes each recruit to complete an obstacle course is recorded as shown in the table. At $\alpha = 0.05$, is there a difference in the times it takes the recruits to complete the course?

Army	15	18	16	17	13	22	24	17	19	21	26	28	Mean = 19.67
Marines	14	9	16	19	10	12	11	8	15	18	25		Mean = 14.27

Solution   **STEP 1**   State the hypotheses, and identify the claim.

H_0: There is no difference in the times it takes the recruits to complete the obstacle course.
H_1: There is a difference in the times it takes recruits to complete the obstacle course (claim).

STEP 2   Find the critical value. Since $\alpha = 0.05$ and this test is a two-tailed test, the z values of $+1.96$ and -1.96 from Table E are used.

STEP 3   Compute the test value.

a. Combine the data from the two samples, arrange the combined data in order, and rank each value. Be sure to indicate the group.

Time	8	9	10	11	12	13	14	15	15	16	16	17
Group	M	M	M	M	M	A	M	A	M	A	M	A
Rank	1	2	3	4	5	6	7	8.5	8.5	10.5	10.5	12.5

Time	17	18	18	19	19	21	22	24	25	26	28
Group	A	M	A	A	M	A	A	A	M	A	A
Rank	12.5	14.5	14.5	16.5	16.5	18	19	20	21	22	23

b. Sum the ranks of the group with the smaller sample size. (*Note:* If both groups have the same sample size, then either one can be used.) In this case, the sample size for the marines is smaller.

$$R = 1 + 2 + 3 + 4 + 5 + 7 + 8.5 + 10.5$$
$$+ \ 14.5 + 16.5 + 21$$
$$= 93$$

c. Substitute in the formulas to find the test value.

$$\mu_R = \frac{n_1(n_1 + n_2 + 1)}{2} = \frac{(11)(11 + 12 + 1)}{2} = 132$$

$$\sigma_R = \sqrt{\frac{n_1 n_2(n_1 + n_2 + 1)}{12}} = \sqrt{\frac{(11)(12)(11 + 12 + 1)}{12}}$$

$$= \sqrt{264} = 16.25$$

$$z = \frac{R - \mu_R}{\sigma_R} = \frac{93 - 132}{16.25} = -2.4$$

STEP 4   Make the decision. The decision is to reject the null hypothesis, since $-2.4 < -1.96$.

STEP 5   Summarize the results. There is enough evidence to support the claim that there is a difference in the times it takes the recruits to complete the course.   ◄

The general procedure for the Wilcoxon rank sum test is outlined in Procedure Table 14.

Procedure Table 14

Wilcoxon Rank Sum Test

Step 1: State the hypotheses, and identify the claim.

Step 2: Find the critical value(s). Use Table E.

Step 3: Compute the test value.

    a. Combine the data from the two samples, arrange the combined data in order, and rank each value.

    b. Sum the ranks of the group with the smaller sample size. (*Note:* If both groups have the same sample size, then either one can be used.)

    c. Use these formulas to find the test value.

$$\mu_R = \frac{n_1(n_1 + n_2 + 1)}{2}$$

$$\sigma_R = \sqrt{\frac{n_1 n_2 (n_1 + n_2 + 1)}{12}}$$

$$z = \frac{R - \mu_R}{\sigma_R}$$

where R is the sum of the ranks of the data in the smaller sample.

Step 4: Make the decision.

Step 5: Summarize the results.

When the samples are dependent, as they would be in a before-after test using the same subjects, the Wilcoxon signed-rank test can be used in place of the t test for dependent samples. Again, this test does not require the condition of normality. Table K is used to find the critical values.

The procedure for this test is shown in the next example.

EXAMPLE 14–5 In a large department store, the owner wishes to see whether the number of shoplifting incidents per day will change if the number of uniformed security officers is doubled. A sample of seven days before the number of officers is increased and seven days after the increase shows the number of shoplifting incidents.

Number of Shoplifting Incidents

Day	Before	After
Monday	7	5
Tuesday	2	3
Wednesday	3	4
Thursday	6	3
Friday	5	1
Saturday	8	6
Monday	12	4

Test the claim, at $\alpha = 0.05$, that there is a difference in the number of shoplifting incidents before and after the increase in security.

Solution **STEP 1** State the hypotheses, and identify the claim.

H_0: There is no difference in the number of shoplifting incidents before and after the increase in security.

H_1: There is a difference in the number of shoplifting incidents before and after the increase in security (claim).

STEP 2 Find the critical value from Table K. Since $n = 7$ and since $\alpha = 0.05$ for this two-tailed test, the critical value is 2. See Figure 14–2.

FIGURE 14–2
Finding the Critical Value in Table K for Example 14–5

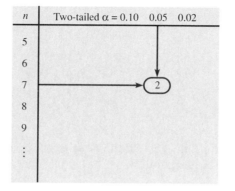

n	Two-tailed $\alpha = 0.10$	0.05	0.02
5			
6			
7		2	
8			
9			
⋮			

STEP 3 Find the test value.

a. Make a table as shown here.

Day	Before, X_B	After, X_A	Difference, $D = X_B - X_A$	Absolute Value, $\lvert D \rvert$	Rank	Signed Rank
Mon.	7	5				
Tues.	2	3				
Wed.	3	4				
Thurs.	6	3				
Fri.	5	1				
Sat.	8	6				
Mon.	12	4				

b. Find the differences (before $-$ after), and place the values in the "Difference" column.

$$7 - 5 = 2 \qquad 6 - 3 = 3 \qquad 8 - 6 = 2$$
$$2 - 3 = -1 \qquad 5 - 1 = 4 \qquad 12 - 4 = 8$$
$$3 - 4 = -1$$

c. Find the absolute value of each difference, and place the results in the "Absolute Value" column. (*Note:* The absolute value of any number except 0 is the positive value of the number. Any differences of 0 should be ignored.)

$$\lvert 2 \rvert = 2 \qquad \lvert 3 \rvert = 3 \qquad \lvert 2 \rvert = 2$$
$$\lvert -1 \rvert = 1 \qquad \lvert 4 \rvert = 4 \qquad \lvert 8 \rvert = 8$$
$$\lvert -1 \rvert = 1$$

d. Rank each absolute value from lowest to highest, and place the rankings in the "Rank" column. In the case of a tie, assign each rank that value plus 0.5.

Value	2	1	1	3	4	2	8
Rank	3.5	1.5	1.5	5	6	3.5	7

e. Give each rank a $+$ or a $-$ sign, according to the sign in the "Difference" column. The completed table is shown here.

Day	Before, X_B	After, X_A	Difference, $D = X_B - X_A$	Absolute Value, $|D|$	Rank	Signed Rank
Mon.	7	5	2	2	3.5	+3.5
Tues.	2	3	−1	1	1.5	−1.5
Wed.	3	4	−1	1	1.5	−1.5
Thurs.	6	3	3	3	5	+5
Fri.	5	1	4	4	6	+6
Sat.	8	6	2	2	3.5	+3.5
Mon.	12	4	8	8	7	+7

f. Find the sum of the positive ranks and the sum of the negative ranks separately.

positive rank sum $(+3.5) + (+5) + (6) + (3.5) + (7) = (+25)$

negative rank sum $(-1.5) + (-1.5)$ $= (-3)$

g. Select the smaller of the absolute values of the sums ($|-3|$), and use this absolute value as the test value w_s. In this case, $w_s = |-3| = 3$.

STEP 4 Make the decision. Reject the null hypothesis if the test value is less than or equal to the critical value. In this case, $3 > 2$; hence, the decision is to not reject the null hypothesis.

STEP 5 Summarize the results. There is not enough evidence to support the claim that there is a difference in the number of shoplifting incidents. Hence, the security increase probably made no difference in the number of shoplifting incidents. ◄

The rationale behind the signed-rank test can be explained by a diet example. If the diet is working, then the majority of the postweights will be smaller than the preweights. When the postweights are subtracted from the preweights, the majority of the signs will be positive, and the sum of the negative ranks will be small. Thus, the sum will probably be smaller than the critical value obtained from Table K, and the null hypothesis will be rejected. On the other hand, if the diet does not work, some people will gain weight, some people will lose weight, and some people will remain about the same in weight. In this case, the sum of the positive ranks and the sum of the negative ranks will be approximately equal and will be about half of the sum of the absolute value of all the ranks. In this case, the smaller of the two sums will still be larger than the critical value obtained from Table K, and the null hypothesis will not be rejected.

When $n \geq 30$, the normal distribution can be used to approximate the Wilcoxon distribution. The same critical values from Table E used for the z test for specific α values are used. The formula is

$$z = \frac{w_s - \dfrac{n(n+1)}{4}}{\sqrt{\dfrac{n(n+1)(2n+1)}{24}}}$$

where

n = number of pairs where the difference is not 0

w_s = smaller sum in absolute value of the signed ranks

The procedure for conducting the Wilcoxon signed-rank test is given in Procedure Table 15.

Procedure Table 15

Wilcoxon Signed-Rank Test

Step 1: State the hypotheses, and identify the claim.

Step 2: Find the critical value from Table K.

Step 3: Compute the test value.

a. Make a table, as shown.

Before, X_B	After, X_A	Difference, $D = X_B - X_A$	Absolute Value, $\lvert D \rvert$	Rank	Signed Rank

b. Find the differences (before − after), and place the values in the "Difference" column.

c. Find the absolute value of each difference, and place the results in the "Absolute Value" column.

d. Rank each absolute value from lowest to highest, and place the rankings in the "Rank" column.

e. Give each rank a + or − sign, according to the sign in the "Difference" column.

f. Find the sum of the positive ranks and the sum of the negative ranks separately.

g. Select the smaller of the absolute values of the sums, and use this absolute value as the test value w_s.

Step 4: Make the decision. Reject the null hypothesis if the test value is less than or equal to the critical value.

Step 5: Summarize the results.

Note: When $n \geq 30$, use Table E and the following test value:

$$z = \frac{w_s - \dfrac{n(n+1)}{4}}{\sqrt{\dfrac{n(n+1)(2n+1)}{24}}}$$

where n = number of pairs where the difference is not 0

w_s = smaller sum in absolute value of the signed ranks.

Speaking of **Statistics**

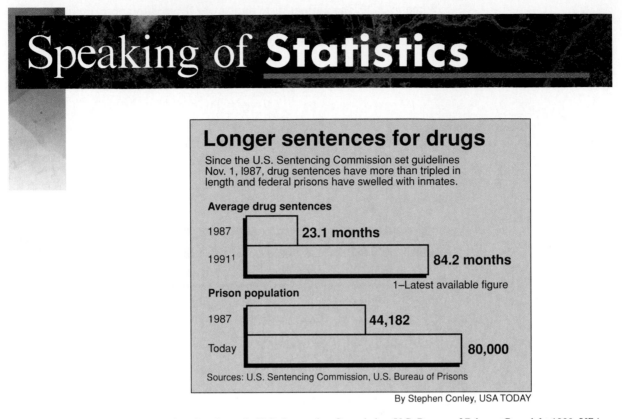

Longer sentences for drugs

Since the U.S. Sentencing Commission set guidelines
Nov. 1, 1987, drug sentences have more than tripled in
length and federal prisons have swelled with inmates.

Average drug sentences

1987 — **23.1 months**

1991[1] — **84.2 months**

1—Latest available figure

Prison population

1987 — **44,182**

Today — **80,000**

Sources: U.S. Sentencing Commission, U.S. Bureau of Prisons

By Stephen Conley, USA TODAY

Source: Based on data from the U.S. Sentencing Commission, U.S. Bureau of Prisons. Copyright 1993, USA TODAY. Reprinted with permission.

This study compares the average drug sentences in 1987 with those in 1991 and compares the prison populations for the two years. Do you think the difference in the averages is significant? If the distributions of the drug sentences were not normal, how could one test to see whether the difference is significant?

EXERCISES

14–36. Explain the difference between the Wilcoxon tests and the sign test.

14–37. What is the difference between the Wilcoxon rank sum test and the Wilcoxon signed-rank test?

14–38. What are the parametric equivalent tests for the Wilcoxon rank sum tests?

14–39. What is the parametric equivalent test for the Wilcoxon signed-rank test?

14–40. In the Wilcoxon signed-rank test, is the smaller or larger sum, in absolute value, of the signed ranks used as the test value?

14–41. What distribution can be used to approximate the Wilcoxon distribution when $n \geq 30$?

For Exercises 14–42 through 14–49 and 14–57 through 14–61, perform the following steps.

a. State the hypotheses, and identify the claim.
b. Find the critical value(s).
c. Compute the test value.
d. Make the decision.
e. Summarize the results.

For Exercises 14–42 through 14–49, use the Wilcoxon rank sum test. Assume that the samples are independent.

14–42. A random sample of men and women in prison was asked to give the length of sentence each received for a certain type of crime. At $\alpha = 0.05$, can one conclude that there is no difference in the sentence received by each gender? The data (given in months) are shown here.

Males	8	12	6	14	22	27	32	24	26	19	15
Females	7	5	2	3	21	26	30	9	4	17	23

Males	13
Females	12 11 16

14–43. A researcher surveyed married women and single women to ascertain whether there was a difference in the number of books each had read during the past year. The data are as shown in the table. At $\alpha = 0.10$, can the researcher conclude that each group read the same number of books?

Married	6	8	7	4	9	12	13	7	10	18	15
Single	2	3	5	11	3	5	11	12	16	4	0	1

14–44. To test the claim that there is no difference in the lifetimes of two brands of microwave ovens, a researcher selects a sample of 11 ovens of each brand. Shown in the table are the lifetimes (in months) of each brand. At $\alpha = 0.01$, can the researcher conclude that there is a difference in the distributions of lifetimes for the two brands?

Brand A	28	36	30	41	38	22	53	46	52	47	29
Brand B	42	26	56	32	43	35	25	51	38	49	41

14–45. Over the past 12 seasons, a statistician kept track of the total number of points scored by North High School and its rival, South High School, in an academic game contest. The data are as shown in the table. At $\alpha = 0.05$, is there a difference in the number of points scored by the teams?

North	60	28	27	42	91	14
South	43	73	48	86	48	10

North	102	24	17	51	50	54
South	90	66	63	53	58	62

14–46. Two groups of employees were given a questionnaire to ascertain their degree of job satisfaction. The scale ranged from 0 to 100. The groups were divided into those who had under five years of work experience and those who had five or more years of experience. The data are as shown in the table. At $\alpha = 0.10$, can one conclude that there is no difference in the job satisfaction of each group, as measured by the questionnaire?

Under 5	78	98	83	86	75	77	72
5 and Over	94	79	82	85	73	66	64

Under 5	68	56	93	97	99	93
5 and Over	59	52	58	63	68	88

14–47. The results of a study of payments for flood damages awarded by insurance companies in two Texas cities are shown here. The data are given in dollars. At $\alpha = 0.05$, is there a difference in the amount of money awarded for flood damages in each city?

City A	563	648	925	602	921	232
City B	869	718	626	453	832	752

City A	953	824	605	601	687	431
City B	769	324	885	927	918	239

14–48. Supervisors were asked to rate the productivity of employees on their jobs. A researcher wishes to see whether married men receive higher ratings than single men. A rating scale of 1 to 50 was used. The data are as shown. At $\alpha = 0.01$, is there evidence to support this claim?

Single Men	48	46	42	50	38	36
Married Men	44	35	41	37	42	43

Single Men	40	31	28	24	49	34
Married Men	29	31	37	32	36	

14–49. A study was conducted to see whether there was a difference in the time it takes employees of a factory to assemble the product. Samples of high school graduates and nongraduates were timed. At $\alpha = 0.05$, is there a difference in the distributions for the two groups in the times needed to assemble the product? The data (in minutes) are as shown.

Graduates	3.6	3.2	4.4	3.0	5.6	6.3	8.2
Nongraduates	2.7	3.8	5.3	1.6	1.9	2.4	2.9

Graduates	7.1	5.8	7.3	6.4	4.2	4.7
Nongraduates	1.7	2.6	2.0	3.1	3.4	3.9

For Exercises 14–50 and 14–51, find the sum of the signed ranks. Assume that the samples are dependent. State which sum is used as the test value.

14–50.

Pretest	18	32	35	37	25	41	52	43	56	62
Posttest	20	21	26	37	29	40	31	37	51	65

14–51.

Pretest	108	97	115	162	156	105	153
Posttest	110	97	103	168	143	112	141

For Exercises 14–52 through 14–56, use Table K to determine whether the null hypothesis should be rejected.

14–52. $w_s = 62$, $n = 21$, $\alpha = 0.05$, two-tailed test

14–53. $w_s = 18$, $n = 15$, $\alpha = 0.02$, two-tailed test

14–54. $w_s = 53$, $n = 25$, $\alpha = 0.05$, one-tailed test

14–55. $w_s = 142$, $n = 28$, $\alpha = 0.05$, one-tailed test

14–56. $w_s = 109$, $n = 27$, $\alpha = 0.025$, one-tailed test

14–57. Seven students were given a pretest to measure math anxiety. They completed a workshop to reduce their anxiety and were then given a posttest. At $\alpha = 0.05$, can one conclude that the workshop reduced their anxiety? The pretest and posttest scores are shown here. (*Note:* A lower score indicates a lower anxiety level.)

Pretest	32	39	43	21	29	35	34
Posttest	30	31	44	21	26	36	28

14–58. In a corporation, male and female workers were matched according to years of experience working for the company. Their salaries were then compared. The data (in thousands of dollars) are shown in the table. At $\alpha = 0.10$, is there a difference in the salaries of the males and females?

Males	18	43	32	27	15	45	21	22
Females	16	38	35	29	15	46	25	28

14–59. A recent survey of marriages recorded the ages of the husbands and wives. At $\alpha = 0.05$, test the claim that there is no difference in the ages of the husbands and wives. The data are as shown in the table.

Husband	56	48	31	35	47	43	26	37
Wife	53	32	32	31	41	43	23	22

14–60. Eight couples are given a questionnaire designed to measure marital compatibility. After completing a workshop, they are given a second questionnaire to see whether there is a difference in their attitudes toward each other. The data are as shown in the table. At $\alpha = 0.10$, is there any difference in the scores of the couples?

Before	43	52	37	29	51	62	57	61
After	48	59	36	29	60	68	59	72

14–61. A movie theater selects a week at random and counts the number of patrons it has. After an offer of free popcorn, the theater again counts the patrons it has during the week. At $\alpha = 0.10$, did the ad campaign make a difference? The data are as shown in the table.

Day	Mon.	Tues.	Wed.	Thurs.	Fri.	Sat.	Sun.
Before	57	99	79	90	112	150	175
After	63	78	81	95	118	160	179

14–5 THE KRUSKAL-WALLIS TEST

The analysis of variance uses the F test to compare the means of three or more populations. The assumptions for the ANOVA test are that the populations are normally distributed and that the population variances are equal. When these assumptions cannot be met, the nonparametric **Kruskal-Wallis test** can be used to compare three or more means. This test is sometimes called the H test.

In this test, each sample size must be five or more. In these situations, the distribution can be approximated by the chi-square distribution with $k - 1$ degrees of freedom. This test also uses ranks. The formula for the test is given next.

Formula for the Kruskal-Wallis Test

$$H = \frac{12}{N(N+1)} \left(\frac{R_1^2}{n_1} + \frac{R_2^2}{n_2} + \cdots + \frac{R_k^2}{n_k} \right) - 3(N+1)$$

where

R_1 = sum of the ranks of sample 1

n_1 = size of sample 1

R_2 = sum of the ranks of sample 2

n_2 = size of sample 2

.

.

.

R_k = sum of the ranks of sample k

n_k = size of sample k

N = $n_1 + n_2 + \cdots + n_k$

k = number of samples

In the Kruskal-Wallis test, one considers all the data values as a group and then ranks them. Next, the ranks are separated, and the H formula is computed. This formula approximates the variance of the ranks. If the samples are from different populations, the sum of the ranks will be different, and the H value will be large; hence, the null hypothesis will be rejected. If the samples are from the same population, the sum of the ranks will be approximately the same, and the H value will be small; therefore, the null hypothesis will not be rejected. This test is always a one-tailed right test. The chi-square table, Table G, should be used for critical values.

The next example illustrates the procedure for conducting the Kruskal-Wallis test.

EXAMPLE 14–6   A researcher tests three different brands of breakfast drinks for the number of milliequivalents of potassium per quart each contains. The following data are obtained.

Brand A	Brand B	Brand C
4.7	5.3	6.3
3.2	6.4	8.2
5.1	7.3	6.2
5.2	6.8	7.1
5.0	7.2	6.6

At $\alpha = 0.05$, test the hypothesis that all brands contain the same amount of potassium.

Solution   **STEP 1**   State the hypotheses, and identify the claim.

H_0: There is no difference in the amount of potassium contained in each brand (claim).

H_1: There is a difference in the amount of potassium contained in each brand.

STEP 2   Find the critical value. Use the chi-square table, Table G, with d.f. $= k - 1$ (k = number of groups). With $\alpha = 0.05$ and d.f. $= 3 - 1 = 2$, the critical value is 5.991.

STEP 3   Compute the test value.

a. Arrange all the data from lowest to highest, and rank each value.

Amount	Brand	Rank
3.2	A	1
4.7	A	2
5.0	A	3
5.1	A	4
5.2	A	5
5.3	B	6
6.2	C	7
6.3	C	8
6.4	B	9
6.6	C	10
6.8	B	11
7.1	C	12
7.2	B	13
7.3	B	14
8.2	C	15

b. Find the sum of the ranks of each brand.

$$\begin{aligned} \text{brand A} \quad & 1 + 2 + 3 + 4 + 5 & = 15 \\ \text{brand B} \quad & 6 + 9 + 11 + 13 + 14 & = 53 \\ \text{brand C} \quad & 7 + 8 + 10 + 12 + 15 & = 52 \end{aligned}$$

c. Substitute in the formula.

$$H = \frac{12}{N(N+1)}\left(\frac{R_1^2}{n_1} + \frac{R_2^2}{n_2} + \frac{R_3^2}{n_3}\right) - 3(N+1)$$

where

$N = 15$ $R_1 = 15$ $R_2 = 53$ $R_3 = 52$
$n_1 = n_2 = n_3 = 5$

Therefore,

$$H = \frac{12}{15(15+1)}\left(\frac{15^2}{5} + \frac{53^2}{5} + \frac{52^2}{5}\right) - 3(15+1) = 9.38$$

STEP 4   Make the decision. Since the test value 9.38 is greater than the critical value 5.991, the decision is to reject the null hypothesis.

STEP 5 Summarize the results. There is enough evidence to reject the claim that there is no difference in the amount of potassium contained in each brand. Hence, all brands do not contain the same amount of potassium. ◀

The procedure for the Kruskal-Wallis test is given in Procedure Table 16.

Procedure Table 16
Kruskal-Wallis Test

Step 1: State the hypotheses, and identify the claim.

Step 2: Find the critical value. Use the chi-square table, Table G, with d.f. = $k - 1$ (k = number of groups).

Step 3: Compute the test value.
a. Arrange the data from lowest to highest, and rank each value.
b. Find the sum of the ranks of each group.
c. Substitute in the formula.

$$H = \frac{12}{N(N + 1)}\left(\frac{R_1^2}{n_1} + \frac{R_2^2}{n_2} + \cdots + \frac{R_k^2}{n_k}\right) - 3(N + 1)$$

where $N = n_1 + n_2 + \cdots + n_k$

R_k = sum of the ranks for the kth group

k = number of groups

Step 4: Make the decision.

Step 5: Summarize the results.

EXERCISES

14–62. What is the parametric test that is equivalent to the Kruskal-Wallis test?

For Exercises 14–63 through 14–73, perform the following steps.

a. State the hypotheses, and identify the claim.
b. Find the critical value.
c. Compute the test value.
d. Make the decision.
e. Summarize the results.

14–63. Samples of four different types of candy bars show the following number of calories for each brand. At $\alpha = 0.05$, is there a difference in the number of calories for each type?

Brand A	Brand B	Brand C	Brand D
327	330	305	452
380	342	298	460
362	356	309	449
371	361	295	447
354	335	303	441
361		307	399
		310	

14–64. A test designed to measure the self-confidence of individuals is given to three different samples of individuals. The scores on the test range from 0 to 50. The data are as shown in the table. At $\alpha = 0.05$, is there a difference in the scores?

Never Married	Married	Widowed
47	50	48
49	46	45
42	42	26
41	44	28
37	39	31
32	43	25

14–65. A large grocery store chain decides to advertise a product by three different methods (one method in each area): radio, television, and newspaper. One week's sales from randomly selected stores in each area are recorded in the table. At $\alpha = 0.10$, is there a difference in sales for each type of advertising?

Radio	Television	Newspaper
$832	$1024	$329
648	996	437
562	1011	561
786	853	329
452	471	382
975		495
		262

14–66. Three different brands of microwave dinners were advertised as low in sodium. Samples of the three different brands show the following milligrams of sodium. At $\alpha = 0.05$, is there a difference in the amount of sodium among the brands?

Brand A	Brand B	Brand C
810	917	893
702	912	790
853	952	603
703	958	744
892	893	623
732		743
713		609
613		

14–67. Three different types of soils are used to grow strawberries. The yields (in quarts) for plots of the same size are as shown in the table. At $\alpha = 0.01$, is there a difference in the yields of the three plots?

Soil A	Soil B	Soil C
32	43	50
38	45	56
31	49	58
40	46	54
39	51	52

14–68. A recent study recorded the number of job offers received by newly graduated chemical engineers at three different colleges. The data are as shown in the table. At $\alpha = 0.05$, is there a difference in the average number of job offers received by the graduates at the three colleges?

College A	College B	College C
6	2	10
8	1	12
7	0	9
5	3	13
6	6	4

14–69. A marketing specialist uses four different colors of boxes to package a product. At randomly selected locations, the number of products sold during a week is recorded by the color of the box. The data are as shown in the table. At $\alpha = 0.10$, does the color of the box make a difference in sales?

Red	Blue	White	Yellow
71	91	105	97
76	88	99	96
82	84	103	82
78	86	98	94
73	93	96	95

14–70. Three different brands of copy machines are used in a large office building. The monthly maintenance cost of each machine is recorded, and the results are as shown in the table. At $\alpha = 0.05$, can one conclude that there is a difference in the monthly maintenance costs?

Type 1	Type 2	Type 3
$56	$63	$82
42	72	81
48	71	79
53	74	77
51	76	55

14–71. In a large city, the number of crimes per week in five different precincts is recorded for five weeks. At $\alpha = 0.01$, is there a difference in the number of crimes? The data are as shown in the table.

Precinct 1	Precinct 2	Precinct 3	Precinct 4	Precinct 5
105	87	74	56	103
108	86	83	43	98
99	91	78	52	94
97	93	74	58	89
92	82	60	62	88

14–72. A recent study examined the number of unemployed people in five cities who are actively seeking employment. They are listed in the following table according to the education each received. At $\alpha = 0.05$, is there a difference in the number of unemployed based on education received?

High School Diploma	College Degree	Postgraduate Degree
49	23	17
43	49	38
51	54	23
108	87	52
68	28	26

14–73. Three different methods of first-aid instruction are given to students. The same examination is given to each class. The data are as shown in the table. At $\alpha = 0.10$, is there a difference in the final examination scores?

Method A	Method B	Method C
98	97	99
100	88	94
95	82	96
92	84	89
86	75	81
76	73	72
71	74	

14–6 SPEARMAN RANK CORRELATION COEFFICIENT AND THE RUNS TEST

Rank Correlation Coefficient

In Chapter 11, the techniques of regression and correlation were explained. In order to determine whether two variables are related, one uses the Pearson product moment correlation coefficient. Its values range from $+1$ to -1. One assumption for testing the hypothesis that $\rho = 0$ for the Pearson coefficient is that the populations from which the samples are obtained are normally distributed. If this requirement cannot be met, the nonparametric equivalent, called the **Spearman rank correlation coefficient** (denoted by r_S), can be used when the data are ranked.

The computations for the rank correlation coefficient are simpler than those for the Pearson coefficient and involve ranking each of the sets of data. The difference in ranks is found, and r_S is computed by using these differences. If both sets of data have the same ranks, r_S will be $+1$. If the sets of data are ranked in exactly the opposite way, r_S will be -1. If there is no relationship between the rankings, r_S will be near 0.

Formula for Computing the Spearman Rank Correlation Coefficient

$$r_S = 1 - \frac{6 \sum d^2}{n(n^2 - 1)}$$

where

d = difference in the ranks
n = number of data pairs

This formula is algebraically equivalent to the formula for r given in Chapter 11, except that ranks are used instead of raw data.

The computational procedure is shown in the next example. For a test of the significance of r_S, Table L is used for values of n up to 30. For larger values, the normal distribution can be used. (See Exercises 14–98 through 14–102.)

EXAMPLE 14–7 Two students were asked to rate eight different textbooks for a specific course. The scale ranged from 0 to 20 points, with 20 being the highest rating and 0 being the lowest rating. Points were assigned for each of several categories, such as reading level, use of illustrations, and use of color. At $\alpha = 0.05$, test the hypothesis that there is a significant linear correlation between the two students' ratings. The data are shown in the table.

Textbook	Student 1's Rating	Student 2's Rating
A	4	4
B	10	6
C	18	20
D	20	14
E	12	16
F	2	8
G	5	11
H	9	7

Solution **STEP 1** State the hypotheses.

$$H_0\text{: } \rho = 0 \qquad \text{and} \qquad H_1\text{: } \rho \neq 0$$

STEP 2 Rank each data set, as shown in the table below.

Textbook	Student 1	Rank	Student 2	Rank
A	4	7	4	8
B	10	4	6	7
C	18	2	20	1
D	20	1	14	3
E	12	3	16	2
F	2	8	8	5
G	5	6	11	4
H	9	5	7	6

Let X_1 be the first student's rankings and X_2 be the second student's rankings.

STEP 3 Subtract the rankings $(X_1 - X_2)$.

$$7 - 8 = -1 \qquad 4 - 7 = -3 \qquad \text{etc.}$$

STEP 4 Square the differences.

$$(-1)^2 = 1 \qquad (-3)^2 = 9 \qquad \text{etc.}$$

STEP 5 Find the sum of the squares.

$$1 + 9 + 1 + 4 + 1 + 9 + 4 + 1 = 30$$

The results can be summarized in a table, as shown here.

X_1	X_2	$d = X_1 - X_2$	d^2
7	8	-1	1
4	7	-3	9
2	1	1	1
1	3	-2	4
3	2	1	1
8	5	3	9
6	4	2	4
5	6	-1	1
			$\Sigma d^2 = 30$

STEP 6   Substitute in the formula to find r_S.

$$r_S = 1 - \frac{6 \Sigma d^2}{n(n^2 - 1)}$$

where $n =$ the number of data pairs. For this problem,

$$r_S = 1 - \frac{(6)(30)}{8(8^2 - 1)} = 1 - \frac{180}{504} = 0.643$$

STEP 7   Find the critical value. Use Table L to find the value for $n = 8$ and $\alpha = 0.05$. It is 0.738. See Figure 14–3.

FIGURE 14–3
Finding the Critical
Value in Table L for
Example 14 7

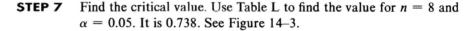

n	$\alpha = 0.10$	$\alpha = 0.05$	$\alpha = 0.02$
5			
6			
7			
8		(0.738)	
9			
⋮			

STEP 8   Make the decision. Do not reject the null hypothesis, since $r_S = 0.643$, which is less than the critical value of 0.738.

STEP 9   Summarize the results. There is not enough evidence to say that there is a correlation between the rankings of the two students.   ◄

Speaking of **Statistics**

Stock permits

The Bureau of Land Management and the U.S. Forest Service together lease 270 million acres to 30,000 livestock ranchers in 16 states. Grazing units[1] issued and average cost per grazing unit:

State	Permits	Cost
Ariz.	1,804,369	$114
Calif.	944,597	$53
Colo.	1,597,434	$75
Idaho	2,747,787	$60
Kan.	120	N/A
Mont.	1,837,335	$76
Neb.	85,334	$140
Nev.	2,743,959	$40
N.M.	2,880,010	$103
N.D.	261,363	$56
Okla.	475	N/A
Ore.	1,442,014	$56
S.D.	95,814	N/A
Utah	2,425,300	$50
Wash.	79,315	N/A
Wyo.	2,594,592	$49

1—One unit is enough forage land to feed one horse or a cow and a calf.
Source: USDA, Interior Department

Source: Based on data from the USDA, Interior Department. Copyright 1993, USA TODAY. Reprinted with permission.

Rank each state according to the number of permits issued and then according to the average cost per unit. Omit the states with the code N/A. Using the Spearman rank correlation coefficient, see whether there is a significant relationship between the ranks.

The procedure for finding and testing the Spearman rank correlation coefficient is given in Procedure Table 17.

Procedure Table 17

Finding and Testing the Spearman Rank Correlation Coefficient

Step 1: State the hypotheses.

Step 2: Rank each data set.

Step 3: Subtract the rankings $(X_1 - X_2)$.

Step 4: Square the differences.

Step 5: Find the sum of the squares.

Step 6: Substitute in the formula.

$$r_S = 1 - \frac{6 \sum d^2}{n(n^2 - 1)}$$

where

$\sum d^2 =$ sum of the squares of the differences of the ranking values

$n =$ number of pairs of data

Step 7: Find the critical value.

Step 8: Make the decision.

Step 9: Summarize the results.

Runs Test   When samples are selected, one assumes that they are selected at random. The question can be asked, "How does one know if a sample is truly random?" Before the answer to this question is given, consider the following situations for a researcher interviewing 20 people for a survey. Let the gender of each person be denoted by M for male and F for female. Suppose the participants were chosen as follows:

situation 1   M M M M M M M M M M F F F F F F F F F F

It does not look like the people were selected at random, since ten men were selected first, followed by ten females.

Consider a different selection:

situation 2   F M F M F M F M F M F M F M F M F M F M

In this case, it seems as if the researcher selected a female, then a male, etc. This selection is probably not random either.

Finally, consider the following selection:

situation 3   F F F M M F M F M M F F M M F F M M M F

This selection looks like it might be random, since there is a mix of men and women and no apparent pattern to their selection.

Rather than trying to guess whether a sample has been selected at random, statisticians have devised a nonparametric test to determine the randomness of a sample. This test is called the **runs test**.

A **run** is a succession of identical letters preceded by or followed by a different letter or no letter at all, such as the beginning or end of the succession.

For example, the first situation presented has 2 runs:

run 1:   M M M M M M M M M

run 2:   F F F F F F F F F F

The second situation has 20 runs. (Each letter constitutes 1 run.) The third situation has 11 runs.

run 1:	F F F	run 5:	F	run 9:	F F
run 2:	M M	run 6:	M M	run 10:	M M M
run 3:	F	run 7:	F F	run 11:	F
run 4:	M	run 8:	M M		

EXAMPLE 14–8   Determine the number of runs in each sequence.

a. F F F M M F F F F M
b. H H H T T T T
c. A A B B A A B B A A B B

Solution   a. There are four runs, as shown.

F F F   M M   F F F F   M

    1   2   3   4

b. There are two runs, as shown.

H H H   T T T T

   1   2

c. There are six runs, as shown.

A A   B B   A A   B B   A A   B B

 1   2   3   4   5   6

The test for randomness considers the number of runs rather than the frequency of the letters. For example, for data to be selected at random, there should not be too few or too many runs, as in situations 1 and 2. The runs test does not consider the questions of how many males or females were selected or how many of each are in a specific run.

In order to determine whether the number of runs is within the random range, one uses Table M. The values are for a two-tailed test with $\alpha = 0.05$. For a sample of 12 males and 8 females, the table values shown in Figure 14–4 mean that any number of runs from 7 to 15 would be considered random. If the number of runs is 6 or less or 16 or more, the sample is probably not random, and the null hypothesis would be rejected.

FIGURE 14–4
Finding the Critical Value in Table M

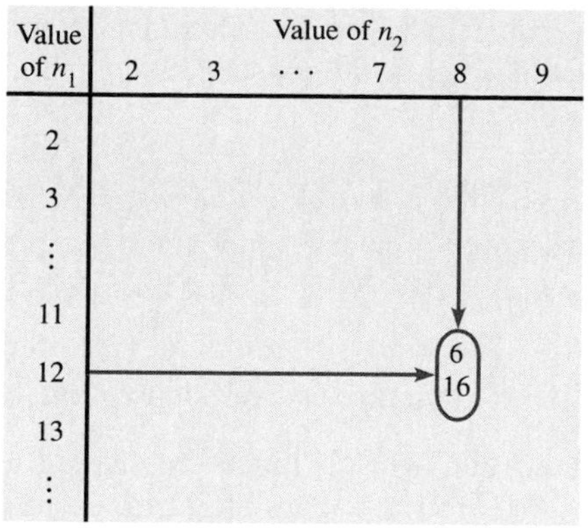

Example 14–9 shows the procedure for conducting the runs test for letters as data; Example 14–10 shows how the runs test can be used for numerical data.

EXAMPLE 14–9 On a commuter train, the conductor wishes to see whether the passengers enter the train at random. He observes the first 25 people, with the following sequence of males (M) and females (F).

F F F M M F F F F M F M M M F F F F M M F F F M M

Test for randomness at $\alpha = 0.05$.

Solution **STEP 1** State the hypotheses, and identify the claim.

H_0: The passengers board the train at random, according to gender (claim).
H_1: The null hypothesis is not true.

STEP 2 Find the number of runs. Arrange the letters according to runs of males and females, as shown.

Run	Gender
1	F F F
2	M M
3	F F F F
4	M
5	F
6	M M M
7	F F F F
8	M M
9	F F F
10	M M

There are 15 females (n_1) and 10 males (n_2).

STEP 3 Find the critical value. Find the number of runs in Table M for $n_1 = 15$, $n_2 = 10$, and $\alpha = 0.05$. The values are 7 and 18.

STEP 4 Make the decision. Compare these critical values with the number of runs. Since the number of runs is 10, and 10 is between 7 and 18, do not reject the null hypothesis.

STEP 5 Summarize the results. There is not enough evidence to reject the hypothesis that the passengers board the train at random according to gender. ◄

EXAMPLE 14–10 Twenty people enrolled in a drug abuse program. Test the claim that the ages of the people, according to the order in which they enroll, occur at random, at $\alpha = 0.05$. The data are 18, 36, 19, 22, 25, 44, 23, 27, 27, 35, 19, 43, 37, 32, 28, 43, 46, 19, 20, 22.

Solution **STEP 1** State the hypotheses, and identify the claim.

H_0: The ages of the people, according to the order in which they enroll in a drug program, occur at random (claim).
H_1: The null hypothesis is not true.

STEP 2 Find the number of runs.

a. Find the median of the data. Arrange the data in ascending order.

18, 19, 19, 19, 20, 22, 22, 23, 25, 27,
27, 28, 32, 35, 36, 37, 43, 43, 44, 46

The median is 27.

b. Replace each number in the original sequence with an A if it is above the median and with a B if it is below the median. Eliminate any numbers that are equal to the median.

B, A, B, B, B, A, B, A, B, A, A, A, A, A, A, B, B, B

c. Arrange the letters according to runs.

Run	Letters
1	B
2	A
3	B B B
4	A
5	B
6	A
7	B
8	A A A A A A
9	B B B

STEP 3 Find the critical value. Table M in the Appendix shows that with $n_1 = 9$, $n_2 = 9$, and $\alpha = 0.05$, the number of runs should be between 5 and 15.

STEP 4 Make the decision. Since there are 9 runs, and 9 falls between 5 and 15, the null hypothesis is not rejected.

STEP 5 Summarize the results. There is not enough evidence to reject the hypothesis that the ages of the people who enroll occur at random. ◄

The procedure for conducting the runs test is given in Procedure Table 18.

Procedure Table 18

The Runs Test

Step 1: State the hypotheses, and identify the claim.

Step 2: Find the number of runs.
Note: When the data are numerical, find the median. Then, compare each data value with the median and classify it as above or below the median.

Step 3: Find the critical value. Use Table M.

Step 4: Make the decision. Compare the actual number of runs with the critical value.

Step 5: Summarize the results.

EXERCISES

For Exercises 14–74 through 14–78, find the critical value from Table L for the rank correlation coefficient, given sample size n and α. Assume that the test is a two-tailed test.

14–74. $n = 30, \alpha = 0.05$

14–75. $n = 14, \alpha = 0.01$

14–76. $n = 28, \alpha = 0.02$

14–77. $n = 10, \alpha = 0.05$

14–78. $n = 9, \alpha = 0.01$

For Exercises 14–79 through 14–88, perform the following steps.

a. Find the Spearman rank correlation coefficient.
b. State the hypotheses.
c. Find the critical value. Use $\alpha = 0.05$.
d. Make the decision.
e. Summarize the results.

14–79. Six actors were ranked according to their performance by the professors of the drama department and by the students in the drama department, with 1 being the highest ranking. The data are shown below. Is there a relationship between the rankings? Use $\alpha = 0.05$.

Actor	A	B	C	D	E	F
Professors	5	4	3	1	2	6
Students	6	4	1	3	2	5

14–80. Eight dogs were treated and ranked on their ease of handling by the veterinarian and her assistant. The data are shown below (1 is the highest ranking). Is there a relationship between the rankings of the two people?

Dog	1	2	3	4	5	6	7	8
Vet	4	5	6	1	7	2	3	8
Assistant	3	5	4	6	8	1	2	7

14–81. Six teaching candidates were interviewed by the dean and then by department members and rated for a position on a scale of 1 (highest ranking) to 20 points. The data are shown below. Is there a relationship between the rankings?

Candidate	A	B	C	D	E	F
Dean	5	20	14	18	12	10
Department	8	16	15	19	10	3

14–82. Six speakers were ranked by the males and females in the audience on the basis of poise, information, and presentation. The data are shown below (1 is the highest ranking). Is there a relationship between the rankings?

Speaker	A	B	C	D	E	F
Male	1	5	2	3	4	6
Female	3	6	2	1	5	4

14–83. Eight 3-minute music videos were ranked by teenagers and their parents on style and clarity, with 1 being the highest ranking. The data are shown on page 526. At $\alpha = 0.05$, is there a relationship between the rankings?

Music Video	1	2	3	4	5	6	7	8
Teenagers	4	6	2	8	1	7	3	5
Parents	1	7	5	4	3	8	2	6

14–84. The sociology department is selecting new textbooks for the next semester. Five instructors and three teaching assistants reviewed seven possible books and assigned each book rating points on a scale of 1 to 12, with 1 being the poorest and 12 being the best. Each book was rated on content, readability, etc. The data are shown below. Is there a relationship between the ratings of the instructors and assistants?

Textbook	1	2	3	4	5	6	7
Instructor	5	8	12	3	11	9	1
Assistant	5	7	10	4	11	8	2

14–85. Six model kitchens were rated for style and convenience by independent interior designers and by potential customers. The scale ran from 1 to 100 points, with 1 being the lowest and 100 being the highest. The data are shown below. At $\alpha = 0.05$, is there a relationship between the two ratings?

Kitchen	A	B	C	D	E	F
Designer	48	76	30	88	61	93
Customer	35	44	28	50	75	85

14–86. Nine tennis players were ranked by sportswriters and by coaches. The data are shown below (1 is the highest ranking). Is there a relationship between the two rankings?

Player	A	B	C	D	E	F	G	H	I
Coach	4	6	5	1	7	2	3	8	9
Writers	7	6	4	3	5	2	1	9	8

14–87. Twelve automobiles were rated for style, performance, driveability, etc., by independent automobile engineers and by potential customers. The scale ran from 1 to 100 points, with 1 being the lowest and 100 being the highest. The data are shown below. Is there a relationship between the two ratings?

Auto	A	B	C	D	E	F
Engineer	81	70	65	54	43	90
Customer	85	75	68	50	52	95

Auto	G	H	I	J	K	L
Engineer	41	88	40	85	82	35
Customer	48	100	44	90	83	20

14–88. Six watercolor paintings at an art show were ranked by professional judges and by the general public. The data are shown in column 2 (1 is the highest ranking). Is there a relationship between the rankings?

Painting	1	2	3	4	5	6
Judge	3	5	1	2	4	6
Public	3	4	2	1	6	5

14–89. A school dentist wishes to test the claim, at $\alpha = 0.05$, that the number of cavities in fourth-grade students is random. Forty students are checked, and the following data give the number of cavities each had. Test for randomness of the values above or below the median.

```
0  4  6  0  6  2  5  3  1  5  1
2  2  1  3  7  3  6  0  2  6  0
2  3  1  5  2  1  3  0  2  3  7
3  1     5     1     1     2  2
```

14–90. A drawing was held each day for a month. A list of winning numbers was categorized as odd or even. Test for randomness, at $\alpha = 0.05$. The data follow.

```
409 872 235 338 472 481 318 129 229
084 291 991 356 212 457 473 834 304
361 301 051 652 405 458 094 633 809
299 712 802
```

14–91. Listed below are the winning numbers for the Pennsylvania State Lotto drawing for April. Classify each as odd or even, and test for randomness, at $\alpha = 0.05$. No drawings were held on weekends.

457, 605, 348, 927, 463, 300, 620, 261, 614, 098, 467, 961, 957, 870, 262, 571, 633, 448, 187, 462, 565, 180, 050

14–92. An irate student believes that the answers to his history professor's final true–false examination are not random. Test claim, at $\alpha = 0.05$. The answers to the questions follow.

T T T F F T T T F F F F F F T
T T F F F T T T F T F F T T F

14–93. A machine manufactures audiocassette cases that are either defective D or acceptable A. The sequence is as follows. At $\alpha = 0.05$, test for randomness.

D A A A A A A D D A D D A A A
D D A A A A A A A D D D A A A

14–94. Twenty people are in line at a fast-food restaurant. At $\alpha = 0.05$, test for randomness of the person being an adult (A) or a child (C). The data are as shown.

A C C A C A A A C A
A C A C A C C C C A

14–95. A supervisor records the number of employees absent over a 30-day period. Test for randomness, at $\alpha = 0.05$.

```
27   6  19  24  18  12  15  17  18  20
 0   9   4  12   3   2   7   7   0   5
32  16  38  31  27  15   5   9   4  10
```

14–96. A ski lodge manager observes the weather for the month of February. If his customers are able to ski, he records S; if weather conditions do not permit his customers to ski, he records N. Test for randomness at $\alpha = 0.05$.

```
S S S S S S N N N N N N N N
N S S S N N S S S S S S S S
```

14–97. The following data are the scores on an IQ exam in the order that the students finished the test. At $\alpha = 0.05$, test for randomness.

101	98	99	110	119	121	118
106	96	88	91	97	92	106
94	93	100	89	86	95	99

When $n \geq 30$, the formula

$$r = \frac{\pm z}{\sqrt{n-1}}$$

can be used to find the critical values for the rank correlation coefficient. For example, if $n = 40$ and $\alpha = 0.05$,

$$r = \frac{\pm 1.96}{\sqrt{40-1}} = \pm 0.314$$

Hence, any r greater than or equal to $+0.314$ or less than or equal to -0.314 is significant.

Find the critical r value for each (assume that the test is a two-tailed test).

*14–98. $n = 50$, $\alpha = 0.05$

*14–99. $n = 30$, $\alpha = 0.01$

*14–100. $n = 35$, $\alpha = 0.02$

*14–101. $n = 60$, $\alpha = 0.10$

*14–102. $n = 40$, $\alpha = 0.01$

14–7 SUMMARY

In many research situations, the assumptions (particularly normality) for the use of parametric statistics cannot be met. Also, some statistical studies do not involve parameters such as means, variances, and proportions. For both situations, statisticians have developed nonparametric statistical methods. These methods are also called distribution-free methods.

There are several advantages to the use of nonparametric methods. The most important one is that no knowledge of the population distributions is required. Other advantages include ease of computation and understanding. The major disadvantage is that they are less efficient than their parametric counterparts when the assumptions for the parametric methods are met. In other words, larger sample sizes are needed to get as accurate results as given by their parametric counterparts.

The following list gives the nonparametric statistical tests presented in this chapter, along with their parametric counterparts.

Nonparametric Test	Parametric Test	Condition
Single-sample sign test	z or t test	One sample
Paired-sample sign test	z or t test	Two dependent samples
Wilcoxon rank sum test	z or t test	Two independent samples
Wilcoxon signed-rank test	t test	Two dependent samples
Kruskal-Wallis test	ANOVA	Three or more independent samples
Spearman rank correlation coefficient	Pearson's correlation coefficient	Relationships between variables
Runs test	None	Randomness

When the assumptions for the parametric tests can be met, they should be used instead of their nonparametric counterparts.

Important Terms

Distribution-free statistics
Kruskal-Wallis test
Nonparametric statistics
Parametric tests
Ranking
Run
Runs test
Sign test
Spearman rank correlation
  coefficient
Wilcoxon rank sum test
Wilcoxon signed-rank test

Important Formulas

Formula for the z test value in the sign test:

$$z = \frac{(X + 0.5) - (n/2)}{\sqrt{n}/2}$$

where

n = sample size (greater than or equal to 26)
X = smaller number of + or − signs

Formula for the Wilcoxon rank sum test:

$$z = \frac{R - \mu_R}{\sigma_R}$$

where

$$\mu_R = \frac{n_1(n_1 + n_2 + 1)}{2}$$

$$\sigma_R = \sqrt{\frac{n_1 n_2(n_1 + n_2 + 1)}{12}}$$

R = sum of the ranks for the smaller sample size (n_1)
n_1 = smaller of the sample sizes
n_2 = larger of the sample sizes

Formula for the Wilcoxon signed-rank test:

$$z = \frac{w_s - \dfrac{n(n + 1)}{4}}{\sqrt{\dfrac{n(n + 1)(2n + 1)}{24}}}$$

where

n = number of pairs where the difference is not 0
w_s = smaller sum in absolute value of the signed ranks

Formula for the Kruskal-Wallis test:

$$H = \frac{12}{N(N + 1)}\left(\frac{R_1^2}{n_1} + \frac{R_2^2}{n_2} + \cdots + \frac{R_k^2}{n_k}\right) - 3(N + 1)$$

where

R_1 = sum of the ranks of sample 1
n_1 = size of sample 1
R_2 = sum of the ranks of sample 2
n_2 = size of sample 2
.
.
.
R_k = sum of the ranks of sample k
n_k = size of sample k
$N = n_1 + n_2 + \cdots + n_k$
k = number of samples

Speaking of **Statistics**

COMPILED BY ELIAS SPORTS BUREAU

TEAM BATTING COMPARISON

Key: BA-batting avg.; OB-on base pct.; AB-at-bats; R-runs; H-hits; TB-total bases; 2B-doubles; 3B-triples; HR-home runs; RBI-runs batted in; SH-sacrifices; SF-sacrifice flies; LOB-left on base; BB-walks; SO-struck out; SB-stolen bases; CS-caught stealing; E-errors.

Team	BA	SLG	OB	AB	R	H	TB	2B	3B	HR	RBI	SH	SF	LOB	BB	SO	SB	CS	E
San Fran.	.278	.429	.342	5356	782	1487	2297	259	31	163	734	99	49	1118	499	887	114	64	99
St. Louis	.275	.400	.344	5340	746	1468	2138	251	34	117	714	56	52	1133	566	834	151	70	152
Philadelphia	.275	.431	.353	5481	858	1506	2362	289	51	155	793	81	48	1233	645	1017	91	30	136
Colorado	.274	.423	.324	5352	741	1467	2264	274	59	135	688	67	51	950	379	907	145	86	163
Chicago	.270	.412	.324	5450	712	1473	2246	249	31	154	683	66	40	1090	424	880	94	41	113
Pittsburgh	.268	.392	.335	5347	678	1433	2094	256	48	103	636	75	50	1153	510	927	87	55	101
Houston	.266	.409	.329	5249	689	1395	2146	280	36	133	631	81	43	1064	476	871	96	58	122
Cincinnati	.265	.399	.324	5353	702	1420	2136	258	28	134	651	60	64	1090	467	984	140	58	119
Atlanta	.260	.406	.329	5310	729	1381	2154	230	27	163	674	71	48	1116	533	901	122	47	102
Los Angeles	.260	.381	.319	5386	645	1399	2053	228	27	124	610	103	47	1117	468	906	124	58	127
Montreal	.256	.384	.326	5286	706	1352	2032	257	33	119	657	97	49	1121	528	828	212	55	153
San Diego	.252	.387	.312	5343	655	1349	2070	231	26	146	609	80	47	1061	427	1013	91	41	157
Florida	.250	.351	.316	5288	571	1322	1855	191	30	94	532	56	43	1140	479	1012	113	55	119
New York	.248	.390	.306	5224	642	1295	2036	215	35	152	603	83	46	973	432	838	78	49	152

TEAM PITCHING COMPARISON

Key: W-wins; L-losses; ERA-earned run avg.; G-games; CG-complete games; SHO-shutouts; SV-saves; IP-innings; H-hits; R-runs; ER-earned runs; HR-home runs; HB-hit batters; BB-walks; SO-strikeouts; WP-wild pitches; BK-balks.

Team	W	L	ERA	CG	SHO	SV	IP	H	R	ER	HR	HB	BB	SO	WP	BK
Atlanta	100	56	3.11	18	16	43	1401	1245	533	484	95	22	459	996	45	9
Los Angeles	79	77	3.48	16	9	35	1418⅔	1350	634	548	99	36	544	997	41	20
Houston	81	75	3.49	17	13	40	1389⅓	1309	606	538	112	40	457	1002	60	11
San Fran.	99	57	3.55	4	9	46	1403⅔	1321	604	554	157	47	420	944	31	18
Montreal	89	67	3.59	8	7	57	1402⅔	1331	668	560	118	44	499	887	45	11
Philadelphia	95	61	3.92	24	11	45	1406⅓	1375	710	619	122	35	547	1065	66	7
Florida	64	92	4.08	4	5	48	1386⅓	1376	689	629	131	31	576	905	80	19
St. Louis	85	71	4.18	5	6	53	1392⅓	1509	726	646	149	43	370	729	40	7
New York	53	103	4.19	15	7	22	1377	1442	737	641	139	49	415	831	32	14
Chicago	81	76	4.19	8	5	53	1406⅔	1475	719	655	148	41	459	874	42	20
San Diego	59	98	4.22	8	6	32	1392⅔	1428	745	653	141	33	539	919	55	13
Cincinnati	71	86	4.52	10	7	36	1390	1465	765	698	155	43	492	964	44	8
Pittsburgh	72	84	4.80	11	4	32	1393⅔	1500	784	744	150	43	470	813	54	11
Colorado	66	91	5.39	9	0	34	1390⅓	1621	936	832	176	41	586	879	78	22

Source: Based on data from Elias Sports Bureau. Copyright 1993, USA TODAY. Reprinted with permission.

Many statistical tests can be performed on sports data. For example, is there a relationship between the rankings of teams according to their earned run averages? Use the Spearman rank correlation coefficient to decide. What other nonparametric tests can be performed on the data?

Formula for the Spearman rank correlation coefficient:

$$r_S = 1 - \frac{6 \Sigma d^2}{n(n^2 - 1)}$$

where

d = difference in the ranks
n = number of data pairs

Review Exercises

14–103. The owner of a shop states that she sells, on average, 400 greeting cards per day. A random sample of 20 days shows the following number of cards sold each day. At $\alpha = 0.10$, is the claim correct? Use the sign test.

371 382 368 405 411 397 206 315 415 457 411
415 397 362 356 419 399 401 382 373

14–104. A tire manufacturer claims that the median lifetime of a certain brand of truck tires is 40,000 miles. A sample of 30 tires shows that 12 lasted longer than 40,000 miles. Test the claim, at $\alpha = 0.05$. Use the sign test.

14–105. A special diet including hormones is fed to adult hogs to see whether they will gain weight. The "before" and "after" weights (in pounds) are given in the table. Use the paired-sample sign test at $\alpha = 0.05$ to determine whether there is a weight gain.

Before	320	432	456	358	371	394	362	359	319
After	333	430	459	362	381	395	367	356	315

14–106. Two groups of bowlers consisting of males and females were asked at what age they first started to bowl. The data are as shown in the table. For the Wilcoxon rank sum test at $\alpha = 0.05$, is there a difference in the ages of the genders?

Males	18	12	24	16	15	19	23	26	22	8	13
Females	9	11	15	10	13	7	8	14	21	22	10

14–107. Samples of students majoring in business and engineering are selected, and the amount (in dollars) each spent on a required textbook for the fall semester is recorded. The data are shown in the table. For the Wilcoxon rank sum test at $\alpha = 0.10$, is there a difference in the amount spent by each group?

Business		Engineering	
48	36	98	73
52	62	72	78
74	50	63	93
63	46	78	88
51	53	55	86
49	58	58	85
		64	

14–108. Twelve automobiles were tested to see how many miles per gallon each one obtained. Under similar driving conditions, the automobiles were tested again using a special additive. The data are as shown in the table. At $\alpha = 0.05$, did the additive improve gas mileage? Use the Wilcoxon signed-rank test.

Before		After	
13.6	22.6	18.3	23.7
18.2	21.9	19.5	20.8
16.1	25.3	18.2	25.3
15.3	28.6	16.7	27.2
19.2	15.2	21.3	17.1
18.8	16.3	17.2	18.5

14–109. The number of sick days taken by seven assembly line workers were recorded for one year. The owners of the company then installed brighter lighting and permitted the employees to take a 10-minute break in the morning and afternoon. The number of sick days taken were recorded for another year. The data are as shown in the table. From the Wilcoxon signed-rank test at $\alpha = 0.05$, can one conclude that the number of sick days taken was reduced?

Before	6	15	18	14	27	17	9
After	8	12	16	9	23	14	15

14–110. Samples of three different types of ropes are tested for breaking strength. The data (in pounds) are as shown in the table. At $\alpha = 0.05$, is there a difference in the breaking strength of the ropes? Use the Kruskal-Wallis test.

Cotton	Nylon	Hemp
230	356	506
432	303	527
505	361	581
487	405	497
451	432	459
380	378	507
462	361	562
531	399	571
366	372	499
372	363	475
453	306	505
488	304	561
462	318	532
467	322	501

14–111. Rats are fed three different diets for one month to see whether diet has any effect on learning. Each rat is then taught to traverse a simple maze. The number of trials it took each rat to learn the correct path is shown in the table. At $\alpha = 0.05$, does diet have any effect on learning? Use the Kruskal-Wallis test.

Diet 1	8	6	12	15	9	7	5	
Diet 2	2	3	6	8	7	4		
Diet 3	9	15	17	8	4	13	18	20

14–112. A statistics instructor wishes to see whether there is a relationship between the number of homework exercises a student completes and his or her exam score. The data are as shown in the table. Using the Spearman rank correlation coefficient, test the hypothesis that there is no relationship, at $\alpha = 0.05$.

Homework Problems	Exam Score
63	85
55	71
58	75
87	98
89	93
52	63
46	72
75	89
105	100

14–113. Six different brands of ice cream were ranked according to taste by third-grade boys and girls. The data are as shown in the table. Using the Spearman rank correlation coefficient, test the hypothesis that there is no relationship in the rankings of boys and girls. Use $\alpha = 0.05$.

Brand	A	B	C	D	E	F
Boys	1	5	4	3	2	6
Girls	2	1	3	6	4	5

14–114. In a recent survey, 24 employees were asked, as they arrived at work, if they drove (D) or used public transportation (P). At $\alpha = 0.05$, test for randomness. The data are as shown.

D D P P P D P D D D P P
P P D D D P P P D D P D

14–115. An instructor wishes to see whether grades of students who finish an exam occur at random. Shown below are the grades of 30 students in the order (from left to right across each row, and then proceed to the next row) that they finished an exam. Test for randomness, at $\alpha = 0.05$.

87	93	82	77	64	98
100	93	88	65	72	73
56	63	85	92	95	91
88	63	72	79	55	53
65	68	54	71	73	72

COMPUTER APPLICATIONS

1. MINITAB can be used for the sign test for the equality of two medians. For example, suppose a new medication is designed to lower diastolic blood pressure in patients. A sample of 10 patients receives the medication. The data are shown in the table. At $\alpha = 0.05$, has the pressure been lowered?

Before	101	106	97	95	100	102	108	111	88	91
After	95	98	93	88	101	97	103	109	82	92

Type the following:

```
MTB  > READ C1 C2
DATA > 101 95
DATA > 106 98
DATA > 97 93
DATA > 95 88
DATA > 100 101
DATA > 102 97
DATA > 108 103
DATA > 111 109
DATA > 88 82
DATA > 91 92
DATA > END
      10 ROWS READ
MTB  > LET C3=C2-C1
MTB  > STEST MEDIAN 0 C3;
SUBC > ALTERNATE -1.
```

The computer will print the following:

```
SIGN TEST OF MEDIAN   0.000 VERSUS L.T.  0.000
      N    BELOW   EQUAL    ABOVE    P-VALUE   MEDIAN
C3   10      8       0        2       0.0547   -5.000
```

As shown, the P-value is 0.0547, which is slightly larger than 0.05; hence, the decision is to not reject the null hypothesis.

Use this program to do Exercise 14–25.

2. MINITAB can be used to test a claim about the median when the data are normal. For example, a coin is tossed 40 times, and 32 heads and 8 tails result. Is the coin balanced? Test at $\alpha = 0.05$. (*Note:* For this test, 1 is used to represent heads and -1 is used to represent tails.)

Type the following:

```
MTB  > SET C1
DATA > 32(1), 8(-1)
DATA > END
MTB  > STEST MEDIAN 0 C1
```

The computer will print the following:

```
SIGN TEST OF MEDIAN 0.000 VERSUS N.E.  0.000
      N    BELOW   EQUAL    ABOVE    P-VALUE   MEDIAN
C1   40      8       0        32      0.0002   1.000
```

The decision is to reject the null hypothesis, since the P-value is 0.0002, which is much smaller than 0.05.

Use this program to do Exercise 14–17.

3. MINITAB will compute the Wilcoxon rank sum test value. This test is equivalent to the Mann-Whitney test used in MINITAB. For example, two classes of college algebra are selected and a mathematics pretest is given to each. At $\alpha = 0.05$, is there a difference in the scores? The data are as shown.

Class A	90	88	96	87	82	73	71	84	95	92	81	63	54	67	63
Class B	72	77	100	98	65	52	78	81	86	84	72	71	68		

Type the following:

```
MTB  > SET C1
DATA > 90 88 96 87 82 73 71 84 95 92 81 63 54 67 63
DATA > END
MTB  > SET C2
DATA > 72 77 100 98 65 52 78 81 86 84 72 71 68
DATA > END
MTB  > MANN-WHITNEY C1 C2
```

The computer will print the following:

```
W = 227.5
CANNOT REJECT AT ALPHA = 0.05
```

Hence, the decision is to not reject the null hypothesis, at $\alpha = 0.05$.

Use this program to do Exercise 14–43.

4. MINITAB will compute the Wilcoxon signed-rank test value for dependent samples. For example, 12 men are given a strength test. Then, they consume 400 international units of vitamin E for three days and are tested again. At $\alpha = 0.05$, test the claim that the vitamin does not affect strength. The data are as shown.

Before	106	214	113	152	187	262	205	193	140	113	175	186
After	111	216	113	162	193	257	216	205	156	114	180	192

Type the following:

```
MTB  >  READ C1 C2
DATA >  106 111
DATA >  214 216
DATA >  113 113
DATA >  152 162
DATA >  187 193
DATA >  262 257
DATA >  205 216
DATA >  193 205
DATA >  140 156
DATA >  113 114
DATA >  175 180
DATA >  186 192
DATA >  END
        12 ROWS READ
MTB  >  LET C3=C2-C1
MTB  >  WTEST C3
```

The computer will print the following:

```
TEST OF MEDIAN = 0.000000 VERSUS MEDIAN N.E. 0.000000
                   N FOR   WILCOXON        ESTIMATED
                 N TEST STATISTIC P-VALUE    MEDIAN
C3              12    11       62.0   0.011     5.500
```

Since this test value is significant only at $\alpha = 0.011$, the null hypothesis is rejected at $\alpha = 0.05$.

Use this program to do Exercise 14–59.

5. The Kruskal-Wallis test can be performed with MINITAB. For example, test the equality of the means of three random samples of new automobiles, at $\alpha = 0.05$. The data are as shown.

Ford	GM	Chrysler
$8,753	$12,510	$15,321
9,261	13,623	9,716
7,318	8,795	14,283
10,645	9,993	14,667
8,134	10,210	15,005
11,328	12,673	
	8,800	

Type the following:

```
MTB  >  SET C1
DATA >  8753 9261 7318 10645 8134 11328
DATA >  END
MTB  >  SET C2
DATA >  12510 13623 8795 9993 10210 12673 8800
DATA >  END
MTB  >  SET C3
DATA >  15321 9716 14283 14667 15005
DATA >  END
MTB  >  STACK C1 C2 C3 C4;
SUBC >  SUBSCRIPTS C5.
MTB  >  KRUSKAL-WALLIS C4 C5
```

The computer will print the following:

```
H = 7.94    d.f. = 2    p = 0.019
```

To get the critical value, type the following:

```
MTB  > INVCDF 0.95;
SUBC > CHISQUARE DF 2.
```

The computer will print the following:

```
0.9500 5.9915
```

Since the test value 7.943 is greater than the critical value 5.9915, the decision is to reject the null hypothesis, at $\alpha = 0.05$.

Use this program to do Exercise 14–73.

6. MINITAB can be used to compute Spearman's rank correlation coefficient. For example, suppose that five speakers are rated on a scale of 1 to 50 points for poise, gestures, speech content, etc. Each of the speakers is rated by a group of speech students and by their instructor. Is there a relationship between the ratings? Use $\alpha = 0.05$. The ratings are as shown.

Speaker	Student Rating	Instructor Rating
A	15	18
B	23	22
C	46	20
D	9	6
E	35	38

MINITAB will rank each rating and then compute the correlation coefficient. Type the following:

```
MTB  > READ C1 C2
DATA > 15 18
DATA > 23 22
DATA > 46 20
DATA > 9 6
DATA > 35 38
DATA > END
MTB  > RANK C1 INTO C3
MTB  > RANK C2 INTO C4
MTB  > CORRELATION C3 C4
```

The computer will print the following:

```
Correlation of C3 and C4 0.7000
```

Note: If the data are given in ranks, then omit the rank statements.

Use this program to do Exercise 14–81.

7. MINITAB can be used to test for randomness in runs. For example, suppose that a researcher wishes to see whether the number of calls a special 24-hour hot line receives is random about the median. Test for randomness at $\alpha = 0.05$. The data are as shown.

32, 18, 12, 54, 63, 17, 5, 63, 27, 9, 56, 58, 47, 9, 14, 16, 18, 50, 42, 37, 35, 29, 18, 3

Type the following:

```
MTB  > SET C1
DATA > 32 18 12 54 63 17 5 63 27 9 56 58
DATA > 47 9 14 16 18 50 42 37 35 29 18 3
DATA > END
MTB  > MEDIAN C1
```

The computer will print the following:

```
MEDIAN 28.000
```

Now, type the following:

```
MTB > RUNS 28 C1
```

The computer will print the following:

```
C1
K = 28.0000
THE OBSERVED NO OF RUNS = 10
THE EXPECTED NO OF RUNS = 13.0000
12 OBSERVATIONS ABOVE K    12 BELOW
THE TEST IS SIGNIFICANT AT 0.2108
CANNOT REJECT AT ALPHA = 0.05
```

Use this program to do Exercise 14–89.

8. MINITAB can be used to test for randomness when the data can be classified into two categories. For example, a coin is flipped 25 times. At $\alpha = 0.05$, test for randomness. The data are as shown.

H H T T T H T T T H H H T H T H T H H H H H H T T T T H H

To use MINITAB, change the data to numbers.

```
MTB > SET C1
DATA > 0 0 1 1 1 0 1 1 1 0 0 1 0 1 0 0 0 0 0 1 1 1 1 0 0
DATA > END
MTB > RUNS C1
```

The computer will print the following:

```
C1
K = 0.4800
THE OBSERVED NO OF RUNS = 11
THE EXPECTED NO OF RUNS = 13.4800
12 OBSERVATIONS ABOVE K    13 BELOW
THIS TEST IS SIGNIFICANT AT 0.3103
CANNOT REJECT AT ALPHA = 0.05
```

Use this program to do Exercise 14–93.

√ DATA ANALYSIS Applying the Concepts

The Data Bank is located in the Appendix.

1. From the Data Bank, choose a sample and use the sign test to test one of the following hypotheses:
   a. For serum cholesterol, test H_0: median = 220 milligram percent (mg%).
   b. For systolic pressure, test H_0: median = 120 millimeters of mercury (mmHg).
   c. For IQ, test H_0: median = 100.
   d. For sodium level, test H_0: median = 140 mEq/l.

2. From the Data Bank, select a sample of subjects. Use the Kruskal-Wallis test to test the sodium levels of smokers and nonsmokers to see whether they are equal.

✎ TEST

Directions: Determine whether the statement is true or false.

1. When the assumption of normality of the population cannot be met, one can use nonparametric statistics.
2. Nonparametric statistics cannot be used to test differences between two means.
3. Nonparametric statistics are more sensitive than their parametric counterparts.
4. Nonparametric statistics use less information than their parametric counterparts.
5. Nonparametric statistics can be used to test hypotheses other than means, proportions, and standard deviations.
6. When the data is nominal or ordinal, parametric methods are preferred over nonparametric methods.
7. To test to see whether a median is equal to a specific value, one would use the runs test.

8. The Wilcoxon rank sum test is used when the samples are dependent.
9. The Kruskal-Wallis test uses the chi-square distribution.
10. The nonparametric counterpart of ANOVA is the Wilcoxon signed-rank test.
11. To test to see whether two rankings are related, one can use the runs test.
12. A test for randomness is the Spearman rank correlation test.
13. Parametric tests are preferred over their nonparametric counterparts if the assumptions can be met.
14. Nonparametric tests are less efficient than their parametric counterparts.

HYPOTHESIS-TESTING SUMMARY 3*

13. Test to see whether the median of a sample is a specific value when $n \geq 26$.

    Example: H_0: median $= 100$

    Use the sign test:

    $$z = \frac{(X + 0.5) - (n/2)}{\sqrt{n}/2}$$

14. Test to see whether two independent samples are obtained from populations that have identical distributions.

    Example: H_0: There is no difference in the ages of the subjects.

    Use the Wilcoxon rank sum test:

    $$z = \frac{R - \mu_R}{\sigma_R}$$

    where

    $$\mu_R = \frac{n_1(n_1 + n_2 + 1)}{2}$$

    $$\sigma_R = \sqrt{\frac{n_1 n_2(n_1 + n_2 + 1)}{12}}$$

15. Test to see whether two dependent samples have identical distributions.

    Example: H_0: There is no difference in the effects of a tranquilizer on the number of hours a person sleeps at night.

Use the Wilcoxon signed-rank test:

$$z = \frac{w_s - \dfrac{n(n + 1)}{4}}{\sqrt{\dfrac{n(n + 1)(2n + 1)}{24}}}$$

when $n \geq 30$.

16. Test to see whether three or more samples come from identical populations.

    Example: H_0: There is no difference in the weights of the three groups.

    Use the Kruskal-Wallis test:

    $$H = \frac{12}{N(N + 1)}\left(\frac{R_1^2}{n_1} + \frac{R_2^2}{n_2} + \cdots + \frac{R_k^2}{n_k}\right) - 3(N + 1)$$

17. Rank correlation coefficient

    $$r_S = 1 - \frac{6 \Sigma d^2}{n(n^2 - 1)}$$

18. Test for randomness.
    Use the runs test.

*This summary is a continuation of Hypothesis-Testing Summary 2.

CHAPTER

15

Sampling and Simulation

15–1 INTRODUCTION

Most people have heard of Gallup, Harris, and Nielsen. These are only a few of the pollsters who gather information about the habits and opinions of the American people. These survey firms, including the U.S. Census Bureau, gather information by selecting samples from well-defined populations. Recall from Chapter 1 that the subjects in the sample should be a subgroup of the subjects in the population. Sampling methods use what are called *random numbers* to select samples.

Random numbers are also used in *simulation techniques*. Instead of studying a real-life situation, which may be costly or dangerous, researchers create a similar situation in a laboratory or with a computer, which simulates the real-life situation. Then, by studying the simulated situation, researchers can gain the necessary information about a real-life situation in a less expensive or safer manner. This chapter will explain some common methods used to obtain samples as well as the techniques used in simulations. Also, included is a section on writing a research report.

15–2 COMMON SAMPLING TECHNIQUES

In Chapter 1, a *population* was defined as all subjects (human or otherwise) under study. Since some populations can be very large, researchers cannot use every single subject, so a sample must be selected. A *sample* is a subgroup of the population. Any subgroup of the population, technically speaking, can be called a sample. However, for researchers to make valid generalizations about population characteristics, the sample must be random.

For a sample to be a random sample, every member of the population must have an equal chance of being selected.

When a sample is chosen at random from a population, it is said to be an **unbiased sample**—i.e., the sample, for the most part, is representative of the population. Conversely, if a sample is selected incorrectly, it may be a biased sample. Samples are said to be **biased samples** when some type of systematic error has been made in the selection of the subjects for the samples.

A sample is used to get information about a population for several reasons:

1. *It saves the researcher time and money.*
2. *It enables the researcher to get information that he or she may not be able to obtain otherwise.* For example, if a person's blood is to be analyzed for cholesterol, a researcher could not analyze every single drop of blood without killing the person. Or if the breaking strength of cables was to be determined, a researcher could not test every single cable manufactured, since the company would not have any cables left to sell.
3. *It enables the researcher to get more detailed information about a particular subject.* If a small number of people are surveyed, the researcher can

conduct in-depth interviews by spending more time with each person, thus getting more information about a subject. This is not to say that the smaller the sample, the better—the opposite is true. In general, larger samples (if correct sampling techniques are used) have a higher probability of being representative of the population.

It would be ideal if the sample were a perfect miniature of the population in all characteristics. This ideal, however, is impossible to achieve, because there are so many human traits (height, weight, IQ, etc.). The best that can be done is to select a sample that will be representative with respect to *some* characteristics, preferably those pertaining to the study. For example, if half of the population subjects are female, then approximately half of the sample subjects should be female. Likewise, other characteristics, such as age, socioeconomic status, and IQ, should be represented proportionately. In order to obtain unbiased samples, statisticians have developed several basic sampling methods. The most common methods are *random, systematic, stratified,* and *cluster sampling.* Each method will be explained in detail in this section.

In addition to the basic methods, there are other methods used to obtain samples. Some of these methods are also explained in this section.

Random Sampling

A random sample is obtained by using methods, such as random numbers. In *random sampling,* the basic requirement is that for a sample of size *n,* all possible samples of this size must have an equal chance of being selected from the population. But before the correct method of obtaining a random sample is explained, a discussion is given of several incorrect methods commonly used by various researchers and agencies to gain information.

One incorrect method commonly used is to ask "the person on the street." News reporters use this technique quite often. Selecting people haphazardly on the street does not meet the requirement for simple random sampling, since all possible samples of a specific size do not have an equal chance of being selected. Many people will be at home or at work when the interview is being conducted and therefore do not have a chance of being selected.

Another incorrect technique is to ask a question either by radio or television and have the listeners or viewers call the station to give their responses or opinions. Again, this sample is not random, since only those who feel strongly for or against the issue may respond.

A third erroneous method is to ask people to respond by mail. Again, only those who are concerned and who have the time are likely to respond.

These methods do not meet the requirement of random sampling, since all possible samples of a specific size do not have an equal chance of being selected. In order to meet this requirement, researchers can use one of two methods. The first method is to number each element of the population and then place the numbers on cards. Place the cards in a hat or fishbowl, mix them, and then select the sample by drawing the cards. When using this procedure, researchers must ensure that the numbers are well mixed. On occasion, when this procedure is used, the numbers are not mixed well, and the numbers chosen for the sample are those that were placed in the bowl last.

The second and preferred way of selecting a random sample is to use random numbers. Random numbers can be obtained by using a calculator, computer, or table. Figure 15–1 shows a table of random numbers generated by a computer.

FIGURE 15–1
Table of Random Numbers

79	41	71	93	60	35	04	67	96	04	79	10	86
26	52	53	13	43	50	92	09	87	21	83	75	17
18	13	41	30	56	20	37	74	49	56	45	46	83
19	82	02	69	34	27	77	34	24	93	16	77	00
14	57	44	30	93	76	32	13	55	29	49	30	77
29	12	18	50	06	33	15	79	50	28	50	45	45
01	27	92	67	93	31	97	55	29	21	64	27	29
55	75	65	68	65	73	07	95	66	43	43	92	16
84	95	95	96	62	30	91	64	74	83	47	89	71
62	62	21	37	82	62	19	44	08	64	34	50	11
66	57	28	69	13	99	74	31	58	19	47	66	89
48	13	69	97	29	01	75	58	05	40	40	18	29
94	31	73	19	75	76	33	18	05	53	04	51	41
00	06	53	98	01	55	08	38	49	42	10	44	38
46	16	44	27	80	15	28	01	64	27	89	03	27
77	49	85	95	62	93	25	39	63	74	54	82	85
81	96	43	27	39	53	85	61	12	90	67	96	02
40	46	15	73	23	75	96	68	13	99	49	64	11

The theory behind random numbers is that each digit, 0 through 9, has an equal probability of occurring. That is, in every sequence of ten digits, each digit has a probability of $\frac{1}{10}$ of occurring. This does not mean that in every sequence of ten digits, one will find each digit. Rather, it means that on the average, each digit will occur once. For example, the digit 2 may occur three times in a sequence of ten digits, but in later sequences, it may not occur at all, thus averaging to a probability of $\frac{1}{10}$.

To obtain a sample by using random numbers, one numbers the elements of the population sequentially, and then each person is selected by using random numbers. This process is shown in Example 15–1.

Random samples can be selected with or without replacement. If the same member of the population cannot be used more than once in the study, then the sample is selected without replacement. That is, once a random number is selected, it cannot be used later.

Note: In the explanations and examples of the sampling procedures, a small population will be used, and small samples will be selected from this population. Small populations are used for illustrative purposes only, because with a small population, the entire population can be included with little difficulty. In real life, however, researchers must usually sample from very large populations, using the procedures shown in this chapter.

EXAMPLE 15–1 Suppose a researcher wants to ask the governors of the 50 states their opinions on capital punishment. Select a random sample of 10 states from the 50.

Speaking of **Statistics**

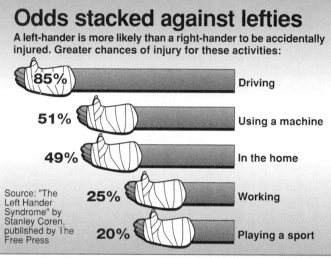

USA SNAPSHOTS®

A look at statistics that shape our lives

Odds stacked against lefties

A left-hander is more likely than a right-hander to be accidentally injured. Greater chances of injury for these activities:

85% Driving

51% Using a machine

49% In the home

25% Working

Source: "The Left Hander Syndrome" by Stanley Coren, published by The Free Press

20% Playing a sport

By Nick Galifianakis, USA TODAY

Source: Based on data in *The Left Hander Syndrome* by Stanley Coren, published by The Free Press. Copyright 1993, USA TODAY. Reprinted with permission.

Many studies have been done comparing left-handed people to right-handed people. Here is one comparing injuries for each group. State some reasons for the differences. For example, why do lefties have 51% more accidents than righties when using a machine?

Solution   **STEP 1**   Number each state from 1 to 50, as shown.

1. Alabama	14. Indiana	27. Nebraska	40. South Carolina
2. Alaska	15. Iowa	28. Nevada	41. South Dakota
3. Arizona	16. Kansas	29. New Hampshire	42. Tennessee
4. Arkansas	17. Kentucky	30. New Jersey	43. Texas
5. California	18. Louisiana	31. New Mexico	44. Utah
6. Colorado	19. Maine	32. New York	45. Vermont
7. Connecticut	20. Maryland	33. North Carolina	46. Virginia
8. Delaware	21. Massachusetts	34. North Dakota	47. Washington
9. Florida	22. Michigan	35. Ohio	48. West Virginia
10. Georgia	23. Minnesota	36. Oklahoma	49. Wisconsin
11. Hawaii	24. Mississippi	37. Oregon	50. Wyoming
12. Idaho	25. Missouri	38. Pennsylvania	
13. Illinois	26. Montana	39. Rhode Island	

STEP 2 Using the random numbers shown in Figure 15–1, find a starting point. To find a starting point, one generally closes one's eyes and places one's finger anywhere on the table. In this case, the first number selected was 27 in the fourth column. Going down the column, and continuing on to the next column, select the first ten numbers. They are 27, 95, 27, 73, 60, 43, 56, 34, 93, and 06. See Figure 15–2.

FIGURE 15–2

Selecting a Starting Point and Ten Digits from the Random Number Table

79	41	71	93	60✓	35	04	67	96	04	79	10	86
26	52	53	13	43✓	50	92	09	87	21	83	75	17
18	13	41	30	56✓	20	37	74	49	56	45	46	83
19	82	02	69	34✓	27	77	34	24	93	16	77	00
14	57	44	30	93✓	76	32	13	55	29	49	30	77
29	12	18	50	06✓	33	15	79	50	28	50	45	45
01	27	92	67	93	31	97	55	29	21	64	27	29
55	75	65	68	65	73	07	95	66	43	43	92	16
84	95	95	96	62	30	91	64	74	83	47	89	71
62	62	21	37	82	62	19	44	08	64	34	50	11
66	57	28	69	13	99	74	31	58	19	47	66	89
48	13	69	97	29	01	75	58	05	40	40	18	29
94	31	73	19	75	76	33	18	05	53	04	51	41
00	06	53	*Start here	01	55	08	38	49	42	10	44	38
46	16	44	⨀27 ✓	80	15	28	01	64	27	89	03	27
77	49	85	95✓	62	93	25	39	63	74	54	82	85
81	96	43	27✓	39	53	85	61	12	90	67	96	02
40	46	15	73✓	23	75	96	68	13	99	49	64	11

Now, refer to the list of states, and identify the state corresponding to each number. The sample consists of the following states:

27	Nebraska	43	Texas
95		56	
27	Nebraska	34	North Dakota
73		93	
60		06	Colorado

STEP 3 Since the numbers 95, 73, 60, 56, and 93 are too large, they are disregarded. And since 27 appears twice, it is also disregarded the second time. Now, one must select six more random numbers between 1 and 50 and omit duplicates, since this sample will be

selected without replacement. This selection is accomplished by continuing down the column and moving over to the next column until a total of ten numbers is selected. The final ten numbers are 27, 43, 34, 06, 13, 29, 01, 39, 23, and 35. See Figure 15–3.

FIGURE 15–3
The Final Ten Numbers Selected

79	41	71	93	60	(35)	04	67	96	04	79	10	86
26	52	53	13	(43)	50	92	09	87	21	83	75	17
18	13	41	30	56	20	37	74	49	56	45	46	83
19	82	02	69	(34)	27	77	34	24	93	16	77	00
14	57	44	30	93	76	32	13	55	29	49	30	77
29	12	18	50	(06)	33	15	79	50	28	50	45	45
01	27	92	67	93	31	97	55	29	21	64	27	29
55	75	65	68	65	73	07	95	66	43	43	92	16
84	95	95	96	62	30	91	64	74	83	47	89	71
62	62	21	37	82	62	19	44	08	64	34	50	11
66	57	28	69	(13)	99	74	31	58	19	47	66	89
48	13	69	97	(29)	01	75	58	05	40	40	18	29
94	31	73	19	75	76	33	18	05	53	04	51	41
00	06	53	98	(01)	55	08	38	49	42	10	44	38
46	16	44	(27)	80	15	28	01	64	27	89	03	27
77	49	85	95	62	93	25	39	63	74	54	82	85
81	96	43	27	(39)	53	85	61	12	90	67	96	02
40	46	15	73	(23)	75	96	68	13	99	49	64	11

These numbers correspond to the following states:

27	Nebraska	29	New Hampshire
43	Texas	01	Alabama
34	North Dakota	39	Rhode Island
06	Colorado	23	Minnesota
13	Illinois	35	Ohio

Thus, the governors of these ten states will constitute the sample. ◄

Random sampling has one limitation. If the population is extremely large, it is time-consuming to number and select the sample elements. Also, notice that the random numbers in the table are two-digit numbers. If three digits are needed, then the first digit from the next column can be used, as shown in Figure 15–4. Table D in the Appendix gives five-digit random numbers.

FIGURE 15–4
Method for Selecting Three-Digit Numbers

79	41	71	93	60	35	04	67	96	04	79	10	86
26	52	53	13	43	50	92	09	87	21	83	75	17
18	13	41	30	56	20	37	74	49	56	45	46	83
19	82	02	69	34	27	77	34	24	93	16	77	00
14	57	44	30	93	76	32	13	55	29	49	30	77
29	12	18	50	06	33	15	79	50	28	50	45	45
01	27	92	67	93	31	97	55	29	21	64	27	29
55	75	65	68	65	73	07	95	66	43	43	92	16
84	95	95	96	62	30	91	64	74	83	47	89	71
62	62	21	37	82	62	19	44	08	64	34	50	11
66	57	28	69	13	99	74	31	58	19	47	66	89
48	13	69	97	29	01	75	58	05	40	40	18	29
94	31	73	19	75	76	33	18	05	53	04	51	41
00	06	53	98	01	55	08	38	49	42	10	44	38
46	16	44	27	80	15	28	01	64	27	89	03	27
77	49	85	95	62	93	25	39	63	74	54	82	85
81	96	43	27	39	53	85	61	12	90	67	96	02
40	46	15	73	23	75	96	68	13	99	49	64	11

Use one column and part of the next column for three digits, i.e., 404.

Systematic Sampling

A systematic sample is a sample obtained by numbering each element in the population and then selecting every third, fifth, tenth, etc., number from the population to be included in the sample.

The procedure of systematic sampling is illustrated in the next example.

EXAMPLE 15–2 Using the population of 50 states in Example 15–1, select a systematic sample of 10 states.

Solution **STEP 1** Number the population units as shown in Example 15–1.

STEP 2 Since there are 50 states, and 10 are to be selected, the rule is to select every fifth state. This rule was determined by dividing 50 by 10 to get 5.

STEP 3 Using the table of random numbers, select the first digit (from 1 to 5) at random. In this case, 4 was selected.

STEP 4   Select every fifth number on the list, starting with 4. The numbers include the following:

1, 2, 3,④, 5, 6, 7, 8,⑨, 10, 11, 12, 13, ⑭, . . .

The selected states are as follows:

4	Arkansas	29	New Hampshire
9	Florida	34	North Dakota
14	Indiana	39	Rhode Island
19	Maine	44	Utah
24	Mississippi	49	Wisconsin

◄

The advantage of systematic sampling is the ease of selecting the sample elements. Also, in many cases, a numbered list of the population units already may exist. For example, the manager of a factory may have a list of employees who work for the company, or there may be an in-house telephone directory.

One disadvantage of systematic sampling that may cause a biased sample is the arrangement of the items on the list. For example, if each unit were arranged, say, as

1. Husband
2. Wife
3. Husband
4. Wife
   etc.

the selection of the starting number could produce a sample of all males or all females, depending on whether the starting number is even or odd and whether the number to be added is even or odd. As another example, if the list were arranged in order of heights of individuals, one would get a different average from two samples if the first were selected by using a small starting number and the second by using a large starting number.

Stratified Sampling

A stratified sample is a sample obtained by dividing the population into subgroups, called *strata*, according to various homogeneous characteristics and then selecting members from each stratum for the sample.

For example, a population may consist of males and females who are smokers or nonsmokers. The researcher will want to include in the sample people from each group—that is, males who smoke, males who do not smoke, females who smoke, and females who do not smoke. In order to accomplish this selection, the researcher divides the population into four subgroups and then selects a random sample from each subgroup. This method ensures that the sample is representative on the basis of the characteristics of gender and smoking. Of course, it may not be representative on the basis of other characteristics.

EXAMPLE 15–3 Using the population of 20 students shown in Figure 15–5, select a sample of 8 students on the basis of gender (male/female) and grade level (freshman/sophomore) by stratification.

FIGURE 15–5
Population of Students
for Example 15–3

1. Ald, Peter	M	Fr	
2. Brown, Danny	M	So	
3. Bear, Theresa	F	Fr	
4. Collins, Carolyn	F	Fr	
5. Carson, Susan	F	Fr	
6. Davis, William	M	Fr	
7. Hogan, Michael	M	Fr	
8. Jones, Lois	F	So	
9. Lutz, Harry	M	So	
10. Lyons, Larry	M	So	

11. Martin, Janice	F	Fr
12. Meloski, Gary	M	Fr
13. Oeler, George	M	So
14. Peters, Michele	F	So
15. Peterson, John	M	Fr
16. Smith, Nancy	F	Fr
17. Thomas, Jeff	M	So
18. Toms, Debbie	F	So
19. Unger, Roberta	F	So
20. Zibert, Mary	F	So

Solution **STEP 1** Divide the population into two subgroups, consisting of males and females, as shown in Figure 15–6.

FIGURE 15–6
Population Divided into
Subgroups by Gender

Males

1. Ald, Peter	M	Fr
2. Brown, Danny	M	So
3. Davis, William	M	Fr
4. Hogan, Michael	M	Fr
5. Lutz, Harry	M	So
6. Lyons, Larry	M	So
7. Meloski, Gary	M	Fr
8. Oeler, George	M	So
9. Peterson, John	M	Fr
10. Thomas, Jeff	M	So

Females

1. Bear, Theresa	F	Fr
2. Collins, Carolyn	F	Fr
3. Carson, Susan	F	Fr
4. Jones, Lois	F	So
5. Martin, Janice	F	Fr
6. Peters, Michele	F	So
7. Smith, Nancy	F	Fr
8. Toms, Debbie	F	So
9. Unger, Roberta	F	So
10. Zibert, Mary	F	So

STEP 2 Divide each subgroup further into two groups of freshmen and sophomores, as shown in Figure 15–7.

FIGURE 15–7
Each Subgroup Divided
into Subgroups by
Grade Level

Group 1

1. Ald, Peter	M	Fr
2. Davis, William	M	Fr
3. Hogan, Michael	M	Fr
4. Meloski, Gary	M	Fr
5. Peterson, John	M	Fr

Group 2

1. Bear, Theresa	F	Fr
2. Collins, Carolyn	F	Fr
3. Carson, Susan	F	Fr
4. Martin, Janice	F	Fr
5. Smith, Nancy	F	Fr

Group 3

1. Brown, Danny	M	So
2. Lutz, Harry	M	So
3. Lyons, Larry	M	So
4. Oeler, George	M	So
5. Thomas, Jeff	M	So

Group 4

1. Jones, Lois	F	So
2. Peters, Michele	F	So
3. Toms, Debbie	F	So
4. Unger, Roberta	F	So
5. Zibert, Mary	F	So

STEP 3 Determine how many students need to be selected from each subgroup to have a proportional representation of each subgroup in the sample. There are 4 groups, and since a total of 8 students are needed for the sample, 2 students must be selected from each subgroup.

STEP 4 Select 2 students from each group by using random numbers. In this case, the random numbers were as follows:

Group 1 Students 5 and 4   Group 2 Students 5 and 2
Group 3 Students 1 and 3   Group 4 Students 3 and 4

The stratified sample then consists of the following people:

Peterson, John	M	Fr	Smith, Nancy	F	Fr
Meloski, Gary	M	Fr	Collins, Carolyn	F	Fr
Brown, Danny	M	So	Toms, Debbie	F	So
Lyons, Larry	M	So	Unger, Roberta	F	So ◄

The major advantage of stratification is that it ensures representation of all subgroups in the population that are important to the study. There are two major drawbacks to stratification, however. First, if there are a large number of variables of interest, the task of dividing a large population into representative subgroups requires a great deal of effort. Second, if the variables are somewhat complex or ambiguous (such as beliefs, attitudes, or prejudices), it is difficult to separate individuals into the subgroups according to these variables.

Cluster Sampling

A cluster sample is a sample obtained by selecting a preexisting or natural group, called a *cluster*, and using the members in the cluster for the sample.

For example, many studies in education use already existing classes, such as the seventh grade in Wilson Junior High School. The voters of a certain electoral district might be surveyed to determine their preferences for a mayoralty candidate in the upcoming election. Or the residents of an entire city block might be polled to ascertain the percentage of households that have two or more incomes. In cluster sampling, researchers may use all units of a cluster if that is feasible, or they may select only a part of a cluster to use as a sample. This selection is done by random methods.

There are three advantages to using a cluster sample instead of other types of samples: (1) A cluster sample can reduce costs; (2) it can simplify fieldwork; and (3) it is convenient. For example, in a dental study involving X-raying fourth-grade students' teeth to see how many cavities each child had, it would be a simple matter to select a single classroom and bring the X-ray equipment to the school to conduct the study. If other sampling methods were used, researchers might have to transport the machine to several different schools in the district to conduct the study or to transport the pupils to the dental office.

The major disadvantage of cluster sampling is that the elements in a cluster may not have the same variations in characteristics as elements selected individually from a population. The reason is that groups of people may be more homogeneous (alike) in specific clusters such as neighborhoods or clubs. For example, the people who live in a certain neighborhood tend to have similar incomes, drive similar automobiles, live in similar houses, and for the most part, have similar habits.

Speaking of **Statistics**

Off-Campus Living a Top Dropout Risk

By Tamara Henry
USA TODAY

Living off campus, working long hours, and having fewer than two campus friends are telltale signs of a potential college dropout, says a study by a University of Maryland psychology professor.

Roger McIntire's survey of 910 students on the College Park, Md., campus identified predictors of whether a college student will drop out. Living off campus is the strongest, poor academic standing one of the weakest. Financial difficulty is another major factor.

"Students consumed by work and travel find the final step of quitting college an easy one—a simple schedule adjustment," concludes McIntire.

The study specified students were at high risk if they:

- Worked more than 21 hours a week.
- Paid more than 30% of their own expenses.
- Commuted eight minutes or more from home to campus.
- Spent less than two hours a week socializing on campus.
- Commuted more than 13 minutes to work from home.
- Had fewer than two friends on campus.

McIntire says campus jobs and affordable housing might help students stay in school.

Source: Copyright 1993, USA TODAY. Reprinted with permission.

Researchers use statistics to identify characteristics of certain types of individuals. In this case, the characteristics of college dropouts have been identified. Once these characteristics are known, a potential dropout can be identified and proper counseling techniques can be used to help the individual stay in school. Speculate about how a study of this type could be conducted.

OTHER TYPES OF SAMPLING TECHNIQUES

In addition to the four basic sampling methods, other methods are sometimes used. In **sequence sampling,** which is used in quality control, successive units taken from production lines are sampled to ensure that the products meet certain standards set by the manufacturing company.

In **double sampling,** a very large population is given a questionnaire to determine those who meet the qualifications for a study. After the questionnaires are reviewed, a second smaller population is defined, and then a sample is selected from this group.

In **multistage sampling,** the researcher uses a combination of sampling methods. For example, suppose a research organization wants to conduct a nationwide survey for a new product being manufactured. A sample can be obtained by using the following combination of methods. First, the researchers divide the states into four or five regions (or clusters). Then, several states from each region are selected at random. Next, the states are divided into various areas by using large cities and small towns. A sample of these areas are then selected. Next, each city and town is divided into districts or wards. Finally, streets in these wards are selected at random, and the families living on these streets are given samples of the product to test and asked to report the results. This hypothetical example illustrates a typical multistage sampling method.

The steps for conducting a sample survey are given in Procedure Table 19.

Procedure Table 19

Conducting a Sample Survey

Step 1: Decide what information is needed.

Step 2: Determine how the data will be collected (phone interview, mail survey, etc.).

Step 3: Select the information-gathering instrument or design the questionnaire if one is not available.

Step 4: Set up a sampling list, if possible.

Step 5: Select the best method for obtaining the sample (random, systematic, stratified, cluster, or other).

Step 6: Conduct the survey and collect the data.

Step 7: Tabulate the data.

Step 8: Conduct the statistical analysis.

Step 9: Report the results.

EXERCISES

15–1. Name the four basic sampling techniques.

15–2. Why are samples used in statistics?

15–3. What is the basic requirement for a sample?

15–4. Why should random numbers be used when one is selecting a random sample?

15–5. List three incorrect ways that are commonly used to obtain a sample.

15–6. What is the principle behind random numbers?

15–7. List the advantages and disadvantages of random sampling.

15–8. List the advantages and disadvantages of systematic sampling.

15–9. List the advantages and disadvantages of stratified sampling.

15–10. List the advantages and disadvantages of cluster sampling.

Using the student survey at Utopia University, shown in Figure 15–8, as the population, complete Exercises 15–11 through 15–15.

FIGURE 15–8

Student Survey at Utopia University for Exercises 15–11 Through 15–15

Student Number	Gender	Class Rank	GPA	Miles Traveled to School	IQ	Major Field	Student Number	Gender	Class Rank	GPA	Miles Traveled to School	IQ	Major Field
1	M	Fr	1.4	1	104	Bio	26	M	Fr	1.1	8	100	Ed
2	M	Fr	2.3	2	95	Ed	27	F	Jr	2.1	3	101	Bus
3	M	So	2.7	6	108	Psy	28	M	Gr	3.7	5	99	Bio
4	F	So	3.2	7	119	Eng	29	M	Se	2.4	8	105	Eng
5	F	Gr	3.8	12	114	Ed	30	M	So	2.1	15	108	Bus
6	M	Jr	4.0	13	91	Psy	31	M	Gr	3.9	2	112	Ed
7	F	Jr	3.0	2	106	Eng	32	F	Jr	2.4	4	111	Psy
8	M	Jr	3.3	6	100	Bio	33	M	Se	2.7	6	107	Eng
9	F	Se	2.7	9	102	Eng	34	F	So	2.5	1	104	Bio
10	F	So	2.3	5	99	Ed	35	M	Se	3.2	3	96	Bus
11	M	Se	1.6	18	100	Bus	36	F	Fr	3.4	7	98	Bio
12	M	Gr	3.2	7	105	Psy	37	M	Gr	3.6	14	105	Ed
13	F	Gr	3.8	3	103	Bus	38	M	Jr	3.8	4	115	Psy
14	F	Se	3.1	5	97	Eng	39	F	Se	2.2	8	113	Eng
15	F	Jr	2.7	5	106	Bio	40	F	So	2.0	8	103	Psy
16	F	Fr	1.4	4	114	Bus	41	F	Fr	2.3	9	103	Eng
17	M	So	3.6	17	102	Ed	42	F	Se	2.5	10	99	Bus
18	M	Fr	2.2	1	101	Psy	43	M	Gr	3.7	13	114	Ed
19	F	Gr	4.0	7	108	Bus	44	M	Fr	3.0	11	121	Bus
20	M	Jr	2.1	4	97	Ed	45	M	Jr	2.1	10	101	Eng
21	F	Fr	2.0	3	113	Bio	46	F	Jr	3.4	2	104	Ed
22	F	So	3.6	4	104	Bio	47	M	So	3.6	9	105	Psy
23	F	Gr	3.3	16	110	Eng	48	M	Se	2.1	1	97	Psy
24	F	Se	2.5	4	99	Psy	49	F	Gr	3.3	12	111	Bio
25	M	So	3.0	5	96	Psy	50	F	Fr	2.2	11	102	Bio

15–11. Using the table of random numbers in Figure 15–1, select ten students and find the sample mean (average) of the GPA, IQ, and distance traveled to school. Compare these sample means with the population means.

15–12. Select a sample of ten students by the systematic method and compute the sample means of the GPA, IQ, and distance traveled to school of this sample. Compare these sample means with the population means.

15–13. Select a cluster of ten students, for example, students 9 through 18, and compute the sample means of the GPA, IQ, and distance traveled to school for this sample. Compare these sample means with the population means.

15–14. Divide the 50 students into subgroups according to class rank. Then, select a sample of 2 students from each rank and compute the mean for the GPA, IQ, and distance traveled to school each day. Compare these sample means with the population means.

15–15. In your opinion, which sampling method(s) provided the best sample to represent the population?

"O.K. You ask me, then I'll ask you."

Source: © 1988, *Scouting Magazine.* Reprinted by permission of Orlando Busino.

Figure 15–9 shows the 50 states and the number of electoral votes each state has in the presidential election. Using this listing as a population, complete Exercises 15–16 through 15–19.

FIGURE 15–9
States and Number of Electoral Votes for Each
(Exercises 15–16 Through 15–19)

1.	Alabama	9	26.	Montana	4
2.	Alaska	3	27.	Nebraska	5
3.	Arizona	7	28.	Nevada	4
4.	Arkansas	6	29.	New Hampshire	4
5.	California	47	30.	New Jersey	16
6.	Colorado	8	31.	New Mexico	5
7.	Connecticut	8	32.	New York	36
8.	Delaware	3	33.	North Carolina	13
9.	Florida	21	34.	North Dakota	3
10.	Georgia	12	35.	Ohio	23
11.	Hawaii	4	36.	Oklahoma	8
12.	Idaho	4	37.	Oregon	7
13.	Illinois	24	38.	Pennsylvania	25
14.	Indiana	12	39.	Rhode Island	4
15.	Iowa	8	40.	South Carolina	8
16.	Kansas	7	41.	South Dakota	3
17.	Kentucky	9	42.	Tennessee	11
18.	Louisiana	10	43.	Texas	29
19.	Maine	4	44.	Utah	5
20.	Maryland	10	45.	Vermont	3
21.	Massachusetts	13	46.	Virginia	12
22.	Michigan	20	47.	Washington	10
23.	Minnesota	10	48.	West Virginia	6
24.	Mississippi	7	49.	Wisconsin	11
25.	Missouri	11	50.	Wyoming	3

15–16. Select a random sample of ten states and find the mean number of electoral votes for this sample. Compare this mean with the population mean.

15–17. Select a systematic sample of ten states and compute the mean number of electoral votes for the sample. Compare this mean with the population mean.

15–18. Divide the 50 states into 5 subgroups by geographic location, using a map of the United States. Each subgroup should include 10 states. The subgroups should be northeast, southeast, central, northwest, and southwest. Select 2 states from each subgroup and find the mean number of electoral votes for the sample. Compare this mean with the population mean.

15–19. Select a cluster of ten states and compute the mean number of electoral votes for the sample. Compare this mean with the population mean.

*15–20. Many research studies described in newspapers and magazines do not report the sample size or the sampling method used. Try to find a research article that gives this information, and state the sampling method that was used and the sample size.

"Now think carefully. The answer you give will represent the opinion of millions of Americans."

Source: Reprinted from *The Saturday Evening Post,* © 1987 BFL&MS, Inc., Indianapolis.

15–3 WRITING THE RESEARCH REPORT

After conducting a statistical study, a researcher must write a final report explaining how the study was conducted and giving the results of the study. The format of research reports, theses, and dissertations vary from school to school; however, they tend to follow the general format explained here.

- *Front materials:* The front materials typically include the following items:

    Title page
    Copyright page
    Acknowledgments
    Table of contents
    Table of appendices
    List of tables
    List of figures

- *Chapter 1: Nature and background of the study.* This chapter should introduce the reader to the nature of the study and present some discussion on the background of the study. It should contain the following information:

    Introduction
    Statement of the problem
    Background of the problem
    Rationale for the study
    Research questions and/or hypotheses
    Assumptions, limitations, and delimitations
    Definitions of terms

- *Chapter 2: Review of literature.* This chapter should explain what has been done in previous research related to the study. It should contain the following information:

    Prior research
    Related literature

- *Chapter 3: Methodology.* This chapter should explain how the study was conducted. It should contain the following information:

    Development of questionnaires, tests, survey instruments, etc.
    Definition of the population
    Sampling methods used
    How the data was collected
    Research design used
    Statistical tests that will be used to analyze the data

- *Chapter 4: Analysis of data.* This chapter should explain the results of the statistical analysis of the data. It should state whether or not the null hypothesis should be rejected. Any statistical tables used to analyze the data should be included here.

- *Chapter 5: Summary, conclusions, and recommendations.* This chapter summarizes the results of the study and explains any conclusions that have resulted from the statistical analysis of the data. The researchers should cite and explain any shortcomings of the study. Recommendations obtained from the study should be included here, and recommendations for further studies should be suggested.

15–4 SIMULATION TECHNIQUES

Many real-life problems can be solved by employing simulation techniques.

A simulation technique uses a probability experiment to mimic a real-life situation.

Instead of studying the actual situation, which might be too costly, too dangerous, or too time-consuming, scientists and researchers create a similar situation but one that is less expensive, less dangerous, or less time-consuming. For example, NASA uses space shuttle flight simulators so that its astronauts can practice flying the shuttle. Most video games use the computer to simulate real-life sports such as boxing, wrestling, and baseball and hockey games.

Simulation techniques go back to ancient times when the game of chess was invented to simulate warfare. Modern techniques date back to the mid 1940s when two physicists, John Von Neuman and Stanislaw Ulam, developed simulation techniques to study the behavior of neutrons in the design of atomic reactors.

Mathematical simulation techniques use probability and random numbers to create conditions similar to those of real-life problems. Also, computers have played an important role in simulation techniques, since they can generate random numbers, perform experiments, tally the outcomes, and compute the probabilities much faster than human beings. The basic simulation technique is called the Monte Carlo method. This topic is discussed in the next section.

15–5 THE MONTE CARLO METHOD

The **Monte Carlo method** is a simulation technique using random numbers. The steps for simulating real-life experiments in the Monte Carlo method follow.

1. List all possible outcomes of the experiment.
2. Determine the probability of each outcome.
3. Set up a correspondence between the outcomes of the experiment and the random numbers.
4. Select random numbers from a table, and conduct the experiment.
5. Repeat the experiment, and tally the outcomes.
6. Compute any statistics, and state the conclusions.

Before examples of the complete simulation technique are given, an explanation is needed for Step 3, "Set up a correspondence between the outcomes of the experiment and the random numbers." Tossing a coin, for instance, can be simulated by using random numbers as follows: Since there are only two outcomes, heads and tails, and since each outcome has a probability of $\frac{1}{2}$, the odd digits, 1, 3, 5, 7, and 9, can be used to represent a head, and the even digits, 0, 2, 4, 6, and 8, can be used to represent a tail.

Suppose a random number, 8631, is selected. This number represents four tosses of a single coin and the results T, T, H, H. This number could also represent one toss of four coins with the same results.

An experiment of rolling a single die can also be simulated by using random numbers. In this case, the digits 1, 2, 3, 4, 5, and 6 can represent the number of spots that appear on the face of the die. The digits 7, 8, 9, and 0 are ignored.

FIGURE 15–10
Spinner with Four
Numbers

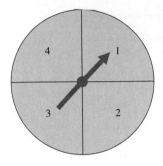

When two dice are rolled, two random digits are needed. For example, the number 26 represents a 2 on the first die and a 6 on the second die. The random number 37 represents a 3 on the first die, but the 7 cannot be used, so another digit must be selected. As another example, a three-digit daily lotto number can be simulated by using three-digit random numbers. Finally, a spinner with four numbers, as shown in Figure 15–10, can be simulated by letting the random numbers 1 and 2 represent 1 on the spinner, 3 and 4 represent 2 on the spinner, 5 and 6 represent 3 on the spinner, and 7 and 8 represent 4 on the spinner, since each number has a probability of 1 out of 4 of being selected. The random numbers 9 and 0 are ignored in this situation.

Many real-life games, such as bowling and baseball, can be simulated by using random numbers, as shown in Figure 15–11.

FIGURE 15–11
Example of Simulation
of a Game

Simulated Bowling Game

Let's use the random digit table to simulate a bowling game. Our game is much simpler than commercial simulation games.

First Ball		Second Ball			
		2-Pin Split		No Split	
Digit	Results	Digit	Results	Digit	Results
1–3	Strike	1	Spare	1–3	Spare
4–5	2-pin split	2–8	Leave one pin	4–6	Leave 1 pin
6–7	9 pins down	9–0	Miss both pins	7–8	*Leave 2 pins
8	8 pins down			9	+Leave 3 pins
9	7 pins down			0	Leave all pins
0	6 pins down				

*If there are fewer than 2 pins, result is a spare.
+If there are fewer than 3 pins, those pins are left.

Here's how to score bowling:
1. There are 10 frames to a **game** or **line**.
2. You roll two balls for each frame, unless you knock all the pins down with the first ball (a **strike**).
3. Your score for a frame is the sum of the pins knocked down by the two balls, if you don't knock down all 10.
4. If you knock all 10 pins down with two balls (a **spare**, shown as ▱), your score is 10 pins plus the number knocked down with the next ball.
5. If you knock all 10 pins down with the first ball (a **strike**, shown as ▨), your score is 10 pins plus the number knocked down with the next **two** balls.
6. A **split** (shown as 0) is when there is a big space between the remaining pins. Place in the circle the number of pins remaining after the second ball.
7. A **miss** is shown as —.

Here is how one person simulated a bowling game using the random digits 7 2 7 4 8 2 2 3 6 1 6 0 4 6 1 5 5, chosen in that order from the table.

Frame

	1	2	3	4	5	6	7	8	9	10
Digit(s)	7/2	7/4	8/2	2	3	6/1	6/0	4/6	1	5/5
Bowling result	9 ▱ 19	9 — 28	8 ▱ 48	▨ 77	▨ 97	9 ▱ 116	9 — 125	8 ① 134	▨ 153	8 ① 162	162

Now you try several.

Frame

	1	2	3	4	5	6	7	8	9	10
Digit(s)										
Bowling result										

	1	2	3	4	5	6	7	8	9	10
Digit(s)										
Bowling result										

If you wish to, you can change the probabilities in the simulation to better reflect *your* actual bowling ability.

EXAMPLE 15-4   Using random numbers, simulate the gender of children born.

Solution   There are only two possibilities, male and female. Since the probability of each outcome is 0.5, the odd digits can be used to represent the male births, and the even digits can be used to represent the female births.   ◄

EXAMPLE 15-5   Using random numbers, simulate the outcomes of a tennis game between two people, Bill and Mike, with the additional condition that Bill is twice as good as Mike.

Solution   Since Bill is twice as good as Mike, he will win approximately two games for every one Mike wins; hence, the probability that Bill wins will be $\frac{2}{3}$ and the probability that Mike wins will be $\frac{1}{3}$. The random digits 1 through 6 can be used to represent a game Bill wins; the random digits 7, 8, and 9 can be used to represent Mike's wins. The digit 0 is disregarded. Suppose they play five games, and the random number 86314 is selected. This number means that Bill won games 2, 3, 4, and 5, and Mike won the first game. The sequence is

8   6   3   1   4
M   B   B   B   B   ◄

More complex problems can be solved by using random numbers, as shown in the next three examples.

EXAMPLE 15-6   A die is rolled until a 6 appears. Using simulation, find the average number of rolls needed. Try the experiment 20 times.

Solution   **STEP 1**   List all possible outcomes. They are 1, 2, 3, 4, 5, 6.

STEP 2   Assign the probabilities. Each outcome has a probability of $\frac{1}{6}$.

STEP 3   Set up a correspondence between the random numbers and the outcome. Random numbers 1 through 6 will be used. The numbers 7, 8, 9, and 0 will be omitted.

STEP 4   Select a block of random numbers and count each digit 1 through 6 until the first 6 is obtained. For example, the block 85723649 means that it takes 4 rolls to get a 6.

8   5   7   2   3   6   4   9
   ↑   ↑   ↑   ↑
   5   2   3   6

STEP 5 Repeat the experiment 19 more times, and tally the data as shown.

Trial	Random Number	Number of Rolls
1	8 5 7 2 3 6	4
2	2 1 0 4 8 0 1 5 1 1 0 1 5 3 6	11
3	2 3 3 6	4
4	2 4 1 3 0 4 8 3 6	7
5	4 2 1 6	4
6	3 7 5 2 0 3 9 8 7 5 8 1 8 3 7 1 6	9
7	7 7 9 2 1 0 6	3
8	9 9 5 6	2
9	9 6	1
10	8 9 5 7 9 1 4 3 4 2 6	7
11	8 5 4 7 5 3 6	5
12	2 8 9 1 8 6	3
13	6	1
14	0 9 4 2 9 9 3 9 6	4
15	1 0 3 6	3
16	0 7 1 1 9 9 7 3 3 6	5
17	5 1 0 8 5 1 2 7 6	6
18	0 2 3 6	3
19	0 1 0 1 1 5 4 0 9 2 3 3 3 6	10
20	5 2 1 6	4
	Total	96

STEP 6 Compute the results and draw a conclusion. In this case, one must find the average

$$\overline{X} = \frac{\Sigma X}{n} = \frac{96}{20} = 4.8$$

Hence, the average is about 5 rolls.

Note: The theoretical average obtained from the expected value formula is 6. If this experiment is done a large number of times, say 1000 times, the results should be closer to the theoretical results. ◄

EXAMPLE 15–7 A person selects a key at random from four keys to open a lock. Only one key fits. If the first key does not fit, the person tries another key; the person keeps trying other keys until one key fits. Find the average of the number of keys a person will have to try to open the lock. Try the experiment 25 times.

Solution Assume that each key is numbered from 1 through 4 and that key 2 fits the lock. Naturally, the person doesn't know this, so she selects the keys at random. For the simulation, one must select a sequence of random digits using only 1 through 4 until the digit 2 is reached. The trials are shown here.

Trial	Random Digit (Key)	Number	Trial	Random Digit (Key)	Number
1	2	1	14	2	1
2	2	1	15	4 2	2
3	1 2	2	16	1 3 2	3
4	1 4 3 2	4	17	1 2	2
5	3 2	2	18	2	1
6	3 1 4 2	4	19	3 4 2	3
7	4 2	2	20	2	1
8	4 3 2	3	21	2	1
9	4 2	2	22	2	1
10	2	1	23	4 2	2
11	4 2	2	24	4 3 1 2	4
12	3 1 2	3	25	3 1 2	3
13	3 1 2	3		Total	54

Next, find the average:

$$\overline{X} = \frac{\Sigma X}{n} = \frac{1 + 1 + \cdots + 3}{25} = \frac{54}{25} = 2.16$$

The theoretical average is 2.5. Again, only 25 repetitions were used; more repetitions should give a result closer to the theoretical average. ◄

EXAMPLE 15-8 A box contains five $1 bills, three $5 bills, and two $10 bills. A person selects a bill at random. What is the expected value of the bill? Perform the experiment 25 times.

Solution **STEP 1** List all possible outcomes. They are $1, $5, and $10.

STEP 2 Assign probabilities to each outcome:

$$P(\$1) = \tfrac{5}{10} \qquad P(\$5) = \tfrac{3}{10} \qquad P(\$10) = \tfrac{2}{10}$$

STEP 3 Set up a correspondence between the random numbers and the outcomes. Use the random numbers 1 through 5 to represent a $1 bill being selected, 6 through 8 to represent a $5 bill being selected, and 9 and 0 to represent a $10 bill being selected.

STEPS 4 and 5 Select 25 random numbers and tally the results.

Number	Results
4 5 8 2 9	$1, $1, $5, $1, $10
2 5 6 4 6	$1, $1, $5, $1, $5
9 1 8 0 3	$10, $1, $5, $10, $1
8 4 0 6 0	$5, $1, $10, $5, $10
9 6 9 4 3	$10, $5, $10, $1, $1

STEP 6 Compute the average:

$$\overline{X} = \frac{\Sigma X}{n} = \frac{\$1 + \$1 + \$5 + \cdots + \$1}{25} = \frac{116}{25} = \$4.64$$

Hence, the average (expected value) is $4.64. ◄

Recall that using the expected value formula, $E(X) = \Sigma\, XP(X)$, gives a theoretical average of

$$E(X) = (0.5)(\$1) + (0.3)(\$5) + (0.2)(\$10) = \$4.00$$

Remember that simulation techniques do not give exact results. The more times the experiment is performed, though, the closer the actual results should be to the theoretical results.

The steps for solving problems using the Monte Carlo method are summarized in Procedure Table 20.

Procedure Table 20

Simulating Experiments Using the Monte Carlo Method

Step 1: List all possible outcomes of the experiment.
Step 2: Determine the probability of each outcome.
Step 3: Set up a correspondence between the outcomes of the experiment and the random numbers.
Step 4: Select random numbers from a table, and conduct the experiment.
Step 5: Repeat the experiment, and tally the outcomes.
Step 6: Compute any statistics and state the conclusions.

EXERCISES

15-21. Define *simulation techniques*.

15-22. Give three examples of simulation techniques.

15-23. Who is responsible for the development of modern-day simulation techniques?

15-24. What role does the computer play in simulation?

15-25. What are the steps in the simulation of an experiment?

15-26. What purpose do random numbers play in simulation?

15-27. What happens when the number of repetitions is increased?

For Exercises 15-28 through 15-33, explain how each experiment can be simulated by using random numbers.

15-28. A spinner contains five equal areas.

15-29. A basketball player makes 60% of her shots.

15-30. A certain brand of batteries manufactured has a 10% defective rate.

15-31. A shooter hits the target 80% of the time.

15-32. Two players match pennies.

15-33. Three players play odd man out. (Three coins are tossed; if all three match, the game is repeated and no one wins. If two players match, the third person wins all three coins.)

For Exercises 15-34 through 15-41, use random numbers to simulate the experiments. The number in parentheses is the number of times the experiment is to be repeated.

15-34. A coin is tossed until four heads are obtained. Find the average number of tosses necessary. (50)

15-35. A die is rolled until all faces appear at least once. Find the average number of tosses. (30)

15-36. A cereal company gives five different prizes, one in each box. They are placed in the boxes at random. Find the average number of boxes a person needs to buy to get all five prizes. (50)

15–37. Two teams are evenly matched. They play a tournament in which the first team to win three games wins the tournament. Find the average number of games the tournament will last. (20)

15–38. To win a certain lotto, a person must spell the word *big*. Sixty percent of the tickets contain the letter b, 30% contain the letter i, and 10% contain the letter g. Find the average number of tickets a person must buy to win the prize. (30)

15–39. Two shooters shoot clay pigeons. Gail has an 80% accuracy rate, and Paul has a 60% accuracy rate. Paul shoots first. The first person who hits the target wins. Find the probability that each wins. (30)

15–40. In Exercise 15–39, find the average number of shots fired. (30)

15–41. A basketball player has a 60% success rate for shooting foul shots. If she gets two shots, find the probability that she will make one or both shots. (50)

*15–42. Select a game such as baseball or football and write a simulation using random numbers.

*15–43. Explain how cards can be used to generate random numbers.

*15–44. Explain how a pair of dice can be used to generate random numbers.

15–6 SUMMARY

In order to obtain information from a large population, researchers select a sample. A sample is a subgroup of the population. Using a sample rather than a population, researchers can save time and money, get more detailed information, and get information that otherwise would be impossible to obtain.

There are four common methods researchers use to obtain samples: random, systematic, stratified, and cluster sampling methods. In random sampling, some type of random method (usually random numbers) is used to obtain the sample. In systematic sampling, the researcher selects every kth person after selecting the first person at random. In stratified sampling, the population is divided into subgroups according to various characteristics, and elements are then selected at random from the subgroups. In cluster sampling, the researcher selects an intact group to use as a sample. When the population is large, multistage sampling (a combination of methods) is used to obtain a subgroup of the population.

At the completion of a statistical study, researchers prepare a research report detailing how the study was conducted and giving the results of the study.

Most sampling methods use random numbers. In addition to being used in sampling, random numbers can be used to simulate many real-life problems or situations. The basic method of simulation is known as the Monte Carlo method. The purpose of simulation is to duplicate situations that are too dangerous, too costly, or too time-consuming to study in real life. Most simulation techniques can be done on the computer, since it can rapidly generate random numbers, count the outcomes, and perform the necessary computations.

Sampling and simulation are two techniques that enable researchers to gain information that might otherwise be unobtainable.

Important Terms

Biased sample	Multistage sampling	Stratified sample
Cluster sample	Random sample	Systematic sample
Double sampling	Sequence sampling	Unbiased sample
Monte Carlo method	Simulation technique	

Review Exercises

Use Figure 15–12 for Exercises 15–45 through 15–52.

FIGURE 15–12
Population for Exercises 15–45 Through 15–52

Individual	Gender	Weight	Systolic Blood Pressure	Individual	Gender	Weight	Systolic Blood Pressure	Individual	Gender	Weight	Systolic Blood Pressure
1	F	122	132	18	F	118	125	35	M	172	116
2	F	128	116	19	F	107	138	36	M	175	123
3	M	183	140	20	M	214	121	37	F	101	114
4	M	165	136	21	F	114	127	38	F	123	113
5	M	192	120	22	M	119	125	39	M	186	145
6	F	116	118	23	F	125	114	40	F	100	119
7	M	206	116	24	M	182	137	41	M	202	135
8	F	131	120	25	F	127	127	42	F	117	121
9	M	155	118	26	F	132	130	43	F	120	130
10	F	106	122	27	M	198	114	44	M	193	125
11	F	103	119	28	F	135	119	45	M	200	115
12	M	169	136	29	M	183	137	46	F	118	132
13	M	173	134	30	F	140	123	47	F	121	143
14	M	195	145	31	M	189	135	48	M	189	128
15	F	107	113	32	M	165	121	49	M	114	118
16	M	201	111	33	M	211	117	50	M	174	138
17	F	114	141	34	F	111	127				

15–45. Select a random sample of ten people and find the mean of the weights of the individuals. Compare this mean with the population mean.

15–46. Select a systematic sample of ten people and compute the mean of the weights of the individuals. Compare this mean with the population mean.

15–47. Divide the individuals into subgroups of males and females. Select five individuals from each group and find the mean of the weights of the individuals. Compare this mean with the population mean.

15–48. Select a cluster of ten people and find the mean of the weights of the individuals. Compare this mean with the population mean.

15–49. Repeat Exercise 15–45 for blood pressure.

15–50. Repeat Exercise 15–46 for blood pressure.

15–51. Repeat Exercise 15–47 for blood pressure.

15–52. Repeat Exercise 15–48 for blood pressure.

For Exercises 15–53 through 15–57, explain how each experiment can be simulated by using random numbers.

15–53. A quarterback completes 40% of his passes.

15–54. An airline overbooks 5% of its flights.

15–55. Two players roll a die, and the one who rolls the higher number wins.

15–56. One player rolls two dice, and one player rolls one die. The lowest number wins.

15–57. Two players play rock, paper, scissors. The rules are as follows:

Paper covers rock. Paper wins.
Rock breaks scissors. Rock wins.
Scissors cuts paper. Scissors wins.

Each person selects either rock, paper, or scissors by random numbers and then compares results.

For Exercises 15–58 through 15–62, use random numbers to simulate the experiments. The number in parentheses is the number of times the experiment is to be repeated.

15–58. A football is placed on the 10-yard line and a team has four downs to score a touchdown. The team can move the ball only 0 to 5 yards per play. Find the average number of times the team will score a touchdown. (30)

15–59. In Exercise 15–58, find the average number of plays it will take to score a touchdown. Ignore the four-downs rule and keep playing until a touchdown is scored. (30)

15–60. Three dice are rolled. Find the average of the sum of the number of spots that will appear. (50)

15–61. A field goal kicker is successful in 70% of his kicks inside the 30-yard line. Find the probability of kicking four field goals in a row. (50)

15–62. A sales representative finds that there is a 20% probability of making a sale over the telephone. For every ten calls, find the probability of making two sales in a row. (40)

COMPUTER APPLICATIONS

1. Write a computer program that will generate and print random numbers. Use n for the number of random numbers desired and p for the number of digits each random number should contain. For example, generate 50 three-digit numbers.

2. MINITAB can be used to generate random numbers. To generate 30 four-digit numbers by using MINITAB, type the following commands:

```
MTB  > RANDOM 30 OBSERVATIONS INTO C1;
SUBC > INTEGERS 0000 TO 9999.
MTB  > PRINT C1
```

The computer will print the following:

```
C1
7957   5499   3778     41   2828   4630   4084   6786   7319   9431   4834
9550   8009   3822   7935   5349   8911   1119   9804   7731   9991   6845
2496   5674   5836   5186   6697   3470   6632   7541
```

The random number 41 would be 0041. To get the computer to print three-digit numbers, type 000 to 999.

Have the computer print 100 three-digit random numbers by using MINITAB.

3. MINITAB can be used to select a random sample of a specific size from a finite population. Sampling is done without replacement. First, enter the population data into C1, and then sample and place the sample values into C2. Finally, print the sample data. Using the results of the student survey at Utopia University in Figure 15–8, enter the IQ values of all 50 students, and then select one sample of 10 students. Compute the mean of the sample, and compare this mean with the population mean.

This problem can be done as follows:

```
MTB  >
MTB  > SET C1
DATA > 104 95 108 119 114 91 106 100 102 99 100
DATA > 105 103 97 106 114 102 101 108 97 113
DATA > 104 110 99 96 100 101 99 105 108 112
DATA > 111 107 104 96 98 105 115 113 103 103
DATA > 99 114 121 101 104 105 97 111 102
DATA > END
MTB  > SAMPLE 10 OBSERVATIONS FROM C1 AND PUT INTO C2
MTB  > PRINT C2
```

The computer will print 30 four-digit random numbers, as shown.

```
C2
    96 98 96 100 103 103 91 110 108 104
```

Repeat this experiment four additional times. For repetitions, you only need to retype the last two lines.

√ DATA ANALYSIS Applying the Concepts

The Data Bank is located in the Appendix.

1. From the Data Bank, choose a variable. Select a random sample of 20 individuals and find the mean of the data.
2. Select a systematic sample of 20 individuals, and using the same variable as in Exercise 1, find the mean.
3. Select a cluster sample of 20 individuals, and using the same variable as in Exercise 1, find the mean of the data.
4. Stratify the data according to marital status and gender, and sample 20 individuals. Compute the mean of the same variable selected in Exercise 1.
5. Compare all four means and decide which one is most appropriate. (*Hint:* Find the population mean.)

✎ TEST

Directions: Determine whether the statement is true or false.

1. When all subjects under study are used, the group is called a sample.
2. When a population is divided into subgroups, with each group having similar characteristics, and then a sample of individuals is selected from each group, the sample was obtained by using the systematic sampling method.
3. In general, when one conducts sampling, the larger the sample, the more representative it will be.
4. When samples are not representative, they are said to be unbiased.
5. When researchers are sampling from large populations, such as adult citizens living in the United States, they use a combination of sampling techniques to ensure representativeness.
6. Interviewing selected people at a local supermarket can be considered an example of random sampling.
7. Random samples are said to be unbiased.
8. When all residents of a street are interviewed for a survey, the sample is said to be stratified.
9. Simulation techniques using random numbers is actually a substitute for performing the actual statistical experiment.
10. When researchers perform simulation experiments, they do not need to use a random number table, since they can make up random numbers.

CHAPTER

16

Quality Control

16–1 INTRODUCTION

After the Industrial Revolution, companies began manufacturing products using assembly lines in factories. Rather than one person's being responsible for a single product from start to finish, many workers helped to complete the product on an assembly line. Each worker was responsible for a single part of the product. As products became more complex, the workers were trained only to perform a specific task in the manufacturing process. For example, in the production of an electric motor, one person would be responsible for winding the armature, another person would be responsible for casting the gears, a third person would be responsible for assembling the motor, etc.

In order to ensure that the manufactured product performed satisfactorily, companies trained some people to inspect the finished product. When poor performance was detected, it was necessary to find the cause and correct it.

During the 1930s and 1940s, Walter A. Shewhart, working for Bell Telephone Laboratories, developed a process for checking each part of the manufacturing process and making any corrections needed before the product was assembled rather than checking the final product, thus saving time and money. This process is known as **quality control,** and it uses small samples and a **control chart,** sometimes called a *statistical process control chart* or a *quality control chart*. The control chart uses means, medians, ranges, or standard deviations of certain measurements of various parts of the product to detect deviations from the norm.

Understanding the quality control process means understanding the variation that can exist when a product is manufactured. For example, when a type of rope is manufactured, the manufacturer might wish the rope to have a breaking strength of, say, 50 pounds. Because it is impossible to manufacture ropes that are exactly the same, there will be a variation in the breaking strength of each rope produced. The manufacturer will then set *limits*—a maximum and a minimum acceptable value—on the breaking strength for the rope. These limits are called the **upper** and **lower control limits.** If the breaking strength is much lower than the desired 50 pounds, the rope cannot be sold and must be destroyed. If the breaking strength is too great, then perhaps too much fiber is being used, thus costing the manufacturer money.

Since it is impossible to test every rope produced, the manufacturer will select several samples and compute the means for the breaking strengths of the ropes. These sample means will be plotted on a graph, and the graph will be analyzed. If one or more means exceeds the upper or lower control limits, the manufacturing process is said to be *out of control*. It is then up to the person in charge to find the problem and correct it. Once the problem is corrected, the process is said to be *in control*.

A control chart for means can be constructed with the following steps.

STEP 1 Several samples of the product or part of the product are selected and tested, and the mean of the attribute being measured for each sample is computed. In the rope example, the means of the samples for the breaking strength of the ropes were computed.

STEP 2 Depending on the type of control chart, a measure of variation such as the range or standard deviation is computed.

STEP 3 From these measures, an upper and a lower control limit is computed. Lines representing these values, along with a central line representing the grand mean ($\overline{X}_{GM}$), are then drawn on a graph.

STEP 4   The means of the samples are plotted on a control chart.

STEP 5   The control chart is analyzed to see whether the manufacturing process is in control or out of control.

STEP 6   If the process is out of control, a probable cause is determined, and an attempt is made to correct the problem.

STEP 7   The procedure is repeated.

Figure 16–1 shows a control chart for ten sample means. Since all the sample means fall within the upper and lower control limits, abbreviated $UCL_{\overline{X}}$ and $LCL_{\overline{X}}$, the manufacturing process is said to be in control.

FIGURE 16–1
Control Chart for a
Process That Is in
Control

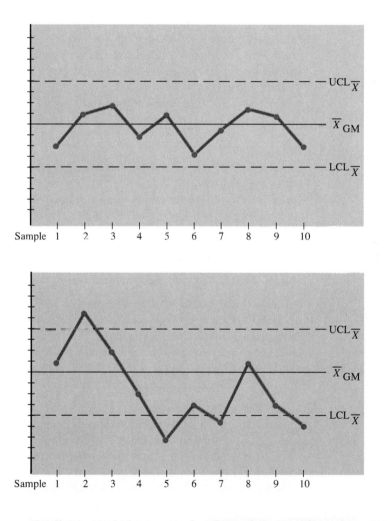

FIGURE 16–2
Control Chart for a
Process That Is Out of
Control

Quality control charts can also show when manufacturing processes are out of control. Figure 16–2 shows a process where several means exceed the upper and lower control limits. When this situation occurs, the process is out of control and should be stopped. The problem should be located and corrected. Then another check should be made. In addition to checking to see whether the sample means fall outside the control limits, an inspection should also look for other unusual situations, such as increasing or decreasing trends in the means. That is, are the means getting consistently larger or smaller in a nonrandom fashion? Other situations that might indicate problems in the manufacturing process are too many means above or below the centerline or too many means near the upper or lower control limits.

Before the method for computing the upper and lower control limits can be explained, the types of variations and the types of control charts must be discussed.

There are two types of variation that can occur in manufacturing processes. The first type is called *chance* or *random variation*. This variation is due to chance causes, such as temperature, dust, humidity, or vibration of machinery. For the most part, this variation occurs without plan, purpose, or pattern and cannot be eliminated entirely. The other type of variation is called *assignable-cause variation*. This variation is not random and is due to defects or problems in the manufacturing process, such as cutting tools becoming dull, operators using the machine incorrectly, or raw materials changing. Assignable-cause variations must be corrected in order to maintain quality in the manufacturing process. These two types of variation are formally defined next.

Chance variations in manufactured items are random and cannot be eliminated entirely.
Assignable-cause variations in manufacturing processes are nonrandom and must be eliminated to maintain the quality of the manufactured product.

There are two basic types of control charts: *variable charts* and *attribute charts*.

Variable charts are used to analyze the quality of a manufacturing process in which measurements, such as lengths, diameters, weights, or breaking strengths, of an item are important.
Attribute charts are used to analyze a manufacturing process in which the product can be classified as acceptable or defective.

Variable and attribute charts are used to help determine answers to questions such as these: Do the seals on a waterproof container leak? Does the key fit the lock? Does the lightbulb fit the socket? If the seals work, the key fits the lock, or the bulb fits the socket, then the items are termed *acceptable;* if not, they are classified as *defective*.

The next section discusses how variable charts are constructed.

16–2 VARIABLE CHARTS

There are several variable charts that can be constructed. The two shown in this section are the X bar chart ($\overline{X}$ chart), also called the mean chart, and the R chart, also called the range chart.

The $\overline{X}$ Chart

The $\overline{X}$ **chart** is constructed by selecting several samples of a specific size and finding the mean and range of each sample. The grand mean is then found and used as a centerline. The upper and lower control limits are found by using the average of the ranges and a value from Table O in the Appendix. Finally, the means for the samples are plotted on the chart and analyzed. The procedure is shown in the next two examples.

EXAMPLE 16-1 A manufacturer of rope selects six samples of five ropes each and tests the breaking strength of each. Construct an $\overline{X}$ chart for the data shown.

Sample	Breaking Strength (Pounds)
1	46 47 45 46 47
2	50 51 52 53 49
3	48 51 50 50 49
4	52 50 49 50 51
5	51 47 46 48 47
6	49 51 50 51 52

Solution **STEP 1** Find the mean $\overline{X}$ and range R for each sample. For sample 1,

$$\overline{X_1} = \frac{\Sigma X}{n} = \frac{46 + 47 + 45 + 46 + 47}{5} = \frac{231}{5} = 46.2$$

$$R_1 = H - L = 47 - 45 = 2$$

where H = highest value and L = lowest value. For sample 2,

$$\overline{X_2} = \frac{50 + 51 + 52 + 53 + 49}{5} = \frac{255}{5} = 51$$

$$R_2 = 53 - 49 = 4$$

For sample 3,

$$\overline{X_3} = \frac{48 + 51 + 50 + 50 + 49}{5} = \frac{248}{5} = 49.6$$

$$R_3 = 51 - 48 = 3$$

For sample 4,

$$\overline{X_4} = \frac{52 + 50 + 49 + 50 + 51}{5} = \frac{252}{5} = 50.4$$

$$R_4 = 52 - 49 = 3$$

For sample 5,

$$\overline{X_5} = \frac{51 + 47 + 46 + 48 + 47}{5} = \frac{239}{5} = 47.8$$

$$R_5 = 51 - 46 = 5$$

For sample 6,

$$\overline{X_6} = \frac{49 + 51 + 50 + 51 + 52}{5} = \frac{253}{5} = 50.6$$

$$R_6 = 52 - 49 = 3$$

STEP 2 Find the grand mean $\overline{X}_{GM}$ (the mean of the sample means) and the mean of the ranges.

$$\overline{X}_{GM} = \frac{\Sigma \overline{X}}{k} = \frac{46.2 + 51 + 49.6 + 50.4 + 47.8 + 50.6}{6} = 49.3$$

$$\overline{R} = \frac{\Sigma R}{k} = \frac{2 + 4 + 3 + 3 + 5 + 3}{6} = \frac{20}{6} = 3.33$$

where k = number of samples.

STEP 3 Compute the $UCL_{\bar{X}}$ and the $LCL_{\bar{X}}$ using these formulas:

$$UCL_{\bar{X}} = \bar{X}_{GM} + A_2 \bar{R}$$
$$LCL_{\bar{X}} = \bar{X}_{GM} - A_2 \bar{R}$$

where A_2 is a constant obtained from Table O in the Appendix. See Figure 16–3. For the chart in Figure 16–3, n is the number of items in each sample. In this case, $n = 5$; hence, $A_2 = 0.577$. Therefore,

$$UCL_{\bar{X}} = \bar{X}_{GM} + A_2\bar{R} = 49.3 + (0.577)(3.33) = 51.22$$
$$LCL_{\bar{X}} = \bar{X}_{GM} - A_2\bar{R} = 49.3 - (0.577)(3.33) = 47.38$$

FIGURE 16–3
Using Table O in the Appendix for Example 16–1

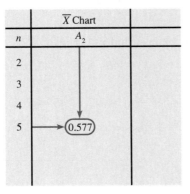

STEP 4 Draw the chart, using the value $\bar{X}_{GM}$ for the centerline, as shown in Figure 16–4, and the calculated values for $UCL_{\bar{X}}$ and $LCL_{\bar{X}}$.

STEP 5 Plot each mean $\bar{X}$ on the chart, and analyze the chart as shown in Figure 16–4. Since there is a mean beyond the control limits, the process is out of control.

FIGURE 16–4
Control Chart for Example 16–1

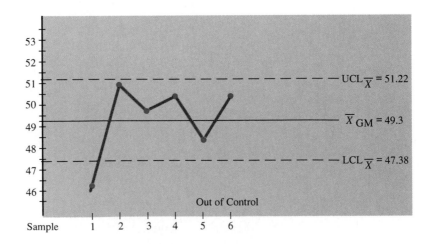

EXAMPLE 16–2   A pressure valve is designed to release at 200 pounds per square inch (psi). Ten samples of six valves each were selected and tested. The mean and range for each sample are shown below. Construct and analyze a quality control chart for the data.

Sample	1	2	3	4	5
$\overline{X}$	200.2	198.9	199.3	201.1	200.8
R	3.1	0.8	2.2	1.0	3.6
Sample	6	7	8	9	10
$\overline{X}$	202.6	201.3	203.7	205.6	206.1
R	2.3	1.4	2.2	2.3	2.4

Solution   **STEP 1**   Since the means and ranges are given, they do not need to be calculated.

STEP 2   Find the grand mean and the mean of the ranges.

$$\overline{X}_{GM} = \frac{\Sigma \overline{X}}{k} = \frac{200.2 + 198.9 + \cdots + 206.1}{10} = \frac{2019.6}{10} = 201.96$$

$$\overline{R} = \frac{\Sigma R}{k} = \frac{3.1 + 0.8 + \cdots + 2.4}{10} = \frac{21.3}{10} = 2.13$$

STEP 3   Find the $UCL_{\overline{X}}$ and the $LCL_{\overline{X}}$ by using the formulas.

$$UCL_{\overline{X}} = \overline{X}_{GM} + A_2\overline{R} \qquad \text{and} \qquad LCL_{\overline{X}} = \overline{X}_{GM} - A_2\overline{R}$$

where A_2 is the constant obtained from Table O in the Appendix and $n = 6$. For this example, $A_2 = 0.483$. Hence,

$$UCL_{\overline{X}} = 201.96 + (0.483)(2.13) = 202.99$$
$$LCL_{\overline{X}} = 201.96 - (0.483)(2.13) = 200.93$$

STEP 4   Draw the chart, with $\overline{X}_{GM}$ as the centerline and using the calculated values of $UCL_{\overline{X}}$ and $LCL_{\overline{X}}$. See Figure 16–5.

STEP 5   Plot each $\overline{X}$ on the chart, and analyze the chart, as shown in Figure 16–5. The process is out of control since there is an increasing trend. Also, three means are beyond the upper control limit, and four means are below the lower limit.

FIGURE 16–5
Control Chart for
Example 16–2

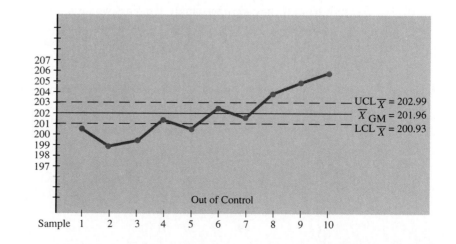

The formulas for the upper and lower control limits are given next.

Formulas for the Upper and Lower Control Limits for the $\overline{X}$ Chart

$$\text{UCL}_{\overline{X}} = \overline{X}_{\text{GM}} + A_2\overline{R} \quad \text{and} \quad \text{LCL}_{\overline{X}} = \overline{X}_{\text{GM}} - A_2\overline{R}$$

where $\overline{X}_{\text{GM}}$ = grand mean of the sample means

A_2 = value obtained from Table O in the Appendix

$\overline{R}$ = mean of the ranges obtained from the samples

The procedure for constructing the $\overline{X}$ quality control chart is given in Procedure Table 21.

Procedure Table 21

Constructing the $\overline{X}$ Chart

Step 1: Find the mean and range for each sample.

Step 2: Find the grand mean $\overline{X}_{\text{GM}}$ and the mean of the ranges $\overline{R}$.

Step 3: Compute the upper and lower control limits, using the formulas

$$\text{UCL}_{\overline{X}} = \overline{X}_{\text{GM}} + A_2\overline{R} \quad \text{and} \quad \text{LCL}_{\overline{X}} = \overline{X}_{\text{GM}} - A_2\overline{R}$$

where A_2 is obtained from Table O in the Appendix.

Step 4: Draw the chart, using the value $\overline{X}_{\text{GM}}$ for the centerline and the calculated values for $\text{UCL}_{\overline{X}}$ and $\text{LCL}_{\overline{X}}$.

Step 5: Plot the means for each sample, and analyze the chart. If any mean exceeds either $\text{UCL}_{\overline{X}}$ or $\text{LCL}_{\overline{X}}$, the process is out of control; otherwise, it is in control.

The R Chart For a product to be manufactured with quality, the ranges of the measurements must also be within specific limits. For example, if a factory were manufacturing pipe fittings, and if some of the pipe diameters were too large, the pipes would leak. If they were too small, the pipes would not fit into the fittings. The **R chart** is used to analyze the ranges of measured values to determine whether a manufacturing process is in or out of control. The next example shows how to construct the R chart.

EXAMPLE 16–3 Using the data for the breaking strengths of rope given in Example 16–1, construct an R chart. The data are repeated below.

Sample	Breaking Strength (Pounds)				
1	46	47	45	46	47
2	50	51	52	53	49
3	48	51	50	50	49
4	52	50	49	50	51
5	51	47	46	48	47
6	49	51	50	51	52

Solution **STEP 1** Find the ranges for each sample.

$$R_1 = 47 - 45 = 2 \qquad R_4 = 52 - 49 = 3$$
$$R_2 = 53 - 49 = 4 \qquad R_5 = 51 - 46 = 5$$
$$R_3 = 51 - 48 = 3 \qquad R_6 = 52 - 49 = 3$$

STEP 2 Find the mean of the ranges, $\overline{R}$.

$$\overline{R} = \frac{\Sigma R}{k} = \frac{2 + 4 + 3 + 3 + 5 + 3}{6} = \frac{20}{6} = 3.33$$

STEP 3 Find the UCL_R and LCL_R by using the following formulas:

$$\text{UCL}_R = D_4\overline{R} \qquad \text{and} \qquad \text{LCL}_R = D_3\overline{R}$$

where the values D_3 and D_4 are obtained from Table O in the Appendix. See Figure 16–6. For a sample size of $n = 5$, $D_3 = 0$ and $D_4 = 2.115$. Hence,

$$\text{UCL}_R = D_4\overline{R} = 2.115(3.33) = 7.04$$
$$\text{LCL}_R = D_3\overline{R} = 0(3.33) = 0$$

FIGURE 16–6
Using Table O in the Appendix for Example 16–3

n	$\overline{X}$ Chart A_2	R Chart D_3	D_4
2			
3			
4			
5		0	2.115
6			
7			

STEP 4 Draw the chart, using the value $\overline{R}$ as the centerline and the calculated values for UCL_R and LCL_R.

STEP 5 Plot the ranges for each sample, and analyze the chart. See Figure 16–7. The process is in control since all ranges fall within the control limits, UCL_R and LCL_R. As with the $\overline{X}$ chart, if any of the ranges fall outside the control limits, the process is out of control.

FIGURE 16–7
Control Chart for Example 16–3

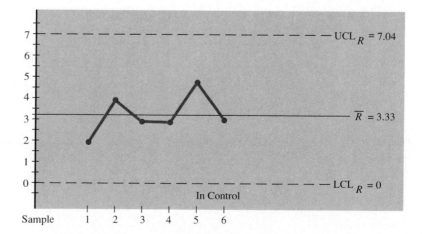

Speaking of **Statistics**

Death Rate a Valid Part of Judging Hospitals

By Leslie Miller
USA TODAY

Hospital mortality rates—often criticized as mainly a product of chance—reflect true quality differences that for some high-risk patients may affect survival, a new study suggests.

The rates, adjusted for risk and published since 1987 by the Health Care Financing Administration, are one of the few objective measures available of hospital quality.

To test them, University of California researchers Harold S. Luft and Patrick S. Romano looked at data from 132,750 patients getting one surgery—coronary artery bypass graft—from 1983 to 1989.

Results, in today's *Journal of the American Medical Association,* suggest rates vary about 50%; some hospitals are consistently lower, others higher.

Luft sees them as "yellow flags, rather than red flags." A high rate alone doesn't mean a hospital is bad, but it's one of many factors that goes into overall quality.

An editorial says rates are useful in quality control, but Luft says doctors and managed care providers can use them to refer high-risk patients "selectively to better hospitals."

Source: Copyright 1993, USA TODAY. Reprinted with permission.

Statisticians study many events that appear to be due to chance. However, after rigorous statistical analyses using quality control techniques, they may find certain patterns that contradict the probabilities of chance events. This news report is an example of such a study. Explain why the death rates for certain hospitals might contradict the probabilities of chance events.

The formulas for the control limits for the range of measurements in a manufacturing process are given next.

Formulas for the Upper and Lower Control Limits for the R Chart
$$\text{UCL}_R = D_4\overline{R} \quad \text{and} \quad \text{LCL}_R = D_3\overline{R}$$
where $\overline{R}$ = mean of the ranges and D_3 and D_4 are values obtained from Table O in the Appendix.

The steps for constructing the R chart are summarized in Procedure Table 22.

Procedure Table 22

Constructing the R Chart

Step 1: Find the ranges of each sample.

Step 2: Find the mean of the ranges, denoted by $\overline{R}$.

Step 3: Find the upper and lower control limits for the ranges by using these formulas:

$$\text{UCL}_R = D_4\overline{R} \qquad \text{and} \qquad \text{LCL}_R = D_3\overline{R}$$

where D_3 and D_4 are obtained from Table O in the Appendix.

Step 4: Draw the chart, using the value $\overline{R}$ as the centerline and the calculated values for UCL_R and LCL_R.

Step 5: Plot the ranges for each sample, and analyze the chart. If any of the ranges exceed UCL_R or LCL_R, the process is out of control; otherwise, it is in control.

EXERCISES

16–1. Explain briefly the purpose of quality control.

16–2. Who was an early pioneer in the development of quality control procedures?

16–3. Name two types of variations that occur in manufacturing processes.

16–4. Name two types of control charts that are used to judge the quality of a manufacturing process.

16–5. A store manager selects seven samples consisting of seven days and records the amount of merchandise returned each day. The data are shown below. Construct and analyze an $\overline{X}$ chart and an R chart for the data.

Sample	Amount for Merchandise Returned						
1	$832	$862	$901	$842	$887	$915	$871
2	746	733	756	801	789	923	706
3	956	899	804	936	939	961	702
4	300	432	331	325	621	431	392
5	802	802	881	795	962	897	856
6	937	892	851	823	821	963	880
7	632	661	705	736	811	821	907

16–6. The sales manager of a large company surveys the sales personnel and asks them to submit the cost of business cards each person purchased last year. Samples of six individuals were selected from four districts. The data are shown in column 2. Construct and analyze an $\overline{X}$ chart and an R chart for the data.

District	Cost of Business Cards					
1	$203	$308	$371	$252	$362	$381
2	327	356	285	371	343	432
3	356	372	386	399	330	282
4	305	506	613	587	561	432

16–7. The weight (in ounces) of five samples of five binoculars is shown below. Construct and analyze an $\overline{X}$ chart and an R chart for the data.

Sample	Weight (Ounces)				
1	32.1	30.6	31.7	32.0	29.2
2	32.2	30.6	32.9	33.1	31.2
3	31.7	32.6	31.9	33.2	34.1
4	32.2	32.6	31.8	33.6	32.4
5	34.2	33.6	34.5	33.7	34.8

16–8. Seven samples of outboard motors are selected and tested over a period of time. The thrust (in pounds) for each motor is shown below. Construct and analyze an $\overline{X}$ chart and an R chart for the data.

Testing Time	Thrust (Pounds)						
9:00 A.M.	17.2	17.1	16.9	16.8	17.1	17.2	17.0
10:00 A.M.	17.2	17.3	17.1	17.0	17.0	17.2	17.3
11:00 A.M.	17.3	17.0	17.0	16.9	16.8	17.2	17.3
12:00 P.M.	17.5	17.8	17.6	17.3	17.4	17.2	17.4
1:00 P.M.	17.6	17.7	17.3	17.8	17.7	17.5	17.8
2:00 P.M.	18.2	18.3	18.0	17.9	18.1	18.3	18.5
3:00 P.M.	17.7	17.9	17.9	17.6	18.0	17.8	17.3

16–9. Eight samples of five rechargeable batteries are charged for 60 minutes and then tested to see how long they will hold the charge. The data (in hours) are shown below. Construct an $\overline{X}$ chart and an R chart for the data. Is the process in control?

Sample	Hours Holding Charge				
1	12.2	12.0	11.9	11.8	12.0
2	12.6	12.3	12.2	11.9	12.1
3	12.0	12.8	12.9	11.2	11.9
4	11.9	11.8	12.1	12.2	12.3
5	12.2	12.3	12.1	12.0	12.2
6	12.0	12.1	12.2	12.0	12.1
7	11.8	12.2	12.1	11.9	12.3
8	11.6	11.9	12.3	11.8	11.8

16–10. Six samples of four electrical extension cords are measured and cut by a machine. The data (in inches) are shown below. Construct and analyze an $\overline{X}$ chart and an R chart for the data.

Sample	Length (Inches)			
1	60.1	60.0	59.8	59.3
2	60.6	60.2	60.3	60.0
3	59.8	59.2	59.3	60.2
4	63.2	62.1	63.4	62.1
5	60.0	60.1	59.8	59.9
6	59.9	90.2	60.0	60.3

16–11. Ten samples, each consisting of five automobile batteries, are tested for strength. The means and ranges of the samples are given below. Construct and analyze an $\overline{X}$ chart and an R chart for the data.

Sample	Mean	R
1	12.2	2.1
2	12.5	1.3
3	12.3	1.5
4	11.9	2.1
5	11.8	1.2
6	11.2	1.1
7	12.1	1.3
8	12.0	1.3
9	12.2	1.8
10	11.8	2.0

16–12. Seven samples each consisting of six rechargeable batteries are selected and tested. The average number of days the battery delivered 1.5 volts is given in the table below. Construct and analyze an $\overline{X}$ chart and an R chart for the data.

Sample	Mean	R
1	59.2	3.0
2	60.1	4.2
3	58.3	3.4
4	60.9	5.8
5	61.2	7.6
6	59.4	4.9
7	60.5	5.4

16–13. Five samples consisting of the times (in minutes) it takes five workers to complete a project are selected. The means and ranges for the samples are as shown below. Construct and analyze an $\overline{X}$ chart and an R chart for the data.

Sample	1	2	3	4	5
Mean	38.6	33.2	35.6	34.9	31.2
R	3.2	5.7	10.8	3.4	2.8

16–14. Six samples, each consisting of ten automobile tires, were selected, and the pressure (in pounds per square inch) in each tire was measured. The tires were filled by a machine on an automobile assembly line. The means and ranges for each sample are as shown below. Construct and analyze an $\overline{X}$ chart and an R chart for the data.

Sample	1	2	3	4	5	6
Mean	32.4	35.1	34.6	32.4	33.4	35.1
R	3.8	2.1	3.7	2.2	3.0	3.1

16–3 ATTRIBUTE CHARTS

Attribute charts are used when manufactured items can be classified as acceptable or defective. For example, when an item is spray-painted, the painted surface can be classified as acceptable or defective. A defective painted surface might contain runs, specks, bubbles, etc. To produce quality products, manufacturers must keep the percentage of defective items to a minimum. To do so, they use two types of charts to measure the quality of attributes: the p bar chart, or $\overline{p}$ chart, and the c bar chart, or $\overline{c}$ chart. Each chart is described in the following subsections.

The $\bar{p}$ Chart   The $\bar{p}$ chart is used to analyze the percentage of defective items per sample. Thus, one must count the number of defective items, find the percentages, and calculate the upper and lower control limits, as shown in the next example.

EXAMPLE 16–4   A machine manufactures ballpoint pens. Five samples of 50 pens each are selected, and the number of defective pens in each sample is recorded, as shown in the table. Construct a $\bar{p}$ chart, and analyze the data.

Sample	Size	No. of Defective Pens
1	50	3
2	50	1
3	50	4
4	50	2
5	50	5

Solution   STEP 1   Find the proportion of defective pens for each sample. For sample 1,

$$p_1 = \frac{X}{n} = \frac{3}{50} = 0.06$$

For sample 2,

$$p_2 = \frac{1}{50} = 0.02$$

For sample 3,

$$p_3 = \frac{4}{50} = 0.08$$

For sample 4,

$$p_4 = \frac{2}{50} = 0.04$$

For sample 5,

$$p_5 = \frac{5}{50} = 0.10$$

STEP 2   Find the mean for the proportions of defective parts.

$$\bar{p} = \frac{\Sigma p}{k} = \frac{0.06 + 0.02 + 0.08 + 0.04 + 0.10}{5} = \frac{0.3}{5} = 0.06$$

where k is the number of samples.

STEP 3   Find the UCL_p and LCL_p by using the following formulas:

$$UCL_p = \bar{p} + 3\sqrt{\frac{\bar{p}(1 - \bar{p})}{n}} \quad \text{and} \quad LCL_p = \bar{p} - 3\sqrt{\frac{\bar{p}(1 - \bar{p})}{n}}$$

where n is the sample size. Thus,

$$UCL_p = 0.06 + 3\sqrt{\frac{0.06(1 - 0.06)}{50}} = 0.06 + 3(0.034) = 0.06 + 0.102$$

$$= 0.162$$

$$LCL_p = 0.06 - 3\sqrt{\frac{0.06(1 - 0.06)}{50}} = 0.06 - 0.102 = -0.042$$

Since there cannot be a negative proportion of defective items, the LCL_p is set equal to zero.

STEP 4 Draw the chart, using $\bar{p}$ for the centerline and the calculated values for UCL_p and LCL_p. See Figure 16–8.

FIGURE 16–8
Control Chart for
Example 16–4

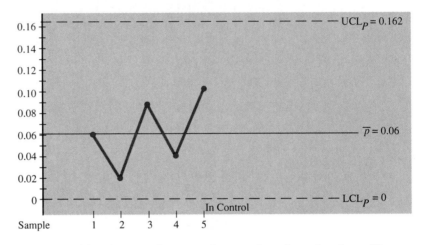

STEP 5 Plot the sample proportions, and analyze the chart. Since none of the values are greater than UCL_p, the process is in control. ◄

The formulas for the upper and lower control limits for the $\bar{p}$ chart are given next.

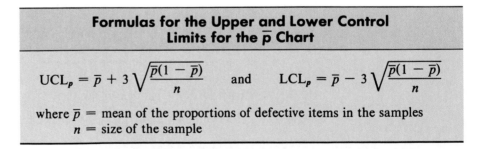

Formulas for the Upper and Lower Control Limits for the $\bar{p}$ Chart

$$UCL_p = \bar{p} + 3\sqrt{\frac{\bar{p}(1 - \bar{p})}{n}} \qquad \text{and} \qquad LCL_p = \bar{p} - 3\sqrt{\frac{\bar{p}(1 - \bar{p})}{n}}$$

where $\bar{p}$ = mean of the proportions of defective items in the samples
n = size of the sample

The steps for constructing the $\bar{p}$ chart are summarized in Procedure Table 23.

Procedure Table 23

Constructing the $\bar{p}$ Chart

Step 1: Find the proportion of defective items for each sample.

Step 2: Find the mean of all the sample proportions.

Step 3: Find the upper and lower control limits by using the following formulas:

$$UCL_p = \bar{p} + 3\sqrt{\frac{\bar{p}(1 - \bar{p})}{n}} \quad \text{and} \quad LCL_p = \bar{p} - 3\sqrt{\frac{\bar{p}(1 - \bar{p})}{n}}$$

where n = sample size

Step 4: Draw the chart, using $\bar{p}$ for the centerline and the calculated values for UCL_p and LCL_p. (*Note:* LCL_p cannot be less than zero.)

Step 5: Plot the sample proportions, and analyze the chart. If any of the values of $\bar{p}$ exceed either UCL_p or LCL_p, the process is out of control; otherwise, it is in control.

The $\bar{c}$ Chart

Another type of attribute chart is called the c bar chart, or $\bar{c}$ **chart**. This chart is used to judge the quality of an item that could have one or more defects per item. For example, a cot might have several defects: The seams of the material might not be properly sewn, the legs might not fit into the frame, and the rubber protectors on the bottom of the legs might not be properly installed. The $\bar{c}$ chart can be used to judge the number of defects per cot.

The next example shows how to construct a $\bar{c}$ chart.

EXAMPLE 16–5

Handheld statistical calculators are manufactured and checked for defects. If a calculator is not defective, it is packaged and shipped to a retail store. Any defective calculators are repaired before they are shipped. Twelve of the defective calculators are checked for the number of defects per calculator. The defects include soldering, lettering, and cracked cases. The number of defects per calculators are 6, 3, 2, 5, 6, 7, 4, 3, 7, 8, 9, and 5. Construct and analyze a $\bar{c}$ chart for the data.

Solution

STEP 1 Find the average number of defects per calculator, $\bar{c}$.

$$\bar{c} = \frac{\Sigma c}{n} = \frac{6 + 3 + 2 + \cdots + 5}{12} = \frac{65}{12} = 5.42$$

STEP 2 Find the UCL_c and LCL_c using the following formulas:

$$UCL_c = \bar{c} + 3\sqrt{\bar{c}} = 5.42 + 3\sqrt{5.42} = 5.42 + 6.98 = 12.40$$
$$LCL_c = \bar{c} - 3\sqrt{\bar{c}} = 5.42 - 3\sqrt{5.42} = 5.42 - 6.98 = -1.56$$

Since the LCL_c cannot be negative, zero is used for its value.

STEP 3 Draw the chart, using $\bar{c}$ as the centerline and the calculated values of UCL_c and LCL_c.

STEP 4 Plot the number of defects on the chart, and analyze the chart, as shown in Figure 16–9. Since all points fall within the lines for UCL_c and LCL_c, the process is in control. That is, in the defective calculators, the number of defects per calculator is not excessive.

FIGURE 16–9
Control Chart for
Example 16–5

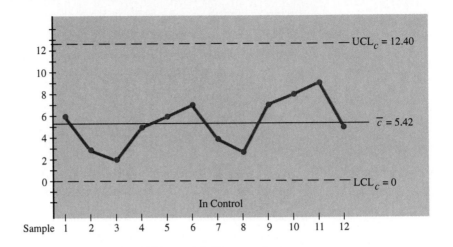

The formulas for UCL_c and LCL_c are given next.

Formulas for the Upper and Lower Control Limits for the $\bar{c}$ Chart

$$UCL_c = \bar{c} + 3\sqrt{\bar{c}} \quad \text{and} \quad LCL_c = \bar{c} - 3\sqrt{\bar{c}}$$

where $\bar{c}$ = mean of the defects per item in the samples

---

The steps for constructing the $\bar{c}$ chart are summarized in Procedure Table 24.

Procedure Table 24

Constructing the $\bar{c}$ Chart

Step 1: Find the average number of defects per item in the samples.

Step 2: Find the upper and lower control limits, using the following formulas:

$$UCL_c = \bar{c} + 3\sqrt{\bar{c}} \quad \text{and} \quad LCL_c = \bar{c} - 3\sqrt{\bar{c}}$$

Step 3: Draw the chart, using $\bar{c}$ as the centerline and the calculated values of UCL_c and LCL_c. (*Note:* LCL_c cannot be less than zero.)

Step 4: Plot the number of defects per item on the chart, and analyze the chart. If any of the numbers exceed UCL_c or LCL_c, the process is out of control; otherwise, it is in control.

Speaking of **Statistics**

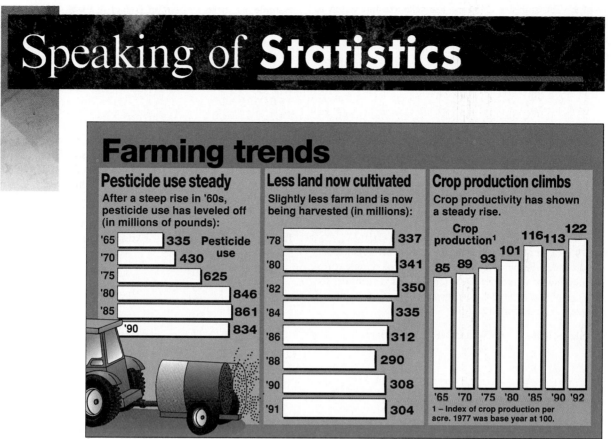

Farming trends

Pesticide use steady

After a steep rise in '60s, pesticide use has leveled off (in millions of pounds):

Year	Pesticide use
'65	335
'70	430
'75	625
'80	846
'85	861
'90	834

Less land now cultivated

Slightly less farm land is now being harvested (in millions):

Year	
'78	337
'80	341
'82	350
'84	335
'86	312
'88	290
'90	308
'91	304

Crop production climbs

Crop productivity has shown a steady rise.

Crop production[1]

85 89 93 101 116 113 122

'65 '70 '75 '80 '85 '90 '92

1 – Index of crop production per acre. 1977 was base year at 100.

Source: USA TODAY research

By Marcy E. Mullins, USA TODAY

Source: Copyright 1993, USA TODAY. Reprinted with permission.

Many statistical studies involve agriculture. This study examines three agricultural variables and how they have changed over the years. Explain how these variables might be related. Cite some ways that quality control techniques might be used in agriculture.

EXERCISES

16-15. Name two types of attribute charts.

16-16. Explain the difference between the two types of attribute charts.

16-17. Samples of a business enterprise are obtained from six geographic locations across the United States, and the number of failures per location are recorded. The data are shown at the right. Construct and analyze a $\bar{p}$ chart for the data.

Sample	Size	No. of Failures
1	100	8
2	100	6
3	100	5
4	100	27
5	100	9
6	100	10

16–18. Ten samples of lamps are selected and tested to see whether they will light when plugged into an electrical circuit. The data are shown below. Construct and analyze a $\bar{p}$ chart for the data.

Sample	Size	No. of Defective Lamps
1	30	6
2	30	8
3	30	2
4	30	3
5	30	10
6	30	5
7	30	6
8	30	7
9	30	4
10	30	20

16–19. Eight samples of sandwich makers are selected and tested to see whether they reach the required temperature. The number of defective sandwich makers is recorded for each sample, and the data are given below. Construct and analyze a $\bar{p}$ chart for the data.

Sample	Size	No. of Defective Products
1	25	5
2	25	4
3	25	0
4	25	6
5	25	5
6	25	3
7	25	2
8	25	3

16–20. Samples of golf balls are selected and screened to see whether there are any defects on the surface of the balls. The number of defective balls recorded for each sample is given below. Construct and analyze a $\bar{p}$ chart for the data.

Sample	Size	No. of Defective Balls
1	10	0
2	10	1
3	10	0
4	10	1
5	10	2

16–21. Samples of O-rings are selected and tested to see whether they will seal properly. The data are shown in column 2. Construct and analyze a $\bar{p}$ chart for the data.

Sample	Size	No. of Defective Rings
1	20	2
2	20	1
3	20	4
4	20	3
5	20	2
6	20	2
7	20	4

16–22. Fire extinguishers are randomly checked to see whether they will function properly when used. Samples are selected and tested over a period of 5 hours. The number of defective extinguishers is given in the table below. Construct and analyze a $\bar{p}$ chart for the data.

Testing Time	Size	No. of Defective Extinguishers
9:00 A.M.	50	8
10:00 A.M.	50	10
11:00 A.M.	50	9
12:00 P.M.	50	8
1:00 P.M.	50	12
2:00 P.M.	50	11

16–23. Samples of toggle switches are selected and tested to see whether they are defective. The data are shown below. Construct and analyze a $\bar{p}$ chart for the data.

Sample	Size	No. of Defective Switches
1	200	10
2	200	8
3	200	9
4	200	32
5	200	15
6	200	35
7	200	38
8	200	10

16–24. T-shirts are manufactured and checked to see whether the logos are properly printed on the shirts. The number of defective logos per sample is shown below. Construct and analyze a $\bar{p}$ chart for the data.

Sample	Size	No. of Defective Logos
1	50	18
2	50	22
3	50	16
4	50	16
5	50	16
6	50	35

16-25. Stereo headphone sets are checked to see whether they work when plugged into a stereo receiver. Eight samples of 30 sets each are selected and checked for defective headphones. The number of defective headphones per sample is shown below. Construct and analyze a $\bar{p}$ chart for the data.

Sample	Size	No. of Defective Headphones
1	30	6
2	30	5
3	30	8
4	30	10
5	30	12
6	30	18
7	30	14
8	30	24

16-26. Plastic collapsible cups are tested to see whether they will leak when opened up and filled with water. Samples are selected and tested over a period of 7 hours. The number of defective cups per sample is shown below. Construct and analyze a $\bar{p}$ chart for the data.

Testing Time	Size	No. of Defective Cups
9:00 A.M.	50	5
10:00 A.M.	50	8
11:00 A.M.	50	6
12:00 P.M.	50	5
1:00 P.M.	50	8
2:00 P.M.	50	9
3:00 P.M.	50	4
4:00 P.M.	50	3

16-27. The number of defects in each of nine exercise bikes is 6, 3, 8, 2, 3, 6, 8, 1, and 3. Construct and analyze a $\bar{c}$ chart for the data.

16-28. The number of defects in five dot matrix printers is 10, 8, 9, 10, and 11. Construct and analyze a $\bar{c}$ chart for the data.

16-29. Samples of wood panels are checked for defects. The number of defects (scratches, blisters, etc.) for each panel is given below. Construct and analyze a $\bar{c}$ chart for the data.

Sample	1	2	3	4	5	6	7	8
No. of Defects	1	2	1	1	1	3	6	2

16-30. Coated aluminum pipes are checked for chips in the coating. Construct and analyze a $\bar{c}$ chart for the data shown below.

Sample	1	2	3	4	5	6	7	8	9	10	11	12
No. of Chips	5	8	13	12	11	10	12	11	8	7	6	3

16-31. Linoleum is checked for defects. Construct and analyze a $\bar{c}$ chart for the number of defects in a 20-foot roll. The data are shown below.

Sample	1	2	3	4	5	6	7	8
No. of Defects	12	6	3	2	1	1	3	4

16-32. Workers are painting a large apartment building. An inspector checks several walls for paint blemishes. The number of blemishes per wall is shown in the table. Construct and analyze a $\bar{c}$ chart for the data.

Sample	1	2	3	4	5	6	7	8	9	10
No. of Blemishes	18	12	15	17	16	16	15	16	18	17

16-33. Bedspreads are checked for defects in the weaving of the material. The number of defects per spread is shown in the table. Construct and analyze a $\bar{c}$ chart for the data.

Spread	1	2	3	4	5	6
No. of Defects	3	1	1	2	3	2

16-34. Sixty-minute cassette tapes are checked for defects. The number of defects per tape for ten tapes is shown in the table. Construct and analyze a $\bar{c}$ chart for the data.

Tape	1	2	3	4	5	6	7	8	9	10
No. of Defects	6	12	9	8	8	6	12	10	11	8

16-4 SUMMARY

In order to ensure and check the quality of a manufactured product, statisticians developed quality control charts. Walter A. Shewhart pioneered this effort in the 1930s and 1940s. Quality control enables manufacturers to cut down on costs and to maintain a high and reliable standard for their manufactured items.

There are two basic types of control charts: variable charts and attribute charts. Variable charts are used to analyze the quality of an item when measurements are important—for example, the weight, length, or diameter of an item. Two types of variable charts are the $\overline{X}$ chart and the R chart. The $\overline{X}$ chart checks the means of the measurements, and the R chart checks the range of the measurements.

Attribute charts are used to analyze a manufacturing process in which an item can be classified as acceptable or defective. For example, an inspection check may test to see whether tires hold air. If they do, the tires are acceptable; if they do not, they are defective. In order to keep defective items to a minimum, manufacturers rely on two types of attribute charts: the $\overline{p}$ chart and the $\overline{c}$ chart. The $\overline{p}$ chart is used to analyze the percentage of defective items per lot or sample, while the $\overline{c}$ chart is used to analyze the number of defects per item in the samples.

Important Terms

Assignable-cause variation
Attribute chart
$\overline{c}$ chart
Chance (or random) variation
Control chart
Lower control limit
$\overline{p}$ chart
Quality control
R chart
Upper control limit
Variable chart
$\overline{X}$ chart

Important Formulas

Formulas for the $\overline{X}$ chart:

$$\overline{X}_{GM} = \frac{\Sigma \overline{X}}{k} \qquad UCL_{\overline{X}} = \overline{X}_{GM} + A_2\overline{R}$$

$$\overline{R} = \frac{\Sigma R}{k} \qquad LCL_{\overline{X}} = \overline{X}_{GM} - A_2\overline{R}$$

where k = number of samples and A_2 is obtained from Table O.

Formulas for the R chart:

$$UCL_R = D_4\overline{R} \qquad LCL_R = D_3\overline{R}$$

where D_3 and D_4 are obtained from Table O.

Formulas for the $\overline{p}$ chart:

$$\overline{p} = \frac{\Sigma p}{k} \qquad UCL_p = \overline{p} + 3\sqrt{\frac{\overline{p}(1-\overline{p})}{n}}$$

$$LCL_p = \overline{p} - 3\sqrt{\frac{\overline{p}(1-\overline{p})}{n}}$$

where

$$k = \text{number of samples} \qquad n = \text{sample size}$$

Formulas for the $\overline{c}$ chart:

$$\overline{c} = \frac{\Sigma c}{n} \qquad UCL_c = \overline{c} + 3\sqrt{\overline{c}} \qquad LCL_c = \overline{c} - 3\sqrt{\overline{c}}$$

where

$$c = \text{number of defects per item} \qquad n = \text{number of items}$$

Review Exercises

16–35. Six samples of emergency room bills were selected. The data are shown here. Construct and analyze an $\overline{X}$ chart and an R chart for the data.

Sample	Emergency Room Bill				
1	$82	$95	$86	$97	$93
2	84	90	99	110	116
3	53	62	43	55	58
4	97	89	90	100	102
5	88	84	87	87	82
6	91	93	95	99	86

16–36. Five samples of shafts for miniature motors were selected and measured. The data (in inches) are shown below. Construct and analyze an $\overline{X}$ chart and an R chart for the data.

Sample	Shaft Diameters (Inches)			
1	1.56	1.54	1.55	1.59
2	1.52	1.55	1.50	1.56
3	1.49	1.48	1.51	1.50
4	1.43	1.50	1.56	1.51
5	1.51	1.56	1.49	1.52

16–37. Eight samples, each consisting of ten remote control door alarms, are selected and tested. The mean and ranges of the effective distances (in feet) for each alarm are given here. Construct and analyze an $\overline{X}$ chart and an R chart for the data.

Sample	1	2	3	4	5	6	7	8
Mean	74.1	72.3	75.6	74.8	76.1	75.3	76.2	71.7
Range	3.1	3.6	2.9	6.7	2.1	3.2	3.1	2.5

16–38. A machine cuts boards to a specific length. Six samples, each consisting of eight boards, are selected and measured. The data (in inches) are shown here. Construct and analyze an $\overline{X}$ chart and an R chart for the data.

Sample	1	2	3	4	5	6
Mean	72.0	71.8	72.2	72.1	71.9	72.2
Range	0.2	0.3	0.2	0.3	0.2	0.2

16–39. Ten samples of alarm clocks are tested to see whether they keep correct time. The number of defective clocks is shown in the table. Construct and analyze a $\overline{p}$ chart for the data.

Sample	Size	Number of Defective Clocks
1	40	5
2	40	4
3	40	6
4	40	3
5	40	5
6	40	3
7	40	3
8	40	4
9	40	3
10	40	5

16–40. Eight samples of water pumps are selected and tested for leaks. Those that leak are considered defective. The data are shown below. Construct and analyze a $\overline{p}$ chart for the data.

Sample	Size	Number of Defective Pumps
1	10	6
2	10	0
3	10	2
4	10	1
5	10	3
6	10	2
7	10	1
8	10	0

16–41. Eight sofas are checked for defects in the material and seams. The number of defects for each is shown below. Construct and analyze a $\overline{c}$ chart for the data.

Sofa	1	2	3	4	5	6	7	8
No. of Defects	10	12	9	7	5	18	6	3

16–42. Ten automobile tires are checked for defects. The number of defects per tire is shown below. Construct and analyze a $\overline{c}$ chart for the data.

Tire	1	2	3	4	5	6	7	8	9	10
No. of Defects	9	10	8	11	10	8	10	9	12	8

COMPUTER APPLICATIONS

1. MINITAB can be used to compute and print quality control charts, including $\overline{X}$ and R charts. The data from Example 16–1 will be used as an example here.

   The data for the $\overline{X}$ chart and the R chart are entered by using the SET C1 command. After the data have been entered, type XBARCHART C1 5 ; . The last number tells MINITAB the size of each sample. Enter the following in the computer:

```
MTB  >  SET  C1
DATA >  46  47  45  46  47

DATA >  50  51  52  53  49
DATA >  48  51  50  50  49
DATA >  52  50  49  50  51
DATA >  51  47  46  48  47
DATA >  49  51  50  51  52
DATA >  END
MTB  >  XBARCHART  C1  5
```

The computer will print the following:

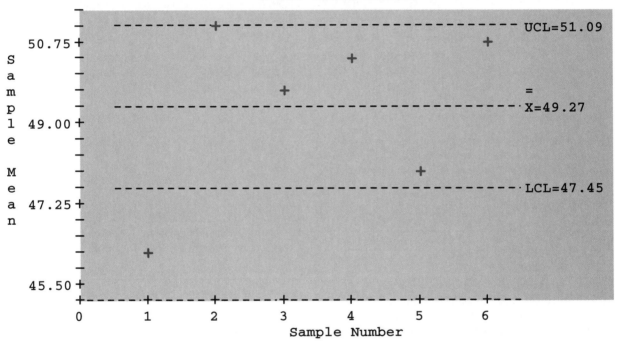

The $UCL_{\bar{x}}$ is 51.09, and the $LCL_{\bar{x}}$ is 47.45. The grand mean is 49.27. Since the mean of sample 1 is outside the lower control limit, the process is out of control. There is a slight difference in the computed values and the MINITAB values due to rounding, and in some cases MINITAB uses different formulas.

To have MINITAB print an R chart, type in the data and the command RCHART C1 5.

```
MTB  >  SET  C1
DATA >  46  47  45  46  47
DATA >  50  51  52  53  49
DATA >  48  51  50  50  49
DATA >  52  50  49  50  51
DATA >  51  47  46  48  47
DATA >  49  51  50  51  52
DATA >  END
MTB  >  RCHART  C1  5
```

The computer will print the following:

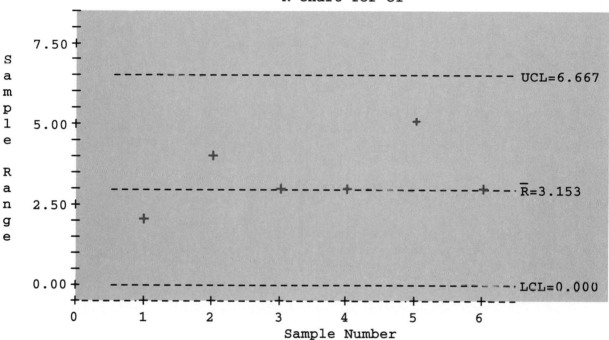

R Chart for C1

The process is in control since all the ranges fall within the control limits. The value for $\bar{R}$ is different from the one obtained in Example 16–1 because MINITAB uses a different formula for its computation.

Use MINITAB to do Exercises 16–5 and 16–61.

2. MINITAB can compute and print a $\bar{p}$ chart. Use the data from Example 16–4. The data are entered by using the READ C1 C2 command. The number of defective items is entered first. The sample size is entered second. The PCHART C1 C2 command tells MINITAB to compute and print a $\bar{p}$ chart. Type in the following:

```
MTB  >  READ  C1  C2
DATA >  3  50
DATA >  1  50
DATA >  4  50
DATA >  2  50
DATA >  5  50
DATA >  END
         5  ROWS  READ
MTB  >  PCHART  C1  C2
```

The computer will print the following:

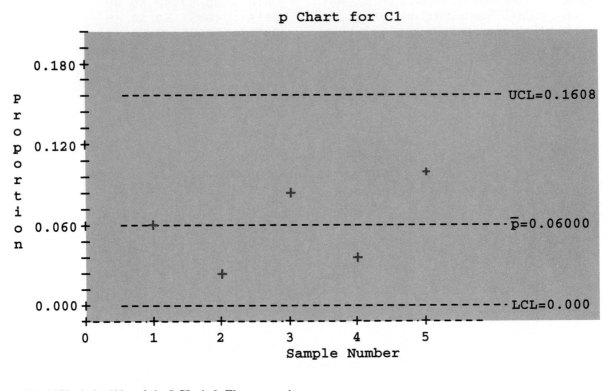

The UCL_p is 0.1608 and the LCL_p is 0. The process is in control. The results vary slightly due to rounding.
Try Exercises 16–17 and 16–18 using MINITAB.

✎ TEST

Directions: Determine whether the statement is true or false.

1. Dr. Walter A. Shewhart was an early pioneer in the development of quality control techniques.
2. Attribute charts are used to analyze a manufacturing process in which measurements are important.
3. The $\bar{p}$ chart is a variable chart.
4. When all the sample means fall within $UCL_{\bar{X}}$ and $LCL_{\bar{X}}$ of an $\bar{X}$ control chart, the process is said to be in control for the means.
5. The R control chart uses the sample means and ranges to compute UCL_R and the LCL_R.
6. In addition to looking for sample means that fall within $UCL_{\bar{X}}$ and $LCL_{\bar{X}}$, one should also look for other irregularities.
7. For variable charts, one need only check the means to determine whether a manufacturing process is out of control.
8. When manufactured items are classified as acceptable or defective, an attribute chart can be used to analyze the process.
9. The $\bar{c}$ control chart uses percentages or proportions in the analysis of the manufacturing process.
10. If a sample of 100 items contains no defective items, it should not be used when constructing a $\bar{p}$ chart.

A–1 FACTORIALS

Definition and Properties of Factorials

The notation called factorial notation is used in probability. *Factorial notation* uses the exclamation point and involves multiplication. For example,

$$5! = 5 \cdot 4 \cdot 3 \cdot 2 \cdot 1 = 120$$
$$4! = 4 \cdot 3 \cdot 2 \cdot 1 = 24$$
$$3! = 3 \cdot 2 \cdot 1 = 6$$
$$2! = 2 \cdot 1 = 2$$
$$1! = 1$$

In general, a factorial is evaluated as follows:

$$n! = n(n - 1)(n - 2) \cdots 3 \cdot 2 \cdot 1$$

Note that the factorial is the product of n factors, with the number decreased by one for each factor.

One property of factorial notation is that it can be stopped at any point by using the exclamation point. For example,

$5! = 5 \cdot 4!$	since	$4! = 4 \cdot 3 \cdot 2 \cdot 1$
$= 5 \cdot 4 \cdot 3!$	since	$3! = 3 \cdot 2 \cdot 1$
$= 5 \cdot 4 \cdot 3 \cdot 2!$	since	$2! = 2 \cdot 1$
$= 5 \cdot 4 \cdot 3 \cdot 2 \cdot 1$		

Thus, $n! = n(n - 1)!$
$$= n(n - 1)(n - 2)!$$
$$= n(n - 1)(n - 2)(n - 3)! \quad \text{etc.}$$

Another property of factorials is

$$0! = 1$$

This fact is needed for formulas.

Operations with Factorials

Factorials cannot be added or subtracted directly. They must be multiplied out; then, the products can be added or subtracted.

EXAMPLE A–1
Evaluate $3! + 4!$.

Solution
$$3! + 4! = (3 \cdot 2 \cdot 1) + (4 \cdot 3 \cdot 2 \cdot 1)$$
$$= 6 + 24 = 30$$

Note: $3! + 4! \neq 7!$, since $7! = 5040$. ◄

EXAMPLE A–2
Evaluate $5! - 3!$.

Solution
$$5! - 3! = (5 \cdot 4 \cdot 3 \cdot 2 \cdot 1) - (3 \cdot 2 \cdot 1)$$
$$= 120 - 6 = 114$$

Note: $5! - 3! \neq 2!$, since $2! = 2$. ◄

Factorials cannot be multiplied directly. Again, one must multiply them out and then multiply the products.

EXAMPLE A–3
Evaluate $3! \cdot 2!$.

Solution
$$3! \cdot 2! = (3 \cdot 2 \cdot 1) \cdot (2 \cdot 1)$$
$$= 6 \cdot 2 = 12$$

Note: $3! \cdot 2! \neq 6!$, since $6! = 720$. ◄

Finally, factorials cannot be divided directly unless they are equal.

EXAMPLE A–4
Evaluate $6! \div 3!$.

Solution
$$\frac{6!}{3!} = \frac{6 \cdot 5 \cdot 4 \cdot 3 \cdot 2 \cdot 1}{3 \cdot 2 \cdot 1} = \frac{720}{6} = 120$$

Note: $\dfrac{6!}{3!} \neq 2!$ since $2! = 2$

But $\dfrac{3!}{3!} = \dfrac{3 \cdot 2 \cdot 1}{3 \cdot 2 \cdot 1} = \dfrac{6}{6} = 1$

In division, one can take some shortcuts, as shown:

$$\frac{6!}{3!} = \frac{6 \cdot 5 \cdot 4 \cdot 3!}{3!} \quad \text{and} \quad \frac{3!}{3!} = 1$$
$$= 6 \cdot 5 \cdot 4 = 120$$

Hence, $\quad \dfrac{8!}{6!} = \dfrac{8 \cdot 7 \cdot 6!}{6!} \quad$ and $\quad \dfrac{6!}{6!} = 1$
$$= 8 \cdot 7 = 56$$

Another shortcut that can be used with factorials is cancellation, after factors have been expanded. For example,

$$\frac{7!}{(4!)(3!)} = \frac{7 \cdot 6 \cdot 5 \cdot 4!}{3 \cdot 2 \cdot 1 \cdot 4!}$$

Now cancel both 4!'s. Then cancel the $3 \cdot 2$ in the denominator with the 6 in the numerator.

$$\frac{7 \cdot \overset{1}{\cancel{6}} \cdot 5 \cdot \overset{1}{\cancel{4!}}}{\underset{1}{\cancel{3}} \cdot \underset{1}{\cancel{2}} \cdot 1 \cdot \underset{1}{\cancel{4!}}} = 7 \cdot 5 = 35 \qquad \blacktriangleleft$$

EXAMPLE A–5
Evaluate $10! \div (6!)(4!)$.

Solution

$$\frac{10!}{(6!)(4!)} = \frac{10 \cdot \overset{3}{\cancel{9}} \cdot \overset{1}{\cancel{8}} \cdot 7 \cdot \overset{1}{\cancel{6!}}}{\underset{1}{\cancel{4}} \cdot \underset{1}{\cancel{3}} \cdot \underset{1}{\cancel{2}} \cdot 1 \cdot \underset{1}{\cancel{6!}}} = 10 \cdot 3 \cdot 7 = 210$$

EXERCISES

Evaluate each expression.

A–1. $9!$ **A–2.** $7!$ **A–3.** $5!$

A–4. $0!$ **A–5.** $1!$ **A–6.** $3!$

A–7. $\dfrac{12!}{9!}$ **A–8.** $\dfrac{10!}{2!}$ **A–9.** $\dfrac{5!}{3!}$

A–10. $\dfrac{11!}{7!}$ **A–11.** $\dfrac{9!}{(4!)(5!)}$ **A–12.** $\dfrac{10!}{(7!)(3!)}$

A–13. $\dfrac{8!}{(4!)(4!)}$ **A–14.** $\dfrac{15!}{(12!)(3!)}$

A–15. $\dfrac{10!}{(10!)(0!)}$ **A–16.** $\dfrac{5!}{(3!)(2!)(1!)}$

A–17. $\dfrac{8!}{(3!)(3!)(2!)}$ **A–18.** $\dfrac{11!}{(7!)(2!)(2!)}$

A–19. $\dfrac{10!}{(3!)(2!)(5!)}$ **A–20.** $\dfrac{6!}{(2!)(2!)(2!)}$

A–2 SUMMATION NOTATION

In mathematics, the symbol Σ (Greek letter sigma) means to add or find the sum. For example, ΣX means to add the numbers represented by the variable X. Thus, when X represents 5, 8, 2, 4, and 6, then ΣX means $5 + 8 + 2 + 4 + 6 = 25$.

Sometimes, a subscript notation is used, such as

$$\sum_{i=1}^{5} X_i$$

This notation means to find the sum of five numbers represented by X, as shown:

$$\sum_{i=1}^{5} X_i = X_1 + X_2 + X_3 + X_4 + X_5$$

When the number of values is not known, the unknown number can be represented by n, such as

$$\sum_{i=1}^{n} X_i = X_1 + X_2 + X_3 + \cdots + X_n$$

There are several important types of summation used in statistics. The notation ΣX^2 means to square each value before summing. For example, if the values of X are 2, 8, 6, 1, and 4, then

$$\Sigma X^2 = 2^2 + 8^2 + 6^2 + 1^2 + 4^2$$
$$= 4 + 64 + 36 + 1 + 16 = 121$$

The notation $(\Sigma X)^2$ means to find the sum of X and then square the answer. For instance, if the values for X are 2, 8, 6, 1, and 4, then

$$(\Sigma X)^2 = (2 + 8 + 6 + 1 + 4)^2 = (21)^2 = 441$$

Another important use of summation notation is in finding the mean (shown in Section 3–2). The mean $\overline{X}$ is defined as

$$\overline{X} = \frac{\Sigma X}{n}$$

For example, to find the mean of 12, 8, 7, 3, and 10, use the formula and substitute the values, as shown:

$$\overline{X} = \frac{\Sigma X}{n} = \frac{12 + 8 + 7 + 3 + 10}{5} = \frac{40}{5} = 8$$

The notation $\Sigma (X - \overline{X})^2$ means to perform the following steps.

STEP 1 Find the mean.

STEP 2 Subtract the mean from each value.

STEP 3 Square the numbers.

STEP 4 Find the sum.

EXAMPLE A–6

Find the value of $\Sigma (X - \overline{X})^2$ for the values 12, 8, 7, 3, and 10 of X.

Solution

STEP 1 Find the mean.

$$\overline{X} = \frac{12 + 8 + 7 + 3 + 10}{5} = \frac{40}{5} = 8$$

STEP 2 Subtract the mean from each value.

$$12 - 8 = 4 \qquad 7 - 8 = -1 \qquad 10 - 8 = 2$$
$$8 - 8 = 0 \qquad 3 - 8 = -5$$

STEP 3 Square the answers.

$$4^2 = 16 \quad 0^2 = 0 \quad (-1)^2 = 1 \quad (-5)^2 = 25 \quad 2^2 = 4$$

STEP 4 Find the sum.

$$16 + 0 + 1 + 25 + 4 = 46 \qquad \blacktriangleleft$$

EXAMPLE A–7

Find $\Sigma (X - \overline{X})^2$ for the following values of X: 5, 7, 2, 1, 3, 6.

Solution

Find the mean:

$$\overline{X} = \frac{5 + 7 + 2 + 1 + 3 + 6}{6} = \frac{24}{6} = 4$$

Then, the steps in Example A–6 can be shortened as follows:

$$
\begin{aligned}
\Sigma (X - \overline{X})^2 &= (5 - 4)^2 + (7 - 4)^2 + (2 - 4)^2 \\
&\quad + (1 - 4)^2 + (3 - 4)^2 + (6 - 4)^2 \\
&= 1^2 + 3^2 + (-2)^2 + (-3)^2 \\
&\quad + (-1)^2 + 2^2 \\
&= 1 + 9 + 4 + 9 + 1 + 4 = 28 \qquad \blacktriangleleft
\end{aligned}
$$

EXERCISES

For each set of values, find ΣX, ΣX^2, $(\Sigma X)^2$, and $\Sigma (X - \overline{X})^2$.

A–21. 9, 17, 32, 16, 8, 2, 9, 7, 3, 18

A–22. 4, 12, 9, 13, 0, 6, 2, 10

A–23. 5, 12, 8, 3, 4

A–24. 6, 2, 18, 30, 31, 42, 16, 5

A–25. 80, 76, 42, 53, 77

A–26. 123, 132, 216, 98, 146, 114

A–27. 53, 72, 81, 42, 63, 71, 73, 85, 98, 55

A–28. 43, 32, 116, 98, 120

A–29. 12, 52, 36, 81, 63, 74

A–30. −9, −12, 18, 0, −2, −15

A–3 THE LINE

Figure A–1 shows the *rectangular coordinate system* or *Cartesian plane*. This figure consists of two axes: the horizontal axis, called the x axis, and the vertical axis, called the y axis. Each of the axes has numerical scales. The point of the intersection of the axes is called the *origin*.

FIGURE A–1

Rectangular Coordinate
System

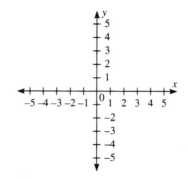

Points can be graphed by using coordinates. For example, the notation for point $P(3, 2)$ means that the x coordinate is 3 and the y coordinate is 2. Hence, P is located at the intersection of $x = 3$ and $y = 2$, as shown in Figure A–2.

FIGURE A–2

Point P and Its
Coordinates

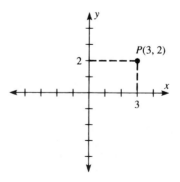

Other points $Q(-5, 2)$, $R(4, +1)$, $S(-3, -4)$ can be plotted as shown in Figure A–3.

When a point lies on the y axis, the x coordinate is 0, as in (0, 6) (0, −3), etc. When a point lies on the x axis, the y coordinate is 0, as in (6, 0) (−8, 0), etc. See Figure A–3.

FIGURE A–3
Points Plotted on the Rectangular Coordinate System

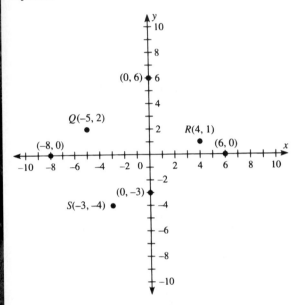

Two points determine a line. There are two properties of a line: its slope and its equation. The *slope m* of a line is determined by the ratio of the rise (called Δy) and the run (called Δx).

$$m = \frac{\text{rise}}{\text{run}} = \frac{\Delta y}{\Delta x}$$

For example, the slope of the line shown in Figure A–4 is $\frac{3}{2}$, or 1.5, since the height Δy is 3 units and the run Δx is 2 units.

The slopes of lines can be positive, negative, or zero, as shown in Figure A–5. A line going uphill from left to right has a positive slope. A line that is horizontal has a slope of zero. And a line going downhill from left to right has a negative slope.

A point a where the line crosses the x axis is called the x *intercept* and has the coordinates $(a, 0)$. A point b where the line crosses the y axis is called the y *intercept* and has the coordinates $(0, b)$. See Figure A–6.

FIGURE A–4
Slope of a Line and Its Rise and Run

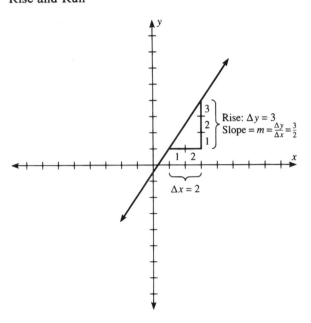

FIGURE A–5
Slopes

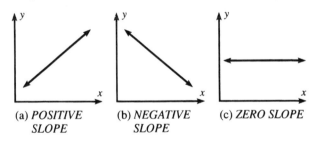

(a) *POSITIVE SLOPE*   (b) *NEGATIVE SLOPE*   (c) *ZERO SLOPE*

FIGURE A–6
A Line and Its x and y Intercepts

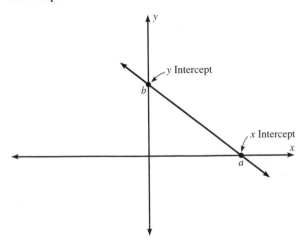

Every line has a unique equation of the form $y = a + bx$. For example, the equations

$$y = 5 + 3x$$
$$y = -8.6 + 3.2x$$
$$y = 5.2 - 6.1x$$

all represent different, unique lines. The number represented by a is the y intercept point; the number represented by the b is the slope. The line whose equation is $y = 3 + 2x$ has a y intercept at 3 and a slope of 2, or 2/1. This line can be shown as in the graph in Figure A–7.

FIGURE A–7
The Line $y = 3 + 2x$

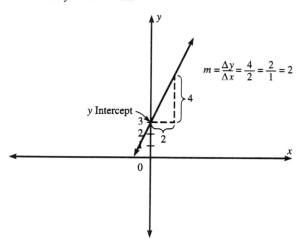

If two points are known, then the graph of the line can be plotted. For example, to find the graph of a line passing through the points $P(2, 1)$ and $Q(3, 5)$, plot the points and connect them as shown in Figure A–8.

FIGURE A–8
The Line Passing
Through (2, 1) and
(3, 5)

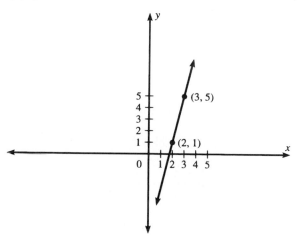

Given the equation of a line, one can graph the line by finding two points and then plotting them.

EXAMPLE A–8
Plot the graph of the line whose equation is $y = 3 + 2x$.

Solution
Select any number as an x value and substitute it in the equation to get the corresponding y value. Let $x = 0$. Then,

$$y = 3 + 2x = 3 + 2(0) = 3$$

Hence, when $x = 0$, then $y = 3$, and the line passes through the point (0, 3).

Now select any other value of x, say $x = 2$.

$$y = 3 + 2x = 3 + 2(2) = 7$$

Hence, a second point is (2, 7).

Then, plot the points and graph the line, as shown in Figure A–9. ◀

FIGURE A–9
Graph of the Line for
Example A–8

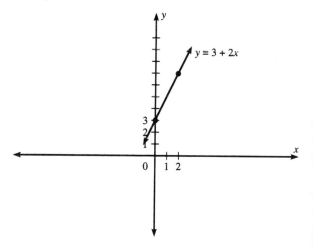

EXERCISES

Plot the lines passing through each set of points.

A–31. $P(3, 2), Q(1, 6)$ **A–32.** $P(0, 5), Q(8, 0)$

A–33. $P(-2, 4), Q(3, 6)$

A–34. $P(-1, -2), Q(-7, 8)$ **A–35.** $P(6, 3), Q(10, 3)$

Find at least two points on each line, and then graph the line containing these points.

A–36. $y = 5 + 2x$ **A–37.** $y = -1 + x$

A–38. $y = 3 + 4x$ **A–39.** $y = -2 - 2x$

A–40. $y = 4 - 3x$

APPENDIX B
Tables

Table A

Factorials

n	$n!$
0	1
1	1
2	2
3	6
4	24
5	120
6	720
7	5,040
8	40,320
9	362,880
10	3,628,800
11	39,916,800
12	479,001,600
13	6,227,020,800
14	87,178,291,200
15	1,307,674,368,000
16	20,922,789,888,000
17	355,687,428,096,000
18	6,402,373,705,728,000
19	121,645,100,408,832,000
20	2,432,902,008,176,640,000

Table B

The Binomial Distribution

n	x	p 0.05	0.1	0.2	0.3	0.4	0.5	0.6	0.7	0.8	0.9	0.95
2	0	0.902	0.810	0.640	0.490	0.360	0.250	0.160	0.090	0.040	0.010	0.002
	1	0.095	0.180	0.320	0.420	0.480	0.500	0.480	0.420	0.320	0.180	0.095
	2	0.002	0.010	0.040	0.090	0.160	0.250	0.360	0.490	0.640	0.810	0.902
3	0	0.857	0.729	0.512	0.343	0.216	0.125	0.064	0.027	0.008	0.001	
	1	0.135	0.243	0.384	0.441	0.432	0.375	0.288	0.189	0.096	0.027	0.007
	2	0.007	0.027	0.096	0.189	0.288	0.375	0.432	0.441	0.384	0.243	0.135
	3		0.001	0.008	0.027	0.064	0.125	0.216	0.343	0.512	0.729	0.857
4	0	0.815	0.656	0.410	0.240	0.130	0.062	0.026	0.008	0.002		
	1	0.171	0.292	0.410	0.412	0.346	0.250	0.154	0.076	0.026	0.004	
	2	0.014	0.049	0.154	0.265	0.346	0.375	0.346	0.265	0.154	0.049	0.014
	3		0.004	0.026	0.076	0.154	0.250	0.346	0.412	0.410	0.292	0.171
	4			0.002	0.008	0.026	0.062	0.130	0.240	0.410	0.656	0.815
5	0	0.774	0.590	0.328	0.168	0.078	0.031	0.010	0.002			
	1	0.204	0.328	0.410	0.360	0.259	0.156	0.077	0.028	0.006		
	2	0.021	0.073	0.205	0.309	0.346	0.312	0.230	0.132	0.051	0.008	0.001
	3	0.001	0.008	0.051	0.132	0.230	0.312	0.346	0.309	0.205	0.073	0.021
	4			0.006	0.028	0.077	0.156	0.259	0.360	0.410	0.328	0.204
	5				0.002	0.010	0.031	0.078	0.168	0.328	0.590	0.774
6	0	0.735	0.531	0.262	0.118	0.047	0.016	0.004	0.001			
	1	0.232	0.354	0.393	0.303	0.187	0.094	0.037	0.010	0.002		
	2	0.031	0.098	0.246	0.324	0.311	0.234	0.138	0.060	0.015	0.001	
	3	0.002	0.015	0.082	0.185	0.276	0.312	0.276	0.185	0.082	0.015	0.002
	4		0.001	0.015	0.060	0.138	0.234	0.311	0.324	0.246	0.098	0.031
	5			0.002	0.010	0.037	0.094	0.187	0.303	0.393	0.354	0.232
	6				0.001	0.004	0.016	0.047	0.118	0.262	0.531	0.735
7	0	0.698	0.478	0.210	0.082	0.028	0.008	0.002				
	1	0.257	0.372	0.367	0.247	0.131	0.055	0.017	0.004			
	2	0.041	0.124	0.275	0.318	0.261	0.164	0.077	0.025	0.004		
	3	0.004	0.023	0.115	0.227	0.290	0.273	0.194	0.097	0.029	0.003	
	4		0.003	0.029	0.097	0.194	0.273	0.290	0.227	0.115	0.023	0.004
	5			0.004	0.025	0.077	0.164	0.261	0.318	0.275	0.124	0.041
	6				0.004	0.017	0.055	0.131	0.247	0.367	0.372	0.257
	7					0.002	0.008	0.028	0.082	0.210	0.478	0.698

Source: John E. Freund, *Modern Elementary Statistics,* 7e, © 1988, pp. 517–521. Reprinted by permission of Prentice-Hall, Inc., Englewood Cliffs, New Jersey.

Note: All values omitted in this table are 0.0005 or less.

Table B

The Binomial Distribution   *Continued*

n	x	0.05	0.1	0.2	0.3	0.4	0.5	0.6	0.7	0.8	0.9	0.95
8	0	0.663	0.430	0.168	0.058	0.017	0.004	0.001				
	1	0.279	0.383	0.336	0.198	0.090	0.031	0.008	0.001			
	2	0.051	0.149	0.294	0.296	0.209	0.109	0.041	0.010	0.001		
	3	0.005	0.033	0.147	0.254	0.279	0.219	0.124	0.047	0.009		
	4		0.005	0.046	0.136	0.232	0.273	0.232	0.136	0.046	0.005	
	5			0.009	0.047	0.124	0.219	0.279	0.254	0.147	0.033	0.005
	6			0.001	0.010	0.041	0.109	0.209	0.296	0.294	0.149	0.051
	7				0.001	0.008	0.031	0.090	0.198	0.336	0.383	0.279
	8					0.001	0.004	0.017	0.058	0.168	0.430	0.663
9	0	0.630	0.387	0.134	0.040	0.010	0.002					
	1	0.299	0.387	0.302	0.156	0.060	0.018	0.004				
	2	0.063	0.172	0.302	0.267	0.161	0.070	0.021	0.004			
	3	0.008	0.045	0.176	0.267	0.251	0.164	0.074	0.021	0.003		
	4	0.001	0.007	0.066	0.172	0.251	0.246	0.167	0.074	0.017	0.001	
	5		0.001	0.017	0.074	0.167	0.246	0.251	0.172	0.066	0.007	0.001
	6			0.003	0.021	0.074	0.164	0.251	0.267	0.176	0.045	0.008
	7				0.004	0.021	0.070	0.161	0.267	0.302	0.172	0.063
	8					0.004	0.018	0.060	0.156	0.302	0.387	0.299
	9						0.002	0.010	0.040	0.134	0.387	0.630
10	0	0.599	0.349	0.107	0.028	0.006	0.001					
	1	0.315	0.387	0.268	0.121	0.040	0.010	0.002				
	2	0.075	0.194	0.302	0.233	0.121	0.044	0.011	0.001			
	3	0.010	0.057	0.201	0.267	0.215	0.117	0.042	0.009	0.001		
	4	0.001	0.011	0.088	0.200	0.251	0.205	0.111	0.037	0.006		
	5		0.001	0.026	0.103	0.201	0.246	0.201	0.103	0.026	0.001	
	6			0.006	0.037	0.111	0.205	0.251	0.200	0.088	0.011	0.001
	7			0.001	0.009	0.042	0.117	0.215	0.267	0.201	0.057	0.010
	8				0.001	0.011	0.044	0.121	0.233	0.302	0.194	0.075
	9					0.002	0.010	0.040	0.121	0.268	0.387	0.315
	10						0.001	0.006	0.028	0.107	0.349	0.599

Table B

The Binomial Distribution *Continued*

n	x	0.05	0.1	0.2	0.3	0.4	0.5	0.6	0.7	0.8	0.9	0.95
11	0	0.569	0.314	0.086	0.020	0.004						
	1	0.329	0.384	0.236	0.093	0.027	0.005	0.001				
	2	0.087	0.213	0.295	0.200	0.089	0.027	0.005	0.001			
	3	0.014	0.071	0.221	0.257	0.177	0.081	0.023	0.004			
	4	0.001	0.016	0.111	0.220	0.236	0.161	0.070	0.017	0.002		
	5		0.002	0.039	0.132	0.221	0.226	0.147	0.057	0.010		
	6			0.010	0.057	0.147	0.226	0.221	0.132	0.039	0.002	
	7			0.002	0.017	0.070	0.161	0.236	0.220	0.111	0.016	0.001
	8				0.004	0.023	0.081	0.177	0.257	0.221	0.071	0.014
	9				0.001	0.005	0.027	0.089	0.200	0.295	0.213	0.087
	10					0.001	0.005	0.027	0.093	0.236	0.384	0.329
	11							0.004	0.020	0.086	0.314	0.569
12	0	0.540	0.282	0.069	0.014	0.002						
	1	0.341	0.377	0.206	0.071	0.017	0.003					
	2	0.099	0.230	0.283	0.168	0.064	0.016	0.002				
	3	0.017	0.085	0.236	0.240	0.142	0.054	0.012	0.001			
	4	0.002	0.021	0.133	0.231	0.213	0.121	0.042	0.008	0.001		
	5		0.004	0.053	0.158	0.227	0.193	0.101	0.029	0.003		
	6			0.016	0.079	0.177	0.226	0.177	0.079	0.016		
	7			0.003	0.029	0.101	0.193	0.227	0.158	0.053	0.004	
	8			0.001	0.008	0.042	0.121	0.213	0.231	0.133	0.021	0.002
	9				0.001	0.012	0.054	0.142	0.240	0.236	0.085	0.017
	10					0.002	0.016	0.064	0.168	0.283	0.230	0.099
	11						0.003	0.017	0.071	0.206	0.377	0.341
	12							0.002	0.014	0.069	0.282	0.540
13	0	0.513	0.254	0.055	0.010	0.001						
	1	0.351	0.367	0.179	0.054	0.011	0.002					
	2	0.111	0.245	0.268	0.139	0.045	0.010	0.001				
	3	0.021	0.100	0.246	0.218	0.111	0.035	0.006	0.001			
	4	0.003	0.028	0.154	0.234	0.184	0.087	0.024	0.003			
	5		0.006	0.069	0.180	0.221	0.157	0.066	0.014	0.001		
	6		0.001	0.023	0.103	0.197	0.209	0.131	0.044	0.006		
	7			0.006	0.044	0.131	0.209	0.197	0.103	0.023	0.001	
	8			0.001	0.014	0.066	0.157	0.221	0.180	0.069	0.006	
	9				0.003	0.024	0.087	0.184	0.234	0.154	0.028	0.003
	10				0.001	0.006	0.035	0.111	0.218	0.246	0.100	0.021
	11					0.001	0.010	0.045	0.139	0.268	0.245	0.111
	12						0.002	0.011	0.054	0.179	0.367	0.351
	13							0.001	0.010	0.055	0.254	0.513

Table B

The Binomial Distribution   *Continued*

n	x	0.05	0.1	0.2	0.3	0.4	0.5	0.6	0.7	0.8	0.9	0.95
14	0	0.488	0.229	0.044	0.007	0.001						
	1	0.359	0.356	0.154	0.041	0.007	0.001					
	2	0.123	0.257	0.250	0.113	0.032	0.006	0.001				
	3	0.026	0.114	0.250	0.194	0.085	0.022	0.003				
	4	0.004	0.035	0.172	0.229	0.155	0.061	0.014	0.001			
	5		0.008	0.086	0.196	0.207	0.122	0.041	0.007			
	6		0.001	0.032	0.126	0.207	0.183	0.092	0.023	0.002		
	7			0.009	0.062	0.157	0.209	0.157	0.062	0.009		
	8			0.002	0.023	0.092	0.183	0.207	0.126	0.032	0.001	
	9				0.007	0.041	0.122	0.207	0.196	0.086	0.008	
	10				0.001	0.014	0.061	0.155	0.229	0.172	0.035	0.004
	11					0.003	0.022	0.085	0.194	0.250	0.114	0.026
	12					0.001	0.006	0.032	0.113	0.250	0.257	0.123
	13						0.001	0.007	0.041	0.154	0.356	0.359
	14							0.001	0.007	0.044	0.229	0.488
15	0	0.463	0.206	0.035	0.005							
	1	0.366	0.343	0.132	0.031	0.005						
	2	0.135	0.267	0.231	0.092	0.022	0.003					
	3	0.031	0.129	0.250	0.170	0.063	0.014	0.002				
	4	0.005	0.043	0.188	0.219	0.127	0.042	0.007	0.001			
	5	0.001	0.010	0.103	0.206	0.186	0.092	0.024	0.003			
	6		0.002	0.043	0.147	0.207	0.153	0.061	0.012	0.001		
	7			0.014	0.081	0.177	0.196	0.118	0.035	0.003		
	8			0.003	0.035	0.118	0.196	0.177	0.081	0.014		
	9			0.001	0.012	0.061	0.153	0.207	0.147	0.043	0.002	
	10				0.003	0.024	0.092	0.186	0.206	0.103	0.010	0.001
	11				0.001	0.007	0.042	0.127	0.219	0.188	0.043	0.005
	12					0.002	0.014	0.063	0.170	0.250	0.129	0.031
	13						0.003	0.022	0.092	0.231	0.267	0.135
	14							0.005	0.031	0.132	0.343	0.366
	15								0.005	0.035	0.206	0.463

Table B

The Binomial Distribution *Continued*

n	x	0.05	0.1	0.2	0.3	0.4	0.5	0.6	0.7	0.8	0.9	0.95
16	0	0.440	0.185	0.028	0.003							
	1	0.371	0.329	0.113	0.023	0.003						
	2	0.146	0.275	0.211	0.073	0.015	0.002					
	3	0.036	0.142	0.246	0.146	0.047	0.009	0.001				
	4	0.006	0.051	0.200	0.204	0.101	0.028	0.004				
	5	0.001	0.014	0.120	0.210	0.162	0.067	0.014	0.001			
	6		0.003	0.055	0.165	0.198	0.122	0.039	0.006			
	7			0.020	0.101	0.189	0.175	0.084	0.019	0.001		
	8			0.006	0.049	0.142	0.196	0.142	0.049	0.006		
	9			0.001	0.019	0.084	0.175	0.189	0.101	0.020		
	10				0.006	0.039	0.122	0.198	0.165	0.055	0.003	
	11				0.001	0.014	0.067	0.162	0.210	0.120	0.014	0.001
	12					0.004	0.028	0.101	0.204	0.200	0.051	0.006
	13					0.001	0.009	0.047	0.146	0.246	0.142	0.036
	14						0.002	0.015	0.073	0.211	0.275	0.146
	15							0.003	0.023	0.113	0.329	0.371
	16								0.003	0.028	0.185	0.440
17	0	0.418	0.167	0.023	0.002							
	1	0.374	0.315	0.096	0.017	0.002						
	2	0.158	0.280	0.191	0.058	0.010	0.001					
	3	0.041	0.156	0.239	0.125	0.034	0.005					
	4	0.008	0.060	0.209	0.187	0.080	0.018	0.002				
	5	0.001	0.017	0.136	0.208	0.138	0.047	0.008	0.001			
	6		0.004	0.068	0.178	0.184	0.094	0.024	0.003			
	7		0.001	0.027	0.120	0.193	0.148	0.057	0.009			
	8			0.008	0.064	0.161	0.185	0.107	0.028	0.002		
	9			0.002	0.028	0.107	0.185	0.161	0.064	0.008		
	10				0.009	0.057	0.148	0.193	0.120	0.027	0.001	
	11				0.003	0.024	0.094	0.184	0.178	0.068	0.004	
	12				0.001	0.008	0.047	0.138	0.208	0.136	0.017	0.001
	13					0.002	0.018	0.080	0.187	0.209	0.060	0.008
	14						0.005	0.034	0.125	0.239	0.156	0.041
	15						0.001	0.010	0.058	0.191	0.280	0.158
	16							0.002	0.017	0.096	0.315	0.374
	17								0.002	0.023	0.167	0.418

Table B

The Binomial Distribution *Continued*

n	x	0.05	0.1	0.2	0.3	0.4	0.5	0.6	0.7	0.8	0.9	0.95
18	0	0.397	0.150	0.018	0.002							
	1	0.376	0.300	0.081	0.013	0.001						
	2	0.168	0.284	0.172	0.046	0.007	0.001					
	3	0.047	0.168	0.230	0.105	0.025	0.003					
	4	0.009	0.070	0.215	0.168	0.061	0.012	0.001				
	5	0.001	0.022	0.151	0.202	0.115	0.033	0.004				
	6		0.005	0.082	0.187	0.166	0.071	0.015	0.001			
	7		0.001	0.035	0.138	0.189	0.121	0.037	0.005			
	8			0.012	0.081	0.173	0.167	0.077	0.015	0.001		
	9			0.003	0.039	0.128	0.185	0.128	0.039	0.003		
	10			0.001	0.015	0.077	0.167	0.173	0.081	0.012		
	11				0.005	0.037	0.121	0.189	0.138	0.035	0.001	
	12				0.001	0.015	0.071	0.166	0.187	0.082	0.005	
	13					0.004	0.033	0.115	0.202	0.151	0.022	0.001
	14					0.001	0.012	0.061	0.168	0.215	0.070	0.009
	15						0.003	0.025	0.105	0.230	0.168	0.047
	16						0.001	0.007	0.046	0.172	0.284	0.168
	17							0.001	0.013	0.081	0.300	0.376
	18								0.002	0.018	0.150	0.397
19	0	0.377	0.135	0.014	0.001							
	1	0.377	0.285	0.068	0.009	0.001						
	2	0.179	0.285	0.154	0.036	0.005						
	3	0.053	0.180	0.218	0.087	0.017	0.002					
	4	0.011	0.080	0.218	0.149	0.047	0.007	0.001				
	5	0.002	0.027	0.164	0.192	0.093	0.022	0.002				
	6		0.007	0.095	0.192	0.145	0.052	0.008	0.001			
	7		0.001	0.044	0.153	0.180	0.096	0.024	0.002			
	8			0.017	0.098	0.180	0.144	0.053	0.008			
	9			0.005	0.051	0.146	0.176	0.098	0.022	0.001		
	10			0.001	0.022	0.098	0.176	0.146	0.051	0.005		
	11				0.008	0.053	0.144	0.180	0.098	0.017		
	12				0.002	0.024	0.096	0.180	0.153	0.044	0.001	
	13				0.001	0.008	0.052	0.145	0.192	0.095	0.007	
	14					0.002	0.022	0.093	0.192	0.164	0.027	0.002
	15					0.001	0.007	0.047	0.149	0.218	0.080	0.011
	16						0.002	0.017	0.087	0.218	0.180	0.053
	17							0.005	0.036	0.154	0.285	0.179
	18							0.001	0.009	0.068	0.285	0.377
	19								0.001	0.014	0.135	0.377

Table B

The Binomial Distribution *Continued*

n	x	0.05	0.1	0.2	0.3	0.4	0.5	0.6	0.7	0.8	0.9	0.95
20	0	0.358	0.122	0.012	0.001							
	1	0.377	0.270	0.058	0.007							
	2	0.189	0.285	0.137	0.028	0.003						
	3	0.060	0.190	0.205	0.072	0.012	0.001					
	4	0.013	0.090	0.218	0.130	0.035	0.005					
	5	0.002	0.032	0.175	0.179	0.075	0.015	0.001				
	6		0.009	0.109	0.192	0.124	0.037	0.005				
	7		0.002	0.055	0.164	0.166	0.074	0.015	0.001			
	8			0.022	0.114	0.180	0.120	0.035	0.004			
	9			0.007	0.065	0.160	0.160	0.071	0.012			
	10			0.002	0.031	0.117	0.176	0.117	0.031	0.002		
	11				0.012	0.071	0.160	0.160	0.065	0.007		
	12				0.004	0.035	0.120	0.180	0.114	0.022		
	13				0.001	0.015	0.074	0.166	0.164	0.055	0.002	
	14					0.005	0.037	0.124	0.192	0.109	0.009	
	15					0.001	0.015	0.075	0.179	0.175	0.032	0.002
	16						0.005	0.035	0.130	0.218	0.090	0.013
	17						0.001	0.012	0.072	0.205	0.190	0.060
	18							0.003	0.028	0.137	0.285	0.189
	19								0.007	0.058	0.270	0.377
	20								0.001	0.012	0.122	0.358

Table C

The Poisson Distribution

x	0.1	0.2	0.3	0.4	0.5	0.6	0.7	0.8	0.9	1.0
0	.9048	.8187	.7408	.6703	.6065	.5488	.4966	.4493	.4066	.3679
1	.0905	.1637	.2222	.2681	.3033	.3293	.3476	.3595	.3659	.3679
2	.0045	.0164	.0333	.0536	.0758	.0988	.1217	.1438	.1647	.1839
3	.0002	.0011	.0033	.0072	.0126	.0198	.0284	.0383	.0494	.0613
4	.0000	.0001	.0003	.0007	.0016	.0030	.0050	.0077	.0111	.0153
5	.0000	.0000	.0000	.0001	.0002	.0004	.0007	.0012	.0020	.0031
6	.0000	.0000	.0000	.0000	.0000	.0000	.0001	.0002	.0003	.0005
7	.0000	.0000	.0000	.0000	.0000	.0000	.0000	.0000	.0000	.0001

Source: From Beyer, W. H., *Handbook of Tables for Probability and Statistics, 2nd Edition,* CRC Press, Boca Raton, Florida, 1986. With permission.

Table C

The Poisson Distribution *Continued*

					λ					
x	1.1	1.2	1.3	1.4	1.5	1.6	1.7	1.8	1.9	2.0
0	.3329	.3012	.2725	.2466	.2231	.2019	.1827	.1653	.1496	.1353
1	.3662	.3614	.3543	.3452	.3347	.3230	.3106	.2975	.2842	.2707
2	.2014	.2169	.2303	.2417	.2510	.2584	.2640	.2678	.2700	.2707
3	.0738	.0867	.0998	.1128	.1255	.1378	.1496	.1607	.1710	.1804
4	.0203	.0260	.0324	.0395	.0471	.0551	.0636	.0723	.0812	.0902
5	.0045	.0062	.0084	.0111	.0141	.0176	.0216	.0260	.0309	.0361
6	.0008	.0012	.0018	.0026	.0035	.0047	.0061	.0078	.0098	.0120
7	.0001	.0002	.0003	.0005	.0008	.0011	.0015	.0020	.0027	.0034
8	.0000	.0000	.0001	.0001	.0001	.0002	.0003	.0005	.0006	.0009
9	.0000	.0000	.0000	.0000	.0000	.0000	.0001	.0001	.0001	.0002

					λ					
x	2.1	2.2	2.3	2.4	2.5	2.6	2.7	2.8	2.9	3.0
0	.1225	.1108	.1003	.0907	.0821	.0743	.0672	.0608	.0550	.0498
1	.2572	.2438	.2306	.2177	.2052	.1931	.1815	.1703	.1596	.1494
2	.2700	.2681	.2652	.2613	.2565	.2510	.2450	.2384	.2314	.2240
3	.1890	.1966	.2033	.2090	.2138	.2176	.2205	.2225	.2237	.2240
4	.0992	.1082	.1169	.1254	.1336	.1414	.1488	.1557	.1622	.1680
5	.0417	.0476	.0538	.0602	.0668	.0735	.0804	.0872	.0940	.1008
6	.0146	.0174	.0206	.0241	.0278	.0319	.0362	.0407	.0455	.0504
7	.0044	.0055	.0068	.0083	.0099	.0118	.0139	.0163	.0188	.0216
8	.0011	.0015	.0019	.0025	.0031	.0038	.0047	.0057	.0068	.0081
9	.0003	.0004	.0005	.0007	.0009	.0011	.0014	.0018	.0022	.0027
10	.0001	.0001	.0001	.0002	.0002	.0003	.0004	.0005	.0006	.0008
11	.0000	.0000	.0000	.0000	.0000	.0001	.0001	.0001	.0002	.0002
12	.0000	.0000	.0000	.0000	.0000	.0000	.0000	.0000	.0000	.0001

					λ					
x	3.1	3.2	3.3	3.4	3.5	3.6	3.7	3.8	3.9	4.0
0	.0450	.0408	.0369	.0334	.0302	.0273	.0247	.0224	.0202	.0183
1	.1397	.1304	.1217	.1135	.1057	.0984	.0915	.0850	.0789	.0733
2	.2165	.2087	.2008	.1929	.1850	.1771	.1692	.1615	.1539	.1465
3	.2237	.2226	.2209	.2186	.2158	.2125	.2087	.2046	.2001	.1954
4	.1734	.1781	.1823	.1858	.1888	.1912	.1931	.1944	.1951	.1954

Table C

The Poisson Distribution *Continued*

x	3.1	3.2	3.3	3.4	3.5	3.6	3.7	3.8	3.9	4.0
					λ					
5	.1075	.1140	.1203	.1264	.1322	.1377	.1429	.1477	.1522	.1563
6	.0555	.0608	.0662	.0716	.0771	.0826	.0881	.0936	.0989	.1042
7	.0246	.0278	.0312	.0348	.0385	.0425	.0466	.0508	.0551	.0595
8	.0095	.0111	.0129	.0148	.0169	.0191	.0215	.0241	.0269	.0298
9	.0033	.0040	.0047	.0056	.0066	.0076	.0089	.0102	.0116	.0132
10	.0010	.0013	.0016	.0019	.0023	.0028	.0033	.0039	.0045	.0053
11	.0003	.0004	.0005	.0006	.0007	.0009	.0011	.0013	.0016	.0019
12	.0001	.0001	.0001	.0002	.0002	.0003	.0003	.0004	.0005	.0006
13	.0000	.0000	.0000	.0000	.0001	.0001	.0001	.0001	.0002	.0002
14	.0000	.0000	.0000	.0000	.0000	.0000	.0000	.0000	.0000	.0001

x	4.1	4.2	4.3	4.4	4.5	4.6	4.7	4.8	4.9	5.0
					λ					
0	.0166	.0150	.0136	.0123	.0111	.0101	.0091	.0082	.0074	.0067
1	.0679	.0630	.0583	.0540	.0500	.0462	.0427	.0395	.0365	.0337
2	.1393	.1323	.1254	.1188	.1125	.1063	.1005	.0948	.0894	.0842
3	.1904	.1852	.1798	.1743	.1687	.1631	.1574	.1517	.1460	.1404
4	.1951	.1944	.1933	.1917	.1898	.1875	.1849	.1820	.1789	.1755
5	.1600	.1633	.1662	.1687	.1708	.1725	.1738	.1747	.1753	.1755
6	.1093	.1143	.1191	.1237	.1281	.1323	.1362	.1398	.1432	.1462
7	.0640	.0686	.0732	.0778	.0824	.0869	.0914	.0959	.1002	.1044
8	.0328	.0360	.0393	.0428	.0463	.0500	.0537	.0575	.0614	.0653
9	.0150	.0168	.0188	.0209	.0232	.0255	.0280	.0307	.0334	.0363
10	.0061	.0071	.0081	.0092	.0104	.0118	.0132	.0147	.0164	.0181
11	.0023	.0027	.0032	.0037	.0043	.0049	.0056	.0064	.0073	.0082
12	.0008	.0009	.0011	.0014	.0016	.0019	.0022	.0026	.0030	.0034
13	.0002	.0003	.0004	.0005	.0006	.0007	.0008	.0009	.0011	.0013
14	.0001	.0001	.0001	.0001	.0002	.0002	.0003	.0003	.0004	.0005
15	.0000	.0000	.0000	.0000	.0001	.0001	.0001	.0001	.0001	.0002

x	5.1	5.2	5.3	5.4	5.5	5.6	5.7	5.8	5.9	6.0
					λ					
0	.0061	.0055	.0050	.0045	.0041	.0037	.0033	.0030	.0027	.0025
1	.0311	.0287	.0265	.0244	.0225	.0207	.0191	.0176	.0162	.0149
2	.0793	.0746	.0701	.0659	.0618	.0580	.0544	.0509	.0477	.0446
3	.1348	.1293	.1239	.1185	.1133	.1082	.1033	.0985	.0938	.0892
4	.1719	.1681	.1641	.1600	.1558	.1515	.1472	.1428	.1383	.1339

Table C

The Poisson Distribution *Continued*

x	λ 5.1	5.2	5.3	5.4	5.5	5.6	5.7	5.8	5.9	6.0
5	.1753	.1748	.1740	.1728	.1714	.1697	.1678	.1656	.1632	.1606
6	.1490	.1515	.1537	.1555	.1571	.1584	.1594	.1601	.1605	.1606
7	.1086	.1125	.1163	.1200	.1234	.1267	.1298	.1326	.1353	.1377
8	.0692	.0731	.0771	.0810	.0849	.0887	.0925	.0962	.0998	.1033
9	.0392	.0423	.0454	.0486	.0519	.0552	.0586	.0620	.0654	.0688
10	.0200	.0220	.0241	.0262	.0285	.0309	.0334	.0359	.0386	.0413
11	.0093	.0104	.0116	.0129	.0143	.0157	.0173	.0190	.0207	.0225
12	.0039	.0045	.0051	.0058	.0065	.0073	.0082	.0092	.0102	.0113
13	.0015	.0018	.0021	.0024	.0028	.0032	.0036	.0041	.0046	.0052
14	.0006	.0007	.0008	.0009	.0011	.0013	.0015	.0017	.0019	.0022
15	.0002	.0002	.0003	.0003	.0004	.0005	.0006	.0007	.0008	.0009
16	.0001	.0001	.0001	.0001	.0001	.0002	.0002	.0002	.0003	.0003
17	.0000	.0000	.0000	.0000	.0000	.0000	.0001	.0001	.0001	.0001

x	λ 6.1	6.2	6.3	6.4	6.5	6.6	6.7	6.8	6.9	7.0
0	.0022	.0020	.0018	.0017	.0015	.0014	.0012	.0011	.0010	.0009
1	.0137	.0126	.0116	.0106	.0098	.0090	.0082	.0076	.0070	.0064
2	.0417	.0390	.0364	.0340	.0318	.0296	.0276	.0258	.0240	.0223
3	.0848	.0806	.0765	.0726	.0688	.0652	.0617	.0584	.0552	.0521
4	.1294	.1249	.1205	.1162	.1118	.1076	.1034	.0992	.0952	.0912
5	.1579	.1549	.1519	.1487	.1454	.1420	.1385	.1349	.1314	.1277
6	.1605	.1601	.1595	.1586	.1575	.1562	.1546	.1529	.1511	.1490
7	.1399	.1418	.1435	.1450	.1462	.1472	.1480	.1486	.1489	.1490
8	.1066	.1099	.1130	.1160	.1188	.1215	.1240	.1263	.1284	.1304
9	.0723	.0757	.0791	.0825	.0858	.0891	.0923	.0954	.0985	.1014
10	.0441	.0469	.0498	.0528	.0558	.0588	.0618	.0649	.0679	.0710
11	.0245	.0265	.0285	.0307	.0330	.0353	.0377	.0401	.0426	.0452
12	.0124	.0137	.0150	.0164	.0179	.0194	.0210	.0227	.0245	.0264
13	.0058	.0065	.0073	.0081	.0089	.0098	.0108	.0119	.0130	.0142
14	.0025	.0029	.0033	.0037	.0041	.0046	.0052	.0058	.0064	.0071
15	.0010	.0012	.0014	.0016	.0018	.0020	.0023	.0026	.0029	.0033
16	.0004	.0005	.0005	.0006	.0007	.0008	.0010	.0011	.0013	.0014
17	.0001	.0002	.0002	.0002	.0003	.0003	.0004	.0004	.0005	.0006
18	.0000	.0001	.0001	.0001	.0001	.0001	.0001	.0002	.0002	.0002
19	.0000	.0000	.0000	.0000	.0000	.0000	.0000	.0001	.0001	.0001

Table C

The Poisson Distribution *Continued*

	λ									
x	*7.1*	*7.2*	*7.3*	*7.4*	*7.5*	*7.6*	*7.7*	*7.8*	*7.9*	*8.0*
0	.0008	.0007	.0007	.0006	.0006	.0005	.0005	.0004	.0004	.0003
1	.0059	.0054	.0049	.0045	.0041	.0038	.0035	.0032	.0029	.0027
2	.0208	.0194	.0180	.0167	.0156	.0145	.0134	.0125	.0116	.0107
3	.0492	.0464	.0438	.0413	.0389	.0366	.0345	.0324	.0305	.0286
4	.0874	.0836	.0799	.0764	.0729	.0696	.0663	.0632	.0602	.0573
5	.1241	.1204	.1167	.1130	.1094	.1057	.1021	.0986	.0951	.0916
6	.1468	.1445	.1420	.1394	.1367	.1339	.1311	.1282	.1252	.1221
7	.1489	.1486	.1481	.1474	.1465	.1454	.1442	.1428	.1413	.1396
8	.1321	.1337	.1351	.1363	.1373	.1382	.1388	.1392	.1395	.1396
9	.1042	.1070	.1096	.1121	.1144	.1167	.1187	.1207	.1224	.1241
10	.0740	.0770	.0800	.0829	.0858	.0887	.0914	.0941	.0967	.0993
11	.0478	.0504	.0531	.0558	.0585	.0613	.0640	.0667	.0695	.0722
12	.0283	.0303	.0323	.0344	.0366	.0388	.0411	.0434	.0457	.0481
13	.0154	.0168	.0181	.0196	.0211	.0227	.0243	.0260	.0278	.0296
14	.0078	.0086	.0095	.0104	.0113	.0123	.0134	.0145	.0157	.0169
15	.0037	.0041	.0046	.0051	.0057	.0062	.0069	.0075	.0083	.0090
16	.0016	.0019	.0021	.0024	.0026	.0030	.0033	.0037	.0041	.0045
17	.0007	.0008	.0009	.0010	.0012	.0013	.0015	.0017	.0019	.0021
18	.0003	.0003	.0004	.0004	.0005	.0006	.0006	.0007	.0008	.0009
19	.0001	.0001	.0001	.0002	.0002	.0002	.0003	.0003	.0003	.0004
20	.0000	.0000	.0001	.0001	.0001	.0001	.0001	.0001	.0001	.0002
21	.0000	.0000	.0000	.0000	.0000	.0000	.0000	.0000	.0001	.0001

	λ									
x	*8.1*	*8.2*	*8.3*	*8.4*	*8.5*	*8.6*	*8.7*	*8.8*	*8.9*	*9.0*
0	.0003	.0003	.0002	.0002	.0002	.0002	.0002	.0002	.0001	.0001
1	.0025	.0023	.0021	.0019	.0017	.0016	.0014	.0013	.0012	.0011
2	.0100	.0092	.0086	.0079	.0074	.0068	.0063	.0058	.0054	.0050
3	.0269	.0252	.0237	.0222	.0208	.0195	.0183	.0171	.0160	.0150
4	.0544	.0517	.0491	.0466	.0443	.0420	.0398	.0377	.0357	.0337
5	.0882	.0849	.0816	.0784	.0752	.0722	.0692	.0663	.0635	.0607
6	.1191	.1160	.1128	.1097	.1066	.1034	.1003	.0972	.0941	.0911
7	.1378	.1358	.1338	.1317	.1294	.1271	.1247	.1222	.1197	.1171
8	.1395	.1392	.1388	.1382	.1375	.1366	.1356	.1344	.1332	.1318
9	.1256	.1269	.1280	.1290	.1299	.1306	.1311	.1315	.1317	.1318

Table C

The Poisson Distribution   *Continued*

x	λ 8.1	8.2	8.3	8.4	8.5	8.6	8.7	8.8	8.9	9.0
10	.1017	.1040	.1063	.1084	.1104	.1123	.1140	.1157	.1172	.1186
11	.0749	.0776	.0802	.0828	.0853	.0878	.0902	.0925	.0948	.0970
12	.0505	.0530	.0555	.0579	.0604	.0629	.0654	.0679	.0703	.0728
13	.0315	.0334	.0354	.0374	.0395	.0416	.0438	.0459	.0481	.0504
14	.0182	.0196	.0210	.0225	.0240	.0256	.0272	.0289	.0306	.0324
15	.0098	.0107	.0116	.0126	.0136	.0147	.0158	.0169	.0182	.0194
16	.0050	.0055	.0060	.0066	.0072	.0079	.0086	.0093	.0101	.0109
17	.0024	.0026	.0029	.0033	.0036	.0040	.0044	.0048	.0053	.0058
18	.0011	.0012	.0014	.0015	.0017	.0019	.0021	.0024	.0026	.0029
19	.0005	.0005	.0006	.0007	.0008	.0009	.0010	.0011	.0012	.0014
20	.0002	.0002	.0002	.0003	.0003	.0004	.0004	.0005	.0005	.0006
21	.0001	.0001	.0001	.0001	.0001	.0002	.0002	.0002	.0002	.0003
22	.0000	.0000	.0000	.0000	.0001	.0001	.0001	.0001	.0001	.0001

x	λ 9.1	9.2	9.3	9.4	9.5	9.6	9.7	9.8	9.9	10
0	.0001	.0001	.0001	.0001	.0001	.0001	.0001	.0001	.0001	.0000
1	.0010	.0009	.0009	.0008	.0007	.0007	.0006	.0005	.0005	.0005
2	.0046	.0043	.0040	.0037	.0034	.0031	.0029	.0027	.0025	.0023
3	.0140	.0131	.0123	.0115	.0107	.0100	.0093	.0087	.0081	.0076
4	.0319	.0302	.0285	.0269	.0254	.0240	.0226	.0213	.0201	.0189
5	.0581	.0555	.0530	.0506	.0483	.0460	.0439	.0418	.0398	.0378
6	.0881	.0851	.0822	.0793	.0764	.0736	.0709	.0682	.0656	.0631
7	.1145	.1118	.1091	.1064	.1037	.1010	.0982	.0955	.0928	.0901
8	.1302	.1286	.1269	.1251	.1232	.1212	.1191	.1170	.1148	.1126
9	.1317	.1315	.1311	.1306	.1300	.1293	.1284	.1274	.1263	.1251
10	.1198	.1210	.1219	.1228	.1235	.1241	.1245	.1249	.1250	.1251
11	.0991	.1012	.1031	.1049	.1067	.1083	.1098	.1112	.1125	.1137
12	.0752	.0776	.0799	.0822	.0844	.0866	.0888	.0908	.0928	.0948
13	.0526	.0549	.0572	.0594	.0617	.0640	.0662	.0685	.0707	.0729
14	.0342	.0361	.0380	.0399	.0419	.0439	.0459	.0479	.0500	.0521
15	.0208	.0221	.0235	.0250	.0265	.0281	.0297	.0313	.0330	.0347
16	.0118	.0127	.0137	.0147	.0157	.0168	.0180	.0192	.0204	.0217
17	.0063	.0069	.0075	.0081	.0088	.0095	.0103	.0111	.0119	.0128
18	.0032	.0035	.0039	.0042	.0046	.0051	.0055	.0060	.0065	.0071
19	.0015	.0017	.0019	.0021	.0023	.0026	.0028	.0031	.0034	.0037

Table C

The Poisson Distribution *Continued*

x	9.1	9.2	9.3	9.4	9.5	9.6	9.7	9.8	9.9	10
					λ					
20	.0007	.0008	.0009	.0010	.0011	.0012	.0014	.0015	.0017	.0019
21	.0003	.0003	.0004	.0004	.0005	.0006	.0006	.0007	.0008	.0009
22	.0001	.0001	.0002	.0002	.0002	.0002	.0003	.0003	.0004	.0004
23	.0000	.0001	.0001	.0001	.0001	.0001	.0001	.0001	.0002	.0002
24	.0000	.0000	.0000	.0000	.0000	.0000	.0000	.0001	.0001	.0001

x	11	12	13	14	15	16	17	18	19	20
					λ					
0	.0000	.0000	.0000	.0000	.0000	.0000	.0000	.0000	.0000	.0000
1	.0002	.0001	.0000	.0000	.0000	.0000	.0000	.0000	.0000	.0000
2	.0010	.0004	.0002	.0001	.0000	.0000	.0000	.0000	.0000	.0000
3	.0037	.0018	.0008	.0004	.0002	.0001	.0000	.0000	.0000	.0000
4	.0102	.0053	.0027	.0013	.0006	.0003	.0001	.0001	.0000	.0000
5	.0224	.0127	.0070	.0037	.0019	.0010	.0005	.0002	.0001	.0001
6	.0411	.0255	.0152	.0087	.0048	.0026	.0014	.0007	.0004	.0002
7	.0646	.0437	.0281	.0174	.0104	.0060	.0034	.0018	.0010	.0005
8	.0888	.0655	.0457	.0304	.0194	.0120	.0072	.0042	.0024	.0013
9	.1085	.0874	.0661	.0473	.0324	.0213	.0135	.0083	.0050	.0029
10	.1194	.1048	.0859	.0663	.0486	.0341	.0230	.0150	.0095	.0058
11	.1194	.1144	.1015	.0844	.0663	.0496	.0355	.0245	.0164	.0106
12	.1094	.1144	.1099	.0984	.0829	.0661	.0504	.0368	.0259	.0176
13	.0926	.1056	.1099	.1060	.0956	.0814	.0658	.0509	.0378	.0271
14	.0728	.0905	.1021	.1060	.1024	.0930	.0800	.0655	.0514	.0387
15	.0534	.0724	.0885	.0989	.1024	.0992	.0906	.0786	.0650	.0516
16	.0367	.0543	.0719	.0866	.0960	.0992	.0963	.0884	.0772	.0646
17	.0237	.0383	.0550	.0713	.0847	.0934	.0963	.0936	.0863	.0760
18	.0145	.0256	.0397	.0554	.0706	.0830	.0909	.0936	.0911	.0844
19	.0084	.0161	.0272	.0409	.0557	.0699	.0814	.0887	.0911	.0888
20	.0046	.0097	.0177	.0286	.0418	.0559	.0692	.0798	.0866	.0888
21	.0024	.0055	.0109	.0191	.0299	.0426	.0560	.0684	.0783	.0846
22	.0012	.0030	.0065	.0121	.0204	.0310	.0433	.0560	.0676	.0769
23	.0006	.0016	.0037	.0074	.0133	.0216	.0320	.0438	.0559	.0669
24	.0003	.0008	.0020	.0043	.0083	.0144	.0226	.0328	.0442	.0557
25	.0001	.0004	.0010	.0024	.0050	.0092	.0154	.0237	.0336	.0446
26	.0000	.0002	.0005	.0013	.0029	.0057	.0101	.0164	.0246	.0343
27	.0000	.0001	.0002	.0007	.0016	.0034	.0063	.0109	.0173	.0254
28	.0000	.0000	.0001	.0003	.0009	.0019	.0038	.0070	.0117	.0181
29	.0000	.0000	.0001	.0002	.0004	.0011	.0023	.0044	.0077	.0125

Table C

The Poisson Distribution *Continued*

	λ									
x	*11*	*12*	*13*	*14*	*15*	*16*	*17*	*18*	*19*	*20*
30	.0000	.0000	.0000	.0001	.0002	.0006	.0013	.0026	.0049	.0083
31	.0000	.0000	.0000	.0000	.0001	.0003	.0007	.0015	.0030	.0054
32	.0000	.0000	.0000	.0000	.0001	.0001	.0004	.0009	.0018	.0034
33	.0000	.0000	.0000	.0000	.0000	.0001	.0002	.0005	.0010	.0020
34	.0000	.0000	.0000	.0000	.0000	.0000	.0001	.0002	.0006	.0012
35	.0000	.0000	.0000	.0000	.0000	.0000	.0000	.0001	.0003	.0007
36	.0000	.0000	.0000	.0000	.0000	.0000	.0000	.0001	.0002	.0004
37	.0000	.0000	.0000	.0000	.0000	.0000	.0000	.0000	.0001	.0002
38	.0000	.0000	.0000	.0000	.0000	.0000	.0000	.0000	.0000	.0001
39	.0000	.0000	.0000	.0000	.0000	.0000	.0000	.0000	.0000	.0001

Table D

Random Numbers

10480	15011	01536	02011	81647	91646	69179	14194	62590	36207	20969	99570	91291	90700
22368	46573	25595	85393	30995	89198	27982	53402	93965	34095	52666	19174	39615	99505
24130	48360	22527	97265	76393	64809	15179	24830	49340	32081	30680	19655	63348	58629
42167	93093	06243	61680	07856	16376	39440	53537	71341	57004	00849	74917	97758	16379
37570	39975	81837	16656	06121	91782	60468	81305	49684	60672	14110	06927	01263	54613
77921	06907	11008	42751	27756	53498	18602	70659	90655	15053	21916	81825	44394	42880
99562	72905	56420	69994	98872	31016	71194	18738	44013	48840	63213	21069	10634	12952
96301	91977	05463	07972	18876	20922	94595	56869	69014	60045	18425	84903	42508	32307
89579	14342	63661	10281	17453	18103	57740	84378	25331	12566	58678	44947	05584	56941
85475	36857	43342	53988	53060	59533	38867	62300	08158	17983	16439	11458	18593	64952
28918	69578	88231	33276	70997	79936	56865	05859	90106	31595	01547	85590	91610	78188
63553	40961	48235	03427	49626	69445	18663	72695	52180	20847	12234	90511	33703	90322
09429	93969	52636	92737	88974	33488	36320	17617	30015	08272	84115	27156	30613	74952
10365	61129	87529	85689	48237	52267	67689	93394	01511	26358	85104	20285	29975	89868
07119	97336	71048	08178	77233	13916	47564	81056	97735	85977	29372	74461	28551	90707
51085	12765	51821	51259	77452	16308	60756	92144	49442	53900	70960	63990	75601	40719
02368	21382	52404	60268	89368	19885	55322	44819	01188	65255	64835	44919	05944	55157
01011	54092	33362	94904	31273	04146	18594	29852	71585	85030	51132	01915	92747	64951
52162	53916	46369	58586	23216	14513	83149	98736	23495	64350	94738	17752	35156	35749
07056	97628	33787	09998	42698	06691	76988	13602	51851	46104	88916	19509	25625	58104

Table D

Random Numbers *Continued*

48663	91245	85828	14346	09172	30168	90229	04734	59193	22178	30421	61666	99904	32812
54164	58492	22421	74103	47070	25306	76468	26384	58151	06646	21524	15227	96909	44592
32639	32363	05597	24200	13363	38005	94342	28728	35806	06912	17012	64161	18296	22851
29334	27001	87637	87308	58731	00256	45834	15398	46557	41135	10367	07684	36188	18510
02488	33062	28834	07351	19731	92420	60952	61280	50001	67658	32586	86679	50720	94953
81525	72295	04839	96423	24878	82651	66566	14778	76797	14780	13300	87074	79666	95725
29676	20591	68086	26432	46901	20849	89768	81536	86645	12659	92259	57102	80428	25280
00742	57392	39064	66432	84673	40027	32832	61362	98947	96067	64760	64584	96096	98253
05366	04213	25669	26422	44407	44048	37937	63904	45766	66134	75470	66520	34693	90449
91921	26418	64117	94305	26766	25940	39972	22209	71500	64568	91402	42416	07844	69618
00582	04711	87917	77341	42206	35126	74087	99547	81817	42607	43808	76655	62028	76630
00725	69884	62797	56170	86324	88072	76222	36086	84637	93161	76038	65855	77919	88006
69011	65797	95876	55293	18988	27354	26575	08625	40801	59920	29841	80150	12777	48501
25976	57948	29888	88604	67917	48708	18912	82271	65424	69774	33611	54262	85963	03547
09763	83473	73577	12908	30883	18317	28290	35797	05998	41688	34952	37888	38917	88050
91567	42595	27958	30134	04024	86385	29880	99730	55536	84855	29080	09250	79656	73211
17955	56349	90999	49127	20044	59931	06115	20542	18059	02008	73708	83517	36103	42791
46503	18584	18845	49618	02304	51038	20655	58727	28168	15475	56942	53389	20562	87338
92157	89634	94824	78171	84610	82834	09922	25417	44137	48413	25555	21246	35509	20468
14577	62765	35605	81263	39667	47358	56873	56307	61607	49518	89656	20103	77490	18062
98427	07523	33362	64270	01638	92477	66969	98420	04880	45585	46565	04102	46880	45709
34914	63976	88720	82765	34476	17032	87589	40836	32427	70002	70663	88863	77775	69348
70060	28277	39475	46473	23219	53416	94970	25832	69975	94884	19661	72828	00102	66794
53976	54914	06990	67245	68350	82948	11398	42878	80287	88267	47363	46634	06541	97809
76072	29515	40980	07391	58745	25774	22987	80059	39911	96189	41151	14222	60697	59583
90725	52210	83974	29992	65831	38857	50490	83765	55657	14361	31720	57375	56228	41546
64364	67412	33339	31926	14883	24413	59744	92351	97473	89286	35931	04110	23726	51900
08962	00358	31662	25388	61642	34072	81249	35648	56891	69352	48373	45578	78547	81788
95012	68379	93526	70765	10593	04542	76463	54328	02349	17247	28865	14777	62730	92277
15664	10493	20492	38391	91132	21999	59516	81652	27195	48223	46751	22923	32261	85653

Source: From Beyer, W. H., *Handbook of Tables for Probability and Statistics, 2nd Edition,* CRC Press, Boca Raton, Florida, 1986. With permission.

Table E

The Standard Normal Distribution

z	.00	.01	.02	.03	.04	.05	.06	.07	.08	.09
0.0	.0000	.0040	.0080	.0120	.0160	.0199	.0239	.0279	.0319	.0359
0.1	.0398	.0438	.0478	.0517	.0557	.0596	.0636	.0675	.0714	.0753
0.2	.0793	.0832	.0871	.0910	.0948	.0987	.1026	.1064	.1103	.1141
0.3	.1179	.1217	.1255	.1293	.1331	.1368	.1406	.1443	.1480	.1517
0.4	.1554	.1591	.1628	.1664	.1700	.1736	.1772	.1808	.1844	.1879
0.5	.1915	.1950	.1985	.2019	.2054	.2088	.2123	.2157	.2190	.2224
0.6	.2257	.2291	.2324	.2357	.2389	.2422	.2454	.2486	.2517	.2549
0.7	.2580	.2611	.2642	.2673	.2704	.2734	.2764	.2794	.2823	.2852
0.8	.2881	.2910	.2939	.2967	.2995	.3023	.3051	.3078	.3106	.3133
0.9	.3159	.3186	.3212	.3238	.3264	.3289	.3315	.3340	.3365	.3389
1.0	.3413	.3438	.3461	.3485	.3508	.3531	.3554	.3577	.3599	.3621
1.1	.3643	.3665	.3686	.3708	.3729	.3749	.3770	.3790	.3810	.3830
1.2	.3849	.3869	.3888	.3907	.3925	.3944	.3962	.3980	.3997	.4015
1.3	.4032	.4049	.4066	.4082	.4099	.4115	.4131	.4147	.4162	.4177
1.4	.4192	.4207	.4222	.4236	.4251	.4265	.4279	.4292	.4306	.4319
1.5	.4332	.4345	.4357	.4370	.4382	.4394	.4406	.4418	.4429	.4441
1.6	.4452	.4463	.4474	.4484	.4495	.4505	.4515	.4525	.4535	.4545
1.7	.4554	.4564	.4573	.4582	.4591	.4599	.4608	.4616	.4625	.4633
1.8	.4641	.4649	.4656	.4664	.4671	.4678	.4686	.4693	.4699	.4706
1.9	.4713	.4719	.4726	.4732	.4738	.4744	.4750	.4756	.4761	.4767
2.0	.4772	.4778	.4783	.4788	.4793	.4798	.4803	.4808	.4812	.4817
2.1	.4821	.4826	.4830	.4834	.4838	.4842	.4846	.4850	.4854	.4857
2.2	.4861	.4864	.4868	.4871	.4875	.4878	.4881	.4884	.4887	.4890
2.3	.4893	.4896	.4898	.4901	.4904	.4906	.4909	.4911	.4913	.4916
2.4	.4918	.4920	.4922	.4925	.4927	.4929	.4931	.4932	.4934	.4936
2.5	.4938	.4940	.4941	.4943	.4945	.4946	.4948	.4949	.4951	.4952
2.6	.4953	.4955	.4956	.4957	.4959	.4960	.4961	.4962	.4963	.4964
2.7	.4965	.4966	.4967	.4968	.4969	.4970	.4971	.4972	.4973	.4974
2.8	.4974	.4975	.4976	.4977	.4977	.4978	.4979	.4979	.4980	.4981
2.9	.4981	.4982	.4982	.4983	.4984	.4984	.4985	.4985	.4986	.4986
3.0	.4987	.4987	.4987	.4988	.4988	.4989	.4989	.4989	.4990	.4990

Source: Frederick Mosteller and Robert E. K. Rourke, *Sturdy Statistics,* Table A–1 (Reading, Mass.: Addison-Wesley, 1973).
Reprinted with permission of the copyright owners.
Note: Use 0.4999 for z values above 3.09.

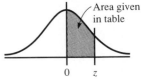

Table F

The *t* Distribution

d.f.	Confidence Intervals	50%	80%	90%	95%	98%	99%
	ONE TAIL, α	0.25	0.10	0.05	0.025	0.01	0.005
	TWO TAILS, α	0.50	0.20	0.10	0.05	0.02	0.01
1		1.000	3.078	6.314	12.706	31.821	63.657
2		.816	1.886	2.920	4.303	6.965	9.925
3		.765	1.638	2.353	3.182	4.541	5.841
4		.741	1.533	2.132	2.776	3.747	4.604
5		.727	1.476	2.015	2.571	3.365	4.032
6		.718	1.440	1.943	2.447	3.143	3.707
7		.711	1.415	1.895	2.365	2.998	3.499
8		.706	1.397	1.860	2.306	2.896	3.355
9		.703	1.383	1.833	2.262	2.821	3.250
10		.700	1.372	1.812	2.228	2.764	3.169
11		.697	1.363	1.796	2.201	2.718	3.106
12		.695	1.356	1.782	2.179	2.681	3.055
13		.694	1.350	1.771	2.160	2.650	3.012
14		.692	1.345	1.761	2.145	2.624	2.977
15		.691	1.341	1.753	2.131	2.602	2.947
16		.690	1.337	1.746	2.120	2.583	2.921
17		.689	1.333	1.740	2.110	2.567	2.898
18		.688	1.330	1.734	2.101	2.552	2.878
19		.688	1.328	1.729	2.093	2.539	2.861
20		.687	1.325	1.725	2.086	2.528	2.845
21		.686	1.323	1.721	2.080	2.518	2.831
22		.686	1.321	1.717	2.074	2.508	2.819
23		.685	1.319	1.714	2.069	2.500	2.807
24		.685	1.318	1.711	2.064	2.492	2.797
25		.684	1.316	1.708	2.060	2.485	2.787
26		.684	1.315	1.706	2.056	2.479	2.779
27		.684	1.314	1.703	2.052	2.473	2.771
28		.683	1.313	1.701	2.048	2.467	2.763
$(z)\ \infty$		.674	1.282[a]	1.645[b]	1.960	2.326[c]	2.576[d]

Source: Adapted from Beyer, W. H., *Handbook of Tables for Probability and Statistics, 2nd Edition,* CRC Press, Boca Raton, Florida, 1986. With permission.

[a]This value has been rounded to 1.28 in the textbook.
[b]This value has been rounded to 1.65 in the textbook.
[c]This value has been rounded to 2.33 in the textbook.
[d]This value has been rounded to 2.58 in the textbook.

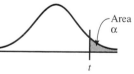

One Tail

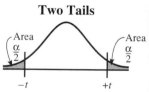

Two Tails

Table G

The Chi-Square Distribution

Degrees of freedom	α									
	0.995	0.99	0.975	0.95	0.90	0.10	0.05	0.025	0.01	0.005
1	—	—	0.001	0.004	0.016	2.706	3.841	5.024	6.635	7.879
2	0.010	0.020	0.051	0.103	0.211	4.605	5.991	7.378	9.210	10.597
3	0.072	0.115	0.216	0.352	0.584	6.251	7.815	9.348	11.345	12.838
4	0.207	0.297	0.484	0.711	1.064	7.779	9.488	11.143	13.277	14.860
5	0.412	0.554	0.831	1.145	1.610	9.236	11.071	12.833	15.086	16.750
6	0.676	0.872	1.237	1.635	2.204	10.645	12.592	14.449	16.812	18.548
7	0.989	1.239	1.690	2.167	2.833	12.017	14.067	16.013	18.475	20.278
8	1.344	1.646	2.180	2.733	3.490	13.362	15.507	17.535	20.090	21.955
9	1.735	2.088	2.700	3.325	4.168	14.684	16.919	19.023	21.666	23.589
10	2.156	2.558	3.247	3.940	4.865	15.987	18.307	20.483	23.209	25.188
11	2.603	3.053	3.816	4.575	5.578	17.275	19.675	21.920	24.725	26.757
12	3.074	3.571	4.404	5.226	6.304	18.549	21.026	23.337	26.217	28.299
13	3.565	4.107	5.009	5.892	7.042	19.812	22.362	24.736	27.688	29.819
14	4.075	4.660	5.629	6.571	7.790	21.064	23.685	26.119	29.141	31.319
15	4.601	5.229	6.262	7.261	8.547	22.307	24.996	27.488	30.578	32.801
16	5.142	5.812	6.908	7.962	9.312	23.542	26.296	28.845	32.000	34.267
17	5.697	6.408	7.564	8.672	10.085	24.769	27.587	30.191	33.409	35.718
18	6.265	7.015	8.231	9.390	10.865	25.989	28.869	31.526	34.805	37.156
19	6.844	7.633	8.907	10.117	11.651	27.204	30.144	32.852	36.191	38.582
20	7.434	8.260	9.591	10.851	12.443	28.412	31.410	34.170	37.566	39.997
21	8.034	8.897	10.283	11.591	13.240	29.615	32.671	35.479	38.932	41.401
22	8.643	9.542	10.982	12.338	14.042	30.813	33.924	36.781	40.289	42.796
23	9.262	10.196	11.689	13.091	14.848	32.007	35.172	38.076	41.638	44.181
24	9.886	10.856	12.401	13.848	15.659	33.196	36.415	39.364	42.980	45.559
25	10.520	11.524	13.120	14.611	16.473	34.382	37.652	40.646	44.314	46.928

Table G

The Chi-Square Distribution *Continued*

Degrees of freedom	α									
	0.995	0.99	0.975	0.95	0.90	0.10	0.05	0.025	0.01	0.005
26	11.160	12.198	13.844	15.379	17.292	35.563	38.885	41.923	45.642	48.290
27	11.808	12.879	14.573	16.151	18.114	36.741	40.113	43.194	46.963	49.645
28	12.461	13.565	15.308	16.928	18.939	37.916	41.337	44.461	48.278	50.993
29	13.121	14.257	16.047	17.708	19.768	39.087	42.557	45.722	49.588	52.336
30	13.787	14.954	16.791	18.493	20.599	40.256	43.773	46.979	50.892	53.672
40	20.707	22.164	24.433	26.509	29.051	51.805	55.758	59.342	63.691	66.766
50	27.991	29.707	32.357	34.764	37.689	63.167	67.505	71.420	76.154	79.490
60	35.534	37.485	40.482	43.188	46.459	74.397	79.082	83.298	88.379	91.952
70	43.275	45.442	48.758	51.739	55.329	85.527	90.531	95.023	100.425	104.215
80	51.172	53.540	57.153	60.391	64.278	96.578	101.879	106.629	112.329	116.321
90	59.196	61.754	65.647	69.126	73.291	107.565	113.145	118.136	124.116	128.299
100	67.328	70.065	74.222	77.929	82.358	118.498	124.342	129.561	135.807	140.169

Source: Donald B. Owen, *Handbook of Statistics Tables,* © 1962, by Addison-Wesley
Publishing Co., Inc., Reading, Massachusetts. Table A–5. Reprinted with permission of the
publisher.

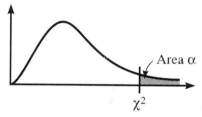

Table H

The F Distribution

d.f.D.: degrees of freedom, denominator	$\alpha = 0.005$ d.f.N.: Degrees of freedom, numerator								
	1	2	3	4	5	6	7	8	9
1	16211	20000	21615	22500	23056	23437	23715	23925	24091
2	198.5	199.0	199.2	199.2	199.3	199.3	199.4	199.4	199.4
3	55.55	49.80	47.47	46.19	45.39	44.84	44.43	44.13	43.88
4	31.33	26.28	24.26	23.15	22.46	21.97	21.62	21.35	21.14
5	22.78	18.31	16.53	15.56	14.94	14.51	14.20	13.96	13.77
6	18.63	14.54	12.92	12.03	11.46	11.07	10.79	10.57	10.39
7	16.24	12.40	10.88	10.05	9.52	9.16	8.89	8.68	8.51
8	14.69	11.04	9.60	8.81	8.30	7.95	7.69	7.50	7.34
9	13.61	10.11	8.72	7.96	7.47	7.13	6.88	6.69	6.54
10	12.83	9.43	8.08	7.34	6.87	6.54	6.30	6.12	5.97
11	12.23	8.91	7.60	6.88	6.42	6.10	5.86	5.68	5.54
12	11.75	8.51	7.23	6.52	6.07	5.76	5.52	5.35	5.20
13	11.37	8.19	6.93	6.23	5.79	5.48	5.25	5.08	4.94
14	11.06	7.92	6.68	6.00	5.56	5.26	5.03	4.86	4.72
15	10.80	7.70	6.48	5.80	5.37	5.07	4.85	4.67	4.54
16	10.58	7.51	6.30	5.64	5.21	4.91	4.69	4.52	4.38
17	10.38	7.35	6.16	5.50	5.07	4.78	4.56	4.39	4.25
18	10.22	7.21	6.03	5.37	4.96	4.66	4.44	4.28	4.14
19	10.07	7.09	5.92	5.27	4.85	4.56	4.34	4.18	4.04
20	9.94	6.99	5.82	5.17	4.76	4.47	4.26	4.09	3.96
21	9.83	6.89	5.73	5.09	4.68	4.39	4.18	4.01	3.88
22	9.73	6.81	5.65	5.02	4.61	4.32	4.11	3.94	3.81
23	9.63	6.73	5.58	4.95	4.54	4.26	4.05	3.88	3.75
24	9.55	6.66	5.52	4.89	4.49	4.20	3.99	3.83	3.69
25	9.48	6.60	5.46	4.84	4.43	4.15	3.94	3.78	3.64
26	9.41	6.54	5.41	4.79	4.38	4.10	3.89	3.73	3.60
27	9.34	6.49	5.36	4.74	4.34	4.06	3.85	3.69	3.56
28	9.28	6.44	5.32	4.70	4.30	4.02	3.81	3.65	3.52
29	9.23	6.40	5.28	4.66	4.26	3.98	3.77	3.61	3.48
30	9.18	6.35	5.24	4.62	4.23	3.95	3.74	3.58	3.45
40	8.83	6.07	4.98	4.37	3.99	3.71	3.51	3.35	3.22
60	8.49	5.79	4.73	4.14	3.76	3.49	3.29	3.13	3.01
120	8.18	5.54	4.50	3.92	3.55	3.28	3.09	2.93	2.81
∞	7.88	5.30	4.28	3.72	3.35	3.09	2.90	2.74	2.62

Source: From Beyer, W. H., *Handbook of Tables for Probability and Statistics, 2nd Edition,* CRC Press, Boca Raton, Florida, 1986. With permission.

10	12	15	20	24	30	40	60	120	∞
24224	24426	24630	24836	24940	25044	25148	25253	25359	25465
199.4	199.4	199.4	199.4	199.5	199.5	199.5	199.5	199.5	199.5
43.69	43.39	43.08	42.78	42.62	42.47	42.31	42.15	41.99	41.83
20.97	20.70	20.44	20.17	20.03	19.89	19.75	19.61	19.47	19.32
13.62	13.38	13.15	12.90	12.78	12.66	12.53	12.40	12.27	12.14
10.25	10.03	9.81	9.59	9.47	9.36	9.24	9.12	9.00	8.88
8.38	8.18	7.97	7.75	7.65	7.53	7.42	7.31	7.19	7.08
7.21	7.01	6.81	6.61	6.50	6.40	6.29	6.18	6.06	5.95
6.42	6.23	6.03	5.83	5.73	5.62	5.52	5.41	5.30	5.19
5.85	5.66	5.47	5.27	5.17	5.07	4.97	4.86	4.75	4.64
5.42	5.24	5.05	4.86	4.76	4.65	4.55	4.44	4.34	4.23
5.09	4.91	4.72	4.53	4.43	4.33	4.23	4.12	4.01	3.90
4.82	4.64	4.46	4.27	4.17	4.07	3.97	3.87	3.76	3.65
4.60	4.43	4.25	4.06	3.96	3.86	3.76	3.66	3.55	3.44
4.42	4.25	4.07	3.88	3.79	3.69	3.58	3.48	3.37	3.26
4.27	4.10	3.92	3.73	3.64	3.54	3.44	3.33	3.22	3.11
4.14	3.97	3.79	3.61	3.51	3.41	3.31	3.21	3.10	2.98
4.03	3.86	3.68	3.50	3.40	3.30	3.20	3.10	2.99	2.87
3.93	3.76	3.59	3.40	3.31	3.21	3.11	3.00	2.89	2.78
3.85	3.68	3.50	3.32	3.22	3.12	3.02	2.92	2.81	2.69
3.77	3.60	3.43	3.24	3.15	3.05	2.95	2.84	2.73	2.61
3.70	3.54	3.36	3.18	3.08	2.98	2.88	2.77	2.66	2.55
3.64	3.47	3.30	3.12	3.02	2.92	2.82	2.71	2.60	2.48
3.59	3.42	3.25	3.06	2.97	2.87	2.77	2.66	2.55	2.43
3.54	3.37	3.20	3.01	2.92	2.82	2.72	2.61	2.50	2.38
3.49	3.33	3.15	2.97	2.87	2.77	2.67	2.56	2.45	2.33
3.45	3.28	3.11	2.93	2.83	2.73	2.63	2.52	2.41	2.25
3.41	3.25	3.07	2.89	2.79	2.69	2.59	2.48	2.37	2.29
3.38	3.21	3.04	2.86	2.76	2.66	2.56	2.45	2.33	2.24
3.34	3.18	3.01	2.82	2.73	2.63	2.52	2.42	2.30	2.18
3.12	2.95	2.78	2.60	2.50	2.40	2.30	2.18	2.06	1.93
2.90	2.74	2.57	2.39	2.29	2.19	2.08	1.96	1.83	1.69
2.71	2.54	2.37	2.19	2.09	1.98	1.87	1.75	1.61	1.43
2.52	2.36	2.19	2.00	1.90	1.79	1.67	1.53	1.36	1.00

Table H

The *F* Distribution *Continued*

d.f.D.: degrees of freedom, denominator	$\alpha = 0.01$								
	d.f.N.: Degrees of freedom, numerator								
	1	*2*	*3*	*4*	*5*	*6*	*7*	*8*	*9*
1	4052	4999.5	5403	5625	5764	5859	5928	5982	6022
2	98.50	99.00	99.17	99.25	99.30	99.33	99.36	99.37	99.39
3	34.12	30.82	29.46	28.71	28.24	27.91	27.67	27.49	27.35
4	21.20	18.00	16.69	15.98	15.52	15.21	14.98	14.80	14.66
5	16.26	13.27	12.06	11.39	10.97	10.67	10.46	10.29	10.16
6	13.75	10.92	9.78	9.15	8.75	8.47	8.26	8.10	7.98
7	12.25	9.55	8.45	7.85	7.46	7.19	6.99	6.84	6.72
8	11.26	8.65	7.59	7.01	6.63	6.37	6.18	6.03	5.91
9	10.56	8.02	6.99	6.42	6.06	5.80	5.61	5.47	5.35
10	10.04	7.56	6.55	5.99	5.64	5.39	5.20	5.06	4.94
11	9.65	7.21	6.22	5.67	5.32	5.07	4.89	4.74	4.63
12	9.33	6.93	5.95	5.41	5.06	4.82	4.64	4.50	4.39
13	9.07	6.70	5.74	5.21	4.86	4.62	4.44	4.30	4.19
14	8.86	6.51	5.56	5.04	4.69	4.46	4.28	4.14	4.03
15	8.68	6.36	5.42	4.89	4.56	4.32	4.14	4.00	3.89
16	8.53	6.23	5.29	4.77	4.44	4.20	4.03	3.89	3.78
17	8.40	6.11	5.18	4.67	4.34	4.10	3.93	3.79	3.68
18	8.29	6.01	5.09	4.58	4.25	4.01	3.84	3.71	3.60
19	8.18	5.93	5.01	4.50	4.17	3.94	3.77	3.63	3.52
20	8.10	5.85	4.94	4.43	4.10	3.87	3.70	3.56	3.46
21	8.02	5.78	4.87	4.37	4.04	3.81	3.64	3.51	3.40
22	7.95	5.72	4.82	4.31	3.99	3.76	3.59	3.45	3.35
23	7.88	5.66	4.76	4.26	3.94	3.71	3.54	3.41	3.30
24	7.82	5.61	4.72	4.22	3.90	3.67	3.50	3.36	3.26
25	7.77	5.57	4.68	4.18	3.85	3.63	3.46	3.32	3.22
26	7.72	5.53	4.64	4.14	3.82	3.59	3.42	3.29	3.18
27	7.68	5.49	4.60	4.11	3.78	3.56	3.39	3.26	3.15
28	7.64	5.45	4.57	4.07	3.75	3.53	3.36	3.23	3.12
29	7.60	5.42	4.54	4.04	3.73	3.50	3.33	3.20	3.09
30	7.56	5.39	4.51	4.02	3.70	3.47	3.30	3.17	3.07
40	7.31	5.18	4.31	3.83	3.51	3.29	3.12	2.99	2.89
60	7.08	4.98	4.13	3.65	3.34	3.12	2.95	2.82	2.72
120	6.85	4.79	3.95	3.48	3.17	2.96	2.79	2.66	2.56
∞	6.63	4.61	3.78	3.32	3.02	2.80	2.64	2.51	2.41

10	12	15	20	24	30	40	60	120	∞
6056	6106	6157	6209	6235	6261	6287	6313	6339	6366
99.40	99.42	99.43	99.45	99.46	99.47	99.47	99.48	99.49	99.50
27.23	27.05	26.87	26.69	26.60	26.50	26.41	26.32	26.22	26.13
14.55	14.37	14.20	14.02	13.93	13.84	13.75	13.65	13.56	13.46
10.05	9.89	9.72	9.55	9.47	9.38	9.29	9.20	9.11	9.02
7.87	7.72	7.56	7.40	7.31	7.23	7.14	7.06	6.97	6.88
6.62	6.47	6.31	6.16	6.07	5.99	5.91	5.82	5.74	5.65
5.81	5.67	5.52	5.36	5.28	5.20	5.12	5.03	4.95	4.86
5.26	5.11	4.96	4.81	4.73	4.65	4.57	4.48	4.40	4.31
4.85	4.71	4.56	4.41	4.33	4.25	4.17	4.08	4.00	3.91
4.54	4.40	4.25	4.10	4.02	3.94	3.86	3.78	3.69	3.60
4.30	4.16	4.01	3.86	3.78	3.70	3.62	3.54	3.45	3.36
4.10	3.96	3.82	3.66	3.59	3.51	3.43	3.34	3.25	3.17
3.94	3.80	3.66	3.51	3.43	3.35	3.27	3.18	3.09	3.00
3.80	3.67	3.52	3.37	3.29	3.21	3.13	3.05	2.96	2.87
3.69	3.55	3.41	3.26	3.18	3.10	3.02	2.93	2.84	2.75
3.59	3.46	3.31	3.16	3.08	3.00	2.92	2.83	2.75	2.65
3.51	3.37	3.23	3.08	3.00	2.92	2.84	2.75	2.66	2.57
3.43	3.30	3.15	3.00	2.92	2.84	2.76	2.67	2.58	2.49
3.37	3.23	3.09	2.94	2.86	2.78	2.69	2.61	2.52	2.42
3.31	3.17	3.03	2.88	2.80	2.72	2.64	2.55	2.46	2.36
3.26	3.12	2.98	2.83	2.75	2.67	2.58	2.50	2.40	2.31
3.21	3.07	2.93	2.78	2.70	2.62	2.54	2.45	2.35	2.26
3.17	3.03	2.89	2.74	2.66	2.58	2.49	2.40	2.31	2.21
3.13	2.99	2.85	2.70	2.62	2.54	2.45	2.36	2.27	2.17
3.09	2.96	2.81	2.66	2.58	2.50	2.42	2.33	2.23	2.13
3.06	2.93	2.78	2.63	2.55	2.47	2.38	2.29	2.20	2.10
3.03	2.90	2.75	2.60	2.52	2.44	2.35	2.26	2.17	2.06
3.00	2.87	2.73	2.57	2.49	2.41	2.33	2.23	2.14	2.03
2.98	2.84	2.70	2.55	2.47	2.39	2.30	2.21	2.11	2.01
2.80	2.66	2.52	2.37	2.29	2.20	2.11	2.02	1.92	1.80
2.63	2.50	2.35	2.20	2.12	2.03	1.94	1.84	1.73	1.60
2.47	2.34	2.19	2.03	1.95	1.86	1.76	1.66	1.53	1.38
2.32	2.18	2.04	1.88	1.79	1.70	1.59	1.47	1.32	1.00

Table H

The F Distribution *Continued*

d.f.D.: degrees of freedom, denominator	$\alpha = 0.025$								
	d.f.N.: Degrees of freedom, numerator								
	1	*2*	*3*	*4*	*5*	*6*	*7*	*8*	*9*
1	647.8	799.5	864.2	899.6	921.8	937.1	948.2	956.7	963.3
2	38.51	39.00	39.17	39.25	39.30	39.33	39.36	39.37	39.39
3	17.44	16.04	15.44	15.10	14.88	14.73	14.62	14.54	14.47
4	12.22	10.65	9.98	9.60	9.36	9.20	9.07	8.98	8.90
5	10.01	8.43	7.76	7.39	7.15	6.98	6.85	6.76	6.68
6	8.81	7.26	6.60	6.23	5.99	5.82	5.70	5.60	5.52
7	8.07	6.54	5.89	5.52	5.29	5.12	4.99	4.90	4.82
8	7.57	6.06	5.42	5.05	4.82	4.65	4.53	4.43	4.36
9	7.21	5.71	5.08	4.72	4.48	4.32	4.20	4.10	4.03
10	6.94	5.46	4.83	4.47	4.24	4.07	3.95	3.85	3.78
11	6.72	5.26	4.63	4.28	4.04	3.88	3.76	3.66	3.59
12	6.55	5.10	4.47	4.12	3.89	3.73	3.61	3.51	3.44
13	6.41	4.97	4.35	4.00	3.77	3.60	3.48	3.39	3.31
14	6.30	4.86	4.24	3.89	3.66	3.50	3.38	3.29	3.21
15	6.20	4.77	4.15	3.80	3.58	3.41	3.29	3.20	3.12
16	6.12	4.69	4.08	3.73	3.50	3.34	3.22	3.12	3.05
17	6.04	4.62	4.01	3.66	3.44	3.28	3.16	3.06	2.98
18	5.98	4.56	3.95	3.61	3.38	3.22	3.10	3.01	2.93
19	5.92	4.51	3.90	3.56	3.33	3.17	3.05	2.96	2.88
20	5.87	4.46	3.86	3.51	3.29	3.13	3.01	2.91	2.84
21	5.83	4.42	3.82	3.48	3.25	3.09	2.97	2.87	2.80
22	5.79	4.38	3.78	3.44	3.22	3.05	2.93	2.84	2.76
23	5.75	4.35	3.75	3.41	3.18	3.02	2.90	2.81	2.73
24	5.72	4.32	3.72	3.38	3.15	2.99	2.87	2.78	2.70
25	5.69	4.29	3.69	3.35	3.13	2.97	2.85	2.75	2.68
26	5.66	4.27	3.67	3.33	3.10	2.94	2.82	2.73	2.65
27	5.63	4.24	3.65	3.31	3.08	2.92	2.80	2.71	2.63
28	5.61	4.22	3.63	3.29	3.06	2.90	2.78	2.69	2.61
29	5.59	4.20	3.61	3.27	3.04	2.88	2.76	2.67	2.59
30	5.57	4.18	3.59	3.25	3.03	2.87	2.75	2.65	2.57
40	5.42	4.05	3.46	3.13	2.90	2.74	2.62	2.53	2.45
60	5.29	3.93	3.34	3.01	2.79	2.63	2.51	2.41	2.33
120	5.15	3.80	3.23	2.89	2.67	2.52	2.39	2.30	2.22
∞	5.02	3.69	3.12	2.79	2.57	2.41	2.29	2.19	2.11

10	12	15	20	24	30	40	60	120	∞
968.6	976.7	984.9	993.1	997.2	1001	1006	1010	1014	1018
39.40	39.41	39.43	39.45	39.46	39.46	39.47	39.48	39.49	39.50
14.42	14.34	14.25	14.17	14.12	14.08	14.04	13.99	13.95	13.90
8.84	8.75	8.66	8.56	8.51	8.46	8.41	8.36	8.31	8.26
6.62	6.52	6.43	6.33	6.28	6.23	6.18	6.12	6.07	6.02
5.46	5.37	5.27	5.17	5.12	5.07	5.01	4.96	4.90	4.85
4.76	4.67	4.57	4.47	4.42	4.36	4.31	4.25	4.20	4.14
4.30	4.20	4.10	4.00	3.95	3.89	3.84	3.78	3.73	3.67
3.96	3.87	3.77	3.67	3.61	3.56	3.51	3.45	3.39	3.33
3.72	3.62	3.52	3.42	3.37	3.31	3.26	3.20	3.14	3.08
3.53	3.43	3.33	3.23	3.17	3.12	3.06	3.00	2.94	2.88
3.37	3.28	3.18	3.07	3.02	2.96	2.91	2.85	2.79	2.72
3.25	3.15	3.05	2.95	2.89	2.84	2.78	2.72	2.66	2.60
3.15	3.05	2.95	2.84	2.79	2.73	2.67	2.61	2.55	2.49
3.06	2.96	2.86	2.76	2.70	2.64	2.59	2.52	2.46	2.40
2.99	2.89	2.79	2.68	2.63	2.57	2.51	2.45	2.38	2.32
2.92	2.82	2.72	2.62	2.56	2.50	2.44	2.38	2.32	2.25
2.87	2.77	2.67	2.56	2.50	2.44	2.38	2.32	2.26	2.19
2.82	2.72	2.62	2.51	2.45	2.39	2.33	2.27	2.20	2.13
2.77	2.68	2.57	2.46	2.41	2.35	2.29	2.22	2.16	2.09
2.73	2.64	2.53	2.42	2.37	2.31	2.25	2.18	2.11	2.04
2.70	2.60	2.50	2.39	2.33	2.27	2.21	2.14	2.08	2.00
2.67	2.57	2.47	2.36	2.30	2.24	2.18	2.11	2.04	1.97
2.64	2.54	2.44	2.33	2.27	2.21	2.15	2.08	2.01	1.94
2.61	2.51	2.41	2.30	2.24	2.18	2.12	2.05	1.98	1.91
2.59	2.49	2.39	2.28	2.22	2.16	2.09	2.03	1.95	1.88
2.57	2.47	2.36	2.25	2.19	2.13	2.07	2.00	1.93	1.85
2.55	2.45	2.34	2.23	2.17	2.11	2.05	1.98	1.91	1.83
2.53	2.43	2.32	2.21	2.15	2.09	2.03	1.96	1.89	1.81
2.51	2.41	2.31	2.20	2.14	2.07	2.01	1.94	1.87	1.79
2.39	2.29	2.18	2.07	2.01	1.94	1.88	1.80	1.72	1.64
2.27	2.17	2.06	1.94	1.88	1.82	1.74	1.67	1.58	1.48
2.16	2.05	1.94	1.82	1.76	1.69	1.61	1.53	1.43	1.31
2.05	1.94	1.83	1.71	1.64	1.57	1.48	1.39	1.27	1.00

Table H

The F Distribution *Continued*

d.f.D.: degrees of freedom, denominator	$\alpha = 0.05$								
	d.f.N.: Degrees of freedom, numerator								
	1	*2*	*3*	*4*	*5*	*6*	*7*	*8*	*9*
1	161.4	199.5	215.7	224.6	230.2	234.0	236.8	238.9	240.5
2	18.51	19.00	19.16	19.25	19.30	19.33	19.35	19.37	19.38
3	10.13	9.55	9.28	9.12	9.01	8.94	8.89	8.85	8.81
4	7.71	6.94	6.59	6.39	6.26	6.16	6.09	6.04	6.00
5	6.61	5.79	5.41	5.19	5.05	4.95	4.88	4.82	4.77
6	5.99	5.14	4.76	4.53	4.39	4.28	4.21	4.15	4.10
7	5.59	4.74	4.35	4.12	3.97	3.87	3.79	3.73	3.68
8	5.32	4.46	4.07	3.84	3.69	3.58	3.50	3.44	3.39
9	5.12	4.26	3.86	3.63	3.48	3.37	3.29	3.23	3.18
10	4.96	4.10	3.71	3.48	3.33	3.22	3.14	3.07	3.02
11	4.84	3.98	3.59	3.36	3.20	3.09	3.01	2.95	2.90
12	4.75	3.89	3.49	3.26	3.11	3.00	2.91	2.85	2.80
13	4.67	3.81	3.41	3.18	3.03	2.92	2.83	2.77	2.71
14	4.60	3.74	3.34	3.11	2.96	2.85	2.76	2.70	2.65
15	4.54	3.68	3.29	3.06	2.90	2.79	2.71	2.64	2.59
16	4.49	3.63	3.24	3.01	2.85	2.74	2.66	2.59	2.54
17	4.45	3.59	3.20	2.96	2.81	2.70	2.61	2.55	2.49
18	4.41	3.55	3.16	2.93	2.77	2.66	2.58	2.51	2.46
19	4.38	3.52	3.13	2.90	2.74	2.63	2.54	2.48	2.42
20	4.35	3.49	3.10	2.87	2.71	2.60	2.51	2.45	2.39
21	4.32	3.47	3.07	2.84	2.68	2.57	2.49	2.42	2.37
22	4.30	3.44	3.05	2.82	2.66	2.55	2.46	2.40	2.34
23	4.28	3.42	3.03	2.80	2.64	2.53	2.44	2.37	2.32
24	4.26	3.40	3.01	2.78	2.62	2.51	2.42	2.36	2.30
25	4.24	3.39	2.99	2.76	2.60	2.49	2.40	2.34	2.28
26	4.23	3.37	2.98	2.74	2.59	2.47	2.39	2.32	2.27
27	4.21	3.35	2.96	2.73	2.57	2.46	2.37	2.31	2.25
28	4.20	3.34	2.95	2.71	2.56	2.45	2.36	2.29	2.24
29	4.18	3.33	2.93	2.70	2.55	2.43	2.35	2.28	2.22
30	4.17	3.32	2.92	2.69	2.53	2.42	2.33	2.27	2.21
40	4.08	3.23	2.84	2.61	2.45	2.34	2.25	2.18	2.12
60	4.00	3.15	2.76	2.53	2.37	2.25	2.17	2.10	2.04
120	3.92	3.07	2.68	2.45	2.29	2.17	2.09	2.02	1.96
∞	3.84	3.00	2.60	2.37	2.21	2.10	2.01	1.94	1.88

10	12	15	20	24	30	40	60	120	∞
241.9	243.9	245.9	248.0	249.1	250.1	251.1	252.2	253.3	254.3
19.40	19.41	19.43	19.45	19.45	19.46	19.47	19.48	19.49	19.50
8.79	8.74	8.70	8.66	8.64	8.62	8.59	8.57	8.55	8.53
5.96	5.91	5.86	5.80	5.77	5.75	5.72	5.69	5.66	5.63
4.74	4.68	4.62	4.56	4.53	4.50	4.46	4.43	4.40	4.36
4.06	4.00	3.94	3.87	3.84	3.81	3.77	3.74	3.70	3.67
3.64	3.57	3.51	3.44	3.41	3.38	3.34	3.30	3.27	3.23
3.35	3.28	3.22	3.15	3.12	3.08	3.04	3.01	2.97	2.93
3.14	3.07	3.01	2.94	2.90	2.86	2.83	2.79	2.75	2.71
2.98	2.91	2.85	2.77	2.74	2.70	2.66	2.62	2.58	2.54
2.85	2.79	2.72	2.65	2.61	2.57	2.53	2.49	2.45	2.40
2.75	2.69	2.62	2.54	2.51	2.47	2.43	2.38	2.34	2.30
2.67	2.60	2.53	2.46	2.42	2.38	2.34	2.30	2.25	2.21
2.60	2.53	2.46	2.39	2.35	2.31	2.27	2.22	2.18	2.13
2.54	2.48	2.40	2.33	2.29	2.25	2.20	2.16	2.11	2.07
2.49	2.42	2.35	2.28	2.24	2.19	2.15	2.11	2.06	2.01
2.45	2.38	2.31	2.23	2.19	2.15	2.10	2.06	2.01	1.96
2.41	2.34	2.27	2.19	2.15	2.11	2.06	2.02	1.97	1.92
2.38	2.31	2.23	2.16	2.11	2.07	2.03	1.98	1.93	1.88
2.35	2.28	2.20	2.12	2.08	2.04	1.99	1.95	1.90	1.84
2.32	2.25	2.18	2.10	2.05	2.01	1.96	1.92	1.87	1.81
2.30	2.23	2.15	2.07	2.03	1.98	1.94	1.89	1.84	1.78
2.27	2.20	2.13	2.05	2.01	1.96	1.91	1.86	1.81	1.76
2.25	2.18	2.11	2.03	1.98	1.94	1.89	1.84	1.79	1.73
2.24	2.16	2.09	2.01	1.96	1.92	1.87	1.82	1.77	1.71
2.22	2.15	2.07	1.99	1.95	1.90	1.85	1.80	1.75	1.69
2.20	2.13	2.06	1.97	1.93	1.88	1.84	1.79	1.73	1.67
2.19	2.12	2.04	1.96	1.91	1.87	1.82	1.77	1.71	1.65
2.18	2.10	2.03	1.94	1.90	1.85	1.81	1.75	1.70	1.64
2.16	2.09	2.01	1.93	1.89	1.84	1.79	1.74	1.68	1.62
2.08	2.00	1.92	1.84	1.79	1.74	1.69	1.64	1.58	1.51
1.99	1.92	1.84	1.75	1.70	1.65	1.59	1.53	1.47	1.39
1.91	1.83	1.75	1.66	1.61	1.55	1.50	1.43	1.35	1.25
1.83	1.75	1.67	1.57	1.52	1.46	1.39	1.32	1.22	1.00

Table H

The *F* Distribution   *Continued*

d.f.D.: degrees of freedom, denominator	$\alpha = 0.10$ d.f.N.: Degrees of freedom, numerator								
	1	2	3	4	5	6	7	8	9
1	39.86	49.50	53.59	55.83	57.24	58.20	58.91	59.44	59.86
2	8.53	9.00	9.16	9.24	9.29	9.33	9.35	9.37	9.38
3	5.54	5.46	5.39	5.34	5.31	5.28	5.27	5.25	5.24
4	4.54	4.32	4.19	4.11	4.05	4.01	3.98	3.95	3.94
5	4.06	3.78	3.62	3.52	3.45	3.40	3.37	3.34	3.32
6	3.78	3.46	3.29	3.18	3.11	3.05	3.01	2.98	2.96
7	3.59	3.26	3.07	2.96	2.88	2.83	2.78	2.75	2.72
8	3.46	3.11	2.92	2.81	2.73	2.67	2.62	2.59	2.56
9	3.36	3.01	2.81	2.69	2.61	2.55	2.51	2.47	2.44
10	3.29	2.92	2.73	2.61	2.52	2.46	2.41	2.38	2.35
11	3.23	2.86	2.66	2.54	2.45	2.39	2.34	2.30	2.27
12	3.18	2.81	2.61	2.48	2.39	2.33	2.28	2.24	2.21
13	3.14	2.76	2.56	2.43	2.35	2.28	2.23	2.20	2.16
14	3.10	2.73	2.52	2.39	2.31	2.24	2.19	2.15	2.12
15	3.07	2.70	2.49	2.36	2.27	2.21	2.16	2.12	2.09
16	3.05	2.67	2.46	2.33	2.24	2.18	2.13	2.09	2.06
17	3.03	2.64	2.44	2.31	2.22	2.15	2.10	2.06	2.03
18	3.01	2.62	2.42	2.29	2.20	2.13	2.08	2.04	2.00
19	2.99	2.61	2.40	2.27	2.18	2.11	2.06	2.02	1.98
20	2.97	2.59	2.38	2.25	2.16	2.09	2.04	2.00	1.96
21	2.96	2.57	2.36	2.23	2.14	2.08	2.02	1.98	1.95
22	2.95	2.56	2.35	2.22	2.13	2.06	2.01	1.97	1.93
23	2.94	2.55	2.34	2.21	2.11	2.05	1.99	1.95	1.92
24	2.93	2.54	2.33	2.19	2.10	2.04	1.98	1.94	1.91
25	2.92	2.53	2.32	2.18	2.09	2.02	1.97	1.93	1.89
26	2.91	2.52	2.31	2.17	2.08	2.01	1.96	1.92	1.88
27	2.90	2.51	2.30	2.17	2.07	2.00	1.95	1.91	1.87
28	2.89	2.50	2.29	2.16	2.06	2.00	1.94	1.90	1.87
29	2.89	2.50	2.28	2.15	2.06	1.99	1.93	1.89	1.86
30	2.88	2.49	2.28	2.14	2.05	1.98	1.93	1.88	1.85
40	2.84	2.44	2.23	2.09	2.00	1.93	1.87	1.83	1.79
60	2.79	2.39	2.18	2.04	1.95	1.87	1.82	1.77	1.74
120	2.75	2.35	2.13	1.99	1.90	1.82	1.77	1.72	1.68
∞	2.71	2.30	2.08	1.94	1.85	1.77	1.72	1.67	1.63

10	12	15	20	24	30	40	60	120	∞
60.19	60.71	61.22	61.74	62.00	62.26	62.53	62.79	63.06	63.33
9.39	9.41	9.42	9.44	9.45	9.46	9.47	9.47	9.48	9.49
5.23	5.22	5.20	5.18	5.18	5.17	5.16	5.15	5.14	5.13
3.92	3.90	3.87	3.84	3.83	3.82	3.80	3.79	3.78	3.76
3.30	3.27	3.24	3.21	3.19	3.17	3.16	3.14	3.12	3.10
2.94	2.90	2.87	2.84	2.82	2.80	2.78	2.76	2.74	2.72
2.70	2.67	2.63	2.59	2.58	2.56	2.54	2.51	2.49	2.47
2.54	2.50	2.46	2.42	2.40	2.38	2.36	2.34	2.32	2.29
2.42	2.38	2.34	2.30	2.28	2.25	2.23	2.21	2.18	2.16
2.32	2.28	2.24	2.20	2.18	2.16	2.13	2.11	2.08	2.06
2.25	2.21	2.17	2.12	2.10	2.08	2.05	2.03	2.00	1.97
2.19	2.15	2.10	2.06	2.04	2.01	1.99	1.96	1.93	1.90
2.14	2.10	2.05	2.01	1.98	1.96	1.93	1.90	1.88	1.85
2.10	2.05	2.01	1.96	1.94	1.91	1.89	1.86	1.83	1.80
2.06	2.02	1.97	1.92	1.90	1.87	1.85	1.82	1.79	1.76
2.03	1.99	1.94	1.89	1.87	1.84	1.81	1.78	1.75	1.72
2.00	1.96	1.91	1.86	1.84	1.81	1.78	1.75	1.72	1.69
1.98	1.93	1.89	1.84	1.81	1.78	1.75	1.72	1.69	1.66
1.96	1.91	1.86	1.81	1.79	1.76	1.73	1.70	1.67	1.63
1.94	1.89	1.84	1.79	1.77	1.74	1.71	1.68	1.64	1.61
1.92	1.87	1.83	1.78	1.75	1.72	1.69	1.66	1.62	1.59
1.90	1.86	1.81	1.76	1.73	1.70	1.67	1.64	1.60	1.57
1.89	1.84	1.80	1.74	1.72	1.69	1.66	1.62	1.59	1.55
1.88	1.83	1.78	1.73	1.70	1.67	1.64	1.61	1.57	1.53
1.87	1.82	1.77	1.72	1.69	1.66	1.63	1.59	1.56	1.52
1.86	1.81	1.76	1.71	1.68	1.65	1.61	1.58	1.54	1.50
1.85	1.80	1.75	1.70	1.67	1.64	1.60	1.57	1.53	1.49
1.84	1.79	1.74	1.69	1.66	1.63	1.59	1.56	1.52	1.48
1.83	1.78	1.73	1.68	1.65	1.62	1.58	1.55	1.51	1.47
1.82	1.77	1.72	1.67	1.64	1.61	1.57	1.54	1.50	1.46
1.76	1.71	1.66	1.61	1.57	1.54	1.51	1.47	1.42	1.38
1.71	1.66	1.60	1.54	1.51	1.48	1.44	1.40	1.35	1.29
1.65	1.60	1.55	1.48	1.45	1.41	1.37	1.32	1.26	1.19
1.60	1.55	1.49	1.42	1.38	1.34	1.30	1.24	1.17	1.00

Table I

Critical Values for the PPMC

Reject H_0: $\rho = 0$ if the absolute value of r is greater than the value given in the table. The values are for a two-tailed test; d.f. $= n - 2$.

d.f.	$\alpha = 0.05$	$\alpha = 0.01$
1	0.999	0.999
2	0.950	0.999
3	0.878	0.959
4	0.811	0.917
5	0.754	0.875
6	0.707	0.834
7	0.666	0.798
8	0.632	0.765
9	0.602	0.735
10	0.576	0.708
11	0.553	0.684
12	0.532	0.661
13	0.514	0.641
14	0.497	0.623
15	0.482	0.606
16	0.468	0.590
17	0.456	0.575
18	0.444	0.561
19	0.433	0.549
20	0.423	0.537
25	0.381	0.487
30	0.349	0.449
35	0.325	0.418
40	0.304	0.393
45	0.288	0.372
50	0.273	0.354
60	0.250	0.325
70	0.232	0.302
80	0.217	0.283
90	0.205	0.267
100	0.195	0.254

Source: From Beyer, W. H., *Handbook of Tables for Probability and Statistics, 2nd Edition,* CRC Press, Boca Raton, Florida, 1986. With permission.

Table J

Critical Values for the Sign Test

Reject the null hypothesis if the smaller number of + or − signs is less than or equal to the value in the table.

n	One-tailed, α = 0.005 Two-tailed, α = 0.01	0.01 0.02	0.025 0.05	0.05 0.10	
8	0	0	0	1	
9	0	0	1	1	
10	0	0	1	1	
11	0	1	1	2	
12	1	1	2	2	
13	1	1	2	3	
14	1	2	3	3	
15	2	2	3	3	
16	2	2	3	4	*Note:* Table J is for one-tailed or two-tailed tests. The term *n* represents the total number of + and − signs. The test value is the number of less frequent signs.
17	2	3	4	4	
18	3	3	4	5	
19	3	4	4	5	
20	3	4	5	5	
21	4	4	5	6	
22	4	5	5	6	
23	4	5	6	7	
24	5	5	6	7	
25	5	6	6	7	

Source: From Beyer, W. H., *Handbook of Tables for Probability and Statistics, 2nd Edition,* CRC Press, Boca Raton, Florida, 1986. With permission.

Table K

Critical Values for the Wilcoxon Signed-Rank Test

Reject the null hypothesis if the test value is less than or equal to the value given in the table.

n	One-tailed, $\alpha = 0.05$ Two-tailed, $\alpha = 0.10$	0.025 0.05	0.01 0.02	0.005 0.01
5	1			
6	2	1		
7	4	2	0	
8	6	4	2	0
9	8	6	3	2
10	11	8	5	3
11	14	11	7	5
12	17	14	10	7
13	21	17	13	10
14	26	21	16	13
15	30	25	20	16
16	36	30	24	19
17	41	35	28	23
18	47	40	33	28
19	54	46	38	32
20	60	52	43	37
21	68	59	49	43
22	75	66	56	49
23	83	73	62	55
24	92	81	69	61
25	101	90	77	68
26	110	98	85	76
27	120	107	93	84
28	130	117	102	92
29	141	127	111	100
30	152	137	120	109

Source: From Beyer, W. H., *Handbook of Tables for Probability and Statistics, 2nd Edition,* CRC Press, Boca Raton, Florida, 1986. With permission.

Table L

Critical Values for the Rank Correlation Coefficient

Reject H_0: $\rho = 0$ if the absolute value of r_S is greater than the value given in the table.

n	$\alpha = 0.10$	$\alpha = 0.05$	$\alpha = 0.02$	$\alpha = 0.01$
5	0.900	—	—	—
6	0.829	0.886	0.943	—
7	0.714	0.786	0.893	0.929
8	0.643	0.738	0.833	0.881
9	0.600	0.700	0.783	0.833
10	0.564	0.648	0.745	0.794
11	0.536	0.618	0.709	0.818
12	0.497	0.591	0.703	0.780
13	0.475	0.566	0.673	0.745
14	0.457	0.545	0.646	0.716
15	0.441	0.525	0.623	0.689
16	0.425	0.507	0.601	0.666
17	0.412	0.490	0.582	0.645
18	0.399	0.476	0.564	0.625
19	0.388	0.462	0.549	0.608
20	0.377	0.450	0.534	0.591
21	0.368	0.438	0.521	0.576
22	0.359	0.428	0.508	0.562
23	0.351	0.418	0.496	0.549
24	0.343	0.409	0.485	0.537
25	0.336	0.400	0.475	0.526
26	0.329	0.392	0.465	0.515
27	0.323	0.385	0.456	0.505
28	0.317	0.377	0.488	0.496
29	0.311	0.370	0.440	0.487
30	0.305	0.364	0.432	0.478

Source: From Beyer, W. H., *Handbook of Tables for Probability and Statistics, 2nd Edition,* CRC Press, Boca Raton, Florida, 1986. With permission.

Table M

Critical Values for the Number of Runs

This table gives the critical values at $\alpha = 0.05$ for a two-tailed test. Reject the null hypothesis if the number of runs is less than or equal to the smaller value or greater than or equal to the larger value.

Value of n_1	Value of n_2																		
	2	3	4	5	6	7	8	9	10	11	12	13	14	15	16	17	18	19	20
2	1	1	1	1	1	1	1	1	1	1	2	2	2	2	2	2	2	2	2
	6	6	6	6	6	6	6	6	6	6	6	6	6	6	6	6	6	6	6
3	1	1	1	1	2	2	2	2	2	2	2	2	2	3	3	3	3	3	3
	6	8	8	8	8	8	8	8	8	8	8	8	8	8	8	8	8	8	8
4	1	1	1	2	2	2	3	3	3	3	3	3	3	3	4	4	4	4	4
	6	8	9	9	9	10	10	10	10	10	10	10	10	10	10	10	10	10	10
5	1	1	2	2	3	3	3	3	3	4	4	4	4	4	4	4	5	5	5
	6	8	9	10	10	11	11	12	12	12	12	12	12	12	12	12	12	12	12
6	1	2	2	3	3	3	3	4	4	4	4	5	5	5	5	5	5	6	6
	6	8	9	10	11	12	12	13	13	13	13	14	14	14	14	14	14	14	14
7	1	2	2	3	3	3	4	4	5	5	5	5	5	6	6	6	6	6	6
	6	8	10	11	12	13	13	14	14	14	14	15	15	15	16	16	16	16	16
8	1	2	3	3	3	4	4	5	5	5	6	6	6	6	6	7	7	7	7
	6	8	10	11	12	13	14	14	15	15	16	16	16	16	17	17	17	17	17
9	1	2	3	3	4	4	5	5	5	6	6	6	7	7	7	7	8	8	8
	6	8	10	12	13	14	14	15	16	16	16	17	17	18	18	18	18	18	18
10	1	2	3	3	4	5	5	5	6	6	7	7	7	7	8	8	8	8	9
	6	8	10	12	13	14	15	16	16	17	17	18	18	18	19	19	19	20	20
11	1	2	3	4	4	5	5	6	6	7	7	7	8	8	8	9	9	9	9
	6	8	10	12	13	14	15	16	17	17	18	19	19	19	20	20	20	21	21
12	2	2	3	4	4	5	6	6	7	7	7	8	8	8	9	9	9	10	10
	6	8	10	12	13	14	16	16	17	18	19	19	20	20	21	21	21	22	22

Table M

Critical Values for the Number of Runs   *Continued*

This table gives the critical values at $\alpha = 0.05$ for a two-tailed test. Reject the null hypothesis if the number of runs is less than or equal to the smaller value or greater than or equal to the larger value.

Value of n_1	Value of n_2																		
	2	3	4	5	6	7	8	9	10	11	12	13	14	15	16	17	18	19	20
13	2	2	3	4	5	5	6	6	7	7	8	8	9	9	9	10	10	10	10
	6	8	10	12	14	15	16	17	18	19	19	20	20	21	21	22	22	23	23
14	2	2	3	4	5	5	6	7	7	8	8	9	9	9	10	10	10	11	11
	6	8	10	12	14	15	16	17	18	19	20	20	21	22	22	23	23	23	24
15	2	3	3	4	5	6	6	7	7	8	8	9	9	10	10	11	11	11	12
	6	8	10	12	14	15	16	18	18	19	20	21	22	22	23	23	24	24	25
16	2	3	4	4	5	6	6	7	8	8	9	9	10	10	11	11	11	12	12
	6	8	10	12	14	16	17	18	19	20	21	21	22	23	23	24	25	25	25
17	2	3	4	4	5	6	7	7	8	9	9	10	10	11	11	11	12	12	13
	6	8	10	12	14	16	17	18	19	20	21	22	23	23	24	25	25	26	26
18	2	3	4	5	5	6	7	8	8	9	9	10	10	11	11	12	12	13	13
	6	8	10	12	14	16	17	18	19	20	21	22	23	24	25	25	26	26	27
19	2	3	4	5	6	6	7	8	8	9	10	10	11	11	12	12	13	13	13
	6	8	10	12	14	16	17	18	20	21	22	23	23	24	25	26	26	27	27
20	2	3	4	5	6	6	7	8	9	9	10	10	11	12	12	13	13	13	14
	6	8	10	12	14	16	17	18	20	21	22	23	24	25	25	26	27	27	28

Source: Adapted from C. Eisenhardt and F. Swed, "Tables for Testing Randomness of Grouping in a Sequence of Alternatives," *The Annals of Statistics,* 14 (1943) 83 86. Reprinted with permission of the Institute of Mathematical Statistics and of the Benjamin/ Cummings Publishing Company, in whose publication, *Elementary Statistics,* 3rd edition (1989), by Mario F. Triola, this table appears.

Table N

Critical Values for the Tukey Test

$\alpha = 0.01$

ν \ k	2	3	4	5	6	7	8	9	10
1	90.03	135.0	164.3	185.6	202.2	215.8	227.2	237.0	245.6
2	14.04	19.02	22.29	24.72	26.63	28.20	29.53	30.68	31.69
3	8.26	10.62	12.17	13.33	14.24	15.00	15.64	16.20	16.69
4	6.51	8.12	9.17	9.96	10.58	11.10	11.55	11.93	12.27
5	5.70	6.98	7.80	8.42	8.91	9.32	9.67	9.97	10.24
6	5.24	6.33	7.03	7.56	7.97	8.32	8.61	8.87	9.10
7	4.95	5.92	6.54	7.01	7.37	7.68	7.94	8.17	8.37
8	4.75	5.64	6.20	6.62	6.96	7.24	7.47	7.68	7.86
9	4.60	5.43	5.96	6.35	6.66	6.91	7.13	7.33	7.49
10	4.48	5.27	5.77	6.14	6.43	6.67	6.87	7.05	7.21
11	4.39	5.15	5.62	5.97	6.25	6.48	6.67	6.84	6.99
12	4.32	5.05	5.50	5.84	6.10	6.32	6.51	6.67	6.81
13	4.26	4.96	5.40	5.73	5.98	6.19	6.37	6.53	6.67
14	4.21	4.89	5.32	5.63	5.88	6.08	6.26	6.41	6.54
15	4.17	4.84	5.25	5.56	5.80	5.99	6.16	6.31	6.44
16	4.13	4.79	5.19	5.49	5.72	5.92	6.08	6.22	6.35
17	4.10	4.74	5.14	5.43	5.66	5.85	6.01	6.15	6.27
18	4.07	4.70	5.09	5.38	5.60	5.79	5.94	6.08	6.20
19	4.05	4.67	5.05	5.33	5.55	5.73	5.89	6.02	6.14
20	4.02	4.64	5.02	5.29	5.51	5.69	5.84	5.97	6.09
24	3.96	4.55	4.91	5.17	5.37	5.54	5.69	5.81	5.92
30	3.89	4.45	4.80	5.05	5.24	5.40	5.54	5.65	5.76
40	3.82	4.37	4.70	4.93	5.11	5.26	5.39	5.50	5.60
60	3.76	4.28	4.59	4.82	4.99	5.13	5.25	5.36	5.45
120	3.70	4.20	4.50	4.71	4.87	5.01	5.12	5.21	5.30
∞	3.64	4.12	4.40	4.60	4.76	4.88	4.99	5.08	5.16

Source: From Beyer, W. H., *Handbook of Tables for Probability and Statistics, 2nd Edition,* CRC Press, Boca Raton, Florida, 1986. With permission.

Table N

Critical Values for the Tukey Test *Continued*

$\alpha = 0.01$

k ν	11	12	13	14	15	16	17	18	19	20
1	253.2	260.0	266.2	271.8	277.0	281.8	286.3	290.4	294.3	298.0
2	32.59	33.40	34.13	34.81	35.43	36.00	36.53	37.03	37.50	37.95
3	17.13	17.53	17.89	18.22	18.52	18.81	19.07	19.32	19.55	19.77
4	12.57	12.84	13.09	13.32	13.53	13.73	13.91	14.08	14.24	14.40
5	10.48	10.70	10.89	11.08	11.24	11.40	11.55	11.68	11.81	11.93
6	9.30	9.48	9.65	9.81	9.95	10.08	10.21	10.32	10.43	10.54
7	8.55	8.71	8.86	9.00	9.12	9.24	9.35	9.46	9.55	9.65
8	8.03	8.18	8.31	8.44	8.55	8.66	8.76	8.85	8.94	9.03
9	7.65	7.78	7.91	8.03	8.13	8.23	8.33	8.41	8.49	8.57
10	7.36	7.49	7.60	7.71	7.81	7.91	7.99	8.08	8.15	8.23
11	7.13	7.25	7.36	7.46	7.56	7.65	7.73	7.81	7.88	7.95
12	6.94	7.06	7.17	7.26	7.36	7.44	7.52	7.59	7.66	7.73
13	6.79	6.90	7.01	7.10	7.19	7.27	7.35	7.42	7.48	7.55
14	6.66	6.77	6.87	6.96	7.05	7.13	7.20	7.27	7.33	7.39
15	6.55	6.66	6.76	6.84	6.93	7.00	7.07	7.14	7.20	7.26
16	6.46	6.56	6.66	6.74	6.82	6.90	6.97	7.03	7.09	7.15
17	6.38	6.48	6.57	6.66	6.73	6.81	6.87	6.94	7.00	7.05
18	6.31	6.41	6.50	6.58	6.65	6.73	6.79	6.85	6.91	6.97
19	6.25	6.34	6.43	6.51	6.58	6.65	6.72	6.78	6.84	6.89
20	6.19	6.28	6.37	6.45	6.52	6.59	6.65	6.71	6.77	6.82
24	6.02	6.11	6.19	6.26	6.33	6.39	6.45	6.51	6.56	6.61
30	5.85	5.93	6.01	6.08	6.14	6.20	6.26	6.31	6.36	6.41
40	5.69	5.76	5.83	5.90	5.96	6.02	6.07	6.12	6.16	6.21
60	5.53	5.60	5.67	5.73	5.78	5.84	5.89	5.93	5.97	6.01
120	5.37	5.44	5.50	5.56	5.61	5.66	5.71	5.75	5.79	5.83
∞	5.23	5.29	5.35	5.40	5.45	5.49	5.54	5.57	5.61	5.65

Table N

Critical Values for the Tukey Test   *Continued*

$\alpha = 0.05$

ν \ k	2	3	4	5	6	7	8	9	10
1	17.97	26.98	32.82	37.08	40.41	43.12	45.40	47.36	49.07
2	6.08	8.33	9.80	10.88	11.74	12.44	13.03	13.54	13.99
3	4.50	5.91	6.82	7.50	8.04	8.48	8.85	9.18	9.46
4	3.93	5.04	5.76	6.29	6.71	7.05	7.35	7.60	7.83
5	3.64	4.60	5.22	5.67	6.03	6.33	6.58	6.80	6.99
6	3.46	4.34	4.90	5.30	5.63	5.90	6.12	6.32	6.49
7	3.34	4.16	4.68	5.06	5.36	5.61	5.82	6.00	6.16
8	3.26	4.04	4.53	4.89	5.17	5.40	5.60	5.77	5.92
9	3.20	3.95	4.41	4.76	5.02	5.24	5.43	5.59	5.74
10	3.15	3.88	4.33	4.65	4.91	5.12	5.30	5.46	5.60
11	3.11	3.82	4.26	4.57	4.82	5.03	5.20	5.35	5.49
12	3.08	3.77	4.20	4.51	4.75	4.95	5.12	5.27	5.39
13	3.06	3.73	4.15	4.45	4.69	4.88	5.05	5.19	5.32
14	3.03	3.70	4.11	4.41	4.64	4.83	4.99	5.13	5.25
15	3.01	3.67	4.08	4.37	4.59	4.78	4.94	5.08	5.20
16	3.00	3.65	4.05	4.33	4.56	4.74	4.90	5.03	5.15
17	2.98	3.63	4.02	4.30	4.52	4.70	4.86	4.99	5.11
18	2.97	3.61	4.00	4.28	4.49	4.67	4.82	4.96	5.07
19	2.96	3.59	3.98	4.25	4.47	4.65	4.79	4.92	5.04
20	2.95	3.58	3.96	4.23	4.45	4.62	4.77	4.90	5.01
24	2.92	3.53	3.90	4.17	4.37	4.54	4.68	4.81	4.92
30	2.89	3.49	3.85	4.10	4.30	4.46	4.60	4.72	4.82
40	2.86	3.44	3.79	4.04	4.23	4.39	4.52	4.63	4.73
60	2.83	3.40	3.74	3.98	4.16	4.31	4.44	4.55	4.65
120	2.80	3.36	3.68	3.92	4.10	4.24	4.36	4.47	4.56
∞	2.77	3.31	3.63	3.86	4.03	4.17	4.29	4.39	4.47

Table N

Critical Values for the Tukey Test   *Continued*

$\alpha = 0.05$

v \ k	11	12	13	14	15	16	17	18	19	20
1	50.59	51.96	53.20	54.33	55.36	56.32	57.22	58.04	58.83	59.56
2	14.39	14.75	15.08	15.38	15.65	15.91	16.14	16.37	16.57	16.77
3	9.72	9.95	10.15	10.35	10.53	10.69	10.84	10.98	11.11	11.24
4	8.03	8.21	8.37	8.52	8.66	8.79	8.91	9.03	9.13	9.23
5	7.17	7.32	7.47	7.60	7.72	7.83	7.93	8.03	8.12	8.21
6	6.65	6.79	6.92	7.03	7.14	7.24	7.34	7.43	7.51	7.59
7	6.30	6.43	6.55	6.66	6.76	6.85	6.94	7.02	7.10	7.17
8	6.05	6.18	6.29	6.39	6.48	6.57	6.65	6.73	6.80	6.87
9	5.87	5.98	6.09	6.19	6.28	6.36	6.44	6.51	6.58	6.64
10	5.72	5.83	5.93	6.03	6.11	6.19	6.27	6.34	6.40	6.47
11	5.61	5.71	5.81	5.90	5.98	6.06	6.13	6.20	6.27	6.33
12	5.51	5.61	5.71	5.80	5.88	5.95	6.02	6.09	6.15	6.21
13	5.43	5.53	5.63	5.71	5.79	5.86	5.93	5.99	6.05	6.11
14	5.36	5.46	5.55	5.64	5.71	5.79	5.85	5.91	5.97	6.03
15	5.31	5.40	5.49	5.57	5.65	5.72	5.78	5.85	5.90	5.96
16	5.26	5.35	5.44	5.52	5.59	5.66	5.73	5.79	5.84	5.90
17	5.21	5.31	5.39	5.47	5.54	5.61	5.67	5.73	5.79	5.84
18	5.17	5.27	5.35	5.43	5.50	5.57	5.63	5.69	5.74	5.79
19	5.14	5.23	5.31	5.39	5.46	5.53	5.59	5.65	5.70	5.75
20	5.11	5.20	5.28	5.36	5.43	5.49	5.55	5.61	5.66	5.71
24	5.01	5.10	5.18	5.25	5.32	5.38	5.44	5.49	5.55	5.59
30	4.92	5.00	5.08	5.15	5.21	5.27	5.33	5.38	5.43	5.47
40	4.82	4.90	4.98	5.04	5.11	5.16	5.22	5.27	5.31	5.36
60	4.73	4.81	4.88	4.94	5.00	5.06	5.11	5.15	5.20	5.24
120	4.64	4.71	4.78	4.84	4.90	4.95	5.00	5.04	5.09	5.13
∞	4.55	4.62	4.68	4.74	4.80	4.85	4.89	4.93	4.97	5.01

Table N

Critical Values for the Tukey Test   *Continued*

$\alpha = 0.10$

ν \ k	2	3	4	5	6	7	8	9	10
1	8.93	13.44	16.36	18.49	20.15	21.51	22.64	23.62	24.48
2	4.13	5.73	6.77	7.54	8.14	8.63	9.05	9.41	9.72
3	3.33	4.47	5.20	5.74	6.16	6.51	6.81	7.06	7.29
4	3.01	3.98	4.59	5.03	5.39	5.68	5.93	6.14	6.33
5	2.85	3.72	4.26	4.66	4.98	5.24	5.46	5.65	5.82
6	2.75	3.56	4.07	4.44	4.73	4.97	5.17	5.34	5.50
7	2.68	3.45	3.93	4.28	4.55	4.78	4.97	5.14	5.28
8	2.63	3.37	3.83	4.17	4.43	4.65	4.83	4.99	5.13
9	2.59	3.32	3.76	4.08	4.34	4.54	4.72	4.87	5.01
10	2.56	3.27	3.70	4.02	4.26	4.47	4.64	4.78	4.91
11	2.54	3.23	3.66	3.96	4.20	4.40	4.57	4.71	4.84
12	2.52	3.20	3.62	3.92	4.16	4.35	4.51	4.65	4.78
13	2.50	3.18	3.59	3.88	4.12	4.30	4.46	4.60	4.72
14	2.49	3.16	3.56	3.85	4.08	4.27	4.42	4.56	4.68
15	2.48	3.14	3.54	3.83	4.05	4.23	4.39	4.52	4.64
16	2.47	3.12	3.52	3.80	4.03	4.21	4.36	4.49	4.61
17	2.46	3.11	3.50	3.78	4.00	4.18	4.33	4.46	4.58
18	2.45	3.10	3.49	3.77	3.98	4.16	4.31	4.44	4.55
19	2.45	3.09	3.47	3.75	3.97	4.14	4.29	4.42	4.53
20	2.44	3.08	3.46	3.74	3.95	4.12	4.27	4.40	4.51
24	2.42	3.05	3.42	3.69	3.90	4.07	4.21	4.34	4.44
30	2.40	3.02	3.39	3.65	3.85	4.02	4.16	4.28	4.38
40	2.38	2.99	3.35	3.60	3.80	3.96	4.10	4.21	4.32
60	2.36	2.96	3.31	3.56	3.75	3.91	4.04	4.16	4.25
120	2.34	2.93	3.28	3.52	3.71	3.86	3.99	4.10	4.19
∞	2.33	2.90	3.24	3.48	3.66	3.81	3.93	4.04	4.13

Table N

Critical Values for the Tukey Test *Continued*

$\alpha = 0.10$

v \ k	11	12	13	14	15	16	17	18	19	20
1	25.24	25.92	26.54	27.10	27.62	28.10	28.54	28.96	29.35	29.71
2	10.01	10.26	10.49	10.70	10.89	11.07	11.24	11.39	11.54	11.68
3	7.49	7.67	7.83	7.98	8.12	8.25	8.37	8.48	8.58	8.68
4	6.49	6.65	6.78	6.91	7.02	7.13	7.23	7.33	7.41	7.50
5	5.97	6.10	6.22	6.34	6.44	6.54	6.63	6.71	6.79	6.86
6	5.64	5.76	5.87	5.98	6.07	6.16	6.25	6.32	6.40	6.47
7	5.41	5.53	5.64	5.74	5.83	5.91	5.99	6.06	6.13	6.19
8	5.25	5.36	5.46	5.56	5.64	5.72	5.80	5.87	5.93	6.00
9	5.13	5.23	5.33	5.42	5.51	5.58	5.66	5.72	5.79	5.85
10	5.03	5.13	5.23	5.32	5.40	5.47	5.54	5.61	5.67	5.73
11	4.95	5.05	5.15	5.23	5.31	5.38	5.45	5.51	5.57	5.63
12	4.89	4.99	5.08	5.16	5.24	5.31	5.37	5.44	5.49	5.55
13	4.83	4.93	5.02	5.10	5.18	5.25	5.31	5.37	5.43	5.48
14	4.79	4.88	4.97	5.05	5.12	5.19	5.26	5.32	5.37	5.43
15	4.75	4.84	4.93	5.01	5.08	5.15	5.21	5.27	5.32	5.38
16	4.71	4.81	4.89	4.97	5.04	5.11	5.17	5.23	5.28	5.33
17	4.68	4.77	4.86	4.93	5.01	5.07	5.13	5.19	5.24	5.30
18	4.65	4.75	4.83	4.90	4.98	5.04	5.10	5.16	5.21	5.26
19	4.63	4.72	4.80	4.88	4.95	5.01	5.07	5.13	5.18	5.23
20	4.61	4.70	4.78	4.85	4.92	4.99	5.05	5.10	5.16	5.20
24	4.54	4.63	4.71	4.78	4.85	4.91	4.97	5.02	5.07	5.12
30	4.47	4.56	4.64	4.71	4.77	4.83	4.89	4.94	4.99	5.03
40	4.41	4.49	4.56	4.63	4.69	4.75	4.81	4.86	4.90	4.95
60	4.34	4.42	4.49	4.56	4.62	4.67	4.73	4.78	4.82	4.86
120	4.28	4.35	4.42	4.48	4.54	4.60	4.65	4.69	4.74	4.78
∞	4.21	4.28	4.35	4.41	4.47	4.52	4.57	4.61	4.65	4.69

Table O

Factors for Computing Control Limits

Number of Observations in Sample, n	$\overline{X}$ Chart Factors for Control Limits	R Chart Factors for Control Limits	
	A_2	D_3	D_4
2	1.880	0	3.267
3	1.023	0	2.575
4	0.729	0	2.282
5	0.577	0	2.115
6	0.483	0	2.004
7	0.419	0.076	1.924
8	0.373	0.136	1.864
9	0.337	0.184	1.816
10	0.308	0.223	1.777
11	0.285	0.256	1.744
12	0.266	0.284	1.716
13	0.249	0.308	1.692
14	0.235	0.329	1.671
15	0.223	0.348	1.652
16	0.212	0.364	1.636
17	0.203	0.379	1.621
18	0.194	0.392	1.608
19	0.187	0.404	1.596
20	0.180	0.414	1.586
21	0.173	0.425	1.575
22	0.167	0.434	1.566
23	0.162	0.443	1.557
24	0.157	0.452	1.548
25	0.153	0.459	1.541

Source: From Beyer, W. H., *Handbook of Tables for Probability and Statistics, 2nd Edition,* CRC Press, Boca Raton, Florida, 1986. With permission.

DATA BANK VALUES

The following list explains the values given for the categories in the Data Bank.

1. "Age" is given in years.
2. "Educational Level" values are defined as follows:

   $$0 = \text{no high school degree} \qquad 2 = \text{college graduate}$$
   $$1 = \text{high school graduate} \qquad 3 = \text{graduate degree}$$

3. "Smoking Status" values are defined as follows:

   0 = does not smoke
   1 = smokes less than one pack per day
   2 = smokes one or more than one pack per day

4. "Exercise" values are defined as follows:

   $$0 = \text{none} \qquad 2 = \text{moderate}$$
   $$1 = \text{light} \qquad 3 = \text{heavy}$$

5. "Weight" is given in pounds.
6. "Serum Cholesterol" is given in milligram percent (mg%).
7. "Systolic Pressure" is given in millimeters of mercury (mmHg).
8. "IQ" is given in standard IQ test score values.
9. "Sodium" is given in milliequivalents per liter (mEq/l).
10. "Gender" is listed as male (M) or female (F).
11. "Marital Status" values are defined as follows:

    $$M = \text{married} \qquad S = \text{single}$$
    $$W = \text{widowed} \qquad D = \text{divorced}$$

Data Bank

ID Number	Age	Educational Level	Smoking Status	Exercise	Weight	Serum Cholesterol	Systolic Pressure	IQ	Sodium	Gender	Marital Status
01	27	2	1	1	120	193	126	118	136	F	M
02	18	1	0	1	145	210	120	105	137	M	S
03	32	2	0	0	118	196	128	115	135	F	M
04	24	2	0	1	162	208	129	108	142	M	M
05	19	1	2	0	106	188	119	106	133	F	S
06	56	1	0	0	143	206	136	111	138	F	W
07	65	1	2	0	160	240	131	99	140	M	W
08	36	2	1	0	215	215	163	106	151	M	D
09	43	1	0	1	127	201	132	111	134	F	M
10	47	1	1	1	132	215	138	109	135	F	D
11	48	3	1	2	196	199	148	115	146	M	D
12	25	2	2	3	109	210	115	114	141	F	S
13	63	0	1	0	170	242	149	101	152	F	D
14	37	2	0	3	187	193	142	109	144	M	M
15	40	0	1	1	234	208	156	98	147	M	M
16	25	1	2	1	199	253	135	103	148	M	S
17	72	0	0	0	143	288	156	103	145	F	M
18	56	1	1	0	156	164	153	99	144	F	D
19	37	2	0	2	142	214	122	110	135	M	M
20	41	1	1	1	123	220	142	108	134	F	M
21	33	2	1	1	165	194	122	112	137	M	S
22	52	1	0	1	157	205	119	106	134	M	D
23	44	2	0	1	121	223	135	116	133	F	M
24	53	1	0	0	131	199	133	121	136	F	M
25	19	1	0	3	128	206	118	122	132	M	S
26	25	1	0	0	143	200	118	103	135	M	M
27	31	2	1	1	152	204	120	119	136	M	M
28	28	2	0	0	119	203	118	116	138	F	M
29	23	1	0	0	111	240	120	105	135	F	S
30	47	2	1	0	149	199	132	123	136	F	M
31	47	2	1	0	179	235	131	113	139	M	M
32	59	1	2	0	206	260	151	99	143	M	W
33	36	2	1	0	191	201	148	118	145	M	D
34	59	0	1	1	156	235	142	100	132	F	W
35	35	1	0	0	122	232	131	106	135	F	M
36	29	2	0	2	175	195	129	121	148	M	M

Data Bank   *Continued*

ID Number	Age	Educational Level	Smoking Status	Exercise	Weight	Serum Cholesterol	Systolic Pressure	IQ	Sodium	Gender	Marital Status
37	43	3	0	3	194	211	138	129	146	M	M
38	44	1	2	0	132	240	130	109	132	F	S
39	63	2	2	1	188	255	156	121	145	M	M
40	36	2	1	1	125	220	126	117	140	F	S
41	21	1	0	1	109	206	114	102	136	F	M
42	31	2	0	2	112	201	116	123	133	F	M
43	57	1	1	1	167	213	141	103	143	M	W
44	20	1	2	3	101	194	110	111	125	F	S
45	24	2	1	3	106	188	113	114	127	F	D
46	42	1	0	1	148	206	136	107	140	M	S
47	55	1	0	0	170	257	152	106	130	F	M
48	23	0	0	1	152	204	116	95	142	M	M
49	32	2	0	0	191	210	132	115	147	M	M
50	28	1	0	1	148	222	135	100	135	M	M
51	67	0	0	0	160	250	141	116	146	F	W
52	22	1	1	1	109	220	121	103	144	F	M
53	19	1	1	1	131	231	117	112	133	M	S
54	25	2	0	2	153	212	121	119	149	M	D
55	41	3	2	2	165	236	130	131	152	M	M
56	24	2	0	3	112	205	118	100	132	F	S
57	32	2	0	1	115	187	115	109	136	F	S
58	50	3	0	1	173	203	136	126	146	M	M
59	32	2	1	0	186	248	119	122	149	M	M
60	26	2	0	1	181	207	123	121	142	M	S
61	36	1	1	0	112	188	117	98	135	F	D
62	40	1	1	0	130	201	121	105	136	F	D
63	19	1	1	1	132	237	115	111	137	M	S
64	37	2	0	2	179	228	141	127	141	F	M
65	65	3	2	1	212	220	158	129	148	M	M
66	21	1	2	2	99	191	117	103	131	F	S
67	25	2	2	1	128	195	120	121	131	F	S
68	68	0	0	0	167	210	142	98	140	M	W
69	18	1	1	2	121	198	123	113	136	F	S

Data Bank   *Concluded*

ID Number	Age	Educational Level	Smoking Status	Exercise	Weight	Serum Cholesterol	Systolic Pressure	IQ	Sodium	Gender	Marital Status
70	26	0	1	1	163	235	128	99	140	M	M
71	45	1	1	1	185	229	125	101	143	M	M
72	44	3	0	0	130	215	128	128	137	F	M
73	50	1	0	0	142	232	135	104	138	F	M
74	63	0	0	0	166	271	143	103	147	F	W
75	48	1	0	3	163	203	131	103	144	M	M
76	27	2	0	3	147	186	118	114	134	M	M
77	31	3	1	1	152	228	116	126	138	M	D
78	28	2	0	2	112	197	120	123	133	F	M
79	36	2	1	2	190	226	123	121	147	M	M
80	43	3	2	0	179	252	127	131	145	M	D
81	21	1	0	1	117	185	116	105	137	F	S
82	32	2	1	0	125	193	123	119	135	F	M
83	29	2	1	0	123	192	131	116	131	F	D
84	49	2	2	1	185	190	129	127	144	M	M
85	24	1	1	1	133	237	121	114	129	M	M
86	36	2	0	2	163	195	115	119	139	M	M
87	34	1	2	0	135	199	133	117	135	F	M
88	36	0	0	1	142	216	138	88	137	F	M
89	29	1	1	1	155	214	120	98	135	M	S
90	42	0	0	2	169	201	123	96	137	M	D
91	41	1	1	1	136	214	133	102	141	F	D
92	29	1	1	0	112	205	120	102	130	F	M
93	43	1	1	0	185	208	127	100	143	M	M
94	61	1	2	0	173	248	142	101	141	M	M
95	21	1	1	3	106	210	111	105	131	F	S
96	56	0	0	0	149	232	142	103	141	F	M
97	63	0	1	0	192	193	163	95	147	M	M
98	74	1	0	0	162	247	151	99	151	F	W
99	35	2	0	1	151	251	147	113	145	F	M
100	28	2	0	3	161	199	129	116	138	M	M

A

alpha the probability of a type I error, represented by the Greek letter α

alternative hypothesis a statistical hypothesis that states a specific difference between a parameter and a specific value or states that there is a difference between two parameters

analysis of variance (ANOVA) a statistical technique used to test a hypothesis concerning the means of three or more populations

ANOVA summary table the table used to summarize the results of an ANOVA test

assignable-cause variation a non-random variation in manufactured items that must be eliminated to maintain the quality of the manufactured product

attribute chart a chart used in quality control to analyze a product and to determine whether it is acceptable or defective

B

bar graph a graph that uses vertical or horizontal bars to represent the frequencies of a distribution

Bayes's theorem a theorem that allows one to compute the revised probability of an event that occurred previously to another event when the events are dependent

beta the probability of a type II error, represented by the Greek letter β

between-group variance a variance estimate using the means of the groups or between the groups in an F test

biased sample a sample for which some type of systematic error has been made in the selection of the subjects for the sample

binomial distribution the outcomes of a binomial experiment and the corresponding probabilities of these outcomes

binomial experiment a probability experiment in which each trial has only two outcomes, there are a fixed number of trials, the outcomes of each trial are independent, and the probability of a success remains the same for each trial

biserial correlation coefficient a correlation coefficient used when variables are continuous but one variable is artificially reduced to two categories

box and whisker plot a graph used to represent a data set when the data set contains a small number of values

C

categorical distribution a frequency distribution used when the data is categorical (nominal)

$\bar{c}$ chart a chart used in quality control to judge the quality of an item that could have one or more defects to determine if a manufacturing process is in or out of control

central limit theorem a theorem that states that as the sample size increases, the shape of the distribution of the sample means taken from the population with mean μ and standard deviation σ will approach a normal distribution; the distribution will have mean μ and a standard deviation σ

chance variation a random variation in manufactured items that cannot be eliminated entirely

Chebyshev's theorem a theorem that states that the proportion of values from a data set that fall within k standard deviations of the mean will be at least $1 - 1/k^2$, where k is a number greater than 1

chi-square distribution a probability distribution obtained from the values of $(n - 1)s^2/\sigma^2$ when random samples are selected from a normally distributed population whose variance is σ^2

classical probability the type of probability that uses sample spaces to determine the numerical probability that an event will happen

class midpoint a value for a class in a frequency distribution obtained by adding the lower and upper class boundaries (or the lower and upper class limits) and dividing by 2

class width the difference between the upper class boundary and the lower class boundary for a class in a frequency distribution

cluster sample a sample obtained by selecting a preexisting or natural group, called a cluster, and using the members in the cluster for the sample

coefficient of determination a measure of the variation of the dependent variable that is explained by the regression line and the independent variable; the ratio of the explained variation to the total variation

coefficient of variation the standard deviation divided by the mean, the result expressed as a percentage

combination a selection of objects without regard to order

complement of an event the set of outcomes in the sample space that are not in the outcomes of the event itself

compound event an event that consists of two or more outcomes or simple events

conditional probability the probability that an event B occurs after an event A has already occurred

confidence interval a specific interval estimate of a parameter determined by using data obtained from a sample and the specific confidence level of the estimate

confidence level the probability that a parameter will fall within the specified interval estimate of the parameter

consistent estimator an estimator whose value approaches the value of the parameter estimated as the sample size increases

contingency table data arranged in table form for the chi-square independence test, with R rows and C columns

continuous variable a variable that can assume all values between any two specific values; a variable obtained by measuring

control chart a chart used to analyze data in quality control

correction for continuity a correction employed when a continuous distribution is used to approximate a discrete distribution

correlation a statistical method used to determine whether a relationship between variables exists

correlation coefficient a statistic or parameter that measures the strength and direction of a relationship between two variables

critical region the range of values of the test value that indicates that there is a significant difference and that the null hypothesis should be rejected in a hypothesis test

critical value a value that separates the critical region from the noncritical region in a hypothesis test

cumulative frequency the sum of the frequencies accumulated up to the upper boundary of a class in a frequency distribution

D

data measurements or observations for a variable

data array a data set that has been ordered

data set a collection of data values

data value or **datum** a value in a data set

decile a location measure of a data value; it divides the distribution into ten groups

degrees of freedom the number of values that are free to vary after a sample statistic has been computed; used when a distribution (such as the t distribution) consists of a family of curves

dependent events events for which the outcome or occurrence of the first event affects the outcome or occurrence of the second event in such a way that the probability is changed

dependent samples samples in which the subjects are paired or matched in some way; i.e., the samples are related

dependent variable a variable in correlation and regression analysis that cannot be controlled or manipulated

descriptive statistics a branch of statistics that consists of the collection, organization, summarization, and presentation of data

discrete variable a variable that assumes values that can be counted

disordinal interaction an interaction between variables in ANOVA, indicated when the graphs of the lines connecting the mean intersect

distribution-free statistics *see* nonparametric statistics

double sampling a sampling method in which a very large population is given a questionnaire to determine those who meet the qualifications for a study; the questionnaire is reviewed, a second smaller population is defined, and a sample is selected from this group

E

empirical probability the type of probability that uses frequency distributions based on observations to determine numerical probabilities of events

empirical rule a rule that states that when a distribution is bell-shaped (normal), approximately 68% of the data values will fall within one standard deviation of the mean; approximately 95% of the data values will fall within two standard deviations of the mean; and approximately 99.7% of the data values will fall within three standard deviations of the mean

equally likely events the events in the sample space that have the same probability of occurring

estimation the process of estimating the value of a parameter from information obtained from a sample

estimator a statistic used to estimate a parameter

event one or more outcomes of a probability experiment

expected frequency the frequency obtained by calculation (as if there were no preference) and used in the chi-square test

expected value the theoretical average of a variable that has a probability distribution

F

factors the independent variables in ANOVA tests

F distribution the sampling distribution of the variances when two independent samples are selected from two normally distributed populations in which the variances are equal and in which the variances s_1^2 and s_2^2 are compared as $s_1^2 \div s_2^2$

finite population correction factor a correction factor used to correct the standard error of the mean when the sample size is greater than 5% of the population size

five-number summary five specific values for a data set that consist of the lowest and highest values, the lower and upper hinges, and the median

frequency the number of values in a specific class of a frequency distribution

frequency distribution an organization of raw data in table form, using classes and frequencies

frequency polygon a graph that displays the data by using lines that connect points plotted for the frequencies at the midpoints of the classes

F test a statistical test used to compare two variances or three or more means

G

goodness-of-fit test a chi-square test used to see whether a frequency distribution fits a specific pattern

grouped frequency distribution a distribution used when the range is large and classes of several units in width are needed

H

histogram a graph that displays the data by using vertical bars of various heights to represent the frequencies of a distribution

hypergeometric distribution the distribution of a variable that has two outcomes when sampling is done without replacement

hypothesis testing a decision-making process for evaluating claims about a population

I

independence test a chi-square test used to test the independence of two variables when data are tabulated in table form in terms of frequencies

independent events events for which the probability of the first occurring does not affect the probability of the second occurring

independent samples samples that are not related

independent variable a variable in correlation and regression analysis that can be controlled or manipulated

inferential statistics a branch of statistics that consists of generalizing from samples to populations, performing hypothesis testing, determining relationships among variables, and making predictions

interaction effect the effect of two or more variables on each other in a two-way ANOVA study

interquartile range $Q_3 - Q_1$

interval estimate a range of values used to estimate a parameter

interval level of measurement a measurement level that ranks data, and precise differences between units of measure do exist

K

Kruskal-Wallis test a nonparametric test used to compare three or more means

L

level a treatment in ANOVA for a variable

level of significance the maximum probability of committing a type I error in hypothesis testing

lower class boundary the lower value of a class in a frequency distribution that has one more decimal place value than the data and ends in a 5 digit

lower class limit the lower value of a class in a frequency distribution that has the same decimal place value as the data

lower control limit the minimum acceptable value of a mean, range, etc., of a sample that is used in a quality control study

lower hinge the median of all values less than or equal to the median when the data set has an odd number of values, or the median of all values less than the median when the data set has an even number of values

M

main effect the effect of the factors or independent variables when there is a nonsignificant interaction effect in a two-way ANOVA study

maximum error of estimate the maximum difference between the point estimate of a parameter and the actual value of the parameter

mean the sum of the values divided by the total number of values

mean square the variance found by dividing the sum of the squares of a variable by the corresponding degrees of freedom; used in ANOVA

measurement scales a type of classification that tells how variables are categorized, counted, or measured; there are four types of scales: nominal, ordinal, interval, and ratio

median the midpoint of a data array

median class the class of a frequency distribution that contains the median

midrange the sum of the lowest and highest data values divided by 2

modal class the class with the largest frequency

mode the value that occurs most often in a data set

Monte Carlo method a simulation technique using random numbers

multinomial distribution a probability distribution for an experiment in which each trial has more than two outcomes

multiple relationship a relationship in which many variables are under study

multistage sampling a sampling technique that uses a combination of sampling methods

mutually exclusive events probability events that cannot occur at the same time

N

negatively skewed distribution a distribution in which the majority of the data values fall to the right of the mean

negative relationship a relationship between variables such that as one variable increases, the other variable decreases, and vice versa

nominal level of measurement a measurement level that classifies data into mutually exclusive (nonoverlapping) exhaustive categories in which no order or ranking can be imposed on the data

noncritical region the range of values of the test value that indicates that the difference was probably due to chance and that the null hypothesis should not be rejected

nonparametric statistics a branch of statistics for use when the population from which the samples are selected is not normally distributed and for use in the test of hypotheses that do not involve specific population parameters

nonrejection region *see* noncritical region

normal distribution a continuous, symmetric, bell-shaped distribution of a variable

null hypothesis a statistical hypothesis that states that there is no difference between a parameter and a specific value or that there is no difference between two parameters

O

observed frequency the actual frequency value obtained from a sample and used in the chi-square test

ogive a graph that represents the cumulative frequencies for the classes in a frequency distribution

one-tailed test a test that indicates that the null hypothesis should be rejected when the test statistic value is in the critical region on one side of the mean

one-way ANOVA a study used to test for differences among means for a single independent variable when there are three or more groups

open-ended distribution a frequency distribution that has no specific beginning value or no specific ending value

ordinal interaction an interaction between variables in ANOVA, indicated when the graphs of the lines connecting the means do not intersect

ordinal level of measurement a measurement level that classifies data into categories that can be ranked; however, precise differences between the ranks do not exist

outcome the result of a single trial of a probability experiment

outlier an extreme value in a data set; it is omitted from a box and whisker plot

P

parameter a characteristic or measure obtained by using all the data values for a specific population

parametric tests statistical tests for population parameters such as means, variances, and proportions that involve assumptions about the populations from which the samples were selected

$\bar{p}$ **chart** a chart used in quality control to analyze the proportion of items in a sample that are defective to determine whether a manufacturing process is in or out of control

Pearson product moment correlation coefficient a statistic used to determine the strength of a relationship when the variables are normally distributed

percentile a location measure of a data value; it divides the distribution into 100 groups

permutation an arrangement of n objects in a specific order

phi correlation coefficient a statistic used to determine the strength of a relationship when both variables are truly dichotomous

pictograph a graph that uses symbols or pictures to represent data

pie graph a circle that is divided into sections or wedges according to the percentage of frequencies in each category of the distribution

point-biserial correlation coefficient a statistic used to determine the strength of a relationship when one variable is a true dichotomous variable and the other variable is continuous and normally distributed

point estimate a specific numerical value estimate of a parameter

Poisson distribution a probability distribution used when n is large and p is small and when the independent variables occur over a period of time

pooled estimate of the variance a weighted average of the variance using the two sample variances and the degrees of freedom of each variance as the weights

population the totality of all subjects possessing certain common characteristics that are being studied

positively skewed distribution a distribution in which the majority of the data values fall to the left of the mean

positive relationship a relationship between two variables such that as one variable increases, the other variable increases; or such that as one variable decreases, the other variable decreases

power of a test the probability of rejecting the null hypothesis when it is false

prediction interval a confidence interval for a predicted value y

probability the chance of an event occurring

probability distribution the values a random variable can assume and the corresponding probabilities of the values

probability experiment a process that leads to well-defined results called outcomes

proportion a part of a whole, represented by a fraction, a decimal, or a percentage

P-value the actual probability of getting the sample mean value if the null hypothesis is true

Q

qualitative variable a variable that can be placed into distinct categories, according to some characteristic or attribute

quality control the analysis of data obtained from small samples of manufactured products to check and correct any defects in the manufacturing process

quantitative variable a variable that is numerical in nature and that can be ordered or ranked

quartile a location measure of a data value; it divides the distribution into four groups

R

random sample a sample obtained by using random or chance methods; a sample for which every member of the population has an equal chance of being selected

random variable a variable whose values are determined by chance

random variation *see* chance variation

range the highest data value minus the lowest data value

ranking the positioning of a data value in a data array according to some rating scale

ratio level of measurement a measurement level that possesses all the characteristics of interval measurement, and there exists a true zero; also, true ratios exist between different units of measure

raw data data collected in original form

R chart a chart used in quality control to analyze the ranges of measured values to determine whether a manufacturing process is in or out of control

regression a statistical method used to describe the nature of the relationship between variables—i.e., a positive or negative, linear or nonlinear relationship

regression line the line of best fit of the data

rejection region *see* critical region

relative frequency graph a graph using proportions instead of raw data as frequencies

relatively efficient estimator an estimator that has the smallest variance from among all the statistics that can be used to estimate a parameter

run a succession of identical letters preceded by or followed by a different letter or no letter at all, such as at the beginning or end of the succession

runs test a nonparametric test used to determine whether data are random

S

sample a subgroup or subset of the population

sample space the set of all possible outcomes of a probability experiment

sampling distribution of sample means a distribution obtained by using the means computed from random samples taken from a population

sampling error the difference between the sample measure and the corresponding population measure due to the fact that the sample is not a perfect representation of the population

scatter plot a graph of the independent and dependent variables in regression and correlation analysis

Scheffé test a test used after ANOVA, if the null hypothesis is rejected, to determine where significant differences in the means lie

sequence sampling a sampling technique used in quality control in which successive units are taken from production lines and tested to see whether they meet the standards set by the manufacturing company

sign test a nonparametric test used to test the value of the median for a specific sample or to test sample means in a comparison of two dependent samples

simple event an outcome that results from a single trial of a probability experiment

simple relationship a relationship in which only two variables are under study

simulation techniques techniques that use probability experiments to mimic real-life situations

Spearman rank correlation coefficient the nonparametric equivalent to the correlation coefficient, used when the data are ranked

standard deviation the square root of the variance

standard error of estimate the standard deviation of the observed y values about the predicted y' values in regression and correlation analysis

standard error of the mean the standard deviation of the sample means for samples taken from the same population

standard normal distribution a normal distribution for which the mean is equal to 0 and the standard deviation is equal to 1

standard score the difference between a data value and the mean divided by the standard deviation

statistic a characteristic or measure obtained by using the data values from a sample

statistical hypothesis a conjecture about a population parameter, which may or may not be true

statistical test a test that uses data obtained from a sample to make a decision about whether or not the null hypothesis should be rejected

stem and leaf plot a data plot that uses part of a data value as the stem and part of the data value as the leaf to form groups or classes

stratified sample a sample obtained by dividing the population into subgroups, called strata, according to various homogeneous characteristics and then selecting members from each stratum for the sample

subjective probability the type of probability that uses a probability value based on an educated guess or estimate, employing opinions and inexact information

sum of squares between groups a statistic computed in the numerator of the fraction used to find the between-group variance in ANOVA

sum of squares within groups a statistic computed in the numerator of the fraction used to find the within-group variance in ANOVA

symmetrical distribution a distribution in which the data values are uniformly distributed about the mean

systematic sample a sample obtained by numbering each element in the population and then selecting every kth number from the population to be included in the sample

T

***t*-distribution** a family of bell-shaped curves based on degrees of freedom; similar to the standard normal distribution with the exception that the variance is greater than 1; used when one is testing small samples and when the population standard deviation is unknown

test value the numerical value obtained from a statistical test, computed from (observed value − expected value) ÷ standard error

tetrachoric correlation coefficient a statistic used to test the strength of a relationship when both variables have been artificially reduced to two categories

time series graph a graph that represents data that occur over a specific period of time

treatment groups the groups used in an ANOVA study

tree diagram a device used to list all possibilities of a sequence of events in a systematic way

***t* test** a statistical test for the mean of a population, used when the population is normally distributed, the population standard deviation is unknown, and the sample size is less than 30

Tukey test a test used to make pairwise comparisons of means in an ANOVA study when samples are the same size

two-tailed test a test that indicates that the null hypothesis should be rejected when the test value is in either of the two critical regions

two-way ANOVA a study used to test the effects of two or more independent variables and the possible interaction between them

type I error the error that occurs if one rejects the null hypothesis when it is true

type II error the error that occurs if one does not reject the null hypothesis when it is false

U

unbiased estimator an estimator whose value approximates the expected value of a population parameter; used for the variance or standard deviation when the sample size is less than 30; an estimator whose expected value or mean must be equal to the mean of the parameter being estimated

unbiased sample a sample chosen at random from the population that is, for the most part, representative of the population

ungrouped frequency distribution a distribution that uses individual data and the range of data is small

upper class boundary the upper value of a class in a frequency distribution that has one more decimal place value than the data and ends in a 5 digit

upper class limit the upper value of a class in a frequency distribution that has the same decimal place value as the data

upper control limit the maximum acceptable value of a mean, range, etc., of a sample that is tested in a quality control study

upper hinge the median of all values greater than or equal to the median when the data set has an odd number of values, or the median of all values greater than the median when the data set has an even number of values

V

variable a characteristic or attribute that can assume different values

variable chart a chart used in quality control to analyze the quality of a product in which measurements, such as lengths, diameters, or weights, are important

variance the average of the squares of the distance each value is from the mean

W

weighted mean the mean found by multiplying each value by its corresponding weight and dividing by the sum of the weights

Wilcoxon rank sum test a nonparametric test used to test independent samples and compare distributions

Wilcoxon signed-rank test a nonparametric test used to test dependent samples and compare distributions

within-group variance a variance estimate using all the sample data for an F test; it is not affected by differences in the means

X

$\overline{X}$ **chart** a chart used in quality control to analyze the means of a measured value to determine whether a manufacturing process is in or out of control

Z

z **distribution** *see* standard normal distribution

z **score** *see* standard score

z **test** a statistical test for means and proportions of a population, used when the population is normally distributed and the population standard deviation is known or the sample size is 30 or more

z **value** same as z score

BIBLIOGRAPHY

Anderson, T. W., and Stanley L. Sclove. *Introductory Statistical Analysis.* Boston: Houghton Mifflin, 1974.

Beyer, William H. *CRC Handbook of Tables for Probability and Statistics,* 2nd ed. Boca Raton, Fla.: CRC Press, 1986.

Chao, Lincoln L. *Introduction to Statistics.* Monterey, Calif.: Brooks/ Cole, 1980.

Daniel, Wayne W., and James C. Terrell. *Business Statistics,* 4th ed. Boston: Houghton Mifflin, 1986.

Edwards, Allan L. *An Introduction to Linear Regression and Correlation,* 2nd ed. New York: Freeman, 1984.

Eves, Howard. *An Introduction to the History of Mathematics,* 3rd ed. New York: Holt, Rinehart and Winston, 1969.

Freund, John E. *Modern Elementary Statistics,* 7th ed. Englewood Cliffs, N.J.: Prentice-Hall, 1988.

Glass, Gene V., and Kenneth D. Hopkins. *Statistical Methods in Education and Psychology,* 2nd ed. Englewood Cliffs, N.J.: Prentice-Hall, 1984.

Guilford, J. P. *Fundamental Statistics in Psychology and Education,* 4th ed. New York: McGraw-Hill, 1965.

Haack, Dennis G. *Statistical Literacy: A Guide to Interpretation.* Boston: Duxbury Press, 1979.

Hartwig, Frederick, with Brian Dearing. *Exploratory Data Analysis.* Newbury Park, Calif.: Sage Publications, 1979.

Hoffman, Mark S., ed. *The World Almanac and Book of Facts, 1993.* New York: Pharos Books, 1992.

Isaac, Stephen, and William B. Michael. *Handbook in Research and Evaluation,* 2nd ed. San Diego: EdITS, 1990.

Johnson, Robert. *Elementary Statistics,* 5th ed. Boston: PWS–Kent, 1988.

Kachigan, Sam Kash. *Statistical Analysis.* New York: Radius Press, 1986.

Khazanie, Ramakant. *Elementary Statistics in a World of Applications,* 3rd ed. Glenview, Ill.: Scott, Foresman, 1990.

Kuzma, Jan W. *Basic Statistics for the Health Sciences.* Mountain View, Calif.: Mayfield, 1984.

Lapham, Lewis H., Michael Pollan, and Eric Ethridge. *The Harper's Index Book.* New York: Henry Holt, 1987.

Lipschultz, Seymour. *Schaum's Outline of Theory and Problems of Probability.* New York: McGraw-Hill, 1968.

Marascuilo, Leonard A., and Maryellen McSweeney. *Nonparametric and Distribution-Free Methods for the Social Sciences.* Monterey, Calif.: Brooks/Cole, 1977.

Mason, Robert D., Douglas A. Lind, and William G. Marchal. *Statistics: An Introduction.* New York: Harcourt Brace Jovanovich, 1988.

Minium, Edward W. *Statistical Reasoning in Psychology and Education.* New York: Wiley, 1970.

Moore, David S. *Statistics: Concepts and Controversies.* San Francisco: Freeman, 1979.

Newmark, Joseph. *Statistics and Probability in Modern Life.* New York: Saunders, 1988.

Pagano, Robert R. *Understanding Statistics,* 3rd ed. New York: West, 1990.

Phillips, John L., Jr. *How to Think About Statistics.* New York: Freeman, 1988.

Reinhardt, Howard E., and Don O. Loftsgaarden. *Elementary Probability and Statistical Reasoning.* Lexington, Mass.: Heath, 1977.

Roscoe, John T. *Fundamental Research Statistics for the Behavioral Sciences,* 2nd ed. New York: Holt, Rinehart and Winston, 1975.

Runyon, Richard P., and Audrey Haber. *Fundamentals of Behavioral Statistics,* 6th ed. New York: Random House, 1988.

Setek, William M., Jr. *Fundamentals of Mathematics.* Beverly Hills, Calif.: Glencoe Press, 1976.

Shulte, Albert P., 1981 yearbook editor, and James R. Smart, general yearbook editor. *Teaching Statistics and Probability, 1981 Yearbook.* Reston, Va.: National Council of Teachers of Mathematics, 1981.

Smith, Gary. *Statistical Reasoning.* Boston: Allyn and Bacon, 1985.

Spiegel, Murray R. *Schaum's Outline of Theory and Problems of Statistics.* New York: McGraw-Hill, 1961.

Triola, Mario F. *Elementary Statistics,* 5th ed. Reading, Mass.: Addison-Wesley, 1992.

Warwick, Donald P., and Charles A. Lininger. *The Sample Survey: Theory and Practice.* New York: McGraw-Hill, 1975.

Weiss, Daniel Evan. *100% American.* New York: Poseidon Press, 1988.

Answers to Odd-Numbered and Selected Even-Numbered Exercises

CHAPTER 1

1–1. Descriptive statistics describes the situation as it is. Inferential statistics goes beyond what is known.

1–3. Answers will vary.

1–5. Samples are used to save time and money when the population is large and when the units must be destroyed to gain information.

1–6. a. Descriptive   b. Descriptive   c. Inferential
d. Inferential   e. Descriptive   f. Inferential
g. Inferential (The entire population was not used.)
h. Inferential   i. Descriptive   j. Inferential

1–7. a. Ratio   b. Ordinal   c. Interval   d. Ratio
e. Ratio   f. Nominal   g. Ratio   h. Ratio
i. Ordinal   j. Ratio

1–8. a. Qualitative   b. Quantitative   c. Qualitative
d. Quantitative   e. Quantitative

1–9. a. Discrete   b. Continuous   c. Discrete
d. Continuous   e. Continuous   f. Discrete
g. Discrete   h. Continuous   i. Continuous
j. Continuous

1–11. Random, systematic, stratified, cluster

1–12. a. Cluster   b. Systematic   c. Random
d. Systematic   e. Stratified

1–13. Jerome Cardan   **1–15.** Chevalier de Méré

1–17. Census taking and record keeping

1–19. Answers will vary.   **1–21.** Answers will vary.

1–23. Answers will vary.

CHAPTER 1 TEST

1. False   2. True   3. False   4. False   5. False
6. True   7. True   8. True   9. True   10. False

All answers are given for the chapter tests.

CHAPTER 2

2–1. To organize data in a meaningful way, to determine the shape of the distribution, to facilitate computational procedures for statistics, to make it easier to draw charts and graphs, to make comparisons among different data sets.

2–3. a. 10.5–15.5, 13, 5
b. 16.5–39.5, 28, 23
c. 292.5–353.5, 323, 61
d. 11.75–14.75, 13.25, 3
e. 3.125–3.935, 3.53, 0.81

2–5. a. Class width is not uniform.
b. Class limits overlap, and class width is not uniform.
c. A class has been omitted.
d. Class width is not uniform.

2–7.

Class	f
15130	5
15131	3
15132	3
15133	7
15134	2
	20

2–9.

	Class	f
0	−0.5–0.5	5
1	0.5–1.5	8
2	1.5–2.5	10
3	2.5–3.5	2
4	3.5–4.5	3
5	4.5–5.5	2
		30

2–11.

Limits	Boundaries	f	cf
26–30	25.5–30.5	5	5
31–35	30.5–35.5	5	10
36–40	35.5–40.5	5	15
41–45	40.5–45.5	9	24
46–50	45.5–50.5	7	31
51–55	50.5–55.5	1	32
56–60	55.5–60.5	2	34
61–65	60.5–65.5	6	40
		40	

2–13.

Limits	Boundaries	f	cf
93– 99	92.5– 99.5	3	3
100–106	99.5–106.5	1	4
107–113	106.5–113.5	2	6
114–120	113.5–120.5	5	11
121–127	120.5–127.5	8	19
128–134	127.5–134.5	5	24
135–141	134.5–141.5	5	29
142–148	141.5–148.5	4	33
149–155	148.5–155.5	1	34
156–162	155.5–162.5	1	35
		35	

2–15.

Limits	Boundaries	f	cf
2.7–3.1	2.65–3.15	1	1
3.2–3.6	3.15–3.65	4	5
3.7–4.1	3.65–4.15	8	13
4.2–4.6	4.15–4.65	13	26
4.7–5.1	4.65–5.15	8	34
5.2–5.6	5.15–5.65	3	37
5.7–6.1	5.65–6.15	3	40
		40	

2–17. 145–205; one possibility

Limits	Boundaries	f	cf
60–80	55–85	4	4
90–110	85–115	2	6
120–140	115–145	8	14
150–170	145–175	4	18
180–200	175–205	1	19
210–230	205–235	1	20
		20	

2–19.

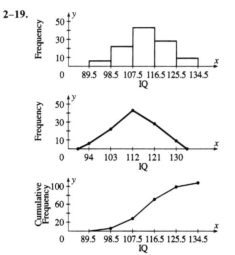

2–21.

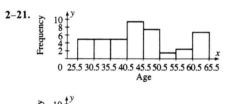
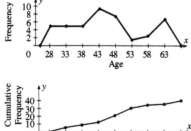
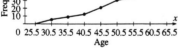

2–23.

2–25.

2–27.

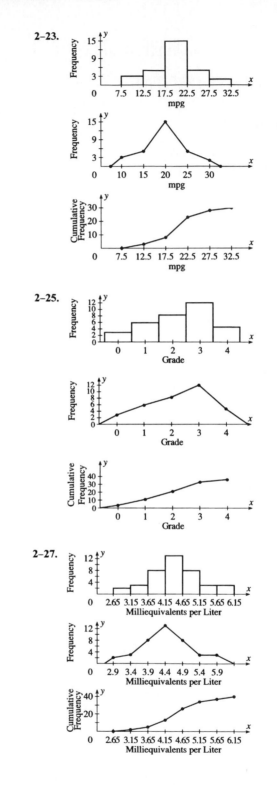

2–29.

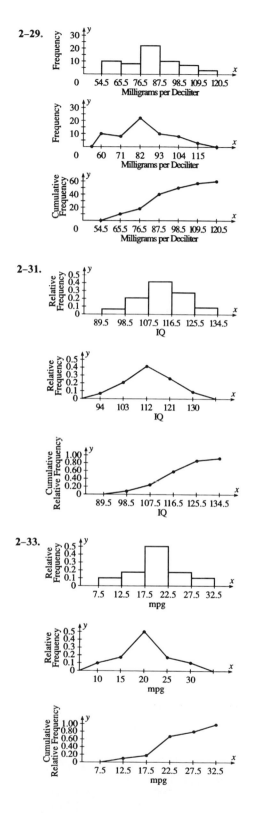

2–31.

2–33.

2–35.

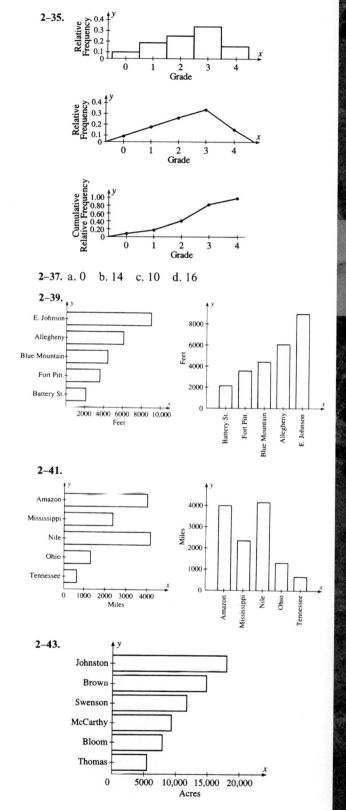

2–37. a. 0   b. 14   c. 10   d. 16

2–39.

2–41.

2–43.

2–45.

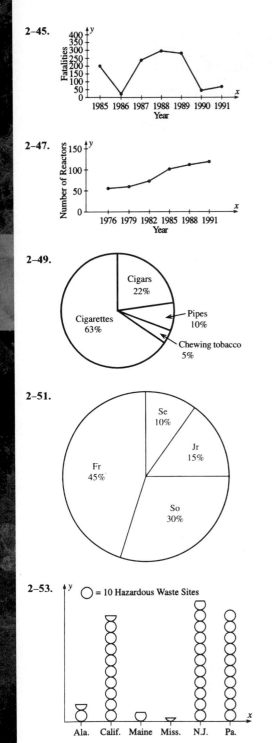

2–47.

2–49.

2–51.

2–53.

2–55.

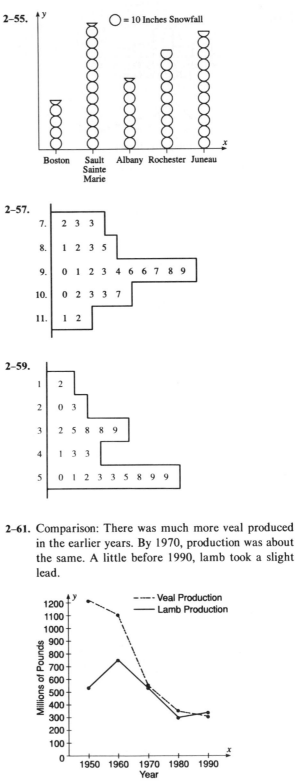

2–57.

2–59.

2–61. Comparison: There was much more veal produced in the earlier years. By 1970, production was about the same. A little before 1990, lamb took a slight lead.

2–63.

Class	f
Newspaper	7
Television	5
Radio	7
Magazine	6
	25

2–65.

Type	f
LP	10
CD	5
CAS	4
45	6
	25

2–67.

Class	f	cf
11	1	1
12	2	3
13	2	5
14	2	7
15	1	8
16	2	10
17	4	14
18	2	16
19	2	18
20	1	19
21	0	19
22	1	20
	20	

2–69.

Limits	Boundaries	f	cf
40–42	39.5–42.5	2	2
43–45	42.5–45.5	7	9
46–48	45.5–48.5	10	19
49–51	48.5–51.5	8	27
52–54	51.5–54.5	3	30
		30	

2–71.

Limits	Boundaries	f	cf
120–128	119.5–128.5	2	2
129–137	128.5–137.5	1	3
138–146	137.5–146.5	4	7
147–155	146.5–155.5	0	7
156–164	155.5–164.5	13	20
165–173	164.5–173.5	10	30

2–73.

2–75.

2–77.

2–79.

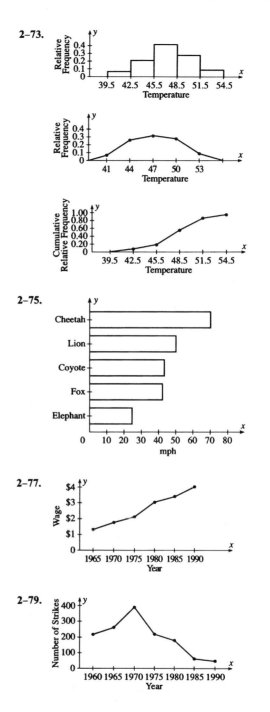

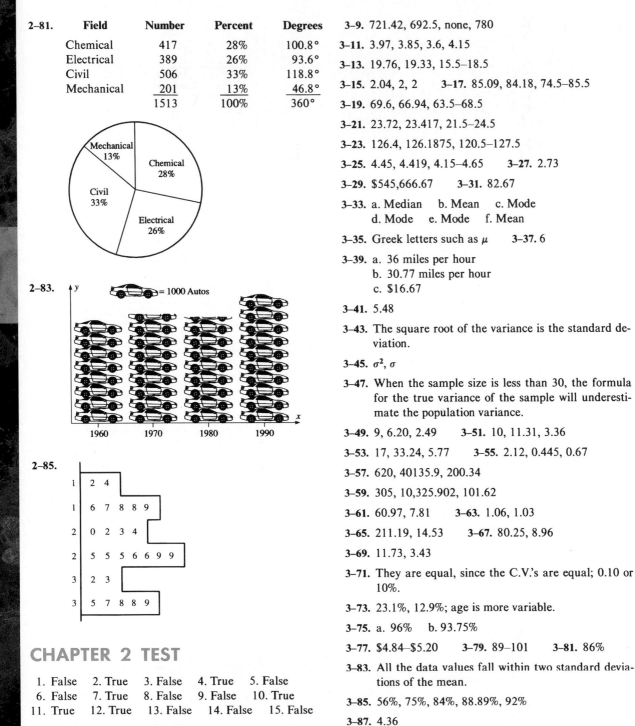

2–81.

Field	Number	Percent	Degrees
Chemical	417	28%	100.8°
Electrical	389	26%	93.6°
Civil	506	33%	118.8°
Mechanical	201	13%	46.8°
	1513	100%	360°

2–83.

2–85.

CHAPTER 2 TEST

1. False 2. True 3. False 4. True 5. False
6. False 7. True 8. False 9. False 10. True
11. True 12. True 13. False 14. False 15. False

CHAPTER 3

3–1. 8.83, 8, 8, 10.5   **3–3.** 43.92, 45.5, 51, 44.5

3–5. 124.29, 121, 121, 124

3–7. 2.78, 2.715, none, 2.91

3–9. 721.42, 692.5, none, 780

3–11. 3.97, 3.85, 3.6, 4.15

3–13. 19.76, 19.33, 15.5–18.5

3–15. 2.04, 2, 2   **3–17.** 85.09, 84.18, 74.5–85.5

3–19. 69.6, 66.94, 63.5–68.5

3–21. 23.72, 23.417, 21.5–24.5

3–23. 126.4, 126.1875, 120.5–127.5

3–25. 4.45, 4.419, 4.15–4.65   **3–27.** 2.73

3–29. $545,666.67   **3–31.** 82.67

3–33. a. Median   b. Mean   c. Mode
d. Mode   e. Mode   f. Mean

3–35. Greek letters such as μ **3–37.** 6

3–39. a. 36 miles per hour
b. 30.77 miles per hour
c. $16.67

3–41. 5.48

3–43. The square root of the variance is the standard deviation.

3–45. σ^2, σ

3–47. When the sample size is less than 30, the formula for the true variance of the sample will underestimate the population variance.

3–49. 9, 6.20, 2.49   **3–51.** 10, 11.31, 3.36

3–53. 17, 33.24, 5.77   **3–55.** 2.12, 0.445, 0.67

3–57. 620, 40135.9, 200.34

3–59. 305, 10,325.902, 101.62

3–61. 60.97, 7.81   **3–63.** 1.06, 1.03

3–65. 211.19, 14.53   **3–67.** 80.25, 8.96

3–69. 11.73, 3.43

3–71. They are equal, since the C.V.'s are equal; 0.10 or 10%.

3–73. 23.1%, 12.9%; age is more variable.

3–75. a. 96%   b. 93.75%

3–77. $4.84–$5.20   **3–79.** 89–101   **3–81.** 86%

3–83. All the data values fall within two standard deviations of the mean.

3–85. 56%, 75%, 84%, 88.89%, 92%

3–87. 4.36

3–89. A z score tells how many standard deviations the data value is above or below the mean.

3–91. A percentile is a relative measurement of position; a percentage is an absolute measure of the part to the total.

3–93. $Q_1 = P_{25}$, $Q_2 = P_{50}$, $Q_3 = P_{75}$

3–95. $D_1 = P_{10}$, $D_2 = P_{20}$, $D_3 = P_{30}$, etc.

3–97. a. 1.00 b. 1.47 c. −0.47 d. 0 e. −1.00

3–99. a. 0.75 b. −1.25 c. 2.25 d. −2 e. −0.5

3–101. a. 1 b. 0.6; exam in part a is higher.

3–103. a. −0.93 b. −0.85 c. −1.4; score in part b is the highest.

3–105. a. 21st b. 58th c. 77th d. 29th

3–106.

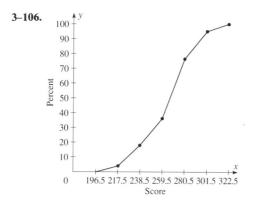

a. 10   b. 28   c. 63   d. 75   e. 92

3–107. a. 235 b. 255 c. 261 d. 275 e. 283

3–108.

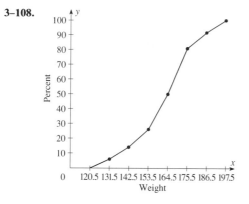

a. 134 b. 148 c. 160 d. 165 e. 170

3–109. a. 2nd b. 13th c. 40th d. 76th e. 96th

3–111. 82   **3–113.** 47   **3–115.** 12

3–117. Negatively skewed

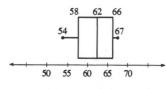

3–119. a. 3 b. 54 c. None d. None e. 145 f. 36

3–121. a. 96 b. 97 c. No mode
d. 96.5 e. 19 f. 42.67 g. 6.53

3–123. a. 3.75   b. 4   c. 4   d. 4   e. 1.11   f. 1.06

3–125. a. 55.5   b. 59.375   c. 57.5–72.5
d. 566.12   e. 23.79

3–127. 1.1   **3–129.** 6   **3–131.** 0.214, 0.417, years

3 133. a.

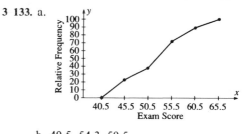

b. 49.5, 54.3, 59.5
c. 31st, 62nd, 89th

3–135. 0.26 and 0.38

3–137. 56%

3–139. 88.89%

CHAPTER 3 TEST

1. False   2. True   3. True   4. True   5. False
6. True   7. False   8. False   9. False   10. False
11. False   12. True   13. False   14. False   15. False
16. True   17. True   18. False   19. True   20. True

CHAPTER 4

4–1. 60   **4–3.** 362,880   **4–5.** 720   **4–7.** 144

4–9. 27,600, 35,152   **4–11.** 720

4–13. 12   **4–15.** 1296   **4–17.** 114   **4–19.** 2^n

4–21. 120   **4–23.** 11   **4–25.** 840

4–27. 151,200   **4–29.** 24   **4–31.** 50,400

4–33. 50,400   **4–35.** 1,860,480   **4–37.** 2520

4–39. 120   **4–41.** 11,441,304,000

4–43. a. 48   b. 60   c. 72   **4–45.** 87

4–47. 22,100   **4–49.** 126, 35   **4–51.** 120

4–53. 462   **4–55.** 166,320   **4–57.** 14,400

4–59. 194,040   **4–61.** 53,130   **4–63.** 126

4–65. 125,970

4–67. a. 4   b. 36   c. 624   d. 3744

4-69.

4-71.

4-73.

4-75.

4-77.

4-79.

4-81.

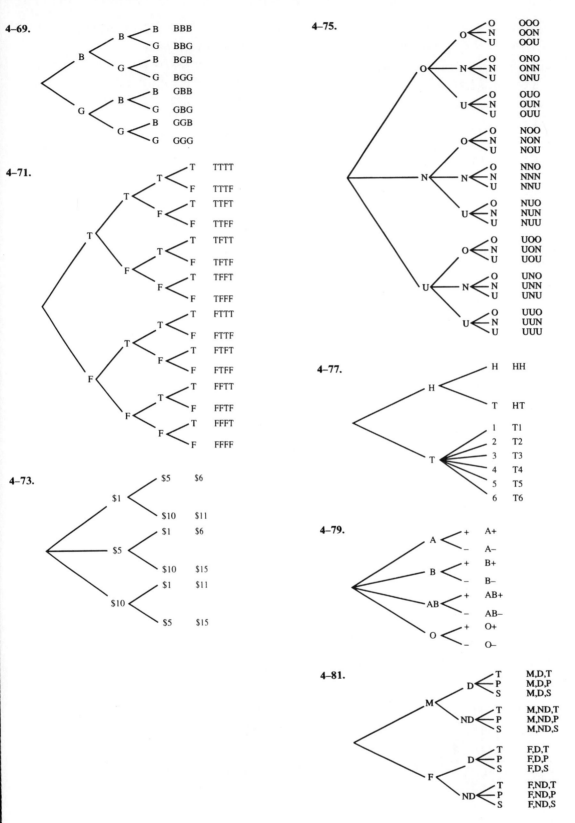

4–83.

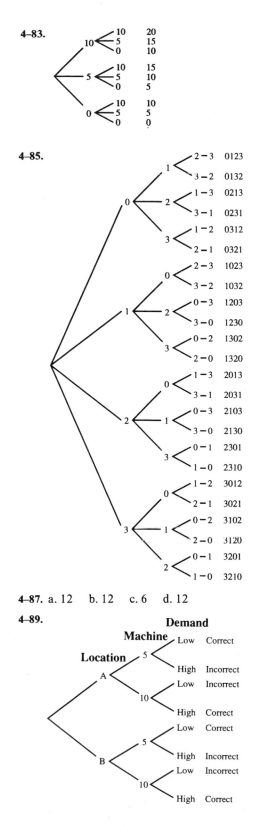

4–85.

4–87. a. 12   b. 12   c. 6   d. 12

4–89.

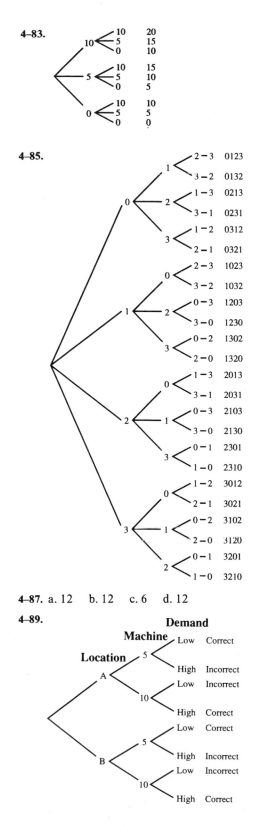

4–91.

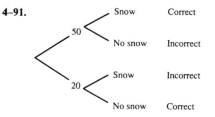

4–93. 20,160   **4–95.** 78   **4–97.** 120

4–99.

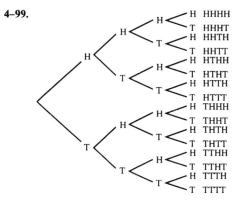

4–101. 40,320   **4–103.** 800

4–105.

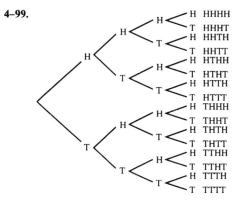

4–107. 720   **4–109.** 24   **4–111.** 25, 20

4–113. 4   **4–115.** 30   **4–117.** 495

4–119. 15,504   **4–121.** 7

CHAPTER 4 TEST

1. True   2. False   3. True   4. False   5. True
6. True   7. True   8. False   9. False   10. True

CHAPTER 5

5–1. A probability experiment is a process that leads to well-defined outcomes.

5–3. An outcome is the result of a single trial of a probability experiment, whereas an event consists of one or more outcomes.

5–5. The range of values is 0 to 1.   **5–7.** 0   **5–9.** 0.4

5–11. a. Empirical   b. Classical   c. Empirical
d. Classical   e. Empirical   f. Empirical
g. Subjective

5–12. a. $\frac{1}{6}$ b. $\frac{1}{2}$ c. $\frac{1}{3}$ d. 1   e. 1
     f. $\frac{5}{6}$ g. $\frac{1}{6}$

5–13. a. $\frac{5}{36}$ b. $\frac{1}{6}$ c. $\frac{2}{9}$ d. $\frac{1}{6}$ e. $\frac{1}{6}$

5–14. a. $\frac{1}{13}$ b. $\frac{1}{4}$ c. $\frac{1}{52}$ d. $\frac{2}{13}$ e. $\frac{4}{13}$
     f. $\frac{4}{13}$ g. $\frac{1}{2}$ h. $\frac{1}{26}$ i. $\frac{7}{13}$ j. $\frac{1}{26}$

5–15. a. $\frac{1}{2}$ b. $\frac{3}{10}$ c. $\frac{7}{10}$ d. $\frac{7}{10}$ e. $\frac{1}{2}$

5–17. $\frac{7}{50}$ **5–19.** 40%

5–21. The sample space is BBB, BBG, BGB, GBB, GGB, GBG, BGG, GGG.
     a. $\frac{1}{8}$ b. $\frac{1}{4}$ c. $\frac{3}{4}$ d. $\frac{3}{4}$

5–23. $\frac{1}{9}$ **5–25.** a. $\frac{9}{19}$ b. $\frac{11}{38}$ c. $\frac{7}{19}$

5–27. 17.543   **5–29.** $\frac{1}{36}$

5–31. a. 20%   b. 43%   c. 47%   d. 40%

5–33. The statement is probably not based on empirical probability.

5–35. Actual outcomes will differ; however, each number should occur approximately $\frac{1}{6}$ of the time.

5–37. Two events are mutually exclusive if they cannot occur at the same time.

5–39. a. No   b. No   c. Yes   d. No
e. No   f. Yes   g. Yes

5–41. $\frac{5}{9}$ **5–43.** $\frac{4}{31}$ **5–45.** 0.53

5–47. a. $\frac{4}{13}$ b. $\frac{1}{2}$ c. $\frac{7}{13}$

5–49. a. $\frac{17}{20}$ b. $\frac{17}{20}$ c. $\frac{3}{5}$

5–51. a. $\frac{8}{59}$ b. $\frac{32}{59}$ c. $\frac{45}{59}$

5–53. a. $\frac{1}{5}$ b. $\frac{3}{5}$ c. $\frac{2}{3}$

5–55. a. 467/1392   b. $\frac{47}{58}$ c. 833/1392

5–57. a. $\frac{1}{20}$ b. $\frac{17}{40}$ c. $\frac{9}{40}$ d. $\frac{5}{8}$

5–59. a. 29/100   b. $\frac{2}{3}$ c. 157/300

5–61. a. $\frac{4}{9}$ b. $\frac{1}{3}$ c. $\frac{5}{12}$

5–63. a. $\frac{1}{36}$ b. $\frac{1}{36}$

5–65. 0.10

5–67. a. Independent   e. Independent
b. Dependent   f. Dependent
c. Dependent   g. Dependent
d. Dependent   h. Independent

5–69. 0.643%   **5–71.** 14.061%   **5–73.** 0.0009

5–75. 1/365   **5–77.** $\frac{1}{27}$ **5–79.** $\frac{1}{15}$ **5–81.** 0.000125

5–83. a. 1/5525   b. 11/850   c. $\frac{2}{17}$

5–85. $\frac{1}{56}$ **5–87.** 35/117   **5–89.** 0.32   **5–91.** 0.0275

5–93. $\frac{2}{7}$ **5–95.** $\frac{2}{9}$ **5–97.** 0.6   **5–99.** 0.9

5–101. 0.5   **5–103.** 0.82

5–105. a. $\frac{18}{47}$ b. $\frac{14}{23}$

5–107. 0.857   **5–109.** 0.653   **5–111.** 0.358

5–113. $\frac{1}{4}$ **5–115.** 0.64

5–117. a. Selecting a nondefective resistor.
b. Selecting a letter from K through Z.
c. Selecting a three-digit number from 334 through 999.
d. Selecting a month that does not begin with J.
e. Selecting a male student in a statistics class.
f. Selecting a day on which it does not snow.
g. Selecting an opponent that the team lost to last year.
h. Selecting a day that is not a national holiday.
i. Selecting a president who is a female.
j. Selecting a black card.

5–119. $\frac{55}{56}$ **5–121.** $\frac{31}{32}$ **5–123.** $\frac{15}{16}$

5–125. 14,498/20,825   **5–127.** 0.266   **5–129.** $\frac{63}{64}$

5–131. 91/216   **5–133.** $\frac{7}{8}$

5–135. a. 1:5, 5:1
b. 1:1, 1:1
c. 1:3, 3:1
d. 1:1, 1:1
e. 1:12, 12:1
f. 1:3, 3:1
g. 1:1, 1:1

5–137. a. 1/2530   b. 38/253   c. 969/2530
d. 114/253

5–139. a. 10/143   b. 60/143   c. 15/1001
d. 160/1001   e. 48/143

5–141. a. $\frac{1}{30}$ b. 1/120   c. $\frac{3}{10}$ d. $\frac{3}{40}$ e. $\frac{3}{20}$

5–143. a. $\frac{14}{55}$ b. $\frac{28}{55}$ c. $\frac{1}{55}$

5–145. a. $\frac{8}{65}$ b. $\frac{1}{35}$ c. $\frac{2}{91}$ d. $\frac{16}{91}$ e. $\frac{4}{13}$

5–147. 400/1001   **5–149.** $\frac{14}{33}$

5–151. a. 4/2,598,960   b. 36/2,598,960
c. 624/2,598,960

5–153. a. $\frac{1}{4}$ b. $\frac{1}{52}$ c. $\frac{4}{13}$ d. $\frac{1}{13}$ e. $\frac{1}{2}$

5–155. If the outcome of the first event in no way affects the outcome of the second event, then the events are said to be independent. Otherwise, they are dependent.

5–157. a. $\frac{9}{35}$ b. $\frac{23}{35}$ c. $\frac{19}{35}$ d. $\frac{19}{35}$

5–159. a. $\frac{1}{4}$ b. $\frac{1}{6}$ c. $\frac{1}{4}$ d. $\frac{1}{4}$ e. 0   f. 1

5–161. 0.24

5–163. 0.015625   **5–165.** a. $\frac{1}{26}$ b. $\frac{1}{4}$ c. $\frac{1}{8}$

5–167. a. 57/104   b. $\frac{10}{13}$ c. $\frac{3}{4}$

5–169. 0.058

5–171. 0.273   **5–173.** 0.0575   **5–175.** a. $\frac{19}{44}$ b. $\frac{1}{4}$

5–177. 0.02   **5–179.** $\frac{31}{32}$ **5–181.** $\frac{12}{55}$ **5–183.** $\frac{2}{7}$

CHAPTER 5 TEST

1. True   2. False   3. False   4. True   5. False
6. False   7. True   8. False   9. True   10. True
11. False   12. False   13. False   14. False   15. False

CHAPTER 6

6–1. A random variable is a variable whose values are determined by chance. Examples will vary.

6–3. The number of commercials a radio station plays during each hour. The number of times a student uses his or her calculator during a mathematics exam. The number of leaves on a specific type of tree.

6–5. A probability distribution is a distribution that consists of the values a random variable can assume along with the corresponding probabilities of these values.

6–7. Yes

6–9. Yes

6–11. No; probability values cannot be greater than 1.

6–13. Discrete   **6–15.** Continuous

6–17. Discrete

6–19.

X	0	1	2	3
P(X)	$\frac{6}{15}$	$\frac{5}{15}$	$\frac{3}{15}$	$\frac{1}{15}$

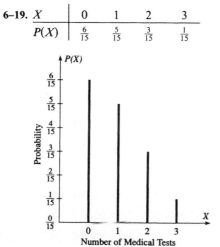

6–21.

X	0	1	2	3	4	5
P(X)	0.75	0.17	0.04	0.025	0.01	0.005

6–23.

X	1	2	3	4	5	6
P(X)	$\frac{1}{2}$	$\frac{1}{6}$	$\frac{1}{12}$	$\frac{1}{12}$	$\frac{1}{12}$	$\frac{1}{12}$

6–25.

X	1	2	3	4	5
P(X)	0.1	0.25	0.25	0.2	0.2

6–27.

X	$1	$5	$10	$20
P(X)	$\frac{3}{7}$	$\frac{2}{7}$	$\frac{1}{7}$	$\frac{1}{7}$

6–29.

X	1	2	3	4
P(X)	$\frac{1}{4}$	$\frac{1}{4}$	$\frac{3}{8}$	$\frac{1}{8}$

6–31.

X	1	2	3
P(X)	$\frac{1}{6}$	$\frac{1}{3}$	$\frac{1}{2}$

Yes

6–33.

X	3	4	7
P(X)	$\frac{3}{6}$	$\frac{4}{6}$	$\frac{7}{6}$

No, the sum of the probabilities is greater than 1, and $P(7) = \frac{7}{6}$, which is also greater than 1.

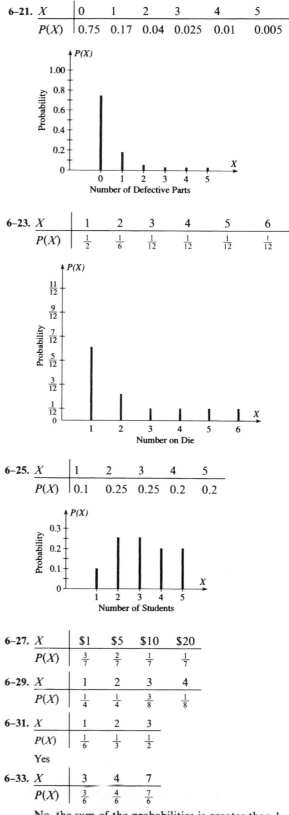

6–35.

X	1	2	4
$P(X)$	$\frac{1}{7}$	$\frac{2}{7}$	$\frac{4}{7}$

Yes

6–37.

X	0	1	2	3
$P(X)$	0.92	0.03	0.03	0.02

0.15, 0.3075, 0.55

6–39. 9.74, 1.5724, 1.254　**6–41.** 1.97, 1.65, 1.28

6–43. 6.59, 1.2619, 1.123　**6–45.** 13.86, 1.3204, 1.1491

6–47. −$1.17

6–49. $0.83; he should pay about $0.83.

6–51. −$1.00　**6–53.** −$0.50, −$0.52

6–55. The probabilities of each are as follows:

Red	$\frac{18}{38}$	0	$\frac{1}{38}$
Black	$\frac{18}{38}$	00	$\frac{1}{38}$
1–18	$\frac{18}{38}$	Any single number	$\frac{1}{38}$
19–36	$\frac{18}{38}$	0 or 00	$\frac{2}{38}$

a. −$0.053　b. −$0.053　c. −$0.053
d. −$0.053　e. −$0.053

6–57. 10.5　**6–59.** Answers will vary.

6–61. Answers will vary.

6–63. a. Yes　b. Yes　c. Yes　d. No　e. Yes
f. Yes　g. Yes　h. Yes　i. No　j. Yes

6–64. a. 0.420　b. 0.346　c. 0.590　d. 0.251
e. 0.000　f. 0.250　g. 0.418　h. 0.176　i. 0.246

6–65. a. 0.0004928　b. 0.131　c. 0.342
d. 0.00743　e. 0.173

6–67. 0.377　**6–69.** 0.267　**6–71.** 0.058

6–73. a. 0.346　b. 0.913　c. 0.663

6–75. a. 0.878　b. 0.201　c. 0.033

6–76. a. 75, 18.75, 4.33　b. 90, 63, 7.94　c. 10, 5, 2.236
d. 8, 1.6, 1.26　e. 100, 90, 9.49　f. 125, 93.75, 9.68
g. 20, 12, 3.46　h. 6, 5, 2.236

6–77. 32, 19.2, 4.38　**6–79.** 10, 9.8, 3.13

6–81. 340, 224.4, 14.98

6–83.

X	0	1	2	3
$P(X)$	0.125	0.375	0.375	0.125

6–85. The binomial distribution is a special case of the multinomial distribution.

6–86. a. 0.135　b. 0.0324　c. 0.0096
d. 0.18　e. 0.0112

6–87. 0.0066　**6–89.** 0.02025　**6–91.** 0.00171

6–93. 0.224

6–95. 0.3033　**6–97.** 0.2642　**6–99.** 0.2205

6–101. 0.0016　**6–103.** 0.252　**6–105.** 0.949

6–107. Yes

6–109. No; the sum of the probabilities is less than 1.

6–111.

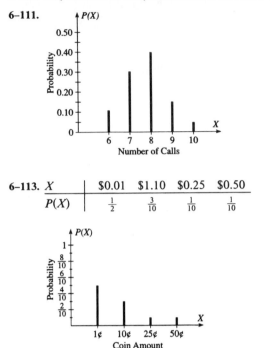

6–113.

X	$0.01	$1.10	$0.25	$0.50
$P(X)$	$\frac{1}{2}$	$\frac{3}{10}$	$\frac{1}{10}$	$\frac{1}{10}$

6–115. 2.05, 1.4075, 1.186　**6–117.** 2.11, 1.4979, 1.22

6–119. $0.182　**6–121.** 0.103　**6–123.** 0.451, 0.129, 0.988

6–125. 5, 4.5, 2.12　**6–127.** 0.00736　**6–129.** 0.0298

6–131. 0.0504　**6–133.** 22,230/105,938 = 0.21

CHAPTER 6 TEST

1. True　2. False　3. True　4. False　5. False
6. True　7. False　8. False　9. False　10. False

CHAPTER 7

7–1. The characteristics of the normal distribution are as follows:
a. It is bell-shaped.
b. It is symmetric about the mean.
c. The mean, median, and mode are equal.
d. It is continuous.
e. It never touches the x axis.
f. The area under the curve is equal to 1.
g. It is unimodal.

7–3. 1, or 100%　**7–5.** 68%, 95%, 99.7%　**7–7.** 0.2123

7–9. 0.4808　**7–11.** 0.4090　**7–13.** 0.0764

7–15. 0.1145　**7–17.** 0.0258　**7–19.** 0.8417

7–21. 0.9846　**7–23.** 0.5714　**7–25.** 0.3015

7–27. 0.2486   **7–29.** 0.4418   **7–31.** 0.0023

7–33. 0.0655   **7–35.** 0.9522   **7–37.** 0.0706

7–39. 0.9222   **7–41.** -1.94 **7–43.** -2.13

7–45. -1.26 **7–47.** -1.86

7–49. a. $z = +1.96$ and $z = -1.96$
b. $z = +1.65$ and $z = -1.65$, approximately
c. $z = +2.58$ and $z = -2.58$, approximately

7–51.

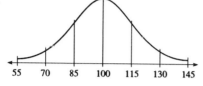

7–53.

x	-2	-1.5	-1	-0.5	0	0.5	1	1.5	2
y	0.05	0.13	0.24	0.35	0.4	0.35	0.24	0.13	0.05

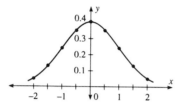

7–55. a. 10.93%   b. 1.13%   **7–57.** a. 0.0427   b. 0.0537

7–59. a. 0.49%   b. 16.24%   c. 74.86%

7–61. a. 0.1525   b. 0.7745   c. 0.1865

7–63. a. 0.9772   b. 0.6915

7–65. a. 0.5691   b. 0.6412

7–67. a. 0.9938   b. 0.1894   c. 0.9637

7–69. a. 0.5987   b. 0.8413   c. 0.2432

7–71. 89.95

7–73. The maximum size is 1927.76 square feet; the minimum size is 1692.24 square feet.

7–75. 66.386 minutes

7–77. The maximum price is $9222; the minimum price is $7290.

7–79. 76.18   **7–81.** $18,840.48   **7–83.** 18.6 months

7–85. a. $\mu = 120, \sigma = 20$ b. $\mu = 15, \sigma = 2.5$
c. $\mu = 30, \sigma = 5$

7–87. There are several mathematical tests that can be used.

7–89. 2.59   **7–91.** $\mu = 45, \sigma = 1.34$

7–93. The distribution is called the sampling distribution of sample means.

7–95. The mean of the sample means is equal to the population mean.

7–97. The distribution will be approximately normal when the sample size is large.

7–99. a. Yes, 0.950   b. Yes, 0.900   c. No
d. No   e. Yes, 0.975   f. No   g. Yes, 0.974
h. No   i. Yes, 0.968   j. No

7–101. 0.2486   **7–103.** 0.2327   **7–105.** 0.0136

7–107. 0.8239   **7–109.** 0.0951   **7–111.** 0.0674

7–113. a. 0.3446   b. 0.0023   c. Yes   d. Very unlikely

7–115. a. 0.3707   b. 0.0475

7–117. a. 0.1815   b. 0.3854
c. Means are less variable than individual data.

7–119. 0.0143   **7–121.** 1.66%   **7–123.** 0.11%

7–125. $\sigma_{\bar{X}} = 1.5, n = 25$

7–127. $\mu = 8, \mu_{\bar{X}} = 8; \sigma = 1.63, \sigma_{\bar{X}} = 1.63/\sqrt{2} = 1.15$

7–128. a. 0.0811   b. 0.0516   c. 0.1052
d. 0.1711   e. 0.2327   f. 0.9988

7–129. a. Yes   b. No   c. No   d. Yes
e. Yes   f. No

7–131. 0.1515   **7–133.** a. 0.0043   b. 0.6331

7–135. 0.1034   **7–137.** 68.44%   **7–139.** 0.9875

7–141. a. 0.4744   b. 0.1443   c. 0.0590
d. 0.8329   e. 0.2139   f. 0.8284
g. 0.0233   h. 0.9131   i. 0.0183
j. 0.9535

7–143. a. 20.39%   b. 0.99%   c. 9.18%

7–145. a. 0.4649   b. 0.0228   c. 0.1894

7–147. a. 0.2611   b. 0.3446   c. 0.4364
d. 0.2601

7–149. Between 92.2 and 107.8

7–151. 1.50%   **7–153.** 0.9871   **7–155.** 0.9678

7–157. 0.0465

CHAPTER 7 TEST

1. False   2. True   3. True   4. True   5. True
6. False   7. False   8. True   9. False   10. True
11. False   12. True   13. False   14. False   15. True
16. False   17. True

CHAPTER 8

8–1. A point estimate of a parameter specifies a specific value, such as $\mu = 87$; an interval estimate specifies a range of values for the parameter, such as $84 < \mu < 90$. The advantage of an interval estimate is that a specific probability (say 95%) can be selected, and one can be 95% confident that the interval contains the parameter that is being estimated.

8–3. The maximum error of estimate is the range of values to the right or left of the statistic in which the parameter may fall.

8–5. A good estimator should be unbiased, consistent, and relatively efficient.

8–7. For one to be able to determine sample size, the maximum error of estimate and the degree of confidence must be specified and the population standard deviation must be known.

8–9. a. 2.58 b. 2.33 c. 1.96 d. 1.65 e. 1.88

8–11. a. $77.03 < \mu < 86.97$
b. $75.46 < \mu < 88.54$
c. The 99% confidence interval is larger because the probability that it contains the mean is larger.

8–13. a. $11.948 < \mu < 13.252$
b. It would be highly unlikely, since this is far larger than 13.252.

8–15. $6.705 < \mu < 7.695$

8–17. a. $36.51 < \mu < 39.49$
b. $35.02 < \mu < 40.98$
c. The confidence interval for part b is larger because the standard deviation is larger.

8–19. $17.421 < \mu < 19.779$

8–21. 45 **8–23.** 25 **8–25.** 5

8–27. W. S. Gosset

8–29. The t distribution should be used when σ is unknown and $n < 30$.

8–30. a. 2.898 b. 2.074 c. 2.624
d. 1.833 e. 2.093

8–31. $14.72 < \mu < 17.28$

8–33. $19.881 < \mu < 22.119$

8–35. $11{,}990.0787 < \mu < 12{,}409.9213$

8–37. $8.656 < \mu < 9.944$

8–39. $17.292 < \mu < 19.768$

8–41. $108.702 < \mu < 121.298$

8–43. $312.07 < \mu < 328.33$

8–45. $\$58{,}197.28 < \mu < \$58{,}240.72$

8–47. a. 0.5, 0.5 b. 0.45, 0.55 c. 0.46, 0.54
d. 0.58, 0.42 e. 0.45, 0.55

8–48. a. 0.12, 0.88 b. 0.29, 0.71 c. 0.65, 0.35
d. 0.53, 0.47 e. 0.67, 0.33

8–49. $0.789 < p < 0.891$ **8–51.** $0.557 < p < 0.743$

8–53. $0.144 < p < 0.296$ **8–55.** $0.045 < p < 0.155$

8–57. $0.337 < p < 0.543$ **8–59.** $0.149 < p < 0.351$

8–61. 3121 **8–63.** 99 **8–65.** 95%

8–67. $69.293 < \mu < 74.707$

8–69. $2493.769 < \mu < 2760.231$

8–71. $9.945 < \mu < 14.055$ **8–73.** 62

8–75. $0.4908 < p < 0.6492$

8–77. $0.545 < p < 0.655$ **8–79.** 8487

CHAPTER 8 TEST

1. False 2. False 3. True 4. True 5. True
6. True 7. False 8. True 9. True 10. False

CHAPTER 9

9–1. The null hypothesis states that there is no difference between a parameter and a specific value or that there is no difference between two parameters. The alternative hypothesis states that there is a specific difference between a parameter and a specific value or that there is a difference between two parameters. Examples will vary.

9–3. A statistical test uses the data obtained from a sample to make a decision about whether or not the null hypothesis should be rejected.

9–5. The critical region is the range of values of the test statistic that indicates that there is a significant difference and that the null hypothesis should be rejected. The noncritical region is the range of values of the test statistic that indicates that the difference was probably due to chance and that the null hypothesis should not be rejected.

9–7. α, β

9–9. A one-tailed test should be used when a specific direction, such as greater than or less than, is being hypothesized; when no direction is specified, a two-tailed test should be used.

9–11. Hypotheses can only be proved true when the entire population is used to compute the test statistic. In most cases, this is impossible.

9–12. a. ± 2.58 b. $+1.65$ c. -2.58 d. -1.28
e. ± 1.96 f. $+1.75$ g. -2.33 h. ± 1.65
i. $+2.05$ j. ± 2.33

9–13. a. H_0: $\mu = 36.3$ (claim), H_1: $\mu \neq 36.3$
b. H_0: $\mu = \$36{,}250$ (claim), H_1: $\mu \neq \$36{,}250$
c. H_0: $\mu \leq 117.3$, H_1: $\mu > 117.3$ (claim)
d. H_0: $\mu \geq 72$, H_1: $\mu < 72$ (claim)
e. H_0: $\mu \geq 100$, H_1: $\mu < 100$ (claim)
f. H_0: $\mu = \$297.75$ (claim), H_1: $\mu \neq \$297.75$
g. H_0: $\mu \leq \$52.98$, H_1: $\mu > \$52.98$ (claim)
h. H_0: $\mu \leq 300$ (claim), H_1: $\mu > 300$
i. H_0: $\mu \geq 3.6$ (claim), H_1: $\mu < 3.6$

9–15. H_0: $\mu \leq 100$ and H_1: $\mu > 100$ (claim); C.V. $= +1.65$; $z = 3.77$, reject. There is enough evidence to support the claim that the IQs of the nursing students are higher than average.

9–17. H_0: $\mu \leq 483$ and H_1: $\mu > 483$ (claim); C.V. $= +1.65$; $z = 0.62$; do not reject. No; there is not enough evidence to support the claim that the course increases the SAT scores.

9–19. H_0: $\mu \geq 80$ (claim) and H_1: $\mu < 80$; C.V. $= -2.33$; $z = -3.56$; reject. No; there is enough evidence to reject the claim that the ski resort averages at least 80 guests per week.

9–21. H_0: $\mu = \$915$ (claim) and H_1: $\mu \neq \$915$; C.V. $= \pm 1.96$; $z = 2.65$; reject. There is enough evidence to reject the agent's claim that the average cost of the trip is $915.

9–23. H_0: $\mu = 36$ (claim) and H_1: $\mu \neq 36$; C.V. $= \pm 2.58$; $z = -3.54$; reject. Yes; there is enough evidence to reject the claim that the average lifetime of the lightbulbs is 36 months.

9–25. H_0: $\mu \geq 240$ and H_1: $\mu < 240$ (claim); C.V. $= -2.33$; $z = -3.87$; reject. Yes; there is enough evidence to support the claim that the medication lowers the cholesterol level.

9–27. H_0: $\mu = \$2126$ (claim) and H_1: $\mu \neq \$2126$; C.V. $= \pm 2.33$; $z = 1.52$; do not reject. There is not enough evidence to reject the claim that the average cost for one year's tuition is $2126.

9–29. $\bar{x} = 5.025$, $s = 3.63$; H_0: $\mu \geq 10$ and H_1: $\mu < 10$ (claim); C.V. $= -1.65$; $z = -8.67$; reject. Yes; there is enough evidence to support the claim that the average number of days missed per year is less than 10.

9–31. H_0: $\mu = 8.65$ (claim) and H_1: $\mu \neq 8.65$; C.V. $= \pm 1.96$; $z = -1.35$; do not reject. Yes; there is not enough evidence to reject the claim that the average hourly wage of the employees is $8.65.

9–33. The t distribution differs from the standard normal distribution in that it is a family of curves and the variance is greater than 1; and as the degrees of freedom increase, the t distribution approaches the standard normal distribution.

9–35. The t test statistic uses s instead of σ when $n < 30$.

9–36. a. $+1.833$ b. ± 1.740 c. -3.365 d. $+2.306$ e. ± 2.145 f. -2.819 g. ± 2.771 h. ± 2.583

9–37. H_0: $\mu = \$850$ (claim) and H_1: $\mu \neq \$850$; d.f. $= 6$; C.V. $= \pm 2.447$; $t = 2.18$; do not reject. There is not enough evidence to reject the claim that the shop sells $850 in baseball equipment.

9–39. H_0: $\mu = 800$ (claim) and H_1: $\mu \neq 800$; C.V. $= \pm 2.262$; d.f. $= 9$; $t = 9.96$; reject. Yes; there is enough evidence to reject the claim that the average rent that small-business establishments pay in Eagle City is $800.

9–41. H_0: $\mu \geq 12$ and H_1: $\mu < 12$ (claim); C.V. $= -2.571$; d.f. $= 5$; $t = -0.47$; do not reject. No; there is not enough evidence to support the claim that the muffler can be changed in less than 12 minutes.

9–43. H_0: $\mu = \$750$ (claim) and H_1: $\mu \neq \$750$; C.V. $= \pm 3.106$; $t = -3.67$; reject. Yes; there is enough evidence to reject the claim that the average rent is $750.

9–45. H_0: $\mu \leq 350$ and H_1: $\mu > 350$ (claim); C.V. $= 1.796$; d.f. $= 11$; $t = 1.732$; do not reject. No; there is not enough evidence to support the claim that the average fine is higher than $350.

9–47. H_0: $\mu \geq 37$ (claim) and H_1: $\mu < 37$; C.V. $= -1.701$; d.f. $= 28$; $t = -1.89$; reject. There is enough evidence to reject the claim that the average household receives at least 37 phone calls per month.

9–49. $\bar{x} = 70.85$; $s = 6.56$; H_0: $\mu = 75$ (claim) and H_1: $\mu \neq 75$; C.V. $= \pm 2.861$; d.f. $= 19$; $t = -2.83$; do not reject. There is not enough evidence to reject the claim that the average score on the real estate exam is 75.

9–51. Answers will vary.   **9–53.** $np \geq 5$ and $nq \geq 5$

9–55. H_0: $p \geq 0.23$ (claim) and H_1: $p < 0.23$; $\mu = 9.2$; $\sigma = 2.66$; C.V. $= -1.65$; $z = -0.83$; do not reject. Yes; there is not enough evidence to reject the claim that at least 23% of the 14-year-old residents own a skateboard.

9–57. H_0: $p \geq 0.40$ (claim) and H_1: $p < 0.40$; $\mu = 32$; $\sigma = 4.38$; C.V. $= -1.28$; $z = -0.457$; do not reject. There is not enough evidence to reject the claim that at least 40% of the arsonists are under 21 years old.

9–59. H_0: $p = 0.21$ (claim) and H_1: $p \neq 0.21$; $\mu = 30.03$; $\sigma = 4.87$; C.V. $= \pm 1.96$; $z = 1.64$; do not reject. There is not enough evidence to reject the claim that 21% of the school's graduates are women.

9–61. H_0: $p \geq 0.15$ (claim) and H_1: $p < 0.15$; $\mu = 12$; $\sigma = 3.19$; C.V. $= -1.65$; $z = -0.94$; do not reject. There is not enough evidence to reject the claim that at least 15% of all eighth-grade students are overweight.

9–63. H_0: $p \leq 0.30$ and H_1: $p > 0.30$ (claim); $\mu = 60$; $\sigma = 6.48$; C.V. $= +1.65$; $z = +1.85$; reject. Yes; there is enough evidence to support the claim that more than 30% of the customers have at least two telephones.

9–65. H_0: $p = 0.18$ (claim) and H_1: $p \neq 0.18$; $\mu = 54$; $\sigma = 6.654$; C.V. $= \pm 1.96$; $z = -0.60$; do not reject. There is not enough evidence to reject the claim that 18% of all high school students smoke at least a pack of cigarettes a day.

9–67. $P(X = 6, p = 0.30, n = 10) = 0.037$; since $0.037 > 0.025$, there is not enough evidence to reject the fashion buyer's claim.

9–69. The *P*-value is the actual probability of getting the sample mean if the null hypothesis is true.

9–70. a. Do not reject.
b. Do not reject.
c. Do not reject.
d. Reject.
e. Reject.

9–71. H_0: $\mu \leq 27.50$ and H_1: $\mu > 27.50$ (claim); $z = 2.55$; *P*-value $= 0.0054$; reject. There is enough evidence to support the claim that the cost of the textbooks is greater than $27.50.

9–73. H_0: $\mu \leq \$875$ and H_1: $\mu > \$875$ (claim); $z = 2.1$; *P*-value $= 0.0179$; do not reject. There is not enough evidence to support the claim that the average cost this year is greater than the average cost last year.

9–75. H_0: $\mu = 800$ (claim) and H_1: $\mu \neq 800$; $z = -2.61$; *P*-value $= 0.0090$; reject. There is enough evidence to reject the null hypothesis that the breaking strength is 800 pounds.

9–77. H_0: $\mu = 65$ (claim) and H_1: $\mu \neq 65$; $z = -1.21$; *P*-value $= 0.2262$; do not reject. There is not enough evidence to reject the hypothesis that the average acreage is 65 acres.

9–79. When the null hypothesis is not rejected, the confidence interval will contain the hypothesized mean. When the null hypothesis is rejected, the confidence interval will not contain the hypothesized mean. The same $\alpha/2$ value must be used in hypothesis testing and in the confidence interval.

9–81. H_0: $\mu = 42$ (claim) and H_1: $\mu \neq 42$; C.V. $= \pm 1.65$; $z = 2.37$; reject; $43.83 < \mu < 52.17$. There is enough evidence to reject the claim that the average is 42. The confidence interval does not contain the hypothesized mean 42.

9–83. H_0: $\mu = 47$ and H_1: $\mu \neq 47$ (claim); C.V. $= \pm 1.65$; $z = -2.26$; reject; $38.35 < \mu < 45.65$. There is enough evidence to support the claim that the mean time has changed. The confidence interval does not contain the hypothesized mean 47.

9–85. H_0: $\mu = 10.8$ (claim) and H_1: $\mu \neq 10.8$; C.V. $= \pm 2.33$; $z = 2.80$; reject; $11.035 < \mu < 13.365$. There is enough evidence to reject the claim that the average time a person spends reading a newspaper is 10.8 minutes. The confidence interval does not contain the hypothesized mean 10.8.

9–87. The power of a test is equal to $1 - \beta$, where β is the probability of a type II error.

9–89. H_0: $\mu = 31.2$ (claim) and H_1: $\mu \neq 31.2$; C.V. $= \pm 1.96$; $z = -6.82$; reject. No; there is enough evidence to reject the hypothesis that the average age is 31.2.

9–91. H_0: $\mu = 14$ (claim) and H_1: $\mu \neq 14$; C.V. $= \pm 1.96$; $z = 4.22$; reject. No; there is enough evidence to reject the claim that the average number of cigarettes a person smokes a day is 14.

9–93. H_0: $\mu = 61.2$ (claim) and H_1: $\mu \neq 61.2$; C.V. $= \pm 2.831$; $t = -4.375$; reject. No; there is enough evidence to reject the claim that the average age is 61.2.

9–95. H_0: $\mu \leq 23.2$ (claim) and H_1: $\mu > 23.2$; C.V. $= 1.74$; d.f. $= 17$; $t = -1.27$; do not reject. No; there is not enough evidence to reject the claim that the average age is less than or equal to 23.2 years.

9–97. H_0: $p \geq 0.3$ (claim) and H_1: $p < 0.3$; $\mu = 18$; $\sigma = 3.55$; C.V. $= -1.65$; $z = -0.85$; do not reject. No; there is not enough evidence to reject the claim that at least 30% of the students are receiving financial aid.

9–99. H_0: $p = 0.80$ (claim) and H_1: $p \neq 0.80$; $\mu = 24$; $\sigma = 2.19$; C.V. $= \pm 2.33$; $z = -1.83$; do not reject. Yes; there is not enough evidence to reject the claim that 80% of the new-home buyers wanted a fireplace.

9–101. H_0: $\mu = 225$ (claim) and H_1: $\mu \neq 225$; $z = 2.36$; *P*-value $= 0.0182$; do not reject. There is not enough evidence to reject the claim that the mean is 225 pounds.

9–103. H_0: $\mu = 35$ (claim) and H_1: $\mu \neq 35$; C.V. $= \pm 1.65$; $z = -3.00$; reject; $32.675 < \mu < 34.325$. No; there is enough evidence to reject the claim that the mean is 35 pounds. Yes, the results agree.

CHAPTER 9 TEST

1. False   2. False   3. True   4. True   5. True
6. True   7. True   8. False   9. False   10. True
11. True   12. False   13. False   14. False   15. True
16. False   17. True   18. True   19. True   20. True

CHAPTER 10

10–1. Testing a single mean involves comparing a sample mean to a specific value such as $\mu = 100$; testing the difference between two means involves comparing the means of two samples, such as $\mu_1 = \mu_2$.

10–3. The populations must be independent of each other, and they must be normally distributed; s_1 and s_2 can be used in place of σ_1 and σ_2 when σ_1 and σ_2 are unknown and both samples are each greater than or equal to 30.

10–5. H_0: $\mu_1 = \mu_2$ and H_1: $\mu_1 \neq \mu_2$ (claim); C.V. $= \pm 2.58$; $z = -2.5$; do not reject; there is not enough evidence to support the claim that there is a difference in the speeds of the two companies.

10–7. H_0: $\mu_1 \leq \mu_2$ and H_1: $\mu_1 > \mu_2$ (claim); C.V. $= +1.65$; $z = 2.56$; reject; yes, there is enough evidence to support the claim that pulse rates of smokers are higher than the pulse rates of nonsmokers.

10–9. H_0: $\mu_1 \leq \mu_2$ and H_1: $\mu_1 > \mu_2$ (claim); C.V. $= 2.05$; $z = 1.43$; do not reject; no, there is not enough evidence to support the claim that taxi drivers have slower reaction times than state police officers.

10–11. H_0: $\mu_1 \geq \mu_2$ and H_1: $\mu_1 < \mu_2$ (claim); C.V. $= -2.05$; $z = -4.42$; reject; yes, there is enough evidence to support the claim that the evening students have higher scores than the day students.

10–13. H_0: $\mu_1 = \mu_2$ and H_1: $\mu_1 \neq \mu_2$ (claim); C.V. $= \pm 1.65$; $z = -20.97$; reject; there is enough evidence to support the claim that there is a difference in the cranking power of the automobile batteries.

10–15. H_0: $\mu_1 \geq \mu_2$ and H_1: $\mu_1 < \mu_2$ (claim); C.V. $= -2.05$; $z = -7.72$; reject; there is enough evidence to support the claim that the Jiffy Meals restaurant has faster service.

10–17. H_0: $\mu_1 - \mu_2 = 8$ (claim) and H_1: $\mu_1 - \mu_2 \neq 8$; C.V. $= \pm 1.96$; $z = -0.73$; do not reject; there is not enough evidence to reject the claim that private school students have an IQ that is 8 points higher than that of students in public schools.

10–19. $-7.300 < \mu_1 - \mu_2 < -1.300$

10–21. $2.822 < \mu_1 - \mu_2 < 5.978$

10–23. H_0: $\mu_1 \leq \mu_2$ and H_1: $\mu_1 > \mu_2$ (claim); C.V. $= 1.753$; d.f. $= 15$; $t = 0.53$; do not reject; there is not enough evidence to support the claim that male nurses earn more than female nurses.

10–25. H_0: $\mu_1 \leq \mu_2$ and H_1: $\mu_1 > \mu_2$ (claim); C.V. $= 2.33$; d.f. $= 36$; $t = 1.26$; do not reject; there is not enough evidence to support the claim that high school girls miss more days of school than high school boys do.

10–27. H_0: $\mu_1 \leq \mu_2$ and H_1: $\mu_1 > \mu_2$ (claim); C.V. $= +2.132$; d.f. $= 4$; $t = 3.54$; reject; yes, there is enough evidence to support the claim that the automobiles using high-octane gasoline get better mileage than the ones using regular gasoline.

10–29. H_0: $\mu_1 \geq \mu_2$ and H_1: $\mu_1 < \mu_2$ (claim); C.V. $= -1.943$; d.f. $= 6$; $t = -3.86$; reject; yes, there is enough evidence to support the claim that the XYZ dishwashers cost more.

10–31. H_0: $\mu_1 \geq \mu_2$ and H_1: $\mu_1 < \mu_2$ (claim); C.V. $= -1.771$; d.f. $= 13$; $t = -0.56$; do not reject; no, there is not enough evidence to support the claim that the Circleville Police issued more speeding tickets.

10–33. Independent samples are not related, whereas dependent samples are.

10–34. a. Dependent
b. Dependent
c. Independent
d. Dependent
e. Independent

10–35. H_0: $\mu_D \leq 0$ and H_1: $\mu_D > 0$ (claim); C.V. $= 1.833$; d.f. $= 9$; $t = 1.25$; do not reject; there is not enough evidence to support the claim that the students missed fewer days after completing the program.

10–37. H_0: $\mu_D = 0$ and H_1: $\mu_D \neq 0$ (claim); C.V. $= \pm 2.201$; d.f. $= 11$; $t = 2.66$; reject; there is enough evidence to support the claim that there was a change in the attitude of the subjects.

10–39. H_0: $\mu_D \geq 0$ and H_1: $\mu_D < 0$ (claim); C.V. $= -1.383$; d.f. $= 9$; $t = -3.40$; reject; there is enough evidence to support the claim that the typing speeds of the secretaries have increased.

10–41. H_0: $\mu_D \geq 0$ and H_1: $\mu_D < 0$ (claim); C.V. $= -2.365$; d.f. $= 7$; $t = 0.765$; do not reject; there is not enough evidence to support the claim that joggers wearing the manufacturer's brand of shoes will jog faster than joggers wearing other brands.

10–43. H_0: $\mu_D \leq 0$ and H_1: $\mu_D > 0$ (claim); C.V. $= 2.262$; d.f. $= 9$; $t = 4.696$; reject; yes, there is enough evidence to support the claim that the blood pressure of the patients has been lowered.

10–45. a. $\hat{p} = \frac{34}{48}$, $\hat{q} = \frac{14}{48}$
b. $\hat{p} = \frac{28}{75}$, $\hat{q} = \frac{47}{75}$
c. $\hat{p} = \frac{50}{100}$, $\hat{q} = \frac{50}{100}$
d. $\hat{p} = \frac{6}{24}$, $\hat{q} = \frac{18}{24}$
e. $\hat{p} = \frac{12}{144}$, $\hat{q} = \frac{132}{144}$

10–46. a. 16   b. 4   c. 4.8   d. 104   e. 30

10–47. a. $\bar{p} = 0.5$, $\bar{q} = 0.5$ b. $\bar{p} = 0.5$, $\bar{q} = 0.5$
c. $\bar{p} = 0.27$, $\bar{q} = 0.73$ d. $\bar{p} = 0.2125$, $\bar{q} = 0.7875$ e. $\bar{p} = 0.216$, $\bar{q} = 0.784$

10–49. $\hat{p}_1 = \frac{5}{50} = 0.1$, $\hat{p}_2 = \frac{15}{80} = 0.1875$; $\bar{p} = 0.15$, $\bar{q} = 0.85$; H_0: $p_1 \geq p_2$ and H_1: $p_1 < p_2$ (claim); C.V. $= -1.65$; $z = -1.36$; do not reject; no, there is not enough evidence to support the claim that the proportion of surgeons who smoke is higher than the proportion of general practitioners who smoke.

10–51. $\hat{p}_1 = \frac{10}{73} = 0.14$, $\hat{p}_2 = \frac{16}{80} = 0.20$; $\bar{p} = 0.17$, $\bar{q} = 0.83$; H_0: $p_1 = p_2$ and H_1: $p_1 \neq p_2$ (claim); C.V. $= \pm 1.96$; $z = -0.99$; do not reject; no, there is not enough evidence to support the claim that there is a difference in the proportions.

10–53. $\hat{p}_1 = 0.15$; $\hat{p}_2 = 0.21$; $\bar{p} = 0.18$, $\bar{q} = 0.82$; H_0: $p_1 = p_2$ and H_1: $p_1 \neq p_2$ (claim); C.V. $= \pm 1.65$; $z = -1.58$; do not reject; no, there is not enough evidence to support the claim that the proportions are different.

10–55. $\hat{p}_1 = 130/200 = 0.65$, $\hat{p}_2 = 63/300 = 0.21$; $\bar{p} = 0.386$, $\bar{q} = 0.614$; H_0: $p_1 \leq p_2$ and H_1: $p_1 > p_2$ (claim); C.V. $= +2.33$; $z = 9.90$; reject; there is enough evidence to support the claim that men are more safety-conscious than women.

10–57. $\hat{p}_1 = 0.30$, $\hat{p}_2 = 0.24$; $X_1 = 30$, $X_2 = 24$; $\bar{p} = 0.27$, $\bar{q} = 0.73$; H_0: $p_1 = p_2$ and H_1: $p_1 \neq p_2$ (claim); C.V.$= \pm 2.33$; $z = 0.96$; do not reject; there is not enough evidence to support the claim that the proportions are different.

10–59. $\hat{p}_1 = \frac{8}{50} = 0.16$, $\hat{p}_2 = \frac{20}{75} = 0.267$; $\bar{p} = 0.224$, $\bar{q} = 0.776$; H_0: $p_1 \geq p_2$ and H_1: $p_1 < p_2$ (claim); C.V. $= -1.65$; $z = -1.41$; do not reject; no, there is not enough evidence to support the claim that the proportions of college freshmen who have their own cars is higher than the proportion of high school seniors who have their own cars.

10–61. $-0.279 < p_1 - p_2 < -0.041$

10–63. H_0: $\mu_1 = \mu_2$ and H_1: $\mu_1 \neq \mu_2$ (claim); C.V. $= \pm 2.58$; $z = -1.77$; do not reject; no, there is not enough evidence to support the claim that the triglyceride levels are different.

10–65. H_0: $\mu_1 = \mu_2$ and H_1: $\mu_1 \neq \mu_2$ (claim); C.V. $= \pm 2.58$; d.f. $= 37$; $t = -14.06$; reject; yes, there is enough evidence to support the claim that there is a difference in the prices of the soups.

10–67. H_0: $\mu_1 = \mu_2$ and H_1: $\mu_1 \neq \mu_2$ (claim); C.V. $= \pm 2.624$; d.f. $= 14$; $t = 6.54$; reject; yes, there is enough evidence to support the claim that there is a difference in the teachers' salaries.

10–69. H_0: $\mu_D \geq 0$ and H_1: $\mu_D < 0$ (claim); C.V. $= -2.821$; d.f. $= 9$; $t = -4.17$; reject; yes, there is enough evidence to support the claim that the tutoring sessions helped to improve the students' vocabulary.

10–71. $\hat{p}_1 = 0.08$, $\hat{p}_2 = 0.09$; $X_1 = 6$, $X_2 = 5.4$; $\bar{p} = 0.084$, $\bar{q} = 0.916$; H_0: $p_1 = p_2$ and H_1: $p_1 \neq p_2$ (claim) C.V. $= \pm 1.96$; $z = -0.208$; do not reject; no, there is not enough evidence to support the claim that there is a difference in the proportions.

CHAPTER 10 TEST

1. True   2. True   3. False   4. False   5. True
6. False   7. True   8. True   9. True

CHAPTER 11

11–1. Two variables are related when a discernable pattern exists between them.

11–3. r, ρ (rho)

11–5. A positive relationship means that as x increases, y increases. A negative relationship means that as x increases, y decreases.

11–7. Answers will vary.

11–9. Pearson product moment correlation coefficient

11–11. There are many other possibilities, such as chance, or relationship to a third variable.

11–13. H_0: $\rho = 0$ and H_1: $\rho \neq 0$; $r = -0.916$; C.V. $= \pm 0.878$; d.f. $= 3$; reject; there is a significant relationship between a person's age and the number of hours a person watches television.

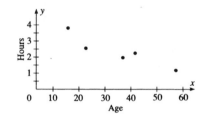

11–15. H_0: $\rho = 0$ and H_1: $\rho \neq 0$; $r = -0.883$; C.V. $= \pm 0.811$; d.f. $= 4$; reject; there is a significant relationship between a person's age and his or her contribution.

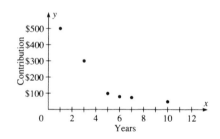

11–17. H_0: $\rho = 0$ and H_1: $\rho \neq 0$; $r = 0.716$; C.V. $= \pm 0.707$; d.f. $= 6$; reject; there is a significant relationship between IQ and GPA.

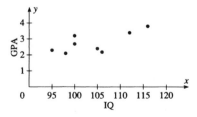

11–19. H_0: $\rho = 0$ and H_1: $\rho \neq 0$; $r = 0.814$; C.V. $= \pm 0.878$; d.f. $= 3$; do not reject; there is not a significant relationship between the variables.

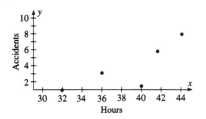

11–21. H_0: $\rho = 0$ and H_1: $\rho \neq 0$; $r = 0.855$; C.V. $= \pm 0.811$; d.f. $= 4$; reject; there is a significant relationship between the number of hours of tutoring a student receives and the final grade of the student.

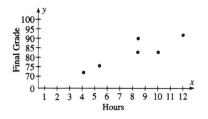

11–23. H_0: $\rho = 0$ and H_1: $\rho \neq 0$; $r = 0.115$; C.V. $= \pm 0.707$; d.f. $= 6$; do not reject; there is not a significant relationship between the number of pounds a mother gains during pregnancy and the birth weight of her baby.

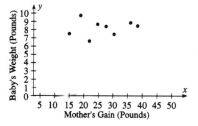

11–25. H_0: $\rho = 0$ and H_1: $\rho \neq 0$; $r = +0.873$; C.V. $= \pm 0.811$; d.f. $= 4$; reject; there is a significant relationship between the IQs of the girls and the boys.

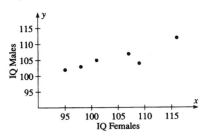

11–27. H_0: $\rho = 0$ and H_1: $\rho \neq 0$; $r = -0.909$; C.V. $= \pm 0.811$; d.f. $= 4$; reject; there is a significant relationship between the years of service and the number of resignations.

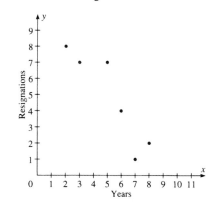

11–29. $r = 1.00$: All values fall in a straight line. $r = 1.00$: The relationship between x and y is the same when the values are interchanged.

11–31. A scatter plot should be drawn and the value of the correlation coefficient should be tested to see whether it is significant.

11–33. $y' = a + bx$

11–35. It is the line that is drawn through the points on the scatter plot such that the sum of the squares of the vertical distances each point is from the line is a minimum.

11–37. When r is positive, b will be positive. When r is negative, b will be negative.

11–39. The closer r is to $+1$ or -1, the more accurate the predicted value will be.

11–41. $y' = 4.45 - 0.0583x$; 2.41 hours

11–43. $y' = 453.176 - 50.439x$; 251.42

11–45. $y' = -3.759 + 0.063x$; 2.793

11–47. Since r is not significant, no regression analysis should be done.

11–49. $y' = 63.580 + 2.39x$; 85.09

11–51. Since r is not significant, no regression analysis should be done.

11–53. $y' = 63.193 + 0.405x$; 105.313

11–55. $y' = 10.770 - 1.149x$; 6.174

11–57. $r = +0.956$; H_0: $\rho = 0$ and H_1: $\rho \neq 0$; C.V. $= \pm 0.754$; d.f. $= 5$; $y' = -10.944 + 1.969x$; when $x = 30$, $y' = 48.126$; reject; there is a significant relationship between the amount of lung damage and the number of years a person has been smoking.

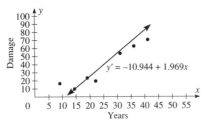

11–59. H_0: $\rho = 0$ and H_1: $\rho \neq 0$; $r = 0.896$; C.V. $= \pm 0.754$; d.f. $= 5$; reject; there is a significant relationship between the income of a person and the amount a person spends on recreation; $y' = -222.493 + 0.355x$.

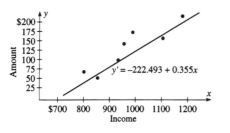

11–61. H_0: $\rho = 0$ and H_1: $\rho \neq 0$; $r = -0.981$; C.V. $= \pm 0.811$; d.f. $= 4$; reject; there is a significant relationship between the number of absences and the final grade; $y' = 96.784 - 2.668x$.

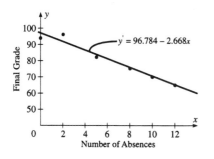

11–63. H_0: $\rho = 0$ and H_1: $\rho \neq 0$; $r = 0.821$; C.V. $= \pm 0.878$; d.f. $= 3$; do not reject; there is no significant relationship between the number of years of experience and monthly sales. Since r is not significant, no regression analysis should be done.

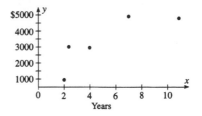

11–65. 453.173, 21.1, -3.7895

11–67. Multiple regression means one dependent variable and two or more independent variables.

11–69. $y' = a + b_1x_1 + b_2x_2 + \cdots + b_nx_n$

11–71. The coefficient of determination is found by squaring the value of the correlation coefficient.

11–73. The coefficient of nondetermination is found by subtracting r^2 from 1.

11–75. When the sample size is large, usually 100 or more.

11–77. 3.48   **11–79.** 575.192   **11–81.** 0.465

11–83. 94.22

11–85. $1.37 < y < 3.33$ **11–87.** $2.256 < y < 7.744$

11–89. H_0: $\rho = 0$ and H_1: $\rho \neq 0$; $r = 0.895$; C.V. $= \pm 0.875$; d.f. $= 5$; reject; there is a significant relationship between the number of hours a student studies and the student's grade point average; $y' = 1.572 + 0.152x$; when $x = 10$, $y' = 3.1$.

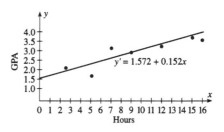

11–91. $r = 0.873$; H_0: $\rho = 0$ and H_1: $\rho \neq 0$; C.V. $= \pm 0.834$; d.f. $= 6$; reject; there is a significant relationship between the mother's age and the number of children she has; $y' = -2.457 + 0.187x$; $y' = 3.9$.

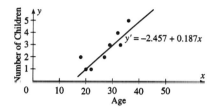

11–93. $r = -0.974$; H_0: $\rho = 0$ and H_1: $\rho \neq 0$; C.V. $= \pm 0.708$; d.f. $= 10$; reject; there is a significant relationship between speed and time; $y' = 14.086 - 0.137x$; $y' = 4.222$.

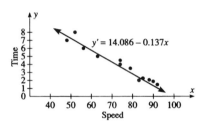

11–95. $r = -0.397$; H_0: $\rho = 0$ and H_1: $\rho \neq 0$; C.V. $= \pm 0.798$; d.f. $= 7$; do not reject; since r is not significant, no regression analysis should be done.

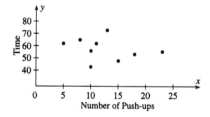

11–97. 116.33 pounds   **11–99.** 0.47

11–101. $3.447 < y < 4.998$

CHAPTER 11 TEST

1. False   2. False   3. True   4. False   5. True
6. True   7. True   8. True   9. False   10. False
11. False   12. True   13. True   14. False   15. True
16. True   17. False   18. False   19. True   20. False

CHAPTER 12

12–1. Three properties of the chi-square distribution are the following:
  a. It is nonnegative.
  b. It is a family of curves based on the degrees of freedom.
  c. The distribution is positively skewed.

12–3. The assumptions are the following:
  a. The sample must be randomly selected.
  b. The population must be normally distributed.
  c. The observations must be independent.

12–4. a. H_0: $\sigma^2 \leq 225$, H_1: $\sigma^2 > 225$; C.V. $= 27.587$; d.f. $= 17$

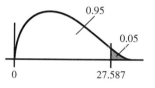

  b. H_0: $\sigma^2 \geq 225$, H_1: $\sigma^2 < 225$; C.V. $= 14.042$; d.f. $= 22$

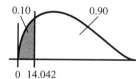

  c. H_0: $\sigma^2 = 225$, H_1: $\sigma^2 \neq 225$; C.V. $= 5.629$, 26.199; d.f. $= 14$

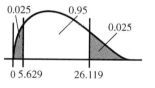

  d. H_0: $\sigma^2 = 225$, H_1: $\sigma^2 \neq 225$; C.V. $= 2.167$, 14.067; d.f. $= 7$

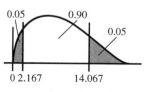

  e. H_0: $\sigma^2 \leq 225$, H_1: $\sigma^2 > 225$; C.V. $= 32.000$, d.f. $= 16$

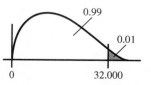

f. $H_0: \sigma^2 \geq 225$, $H_1: \sigma^2 < 225$; C.V. = 8.907; d.f. = 19

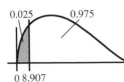

g. $H_0: \sigma^2 = 225$, $H_1: \sigma^2 \neq 225$; C.V. = 3.074, 28.299; d.f. = 12

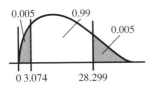

h. $H_0: \sigma^2 \geq 225$, $H_1: \sigma^2 < 225$; C.V. = 15.308; d.f. = 28

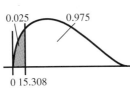

12–5. $H_0: \sigma \leq 1.5$ (claim) and $H_1: \sigma > 1.5$; C.V. = 36.415; $\alpha = 0.05$; d.f. = 24; $\chi^2 = 34.6$; do not reject; yes, there is not enough evidence to reject the claim that the standard deviation is less than or equal to 1.5 months.

12–7. $H_0: \sigma^2 \leq 2.8$ and $H_1: \sigma^2 > 2.8$ (claim); C.V. = 24.725; d.f. = 11; $\chi^2 = 11.786$; do not reject; there is not enough evidence to support the claim that the population variance is greater than 2.8.

12–9. $H_0: \sigma \leq 1.8$ and $H_1: \sigma^2 > 1.8$ (claim); C.V. = 67.505; $\alpha = 0.05$; d.f. = 59; $\chi^2 = 68.83$; reject; yes, there is enough evidence to support the claim that the variance is greater than 1.8.

12–11. $0.251 < \sigma < 0.514$ **12–13.** $0.482 < \sigma^2 < 21.549$

12–15. $5.083 < \sigma^2 < 18.323$

12–17. $H_0: \sigma \leq 0.03$ (claim) and $H_1: \sigma > 0.03$; C.V. = 14.067; $\alpha = 0.05$; d.f. = 7; $\chi^2 = 14.381$; reject; no, there is enough evidence to reject the claim that the standard deviation is less than or equal to 0.03 ounce.

12–19. The variance test compares a sample variance with a hypothesized population variance; the goodness-of-fit test compares a distribution obtained from a sample with a hypothesized distribution.

12–21. The expected values are computed on the basis of what the null hypothesis states about the distribution.

12–23. H_0: The number of accidents is equally distributed throughout the week (claim). H_1: The number of accidents is not equally distributed throughout the week. C.V. = 12.592; d.f. = 6; $\chi^2 = 28.887$; reject; there is enough evidence to reject the claim that the number of accidents is equally distributed during the week.

12–25. H_0: The flavors are selected with equal frequency (claim). H_1: The flavors are not selected with equal frequency. C.V. = 11.345; d.f. = 3; $\chi^2 = 13.266$; reject; no, there is enough evidence to reject the claim that the flavors are selected with equal frequency.

12–27. H_0: The blood types of individuals are 28% type A, 32% type O, 25% type B, and 15% type AB (claim). H_1: The null hypothesis is not true. C.V. = 6.251; d.f. = 3; $\chi^2 = 2.72$; do not reject; there is not enough evidence to reject the examiner's hypothesis.

12–29. H_0: The number of calls received is distributed as 30% on Saturday, 30% on Sunday, and 8% on each of the weekdays (claim). H_1: The distribution is different from that stated in the null hypothesis. C.V. = 12.592; d.f. = 6; $\chi^2 = 4.676$; do not reject; there is not enough evidence to reject the director's claim.

12–31. H_0: The customers show no preference for the items (claim). H_1: The customers have a preference. C.V. = 9.210; d.f. = 2; $\chi^2 = 5.607$; do not reject; there is not enough evidence to reject the claim that the customers show no preference.

12–33. H_0: The distribution of college majors is as follows: 40% business, 25% computer science, 15% science, 10% social science, 5% liberal arts, and 5% general studies (claim). H_1: The null hypothesis is not true. C.V. = 9.236; d.f. = 5; $\chi^2 = 5.613$; do not reject; there is not enough evidence to reject the dean's hypothesis.

12–35. H_0: 50% of customers purchase word-processing programs, 25% purchase spreadsheet programs, and 25% purchase data base programs (claim). H_1: The null hypothesis is not true. C.V. = 5.991; d.f. = 2; $\chi^2 = 0.6$; do not reject; there is not enough evidence to reject the store owner's assumption.

12–37. Answers will vary.

12–39. d.f. = (rows − 1)(columns − 1)

12–41. H_0: The variables are independent (or not related). H_1: The variables are dependent (or related).

12–43. The expected values are computed as (row total × column total) ÷ grand total.

12–45. H_0: Milk consumption of individuals is independent of their age. H_1: Milk consumption is dependent upon age (claim). C.V. = 10.645; d.f. = 6; $\chi^2 = 13.365$; reject; yes, there is enough evidence to support the claim that milk consumption is dependent upon the age of the individual.

12–47. H_0: The amount of liability a person carries is independent of a person's age (claim). H_1: The amount of liability is dependent on the age of the person. C.V. = 12.592; d.f. = 6; χ^2 = 11.878; do not reject; there is not enough evidence to reject the claim that the amount of coverage is independent of the age of the driver.

12–49. H_0: The number of plants sold is independent of the location of the stand. H_1: The number of plants sold is dependent upon the location of the stand (claim). C.V. = 13.277; d.f. = 4; χ^2 = 28.588; reject; yes, there is enough evidence to support the claim that the number of plants sold is dependent upon the location of the stand.

12–51. H_0: The student's rating of the instructor is independent of the type of degree the instructor has. H_1: The rating is dependent on the type of degree the instructor has (claim). C.V. = 7.779; d.f. = 4; χ^2 = 19.507; reject; yes, there is enough evidence to support the claim that the degree of the instructor is related to the students' opinions about instructors' effectiveness.

12–53. H_0: The type of video rented by a person is independent of the person's age. H_1: The type of video rented by a person is dependent on the person's age (claim). C.V. = 13.362; d.f. = 8; χ^2 = 46.733; reject; yes, there is enough evidence to support the claim that the type of movie selected is related to the age of the customer.

12–55. χ^2 = 1.075

12–57. H_0: σ = 3.4 (claim) and H_1: $\sigma \neq$ 3.4; C.V. = 11.689 and 38.076; d.f. = 23; χ^2 = 35.1; do not reject; no, there is not enough evidence to reject the claim that the standard deviation is 3.4 minutes.

12–59. H_0: σ = 14 (claim) and H_1: $\sigma \neq$ 14; C.V. = 66.766 and 20.707; d.f. = 49; χ^2 = 81; reject; no, there is enough evidence to reject the claim that the standard deviation is 2 weeks.

12–61. H_0: $\sigma \geq$ 4.3 (claim) and H_1: $\sigma <$ 4.3; C.V. = 10.117; d.f. = 19; χ^2 = 6.95; reject; yes, there is enough evidence to reject the claim that the standard deviation is greater than or equal to 4.3 miles per gallon.

12–63. H_0: σ = 18 (claim) and H_1: $\sigma \neq$ 18; C.V. = 11.143 and 0.484; d.f. = 4; χ^2 = 5.44; do not reject; there is not enough evidence to reject the claim that the standard deviation is 21 minutes.

12–65. H_0: The number of sales is equally distributed over five regions (claim). H_1: The null hypothesis is not true. C.V. = 9.488; d.f. = 4; χ^2 = 87.14; reject; no, there is enough evidence to reject the claim that the number of items sold in each region is the same.

12–67. H_0: 50% of the customers order beer, 35% order a cocktail, and 15% order a nonalcoholic drink (claim). H_1: The distribution is not the same as stated in the null hypothesis. C.V. = 5.991; d.f. = 2; χ^2 = 5.647; do not reject; there is not enough evidence to reject the bartender's claim that 50% of the customers order beer, 35% order a cocktail, and 15% order a nonalcoholic drink.

12–69. H_0: The subject selected by a student is independent of the gender of the student (claim). H_1: The subject selected is dependent on the gender of the student. C.V. = 5.991; d.f. = 2; χ^2 = 114.373; reject; no, there is enough evidence to reject the claim that the subject selected is independent of gender.

12–71. H_0: The condiment preference is independent of the gender of the purchaser (claim). H_1: The condiment preference is dependent on the gender of the individual. C.V. = 4.605; d.f. = 2; χ^2 = 3.05; do not reject; there is not enough evidence to reject the claim that the condiment chosen is independent of the gender of the individual.

CHAPTER 12 TEST

1. True   2. False   3. False   4. True   5. False
6. False   7. True   8. True   9. True   10. True

CHAPTER 13

13–1. $F = s_1^2/s_2^2$

13–3. The larger variance is placed in the numerator of the formula.

13–5. One degree of freedom is used for the variance associated with the numerator, and one is used for the variance associated with the denominator.

13–7. a. d.f.N. = 15, d.f.D. = 22; C.V. = 3.36
b. d.f.N. = 24, d.f.D. = 13; C.V. = 3.59
c. d.f.N. = 45, d.f.D. = 29; C.V. = 2.03
d. d.f.N. = 20, d.f.D. = 16; C.V. = 2.28
e. d.f.N. = 10, d.f.D. = 10; C.V. = 2.98

13–9. H_0: $\sigma_1^2 = \sigma_2^2$ (claim) and H_1: $\sigma_1^2 \neq \sigma_2^2$; C.V. = 2.85; α = 0.10; d.f.N. = 11, d.f.D. = 11; F = 1.39; do not reject; there is not enough evidence to reject the claim that the variances are equal.

13–11. H_0: $\sigma_1 = \sigma_2$ and H_1: $\sigma_1 \neq \sigma_2$ (claim); C.V. = 2.24; α = 0.05; d.f.N. = 30, d.f.D. = 23; F = 3.178; reject; yes, there is enough evidence to support the claim that the standard deviations are not equal.

13–13. H_0: $\sigma_1 = \sigma_2$ (claim) and H_1: $\sigma_1 \neq \sigma_2$; C.V. = 5.05; α = 0.05; d.f.N. = 4, d.f.D. = 8; F = 5.22; reject; no, there is enough evidence to reject the claim that the standard deviations are equal.

13–15. H_0: $\sigma_1^2 \le \sigma_2^2$ and H_1: $\sigma_1^2 > \sigma_2^2$ (claim); C.V. = 1.84; $\alpha = 0.10$; d.f.N. = 25, d.f.D. = 17; $F = 1.47$; do not reject; no, there is not enough evidence to support the claim that the variation in the number of years of teaching experience of senior high school teachers is greater than the variation in the number of years of teaching experience of elementary school teachers.

13–17. H_0: $\sigma_1 = \sigma_2$ (claim) and H_1: $\sigma_1 \neq \sigma_2$; C.V. = 2.11; $\alpha = 0.10$; d.f.N. = 17, d.f.D. = 24; $F = 1.79$; do not reject; there is not enough evidence to reject the claim that the standard deviations are equal.

13–19. H_0: $\sigma_1^2 \le \sigma_2^2$ and H_1: $\sigma_1^2 > \sigma_2^2$ (claim); C.V. = 1.90; $\alpha = 0.05$; d.f.N. = 29, d.f.D. = 29; $F = 2.91$; reject; yes, there is enough evidence to support the claim that the variation in the salaries of the elementary school teachers is greater than the variation in the salaries of the secondary school teachers.

13–21. Brand A: $s_1^2 = 86.61$, $n_1 = 12$; brand B: $s_2^2 = 132.08$, $n_2 = 12$. H_0: $\sigma_1^2 = \sigma_2^2$ (claim) and H_1: $\sigma_1^2 \neq \sigma_2^2$; C.V. = 3.53; $\alpha = 0.05$; d.f.N. = 11, d.f.D. = 11; $F = 1.52$; do not reject; there is not enough evidence to reject the claim that the variances are equal.

13–23. a. Comparing two means at a time ignores all other means.
b. The probability of a type I error is larger than α when multiple t tests are used.
c. The more sample means, the more t tests are needed.

13–25. The between-group variance estimates the population variance using the means. The within-group variance estimates the population variance using all the data values.

13–27. H_0: $\mu_1 = \mu_2 = \cdots = \mu_n$ and H_1: At least one mean is different from the others.

13–29. H_0: $\mu_1 = \mu_2 = \mu_3$ and H_1: at least one mean is different (claim). C.V. = 3.63; $\alpha = 0.05$; d.f.N. = 2, d.f.D. = 16; $F = 5.96$; reject. Scheffé test: $F' = 7.26$; $\overline{X}_1 = 28.83$, $\overline{X}_2 = 46.43$, $\overline{X}_3 = 31.5$; $\overline{X}_1$ vs. $\overline{X}_2$: $F_S = 10.02$; $\overline{X}_1$ vs. $\overline{X}_3$: $F_S = 0.21$; $\overline{X}_2$ vs. $\overline{X}_3$: $F_S = 7.21$. There is a significant difference between $\overline{X}_1$ and $\overline{X}_2$.

13–31. H_0: $\mu_1 = \mu_2 = \mu_3$ and H_1: at least one mean is different (claim). C.V. = 2.81; $\alpha = 0.10$; d.f.N. = 2, d.f.D. = 12; $F = 8.448$; reject. Tukey test: C.V. = 3.20; $\overline{X}_1 = 2.48$, $\overline{X}_2 = 3.34$, $\overline{X}_3 = 2.12$; $\overline{X}_1$ vs. $\overline{X}_2$: $q = -4$; $\overline{X}_1$ vs. $\overline{X}_3$: $q = 1.67$; $\overline{X}_2$ vs. $\overline{X}_3$: $q = 5.67$. There is a significant difference between $\overline{X}_1$ and $\overline{X}_2$ and between $\overline{X}_2$ and $\overline{X}_3$.

13–33. H_0: $\mu_1 = \mu_2 = \mu_3$ and H_1: at least one mean is different. C.V. = 3.98; $\alpha = 0.05$; d.f.N. = 2, d.f.D. = 11; $F = 7.75$; reject. Scheffé test: $F' = 7.96$; $\overline{X}_1$ vs.

$\overline{X}_2$: $F = 9.811$; $\overline{X}_1$ vs. $\overline{X}_3$: $F = 0.077$; $\overline{X}_2$ vs. $\overline{X}_3$: $F = 11.801$. There is a significant difference between $\overline{X}_1$ and $\overline{X}_2$ and between $\overline{X}_2$ and $\overline{X}_3$.

13–35. H_0: $\mu_1 = \mu_2 = \mu_3 = \mu_4$ and H_1: at least one mean is different. C.V. = 3.01; $\alpha = 0.05$; d.f.N. = 3, d.f.D. = 24; $F = 6.974$; reject. Tukey test: C.V. = 3.90; $\overline{X}_1 = 3.286$, $\overline{X}_2 = 4.714$, $\overline{X}_3 = 11.714$, $\overline{X}_4 = 4.286$; $\overline{X}_1$ vs. $\overline{X}_2$: $q = -0.98$; $\overline{X}_1$ vs. $\overline{X}_3$: $q = -5.78$; $\overline{X}_1$ vs. $\overline{X}_4$: $q = -0.69$; $\overline{X}_2$ vs. $\overline{X}_3$: $q = -4.80$; $\overline{X}_2$ vs. $\overline{X}_4$: $q = 0.29$; $\overline{X}_3$ vs. $\overline{X}_4$: $q = 5.09$. There is a significant difference between $\overline{X}_1$ and $\overline{X}_3$, between $\overline{X}_2$ and $\overline{X}_3$, and between $\overline{X}_3$ and $\overline{X}_4$.

13–37. H_0: $\mu_1 = \mu_2 = \mu_3$ and H_1: at least one mean is different (claim). C.V. = 2.81; $\alpha = 0.10$; d.f.N. = 2, d.f.D. = 12; $F = 9.16$; reject. Tukey test: C.V. = 3.20; $\overline{X}_1 = 157.4$, $\overline{X}_2 = 94.6$, $\overline{X}_3 = 97.6$; $\overline{X}_1$ vs. $\overline{X}_2$: $q = 5.37$; $\overline{X}_1$ vs. $\overline{X}_3$: $q = 5.11$; $\overline{X}_2$ vs. $\overline{X}_3$: $q = -0.26$. There is a significant difference between $\overline{X}_1$ and $\overline{X}_2$ and between $\overline{X}_1$ and $\overline{X}_3$.

13–39. H_0: $\mu_1 = \mu_2 = \mu_3$ and H_1: at least one mean is different. C.V. = 3.89; $\alpha = 0.05$; d.f.N. = 2, d.f.D. = 12; $F = 5.877$; reject. Tukey test: C.V. = 3.77; $\overline{X}_1 = 17.6$, $\overline{X}_2 = 13.6$, $\overline{X}_3 = 8.6$; $\overline{X}_1$ vs. $\overline{X}_2$: $q = 2.15$; $\overline{X}_1$ vs. $\overline{X}_3$: $q = 4.84$; $\overline{X}_2$ vs. $\overline{X}_3$: $q = 2.69$. There is a significant difference between $\overline{X}_1$ and $\overline{X}_3$.

13–41. The two-way ANOVA allows the researcher to test the effects of two independent variables and a possible interaction effect. The one-way ANOVA can test the effects of one independent variable only.

13–43. Factor

13–45. The F test value is computed by dividing the mean square for the variable by the mean square for the within (error) term.

13–47. a. d.f.N. = $(5 - 1) = 4$
b. d.f.N. = $(4 - 1) = 3$
c. d.f.N. = $(5 - 1)(4 - 1) = 12$
d. d.f.D. = $5 \cdot 4(6 - 1) = 100$

13–49. The main effects can be interpreted independently when the interaction effect is not significant or the interaction is an ordinal interaction.

13–51. H_0: There is no interaction effect between the type of ad and the medium on the effectiveness of the ad.
H_1: There is an interaction effect between the type of ad and the medium on the effectiveness of the ad.
H_0: There is no difference between the means of the ratings for the type of ad used.
H_1: There is a difference between the means of the ratings for the type of ad used.
H_0: There is no difference between the means of the ratings for the medium used.
H_1: There is a difference between the means of the ratings for the medium used.

ANOVA Summary Table

Source	SS	d.f.	MS	F
Type	10.563	1	10.563	1.913
Medium	175.563	1	175.563	31.800
Interaction	0.063	1	0.063	0.011
Within	66.250	12	5.512	
Total	252.439	15		

The critical value at $\alpha = 0.01$ with d.f.N. = 1 and d.f.D. = 12 is 9.33 for F_A, F_B, and $F_{A \times B}$. Since the F test value for the medium, 31.800, is greater than the critical value, 9.33, the decision is to reject the null hypothesis for the medium. It can be concluded that there is a significant difference in the ratings for the medium used. Since each of 1.913 and 0.011 is less than the C.V. of 9.33, do not reject the null hypothesis for type and interaction.

13–53. H_0: There is no interaction effect between the subcontractors and the types of homes they build on the times it takes to build the homes.
H_1: There is an interaction effect between the subcontractors and the types of homes they build on the times it takes to build the homes.
H_0: There is no difference in the means of the times it takes the subcontractors to build the homes.
H_1: There is a difference in the means of the times it takes the subcontractors to build the homes.
H_0: There is no difference among the means of the times for the types of homes built.
H_1: There is a difference among the means of the times for the types of homes built.

ANOVA Summary Table

Source	SS	d.f.	MS	F
Subcontractor	1672.553	1	1672.553	122.084
Home type	444.867	2	222.434	16.236
Interaction	313.267	2	156.634	11.433
Within	328.8	24	13.700	
Total	2759.487	29		

The critical values at $\alpha = 0.05$: for the subcontractor with d.f.N. = 1 and d.f.D. = 24, C.V. = 4.26; for the home type and interaction with d.f.N. = 2 and d.f.D. = 24, C.V. = 3.40. All F test values exceed the critical values and all of the null hypotheses are rejected. Since there is a significant interaction effect, the means of the cells must be computed and graphed to determine the type of interaction.
Cell means:

	Home Type		
Subcontractor	**I**	**II**	**III**
A	28	31.4	44.4
B	18.6	20.000	20.400

Since all of the three means for the home types for subcontractor A are greater than the three means for subcontractor B and the differences are not equal, there is an ordinal interaction. Hence, it can be concluded that there is a difference in means for the subcontractors and home types; also an interaction effect is present.

13–55. H_0: There is no interaction effect between the type of paint and the geographic location on the lifetimes of the paint.
H_1: There is an interaction effect between the type of paint and the geographic location on the lifetimes of the paint.
H_0: There is no difference between the means of the lifetimes of the two types of paints.
H_1: There is a difference between the means of the lifetimes of the two types of paints.
H_0: There is no difference among the means of the lifetimes of the paints used in different geographic locations.
H_1: There is a difference in the means of the lifetimes of the paints used in different geographic locations.

ANOVA Summary Table

Source	SS	d.f.	MS	F
Paint type	12.1	1	12.1	0.166
Location	2501	3	833.667	11.465
Interaction	268.1	3	89.367	1.229
Within	2326.8	32	72.713	
Total	5108	39		

The critical values for $\alpha = 0.01$: for the paint type with d.f.N. = 1 and d.f.D. = 32 (use 30), C.V. = 7.56; for the location and interaction with d.f.N. = 3 and d.f.D. = 32 (use 30), C.V. = 4.51.
Since the only F test value that exceeds the critical value is the one for the location, it can be concluded that there is a difference in the means for the geographic locations.

13–57. H_0: $\sigma_1^2 = \sigma_2^2$ and H_1: $\sigma_1^2 \neq \sigma_2^2$ (claim); C.V. = 2.86; $\alpha = 0.05$; d.f.N. = 17, d.f.D. = 15; $F = 2.24$; do not reject; no, there is not enough evidence to support the claim that the variances are different.

13–59. H_0: $\sigma_1^2 = \sigma_2^2$ and H_1: $\sigma_1^2 \neq \sigma_2^2$ (claim); C.V. = 3.18; $\alpha = 0.05$; d.f.N. = 9, d.f.D. = 9; $F = 5.06$; reject; there is enough evidence to support the claim that the variance of the number of speeding tickets issued on Route 19 is greater than the variance of the number of speeding tickets issued on Route 22.

13–61. $H_0: \sigma_1^2 \leq \sigma_2^2$ and $H_1: \sigma_1^2 > \sigma_2^2$ (claim); C.V. = 1.47; $\alpha = 0.10$; d.f.N. = 64, d.f.D. = 41; $F = 2.32$; reject; there is enough evidence to support the claim that the variation in the number of days factory workers miss per year due to illness is greater than the variation in the number of days hospital workers miss per year.

13–63. $H_0: \mu_1 = \mu_2 = \mu_3$ (claim) and H_1: at least one mean is different. C.V. = 6.01; $\alpha = 0.01$; d.f.N. = 2, d.f.D. = 18; $F = 27.02$; reject. Tukey test: C.V. = 4.70; $\overline{X}_1 = 197.57$, $\overline{X}_2 = 295.57$, $\overline{X}_3 = 103.14$; $\overline{X}_1$ vs. $\overline{X}_2$: $q = -5.29$; $\overline{X}_1$ vs. $\overline{X}_3$: $q = 5.10$; $\overline{X}_2$ vs. $\overline{X}_3$: $q = 10.40$. There is a significant difference between any two means.

13–65. $H_0: \mu_1 = \mu_2 = \mu_3$ and H_1: at least one mean is different. C.V. = 3.89; $\alpha = 0.05$; d.f.N. = 2, d.f.D. = 12; $F = 6.141$; reject. Scheffé test: $F' = 7.78$; $\overline{X}_1$ vs. $\overline{X}_2$: $F = 12.167$; $\overline{X}_1$ vs. $\overline{X}_3$: $F = 2.849$; $\overline{X}_2$ vs. $\overline{X}_3$: $F = 2.509$. There is a significant difference between $\overline{X}_1$ and $\overline{X}_2$.

13–67. $H_0: \mu_1 = \mu_2 = \mu_3$ and H_1: at least one mean is different. C.V. = 3.89; $\alpha = 0.05$; d.f.N. = 2, d.f.D. = 12; $F = 3.673$; do not reject; there is not enough evidence to conclude that there is a difference in the means.

13–69. H_0: There is no interaction effect between the ages of the students and the subjects on the anxiety scores of the students.

H_1: There is an interaction effect between the ages of the students and the subjects on the anxiety scores of the students.

H_0: There is no difference between the means of the anxiety scores of the students in the two age groups.
H_1: There is a difference between the means of the anxiety scores of the students in the two age groups.
H_0: There is no difference between the means of the anxiety scores of the students in the two subjects.
H_1: There is a difference between the means of the anxiety scores of the students in the two subjects.

ANOVA Summary Table

Source	SS	d.f.	MS	F
Age	2376.2	1	2376.2	49.816
Class	105.8	1	105.8	2.218
Interaction	2645	1	2645	55.451
Within	763.2	16	47.7	
Total	5890.2	19		

At $\alpha = 0.10$, d.f.N. = 1 and d.f.D. = 16; the critical value is 3.05. Since the interaction hypothesis is rejected, it is necessary to calculate and graph the cell means. *Note:* H_0 for age is also rejected. Cell means:

Age	Class	
	Calculus	Statistics
Under 20	53.6	26.00
20 and over	52.4	70.800

The graph is shown below.

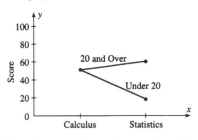

There is a significant interaction effect between the age of the student and the subject on the anxiety test score.

CHAPTER 13 TEST

1. True 2. False 3. False 4. False 5. False
6. True 7. False 8. True 9. True 10. False
11. True 12. True

CHAPTER 14

14–1. *Nonparametric* means hypotheses other than those using population parameters can be tested; whereas *distribution-free* means no assumptions about the population distributions have to be satisfied.

14–3. Nonparametric methods have the following advantages:
a. They can be used to test population parameters when the variable is not normally distributed.
b. They can be used when data is nominal or ordinal.
c. They can be used to test hypotheses other than those involving population parameters.
d. The computations are easier in some cases than the computations of the parametric counterparts.
e. They are easier to understand.

The disadvantages are as follows:
a. They are less sensitive than their parametric counterparts.
b. They tend to use less information than their parametric counterparts.
c. They are less efficient than their parametric counterparts.

14–5.

Data	21	31	34	41	41	61	65	72
Rank	1	2	3	4.5	4.5	6	7	8

14–7.

Data	3	5	5	6	7	8	8	9	12	14	15	17
Rank	1	2.5	2.5	4	5	6.5	6.5	8	9	10	11	12

14–9.

Data	187	190	190	236	321	532	673
Rank	1	2.5	2.5	4	5	6	7

14–11. The sign test uses only $+$ or $-$ signs.

14–13. The smaller number of $+$ or $-$ signs.

14–15. H_0: median $= 2.8$ (claim) and H_1: median $\neq 2.8$; test value $= 8$; C.V. $= 4$; do not reject; there is not enough evidence to reject the claim that the median is 2.8.

14–17. H_0: median $= \$325$ (claim) and H_1: median $\neq \$325$; C.V. $= 2$; test value $= 3$; do not reject; there is not enough evidence to reject the claim that the median rent is $325.

14–19. H_0: median $= 6.25$ (claim) and H_1: median $\neq 6.25$; C.V. $= \pm 1.96$; $z = -0.34$; do not reject; there is not enough evidence to reject the claim that the median is 6.25 months.

14–21. H_0: median $= 25$ (claim) and H_1: median $\neq 25$; C.V. $= \pm 2.33$; $z = -0.99$; do not reject; there is not enough evidence to reject the claim that 50% of the students favor single-room dormitories.

14–23. H_0: median $= 50$ (claim) and H_1: median $\neq 50$; C.V. $= \pm 1.65$; $z = -2.3$; reject; there is enough evidence to reject the claim that 50% of the students are against extending the school year.

14–25. H_0: The medication has no effect on weight loss. H_1: The medication affects weight loss (claim). C.V. $= 1$; test value $= 1$; reject; there is enough evidence to support the claim that the medication affects weight loss.

14–27. H_0: Reasoning ability will not be affected by the course. H_1: Reasoning ability increased after the course. C.V. $= 1$; test value $= 1$; reject; there is enough evidence to support the claim that reasoning ability has increased after the course.

14–29. H_0: Alcohol has no effect on a person's IQ test score. H_1: Alcohol does affect a person's IQ test score. C.V. $= 1$; test value $= 3$; do not reject; there is not enough evidence to reject the claim that alcohol has no effect on a person's IQ score.

14–31. $6 \leq$ median ≤ 22 **14–33.** $4.7 \leq$ median ≤ 9.3

14–35. $17 \leq$ median ≤ 33

14–37. The rank sum test is used for independent samples, and the signed-rank test is used for dependent samples.

14–39. The t test for dependent samples

14–41. The standard normal distribution

14–43. H_0: There is no difference in the number of books each group read (claim). H_1: There is a difference in the number of books each group read. C.V. $= \pm 1.65$; $z = +1.97$; reject; there is enough evidence to reject the claim that there is no difference in the number of books read by each group.

14–45. H_0: There is no difference in the number of points each team scored. H_1: There is a difference in the number of points each team scored (claim). C.V. $= \pm 1.96$; $z = 1.33$; do not reject; there is not enough evidence to support the claim that the number of points each team scored is different.

14–47. H_0: There is no difference in the amount of money awarded to each city. H_1: There is a difference in the amount of money awarded to each city (claim). C.V. $= \pm 1.96$; $z = -0.520$; do not reject; there is not enough evidence to support the claim that there is a difference in the amount of money awarded to the cities.

14–49. H_0: There is no difference in the times needed to assemble the product. H_1: There is a difference in the times needed to assemble the product (claim). C.V. $= \pm 1.96$; $z = +3.56$; reject; there is enough evidence to support the claim that there is a difference in the productivity of the two groups.

14–51. Sum of $-$ ranks is -6; sum of $+$ ranks is $+15$.

14–53. C.V. $= 20$; reject

14–55. C.V. $= 130$; do not reject

14–57. H_0: The workshop did not reduce anxiety. H_1: The workshop reduced anxiety (claim). C.V. $= 2$; $T = 3$; do not reject; there is not enough evidence to support the claim that the workshop reduced anxiety.

14–59. H_0: There is no difference in the ages of husbands and wives (claim). H_1: There is a difference in the ages of husbands and wives. C.V. $= 2$; $T = 1$; reject; there is enough evidence to reject the claim that there is no difference in the ages of husbands and wives.

14–61. H_0: There is no difference in the number of patrons purchasing tickets. H_1: There is a difference in the number of patrons purchasing tickets (claim). C.V. $= 4$; $T = +7$; do not reject; there is not enough evidence to support the claim that the ad campaign made a difference in the number of patrons.

14–63. H_0: There is no difference in the number of calories each brand contains. H_1: There is a difference in the number of calories each brand contains (claim). C.V. $= 7.815$; $H = 20.3$; reject; there is enough evidence to support the claim that the number of calories each brand contains is different.

14–65. H_0: There is no difference in the sales of the stores. H_1: There is a difference in the sales of the stores (claim). C.V. = 4.605; H = 10.8; reject; there is enough evidence to support the claim that there is a difference in sales.

14–67. H_0: There is no difference in the yields of the three plots. H_1: There is a difference in the yields of the three plots (claim). C.V. = 9.210; H = 12.020; reject; there is enough evidence to support the claim that the yields of the three plots are different.

14–69. H_0: The color of the box did not make a difference in the number of products sold. H_1: The color of the box did make a difference in the number of products sold (claim). C.V. = 6.251; H = 15.959; reject; there is enough evidence to support the claim that the color of the box makes a difference.

14–71. H_0: There is no difference in the number of crimes in the five precincts. H_1: There is a difference in the number of crimes in the five precincts (claim). C.V. = 13.277; H = 20.753; reject; there is enough evidence to support the claim that there is a difference in the number of crimes in the five precincts.

14–73. H_0: There is no difference in the final exam scores of the three groups. H_1: There is a difference in the final exam scores of the three groups (claim). C.V. = 4.605; H = 1.710; do not reject; there is not enough evidence to support the claim that there is a difference in the final exam scores of the three groups.

14–75. 0.716   **14–77.** 0.648

14–79. r_S = 0.714; H_0: ρ = 0 and H_1: $\rho \neq 0$; C.V. = ±0.886; do not reject; there is no significant relationship between the rankings.

14–81. r_S = 0.8857; H_0: ρ = 0 and H_1: $\rho \neq 0$; C.V. = 0.886; do not reject; there is no significant relationship between the rankings.

14–83. r_S = 0.857; H_0: ρ = 0 and H_1: $\rho \neq 0$; C.V. = ±0.738; reject; there is a significant relationship between the rankings.

14–85. r_S = 0.829; H_0: ρ = 0 and H_1: $\rho \neq 0$; C.V. = ±0.886; do not reject; there is no significant relationship between the rankings.

14–87. r_S = 0.979; H_0: ρ = 0 and H_1: $\rho \neq 0$; C.V. = ±0.591; reject; there is a significant relationship between the rankings.

14–89. H_0: The number of cavities in a person occurs at random. H_1: The null hypothesis is not true. There are 12 runs; the expected number of runs is between 10 and 22; therefore, do not reject the null hypothesis; the number of cavities in a person occurs at random.

14–91. H_0: The Lotto numbers occur at random. H_1: The null hypothesis is not true. There are 14 runs, and this value is between 7 and 18; hence, do not reject the null hypothesis; the Lotto occurs at random.

14–93. H_0: The number of defective cassette cases manufactured by the machine occurs at random. H_1: The null hypothesis is not true. There are 10 runs; and since this value is between 9 and 20, do not reject the null hypothesis; the number of defective cassette cases manufactured by a machine occurs at random.

14–95. H_0: The number of absences of employees occurs at random over a 30-day period. H_1: The null hypothesis is not true. There are only 6 runs, and this value does not fall within the 9-to-21 range; hence, the null hypothesis is rejected; the absences do not occur at random.

14–97. H_0: The IQs of students who complete the tests are random. H_1: The null hypothesis is not true. Do not reject, since there are 7 runs, and this value is within the 6-to-16 range; the IQs occur at random.

14–99. ±0.479   **14–101.** ±0.215

14–103. H_0: median = 400 and H_1: median ≠ 400; C.V. = 5; test value = 8; do not reject; there is not enough evidence to reject the claim that the median of the sales is 400 cards.

14–105. H_0: The special diet has no effect on weight. H_1: The diet increases weight. C.V. = 1; test value = 3; do not reject; there is not enough evidence to support the claim that there was an increase in weight.

14–107. H_0: There is no difference in the amount of money each group spent for the textbook. H_1: There is a difference in the amount of money each group spent for the textbook (claim). C.V. = ±1.65; z = −3.59; reject; there is enough evidence to support the claim that there is a difference in the amount of money each group spent on the textbooks.

14–109. H_0: The number of sick days workers used was not reduced. H_1: The number of sick days workers used was reduced. C.V. = 4; T = 8.5; do not reject; there is not enough evidence to support the claim that the number of sick days was reduced.

14–111. H_0: The diet has no effect on learning. H_1: The diet affects learning (claim). C.V. = 5.991; H = 8.5; reject; there is enough evidence to support the claim that the diets do affect learning.

14–113. r_S = 0.086; H_0: ρ = 0 and H_1: $\rho \neq 0$; C.V. = ±0.886; do not reject; there is no significant relationship in the rankings of the boys and girls.

14–115. H_0: The grades of students who finish the exam occur at random. H_1: The null hypothesis is not true. Since there are 8 runs, and this value does not fall between the 9-to-21 interval, the null hypothesis is rejected; the grades do not occur at random.

CHAPTER 14 TEST

1. True   2. False   3. False   4. True   5. True
6. False   7. False   8. False   9. True   10. False
11. False   12. False   13. True   14. True

CHAPTER 15

15–1. Random, systematic, stratified, cluster

15–3. A sample must be randomly selected.

15–5. Talking to people on the street, calling people on the phone, and asking one's friends are three incorrect ways of obtaining a sample.

15–7. Random sampling has the advantage that each unit of the population has an equal chance of being selected. One disadvantage is that the units of the population must be numbered; and if the population is large, this could be somewhat time-consuming.

15–9. An advantage of stratified sampling is that it ensures representation for the groups used in stratification; however, it is virtually impossible to stratify the population so that all groups are represented.

15–11 through 15–19.   Answers will vary.

15–21. Simulation involves setting up probability experiments that mimic the behavior of real-life events.

15–23. John Von Neuman and Stanislaw Ulam

15–25. The steps are as follows:
  a. List all possible outcomes.
  b. Determine the probability of each outcome.
  c. Set up a correspondence between the outcomes and the random numbers.
  d. Conduct the experiment by using random numbers.
  e. Repeat the experiment and tally the outcomes.
  f. Compute any statistics, and state the conclusions.

15–27. When the repetitions increase, there is a higher probability that the simulation will yield more precise answers.

15–29. Use random numbers 1 through 6 to represent making a shot, and use 7 through 9 and 0 to represent a miss.

15–31. Use random numbers 1 through 8 to represent a hit, and use 9 and 0 to represent a miss.

15–33. Let an odd number represent heads and an even number represent tails; then, each person selects a digit at random.

15–35 through 15–51.   Answers will vary.

15–53. Use the digits 1 through 4 to represent a completed pass, and use the remaining digits to represent an incomplete pass.

15–55. Select two digits between 1 and 6 to represent the dice.

15–57. Let the digits 1–3 represent rock; let 4–6 represent paper; let 7–9 represent scissors; and omit 0.

15–59 through 15–61.   Answers will vary.

CHAPTER 15 TEST

1. False   2. False   3. True   4. False   5. True
6. False   7. True   8. False   9. True   10. False

CHAPTER 16

16–1. The purpose of quality control is to ensure that a manufactured product meets certain acceptable standards.

16–3. Two types of variation in manufacturing processes are chance variation and assignable-cause variation.

16–5. $UCL_{\bar{X}} = 863.71$; $LCL_{\bar{X}} = 688.45$; $\bar{X}_{GM} = 776.08$; $UCL_R = 402.39$; $LCL_R = 15.89$; $\bar{R} = 209.14$

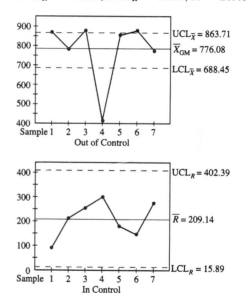

16–7. $\text{UCL}_{\overline{X}} = 33.746$; $\text{LCL}_{\overline{X}} = 31.254$; $\overline{X}_{\text{GM}} = 32.5$; $\text{UCL}_R = 4.568$; $\text{LCL}_R = 0$; $\overline{R} = 2.16$

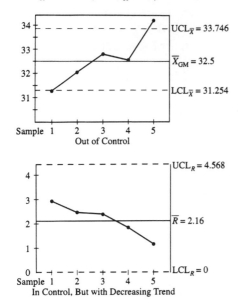

Out of Control

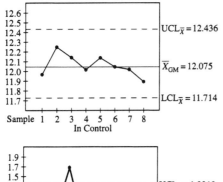

In Control, But with Decreasing Trend

16–11. $\text{UCL}_{\overline{X}} = 12.91$; $\text{LCL}_{\overline{X}} = 11.09$; $\overline{X}_{\text{GM}} = 12$; $\text{UCL}_R = 3.32$; $\text{LCL}_R = 0$; $\overline{R} = 1.57$

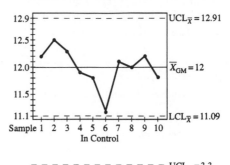

In Control

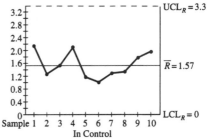

In Control

16–9. $\text{UCL}_{\overline{X}} = 12.436$; $\text{LCL}_{\overline{X}} = 11.714$; $\overline{X}_{\text{GM}} = 12.075$; $\text{UCL}_R = 1.3219$; $\text{LCL}_R = 0$; $\overline{R} = 0.625$

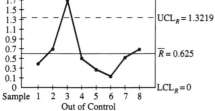

16–13. $\text{UCL}_{\overline{X}} = 37.69$; $\text{LCL}_{\overline{X}} = 31.71$; $\overline{X}_{\text{GM}} = 34.7$; $\text{UCL}_R = 10.9557$; $\text{LCL}_R = 0$; $\overline{R} = 5.18$

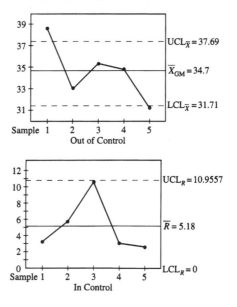

16–15. $\bar{p}$ charts and $\bar{c}$ charts

16–17. $UCL_p = 0.198$; $LCL_p = 0.018$; $\bar{p} = 0.108$

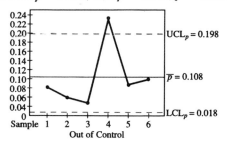

16–19. $UCL_p = 0.348$; $LCL_p = 0$; $\bar{p} = 0.14$

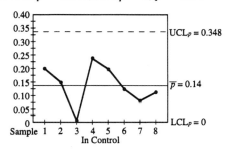

16–21. $UCL_p = 0.355$; $LCL_p = 0$; $\bar{p} = 0.13$

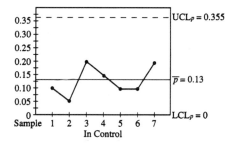

16–23. $UCL_p = 0.161$; $LCL_p = 0.035$; $\bar{p} = 0.098$

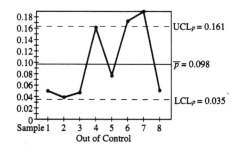

16–25. $UCL_p = 0.675$; $LCL_p = 0.135$; $\bar{p} = 0.405$

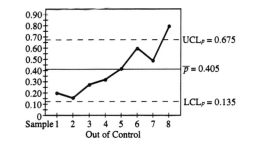

16–27. $UCL_c = 10.76$; $LCL_c = 0$; $\bar{c} = 4.44$

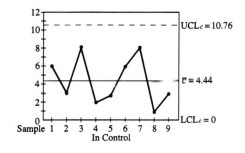

16–29. $UCL_c = 6.498$; $LCL_c = 0$; $\bar{c} = 2.125$

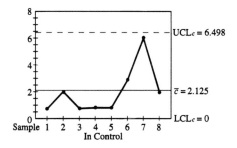

16–31. $UCL_c = 10$; $LCL_c = 0$; $\bar{c} = 4$

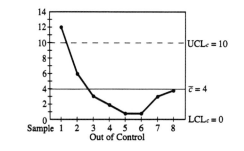

16–33. $UCL_c = 6.24$; $LCL_c = 0$; $\bar{c} = 2$

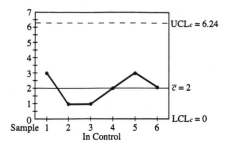

16–35. $UCL_{\bar{X}} = 95.84$; $LCL_{\bar{X}} = 77.02$; $\overline{X}_{GM} = 86.43$; $UCL_R = 34.4745$; $LCL_R = 0$; $\overline{R} = 16.3$

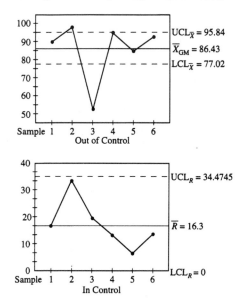

16–37. $UCL_{\bar{X}} = 75.56$; $LCL_{\bar{X}} = 73.46$; $\overline{X}_{GM} = 74.51$; $UCL_R = 6.0418$; $LCL_R = 0.7582$; $\overline{R} = 3.4$

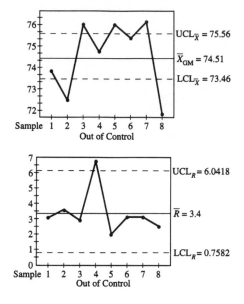

16–39. $UCL_p = 0.247$; $LCL_p = 0$; $\bar{p} = 0.1025$

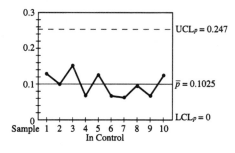

16–41. $UCL_c = 17.62$; $LCL_c = 0$; $\bar{c} = 8.75$

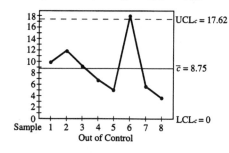

CHAPTER 16 TEST

1. True   2. False   3. False   4. True   5. False
6. True   7. False   8. True   9. False   10. False

APPENDIX A

A–1. 362,880   **A–3.** 120   **A–5.** 1   **A–7.** 1320
A–9. 20   **A–11.** 126   **A–13.** 70   **A–15.** 1   **A–17.** 560
A–19. 2520   **A–21.** 121, 2181, 14,641, 716.9
A–23. 32, 258, 1024, 53.2
A–25. 328, 22,678, 107,584, 1161.2
A–27. 693, 50,511, 480,249, 2486.1
A–29. 318, 20,150, 101,124, 3296

A–31.

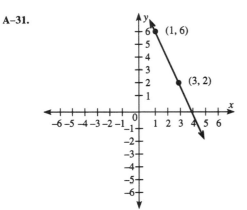

A–33.

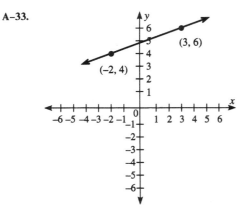

A–35.

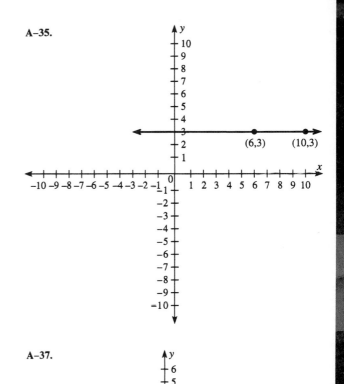

A–37.

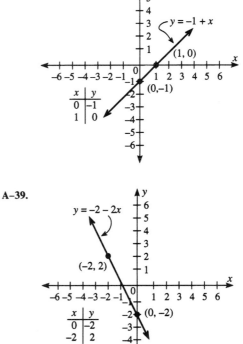

A–39.

INDEX

GLOSSARY OF SYMBOLS

a	y intercept of a line	MD	Median
α	Probability of a type I error	MR	Midrange
b	Slope of a line	MS_B	Mean square between groups
β	Probability of a type II error	MS_W	Mean square within groups (error)
C	Column frequency	n	Sample size
c	Count; number of defects per item	N	Population size
cf	Cumulative frequency	$n(E)$	Number of ways E can occur
$_nC_r$	Number of combinations of n objects taken r at a time	$n(S)$	Number of outcomes in the sample space
		O	Observed frequency
C.V.	Critical value	P	Percentile
CVar	Coefficient of variation	p	Probability; population proportion
D	Difference; decile	$\hat{p}$	Sample proportion
D	Mean of the differences	$\bar{p}$	Weighted estimate of p
d.f.	Degrees of freedom	$P(B\mid A)$	Conditional probability
d.f.N.	Degrees of freedom, numerator	$P(E)$	Probability of an event E
d.f.D.	Degrees of freedom, denominator	$P(\overline{E})$	Probability of the complement of E
E	Event; expected frequency; maximum error of estimate	$_nP_r$	Number of permutations of n objects taken r at a time
$\overline{E}$	Complement of an event	π	Pi = 3.14
e	Euler's constant = 2.7184	Q	Quartile
$E(x)$	Expected value	q	$1 - p$; test value for Tukey test
f	Frequency	$\hat{q}$	$1 - \hat{p}$
F	F test value; failure	$\bar{q}$	$1 - p$
F'	Critical value for the Scheffé test	R	Range; rank sum